Manufacturing Technology

Materials, Processes, and Equipment

Manufacturing Technology

Materials, Processes, and Equipment

Helmi A. Youssef

Hassan A. El-Hofy

Mahmoud H. Ahmed

CRC Press
Taylor & Francis Group
Boca Raton   London   New York

CRC Press is an imprint of the
Taylor & Francis Group, an **informa** business

CRC Press
Taylor & Francis Group
6000 Broken Sound Parkway NW, Suite 300
Boca Raton, FL 33487-2742

First issued in paperback 2017

ISBN 13: 978-1-138-07213-8 (pbk)
ISBN 13: 978-1-4398-1085-9 (hbk)

Library of Congress Cataloging-in-Publication Data

Youssef, Helmi A.
    Manufacturing technology : materials, processes, and equipment / authors: Helmi A. Youssef, Hassan A. El-Hofy, and Mahmoud H. Ahmed.
        p. cm.
    Includes bibliographical references and index.
    ISBN 978-1-4398-1085-9 (hardback)
    1. Manufacturing processes. I. El-Hofy, Hassan. II. Ahmed, Mahmoud H. III. Title.

TS183.Y68 2011
670.42'7--dc22
                                                                                2011007861

Visit the Taylor & Francis Web site at
http://www.taylorandfrancis.com

and the CRC Press Web site at
http://www.crcpress.com

Little Drops of Water,
Little Grains of Sand,
Make the Mighty Ocean,
and the Pleasant Land

Julia Carney, 1845

Dedicated to our little angels
H. Youssef: To Youssef, Nour, Anorine, Fayrouz, and Yousra
H. El-Hofy: To Omar, Youssef, and Zaina
M. Ahmed: To Nada

Contents

Preface

This textbook provides comprehensive knowledge and insight into various aspects of manufacturing technology, processes, materials, tooling, and equipment. Its main objective is to introduce the grand spectrum of manufacturing technology to individuals who will be involved in the design and manufacturing of finished products, to provide them with the basic information on manufacturing technologies. The text material is presented mainly in descriptive manner, where the emphasis is on the fundamentals of the process, its capabilities, typical applications, advantages, and limitations. Mathematical modeling and equations are used only when they enhance the basic understanding of with which the material deals.

The book has been written specifically for undergraduates in mechanical, industrial, manufacturing, and materials engineering disciplines of the second to fourth levels to cover complete courses of manufacturing technology taught in engineering colleges and institutions all over the world. It also covers the needs of production and manufacturing engineers and technologists participating in related industries where it is expected to be part of their professional library. Additionally, the book can be used by students in other disciplines concerned with design and manufacturing, such as automotive and aerospace engineering.

The book is a fundamental textbook that covers all of the manufacturing processes, materials, and equipment used to convert raw materials to a final product. It presents the materials used in manufacturing processes. The book also presents heat treatment processes, smelting of metals, and other technological processes such as casting, forming, powder metallurgy, joining, and surface technology. Manufacturing processes for polymers, ceramics, and composites are also covered.

The book also covers traditional, nontraditional, and advanced manufacturing technologies and applications. It sheds light on modern manufacturing technologies. In this regard, numerical control, industrial robots, and hexapods are covered. Product quality control, automation in manufacturing, and health, safety, and environmental aspects in manufacturing are also discussed.

The book is written in 26 chapters and 3 appendices having the following titles:

1. Introduction to Manufacturing Technology
2. Properties of Engineering Materials
3. Structure of Metals and Alloys
4. Engineering Materials and Their Applications
5. Heat Treatment of Metals and Alloys
6. Smelting of Metallic Materials
7. Casting of Metallic Materials
8. Fundamentals of Metal Forming
9. Bulk Forming of Metallic Materials
10. Sheet Metal Forming Processes
11. High-Velocity Forming and High-Energy-Rate Forming
12. Powder Metallurgy and Processing of Ceramic Materials
13. Polymeric Materials and Their Processing
14. Composite Materials and Their Fabrication Processes
15. Fundamentals of Traditional Machining Processes
16. Machine Tools for Traditional Machining
17. Fundamentals of Nontraditional Machining Processes
18. Numerical Control of Machine Tools
19. Industrial Robots and Hexapods

OUTLINE OF THE BOOK

The following topics are covered by individual chapters of the book:

Chapter 1 presents a general classification showing the importance of manufacturing technology, its attainable accuracy, and economical considerations.

Chapter 2 presents the physical, mechanical, and fabricating properties of engineering materials.

Chapter 3 presents the structures of metals and alloys. The lattice structures and imperfections (point, line, surface, and volume defects) are presented. The solidification of pure metals and alloys along with the related phase diagrams and lever rule are described. The iron–carbon phase diagram is also considered.

Engineering materials are classified and discussed in Chapter 4, which deals mainly with ferrous and nonferrous metals and alloys. Steels and alloy steels are classified according to AISI and DIN standards. They are categorized according to their applications. The production of common types of cast iron is presented. Nonferrous metals and alloys, in addition to superalloys, refractory and noble metals, are surveyed. Newly developed materials, such as nanomaterials, metal foams, amorphous alloys, and shape memory alloys (SMAs) are also presented. Non-metallic materials such as polymers, ceramics, and composites are discussed in subsequent chapters.

Chapter 5 covers the basics of heat treatment operations of metals and alloys. The heat treatment of steels and alloy steels, such as annealing, normalizing, tempering, austempering, martempering, hardening, and surface hardening techniques are described. Also, the heat treatment operations of cast iron and the precipitation hardening of nonferrous alloys are considered.

Chapter 6 deals with smelting, extraction, and refining technologies of metals. Smelting and refining of ferrous metals using blast furnaces and steel refining furnaces are described. The technology of continuous casting is presented. Smelting and extraction processes of some important nonferrous metals are also covered.

Chapter 7 presents casting of metallic materials. It covers a wide variety of processes adopted by industry, including sand casting using different molding techniques, investment casting, permanent mold casting, and centrifugal casting. The furnaces used in foundries are also introduced. The basic relations controlling the process variables are presented as an introduction for process modeling.

Chapter 8 presents both the mechanical and metallurgical fundamentals of the plastic deformation of metals as a prelude to the next two chapters dealing with bulk forming and sheet forming of metallic materials. Mechanical fundamentals cover the analysis of the limiting forces, stresses, strains, and strain rates involved in the metal flow during the process as well as the associated work hardening. Metallurgical aspects include the effect of temperature and plastic deformation on the crystal structure of the material, and accordingly, on the resulting mechanical properties. The effects of friction and lubrication on the forming process are also included.

Chapter 9 introduces plastic forming processes for bulk metallic materials, including forging, rolling, extrusion, and drawing. The different techniques adopted in these processes are discussed, emphasizing the applications, advantages and limitations of each, as well as the utilized equipment and tooling. Modeling of these processes is also included by using simple mathematical relations based on the element equilibrium approach to train students on estimating the limiting stresses, strains, strain rates, forces, and consumed power. These calculations are applied in designing the required dies and selecting the right capacity for equipment. Solved examples are given to support appreciation of how to apply the driven and commonly used mathematical relations.

Chapter 10 presents sheet metal forming processes, including shearing, bending, deep drawing, spinning, and superplastic forming of sheet blanks. Similar to Chapter 9, a variety of techniques involved in these processes are discussed, stressing the applications, tooling, and equipment as well as the advantages and limitations of each. The limiting characteristics of sheet metals such as the spring back, anisotropy, limiting-drawing ratio, and residual stresses are discussed to a reasonable depth to convey the limits of sheet processing. The chapter ends with an introduction of the technological characteristic of sheet formability, the standard tests for its measurement, and the forming limit diagrams for determining the limiting conditions for sheet forming applications.

Chapter 11 introduces the nonconventional family of forming processes based on developing high-velocity forming hammers (high-velocity forming, HVF) or adopting other physical techniques for sudden release of energy (high–energy rate forming, HERF). The presented processes include high velocity forming, explosive forming, electromagnetic forming, and electrohydraulic forming. The principles, equipment and tooling, applications, advantages and limitations are discussed for each process. The basic mathematical relations serving the objectives of process understanding are presented without details that are beyond the scope of the book.

Chapter 12 presents powder metallurgy as an alternative manufacturing technique that was first developed to form high precision (net shape) metallic products. The procedure involves producing metals first in the form of powder using different methods, and then consolidating powders into a solid form by the application of pressure (compaction), and heat (sintering) at a temperature below the melting point of the main constituent. Powder metallurgy was later adopted as a major process for manufacturing advanced ceramic products using the same steps. Therefore, it is crucial in this chapter to introduce ceramic materials and discuss the methods used in their manufacture, including powder metallurgy procedures.

Chapter 13 presents polymeric materials and their processing techniques. The definition and polymerization reactions are first introduced. Due to the different nature and thermomechanical behavior of polymers compared to metals, their structure and properties are discussed for the three classes of polymeric materials, namely, thermoplastics, thermosettings and elastomers. The presented manufacturing processes include extrusion, calendaring, spinning, injection molding, compression molding, transfer molding, thermoforming, casting, foaming, and joining of plastics.

Chapter 14 introduces composite materials, their classes and the main characteristics of their constituents, emphasizing their applications. Then their major manufacturing processes are presented. These processes include the different types of molding, prepeg fabrication, filament winding, and pultrusion.

The fundamentals of traditional machining are presented in Chapter 15. The mechanics and kinematics of chipping processes are investigated. Tool geometry, tool material, tool life, machining economy, and machinability are discussed. This chapter also considers the thermal aspects of chipping operations. The basics of abrasion processes, such as grinding, honing, superfinishing, and lapping along with mechanics and kinematics of the grinding operation are considered.

Chapter 16 presents the general purpose machine tools used for cutting cylindrical surfaces such as lathe, boring, and drilling machines. Additionally machine tools used for cutting flat surfaces such as shapers, planers, and slotters, and milling and broaching machines are given. It presents surface and cylindrical grinding machines as well as honing, superfinishing, and lapping machines normally used for microfinishing operations. Machines used for thread and gear cutting and finishing are also described.

Chapter 17 describes the fundamentals of nontraditional machining processes. Mechanical nontraditional machining is represented by jet machining and ultrasonic machining (USM). Chemical milling, spray etching, electrochemical machining (ECM), and electrochemical grinding (ECG) are also described, along with the thermal nontraditional processes represented by electrodischarge machining (EDM), electrodischarge wire cutting (EDWC), ED-milling, laser beam machining (LBM), electron beam machining (EBM), and plasma beam machining (PBM). Related machines and equipment, their elements, and accuracies are given.

Chapter 18 provides the CNC concepts, movements in CNC systems, control of CNC machine tools, types of CNC machines, and features of CNC systems. It provides the fundamentals of part programming using manual, computer-assisted, and CAD/CAM methods.

Chapter 19 introduces robotics and hexapods and their role in manufacturing technology. The basic elements of industrial robots and hexapods are examined. The economical features and characteristics of robots and hexapods are presented.

Chapter 20 presents the different surface technologies, which include smoothing, cleaning, protection, deburring, and roll burnishing and ballizing. The surface protection methods are sacrificial and direct protection. The latter comprises conversion coating, electroplating, organic coatings, vaporized metal coating (PVD and CVD), metalizing and cladding, all of which are presented in this chapter.

Chapter 21 covers fusion welding operations such as gas welding, thermit welding, and methods of electric arc welding. Methods of resistance spot, projection, seam, and flash welding are introduced. High energy beam welding by electron and laser beams are given. The chapter covers the metallurgy of welded joints, welding defects, and welding quality control. Solid state and solid-liquid state welding techniques, in addition to welding of plastics and mechanical joining, are also explained.

Chapter 22 describes the concept of near net shape manufacturing, microfabrication technology, and nanotechnology. Semiconductor device fabrication, testing, assembly, and packaging are explained. It also introduces the concept of sustainable and green manufacturing.

Function, material, process, and shape interaction are given in Chapter 23. Manufacturing process capabilities and their selection through process information maps and elimination and ranking strategy are given. Design for manufacturing is presented by setting design rules and recommendations for many manufacturing technologies such as casting, sheet metal forming, die forging, machining and welding, as well as assembly operations and design for environment.

Chapter 24 introduces the principles of statistical quality control, control charts, control limits and specifications, process capability as well as acceptance sampling. Concepts of total quality control and the ISO 9000 Standard are also covered. Dimensional control, interchangeability, tolerance and fit are discussed. Measuring tools and equipment commonly used for measuring quality characteristics such as limit gauges and those used for dimensional, angular, and geometric measurements are described. Surface measurements, nondestructive testing and inspection, and destructive testing are introduced.

In Chapter 25, automation in manufacturing technology is presented. The difference between automation and mechanization is clearly defined. The necessity of adopting automation in manufacturing is discussed. Automation is realized in the form of MSs, FMSs, CAD, CAM, CAPP, CIM, lean production, AC, AI, CNC, robotics, and hexapods.

Chapter 26 covers the issues of health and safety at work, different sources of manufacturing hazards, and necessary personal protective equipment. It focuses on the different hazards associated with several manufacturing processes such as melting of metals, sand casting, welding, metal forming operations, and machining by traditional and nontraditional techniques.

Many solved examples are introduced in the text to make students aware of the importance of the relevant topics. At the end of each chapter, review questions and problems are provided. Individuals desiring additional information on specific items of the book are directed to the various references listed at the end of each chapter.

ADVANTAGES OF THE BOOK

- Introduces new trends in surface hardening technology
- Introduces the newly developed materials such as nanomaterials, memory shape materials, amorphous alloys, metal foams, advanced ceramics, and composite materials. It also introduces smart materials and strategic engineering materials
- Provides selection guidelines for engineering materials and manufacturing processes
- Presents the principles of design for manufacturing
- Presents the principles of CNC, robotics, and hexapods as well as their application in manufacturing technology
- Covers the fundamentals of traditional and nontraditional machining processes
- Presents nontraditional forming processes such as high velocity and high energy rate forming
- Presents the technologies of surface treatment
- Presents the ultrasonic technology and its applications in manufacturing industries.
- Introduces the safe technologies for manufacturing toxic materials such as beryllium and asbestos
- Provides the new trends in centrifugal casting
- Presents nano- and micromachining technologies
- Explains the concept and environmental aspects of manufacturing and clean factories
- Explains the principles of near-net shape and net-shape processing and rapid prototyping
- Presents the surface characteristics due to manufacturing processes
- Introduces the recently developed advanced ceramics and their latest processing techniques
- Presents novel composite materials and their fabrication
- Introduces the superplastic phenomenon in some metallic alloys and the related superplastic forming processes
- Cover the manufacturing topic in a simple and descriptive way.
- Covers new topics that are not mentioned in earlier books.
- Presents engineering materials, processes, tools, and equipment used in manufacturing.
- Covers the basic as well as the most recent advanced manufacturing technologies.

Acknowledgments

Many individuals have contributed to the development of this textbook. It is a pleasure to express our deep gratitude to Professor E. M. Abdel-Rasoul, Mansoura University, Egypt, for supplying valuable materials during the preparation of this book. The assistance of Nagham Elberishi, Production Engineering Department, Alexandria University, and Saeid Teileb of the Lord Alexandria Razor Company for their valuable AutoCAD drawings is greatly appreciated. Thanks and apologies to others whose contributions have been overlooked.

We appreciate very much permission from many publishers to reproduce illustrations and tabulated data from a number of authors as well as the courtesy of many industrial companies that provided photographs and drawings of their products to be included in the book. Their generous cooperation is a mark of sincere interest in enhancing the level of engineering education. The credits for all such great help are provided in the captions under the corresponding illustrations.

We would like to acknowledge with thanks the dedication and continued help of the editorial board and production staff of CRC Press for their effort in ensuring that the book is as well designed as possible.

Last, but not least, we sincerely appreciate the support, great patience, encouragement, and enthusiasm of our families during the preparation of the manuscript.

Authors

Helmi A. Youssef earned his BSc with honors in production engineering from Alexandria University in 1960. He then completed his scientific degree in Carolo-Wilhelmina during the period 1961–1967. In June of 1964, he acquired his Dipl-Ing, then in December of 1967, he completed his Dr-Ing in the domain of nontraditional machining. In 1968, he returned to Alexandria University's Production Engineering Department as an assistant professor. In 1973, he was promoted to associate, and in 1978, to full professor. In the period of 1995–1998, Professor Youssef was the chairman of the Production Engineering Department at Alexandria University. Since 1989, he has been a member of the scientific committee for promotion of professors in Egyptian universities.

His experience extends to include topics related to machining technology such as theories of metal cutting, machine tools, automatics, gear cutting, tool design, jigs and fixtures, NC and CNC machines, automation in production technology, and theories and technologies of nontraditional machining.

Based on several research and educational laboratories that he has built, Professor Youssef founded his own scientific school for both traditional and non-traditional machining technologies. In the early 1970s, he established the first NTM research laboratory in Alexandria University, and perhaps in the whole region. Since then, he has carried out intensive research in his fields of specialization and supervised many PhD and MSc theses.

Between 1975 and 1995, Professor Youssef was a visiting professor in Arabic universities, such as El-Fateh University in Tripoli, the Technical University in Baghdad, King Saud University (KSU) in Riyadh, and Beirut Arab University (BAU) in Beirut. In addition to his teaching activities in these universities, he established laboratories and supervised many MSc theses. Moreover, he was a visiting professor in different academic institutions in Egypt and abroad.

Professor Youssef has organized and participated in a number international conferences. He has published many scientific papers in specialized journals, and has authored books in his fields of specialization, two of which are single-authored. One book, which he coauthored on machining technology, was published in 2008 by CRC Press. Currently, he is an emeritus professor in PED, Alexandria University.

Hassan A. El-Hofy received a BSc in production engineering from Alexandria University (Egypt) in 1976 and then served as a teaching assistant in the same department. He received an MSc in production engineering from Alexandria University in 1979 under the supervision of Professor H. Youssef. Professor El-Hofy has had a successful university career in education, training, and research. Following his MSc, he worked as an assistant lecturer until October of 1980, when he went to Aberdeen University in Scotland and began his PhD work with Professor J. McGeough in hybrid machining processes. He won the Overseas Research Student (ORS) Award during pursuit of his doctoral degree, which was completed in 1985. He then went back to Alexandria University and resumed his work as an assistant professor. In 1990, he was promoted to an associate professor. He was on leave as a visiting professor for Al-Fateh University in Tripoli between 1989 and 1994.

In July of 1994, Professor El-Hofy returned to Alexandria University, and in November of 1997, he was promoted to a full professor. In September of 2000, he was selected to work as a professor in the University of Qatar. He chaired the accreditation committee for the Mechanical Engineering program toward ABET Substantial Equivalency Recognition that was granted to the College of Engineering programs in 2005. Due to his role, he received the Qatar University Award and a certificate of appreciation. Professor El-Hofy wrote his first book, *Advanced Machining Processes: Nontraditional and Hybrid Processes*, which was published by McGraw-Hill Company on March 1, 2005. His second book, *Fundamentals of Machining Processes*, was published in September of 2007 by CRC, Taylor & Francis. He coauthored his third book *Machining Technology–Machine Tools and Operations* with Professor Youssef, published by CRC in 2008. He has published over 50 scientific and technical papers and supervised many graduate students in the area of machining by nontraditional methods. He is a consulting editor to many international journals and a regular participant in international conferences.

Between August 2007 and August 2010, he was a professor and chairman of the Production Engineering Department of Alexandria University, College of Engineering, where he taught several machining and related courses.

Mahmoud H. Ahmed received a BSc in production engineering from Alexandria University in 1970, with First Degree of Honors. He was assigned as an instructor in the same department, where he obtained an MSc in 1973. Accordingly, he was promoted to the position of assistant lecturer. In 1974, he was granted a scholarship from the University of Birmingham, United Kingdom, to study for the PhD degree in the Department of Mechanical Engineering. He pursued his research in the field of shearing of metals until he obtained his PhD in 1978. During that period, he contributed to the teaching effort in the department as a teaching assistant.

In 1978, he returned to his homeland and resumed work at Alexandria University as an assistant professor. He left for the United Arab Emirates, on secondment, to work for the UAE University over the period of five years, from 1982 to 1987. He was promoted to the position of associate professor at Alexandria University in 1986 while working at the UAE University and returned back home one year later. He was assigned, for another secondment, to King Abdul Aziz University in Saudi Arabia starting in 1997. He returned home again to Alexandria University in 2002, where he has been working since. In addition to the long-term secondments, Professor Ahmed worked as a part-time visiting professor for the Arab University of Beirut, Lebanon (for 3 months in 1980/1981 and for 2 months in 1981/1982), Qatar University (for one semester, in 1995), and many Egyptian universities, including The Arab Academy for Science and Technology-Alexandria (1995–1997, 2002–2007), El-Mansoura University (1978–1981), Kima High Institute of Technology-Aswan (1979–1982), High Institute of Public Health-Alexandria (1980–1982), El-Minia University (1978–1982), and El-Menoufia University (1981–1982).

During his career, Professor Ahmed has constructed and taught numerous graduate and undergraduate courses in the general fields of materials and manufacturing: failure analysis, material selection, finite element analysis, fracture mechanics, non-destructive testing, advanced manufacturing processes, theory of plasticity, solid mechanics, die design, metal forming, metal cutting, nonconventional machining, welding technology, engineering materials, and manufacturing technology, to name a few. Professor Ahmed took part in establishing and developing laboratories in the same fields, including the Material Technology Laboratory (Alexandria University), Material Testing, Forming Machines, CNC Machining, Metrology, and Electroplating Laboratories (UAE University), as well as Nonconventional Machining and CNC Machining Laboratories (King Abdul Aziz University, KSA). Over the years, Professor Ahmed has supervised numerous MSc and PhD degrees, covering the areas of electrodischarge machining, failure of welded joints, extrusion of fluted sections, plasma cutting, ultrasonic machining, pulsed current MIG welding, compression of tubular sections, forward tube spinning, characterization of engineering materials using nodal analysis, selection of nontraditional machining processes, and thermomechanical rolling. Professor Ahmed has a record of publishing in national and international conferences and reputable journals. He has also contributed to the development and improvement of industrial activities within Alexandria through consultations related to solving design and manufacturing problems, material and product inspection, failure analysis, plant layouts, feasibility and opportunity studies, as well as running crash and training courses in the relevant fields of interest.

1 Introduction to Manufacturing Technology

1.1 IMPORTANCE OF MANUFACTURING TECHNOLOGY

The word *manufacture* was derived from its Latin origin: *manu factus*, i.e., to make by hand. Today, however, manufacturing is mainly done by machinery. Manufacturing technology is the largest sector of modern industry and embraces many branches of industrial activities, such as the food, drug, machinery, weapon, textile, leather, and shoe industries. These technologies are difficult to be contained in one book. To be limited, this book must be confined to technologies that are basic to all industries.

Manufacturing is the art of processing materials. It involves the use of machines, tools, and labor to convert the raw materials, usually supplied in simple or shapeless forms, into finished products with specific shape, structure, and properties designed to fulfill specific consumer needs. It includes human activities ranging from handicraft to the use of high technology. Manufacturing is most commonly applied to industrial production, where raw materials are transformed into finished goods on a large scale. In a free market economy, manufacturing is usually directed toward the mass production of products for sale to consumers at a profit. Some industries, such as semiconductor and steel manufacturers, use the term *fabrication* instead.

Manufacturing technology provides the tools that enable production of all manufactured goods. These master tools of industry magnify the effort of individual workers and give an industrial nation the power to turn raw materials into the affordable, quality goods essential to today's society. Manufacturing technology provides the productive tools that power a growing, stable economy, and raises the standard of living. These tools make possible modern communications, affordable agricultural products, efficient transportation, innovative medical procedures, space exploration, and the everyday conveniences we take for granted.

Manufacturing is a value-adding activity, where the conversion of materials into products adds value to the original material. A well-designed manufacturing system is achieved through minimizing waste and maximizing efficiency. These goals are achieved if the manufacturing processes are properly selected and arranged to permit smooth and controlled flow of material through the factory and also by providing an optimal degree of product flexibility.

Production tools include machine tools and other related equipment and their accessories and tooling. Machine tools are nonportable, power-driven manufacturing machinery and systems used to perform specific operations on man-made materials to produce durable goods or components. Related technologies include computer-aided design (CAD) and computer-aided manufacturing (CAM) as well as assembly and test systems to create a final product or subassembly.

Advanced manufacturing involves the industries that increasingly integrate new innovative technologies into both products and processes. In other words, advanced manufacturing entails rapid transfer of science and technology into manufacturing products and processes. Examples of major manufacturers in North America include General Motors Corporation, General Electric, and Pfizer. Examples in Europe include Volkswagen Group, Siemens, and Michelin. Examples in Asia include Toyota, Samsung, and Bridgestone.

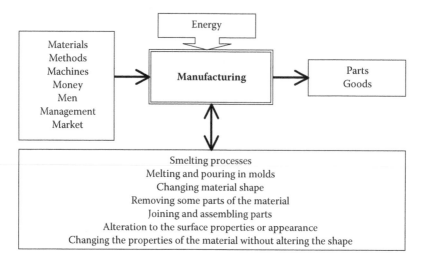

FIGURE 1.1   Manufacturing system.

Figure 1.1 shows typical manufacturing system components. The major activities required for changing the raw material into a usable form of products include the following:

1. Smelting processes
2. Melting and pouring in mold
3. Changing material shape
4. Removing parts of the material
5. Joining and assembling of parts
6. Altering the surface properties or appearance
7. Changing material properties without altering shape

In order to make sure that parts are manufactured according to the design requirements, measurements and inspection are necessary. As a result of manufacturing operations, variations in product quality are inevitable, and can be attributed to either natural or assignable causes. Variations due to natural causes are inherent in the manufacturing process, are difficult to identify, uneconomical to eliminate, and do not follow statistical laws. On the other hand, assignable variations are due to individual causes that can be identified and eliminated.

In this respect, quality control (QC) is a set of procedures intended to ensure that a manufactured product adheres to a defined set of quality criteria, or meets the requirements of the customer. Total quality control (TQC) is a system approach where the management and workers make an integrated effort to manufacture high quality parts consistently. Total quality management (TQM) is a management approach for an organization, centered on quality, based on the participation of all its members, and aimed at long-term success through customer satisfaction, and benefits all members of the organization and society.

1.2   SELECTION OF MATERIALS AND MANUFACTURING PROCESSES

Materials are substances whose properties make them useful in structures, devices, or products. They include metals, ceramics, glasses, semiconductors, dielectric, fibers, wood, sand, stone, and many composites. The number of available materials is almost infinite. The various compositions of steel alone run into the thousands. There are more than 10,000 varieties of glass, and plastics are equally great. In addition, several hundred new varieties of materials appear on the market each month. This means that the selection of an appropriate engineering material to manufacture a product is a tedious task, since each product requires a material with specific characteristics. Therefore, the manufacturing engineer must have sufficient knowledge to select the optimum material for each application, taking into consideration material availability, processing requirements, economical and environmental aspects, in addition to the design and safety considerations which match the part to the system as a whole.

1.2.1 Selection of Materials for Manufacturing

Each material has its own characteristics and advantages to be manufactured using a specific manufacturing process. Manufacturing characteristics of materials (or manufacturability) typically include casability, workability, formability, machinability, grindability, weldability, and ability to undergo heat treatment. The selection of materials for manufacturability has become easier and faster because of the availability of computerized databases. The following aspects should be carefully considered when selecting a material for manufacturing.

- The selected material can be replaced by others that are less expensive.
- The material has properties that unnecessarily exceed the minimum requirements of the product.
- The selected material has appropriate manufacturability to be processed by the selected process.
- The material to be ordered for manufacturing a product is available in the required size, dimensions, tolerances, and surface finish.
- The material supply is reliable without significant price increase, and supplied with the required quantity in the desired time without delay.

1.2.2 Selection of Manufacturing Processes

The selection of a manufacturing process is governed by some important considerations, such as the properties of material processed, geometrical configurations and size of the product and its accuracy and surface quality, the functional requirements of the product, the batch size, level of automation, and the manufacturing cost.

The following aspects should be considered when selecting a manufacturing process.

- All alternative manufacturing processes have been investigated. A part can often be made by several techniques.
- The part is formed to the final dimensions without additional processes.
- The tooling required is available in the plant, specially manufactured, or purchased as a standard item.
- The scrap produced is minimized and recyclable.
- The process parameters are optimized.
- The inspection tools required are available.
- The inspection techniques and quality control are being implemented properly.
- All components of the product are manufactured in the plant, or some of them are available as standard items from external sources.

1.2.3 Classification of Manufacturing Processes

Basic manufacturing processes can be classified as shown in Figure 1.2

1. Smelting and casting processes
2. Plastic forming processes for metallic materials

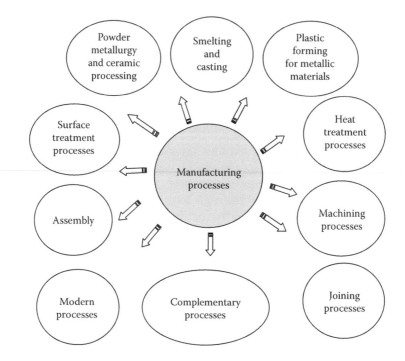

FIGURE 1.2 Classification of manufacturing processes.

3. Powder metallurgy and ceramic processing
4. Machining processes
5. Joining processes
6. Surface treatment processes
7. Heat treatment processes
8. Assembly
9. Others

1. Smelting and casting processes: Casting is the oldest known manufacturing process for manufacturing of metallic products. It involves heating the metal to a temperature higher than the melting point to change it to the liquid state. When the molten metal is poured into a mold having a predetermined shape, it fills the mold cavity. Then the metal is left to cool down inside the mold where it solidifies, keeping the geometrical configurations of the mold cavity. The process is called *smelting*, or *primary casting*, when the objective is to produce a block of metal or alloy with a required chemical composition at a required level of purity. In this case, the block usually has a simple shape such as ingots or slabs. When the objective is to get a required product shape, the process is known as *secondary casting* or simply *casting*, and the product itself is known as a casting.

The mold represents the main element of a casting process. It should be suitably constructed to provide the required geometrical configurations of the shape to be produced, and to allow for extraction of the product after solidification. Meanwhile, it should provide a gating system for leading the molten metal to the mold cavity at a convenient flow rate and to get rid of the air in the gap and the gases entrapped in the melt. The mold material and structure should withstand the high temperature of the molten metal and the thermal gradient during cooling.

Casting processes are generally classified into two major categories, expendable mold casting and permanent mold casting. The first category, based on crushing the mold after casting to extract the product, is subclassified to sand casting, with different techniques of sand molding, shell molding, vacuum molding, plaster molding, ceramic molding, investment casting, and evaporative mold casting. The second category of permanent mold casting is subclassified to die casting processes, where the mold is made of die halves that can be opened to extract the product, and centrifugal casting using rotating molds.

The furnaces used for melting are also numerous, depending on the type and melting point of the metallic material, the amount of metal to be melted, the purity or degree of control of the material constituents, and the production type, whether being continuous or in batches. Furnaces used in casting shops include the cupola, crucible furnace, electric arc furnace, and induction furnace.

2. Plastic forming processes for metallic materials: Plastic forming processes involve applying mechanical force to a metallic material beyond the yield stress limit to reach the plastic stress–strain conditions, causing permanent deformation of the material. The force is usually applied through a specially designed tooling system that ensures that the material is deformed to the required shape. Plastic deformation is usually associated with changes in the mechanical properties and crystal structure of the material. Accordingly, the objective of plastic forming processes is not only attaining the required shape, but also accomplishing the required mechanical properties of the product. Therefore, it is a common practice in metal manufacturing industries that all products of smelting processes (ingots or slabs) should go through a series of primary plastic forming processes to reach intermediate shapes such as billets, standard sections, plates, sheets, rods, or wires with the required dimensions and mechanical properties. These intermediate products can be used directly as parts of products, or they may go through further plastic forming processes (known as secondary processes) to reach a final product.

To study metal forming processes, it is essential to first study the effect of forces and stresses on the deformation, strain, and strain rate of the metallic material in both the elastic and plastic ranges. The associated strain hardening and the limiting stresses that cause failure are also essential. The effects of these stresses and strains on the crystal structure of the material and its mechanical properties should also be considered. The effect of temperature on these parameters should also be studied, as it is known that heating the metal above the recrystallization temperature improves metal flow in the plastic range and reduces the required forces. Heating is only performed in processes involved in large strains (hot forming processes), since the heat affects the tool life and leads to a relatively soft product.

The majority of forming processes deal with products that have a large thickness compared with other dimensions. These are known as bulk forming processes. They include four major types of processes, which are forging processes, rolling processes, extrusion processes, and rod, wire, and tube drawing processes. In forging processes, a metal billet or block is compressed between die halves, using a press or a hammer. When the die halves are flat or simple shaped, the process is known as open die forging, whereas forging in dies comprising the full geometrical configurations of the product when fully closed is known as closed die forging. Further, there are special forming processes including rotary swaging, radial forging, orbital forging, and coining. Rolling processes are based on passing the material between two rotating rolls to squeeze it to the required shape. Rolls may be cylindrical to produce flat sections such as plates, sheets, or foils. Formed or profiled rolls are used to produce profiles such as the standard sections (rods, bars, tubes, I, T, L, and U sections). Extrusion processes include compression of metal billets into a container with a die of the required shape at the end. The compressed metal passes through this die, producing long rods, bars, tubes, or sections with any required configurations. There are also special extrusion processes such as hydraulic extrusion, porthole die extrusion, as well as cold and impact extrusion processes. In rod, wire, and tube drawing, a previously formed section is pulled through a die under tensile force to reduce the section size.

Sheet metal forming processes are presented due to the special characteristics of sheets that have large area-to-thickness ratio. All sheets are manufactured by flat rolling processes, which may generate anisotropic characteristics. This anisotropy may lead to thinning of the sheet blank during forming, causing failure. Further, there are other limiting characteristics in sheet forming processes such as springback and residual stresses due to bending, and wrinkling and drawing limit during deep drawing. These characteristics necessitate special design considerations during the design of sheet forming tools. Sheet forming processes have been highly developed during the past few decades to cope with the vast expansion of related industries such as automotive, aerospace, electronics, household appliances, and utensils. These processes can be generally classified to shearing processes, bending processes, deep drawing, spinning, and superplastic forming processes. In shearing processes, high shearing stresses are applied on a specific contour on the sheet surface, leading to separation along the thickness of the contour. Bending of sheet blanks and strips produces a wide variety of shapes. Deep drawing forms sheet blanks into axial symmetric cup shapes. Spinning processes form sheet blanks into cup shapes or increases the length (and strength) of tubular parts. Superplastic deformation is applied to special alloys capable of extremely large elongation under special forming conditions. All of these processes are presented, emphasizing the advantages, limitations, and design considerations for each process. The limits of sheet formability and the standard tests for this basic characteristic are also discussed.

In the past 50 years, it has been established that increasing the speed of deformation during plastic forming improves formability and uniformity of strain distribution, and reduces the cost of tooling and some process defects. This has led to the development of nontraditional forming methods known as high velocity forming (HVF), and high energy rate forming (HERF). In HVF, the forming speed is increased by development of high speed hammers such as compressed air (pneumatic) hammers, compressed gas hammers, and gas combustion hammers. These hammers increased the speed from 3 to 5 m/s in conventional forming, up to 60 m/s. HERF processes are based on generating high energy at very short times through detonation of an explosive charge (explosive forming), interaction of magnetic fields (electromagnetic forming, or EMF), and sudden electric charge in a fluid (electrohydraulic forming, or EHF). These nontraditional forming processes are commonly applied for sheet, plate, and tubular shapes often with a large size that could be very costly, or even impossible to form using conventional forming processes.

3. Powder metallurgy and ceramic processing: Powder metallurgy (PM) is a processing technology for manufacturing metallic materials by producing them first in the form of powder, and then consolidating them into a solid form by compaction (applying pressure), and sintering (heating below the melting point). The process started in ancient times as an alternative method for producing metals with a high melting point, which were difficult to melt. However, it has been developed to be a widely used, cost effective processing technique for producing high precision (net shape) metallic materials, even those with a low melting point. It has further extended to produce a broad range of the recently developed advanced ceramics such as oxides, carbides, nitrides, and borides. The process can also produce alloys with immiscible phases when applying solid solution melting techniques, and conventionally impossible material combinations such as graded structures and composite metals. PM products offer higher strength and corrosion resistance, higher material saving, higher productivity, and unmatched design flexibility. The adoption of PM processing has led to the development of modern processing techniques, including chemical vapor deposition (CVD), directed metal oxidation, reaction bonding, sol-gel processing (converting a solution of metal compounds or a suspension of fine particles into a viscous mass known as gel), polymer pyrolysis, and melt casting. Details of PM processing techniques and the recently developed ceramic processing techniques are presented.

4. Machining processes: The term *machining* means all methods of shaping parts by removing selected volumes of work material called machining allowance. In other words, it is the controlled removal of unwanted material, so that what is left is an item having the required geometry. Machining is generally used as a final finishing operation for the products roughly produced by other manufacturing processes like casting and metal forming, before these products are ready for use. However, machining processes are generally hampered with the following disadvantages as compared with casting and forming processes.

- They need a longer time.
- They are energy, capital, and labor consuming.

- They need highly qualified personnel.
- They necessitate sophisticated measuring tools.

The machining processes are generally classified into traditional (TMPs) and nontraditional (NTMPs) machining processes (Figure 1.3).

A. Traditional machining processes, in which chips are formed by the interaction of a cutting tool with the material being machined. These processes employ traditional tools of a basic wedge form to penetrate into the workpiece. These tools must be harder than the material to be machined.

TMPs comprise two categories
1. Cutting: chipping processes that use tools of definite geometry such as turning, planing, drilling, milling, broaching, etc.
2. Abrasion processes that use tools of nondefinite geometry such as grinding, honing, lapping, etc.

B. Nontraditional machining processes, in which the machining energy is utilized in its direct form. These processes are less familiar, and are desired to meet the increasingly difficult demands for which TMPs cannot be used.

NTMPs also comprise two categories.
- Abrasion processes, where the mechanical energy is used. These processes include USM, WJM, AJM, AWJM, etc.
- Erosion processes using chemical and electrochemical energy such as CHM, ECM, etc. or using thermal energy such as EDM, LBM, EBM, PBM, etc.

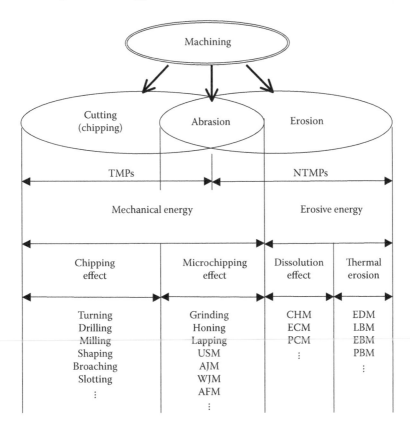

FIGURE 1.3   Classification of machining processes to traditional and nontraditional machining.

5. Joining processes: Metal joining is frequently used for manufacturing products through temporary joining of metal parts, which allows the parts to be disassembled for maintenance and repair purposes. On the other hand, permanent joining of materials can be achieved using fusion welding, which melts the ends of the joint so that the molten metal fuses, forming the welding joint when solidified (Figure 1.4). Some of these methods use chemical reactions for generating the heat necessary for melting the ends of the joint, as in gas and thermite welding methods. Electric arc is another source of heat required in arc welding, whereas resistance as a source of heat is used in resistance welding. High energy beam welding employs laser beam or electron beam in fine welding applications. Solid state welding without metal fusion, solid–liquid state and plastics welding cover a wide range of industrial application. The metallurgy of welded joints, welding defects, and nondestructive testing methods provide the manufacturing engineer with quality control applications for welding technology.

6. Surface treatment processes: After a component is manufactured, its surfaces, totally or partially, may have to be treated in order to impart desired characteristics, using one or more surface treatment techniques, based on mechanical, chemical, thermal, or physical methods. These techniques include

- Smoothing: burnishing and electroplating
- Cleaning: peening, tumbling, ultrasonic, and chemical cleaning
- Protection: sacrificial (galvanic), and direct protection such as coating, metal spraying, and cladding
- Hardening: ballizing and roll burnishing
- Deburring: rounding sharp-edged surfaces

7. Heat treatment processes: The microstructure of metals and alloys developed during smelting or other manufacturing processes can be modified by heat treatment techniques, which involve controlled heating and cooling of alloys at various rates. These techniques induce phase transformation that greatly influences the mechanical properties of alloys. Heat treatment techniques depend mainly on the composition and microstructure of the treated alloy, degree of prior cold work, and the final mechanical properties (strength, ductility, hardness) desired, to cope with the next manufacturing process intended to be performed. The processes of recovery, recrystallization, and grain

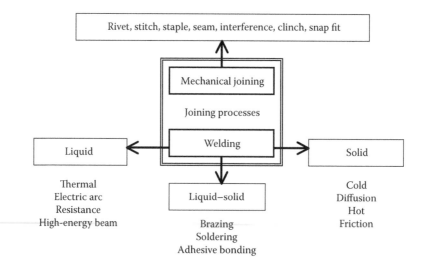

FIGURE 1.4 Joining processes.

growth, along with the basic heat treatment processes such as annealing, normalizing, hardening, surface hardening, etc., are common examples of heat treatment processes.

8. Assembly: Products are designed and manufactured with ease of assembly in mind. In this respect, a product containing fewer parts takes less time to assemble, thereby reducing the assembly costs. Similarly, parts that are easier to grasp, move, orient, and insert reduce assembly time and costs. Therefore, the major cost benefit of the application of design for assembly is achieved through the reduction of the number of parts in an assembly. Assembly methods include manual assembly, where workers manually assemble the product or its components using hand tools. Although it is a flexible and adaptable method, it is limited to small production volume and is associated with high labor costs. Fixed (hard automation/Detroit-type) assembly uses machines and feeders that assemble a specific product. This machinery requires a large capital investment which is justified by the large production volume. Soft automation (robotic) assembly incorporates a single robot, or a multistation robotic assembly cell with all activities simultaneously controlled and coordinated by a programmable logic controller (PLC) or a computer. Although this type of assembly requires large capital costs, its flexibility helps offset the expenses across many different products.

9. Modern manufacturing processes: Modern manufacturing technologies include the following.

1. Near net shape is a manufacturing technique that produces items that are very close to their final (net) shape. It reduces the need for traditional finishing operations and eliminates about two-thirds of the production costs.

2. Due to the high capability of machining processes in terms of accuracy and surface quality, as compared with other manufacturing processes, microfabrication is restricted to micromachining operations. Their applications include semiconductor devices, compact electrical circuits, and integrated circuit packages containing devices of microdimensions.

3. Nanotechnology or molecular manufacturing is the science of manufacturing materials and machines at the nanometer, or atomic/molecular scale. Parts can be made by moving atoms and molecules into the desired arrangements until the required product is finished. It is a clean and cheap production method for the finest computer processors and food.

4. Semiconductor device fabrication is used to create chips and integrated circuits for electronic devices. It consists of a series of steps that deposit special layers of materials on wafers in precise sequence, amounts, and patterns. Accordingly, photographic and chemical processing techniques gradually create electronic circuits on a wafer made of semiconducting material such as silicon.

5. Sustainable/green manufacturing is the production of parts that use nonpolluting processes, conserve energy and natural resources, and are economically sound and safe for employees, communities, and consumers. It reduces cost, enhances process efficiency, and develops new eco-friendly products.

10. Complementary processes: In addition to the previously mentioned basic processes, there are some complementary activities which are very essential to finishing the manufacturing duties. These activities include

1. Inspection, which determines whether the desired quality has been achieved.

2. Testing, which determines if the product is functioning according to the designer's objectives. It includes service tests, destructive and nondestructive testing, and so on.

3. Material and product handling in the factory. It includes loading, positioning, and unloading. Automatic handling in today's manufacturing is essential. In this respect, there is a trend to replace forklift trucks with industrial robots and conveyors.

4. Packaging, which prepares the products for delivery to the customer. Weighing, filling, sealing, and labeling are packaging activities that are highly automated in many industries.

5. Storage is another manufacturing activity in which nothing happens to the product or raw material except the passage of time. However, it is sometimes costly. Therefore, it is a good practice to keep the storage time as short as possible, and to eliminate the storage during processing.

1.3 COMPANY ORGANIZATION

Referring to Figure 1.5, the manufacturing concerns at the company level include finance, personnel, sales and marketing, research and development, and product and manufacturing departments. At the plant level, all aspects of production are under the direction of the plant manager. These aspects include the following groups.

1. Process engineering group: Typical tasks of this group include evaluation of manufacturing feasibility and cost, selection of optimum processes and process sequence, production equipment, tooling, jigs and fixtures, material handling methods and equipment, and plant layout. One of its main jobs is writing any computer-controlled machine tool programs that may be required. This group also issues specifications for auxiliary and finishing processes, treating individual parts, and finished assemblies. Cost and value-analysis is an important task of this group. This group cooperates closely with research and development specialists and delegates its members to concurrent engineering (CE) teams. In the present environment, personnel of this group must understand processes and their control and must be versed in materials, mechanics, electronics, computers, and system analysis.

2. Methods and work standard groups: Its main task is the provision of time and motion studies and norms. Sometimes cooperation with the process engineering group is necessary.

3. Inspection and quality control group: Due to its high vitality, its data must be directly reported to high levels of management, while also continuously integrated into production.

4. Production planning and control group: It determines the economical lot sizes and establishes schedules for manufacturing and assembly. It performs computer-aided process planning (CAPP) and material requirement planning (MRP) based on the master production schedule, the bill of materials and inventory records of raw materials, purchased

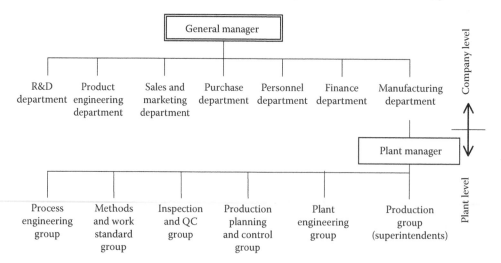

FIGURE 1.5 Company organization chart according to North American practice.

components, parts to be delivered, in-process materials and parts, finished products, tools and maintenance supplies.

5. Plant engineering group: It is responsible for preventive maintenance of equipment, replacement of machinery, and provision of services (power, heating, etc.). Since downtime results in large economic losses, maintenance of computerized equipment and robotics is particularly critical. Total production maintenance (TPM) keeps the equipment running at all scheduled times.

6. Production group: It is headed by the manufacturing superintendent, with the assistance of foremen. Sometimes the several levels of supervision are eliminated by highly trained technicians provided with broad responsibilities and authorities.

1.4 ROLE OF COMPUTERS IN MODERN MANUFACTURING TECHNOLOGY

Since the second industrial revolution in the 1940s, automation has been accelerated because of rapid advances of control systems for machines and computer technology. Computers are now used in many applications, including control and optimization of manufacturing processes, material handling, assembly, automated inspection, inventory control, and several management activities. Beginning with computer graphics, CAD, and CAM, the use of computers has been extended to computer-integrated manufacturing (CIM). In CIM, the traditional separate functions of product design, research and development, production, assembly, inspection, and quality control (QC) are limited. CIM should be considered as a methodology and a goal rather than merely an assemblage of equipment and computers. CIM is not applicable for small- and medium-size companies. The major applications of computers in manufacturing covered in the present book include adaptive control (AC), computer numerical control (CNC), industrial robotics (IR), hexapods, automated handling, computer-aided process planning (CAPP), group technology (GT), just-in-time production (JIT), flexible manufacturing systems (FMSs), cellular manufacturing, expert systems, and artificial intelligence (AI).

1.5 PLANNING FOR MANUFACTURING

There is a close interdependent relationship between the design of a product, selection of material, and selection of the processes and equipment. Figure 1.6 shows the steps of manufacturing a part from the original idea. Because of the highly competitive nature of manufacturing processes, the question of finding ways to reduce cost is ever present. A good starting point for cost reduction is in the design stage of the product. The design engineer should always keep in mind the possible alternatives available in making his design. Unfortunately, designers often consider that their job is to design the product for performance, appearance, and reliability, and that it is the manufacturing engineer's job to produce whatever has been designed. Of course, there is often a natural reluctance to change a proven design for the sake of a reduction in manufacturing cost. As a subject, design for manufacturing hardly exists as compared with design for strength. Successful design is ensured through high product quality, while minimizing the manufacturing cost.

1.6 ACCURACY OF MANUFACTURING PROCESSES

When producing any component, it is necessary to satisfy the surface technological requirements in terms of high product accuracy, good surface finish, and a minimum of drawbacks that may arise as a result of the manufacturing process. The nature of the surface layer has a strong influence on the mechanical properties of the part. Some primary metal forming and casting processes produce surfaces that require further machining operations to control their dimensions at closer tolerances. The surface roughness is considered as an important factor in contact-to-contact surfaces and for functional properties such as wear resistance and fatigue strength of a part. According to the surface roughness required by the design specifications, the optimum manufacturing method can be selected.

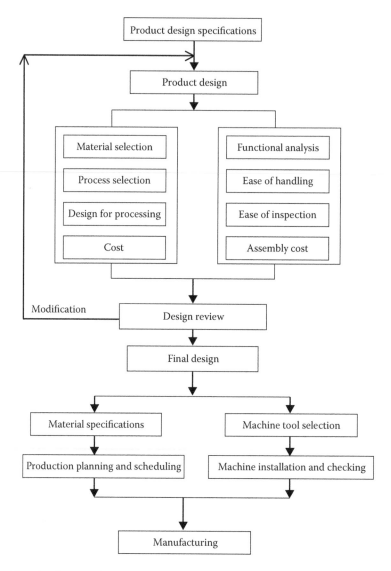

FIGURE 1.6   Planning for manufacturing.

Each manufacturing process is capable of producing certain surface finish and tolerance range without extra cost. For proper functioning and ease of manufacturing the part, it is necessary to specify both maximum and minimum surface roughness values. Additionally, the specified tolerances also should be within the range obtained by the selected manufacturing process, so as to avoid further finishing operations and rise in cost. Therefore, the use of unnecessarily tight tolerance and fine surface finish specifications is a major source of excessive manufacturing costs. The selected manufacturing process must be capable of producing parts within the required tolerance. Tolerances are important for parts that need to be assembled or mated together. Free functional surfaces do not need close tolerance control.

1.7   ECONOMICAL AND ENVIRONMENTAL CONSIDERATIONS

Emerging technologies have provided some new growth in advanced manufacturing employment opportunities in the United States. Manufacturing provides important material support for national infrastructure and for national defense.

On the other hand, most manufacturing may involve significant social and environmental costs. The clean-up costs of hazardous waste, for example, may outweigh the benefits of a product that creates it. Hazardous materials may expose workers to health risks. Developed countries regulate manufacturing activity with labor laws and environmental laws. In the United States, manufacturers are subject to regulations by the Occupational Safety and Health Administration and the U.S. Environmental Protection Agency. In Europe, pollution taxes to offset environmental costs are another form of regulation on manufacturing activity. Labor unions and craft guilds have played an historic role in negotiation of worker rights and wages. Environment laws and labor protections that are available in developed nations may not be available in the third world. Tort law and product liability impose additional costs on manufacturing.

Design for environment (DFE) is a product design approach adopted for reducing the impact of products on the environment during their manufacture, which require the use of highly polluting processes and the consumption of large quantities of raw materials. Products can also have an adverse effect through the consumption of large amounts of energy during manufacturing, use, and disposal. During the product life cycle, many events create pollution, and there are many opportunities for recycling, remanufacturing, and reuse, thus reducing the environmental impact. Manufacturing products that impact the environment less becomes a market advantage. There are three major elements of design for the environment, which include design for environmental manufacturing through the use of nontoxic processes and materials, minimizing the energy utilization, emissions, waste, scrap, and by-products, and design for environmental packaging. Design for disposal and recyclability should also be considered.

1.8　HEALTH AND SAFETY ASPECTS

The manufacturing industry transforms raw materials into finished or semifinished products. Such operations often rely on the integration of numerous components, which frequently results in the generation of toxic or hazardous wastes. A *hazard* is any source of potential damage, harm, or adverse health effects under certain conditions at work. Basically, a hazard can cause harm or adverse effects to individuals (health effects) or to organizations in the form of property or equipment losses. Manufacturing hazards affect people, machines, systems, and other inhabitants of the environment. For example, dust, radiation, temperature, and other environmental factors are hazardous to people and equipment. It is therefore necessary to protect workers within a plant from dangerous environmental conditions, and the general public from the unsafe conditions created by the manufacturing operation or products of the plant.

1.9　REVIEW QUESTIONS

1. Explain the importance of manufacturing technology in raising the standard of living.
2. What are the major types of manufacturing processes?
3. Explain how the company organization at different levels participates in the success of the manufacturing process.
4. Describe how computers are used in different manufacturing-related activities.
5. What is meant by process capability related to accuracy of manufactured parts?
6. Explain the following terms:
   - Design for manufacturability
   - Design for environment
   - Design for assembly
7. What are the main functions of the following plant groups:
   - Process engineering
   - Plant engineering
   - Inspection and control
   - Production (superintendents)

BIBLIOGRAPHY

Degarmo, E. O., Black, J. T., and Kohser, R. A. 1997. *Materials and Processes in Manufacturing*, 8th ed. Upper Saddle River, NJ: Prentice-Hall.

Kalpakjian, S., and Schmid, S. R. 2003. *Manufacturing Processes for Engineering Materials,* 4th ed. Upper Saddle River, NJ: Pearson Education.

Wright, P. K. 2001. *21st Century Manufacturing.* Upper Saddle River, NJ: Prentice-Hall.

2 Properties of Engineering Materials

2.1 INTRODUCTION

Every engineer should be vitally concerned with the materials available to him. Whether the product under consideration is a bridge, a computer, a machine tool, or an automobile exhaust system, he must have an intimate knowledge of the properties and characteristics of the material from which the product is to be designed to function satisfactorily. In making his choices, the designer must take into account such properties as strength, rigidity, hardness, toughness, electrical and/or thermal conductivity, density, and others. Furthermore, he must consider the material behavior during processing, such as formability, machinability, electrical and thermal stability, as well as cost and availability. Finally, the engineer must know which of the properties are significant, how these properties are determined, and what restrictions should be placed on their use.

The following discussion of the properties of engineering materials will be introductory only. Its purpose will be to identify the more common properties in order to compare between different materials or different structures of the same material. Three general types of properties are physical, mechanical, and manufacturing (fabricating).

2.2 PHYSICAL PROPERTIES

A common means of distinguishing one material from another is by evaluating physical properties. Physical properties of particular interest are density, thermal properties, electrical and magnetic properties, optical properties, chemical properties, and resistance to oxidation and corrosion. For some engineering applications, the physical properties of an engineering material may be even more important than mechanical properties. For this reason, it is essential to discuss some important physical properties briefly.

2.2.1 Density

The density of a material is the mass per unit volume. It depends on the atomic weight, atomic radius, and the packing of atoms in the materials unit cell. The SI unit of mass is the gram or megagram (Mg = metric ton). Thus, density is in units of Mg/m^3 (= g/cm^3), or in the U.S. conventional system, lb/in^3 (1 kg = 2.2 lb). The density may be expressed relative to water density. It is then called specific gravity; thus, it is dimensionless.

Density is one of the material selection factors. If an Al alloy has the same strength as the steel, it will have three times the strength-to-weight ratio. This explains why airplanes and modern automotive bodies are made mostly of Al alloys. In high-speed equipment such as printing and textile machinery, Mg alloys are frequently used.

2.2.2 Thermal Properties

Thermal properties such as melting point, thermal expansion, specific heat, thermal conductivity, and latent heat of fusion are important in both manufacturing and service. Their values are found in materials handbooks. A selection of thermal properties is given for some common engineering materials in Table 2.1.

1. Melting point: This is an important parameter, especially in manufacturing operations such as casting, welding, and thermal nontraditional machining processes. The castability, weldability, and machinability improve with decreased melting point of the engineering material.

   The melting points of pure metals are constant, but alloys melt over a range of temperatures governed by the chemical composition. Among low melting point metals are zinc, tin, lead, and bismuth, while high melting point ones are tungsten, molybdenum titanium, and others which are commonly known as refractory metals.

TABLE 2.1
Physical Properties of Engineering Materials at Room Temperature

	Density (kg/m³)	Melting Point (°C)	Specific Heat (J/kg K)	Thermal Conductivity (W/mK)	Coefficient of Thermal Expansion (μm/m °C)
Metals					
Aluminum	2700	660	900	222	23.6
Aluminum alloys	2630–2820	476–654	880–920	121–239	23.0–23.6
Copper	8970	1082	385	393	16.5
Copper alloys	7470–8940	885–1260	337–435	29–234	16.5–20
Gold	19,300	1063	129	317	19.3
Iron	7860	1537	460	74	11.5
Steels	6920–9130	1371–1532	448–502	15–52	11.7–17.3
Lead	11,350	327	130	35	29.4
Lead alloys	8850–11,350	182–326	126–188	24–46	27.1–31.1
Magnesium	1745	650	1025	154	26.0
Magnesium alloys	1770–1780	610–621	1046	75–138	26.0
Nickel	8910	1453	440	92	13.3
Nickel alloys	7750–8850	1110–1454	381–544	12–63	12.7–8.4
Silver	10,500	961	235	429	19.3
Tantalum alloys	16,600	2996	142	54	6.5
Titanium	4510	1668	519	17	8.35
Titanium alloys	4430–4700	1549–1649	502–544	8–12	8.1–9.5
Tungsten	19,290	3410	138	166	4.5
304 stainless steel	—	—	—	15	16.5
410 stainless steel	—	—	—	24	10.0
Invar (Fe-36Ni)	—	—	—	11	0.3–0.6
Kovar (Fe-28Ni-18 Co)	—	—	—	16.7	4.4
Nonmetallic					
Ceramics	2300–5500	—	750–950	10–17	5.5–13.5
Glasses	2400–2700	580–1540	500–850	0.6–1.7	4.6–70
Graphite	1900–2200	—	840	5–10	7.86
Plastics	900–2000	110–330	1000–2000	0.1–0.4	72–200
Wood	400–700	—	2400–2800	0.1–0.4	2–60

Source: Kalpakjian, S., and Schmid, S. R., *Manufacturing Processes for Engineering Materials*, 4th ed., Pearson Education, Prentice-Hall, Upper Saddle River, NJ, 2003.

2. Thermal expansion: This is of prime importance for composites. A large difference of thermal expansion between constituents leads to high stresses and even cracking. Differential expansion leads to warping in the ductile constituents of a composite, and fracture in brittle ones.

   Another area in which differential expansion should be avoided is in the moving elements in machinery that require proper fit (shrink or clearance) for proper functioning. Dimensional changes resulting from differential thermal expansion can be significant in measuring instruments and limit gauges. To counteract some of the problems associated with thermal expansion, some low expansion Fe–Ni alloys have been developed. Typical alloys are Invar (Fe-36% Ni) and Kovar (Fe-28% Ni-18% Cu). These alloys possess good thermal fatigue resistance.

3. Specific heat: This is the energy required to raise the temperature of unit mass of a material by one degree kelvin or celsius (J/kg $^\circ$K). In machining and forming, the lower the specific heat of a material, the higher the temperature will rise. Excessive high temperatures have considerable effects on accuracy and finish, cause tool and die wear, and lead to adverse metallurgical changes of the material. Sometimes, the term heat capacity is used to represent the specific heat times the mass of the material in joules per kelvin.

4. Thermal conductivity: This indicates the rate with which heat flows through the material cross section. In general, metals and alloys have high thermal conductivity, whereas nonmetallic materials have poor conductivity. While alloying elements have minor effects on the specific heat of a metal, they exert a significant effect on the thermal conductivity of that metal (Table 2.1). When heat is generated by plastic deformation, machining, or friction, it should be conducted away rapidly to avoid severe rise of temperatures. The well-known difficulty in machining Ti is due to its low thermal conductivity, resulting in high thermal gradients, which causes inhomogeneous deformation of the product, and thermal failure of cutting tools.

2.2.3 ELECTRICAL AND MAGNETIC PROPERTIES

Metals consist of positively charged ions bonded by freely moving electrons. Thus, they are conductors with resistivities in the order of 20 nΩ m. Conductivity is the reciprocal of resistivity. It is the ability of metals to conduct electric current. Alloying elements (and crystal imperfections) make the passage of electrons more difficult and hence alloys are less conductive to electricity than pure metals. Electrical conductivity is important in certain processes, such as magnetic pulse forming, EDM and ECM, where the workpiece must be electrically conductive.

At some critical temperatures, some materials become superconductive, such that their resistivity drops to zero. Metallic NbTi and Nb_3Sn, cooled by liquid helium (boiling at 4°K), were the first subconductors in practical applications to power highly stable magnets for high energy accelerators. Insulators are materials in which all (or virtually all) electrons are tied down in covalent, ionic, or molecular bonds. A large energy is required to break loose an electron (resistivity > 10 Ω m). They lose their insulating quality only at some critical field intensity (dielectric strength). Semiconductors are solids that are normally insulators but become conductors when an electric field is applied. They form the basis of semiconductor industries needed in the manufacturing of integrated circuits.

Some materials (ceramics and quartz crystals) exhibit piezoelectric effect. If subjected to mechanical stress, these crystals generate a potential difference. Thus, they can be used as a force transducer. In the reverse mode, an applied potential difference causes a dimensional change which can be implemented in ultrasonic transducers and sonar detectors. Other materials such as pure Ni and some iron–nickel alloys expand or contract when subjected to a magnetic field. This phenomenon is called magnetostrictive effect, which is one of the principles on which ultrasonic machining is based.

2.2.4 OPTICAL PROPERTIES

The optical behavior of a solid material is a function of its interactions with electromagnetic radiation, having wave lengths within the visible region of the spectrum. Possible interactive phenomena include refraction, reflection, absorption, and transmission of incident light. Metals appear opaque as a result of absorption and then reemission of light radiation taking place within a thin outer surface layer. The perceived color of a metal is determined by the spectral composition of the reflected light. Nonmetallic materials are either intrinsically transparent or opaque. Opacity results in relatively narrow band gap materials as a result of absorption. Transparent nonmetals have band gaps of greater than about 3 eV.

The manufacturing processes are controlled to finish parts with desired optical attributes. A very smooth finish reflects light at the same angle of incidence (specular reflection), whereas a rough surface, which is sometimes needed (matte finish), reflects light randomly (diffused reflections).

2.2.5 CHEMICAL AND CORROSION PROPERTIES

Corrosion resistance is an important aspect in material selection, especially for applications in chemical, food, and petroleum industries. Many fabricated structures are designed to survive for prolonged periods while being exposed to the surrounding environment. Hence, they deteriorate or corrode by chemical and electrochemical action. Their deterioration is governed by material selection but is also affected by the precautions and protection measures considered during manufacturing. For example, residual stresses can lead to accelerated corrosion. Steel screws corrode when used for joining brass sheets. Stainless steels lose their corrosion resistance if slowly cooled from welding temperatures. Most automobile bodies are protected from corrosion if zinc-plated sheets are used. In most cases, special measures must be taken by using protective coatings.

2.3 MECHANICAL PROPERTIES

The mechanical properties exert a decisive influence in determining the suitability of one or another material for specific constructional or processing application. The mechanical behavior of a material (static or dynamic) reflects the relationship between its response (or deformation) and the applied load. It is important to characterize the behavior of materials under such conditions. The test results can then be used to select materials even when the service conditions differ, where the results are used qualitatively to compare various candidate materials. Important mechanical properties are strength, hardness, ductility, and stiffness.

2.3.1 STRENGTH

Deformation occurs when forces are applied to a material. The amount of deformation per unit length is known as strain and the force per unit area as stress. Energy is absorbed by a material during deformation because a force has acted along the deformation distance. Strength is a measure of the level of stress required to make a material fail. Therefore, strength means the ability of a part made of the material to resist externally applied forces.

If a load is static or changes relatively slowly and is applied uniformly to a cross section of a member, the mechanical behavior may be represented by a simple stress–strain test. These tests are conducted on metals at room temperature. There are three principal ways in which a load may be applied; namely, tension, compression, and shear. In engineering practice, many loads are torsional rather than pure shear. The uniaxial tensile test is the most common, since it provides information about a variety of strength properties. In this test, the material behaves either elastically or plastically. Such behaviors will be dealt with in a detailed manner later on in Chapter 8. However, some simple and basic relationships will be given here to explain the concept of strength by considering the engineering stress–strain curve of low-carbon steel (Figure 2.1).

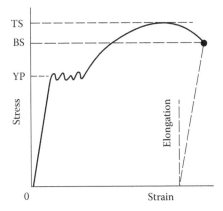

FIGURE 2.1 Engineering stress–strain curve of low-carbon steel specimen.

Elastic range: It can be noted that the initial response of Figure 2.1 is linear up to a certain stress. The stress at which this proportionality exists is known as the *proportional limit* (*yield point YP, or elastic limit*). Up to this stress, the material is elastic and obeys Hook's law. The ratio of stress to strain in this region is known as *Young's modulus* or the modulus of elasticity E. This is an important property of a given engineering material, and it is of considerable design importance. The designer must be fully aware of this property, since it is a measure of interatomic bonding forces, and hence it is a measure of *stiffness* and *rigidity*. It indicates the ability of a material to resist deflection or stretching.

$$E = \frac{\sigma}{\varepsilon} \quad \frac{N/m^2}{m/m} \text{ or Pascal} \qquad (2.1)$$

Up to the proportional (elastic) limit, if the load is removed, the specimen will return to its original length (i.e., the elastic strain is recovered). Table 2.2 lists the moduli of elasticity of several engineering materials in Pascal (N/m^2).

The amount of strain energy that a unit volume of material can absorb while in the elastic range is called the modulus of resilience U_r ($N\,m/m^3 = N/m^2$). It is the area under the stress–strain curve up to the elastic limit (yield point YP).

$$U_r = \varepsilon_y \sigma d\varepsilon = \frac{1}{2}\sigma_y \varepsilon_y = \frac{\sigma_y^2}{2E} \qquad (2.2)$$

Plastic range: The elongation beyond the elastic limit becomes unrecoverable and is known as *plastic deformation*. When the load is removed, the specimen retains a permanent change in shape.

TABLE 2.2
Moduli of Elasticity of Some Engineering Materials

Material	E (Pa)	Material	E (Pa)
Aluminum	7×10^{10}	Window glass	7×10^{10}
Copper	11×10^{10}	Polyethylene	$1{-}14 \times 10^{8}$
Steel	21×10^{10}	Rubber	$4{-}80 \times 10^{6}$
Cast iron	10×10^{10}		

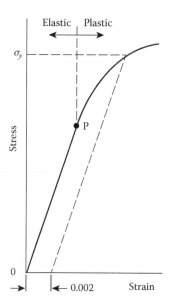

FIGURE 2.2 Stress–strain curve of a material that does not have a well-defined yield point, showing the offset method for determining the yield stress.

In design applications, elastic response is of interest, since allowable stresses employed in machine and structure design are usually kept below the elastic limits to avoid plastic deformation. On the other hand, in manufacturing, plastic deformation is often used to shape a product, and the applied stresses must be sufficient to induce the necessary plastic flow.

When the elastic limit is exceeded, considerable increases in strain do not require proportionate increases in stress. For low-carbon steels (Figure 2.2), a distinct point is significant; that is, the yielding point, $YP(\sigma_y)$. Most materials, however, do not have a well-defined yield point P, but exhibit nondistinct transition from elastic to plastic regime (e.g., Al) (Figure 2.2), and the elastic limit is therefore defined through the use of the offset yield strength, or the value of stress that will produce a given amount of permanent strain. For most materials, the amount of the offset strain is set at 0.2%, but values of 0.1%, or even 0.02% may be specified when small amounts of plastic deformation could lead to failure.

If plastic deformation is continued (Figure 2.1), the material acquires an increased ability to bear load. That means the material is getting stronger with increased deformation (strain hardened). The highest stress is known as the ultimate tensile strength (UTS) of the material. At that point, localized reduction in cross-sectional area (necking) starts to occur. If straining is continued far enough, the tensile specimen will ultimately fracture, and the stress at which fracture occurs is known as breaking strength (BS). For relatively ductile materials, the breaking strength is always less than the ultimate tensile strength, and necking precedes fracture. For brittle materials, fracture usually terminates the stress–strain curve before necking and possibly before the onset of plastic flow (Figure 2.3).

2.3.2 DUCTILITY AND BRITTLENESS

Ductility is a measure of plastic deformation before final failure and may be expressed as either elongation or reduction in area at the point of fracture, as will also be discussed later on in Chapter 8. When materials fail with little or no ductility, they are said to be brittle. *Brittleness*, therefore, is the opposite of ductility and should not be confused with a lack of strength. A brittle material is simply one that lacks significant ductility. Brittle materials are generally considered to be those having a fracture strain of less than 5%.

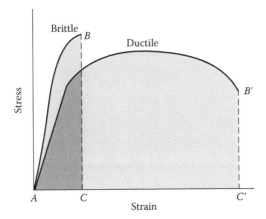

FIGURE 2.3 Stress–strain curves for brittle and ductile materials loaded to fracture (with brittle material fracture usually terminated before necking).

2.3.3 Toughness

This is a measure of the energy required to break a material, whereas strength is a measure of the stress required to produce fracture. This contrast is especially important in view of the fact that toughness is often inaccurately called "impact strength." The modulus of toughness is the energy per unit volume expressed in J/m^3; it is the area under the stress–strain curve. A ductile material with the same strength as a nonductile material will require much more energy for breaking and thus will be tougher. Figure 2.3 demonstrates the stress–strain curves for brittle and tough material types. Even though the brittle material has higher yield tensile strengths, it has a lower toughness than the ductile one due to the lack of ductility. Standardized Charpy or Izod tests are two of several procedures used to measure toughness. They differ in the shape of the test specimen and method of applying the energy (Figure 2.4).

2.3.4 Hardness

Another important mechanical property to consider is *hardness*, which is a measure of material resistance to localized plastic deformation. Therefore, both tensile strength and hardness are indicators of a metal's resistance to plastic deformation. Consequently, they are roughly proportional, as shown in Figure 2.5, for tensile strength (TS) as a function of the Brinell hardness number (HB) for cast iron, steel, and brass. The same proportionality relationship does not hold for all

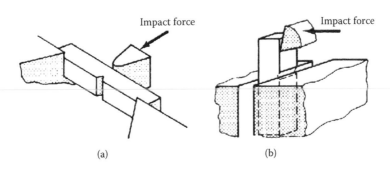

FIGURE 2.4 Test specimens and methods of applying energy for (a) Charpy and (b) Izod impact tests.

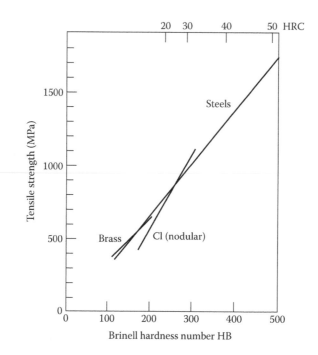

FIGURE 2.5 Relationship between hardness (HB and HRC) and tensile strength for steel, brass, and nodular cast iron.

metals, as Figure 2.5 indicates. As a rule of thumb, for most steels, the HB and TS are related according to:

$$TS \text{ (MPa)} = 3.5 \times HB \qquad (2.3)$$

Over the years, various techniques have been developed in which a small indenter is forced into the surface of the material to be tested, under controlled conditions of load and rate of application. The depth or size of the resulting indentation is measured, which in turn is related to a hardness number; the softer the material, the larger and deeper the indentation, and the lower the hardness index number. Among the most common standardized hardness tests are Brinell, Rockwell, Vickers, and Knoop (Figure 2.6). Measured hardness values are only relative (rather than absolute), and care should be exercised when comparing values determined by different techniques.

Hardness tests are performed more frequently for several reasons.

1. They are simple and inexpensive.
2. The test is nondestructive.
3. Other mechanical properties often may be estimated from hardness data, such as yield and tensile strengths.

Hardness Techniques:

1. Brinell hardness HB: In the Brinell test, a steel or WC ball 10 mm in diameter is pressed against the surface with a load of 500, 1500, and 3000 kg, which is maintained for 5–10 seconds. HB is defined as the ratio of the load F to the curved area of indentation, kg/mm^2 as shown in Figure 2.6. WC balls are generally recommended for HB > 500. The Brinell test is generally suitable for materials of low to medium hardness.

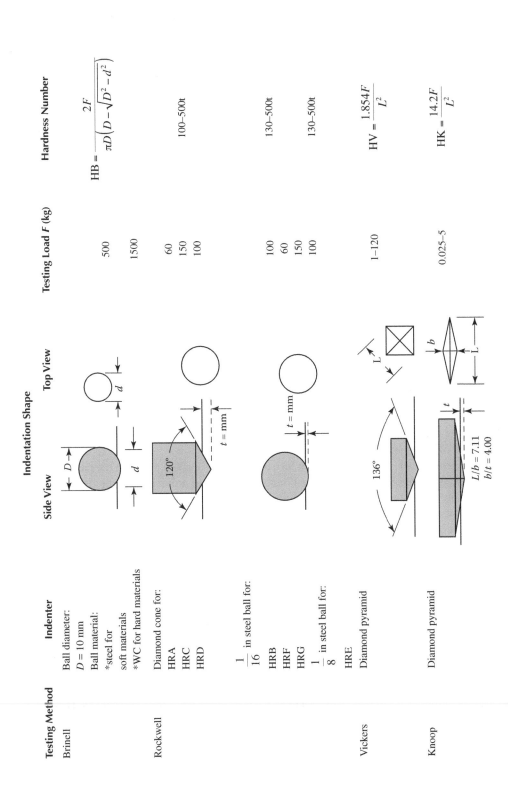

FIGURE 2.6 Hardness testing methods. (Adapted from William G. Moffatt, *Structure and Properties of Materials, Volume 1,* John Wiley & Sons, New York, 1965.)

2. Rockwell hardness HR: This is the most common, because it is simple and requires no special skills. In the Rockwell test, the depth of penetration is measured. The indenter is pressed on the surface, first with a minor load (10 kg). The indicator on the dial of the tester is then set to zero. Finally, a major load is applied to the indenter. Several different scales may be utilized with different indenter geometries and different loads to permit testing of virtually all metals and alloys from the hardest to the softest. Some of the more common Rockwell hardness scales and relevant indenters are listed in Figure 2.6. Rockwell superficial hardness tests, which use lighter loads (15, 30, or 45 kg) are also available.

3. Vickers hardness HV: The Vickers test uses a pyramid-shaped diamond indenter (Figure 2.6), with loads ranging from 1 to 120 kg. The impressions are typically less than 0.5 mm diagonally. The Vickers test gives the same hardness number regardless of the load. HV is suitable for materials with a wide range of hardness, including very hard steels.

4. Knoop hardness HK: The Knoop test uses a diamond indenter in the shape of an elongated pyramid (Figure 2.6), with loads ranging from 25 g to 5 kg. HK is referred to as micro-hardness, because of light loads used. Hence, it is used for very small or thin specimens and for hard and brittle materials, such as carbides and glass. It is also used for measuring the hardness of individual grains in a metal. The surface preparation is very important, since the size of the indentation is ranging from 10 to 100 μm. Because HK depends on the applied load, test results should always cite the load applied.

5. Mohs hardness HM: This is based on the capability of one material to scratch another. HM is scaled 1 for talc and 10 for diamond, and is generally used by geologists and mineralogists. Although the Mohs scale is qualitative, good correlation is observed with Knoop hardness.

6. Scleroscope number: The scleroscope is an apparatus tipped with a diamond indenter enclosed in a glass tube. The indenter is dropped on the specimen from a certain height. The rebound of the indenter is a measure of the hardness. The higher the rebound, the harder the specimen. Because the scleroscope is portable, it is useful for measuring the hardness of large objects.

7. Durometer number: A durometer is an instrument that is used to measure the hardness of soft and elastic materials such as elastomers, rubbers, and polymers. It can also be used to evaluate the strength of foundry molding sands. The spring-loaded indenter is rapidly pressed against the surface with a constant load, and the penetration depth is measured after 1 second. The hardness is inversely related to that depth. The test may be performed by a blunt indenter and a load of 1 kg for softer materials or by a sharp indenter and a load of 5 kg for harder materials. The hardness numbers in both tests range up to 100.

Table 2.3 presents a comparison of hardness values for plain carbon and low-alloy steels. It may be noted that for HRC above 20, HB values are approximately 10 times the Rockwell number. Also for HB below 320, the HV and HB values agree quite closely. Since the relationships between various tests vary with material, mechanical processing and heat treatment (Table 2.3) should be used with caution. The table also shows a correlation between tensile strength and hardness for steel.

Task: A large rolling-mill roll was supposed to be heat treated to a hardness of HRC 56. How could you check if this has indeed been done?

The roll is too large to be placed on a Rockwell hardness tester and no specimen can be cut from it. Therefore, a scleroscope must be used. It gives a reading of 78. Referring to the conversion table (Table 2.3), this corresponds to HRC 58, which means the heat treatment has indeed been carried out.

TABLE 2.3
Hardness Conversion Table

Brinell Number	Vickers Number	Rockwell Number		Scleroscope Number	Tensile Strength (MPa)
		C	B		
	940	68		97	2537
757	860	66		92	2427
722	800	64		88	2324
686	745	62		84	2234
660	700	60		81	2144
615	655	58		78	2055
559	595	55		73	1903
500	545	52		69	1765
475	510	50		67	1703
452	485	48		65	1641
431	459	46		62	1462
410	435	44		58	1407
390	412	42		56	1351
370	392	40		53	1303
350	370	38	110	51	1213
341	350	36	109	48	1138
321	327	34	108	45	1069
302	305	32	107	43	1007
285	287	30	105	40	951
277	279	28	104	39	924
262	263	26	103	37	883
248	248	24	102	36	841
228	240	20	98	34	800
210	222	17	96	32	738
202	213	14	94	30	683
192	202	12	92	29	655
183	192	9	90	28	627
174	182	7	88	26	600
166	175	4	86	25	572
159	167	2	84	24	552
153	162		82	23	524
148	156		80	22	510
140	148		78	22	490
135	142		76	21	469
131	137		74	20	455
126	132		72	20	441
121	121		70		427
112	114		66		

(Brinell column annotation: Steel ball ↓ • ↑ WC ball)

2.3.5 FATIGUE

The term *fatigue* refers to a fracture arising from cyclic stresses. Commonly, these stresses alternate between tension and compression, as in a loaded rotating shaft. Fatigue can also occur as the result of fluctuations due to cyclic stresses of the same sign, as in leaf springs or similarly loaded components of a car. Even though each individual loading event is insufficient to cause permanent deformation, the repeated application of stress can lead to a fatigue failure. Fatigue is the result of cumulative damage caused by stresses much smaller than the tensile strength.

FIGURE 2.7   Fatigue failure in a ductile steel shaft. (Courtesy of H. Mindlin, Battelle Memorial Institute.)

Cyclic fatigue is a characteristic of ductile materials. Even so, the final fracture is rapid (Figure 2.7). The fracture surface tends to exhibit two distinct regions: A smooth, relatively flat region where the crack was propagating by cyclic fatigue, and a coarse, ragged region of ductile overload tearing. The number of stress cycles before fracture is a function of applied stress (Figure 2.8). Fortunately, many metals have an endurance limit (fatigue limit), below which the metal is not subject to fatigue failure. Table 2.4 shows the approximate ratio of the endurance limit σ_{ind} to the ultimate tensile strength (UTS) for some important engineering metals. For many steels, this ratio is about 0.5, and for nonferrous metal, the ratio is significantly lower.

Fatigue failure begins with the generation of small cracks, invisible to the naked eye. Fatigue cracks usually start at the surface because

- Bending or torsion will cause the highest stresses to occur at the outer fiber.
- Surface irregularities introduce stress concentrations.

As a result, the endurance limit is very sensitive to surface finish (Table 2.5).

Fatigue strength is reduced if the surface is rough, especially in high-strength materials that are less ductile. Shot peening and other procedures such as nitriding and case hardening, which introduce surface compression, tend to raise the endurance limit. Furthermore, fatigue lifetime can be shortened considerably by an even mildly corrosive environment. The endurance limit in dry air is higher than in humid industrial air and still higher in a vacuum. A special form of fatigue failure occurs when certain materials are exposed to a chemically aggressive environment. Surface

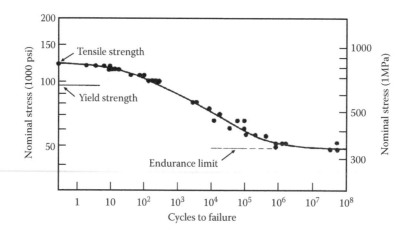

FIGURE 2.8   S–N curve (material: SAE 4140 normalized steel).

TABLE 2.4
Ratio of Endurance Limit to Tensile Strength for Some Important Metals

Metal	Ratio $\dfrac{\sigma_{ind}}{UTS}$
Aluminum	0.38
Beryllium copper (heat treated)	0.29
Cu (hard)	0.33
Steel	
AISI 1035	0.64
AISI 4140 (normalized)	0.54
Wrought iron	0.63

cracks form, and in combination with the applied stress, lead to stress-corrosion cracking. If there are residual stresses on the surface of the part, cracks develop in a part even in absence of external loads. Repeated loading at elevated temperatures may lead to thermal fatigue. This is particularly troublesome in forging tools, cutting tools, casting dies, and glass molds, because cracking of the surface is reproduced on the surface of the part.

As a conclusion, fatigue failures probably account for nearly 90% of all mechanical failures. Accordingly, it is important to know how materials will respond to fatigue conditions. Finally, any design factor that concentrates stresses can lead to premature fatigue failure. We have already seen that the surface finish of the part is critical. Keyways and other notches are also critical. It is important to emphasize the role of generous fillets to be provided in designed parts (Figure 2.9). In fact, in many cases, a component may be made stronger by removing material if such removal will increase the radius of curvature at the point of stress concentration.

2.3.6 CREEP

At elevated temperatures, even low stresses (below yield) produce a slow plastic deformation called creep. It is a phenomenon of metals and certain nonmetallic materials, such as thermoplastics and rubbers, and can occur at any temperature. Lead, for example, creeps under constant tensile load at room temperature. The thickness of window glass in old houses has been found to be greater at the bottom than at the top, with the glass having undergone creep by its own weight over many years.

TABLE 2.5
Surface Roughness R_a versus Endurance Limit

Type of Finish	R_a (μ_m)	σ_{ind} (MPa)
Cylindrical grinding	0.4–0.6	630
Machine lapped	0.3–0.5	720
Surface grinding	0.2–0.3	770
Superfinishing	0.01–0.05	805

Source: Garwood, M. F., H. H. Zurburg, and M. A. Erickson, *Correlation of Laboratory and Service Performance*, ASM, Metal Park, OH, 1951.

Note: Material: 4063 steel quenched and tempered to 44 HRC.

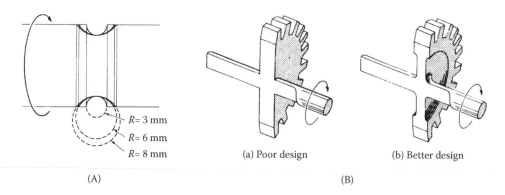

$R= 3$ mm
$R= 6$ mm
$R= 8$ mm

(A)

(a) Poor design   (b) Better design

(B)

FIGURE 2.9 Recommended design of parts for fatigue. (A) Provision of large radius for less stress concentration (adapted from Garwood, M. F., Zurburg, H. H., and Erickson, M. A., *Correlation of Laboratory and Service Performance*, ASM, Metal Park, OH, 1951) and (B) Generous fillets (adapted from Van Vlack, L. H., *A Textbook of Materials Technology*, Addison-Wesley, 1973).

A typical creep characteristic is shown in Figure 2.10. Like the deformation behaviors described before, there is an initial elastic strain ε_0. This is followed in turn by primary creep (stage 1), secondary or steady-state creep (stage 2), and tertiary creep (stage 3) before ultimate rapture. The primary creep is partially inelastic and is recoverable if the stress is removed. The steady-state creep is of greater engineering importance because it leads to extensive irrecoverable strain. During this stage, the creep rate $d\varepsilon/dt$ is commonly reported. Higher temperatures or higher stresses increase the creep rate. The accelerated creep (stage 3) is a result of an increase in the true stress, either because of necking or because of internal cracking.

Although the strain rate may be low, high temperature service applications are often such that materials are exposed to stresses for long periods (e.g., boiler tubes). For parts expected to give long service, design is based on the stress that produces a linear creep rate $d\varepsilon/dt$ of 1% per 100.000 h for steam and gas turbine components. To accelerate creep testing and to obtain design data, tests are conducted at higher stresses and higher temperatures to exhibit higher creep rates and shorter rapture times. Fine-grained materials have a faster creep rate at high temperature applications, and therefore they should be avoided. Generally, resistance to creep increases with the melting point of a material. This fact is a guideline in selecting materials for design purposes. Stainless steels, superalloys, and refractory metals and alloys are used in applications when creep resistance is required.

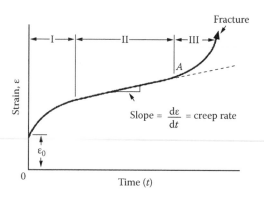

FIGURE 2.10 Stages of creep for metallic materials.

2.4 MANUFACTURING (FABRICATING) PROPERTIES

Manufacturing characteristics of materials determine their suitability for various manufacturing processes. They typically include castability, formability, machinability, grindability, weldability, and heat treatability to perform individual components of specific shapes and dimensions. Such properties are crucial to proper selection of materials. They actually refer to the way a given material responds to a specific processing technique, and are therefore quite difficult to define.

2.4.1 Castability

Castability is generally used to describe the ease with which a metal can be cast to obtain a casting with a good quality. This term incorporates, in addition to the technological concept of fluidity, aspects that define the ease of producing a casting under average foundry conditions. Fluidity means the ability of a metal to fill all cavities of a mold during casting.

An alloy therefore is regarded as highly castable when it not only has high fluidity but is also more tolerant to the design of the fluid supply system. It is less sensitive to wall thickness variations, and in general will produce castings of acceptable quality with fewer labor skills.

2.4.2 Ductility, Malleability, Workability, and Formability

Ductility, *malleability*, *workability*, and *formability* describe the material suitability for plastic forming. However, materials can behave differently at different temperatures. A material with good "hot formability" may behave poorly when deformed at room temperature. Some materials that flow nicely at low deformation speeds behave in a brittle manner when loaded at rapid rates. Thus, formability needs to be evaluated for a specific combination of material, process, and process conditions, and the results can be extrapolated to other processes or process conditions. Sheet formability and its measurements are presented in Chapter 10.

2.4.3 Machinability

Machinability is the ability of a material to be shaped by cutting tools. It depends not only on the material being machined, but also on the specific machining process and its aspects. Machinability ratings are generally based on relative tool life data. In some cases, one may be interested in how easy or fast a metal can be cut, irrespective of the resulting tool wear or surface finish. In other applications, surface finish and accuracy may be of prime importance. For some processes, the formation of fine disposable chips may be a desirable feature. Therefore, the term *machinability* means different things to different people and frequently involves multiple properties of a material acting in unison.

2.4.4 Grindability

Grindability of a material, like *machinability*, is difficult to precisely define. It is a general indication of how easy it is to grind a material, including considerations such as the quality of the surface produced, surface integrity, accuracy, wear of the grinding wheel, cycle time, and overall economics. The grindability of a material can be enhanced greatly by the proper selection of process parameters, the grinding wheel marking and grinding fluids, as well as machine characteristics and fixturing devices.

2.4.5 Weldability

Weldability is the ability of separate elements to be firmly joined in the zone of close contact under the action of local heating up to the plastic or molten state. Weldability involves a large number of variables, making the term extremely nebulous. One process might produce excellent results when applied to a given material, while another may produce a dismal failure. Material characteristics, such as alloying elements, impurities inclusions, grain structure, the processing history of the base metal, and filler are important. Other factors that influence weldability are mechanical and physical properties such as strength, toughness, ductility, notch sensitivity, modulus of elasticity, specific heat, melting point, thermal expansion, surface tension characteristics of molten metal, and corrosion.

2.4.6 Heat Treatability

Heat treatability designates the capability of an alloy to be heat-treated to achieve desired properties. Hardenability is a measure of depth to which a specific ferrous alloy may be hardened by the formation of martensite upon quenching from a temperature above the upper critical temperature.

2.5 REVIEW QUESTIONS

1. A steel bar 12.5 mm in diameter supports a load of 7000 kg. What is the stress placed on the bar? If this bar has $E = 205{,}000$ MPa, how much could the bar be strained with this load? If this bar supports a maximum load of 11,800 kg without plastic deformation, what is its yield strength?
2. Monel (70 Ni-30 Cu) has $E = 180{,}000$ MPa and a yield strength of 450 MPa. How much load can be supported by a rod 18 mm in diameter without yielding? If maximum total elongation of 2.5 mm is allowable in a 2.1-m bar, how large a load could be applied to this rod?
3. A cylindrical rod of steel ($E = 207$ GPa) with a yield strength of 310 MPa is subjected to a load of 11,100 N. If the length of the rod is 500 mm, what must be the diameter to allow an elongation of 0.38 mm?
4. A cold drawn steel bar has a Brinell hardness of HB = 190. What would be its tensile strength?
5. Define: Young's modulus, elastic limit, and ultimate strength.
6. Take the cord of a household appliance, such as a mixer. List the materials used and the probable reasons for their selection.
7. Why is there not a single standard means of assessing characteristics such as machinability, formability, or weldability?
8. What are the three primary thermal properties of a material? What do they measure, and what are their SI units?
9. How would you measure the hardness of a very large object?
10. Which hardness tests would you use for very thin strips of material such as aluminum foil, and why?
11. Explain clearly the difference between resilience and toughness. Give an application for which each is important.

2.6 PROBLEMS

1. What is the stress in a wire whose diameter is 1 mm, which supports a load of 100 Newton?
2. Suppose that the wire in the above problem is copper and is 30 m long. How much will it stretch? Repeat for steel and aluminum wires.
3. Estimate the HB hardness of an alloy steel of ultimate tensile strength TS = 1500 MPa using Equation 2.3. Use the hardness conversion table to determine the HRC hardness.

4. A constructional component made of Al alloy represented as a bar of diameter 20 mm and length 400 mm is subjected to pure tension. Calculate
   a. Extension of the bar if the imposed load = 8 kN.
   b. The load at which the bar suffers permanent deformation.
   c. Maximum load the bar can withstand before fracture.
   Assume the following mechanical properties: E = 70 GPa, YS = 496 MPa, and TS = 560 MPa.

BIBLIOGRAPHY

Callister, Jr., W. D. 1997. *Material Science and Engineering—An Introduction*, 4th ed. New York: John Wiley & Sons.

Degarmo, E. O., Black, J. T., and Kohser, R. A. 1997. *Materials and Processes in Manufacturing,* 8th ed. Upper Saddle River, NJ: Prentice-Hall.

Garwood, M. F., Zurburg, H. H., and Erickson, M. A. 1951. *Correlation of Laboratory and Service Performance.* Metal Park, OH: ASM.

Kalpakjian, S., and Schmid, S. R. 2003. *Manufacturing Processes for Engineering Materials*, 4th ed. Upper Saddle River, NJ: Pearson Education, Prentice-Hall.

3 Structure of Metals and Alloys

3.1 INTRODUCTION

The structure of materials influences their behavior and properties. Therefore, understanding this structure helps to make appropriate selection of these materials for specific applications. Depending on the manner of atomic grouping, materials are classified as having molecular, crystal, or amorphous structures. In molecular structures, atoms are held together by primary bonds. They have only weak attraction. Typical examples of molecular structure include O_2, H_2O, and C_2H_4 (ethylene). Each molecule is free to act independently, so these materials possess relatively low melting and boiling points, since their molecules can move easily with respect to each other. These materials tend to be weak. Solid metals and alloys and most minerals have crystalline structure, where atoms are arranged in a regular geometric array as a lattice of a unit building block that is repetitive throughout the space. In an amorphous structure, such as glass, atoms have a certain degree of local order but lack the periodically ordered arrangement of the crystalline solid.

3.2 LATTICE STRUCTURE OF METALS

Metals and alloys are an extremely important class of materials, since they are frequently processed to manufacture tools, machinery, and many metallic products. Metals are characterized by the metallic bond in three dimensions, offering them their distinguishing characteristics of strength, good electrical and thermal conductivity, luster, the ability to be plastically deformed to a fair degree without fracturing, and a relatively high density compared with nonmetallic materials. When metals and alloys solidify from their molten state, they assume a crystalline structure in which atoms arrange themselves in a geometric lattice.

3.2.1 SPACE LATTICES

There are three basic types of crystal structures (lattice cells) found in nearly all commercially important metals. The shape and dimensions of the lattice in a metal play an important role in the control of mechanical properties. X-ray diffraction measurements determine the parameters (dimensions) of lattice cells.

The three basic types of lattice cells are

a. Body-centered cubic (bcc): As the term implies, an atom is centered in the cubic structure, equidistant from the eight corner atoms. This central atom is completely contained within the unit cell boundaries, hence the name "body centered." To assist in perception of this model, the centered atom is shown shaded in Figure 3.1a, while an isometric of the cell is shown in Figure 3.1b. The lattice parameter a_{bcc} for all sides of the cube is shown in the figure.

The effective number of atoms/unit cell of bcc can then be calculated as

$$\text{Corners:} \quad \frac{1}{8} \times 8 = 1 \text{ atom}$$

$$\text{Center:} \quad = 1 \text{ atom}$$

$$\text{Total} \quad = 2 \text{ atom/unit cell}$$

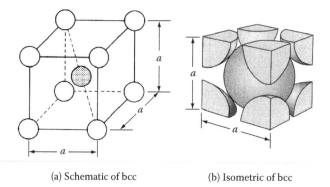

(a) Schematic of bcc   (b) Isometric of bcc

FIGURE 3.1   Body-centered cubic lattice.

Figure 3.2 visualizes the trigonometric representation of the main diagonal through the bcc unit cell, from which

$$B = \sqrt{2}\, a_{bcc}, C = \sqrt{3}\, a_{bcc} = 4r$$

where r is the atomic radius and a_{bcc} is the cell parameter.

Therefore,

$$a_{bcc} = \frac{4}{\sqrt{3}} r$$

Atomic packing factor (APF):

$$APF = \frac{\text{no. of effective atoms (volume/atom)}}{\text{volume of unit cell}}$$

$$(APF)_{bcc} = \frac{2 \times \frac{4}{3}\pi r^3}{\left(\frac{4}{\sqrt{3}} r\right)^3} = 0.68 \tag{3.1}$$

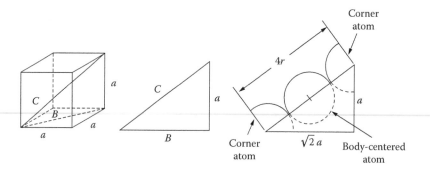

FIGURE 3.2   Trigonometric relationships of bcc lattice.

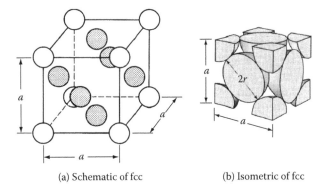

(a) Schematic of fcc   (b) Isometric of fcc

FIGURE 3.3 Face-centered cubic lattice.

This lattice arrangement is found in metals such as Fe (α), Cr, W, Ti, Ta, V, and so on.

b. Face-centered cubic (fcc): A unit cell of the fcc is shown schematically in Figure 3.3a, and its isometric shape is shown in Figure 3.3b. The lattice parameter is indicated by a_{fcc}. The effective number of atoms/unit cell of fcc can be calculated as

$$\text{Corners:} \quad \frac{1}{8} \times 8 = 1 \text{ atom}$$

$$\text{Faces:} \quad = \frac{1}{2} \times 6 = 3 \text{ atom}$$

$$\text{Total} \quad = 4 \text{ atom/unit cell}$$

Figure 3.4 visualizes the trigonometric relationships in this case, from which

$$4r = \sqrt{2}\, a_{fcc}, \, a_{fcc} = \frac{4}{\sqrt{2}} r$$

where r is the atomic radius and a_{fcc} is the cell parameter.

Therefore,

$$(\text{APF})_{fcc} = \frac{4 \times \frac{4}{3}\pi r^3}{\frac{4}{\sqrt{2}}} = 0.74 \tag{3.2}$$

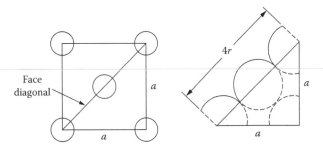

Face
diagonal

FIGURE 3.4 Trigonometric relationships of fcc lattice.

This arrangement is found in most ductile metals such as Fe (γ), Al, Cu, Ni, Au, Ag, Pb, Pt, and so on.

c. Hexagonal close-packed (hcp): As the term "close-packed" implies, this structure displays a high-density packing arrangement of atoms, similar to the fcc. The hcp unit cell is illustrated in Figure 3.5. Inspection of this lattice reveals seven atoms in each basal plane, plus three atoms positioned between the basal planes on axes 120° apart. The hcp lattice parameters are a_{hcp} (hexagon side of basal plane) and c (the height of the lattice). The effective number of atoms/unit cell of hcp is therefore,

$$\text{Corners:} \quad \frac{1}{6} \times 6 = 1 \text{ atom}$$

$$\frac{1}{6} \times 6 = 1 \text{ atom}$$

$$\text{Faces:} \quad \frac{1}{2} \times 2 = 1 \text{ atom}$$

$$\text{Body:} \qquad \quad = 3 \text{ atom}$$

$$\text{Total} \quad = 6 \text{ atoms/unit cell}$$

Atomic packing factor $(APF)_{hcp}$: a relationship exists between the lattice parameters c and a_{hcp}, assuming atoms of perfect spherical shape.

$$c = 1.633 \, a_{hcp}$$

Accordingly, the unit-cell lattice volume V_{hcp} is calculated as

$$V_{hcp} = \frac{3}{2}\sqrt{3} \, a_{hcp}^{2} \times c$$

$$= \frac{3}{2}\sqrt{3} \times 1.633 \, a_{hcp}^{3}$$

and $a_{hcp} = 2r$ (r = atomic radius)

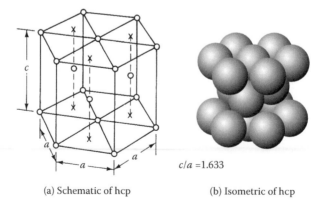

(a) Schematic of hcp   (b) Isometric of hcp

FIGURE 3.5   Hexagon close-packed lattice.

TABLE 3.1
Lattice Parameters Ratio c/a_{hcp} of Different Metals

Metal	Ratio c/a_{hcp}	Remarks	
Cd	1.886 ⎫	Prolated spheroids	◯
Zn	1.856 ⎭		
Sphere	1.633 →	Perfect sphere	◯
Mg	1.624		
Zr	1.590	Slightly compressed	⬯
Ti	1.588		
Be	1.586		

Therefore,

$$(APF)_{hcp} = \frac{6 \times \frac{4}{3}\pi r^3}{\frac{3}{2}\sqrt{3} \times 1.633(2r)^3} = 0.74 \tag{3.3}$$

which is the same value as that of the fcc lattice.

Metals having the hcp structure tend to have poor formability (less ductile), like Zn, Mg, Zr, Ti, Cd, and so on. They often require special processing techniques. Measurements of parameters a_{hcp} and c carried out on different metals of this group have revealed that the ratio c/a_{hcp} differs from metal to metal, with Table 3.1 indicating that the atoms are not perfect spheres, but are prolated or compressed to attain ellipsoidal shape, leading to slight deviation from the above theoretically calculated value of $(APF)_{hcp}$.

Table 3.2 summarizes and compares the data of different crystal structures of metals. As a conclusion, the fcc space lattice is in general more ductile and malleable than the bcc type. The bcc is usually the harder and stronger of the two. The hcp type lacks ductility and accepts little cold-working without failure. Of course, there are exceptions to these rules.

TABLE 3.2
Summary of Crystal Structures Data

Lattice Structure	Lattice Parameter Relationship	No. of Atoms per Unit Cell	APF	Typical Metals
bcc	$a_{bcc} = \dfrac{4r}{\sqrt{3}}$	2	0.68	Fe (α), Cr, Mn, W, Ti, Ta, V, etc.
fcc	$a_{fcc} = \dfrac{4r}{\sqrt{2}}$	4	0.74	Fe (γ), Al, Cu, Ni, Pb, Au, Ag, Pt, etc.
hcp	$a_{hcp} = 2r$ $c/a_{hcp} = 1.633$	6	0.74	Be, Cd, Mg, Zn, Zr, etc.

Note: r = atomic radius, a = cube or hexagon side, c = hexagon height.

TABLE 3.3
Examples of Allotropic Metals and Their Allotropic Transformation Ranges

Allotropic Metals	Temperature (°C)
Iron : Fe	
bcc (α)	<906
fcc (γ)	906–1403
bcc (δ)	1403–1535 (melting point)
Cobalt : Co	
hcp	<420
fcc	420–1495 (melting point)
Tin : Sn	
bcc (α) gray	<13
hcp (β) white	13–232 (melting point)
Titanium : Ti	
hcp (α)	<879
bcc (β)	879–1727 (melting point)

3.2.2 ALLOTROPIC (POLYMORPHIC) CRYSTALS (CHANGES IN CRYSTAL STRUCTURE)

Many metals exist in only one lattice form. Some, however, can exist in the solid state in two or more lattice forms, depending on temperature and sometimes on pressure. Such metals are said to be allotropic or polymorphic (Fe, Co, Sn, Mn, Cr, Ti, etc.). The most notable well-known example of such a metal is iron, where allotropic transformation makes it possible for heat treatment methods to produce a wide range of mechanical properties. Because of its allotropy, iron has become the most important alloy used in many applications. Table 3.3 lists some common examples of allotropic metals and the temperature ranges in which allotropic transformation takes place.

3.2.3 EFFECT OF LATTICE STRUCTURE ON DENSITY OF METALS

Density is the mass/unit volume. This concept can be applied to the unit cell. Accordingly, the density of a metal is the weight of atoms per unit cell, divided by the volume of that particular unit cell. This definition is valid in the case of crystals with no imperfections (perfect space lattice). Unfortunately, this condition is difficult to obtain in most engineering materials. The periodicity of the lattice is frequently interrupted by crystalline imperfections which take up space or volume, but add no mass. Therefore, the theoretical density based on the knowledge of crystal structure will be slightly greater than actual densities. Therefore, the theoretical density of a metal will be

$$D = \frac{(\text{no. of atoms/unit cell}) \times (\text{atomic weight/Avogadro's no.})}{\text{volume of unit cell}} \tag{3.4}$$

The Avogadro's number $(A_0) = 6.02 \times 10^{23}$ amu/g.

Example 1

Aluminum has an fcc crystal structure. Its atomic weight equals 26.98 amu. The approximate atom radius equals 1.431 Å (Å = 10^{-10} m) (Table 3.4). Determine the weight density of aluminum.

TABLE 3.4
Properties of Selected Elements

Element	Atomic Mass (amu)	Density (g/cm³)	Crystal Structure	Atomic Radius (Å)	Most Common Valence	Lattice Parameters a (Å)	Lattice Parameters c (Å)	Melting Point (°C)
Aluminum (Al)	26.98	2.70	fcc	1.431	3	4.05	—	660
Beryllium (Be)	9.01	1.85	hcp	1.143	2	2.29	3.58	1277
Cadmium (Cd)	112.41	8.65	hcp	1.489	2	2.98	5.62	321
Carbon (C)	12.01	2.25	hex	0.770	4	—	—	3500
Chromium (Cr)	52.01	7.19	bcc	1.249	3	2.88	—	1875
Cobalt (Co)	58.94	8.85	hcp	1.245	2	2.51	4.07	1495
Copper (Cu)	63.54	8.96	fcc	1.278	1	3.61	—	1083
Gold (Au)	197.00	19.32	fcc	1.441	1	4.08	—	1063
Iron (Fe)	55.85	7.87	bcc	1.241	2	2.87	—	1537
			fcc	1.269	3	3.66	—	—
Lead (Pb)	207.21	11.36	fcc	1.750	2	4.95	—	327
Magnesium (Mg)	24.32	1.74	hcp	1.604	2	3.21	5.21	650
Nickel (Ni)	58.71	8.90	fcc	1.246	2	3.52	—	1453
Platinum (Pt)	195.09	21.45	fcc	1.388	2	3.92	—	1769
Silver (Ag)	107.88	10.49	fcc	1.444	1	4.09	—	961
Tantalum (Ta)	180.95	16.60	bcc	1.430	5	3.31	—	2996
Tin (Sn)	118.70	7.30	Tetra	1.509	4	—	—	232
Titanium (Ti)	49.90	4.51	hcp	1.475	4	2.95	4.67	1668
Tungsten (W)	183.86	19.30	bcc	1.367	4	3.16	—	3410
Vanadium (V)	50.95	6.10	bcc	1.316	4	3.02	—	1900
Zinc (Zn)	65.38	7.13	hcp	1.332	2	2.66	4.95	420

Solution

$$D = \frac{(\text{no. of atoms/unit cell}) \times (\text{atomic weight}/A_o)}{\text{volume of unit cell}}$$

$$D_{Al} = \frac{(4 \text{ atoms})(26.98 \text{ amu}/6.02 \times 10^{23} \text{ amu/g})}{a^3}$$

$$a_{fcc} = \frac{4r}{\sqrt{2}} = \frac{4 \times 1.431 \times 10^{-8}}{\sqrt{2}} \text{cm} = 4.047 \times 10^{-8} \text{cm}$$

$$a_{fcc}^3 = 66.314 \times 10^{-24} \text{ cm}^3$$

Therefore,

$$D_{Al} = \frac{4 \times 26.98}{66.3314 \times 10^{-24} \times 6.02 \times 10^{23}}$$

$$= 2.703 \text{ g/cm}^3$$

The actual density of Al, as given in Table 2.4, is only 2.7 (slightly less than that theoretically calculated).

3.3 IMPERFECTIONS IN LATTICE STRUCTURE

Actual crystalline structures contain a large number of imperfections. The theoretical (cohesive) strength of metals as based on the sliding of atomic planes over one another predicts yield strengths on the order of 20,000 MPa. Actual strengths are typically 100 to 150 times less than this value. Crystalline imperfections can account for the discrepancies between theory and reality. The theoretical analysis is based on the assumption to displace all atoms involved at the same time, which is not altogether correct since atoms are displaced sequentially.

Investigations have indicated that atomic movements emanate from, and are affected by, imperfections in the crystals. The number and distribution of these imperfections has a great effect on the properties of metals. The mechanical properties of metals such as yield and ultimate strength, fracture strength, ductility, and hardness are strongly affected by lattice imperfections. These properties are known as structure-sensitive properties. On the other hand, physical properties such as density, elastic constants, melting temperature, thermal conductivity, thermal expansion, specific heat, reflectivity, emissivity, electrical conductivity, and electrochemical potential, are not sensitive to these imperfections and are thus known as structure-insensitive properties. There are several kinds of crystalline imperfections which may be categorized on the basis of their geometry and size, such as point defects, line defects (dislocations), surface defects (planar), and volume or bulk defects. Each type of imperfection has some influence on the properties of a material.

3.3.1 POINT DEFECTS

This category is referred to as point defects because they involve single sites in the space lattice (Figure 3.6). Although these imperfections in the space lattice necessarily cover some finite volume, their extent is limited to a localized disturbance. Therefore, they are treated as though they behave like a point.

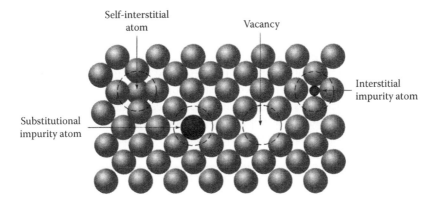

FIGURE 3.6   Point defects in single crystals. (Courtesy of Dr. William G. Moffatt.)

Types of Point Defects:

- *Vacancy:* This is simply a vacant lattice site, where an atom (or an ion) is missing from its normal position in the crystal, as illustrated schematically in Figure 3.6. Vacancies may occur in other forms (di-vacancies), which is simply a situation where two neighboring atoms are missing.
- *Self-interstitial atom:* This is an extra atom in the lattice (Figure 3.6).
- *Interstitial impurity atom:* This is a smaller atom occupying an interstitial site in the lattice of relatively larger atoms (Figure 3.6). Interstitial atoms fit into the interstices between the normal atom structure, and produce a local disturbance in the lattice, depending on their size, the size of host atoms, valence effects, and the structure of the host lattice. Such a structure exists when carbon atoms (much smaller) are introduced into iron atoms and they squeeze into the structure (Figure 3.7a).
- *Substitutional impurity atom:* In this type, the impurity atom is larger than its host. However, this is not always the case, and substitutional atoms may be smaller than their hosts. Depending on the particular geometrical and chemical differential, a localized disturbance is created in the lattice (Figure 3.7b). This perturbation affects the mechanical and physical properties of the crystal. In spite of the connotation of "defect," the impurity atoms can be extremely beneficial to certain classes of engineering materials which should be strengthened by adding substitutional atoms of different spaces to host metals and alloys.

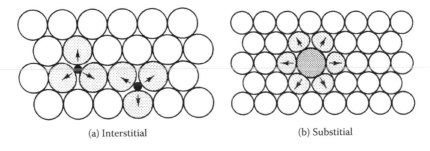

(a) Interstitial   (b) Substial

FIGURE 3.7   Interstitial and substantial point defects.

3.3.2 LINE DEFECTS (DISLOCATIONS)

This category of imperfections consists of defects which are considerably larger than the point defects, in that they extend many atom distances through the crystal. Such imperfections are often referred to as dislocation. A dislocation is an interruption of the periodicity of a lattice, which puts the crystal out of order. Dislocation can be created during crystallization (mismatch in crystallographic orientation) or when the structure is subjected to plastic deformation. An annealed metal can typically have dislocation densities on the order of 10^6 to 10^8 dislocations/cm^2. Heavily cold-worked metals (severely deformed at room temperature) can exhibit as many as 10^{12} dislocations/cm^2. That means, the density of dislocations resulting only due to cold working is as much as 10^6 dislocations/cm^2.

Types of Dislocations: Two types of dislocations can be present in crystals, and they are referred to as edge and screw dislocations.

> Edge dislocation: A slip plain containing a dislocation requires lower shear stress to cause slip than does a plain in a perfect lattice (Figure 3.8a). An analogy used to describe the movement of an edge dislocation (Figure 3.8b) is that of moving a large carpet by forming a hump at one end and moving the hump forward toward the other end. The force required to move the carpet in this way is much less than that required for sliding the whole carpet along the floor.
>
> Screw dislocation: This differs from the edge type in that the displacement of atoms is parallel to the line of dislocation rather than normal to it. It is so named because the atomic planes form a spiral ramp (Figure 3.9).

In real crystals, there can be combinations of edge and screw dislocations. Figure 3.10 illustrates a high magnification transmission electron micrograph of a Ti alloy in which dark lines of dislocations are observed.

3.3.3 SURFACE OR PLANAR DEFECTS (GRAIN BOUNDARIES)

For a crystalline solid, when the periodic and repeated arrangement of atoms is perfect, the result is a single crystal or grain. Most crystalline solids and engineering materials are composed of a collection of many crystals or grains. Such materials are termed polycrystalline. Various stages in solidification (crystallization) of polycrystalline structures are represented schematically in Figure

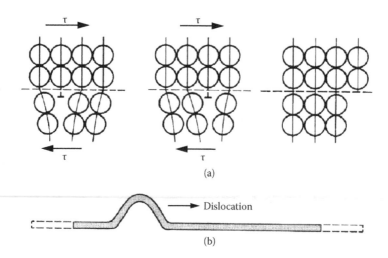

FIGURE 3.8   Edge dislocation and carpet analogy.

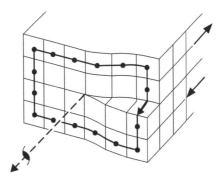

FIGURE 3.9 Screw dislocation.

3.11. Initially, small nuclei form at various positions (Figure 3.11a). Grains of random crystallographic orientations (Figure 3.11b–d), as indicated by the square grits, form. The small grains grow by successive addition from the surrounding liquid of atoms to the structure of each forming dendritic structure. The extremities of adjacent grains impinge on one another as the solidification process approaches completion (Figure 3.11d). According to this figure, the crystallographic orientation varies from grain to grain, resulting in the surface or planar imperfection. Atomic mismatch exists at the grain boundary.

Mechanical and other properties of a single crystal (grain) are anisotropic (exhibiting different values of properties in different crystallographic directions). In contrast, a polycrystalline structure consisting of a large number of randomly oriented grains is isotropic (has the same properties in all directions). The sizes and diverse spatial orientation of grains in a metal largely affect its properties. A fine-grained metal is likely to have a better distribution of grains oriented to respond to stresses in any direction (isotropic) than a coarse-grained material. Of more importance, fine grains present more grain boundaries to inhibit the propagation of dislocations. For these reasons, a fine-grained metal as a rule has a greater yield and ultimate strength, hardness, fatigue strength, and resistance to impact.

Dislocations travel through a crystal on many planes until they reach grain boundaries or imperfections in the lattice that stop them. Cold-worked metal is said to strain harden because a higher stress is necessary to move the entangled and crowded dislocations. As dislocations are piled up under higher and higher stress, they are forced to combine into small cracks that ultimately grow to fractures in the metal.

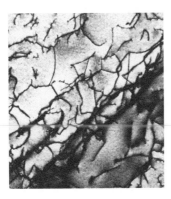

FIGURE 3.10 Dark lines of dislocations of Ti alloy. (Courtesy of Plichta, M. R., Michigan Technological University.)

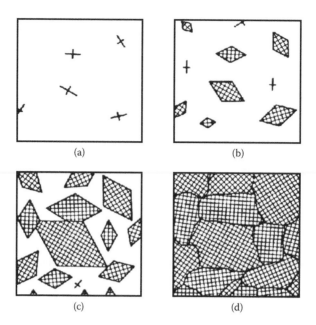

FIGURE 3.11   Solidification of polycrystalline structure and formation of grain boundaries. (From Smith, C., *The Science of Engineering Materials*, Prentice-Hall, Englewood Cliffs, NJ, 1977. With permission.)

Anything that interferes with the flow of dislocations across a grain makes it harder. That may be a distortion of the space lattice or the presence of a foreign material. The former mechanism is referred to as solid solution hardening, whereas the latter is called dispersion hardening. Heat treatment process is often used to control those conditions and thus the properties of the metal. Alloying of metals is another way of controlling lattice conditions.

The space lattice of an alloy is commonly distorted because atoms of different metals are of different sizes and exert different atomic forces. Added atoms may or may not replace atoms of the parent metal in the space lattice, but in any case, they do cause lattice distortions.

The number and size of grains (crystals) developed in a unit volume of a metal depends on the rate of nucleation that takes place, and the rate at which these crystals grow affects the size of grains developed. Rapid cooling generally produces smaller grains, and slow cooling produces coarse grains.

Micrography of Polycrystalline Structures: Since adjacent grains have different orientations, the grain boundary is a zone of disorder. To reveal grain boundaries, the solidified surface may be ground, polished, and etched with a suitable reagent. Because of the higher chemical energy of atoms on the grain boundary, they dissolve at a greater rate than those within the grains. The grooves appear as a dark line when viewed under a microscope because they reflect light at an angle different from that of the grains themselves (Figure 3.12). Figure 3.13 shows schematically, three polished and etched grains of polycrystalline specimen of grains having different orientations. The luster or texture of each depends on its reflectance properties.

When the microstructure of a two-phase alloy is to be examined, an etchant is chosen that produces a different texture for each plane so that the different phases may be distinguished from each other. Microscopic investigations of crystalline structure are only possible with optical and electronic microscopes, usually in conjunction with appropriate photographic equipment. Transmissive and reflective modes are possible for both microscopes.

Determination of Grain Size: The grain size is frequently determined using two methods of photographic techniques, with charts proposed by ASTM (E112).

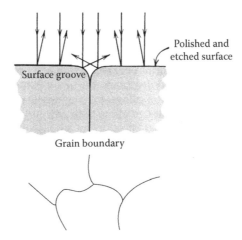

FIGURE 3.12   Photograph of α-ferrite showing grain boundaries and surface groove produced by etching. (Adapted from Callister, W. D., *Material Science and Engineering—An Introduction*, 4th ed., John Wiley, 1997.)

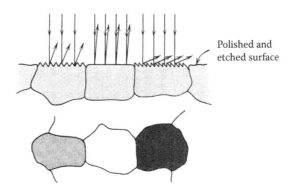

FIGURE 3.13   Schematic of three polished and etched grains of a polycrystalline specimen having grains of different orientations. (Adapted from Callister, W. D., *Material Science and Engineering—An Introduction*, 4th ed., John Wiley, 1997.)

3.3.4   Volume or Bulk Defects

These are other defects existing in all solid materials that are much larger than those defects discussed before. These include pores, cracks, foreign inclusions (nonmetallic elements such as oxides, sulfides, and silicates), and other phases. They are mainly introduced during processing and fabrication steps.

3.4   SOLIDIFICATION OF METALS AND ALLOYS

Casting is a solidification process. Many structural features that control product quality are set during solidification. Many casting defects such as porosity and shrinkage are solidification phenomena, which can be reduced by controlling the solidification process. Cooling curves provide one of the most important tools for studying the solidification process. By inserting thermocouples into the

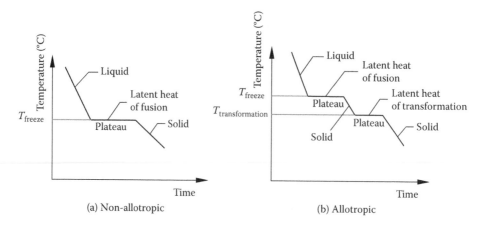

FIGURE 3.14 Solidification of pure metal.

molten metal and monitoring the temperature versus time, one can obtain valuable insight into what is happening during solidification.

3.4.1 Solidification of Pure Metals and Alloys

Pure metals have a clearly defined melting or freezing point, and solidification takes place at a constant temperature. When the temperature of a molten pure metal is reduced to the freezing phase, the latent heat of fusion is given off while the temperature remains constant (plateau of Figure 3.14a). At the end of this isothermal stage, solidification is complete, and the solid metal cools to room temperature. The process is reversible and the same latent heat is absorbed during the melting process. In allotropic pure metals, another kind of change of state involves a relocation of the atoms in a solid metal, causing an allotropic transformation. The space lattice changes at a specific temperature as heat is added or taken away. The heat given off or absorbed is called latent heat of transformation (Figure 3.14b).

On the other hand, alloys solidify over a range of temperatures (Figure 3.15a). Solidification begins when the temperature drops below the liquidus (points x) and completes when the temperature reaches the solidus (points y) (Figure 3.16). Within this temperature range, the alloy is in a mushy or pasty state. Its composition and state are described by the relevant alloy's phase diagram

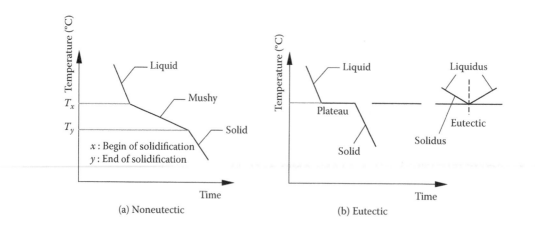

FIGURE 3.15 Solidification of an alloy.

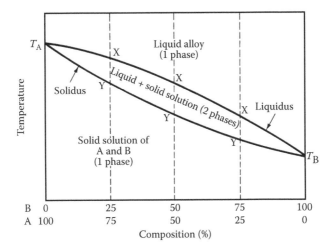

FIGURE 3.16 Hypothetical alloy system of solid state solubility (type 1).

(Figure 3.16). If the alloy has a eutectic or eutectoid composition, it solidifies isothermally as if it is pure metal (Figure 3.15b).

3.4.2 Solid Solutions

A solid solution is one in which two or more elements are soluble in a solid state, forming a single homogeneous solid phase in which the alloying elements are uniformly distributed throughout the solid mass. An alloy consists of two or more metals, or at least one metal and a nonmetal, mixed intimately by diffusion. Diffusion is the movement of atoms of one material among the atoms of another material. A certain resistance exists to the movement of the solute atoms in a solid, and it must be overcome for diffusion to occur. The necessary energy that just overcomes such resistance is called the activation energy. More energy just increases the rate of diffusion. Diffusion is the most important mechanism not only in solidification, but also in many other phases of manufacture. It is greatly accelerated by high temperatures. Diffusion necessitates time, so it is reduced if sufficient time is not available. There are two forms of solid solutions.

1. Substitutional solid solutions: Each atom of the solute replaces an atom of the solvent in its space lattice. The favorable conditions for this form are
   - Size (diameter) of the solute atom is similar to that of the solvent (differs by no more than 15%).
   - Space lattices of solvent and solute are similar.
   - Proximity of the electromotive series (same valency).

   An example of this form is the brass (Cu–Zn alloy), in which Zn is the solute atom.
2. Interstitial solid solutions: Atoms of the solute take position interstitially in the spaces of the lattice of solvent. The favorable conditions for this form are
   - The size (diameter) of the solute atom is less than 60% of that of the solvent.
   - The solvent metal is polyvalent.
   - The proximity of the electromotive series.

An important example of this form is steel, an alloy of iron and carbon, in which carbon atoms are present interstitially in solvent atoms. The properties of steels can be varied over a wide range by controlling the amount of carbon in iron (up to a maximum of 2% carbon). Accordingly, steel is versatile, inexpensive, and useful for many applications.

3.4.3 PHASE DIAGRAMS

Phase diagrams are also called equilibrium or constitutional diagrams. Equilibrium or phase diagrams show the various phases in which an alloy exists, depending on temperature and composition. These diagrams also delineate the conditions under which phases coexist in equilibrium (solubility lines).

There are three standard types of equilibrium diagrams. Only simple examples of each will be considered to illustrate the principles.

Type 1: Equilibrium diagram for metals completely soluble in both the liquid and solid states (solid state solubility), for example, Sb–Bi and Cu–Ni alloys.

Figure 3.16 visualizes the phase diagram for the hypothetical alloy system of A and B, the melting points T_A and T_B, and the respective T_X (beginning of solidification) and T_Y (completion of solidification) are plotted corresponding to the different compositions of the alloy.

Three regions are outlined on Figure 3.16. The uppermost portion labeled liquid alloy single phase is a liquid solution of A and B for all compositions. The central, mushy region contains two phases, liquid alloy plus a solid solution of the components. This is the region of solidification. The lower region is completely solid (solid solution A and B). It is a single phase. An interesting feature of this phase diagram is the narrow two-phase central region (transmission zone). Its upper boundary is the locus of T_X-points (liquidus line). The lower boundary is the locus of T_Y-points (solidus line).

The Inverse Lever Arm Rule:

This is an analytical technique for determining the proportions of solid and liquid contents coexisting in the transition 2-phase region (not applicable for single phases). This technique is applicable at any temperature and composition. It is also valid for other types of equilibrium diagrams. This rule is based on the simple analogy for the equilibrium of levers (Figure 3.17). For a selected temperature T and composition of alloy C_o, the lever is balanced if

$$\text{solid} \times \ell_1 = \text{liquid} \times \ell_2 \tag{3.5}$$

$$\frac{\text{solid}}{\text{solid} + \text{liquid}} = \frac{\ell_2}{\ell_1 + \ell_2} \tag{3.6}$$

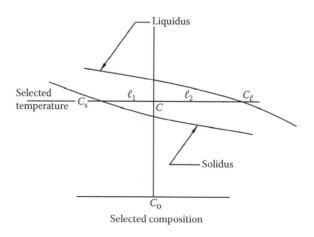

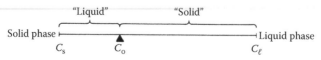

FIGURE 3.17   Inverse-lever rule.

Therefore,

$$solid = \frac{\ell_2}{\ell_1 + \ell_2} = \frac{C_\ell - C_o}{C_\ell - C_s} \times 100 \qquad (3.7)$$

and

$$liquid = \frac{\ell_1}{\ell_1 + \ell_2} = \frac{C_o - C_s}{C_\ell - C_s} \times 100 \qquad (3.8)$$

Example 2

Referring to the Cu–Ni phase diagram (Figure 3.18),

a. For an alloy 70% Cu, 30% Ni, determine the amount of solid solution (x-phase) that has formed at a temperature of 1200°C.
b. How much liquid exists at the same temperature?

Solution:

a. Construct the lever going through 1200°C, with its fulcrum Co = 70% Cu. Intersection of lever with solidus C_s = 62% and liquidus C_ℓ = 78%, respectively.
   Referring to Figure 3.18,

$$solid\ phase = \frac{C_\ell - C_o}{C_\ell + C_s} \times 100$$

$$= \frac{78 - 70}{78 - 62} \times 100 = 50\% \text{ (solid solution)}$$

$$liquid\ phase = \frac{C_o - C_s}{C_\ell - C_s} \times 100$$

$$= \frac{70 - 62}{78 - 62} \times 100 = 50\%$$

Type 2: Equilibrium diagram for metals soluble in the liquid state but insoluble in the solid state (eutectic alloy), for example, Bi–Cd alloy (Figure 3.19).

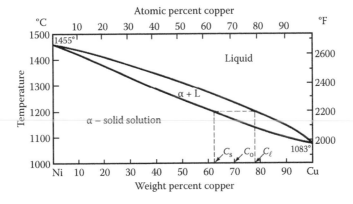

FIGURE 3.18   Cu–Ni phase diagram, solid state solubility (type 1).

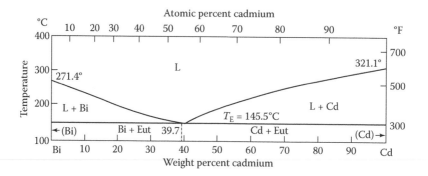

FIGURE 3.19   Bi–Cd phase diagram, insoluble in the solid state (type 2).

In this phase diagram, below the liquidus, a two-phase region exists for all compositions. The solid phase above the eutectic temperature ($T_E = 145.5$) is either pure Cd or pure Bi. The lowest temperature that the alloy is still liquid is referred to as eutectic point, at which the reaction is given by

$$\text{Liquid} \xleftarrow[\text{heating}]{\text{cooling}} \text{solid Bi} + \text{solid Cd} \tag{3.9}$$

In this case, the solids are Bi and Cd in the ratio 60% Bi and 40% Cd. Below T_E, all compositions are completely solid. However, because of mutual insolubility of Bi and Cd, the product formed is not a solid solution. Rather, the solid formed contains the eutectic mixture plus either free Cd or Bi (Figure 3.19). At the eutectic temperature T_E, the composition of the remaining liquid contains 60% Bi–40% Cd, regardless of what C_o was started with. The lever rule is applicable to predict the constituents coexisting at room temperature, as illustrated in the following example.

Example 3

Referring to Figure 3.19, determine the constituent coexisting at room temperature for the 70% Cd alloy.

Solution

Below the eutectic temperature, no other phase changes take place in this type of alloy. Therefore, at room temperature, the constituent of this alloy are (Equation 3.7):

$$\text{Solid Cd} = \frac{70 - 40}{100 - 40} \times 100$$

$$= 50\%$$

Eutectic   solid Cd

|———————————|————————————|

40   70   100

Therefore,

$$\text{Eutectic} = 100\% - 50\% \text{ (balance of mass)} = 50\%$$

Type 3: Equilibrium diagram for metals completely soluble in the liquid state, but partially soluble in the solid state (partial solid solubility), for example, Pd–Sn, and Cu–Ag alloys. This diagram (Figure 3.20) is a combination of type 1 and type 2.

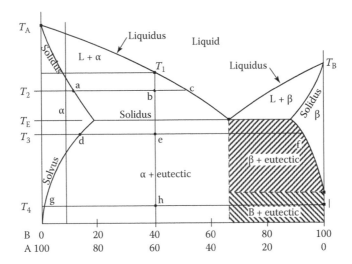

FIGURE 3.20 Phase diagram for hypothetical binary system that exhibits partial solid solution (type 3).

It is worthwhile to emphasize three regions which compose the solid state of this type of alloy

1. α Solid solution: atoms of B (solute) dispersed in A (solvent).
2. β Solid solution: atoms of A (solute) dispersed in B (solvent).
3. Eutectic mixture of α and β. The eutectic of this binary system is 35% A–65% B.

The essential difference in type 3 is the solid solution which forms on either side of the eutectic region (20 ≤ % B ≤ 88). The partial solubility characteristics of these components in the solid state result in the regions bounded by the solidus and solvus lines. The solvus indicates the maximum solubility of the components in their respective solid solutions α and β.

An alloy of 60% A and 40% B acts like type 2 alloy. At temperature T_1 and above, the alloy is single-phase liquid solution. At T_2, it is composed of a liquid solution and solid solution. The proportion and composition of the constituents are found from the lever a–b–c. At T_3, the lever d–e–f indicates a mixture of α solid solution and β solid solution. At T_4, the lever g–h–i shows the proportions of α solid solution and pure metal B, the β solid solution does not exist at that temperature. The alloy 90% A–10% B (Figure 3.20) is comparable to the type 1 alloy. This alloy comprises a mixture of liquid solution and α solid solution at temperatures between liquidus and solidus lines. The proportions and compositions of each in this two-phase region are depicted by the use of the inverse-lever rule. Only the α solid solution exists at temperatures between the solidus and solvus lines (Figure 3.20), and its composition is the basic composition of the alloy. The lever rule is not applicable in any single-phase region, and thus not in this α region. Below the solvus at T_4, both α solid solution and pure metal B exist, where the lever rule applies in this two-phase area. Real type 3 systems are illustrated by the Pb–Sn phase diagram (Figure 3.21) and also the Cu–Ag phase diagram (Figure 3.22).

Example 4

Referring to Figure 3.21, illustrating the lead-tin phase diagram ($C_E = 61.9\%$, $T_E = 183°C$). For the alloys 10% Pb and 30% Pb, determine

1. At 350°C, what are the quantities of Pb and Sn existing?
2. At 200°C, what are the amounts of phases existing?
3. At 100°C, what are the amounts of phases existing?
4. At 0°C, what are the amounts of phases existing?

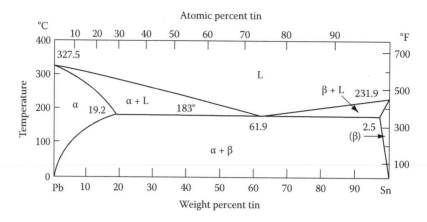

FIGURE 3.21   Pb–Sn phase diagram, partially soluble in the solid state (type 3).

Solution

The answer is given in the following table

Item	10% Pb	30% Pb
1. At 350°C	One-phase liquid region, consisting of: 10% Pb–90% Sn	One-phase liquid region, consisting of: 30% Pb–70% Sn
2. At 200°C	One-phase α solid solution, consisting of: 10% Pb–90% Sn	Two-phase, consisting of α + L
		$\alpha \vdash\!\!\!\!\!-\!\!\!\!-\!\!\!-\!\!+\!\!\!-\!\!\!\!-\!\!\dashv L$ $\qquad$ 18 $\qquad$ 30 $\qquad$ 58
		$\alpha = \dfrac{58 - 30}{58 - 18} \times 100 = 70\%$
		$L = 30\%$
3. At 100°C	Two-phase, consisting of α + β	Two-phase, consisting of α + β.
	$\alpha \vdash\!\!\!-\!\!\!-\!\!+\!\!\!-\!\!\!-\!\!\dashv \beta$ $\quad$ 5 $\qquad$ 10 $\qquad$ 98	$\alpha \vdash\!\!\!-\!\!\!-\!\!+\!\!\!-\!\!\!-\!\!\dashv \beta$ $\quad$ 5 $\qquad$ 30 $\qquad$ 98
	$\alpha = \dfrac{98 - 10}{98 - 5} \times 100 = 94.6\%$	$\alpha = \dfrac{98 - 30}{98 - 5} \times 100 = 73\%$
	$\beta = 5.4\%$	$\beta = 27\%$
At 0°C	Two-phase, consisting of α + β	Two-phase, consisting of α + β.
	$\alpha \vdash\!\!\!-\!\!\!-\!\!+\!\!\!-\!\!\!-\!\!\dashv \beta$ $\quad$ 0 $\qquad$ 10 $\qquad$ 100	$\alpha \vdash\!\!\!-\!\!\!-\!\!+\!\!\!-\!\!\!-\!\!\dashv \beta$ $\quad$ 0 $\qquad$ 30 $\qquad$ 100
	$\alpha = \dfrac{100 - 10}{100 - 0} \times 100 = 90\%$	$\alpha = \dfrac{100 - 30}{100 - 0} \times 100 = 70\%$
	$\beta = 10\%$	$\beta = 30\%$

3.4.4   Iron–Carbon Phase Diagram

A more detailed attention to the Fe–C phase diagram should be given because it is basic to ferrous metals, one of the most versatile and widely used engineering materials. The diagram most frequently encountered, however, is not the full iron–carbon diagram but the iron–iron carbide diagram shown in Figure 3.23. Here, the intermetallic compound Fe_3C is used to terminate the carbon range at 6.7% C.

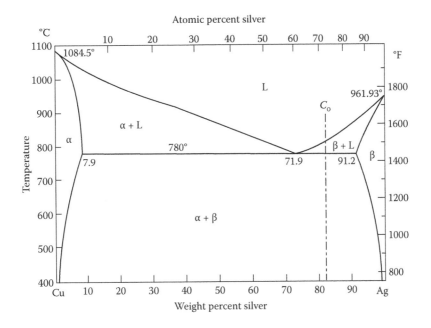

FIGURE 3.22 Cu–Ag phase diagram, partially soluble in the solid state (type 3).

There are four single phases within the Fe–Fe$_3$C diagram. Three of these occur in pure iron, and the fourth is the carbide intermetallic at 6.67%. On heating, pure iron changes its crystalline structure from bcc to fcc at 910°C. Iron is unique among materials in that it reverts to bcc at 1400°C. Referring to Figure 3.23, it is convenient to assign some basic names of phases for discussion.

Ferrite (α-Iron): Pure iron at room temperature is called either α-iron or ferrite. Ferrite is quite soft and ductile. Its tensile strength is less than 300 MPa. It is a ferromagnetic material at

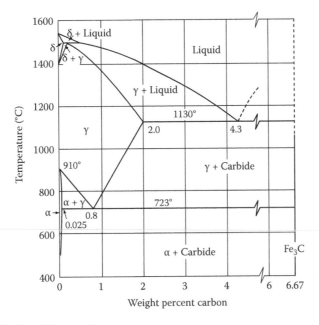

FIGURE 3.23 Fe–Fe$_3$C equilibrium diagram.

temperatures under 767°C (Curie point). Because this transmission is not associated with any change in phase, it does not appear in the equilibrium phase diagram. Since ferrite has a bcc structure, the interatomic spaces are small, and hence cannot readily accommodate even a small carbon atom. Therefore, solubility of carbon in ferrite is very low. It can only hold 0.02% C in solid solution. The carbon atom is too small for substitutional solid solution, and too large for extensive interstitial solid solution.

Austenite (γ-Iron): The face-centered modification of iron is called austenite (γ-iron), in honor of the famed metallurgist Sir Robert Austen of England. It is the stable form of pure iron, at temperature between 910°C and 1400°C. It is difficult to make a direct comparison between austenite and ferrite, regarding the mechanical properties, because they must be compared at different temperatures. However, at its stable temperature, austenite is soft and ductile, and consequently is well suited to fabrication processes. Many steel forging and rolling operations are performed at temperatures of 1100°C or above, where the iron is fcc. Austenite is totally paramagnetic. The fcc structure of iron has larger interatomic spacing than does ferrite, so that they are able to accommodate the carbon atoms in solution. The solubility is limited to 2% C (Figure 3.23). By definition, steels contain less than 2% C. Steels may have their carbon completely dissolved in austenite at high temperatures. Most heat treatments of steel begin with the single-phase austenite structure.

δ-Iron: Above 1400°C, austenite is no longer the most stable form of iron, and the crystal structure changes back to a bcc phase called δ-iron. This iron is the same phase as α-iron except for its temperature range, and so it is commonly called δ-ferrite. The solubility of carbon in δ-ferrite is small, but it is appreciably larger than α-ferrite, because of higher temperature.

Cementite (Iron Carbide): Iron carbide has a chemical composition of Fe_3C. This does not mean that iron carbide forms molecules of Fe_3C, but that the crystal lattice contains iron and carbon atoms in a 3:1 ratio. The compound Fe_3C has an orthorhombic unit cell with 12 iron atoms and 4 carbon atoms per cell, and thus has a carbon content of 6.7%. As compared with austenite and ferrite, cementite, as most intermetallics, is very hard and brittle. Its presence with ferrite in steel greatly increases the strength of the steel. However, since pure iron carbide lacks ductility, it cannot adjust to stress concentrations and tends to embrittle steel. Therefore, alloys with excessive amounts of cementite are highly undesirable.

Example 5

Based on the unit cell of iron carbide Fe_3C, determine the weight percent of carbon in Fe_3C intermetallic compound.

Solution

The Fe_3C cell contains 4 carbon atoms and 12 iron atoms. Referring to Equation 3.4, the weight of carbon per unit cell W_c, and the weight of iron per unit cell W_{Fe} can be calculated.

$$W_c = \frac{(4 \text{ atoms})12 \text{ amu/atom}}{6.02 \times 10^{23} \text{ amu/g}} = 8 \times 10^{-23} \text{ g/unit cell}$$

$$W_{Fe} = \frac{(12 \text{ atoms}) \times 55.8 \text{ amu/atom}}{6.02 \times 10^{23} \text{ amu/g}} = 111.2 \times 10^{-23} \text{ g/unit cell}$$

Total $W_t = W_C + W_{Fe} = 119.2 \times 10^{-23}$ g/unit cell

$$\% \text{ age of C} = \frac{8 \times 10^{-23}}{119.2 \times 10^{-23}} \times 100 = 6.7\%$$

The Three Phase Reactions (Peritectic, Eutectic, and Eutectoid)

Three distinct phase reactions can be identified (Figure 3.23).

1. Peritectic: At 1495°C, a peritectic reaction occurs for alloys with a low weight percentage of carbon. The peritectic reaction rarely assumes any engineering significance.
2. Eutectic (ledeburite): Eutectic reaction is observed at 4.3% carbon, and eutectic temperature of 1130°C. This transformation is identical, in principle, to the eutectic reactions studied before. Inspection of the Fe–Fe_3C phase diagram shows that the eutectic product, also referred to as ledeburite, is a mixture of austenite and iron–carbide Fe_3C, and this transformation occurs outside the composition of steels. All alloys containing more than 2% carbon experience the eutectic reaction, and are generally classified as cast irons.
3. Eutectoid (pearlite): This reaction occurs under equilibrium conditions at a temperature of 723°C and composition of about 0.8% C (Figure 3.23). The eutectoid reaction on cooling involves the simultaneous formation of ferrite and carbide from austenite of eutectoid composition. There is nearly 12% carbide and 88% ferrite in the resulting mixture. Since carbide and ferrite form simultaneously, they are intimately composed of alternate layers of ferrite (white) and carbide (dark). The resulting microstructure, called pearlite, is very important in iron and steel technology.

Example 6

Referring to the Fe–Fe_3C phase diagram given in Figure 3.23, determine the proportions of ferrite α and iron carbide Fe_3C in pearlite (0.8% C), using the inverse-lever rule.

Solution

These phases can be quantitatively calculated from the inverse-lever rule.

$$\alpha \vdash\!\!\!\!-\!\!\!-\!\!\!-\!\!\!+\!\!\!-\!\!\!-\!\!\!-\!\!\dashv Fe_3C$$
$$0.025 \qquad 0.8 \qquad C\% \rightarrow 6.7$$

$$\alpha = \frac{6.7 - 0.8}{6.7 - 0.025} \times 100 = 88.4\%$$

From the balance,

$$Fe_3C = 100 - 88.4 = 11.6\%$$

These values are approximately,

$$\text{Eutectoid } \alpha = 88\%$$

$$\text{Eutectoid } Fe_3C = 12\%$$

3.5 REVIEW QUESTIONS

1. What are important differences between solidification in pure metals and alloys?
2. What are disadvantages of grain growth?
3. What is an allotropic material?
4. What are the three most common crystal structures found in metals?

5. What is a dislocation? How do dislocations affect the mechanical properties of metals?
6. A silver solder brazing alloy contains 72% Ag and 28% Cu. Determine the percentages of the phases of this alloy if it is slowly cooled to room temperature.
7. Microscopic examination of a Cu–Ag alloy reveals a structure containing approximately 70% eutectic and 30% β solid solution. Estimate the nominal composition of this alloy.
8. Why is it difficult to compare between the mechanical properties of austenite and ferrite?
9. Why is austenite capable of accommodating carbon atoms, and ferrite is not?
10. Calculate the lattice parameter for fcc aluminum ($r = 1.431$ Å).
11. Ni forms an fcc structure and has an atomic radius of 1.246 Å. Calculate the volume of its unit cell.
12. Iron experiences an allotropic transformation at 910°C. Calculate the difference that this transformation produces in volume of unit cell when iron is heated just above this temperature, assuming that the temperature is not a factor.
13. Assuming that APF for hcp lattice is 0.74, calculate the theoretical density of Mg ($r = 1.610$ Å).
14. Be exhibits an hcp structure and atomic radius $r = 1.14$ Å. If $c/a = 1.586$, determine its unit cell volume, and theoretical density.
15. Sterling silver consists of 92.5% Ag, and 7.5% Cu (Figure 3.22). Calculate the percentage of phases that would be formed by cooling this alloy at room temperature under near-equilibrium conditions.

3.6   PROBLEMS

1. Determine the weight density of Cu in g/cm³, provided that
   Atomic weight of Cu = 63.54
   Atomic radius of Cu = 1.278 Å
   Avogadro's number = 6.02×10^{23}
2. Ti exhibits an hcp crystal structure, and atomic weight of 47.9. If $c = 1.588\, a_{hcp}$, $a_{hcp} = 2r$, $r = 1.475$ Å, calculate the weight density of Ti.
3. The phase diagram of Ag–Cu system is given ($C_E = 71.9\%$, $T_E = 780$°C). For an alloy of $C_o = 82\%$ Ag, determine
   a. At 900°C, what are the quantities of Ag and Cu in existence?
   b. At 800°C, what phases exist?
   c. At 400°C, what phases exist?
   d. At 400°C, what is the amount of eutectic constituent?
   e. At 400°C, what individual phases coexist?

BIBLIOGRAPHY

Callister, W. D. 1996. *Material Science and Engineering—An Introduction*, 4th ed. New York: John Wiley & Sons.
Degarmo, E. P., Black, J. T. and Kosher, R. A. 1997. *Materials and Processes in Manufacturing*, 8th ed. Upper Saddle River, NJ: Prentice-Hall.
Dibenedetto, A. T. 1967. *The Structure and Properties of Materials*. New York: McGraw-Hill.
Doyle, L. E., Keyser, C. A., Leach, J. L., Schrader, G. F., and Singer, M. B. 1985. *Manufacturing Processes and Materials for Engineers*, 3rd ed. Upper Saddle River, NJ: Prentice-Hall.
Kalpakjian, S., and Schmid, S. R. 2003. *Manufacturing Processes for Engineering Materials*, 4th ed. Upper Saddle River, NJ: Prentice-Hall.
Schey, J. A., 2000. *Introduction to Manufacturing Processes*, 3rd ed. New York: McGraw-Hill.
Thornton, P. A., and Colangelo, V. J. 1985. *Fundamentals of Engineering Materials*. Upper Saddle River, NJ: Prentice-Hall.
Van Vlack, L. H. 1973. *Material Science for Engineers*, 4th ed. Reading, MA: Addison-Wesley.

4 Engineering Materials and Their Applications

4.1 INTRODUCTION

Materials are all about us. They have been so intimately related to the emergence and ascent of man that they have given names to the Stone, Bronze, and Iron Ages of civilization. Naturally existing and man-made materials have become an integral part of our lives. The development of materials used for engineering purposes has experienced unprecedented growth over the past few decades, and engineering materials have truly become an essential part of modern science and technology. This integral relationship is evident in every product and every industry that one can think of.

A wide spectrum of engineering materials is available today. A proper selection has to be made to suit the desired application. Metals are specified when strength, toughness and durability are the primary requirements. Ceramics are generally limited to low-value applications where heat or chemical resistance is required, and when acting loads are compressive. Glass is used for its optical transparency, plastics are appointed to applications where low cost and light weight are attractive features, and semiconductors are used in integrated circuits necessary for computer industries. By careful selection of engineering materials, it is possible to achieve an optimum blend of properties by using a variety of metals, organic materials, ceramics, and so on.

For that reason, the designer sometimes should consult materials specialists when selecting materials that will be used to convert the design into reality. Without this knowledge, a material is simply a *black box* and the designer will have no conception of materials limitations or of possible modifications in materials selection and design.

The selection of a material for a particular application involves consideration of some factors like service requirements (strength, manner of load application, wear, corrosion resistance, electrical properties, aesthetic consideration, and so on), manufacturing requirements (machinability, formability, finish desired, need for heat treatment, etc.) and cost of raw material.

Often, one may have two or three possible options in selection of an appropriate material. The final decision then should be based on preference and experience of the designer and user, and important considerations like ease of repair, possible occurrence of fault, availability of repair facilities, useful life, and so on.

Tens of thousands of different materials have evolved with rather specialized characteristics which meet the needs of our modern complex society. These include metals and alloys, ceramics, glasses, plastics, elastomers, concrete, composites, asbestos, wood, papers, and others.

The development of many technologies that make our existence so comfortable has been intimately associated with the accessibility of suitable materials. The automotive industries alone now consume approximately 60 million tons of engineering materials worldwide every year, primarily steel, cast iron, aluminum, copper, glass, lead, polymers, rubber, and zinc. According to U.S. automakers, it is claimed that there is increased use of lighter-weight materials and high-strength steels as well as plastics and composites. A variety of steels are used in a car. For example, the thin sheet steel of a fender must have different properties from the steel of transmission gears. The steel of a fender must be soft and ductile so that it may be rolled to a thickness of only 1 mm without cracking. The steel of a gear must be strong and tough, with a hard surface, so that its teeth will not snap

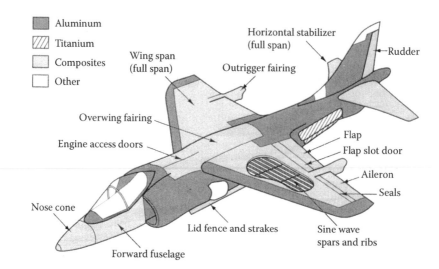

FIGURE 4.1 Materials spectrum in the U.S. Marine Corps' AV-8B Harrier II fighter plane, built by Douglas Corp., 1985, depicting extensive use of composites (26%).

off or wear. In the United States, there is now an increasing tendency to manufacture smaller cars with an average weight of less than 1400 kg, composed of 53% steel, 12% CI, 10% polymers, 6% Al, and 19% other materials. Previously, bicycle frames were constructed exclusively from welded steel tubing. Now, several companies offer frames in a wide range of engineering materials, including Al alloys, Ti, and various fiber-reinforced composites. The lightest carbon fiber frame now weighs only 1.2 kg.

Aerospace applications frequently require light weight, stiffness, and fatigue resistance. Figure 4.1 illustrates a schematic of the U.S. Marine Corps' AV-8B Harrier II fighter plane (built by Douglas Corp., ca. 1985), depicting the extensive use of composite materials, accounting for 26% of the plane's weight. The composites are primarily graphite fiber-reinforced epoxy. This plane can take off on a 500-m runway and make vertical landings. Today composites may account for 65% of the weight of planes of supersonic designs. An added benefit when using composites in military aircraft is reduced detectability by radar.

In the recent future, a hypersonic airplane X-30 (NASP) is intended to be manufactured in the United States. It can take off from a runway, fly directly into space, and return to land on another runway. The incredible speeds (25 mach) will generate elevated temperatures at leading edges of the body near to 1800°C. Therefore, radically different aerospace composite materials will be required to provide the properties of high strength, light weight, and stability at these highly elevated temperatures.

4.2 CLASSIFICATION OF ENGINEERING MATERIALS

Engineering materials used in manufacturing today can be broadly classified into three main groups (Figure 4.2).

1. Metallic materials and alloys: These are further classified into ferrous (steels and cast irons) and nonferrous. Nonferrous metals are generally more expensive than ferrous metals. These are subdivided into light nonferrous metals such as Al, Zn, and so on ($\rho < 5$ g/cm^3), and heavy nonferrous metals such as Cu, W, and so on ($\rho > 5$ g/cm^3). A metallic alloy is composed of a base metal (ferrous or nonferrous, usually 50% content), and one or more

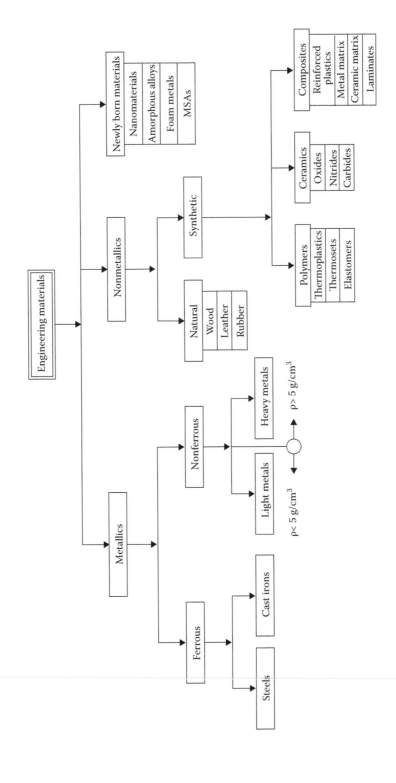

FIGURE 4.2   Classification of engineering materials.

alloying elements. In engineering applications, pure metals are used less frequently than alloys, since they are relatively weak and not readily available in most cases. Their use is generally confined to application in which the material is required to possess a high thermal and electrical conductivity and a high melting point. Alloys have numerous advantages over pure metals. They are characterized by greater strength, a lower melting point, the ability to change their properties according to their chemical composition, and they are easily controlled by heat treatment. They display higher fluidity in the molten state and a lower shrinkage.

2. Nonmetallic materials: Because of their wide range of attractive properties, nonmetallic materials have always played a significant role in manufacturing. Wood has been a key engineering material down through the centuries to manufacture a broad spectrum of furniture and similar products. Stones and rocks are used as constructional materials for buildings, and clay products can be traced to antiquity. Also, leather was previously used for fenders in early automobiles.

The materials family has expanded greatly from the natural materials described above, and now includes a list of polymers, elastomers, ceramics, and composites, which are manufactured, so a wide variety of properties can be obtained. The design requirements of engineering products dictate a corresponding development of nonmetallic materials in the recent future. Important nonmetallic engineering materials include

A. Polymers: Polymers are substances composed of a large number of molecules, joined together in chain-like fashion. The majority of engineering polymers are based on hydrocarbons. Polymers are noted for their low density and their use as insulators, both thermal and electrical. They are poor reflectors of light, tending to be transparent or translucent. Some of them are flexible and deformable. In addition, polymers typically resist atmospheric and other forms of corrosion. Therefore, they can eliminate potentially dangerous corrosion that would ordinarily occur in metallic materials.

By learning more of their architecture, scientists and engineers have been able to produce plastics having the desired properties for an ever-expanding set of technical applications. For many engineering applications, polymers consist of a large number of synthetic plastics in addition to many natural polymers such as wood and rubber. We are probably familiar with many of the synthetics, such as thermoplastics and thermosets. Some polymers display good compatibility with human tissue, which makes them excellent materials for surgical implantation and other biomedical applications.

B. Ceramics: These are compounds which contain metallic and nonmetallic elements. Many ceramic materials are now available for industrial applications. They range from the cement of concrete, to glass, to spark plug materials. Some properties of ceramics are significantly better than those of metals, particularly their hardness, wear resistance, and thermal and electrical resistance. Therefore, they are becoming increasingly important in heat engines, as well as in tool and die manufacturing. More recent applications of ceramics are in automobile components such as exhaust-port liners and coated pistons with the desirable properties of strength and corrosion resistance at high operating temperatures. Grain size has a major influence on the strength and other properties of ceramics; the finer the grain size, the higher strength and toughness. Hence, *fine ceramics* acquire vital importance in specific engineering applications. Another category is traditional or coarse ceramics, which are used in manufacturing of products such as whiteware, dinnerware, tiles, bricks, and pottery.

C. Composites: A number of composite materials that consist of more than one material type have been engineered. Reinforced plastics, metal matrix, and ceramic matrix composites and honeycomb structures are familiar examples of composites. A composite is designed to display a combination of the best characteristics of each of the component materials. Reinforced plastics acquire strength from glass fibers, and flexibility from the polymers. Nonmetallic materials including ceramics, polymers, and composites will be separately considered in a detailed manner along with their processing techniques in Chapters 12, 13, and 14, respectively.

3. Newly born engineering materials: These will be dealt with at the end of this chapter.

4.3 FERROUS METALS AND ALLOYS

Ferrous alloys are among the most useful and applicable types of metals. They contain iron as their base metal and are variously classified into steels and cast irons.

4.3.1 STEELS AND STEEL ALLOYS

Steels and alloy steels are among the most commonly used metallic materials, and have a wide variety of applications. Several alloying elements are added to steels to impart various properties such as hardenability, strength, wear resistance, heat resistance, corrosion resistance, workability, weldability, and machinability. Generally, the higher the percentages of these elements, the higher the particular properties that steels impart. Before dealing with the types of steels and steel alloys, it is important at first to be acquainted with important systems of steel designation.

4.3.1.1 Steel Designation

Many systems of steel designation are adopted by different industrial countries. The American designation suggested some standards, which are suitable for certain industrial applications. These are provided by the American Iron and Steel Institute (AISI), the Society of Automotive Engineers (SAE), the American Society of Metals (ASM), the American Society of Testing Materials (ASTM), and the Aerospace Materials Specifications (AMS). Other systems are adopted in Germany (DIN: Deutsches Institut fuer Normung), France (AFNOR: Association Française de Normalisation), Russia (GOST: Goostandart), Japan (JIS: Japanese Industrial Standards), India (BIS: Bureau of Indian Standard), and Sweden (SS: Sweden Standards).

Two important systems of designation will be considered. These are the American AISI/SAE and the German DIN designation systems.

1. AISI/SAE designation system: A designation system was started by the Society of Automotive Engineers (SAE) to provide some standardization for steels used in the automotive industry. It was later adopted and expanded by the American Iron and Steel Institute (AISI). Both plain carbon and low-alloy steels are classified through a four-digit numbering scheme, in which the last two digits (or three in some cases) indicate the carbon content in hundredths of a percent. For example, a 1040 steel has 0.40% carbon (± a small workable range). The first two digits identify the type of element (or elements) that has been added to the steel and its percentage (Tables 4.1 and 4.2; Figure 4.3). The classification 10xx is reserved for plain carbon steels with only a minimum amount of other alloying elements, and 1112 is resulfurized steel with a carbon content of 0.12%.

   According to AISI designation, a C1018 steel is a plain carbon steel grade (10), containing an average carbon content of 0.18%, and was produced by the basic open-hearth process. An E 52100 steel (ball bearing steel), is a high chromium steel containing approximately 1.45% Cr and 1.00% C. The E denotes that the steel was melted in an electric furnace. Composition and processing of steels are controlled in a manner that makes them suitable for different applications. They are available in various shapes: plates, bars, sheets, wire, casting, and forgings (Figure 4.3). Many commercial steels are not included in the classification scheme listed in Table 4.5. It does not include stainless steels and modifications to the standard grades which many users specify in order to achieve certain desirable

TABLE 4.1
Nomenclature for AISI-SAE Steels

AISI or SAE Number	Composition
10 xx	Plain carbon steels
11 xx	Plain carbon (resulfurized for machinability)
12 xx	Plain carbon (S+P for machinability, free cutting steel)
13 xx	Manganese (1.5%–2.0%)
23 xx	Nickel (3.35%–3.75%)
25 xx	Nickel (4.75%–5.25%)
31 xx	Nickel (1.10%–1.40%), chromium (0.55%–0.90%)
33 xx	Nickel (3.25%–3.75%), chromium (1.40%–1.75%)
40 xx	Molybdenum (0.20%–0.40%)
41 xx	Chromium (0.40%–1.20%), molybdenum (0.08%–0.25%)
43 xx	Nickel (1.65%–2.00%), chromium (0.40%–0.90%), molybdenum (0.20%–0.30%)
46 xx	Nickel (1.40%–2.00%), molybdenum (0.15%–0.30%)
48 xx	Nickel (3.25%–3.75%), molybdenum (0.20%–0.30%)
51 xx	Chromium (0.70%–1.20%)
61 xx	Chromium (0.70%–1.10%), vanadium (0.10%)
81 xx	Nickel (0.20%–0.40%), chromium (0.30%–0.55%), molybdenum (0.08%–0.15%)
86 xx	Nickel (0.30%–0.70%), chromium (0.40%–0.85%), molybdenum (0.08%–0.25%)
87 xx	Nickel (0.40%–0.70%), chromium (0.40%–0.60%), molybdenum (0.2%–0.3%)
92 xx	Silicon (1.80%–2.20%)

xx, Carbon content, 0.xx%.

All steels contain about 0.5 Mn, unless otherwise stated.

Leaded, plain carbon machinable steels are identified by the letter L inserted between the second and third numerates (10L20).

TABLE 4.2
Major AISI Classifications of Steels Concerning Alloying Elements

AISI Designation	Type
1xxx	Carbon steels
2xxx	Ni steels
3xxx	Ni–Cr steels
4xxx	Mo steels
5xxx	Cr steels
6xxx	Cr–V steels
7xxx	W–Cr steels
8xxx	Ni–Cr–Mo steels
9xxx	Si steels

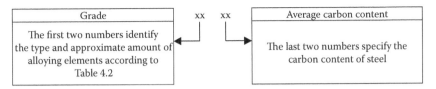

Grade	xx xx	Average carbon content
The first two numbers identify the type and approximate amount of alloying elements according to Table 4.2		The last two numbers specify the carbon content of steel

AISI-prefixes
In case of AISI, the following letters precede the steel
designation to indicate the melting practice

A	Basic open-hearth alloy steel
B	Acid Bessemer carbon steel
C	Basic open-hearth carbon steel
D	Acid open-hearth carbon steel
E	Electric furnace steel

Steel forms available

HB	Hot-rolled bars
CB	Cold-rolled bars
B	Bars
R	Rods
W	Wire

FIGURE 4.3 AISI and SAE designation of plain and low-alloy steels using a four-digit system.

properties. The last mentioned have more specialized applications and may not be stocked as a regular warehouse item.

2. DIN designation system: The self-explanatory Table 4.3 summarizes the German designation system for plain carbon and alloy steels (including high-alloy steels such as stainless steels, tool steels, and heat-resisting steels).

4.3.1.2 Steels and Typical Applications

Steels are ferrous-based alloys, containing significant amounts of one or more alloying elements, which have been either intentionally added or retained during the refining process. According to their chemical composition, they are classified as plain-carbon steels, the mechanical properties of which are determined by the carbon content, or as alloy steels, the mechanical properties of which are influenced by such alloying elements as W, Cr, Ni, Mo, V, Si, Mn, etc. Furthermore, these properties are influenced by the heat treatment methods.

According to the field of application, distinction is made between the following types of steels.

a. Plain carbon steels: Basically, plane carbon steel consists of iron containing small amounts of carbon, where the carbon content can vary from 0.08% to approximately 2.0%. It may also contain limited amounts of Mn (1.6% max.), Si (0.6% max.) and Cu (0.6% max.), and sulfur and phosphorus, which should not exceed 0.05%. The standard carbon steels are the types such as 1005 through 1095 (Table 4.1). Carbon steels in the lower end of this range are generally used in applications where a soft, deformable material is needed. This includes structural shapes, rivets, nails, wire, and pipe. Steels with higher carbon contents (>0.3%) are hardenable by quenching, and therefore find application when greater strength is required.

Medium carbon steels (0.3%–0.8%) can be quenched in water, brine, or oil to form martensite or biainite. Thus, the best balance of properties is obtained at these carbon levels, where a good compromise between toughness and ductility with strength and hardness is achieved. For this reason applications of the midcarbon range include gears, shafts, axles,

TABLE 4.3
German Designation System for Plain Carbon and Alloyed Steels, Summarized According to DIN 17006

1. Plain Carbon Steel

Nontreatable (Structural)	Treatable Steel		Tool Steels
	Basic	Quality	
Ex.: St 42	**Ex.: C 15**	**Ex.: CK 15**	**Ex.: C 100 W2**
St. (notation) is followed by the tensile strength in kg/mm². These steels are used in case of nonspecial requirements.	C notation is followed by the carbon content in hundredths of a percent. K used for quality or refined (S+P<0.04%).		C notation is followed by the carbon content in hundredths of a percent. W abbreviation of Werkzeug (tool). 2 a quality number (1, 2, 3 are quality numbers).

2. Alloy Steels

Low-alloy steel (alloying elements <10%)

Example: Cr–Mo low-alloy steel, 24 Cr Mo 5 5, C = 0.24%, Cr=1.25%, Mo = 0.5%

x x □ □ □ □ x x x x

- Carb. content
- 1st alloying element
- 2nd alloying element
- 3rd alloying element
- 1st alloying element content
- 2nd alloying element content
- 3rd alloying element content

High-alloy steel (alloying elements >10%)

Example: Cr–Ni stainless X10 Cr Ni 18 8, C = 0.1%, Cr = 18%, Ni 8%

Abbreviation of high alloy steel

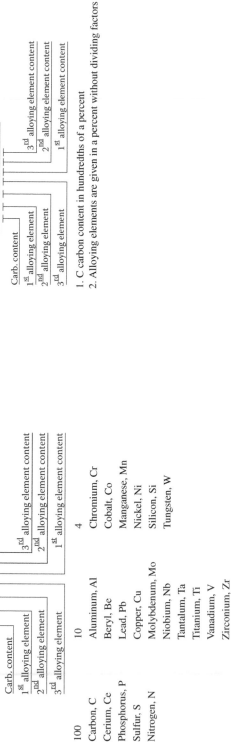

X xx □ □ □ □ x x x x

- Carb. content
- 1st alloying element
- 2nd alloying element
- 3rd alloying element
- 1st alloying element content
- 2nd alloying element content
- 3rd alloying element content

1. C carbon content in hundredths of a percent
2. Alloying elements are given in a percent without dividing factors

100	10	4
Carbon, C	Aluminum, Al	Chromium, Cr
Cerium, Ce	Beryl, Be	Cobalt, Co
Phosphorus, P	Lead, Pb	Manganese, Mn
Sulfur, S	Copper, Cu	Nickel, Ni
Nitrogen, N	Molybdenum, Mo	Silicon, Si
	Niobium, Nb	Tungsten, W
	Tantalum, Ta	
	Titanium, Ti	
	Vanadium, V	
	Zirconium, Zr	

rods, and a multitude of machine parts. The toughness and formability of high-carbon steels (<0.8% C) are quite low, but the hardness and wear resistance are high. High-carbon steels find application in hammers, chisels, drills, punches, files, cutters, knives, saws, wire, and dies for all purposes. Properties that are typically developed in plain-carbon steels are given in Table 4.4.

b. Alloy steels: Alloy steels contain appreciable quantities of alloying elements in addition to carbon. They include the following categories.

1. High-strength low-alloy structural steels (HSLA): HSLA steels contain insufficient carbon and alloying elements to be hardened effectively by quenching to martensite, and hence they rely on chemical composition rather than heat treatment to develop the desired mechanical properties in the as-rolled condition. HSLAs find applications in large welded structures, where the size precludes subsequent heat treatment. These applications require high yield strength, good weldability, acceptable corrosion resistance and limited ductility, while hardenability is of no concern.

These steels contain slightly more phosphorus and silicon than carbon steels, thereby strengthening the ferrite network, and raising the yield strength to about 50% above that of carbon structural steels. HSLA steels can be made corrosion resistant (8 times as compared with that of plain carbon steel) by proper addition of P, Cu, Si, Cr, V and Mo. For this reason, HSLAs are known as *weathering steels*. Because of their high yield strength, weight saving of 20% to 30% can often be achieved with no sacrifice of strength and safety.

Rolled and welded HSLA steels are now being used in automobiles and railway cars to save weight, and for bridges, towers, pressure vessels, and building structures to save painting. Because of their low-alloy content (0.1%–0.25% C and ca 1.5%, other alloying elements) and high volume application, their cost is often little more than that of ordinary carbon structural steels. Another class of low-alloy but high-strength steels is the dual-phase type, having a microstructure of islands of high-carbon martensite–austenite in a matrix of much softer ferrite. They have a high work-hardening characteristic, which improves their ductility and formability. Dual-phase steels were developed in the late 1960s and have now become important sheet metal material in the automobile industry.

TABLE 4.4
Typical Properties for Plain Carbon Steels

AISI Grade[a]	TS (MPa)	YS (MPa)	Reduction of Area (%)	Elongation (%)	HBN
			Water quench		
1030	607	473	69	28	179
1040	744	542	63	23	217
1050	905	636	55	20	262
1095	1138	707	41	16	311
			Oil quench		
1060	945	592	48	18	269
1080	1145	714	38	15	331
1137	745	523	56	21	223
1141	760	520	57	22	217
1144	749	502	46	19	223

Source: Adapted from Modern Steels and their Properties, HB.3310, Bethlehem Steel, PA, 1967.

[a] 25 mm diameter bars tempered at 538°C.

2. Quenched and tempered low-carbon constructional alloy steels: They are also known as low-carbon martensites, which tend to have slightly higher carbon and alloy contents than HSLA. Additionally, they contain alloying elements such as B, V, and Mo, all of which contribute to hardenability, and V, Mo, and Ti, which form persistent carbides that resist softening upon tempering. Such steels are used for applications like pressure vessels, mining equipment, and earthmovers.

3. AISI-SAE alloy steels: With the exception of low, plain carbon steels, the AISI-SAE steels are always used in the heat-treated condition. Although the low-carbon, plain-carbon steels are sometimes used in the unhardened condition, they are also sometimes surface-hardened by carburizing. The low-carbon AISI-SAE alloy steels are used for carburizing if somewhat better core properties and a greater depth of hardening is required. The steels containing more than 0.25% carbon are used in the quenched and tempered condition. Higher carbon contents, up to about 0.6%, provide tempered martensite of greater hardness. Higher alloy contents serve to increase the depth to which hard martensite will form beneath the surface. In other words, the maximum hardness achievable is a primary function of carbon content and the depth is a primary function of alloy contents, particularly with Mn, Mo, Cr, and Ni.

The AISI-SAE steels are generally used for applications such as carburized or through-hardened parts. Through-hardening alloy steels, such as 4130, 4340, 8640, and 9260, are widely used in applications such as connecting rods, springs, torsion elements, and ordnance parts, and 52100 is used mainly for ball and roller bearings. Surface-hardening grades, such as 1320, 4620, and 8620, are used in shafts and gears as well as in many other parts that require surface wear and abrasion resistance, combined with tough interior. The most commonly used carbon and alloy steels according to AISI designation (Table 4.1) and their applications are abstracted in Table 4.5.

The increasing use of high-speed machining, particularly on automated machine tools, has spurred the use of free-machining steels 11xx and 12xx (also known as

TABLE 4.5
Abstracted Carbon and Alloy Steel Grades According to AISI Designation and Their Applications

AISI Grade	Application
1010	Automobile bodies
1040, 4140	Axles
1035, 4042, 4815	Bolts
1020, 1040	Camshafts
1040, 3140, 4340	Connecting rods
1045, 1145, 3135, 3140	Forged crankshafts
1060	Lock washers
1080	Railroad rails and wheels
1095, 4063, 6150	Springs (coil)
1085, 4063, 6150, 9260	Springs (leaf)
1040	Tubing
1045, 1055	Wire
1085	Wire (music)
4140, 8740	Aircraft forgings, tubing, fittings
52100	Ball bearings and races
3135, 3140	Transmission chains
4023	Differential gears
4027, 4032	Gears (cars and trucks)

automatic steels). They are basically carbon steels that have been modified by alloy additions to enhance machinability. Sulfur, lead, bismuth, selenium, tellurium, and phosphorus are added for this purpose. Sulfur (0.08%–0.33%) combines with manganese (0.7%–1.6%) to form soft MnS inclusions. These act as discontinuities in the structure, serving as sites to form broken chips. These provide a built-in lubricant that prevents the formation of built-up edge (BUE). Insoluble lead particles serve the same purpose. A better free-machining agent is bismuth. It is environmentally acceptable and is uniformly dispersed since its density is closer to that of Fe. The tool life is noticeably extended when using this type of free cutting steel.

4. Alloy tool and die steels: Tool and die steels are basically high-carbon steels modified by alloy additions to provide the desired balance of toughness and wear. They are commonly used in forming and machining of metals, and thus they are characterized by
   - High strength
   - High impact toughness
   - High wear resistance at room and elevated temperatures
   Table 4.6 lists the seven basic types of tool and die steels.

5. Stainless steels: Stainless steels are characterized primarily by their corrosion resistance, high strength and ductility, and high chromium content. They are so called because in the presence of oxygen (air), they develop a thin, hard, adherent, and stable film of chromium oxide Cr_2O_3 that protects the metal from corrosion. The protective film builds up again if the surface is scratched. For passivation to occur, the minimum Cr content should be 11% to 12%, approximately. There are three basic classes of stainless steels, namely austenitic, ferritic, and martensic. The names reflect the microstructure of which the steel is normally composed. The designation scheme according to AISI and the typical alloying elements of these basic classes are presented in Tables 4.7 and 4.8.

   The basic classes and their main characteristics are briefly described as follows
   a. Austenitic stainless steel (200, 300 series): They are generally composed of Cr, Ni, and Mn in iron. Austenitic stainless steels are nonmagnetic and have excellent

TABLE 4.6
AISI-SAE Classification of Tool and Die Steels

Type	AISI-SAE Grade	Features	
1. Water hardening	W	High plain-carbon steel, must be quenched in water, brittle, undesired softness if tempered above 150°C, least expensive types, used for small tooling not subject to severe usage or elevated temperatures.	
2. Cold work	O	Oil-hardening medium alloy	These grades, characterized by less
	A	Air-hardening medium alloy	distortion and wear resistance. Used for
	D	High-C, high-Cr (10%–18%)	cold-working operations.
3. Hot work	H	H1–H19: Cr-based, H20–H39: W-based coupled with Cr, H40–H59: Mo-based	These grades are resistant to wear and cracking. Cr, W, Mo, and V, designed for use at elevated temperatures.
4. Shock resisting	S	0.5% C, high impact toughness. Properties depend on alloying elements, designed for both hot and cold applications.	
5. Plastic mold	P	P1–P19 (low carbon) P20–P39 (others)	Used for Zn-die casting and plastic injection molding dies
6. Special purpose	L	Low alloy for extreme toughness	
	F	C–W water hardenable, substantially more wear resistant than plain carbon steel.	
7. HSS (see CH15)	M	Mo-based (95% of all HSS types)	
	T	W-based (only 5%)	

TABLE 4.7
AISI Designation Scheme of Stainless Steels

Series	Alloying Elements	Class (Microstructure)
200	Cr, Ni, Mn, or N_2	Austenitic
300	Cr and Ni	Austenitic
400	Cr only	Ferritic or martensitic
500	Low Cr (<12%)	Martensitic

corrosion resistance, but are susceptible to stress corrosion cracking. They are hardened by cold working, and are the most ductile of all types of stainless steels. They can be formed easily; however, with increasing cold work, their formability is reduced. Austenitic steels are used in a wide variety of applications, such as kitchenware, fittings, furnaces, heat exchangers, and components for several chemical environments. They cost about 10 times as much as ordinary steels.

b. Ferritic stainless steels (400 series): They have high Cr content (up to 27%) and no nickel. Ferritic steels are magnetic and have good corrosion resistance but have lower ductility than austenitic stainless steels because of the bcc crystal structure. They harden by cold working, and are not heat treatable; it is impossible to harden them by quenching and tempering. They are readily weldable, since there is no possibility for martensite formation. Ferritic stainless steels are generally used for nonstructural applications such as kitchen equipment and automotive trim. They are the cheapest, and that is why they should be given the first consideration when a stainless steel alloy is required.

c. Martensitic stainless steels (400, 500 series): They do not contain nickel and are hardenable by heat treatment. Their Cr content may be as high as 18%, and they contain between 0.15% and 0.75% carbon. Martensitic steels are magnetic and possess high strength, hardness, fatigue resistance, and good ductility, but moderate corrosion resistance. Martensitic stainless steels are used for cutlery, razor blades, surgical tooling, instruments, valves, turbine parts, oil well equipment and springs.

Additionally, there are also two important types of stainless steels of special interest. They are

d. Precipitation-hardenable PH stainless steels: They contain Cr and Ni, along with Cu, Al, Ti, or Mo. They have good corrosion resistance, good ductility, and high strength at elevated temperatures. Basically, these are made of martensitic, semi-austenitic, or austenitic steels by varying the Cr/Ni ratio. Since Cr is a ferritic stabilizer and Ni is an austenite stabilizer, lowering the ratio of Cr to Ni tends to

TABLE 4.8
Typical Composition of the Basic Classes of Stainless Steels

Alloying Element	Ferritic	Martensitic	Austenitic
C	0.08–0.2	0.15–1.2	0.03–0.25
Cr	11–27	11.5–18	16–26
Mn	1–1.5	1	2
Mo	—	—	Some classes
Ni	—	—	3.5–22
Si	1	1	1–2
Ti	—	—	Some classes

promote an austenitic condition, while raising this ratio promotes transformation to martensite. Small additions of Al, Ti, Mo, and Cu lead to precipitation of inter-metallic compounds during heat treatment. PH stainless steels are mainly used in aircraft, aerospace, and structural components such as valves, bearings, gears, mandrels, and valve seats.

    e. Duplex stainless steels: These contain between 21% and 25% C, and between 5% and 7% Ni and are water quenched from hot working temperature between 1000°C and 1050°C, to produce a balanced microstructure (mixture of austenitic and fer-ritic grades, 50% each). Duplex stainless steels are characterized by good strength and higher resistance to corrosion in most environments. Their stress corrosion cracking is superior than that of the 300 series austenitic steels. Water treatment plants and heat exchanger components are some of their typical applications.

6. Maraging steels: The term *maraging* is derived from a combination of strengthening due to martensite transformation followed by age hardening. Maraging steel consists basically of an extra-low-carbon (<0.03% C), iron-based alloy to which a high per-cent of Ni (17%–26%) has been added, plus significant amounts of Co, Mo, and Ti. This steel attains super high strength accompanied by good levels of ductility and toughness. It can be hot-worked at temperatures between 760°C and 1250°C. When air cooled from 815°C, the structure is soft and tough, and low carbon martensite of 30 HRC is obtained. It can be easily machined, and because of its low strain hardening property, it can be cold worked to a high degree. Aging at 480°C for 3–6 h followed by air cooling raises the hardness to about 52 HRC.

    Maraging steels are very useful in applications where ultra-high strength and toughness are required. They are quite expensive due to the large amount of alloying additions (>30%), and therefore, they should only be specified when their outstand-ing properties are required. These steels are particularly useful in the manufacture of space-vehicle cases, hydrofoil struts, and extrusion press rams.

7. Hadfield austenitic manganese steels: Hadfield steel is an alloy of work-hardening char-acteristic that imparts good abrasion and wear resistance. It contains about 1% C and 12% Mn and thus exhibits a homogeneous austenitic structure at room temperature (by quenching from 1040°C). It is mainly used for power shovel buckets, railway switch frogs, rock crusher jaws, and similar applications where service conditions result in work hardening and hence improvement in properties.

8. Heat-resisting steels: In order to be useful at elevated temperatures, alloys (ferrous and nonferrous) should possess good strength and resistance to creep at elevated tem-perature, corrosion, and scaling. Moreover, they should not undergo crystallographic changes when exposed to operating conditions. The stability of protective scale cov-erings under condition of thermal shock is important. Particularly, austenitic steels perform well at elevated temperatures, and they give better resistance to scaling if Si or Al is added. The best performance at high temperatures is provided not by steels but by alloys in which elements such as Cr, Ni, Co, Mo, and Fe are present, with no one element present in amounts greater than 50%. These heat resisting alloys can hardly be called steels.

9. Magnetic alloys: These comprise two classes.

    – Permanent magnets: The best permanent magnets are not steels, although steels of high carbon alloys containing Cr, W, Mo, and Co are sometimes used. They are first hardened and then permanently magnetized. They should not be tempered because this destroys the magnetic properties.

    – Temporary magnets: These are soft steels, which are readily magnetized and demagnetized by alternating current. The magnetic performance is impaired by any composition change or treatment which hardens them. This includes even cold working associated with blanking, stamping, or stacking. Hardness makes steels more permanent, i.e., difficult to magnetize and demagnetize in an AC field, resulting in power losses (hysteresis) and eddy losses. The latter is minimized by using insulated laminates in an assembly. The sheet steel used for the highest quality temporary magnet is Si-bearing steel (4% Si) of very low carbon content. Si is used because it increases effectively the electrical resistance and provides the best magnetic properties.

4.3.2 CAST IRONS

Cast iron (CI) is one of the most important engineering materials. Machine tool beds, cylinder blocks, cylinder liners, piston rings, gears, and many other parts are made of CI. The term CI refers to the family of ferrous alloys composed of iron, carbon (ranging from 2% to 4.3%), and silicon (up to 3%).

4.3.2.1 General Characteristics

The characteristics that make CI a valuable engineering material are its

- Cheapness
- Castability
- Excellent machinability
- Fair wear resistance
- High damping capacity
- Fair sliding property
- Lack of sensitivity of surface finish quality
- Fair mechanical properties (reasonable tensile strength ranging from 130 to 400 MPa, associated with a very high compressive strength)

The newly developed irons formed by alloying with sometimes expensive elements or by melting and casting methods are becoming competitors to steel.

Referring to the equilibrium phase diagram (Figure 3.23), cast irons have less melting temperatures as compared to those of steels. That is why casting is a suitable process for cast irons (>2% C).

While carbon in ordinary steels exists as cementite Fe_3C, it occurs in CI in two forms

1. Stable form: graphite C
2. Unstable form: cementite Fe_3C

Graphite is gray, soft, and occupies a large volume, hence counteracting shrinkage, while cementite is intensively hard, with a density of the same order as iron. The relative amounts, shape, and distribution of these two forms of carbon largely control the general properties of the iron.

4.3.2.2 Common Types of Cast Irons

Cast irons are properly classified (Figure 4.4) according to their structure into

1. White cast iron: Of the several varieties of cast irons, white cast iron is the only one whose properties and structure are indicated by the iron–iron carbide diagram (Figure 3.23).

    Due to the rapid cooling, white CI has essentially all of its carbon in the form of iron carbide (Fe_3C). It receives its name from the white surface that appears when the material is fractured. By virtue of the large amounts of iron carbide, white CI is very hard and brittle. It finds applications when high resistance to wear and abrasion is required. Its low ductility and impact strength render it unsuitable for structural parts, and it is not machinable. White CI is important as an intermediate step in the production of malleable iron castings.

2. Malleable cast iron (malleabilizing by annealing): It is made by heating the white CI to 927°C for about 50 h, followed by slow cooling. The castings are packed in a neutral slag or scale during heating and cooling. Decomposition of cementite is promoted by the long heating period (extended annealing). In malleable iron castings, cementite which existed in both the matrix and in the pearlite of the white CI is decomposed to graphite and ferrite (or graphite and pearlite) (Figure 4.4). Malleable CI has good machinability and is more ductile than white and gray CI. Its mechanical properties compare favorably with those of low-carbon steels.

3. Gray cast iron: The structure of gray CI can be predicted better from the Fe–C diagram. However, satisfactory approximations can be made using the Fe–Fe_3C diagram (Figure 3.23). The formation of gray CIs, i.e., the promotion of graphitization, is accomplished most readily by
   • Increasing the carbon content
   • Increasing Si and Ni contents (graphitizers)
   • Decreasing the cooling rate

   The cooling of a gray CI containing, for example, about 4% C, involves formation of primary austenite dendrites. At about 1135°C, there is a eutectic form in which the matrix is also austenite. Throughout the matrix, flakes of graphite are forming, whose size and distribution may vary quite widely. As the temperature drops, austenite continues to precipitate graphite, and at approximately 723°C, the austenitic–pearlitic transformation occurs. If cooling is sufficiently slow, a partial or complete decomposition of austenite into ferrite and graphite may occur. Thus, the final structure depends on the cooling rate. Three possibilities may occur (Figure 4.4)
   • Fairly rapid cooling: graphite flakes embedded in pearlitic matrix
   • Intermediate cooling: graphite flakes embedded in pearlite and ferrite
   • Extremely slow cooling: graphite embedded in ferritic matrix

   The intermediate cooling provides the best structures, as it provides strength combined with machinability. The graphite flakes are actually small stress raisers, and hence, gray CIs are extremely brittle. However, the presence of carbon in the form of graphite accounts for the two outstanding characteristics of gray CI: superior machinability and damping

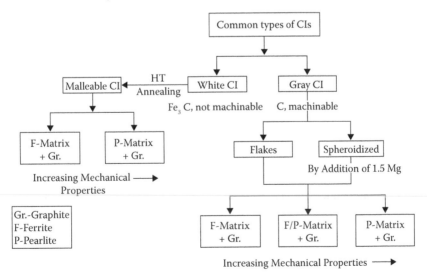

FIGURE 4.4 Grades of most common types of cast irons.

capacity. By controlling the composition and cooling rates, it is possible to make castings in which both gray and white CI structures are present in the same casting. Thin sections cool fast enough to produce the chilled structure of white CI, while thick sections cool more slowly, developing a gray CI structure. Introduction of chill blocks in the mold may produce local areas of white CI. This is useful in casting machine bases and frames, where hardness in wearing surfaces is to be coupled with the ability to absorb vibrations.

4. Spheroidal (ductile) cast iron (Mg addition during casting): It contains spherical graphite in which the graphite flakes of the gray CI form into balls, as do cabbage leaves. Ductile iron is so named because, in the as-cast form, it exhibits measurable ductility. It is obtained from ordinary gray CI by a ladle, added to the liquid metal (at 1400°C) of Mg in an amount of 0.3%–1.5%. The larger the casting, the more Mg is added. Mg causes the graphite, which precipitates in the structure of CI during solidification, to take a strictly spheroidal form instead of the flakes found in gray CI. Graphite in spheroidized form weakens the matrix to a lesser extent, thus providing for higher mechanical properties. Mg not only changes the form of the graphite, but also increases the strength of the matrix. High-ductility spheroidized CIs have structures composed of
   • Pearlite + spheroidal graphite
   • Pearlite + ferrite + spheroidal graphite
   • Ferrite + spheroidal graphite

   Spheroidizable and malleable CIs have almost similar microstructures; however, spheroidizable gray CI has superior mechanical properties than malleable CI. Moreover, malleable CI necessitates prolonged time in treatment. For these reasons, malleable iron has been gradually replaced by the spheroidized type.

5. Alloy cast irons: The chief purposes of alloying are to increase strength and resistance to wear, heat, and corrosion. High-alloy CI contains over 3% alloying contents and it is commercially classified separately. It may be a type of white, gray, or ductile CI. The matrix may be pearlitic, sorbitic, martensitic, or austenitic. A combination of alloying elements is used so that a balance is attained between carbide formers and graphitizers. The most commonly added elements are Ni, Cr, Cu, Mo. Ni tends to produce gray CI; it is less powerful than Si in this respect. It prevents chilling in thin sections. Cr is a carbide former, and acts in the opposite way to Ni; however, it exerts a grain refining action. Ni and Cr, singly or together, are commonly found in motor cylinder blocks. Mo nodules the graphite, which promotes fine pearlite, but CI is best used with elements such as Ni or Cr.

   The outstanding characteristics of alloyed CIs as compared to ordinary CIs are
   1. Marked resistance to corrosion and heat
   2. Nonmagnetic
   3. High electrical resistance
   4. High coefficient of thermal expansion

   Finally, addition of 20%–24% Al lends CI high thermal resistance (up to 950°C), which is used in furnace connections. High-silicon CI (5%–14% Si) lends both high thermal (up to 900°C) and corrosion resistance.

4.4 NONFERROUS METALS AND ALLOYS

Nonferrous metals and alloys are playing an increasingly important role in modern manufacturing technology. Because of their wide variation in numbers and properties, they provide a limitless range of properties for various applications. They are often more costly than ferrous metals. However, they often possess a combination of properties such as

• Corrosion resistance without necessity of the special and expensive alloying elements used in steels

- Good manufacturing properties
- High electrical and thermal conductivity
- Light weight
- Strength at elevated temperatures
- Color

Usually, the strength and stiffness of nonferrous alloys are inferior to those of steels, a fact that places them at a distinct disadvantage when carrying capacity is a necessary demand. Nonferrous alloys often have lower melting points, making them easy to cast. Their high ductility coupled with lower yield points provides high formability and easy cold working features. Their weldability is sometimes inferior to that of low-carbon steels. Savings due to good manufacturability can often overcome their higher material cost, and they are favored in place of steels. Nonferrous alloys are much more commonly used than pure metals, when mechanical behavior is the primary criterion of concern. In addition, alloying sometimes produces compositions whose mechanical properties can be further enhanced by heat treatment. Heat treatment is applied to pure metals only to remove the effects of cold working; they cannot be hardened by heat treatment. Ductility is often reduced by alloying. However, most alloys can be machined, formed, and welded with relative ease.

Typical applications of nonferrous metals and alloys include

- Aluminum for cooking utensils and aircraft bodies
- Cu wire for electricity and Cu tubing for heat exchangers
- Titanium for jet engine turbine blades and prosthetic joints
- Tantalum for rocket engines

A Boeing turbofan jet engine typically contains an alloy of the following content: 38% Ti, 37% Ni, 12% Cr, 6% Co, 5% Al, 1% Nb, and 0.02% Ta.

Important nonferrous metals and alloys will be dealt with in the following sections.

4.4.1 ALUMINUM AND ALUMINUM ALLOYS

Pure aluminum is known for high strength-to-weight ratio, excellent electrical and thermal conductivity, corrosion resistance, nontoxicity, light reflectivity, low specific gravity, nonmagnetism, and softness and ductility. Joining of Al alloys is accomplished successfully by a variety of welding, brazing, and soldering methods. Inert-gas-shielded arc welding is popular. Soldering is difficult and not generally recommended unless the alloy has been tin coated. Al alloys can be mechanically worked, either cold or hot, since they are relatively soft and ductile. Although Al alloys can be electroplated, they are more difficult to electroplate than ferrous, copper, and zinc alloys.

Typical applications of Al and Al alloys

1. Cooking utensils
2. Their softness and ductility, coupled with corrosion resistance and nontoxic nature, have resulted in their use in aluminum cans, foils, and packing material.
3. Containers, building windows, and other types of construction
4. Transportation (aircraft, aerospace applications, busses, cylinder heads for automobiles, railroad cars, and marine craft)
5. Economical conductors and high-voltage transmission wiring
6. Frames of portable tools
7. Radiator fin material for automotive and air conditioning units, due to their good thermal conductivity
8. Sheet metal light reflector, and as a coating for high-grade optical reflectors
9. Furniture

Some typical Al alloys

- Duralumin: It is an Al alloy, containing
  3.5%–4% Cu; 0.4%–0.7% Mn and Mg (each);
  Fe or Si <0.7%; Al (rest).
    This alloy is machinable and can be heat treated to increase its tensile strength without sacrificing ductility. Its strength is comparable to that of steel, but has only about one third of its weight, so it is frequently used to fabricate parts of aircrafts and automobiles.
- Al–Si alloys: These have excellent castability and resistance to corrosion. They are easy to cast in thin or thick sections, but are difficult to machine.
- Al–Mg alloys: These are superior to all other aluminum casting alloys, with respect to resistance to corrosion and machinability. Additionally, they have high mechanical strength and ductility.

4.4.2 MAGNESIUM AND MAGNESIUM ALLOYS

Magnesium is the lightest engineering metal ($\gamma = 1.74$). It is frequently used as an alloying element in various nonferrous metals. Magnesium and magnesium alloys are generally used in applications where light weight is of primary importance. Mg alloys are weaker and more brittle than Al alloys. The principal alloying additions are Al and Zn. Aluminum, zinc, and zirconium promote precipitation hardening; Zn is a solid solution hardener, Mn and Zn improve corrosion resistance, and tin improves castability.

Mg alloys are among the most machinable of the metallic materials, since they have the ideal combination of softness and brittleness for good machinability. Their brittleness makes them difficult to cold work. However, they are readily hot worked and die cast. Strength drops rapidly when temperature exceeds 100°C, so magnesium alloys should not be considered for elevated temperature service. Mg alloys are mainly used when weight saving and vibration damping represent a special requirement, such as in printing and textile machinery, to minimize inertia forces in high speed components. Other typical applications include aircraft die castings, missile components, material handling equipment, portable power tools (drills and saws), ladders, bicycles, sporting goods, and general lightweight components.

Mg alloys are highly combustible when in a finely divided form such as powder and fine chips during machining and grinding. This hazard should not be ignored. When the alloy is heated above 400°C, a noncombustible, oxygen-free atmosphere is required to suppress burning. Casting operations often require additional precautions. As finished parts or castings, however, Mg alloys present no real fire hazard.

4.4.3 ZINC AND ZINC ALLOYS

The outstanding characteristics of zinc are its corrosion resistance to ordinary atmospheric conditions and the fact that it is above iron in the galvanic series. Because of these characteristics, the major use of Zn is as a galvanizing coating applied to steel wire and sheets. This coating provides excellent corrosion resistance, even when the surface is rough and badly scratched. Moreover, Zn is used in sacrificial protection (Chapter 20). Galvanizing accounts for about 35% of all zinc used.

Zn is used as a base metal for a variety of die-casting alloys. Unfortunately, pure Zn is almost as heavy as steel and is also rather weak and brittle. The most important additions for Zn are Mg and Al, which act respectively to improve corrosion resistance and strength in Zn-based die-casting alloys. Copper is sometimes added to stabilize die-casting dimensions. Superplastic zinc alloys containing about 22% Al can be easily formed into intricate shapes and then heat treated to increase the yield point from 250 to 350 MPa.

4.4.4 LEAD, TIN, AND LEAD–TIN ALLOYS (WHITE METALS)

Lead (Pb, Latin *plumbum*) has properties of high density, good resistance to corrosion, softness, low strength, high ductility, and formability. As a pure metal, its principal uses include acid storage batteries, cable cladding, sound and vibration damping, and radiation shielding against X-rays. Lead is also an alloying element in solders, steels, and Cu alloys, and promotes corrosion resistance and machinability. Because of its toxicity, environmental contamination by lead is a significant concern.

Tin (Sn, Latin *stannum*), although used in small amounts, is an important metal. It is a silvery white, lustrous metal, extensively used as a protective coating on steel for making containers for food and similar products. Inside a sealed can, the steel is cathodic and is protected by tin (anode), so that the steel does not corrode. Tin coating improves the performance in deep drawing and other press working operations.

Lead- or tin-based alloys are also known as white metals. Therefore, light metals are low melting alloys in which lead, tin, or antimony predominate depending on the composition. There are two main classes of white metals, which are bearing alloys and fusible alloys (solders).

1. Bearing alloys (babbits): Babbits are quite readily cast in the shop. They are of two categories.
   a. Tin babbit is one of the best bearing materials and is composed of 84% tin, 8% copper, and 8% antimony.
   b. Lead babbit is a more widely used bearing because of the high cost of tin. It is quite adequate for low speeds and moderate loads and is composed of 84.5% lead, 5% tin, 10% antimony, and 0.5% copper.

      In lead babbits, the shaft rides on the harder cube-shaped particles (formed by the combination of antimony and tin) with little friction, while the softer lead matrix acts as a cushion that can distort sufficiently to compensate for misalignment and assure a proper fit between the two surfaces.
2. Soft solders: They are basically Pb–Sn alloys with minimum melting point of the eutectic composition of 62% (optimum tin content). The high cost of tin has forced many users to specify solders of higher lead content.

4.4.5 COPPER AND COPPER ALLOYS

Pure copper is used for electric wire, bars, bus-bars, and tubing for its outstanding electrical and thermal conductivity, corrosion resistance, and solderability. It can be hardened and strengthened by cold working and by solid solution alloying with Zn, Sn, Al, Si, Mn, and Ni. The ductility of many solid solution alloys is greater than that of Cu. Copper forms a precipitation-hardenable alloy when small amounts of beryllium are added. The mechanical properties of Cu alloys at room temperature are intermediate between those of Al alloys and steel. The elevated temperature properties are superior to those of Al alloys. However, they are not outstanding. Copper alloys lend themselves to the whole spectrum of fabrication processing, including casting, machining, and welding.

Cu alloys are known for attractive appearance and corrosion resistance, accounting for large scale use in different applications, such as electrical and electronic components, springs, cartridges for small armaments, plumbing heat exchangers, and marine hardware as well as some consumer goods such as jewelry and decorative objectives.

Copper-based alloys

1. Cu–Zn alloys (brasses)
   - α-Phase brass with up to 36% Zn, includes ductile, easily worked composition. Within this range, cartridge brass (70% Cu, 30% Zn) offers the best combination of strength and ductility and is quite popular for deep drawing (Figure 4.5).

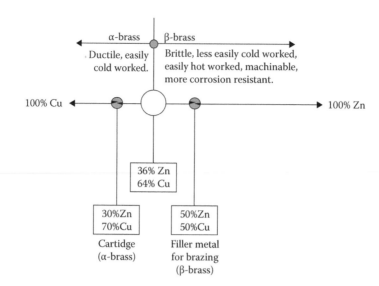

FIGURE 4.5 Different types of brasses.

- β-Phase brass appears with a larger proportion of Zn (>36%). It is more brittle and less easily cold worked, but more corrosion resistant, machinable, and easily hot worked (Figure 4.5). When high machinability is required, as with automatic screw machine stock, 2%–3% Pb is added to brass to ensure the formation of free-breaking chips. A β-brass (50% Cu, 50% Zn) is often used as filler metal in brazing. It is an effective material for brazing steel, cast iron, brasses, and Cu, and produces joints that are nearly as strong as those obtained by welding. Many β-brass casting alloys also contain lead, which improves fluidity. Colors vary in various compositions from red brass to yellow brass. Brasses cost from 8 to 14 times as much as the same mass of carbon steels.

2. Cu–Sn alloys (bronzes): Most bronzes are alloys of copper with 12% tin. These alloys are characterized by their toughness, wear and corrosion resistance, and good strength. Typical applications are bearings, gears, worm gears and pump parts. Bronzes can also be produced by mixing powder of copper and tin, and processing by powder metallurgy technique. The resulting porous product can be used as a filter for high temperature or corrosion media, or infiltrated with oil to produce self-lubricating bearings. Phosphorous bronze alloys have good strength and hardness for applications such as springs and bearings.

3. Precipitation-hardenable Cu-based alloys: These are bronzes with additions of alloying elements (Al, Si, or Be) to produce precipitation-hardenable Cu-based alloys.
   - Al-bronze alloys: They are characterized by high strength and excellent corrosion resistance, and are often considered as cost-effective alternatives to stainless steel and Ni-based alloys. These alloys are ductile if Al is less than 8%. When Al exceeds 9%, however, the ductility drops and the hardness approaches that of steel. Still higher aluminum contents result in brittle but wear-resisting alloy. Typical applications of Al-bronze include marine hardware, power shafts, journal bearings, pumps, and valves.
   - Si-bronze alloys: They contain up to 4% Si, 1.5% Zn (higher Zn contents are used if the material is cast). These alloys are characterized by high strength, formability, machinability, weldability, and corrosion resistance. Si bronzes are used in applications such as boilers, tanks, and stoves.
   - Be-bronze alloys: Beryllium bronzes are quite expensive because they contain up to 2% Be. This alloy has the strength of steel; however, they are nonsparking, nonmagnetic,

and conductive. The toxicity of beryllium has somewhat reduced its popularity and has created a demand for alternative alloys of similar properties.

4. Cu–Ni alloys:
   - Cupronickels (2%–30% Ni)
     The key features of cupronickels include high thermal conductivity and high temperature strength coupled with corrosion resistance and high resistance to stress corrosion cracking, which make cupronickels a good choice for heat exchangers, desalination stations, and coins.
   - Nickel silvers or German silvers (10%–30% Ni and at least 5% Zn)
     These contain no silver. Their silvery luster makes them attractive for ornamental applications and musical instruments.
   - Constantan (45% Ni, 55%)
     This is of high electric resistance, and is used in thermocouples.

4.4.6 NICKEL AND NICKEL ALLOYS

Ni is a major alloying element that imparts strength, toughness, and corrosion resistance to metals. Almost 60% of all nickel produced is used as alloying element in steel and iron, particularly the austenitic stainless steels. Ni plating consumes the second largest quantity, followed by the high-Ni alloys, which account for about 15% of the total nickel production. Pure Ni is also used as a catalyst for chemical reactions, and sometimes as a binder for cutting carbides. Ni-based alloys are most noted for their outstanding strength and corrosion resistance, particularly at high temperatures. They are generally difficult to cast but can be forged and hot worked. Welding can be performed with little difficulty.

Ni-based alloys (nichromes) containing Cr or Cr and Fe, due to their high resistivity and resistance to corrosion at red heat, are used in electric resistance heater elements.

The most important nichromes are described. They have the following commercial names.

1. Inconel (80% Ni, 14% Cr, and 6% Fe): It has high mechanical properties coupled with corrosion and heat resisting properties. It is highly resistant to oxidation even at temperatures up to 900°C. It can be rolled, cast, welded, brazed, and soldered. Inconel is used in the food processing industry, especially in milk and milk products.
2. Monels (67% Ni, 30% Cu, and small additions of Fe, Mn, Si, and C): Monels have tensile strength ranging from 500 to 1200 MPa. They are quite similar to stainless steels in corrosion resistance, appearance, and properties. They are difficult to machine, but welding does not cause them to corrode, as happens with austenitic steels. Monels are used in the food processing industry. They are particularly resistant to salt water, sulfuric acid, and superheated steam. For the latter reason, Monels have been used for steam turbine blades.
3. Invar (Invariable): This is a Ni–Fe alloy (36% Ni, 0.2% C, and 0.5% Mn) of relatively low sensitivity to temperature. The coefficient of expansion is practically zero between 0°C and 100°C, and hence invariable. It is extensively used in clocks and differential expansion regulators. It is also a Ni-Fe alloy
4. Kovar: It is also a Ni–Fe alloy (28% Ni–18% Co): This is a heat-resisting iron alloy.
5. Permalloy (78.5% Ni and 21.5 Fe): This is a carbon-free binary alloy system. It has a high permeability in small magnetic fields. It is used in transformers working with very high frequency currents.

4.4.7 TITANIUM AND TITANIUM ALLOYS

Unalloyed titanium contains up to 1% of various alloying elements. It is useful mainly because of its corrosion resistance and low specific gravity ($\gamma = 4.5$). It is as strong as most Cu- and Al-based alloys and low-carbon steels but is weaker than alloyed titanium. It is, however, most ductile. Pure titanium is used because of its corrosion resistance in chemical piping, valves, tanks, and pros-thetic elements. In airplanes, it is used for applications where high temperature rather than strength requirements are of greater importance. These include fire walls, tail pipes, and jet compressor cases. Titanium alloys are used as substitutes for stainless steels, particularly the austenitic grades, for their favorable characteristics concerning corrosion resistance and strength. For these reasons, about 90% of Ti alloys produced are used in aerospace components.

Ti alloys have three different structures. These are

- α, which has an hcp structure and cannot be hardened by heat treatment.
- β, which has a bcc structure and can be age hardened.
- α–β mixture, which comprises more than 60% of Ti alloys. It is hardenable by heat treatment.

Al acts as α-stabilizer, dissolves in α-phase, and hence strengthens this alloy. For this reason, the presence of Al is important in α or α–β alloys, which are used at temperatures above 370°C. Al-free alloys lose strength rapidly for temperatures above 370°C. Most other alloying elements act as β sta-bilizers. Vanadium and chromium dissolve in β-phase, making it stronger than α-phase. α–β alloys contain Al and β-stabilizers. They can be quenched to a metastable structure containing more than equilibrium amounts of β; upon aging, α precipitates and the α–β alloys are strengthened.

Manufacturability of Ti alloys is much more difficult than that of Al- and Cu-based alloys. Machinability is about the same as for stainless steels, i.e., difficult. Arc welding can be done only if good shielding is provided by He and Ar to protect Ti from O_2 and N_2 in the air to avoid embrittle-ment tendencies. Forming dies made of Ti alloys suffer from wearing and galling.

Ti alloys are used for applications such as aircraft compressor blades, missile fuel tanks, struc-tural parts operating for short times up to 600°C, and autoclaves and process equipment operating up to 480°C. Low specific gravity, good strength even at high temperatures, and corrosion resis-tance justify the use of these materials, which cost about five times as much as stainless steels.

4.4.8 Superalloys

From the previous discussion, Ti and Ti alloys have been found to be useful in providing strength at elevated temperatures of not more than 480°C. Jet engines, gas turbines, rockets, and nuclear appli-cations often require materials that possess high strength, creep resistance, oxidation, and corrosion resistance and fatigue resistance at temperatures up to 1100°C. One class of materials offering these properties is the superalloys, which are commonly Ni-based alloys that are designed to provide good mechanical properties at extremely high temperatures. Unfortunately, their density is greater than that of iron, so their use is often at the expense of additional weight.

Superalloys may be Ni-based (Hastelloy, Waspaloy, René, Inconel, etc.), Fe–Ni-based (Illium, Incoloy, etc.), or Co-based (Hynes, MAR, etc.); most are precipitation-hardenable. Most of the superalloys are very difficult to form or machine, so methods such as EDM, ECM, or USM are often utilized, or the products are made to final shape by investment casting or powder metallurgy. Because of their ingredients, these alloys are quite expensive, which limits their use to small or criti-cal parts where the cost is not a determining factor.

4.4.9 Refractory Metals

These are increasingly important as high temperature materials when temperature approaches about 1000°C and more. Some engineering applications necessitate higher temperature limits than those provided by superalloys. For example, the exhaust temperatures of today's jet engines exceed 1450°C. In such cases, refractory metals are used. The high-temperature strength of refractory met-als is due to their high melting points. These metals along with their melting points and key proper-ties are listed in Table 4.9.

TABLE 4.9
Physical and Mechanical Properties of Some Refractory Metals

Refractory Metal	Melting Point (°C)	Density (g/cm³)	Thermal Conductivity $\dfrac{cal}{cm^2 \cdot s} \Big/ \dfrac{C°}{cm}$	E at 25°C (MPa)	TSᵃ (MPa) at 25°C	at 1000°C
Molybdenum	2620	10.20	0.35	316.900	827–1378	324–560
Niobium	2470	8.66	0.12	103.400	207–314	180–380
Tantalum	3030	16.60	0.13	186.000	241–482	130–250
Tungsten	3410	19.30	0.40	406.000	689–3445	140–750
Rhenium	3180	21.00	0.17	464.000	1929	N.A.

ᵃ Ranges of TS are given because TS of these metals vary significantly with their form and processing history.

Important refractory metals are

1. Tungsten: It has the highest melting temperature of metals. It is extremely strong. As a pure metal, it is used as filament wire in light bulbs. Tungsten is an important element in tool and die steels, imparting strength and hardness at elevated temperatures. WC with Co as a binder is one of the tool and die materials.
2. Tantalum and molybdenum: They are also refractory metals. Both are used in electronic tubes, and as alloying additions to steels and carbide tools. Ta has a high density and poor resistance to chemicals at temperatures above 150°C. It is used for surgical implants. A variety of Ta-based alloys is available in many forms for use in missiles and aircrafts. Mo is useful as an electrical contact material.
3. Niobium (columbium): Possesses good ductility and formability, and it has better resistance to oxidation than other refractory metals. Nb-based alloys are used in rockets, missiles, and nuclear applications. Nb is also an alloying element in various steel alloys and superalloys.
4. Beryllium: This is an expensive light metal; it is only slightly denser (1.85 g/cm³) than Mg, but of much higher melting point (1277°C). Beryllium costs 1200 times as much as carbon steel. It is used in applications where weight saving is important enough to justify its cost. It has the highest strength-to-weight ratio among metals. Be foils are used for X-ray windows and counter tubes, since it has a very low absorption coefficient for short-wave radiant energy. It is used in atomic energy installations as a neutron reflector and modulator. It is also used in rocket nozzles, space and missile structures, aircraft disc brakes, precision instruments, and mirrors. Newly developed Be alloys allow investment casting of thin-walled parts for aerospace applications. Be and Be oxide are toxic and should not be inhaled.
5. Germanium: It is a semiconductor used in rectifiers and transistors. Si is less expensive and performs the same function.
6. Zirconium: It is silvery in appearance and has good strength and ductility at elevated temperature, and good corrosion resistance, because of its adherent oxide film. Zr is used in electronic components and nuclear power reactor applications.

4.4.10 Noble Metals

The most important noble metals are

1. Platinum is a silvery white metal of an fcc structure and density of 21.5 g/cm³. It is produced as a byproduct of nickel refining. Pt is malleable and ductile and resistant to corrosion

even at elevated temperatures. When pure, it is used for decorative purposes, in crucibles, electrical contacts, spark plugs, filaments, and nozzles, as catalysts for chemical reactions, and recently in automobile pollution control devices to reduce noxious emissions. Other applications include jewelry and dental work.

2. Gold is a soft, lustrous yellow metal of density 19.3 g/cm³ and of an fcc structure. It has been known to man since antiquity and has been used in jewelry and for decoration before recorded history. Pure gold is 24 karats and 18-karat gold contains 75% pure gold. Most gold jewelry is made from 14- to 18-karat gold alloyed with Cu because the pure metal is much too soft and does not exhibit adequate hardness and wear resistance. It can be drawn into wire finer than a human hair, and a single troy ounce (31 g) of pure gold can be hammered into a sheet measuring 27 m². Gold is highly corrosion resistant and is not attacked by corrosive media and even many acids. It is for this reason that it is often found free and uncombined in nature. Gold possesses good thermal and electrical conductivity and reflectivity. For these reasons, it is used in electroplating to provide thermal protection on satellites and other aerospace applications. Gold is also used in electrical contacts, coins, jewelry, and dental restorations.

3. Silver is a brilliant white metal with an fcc structure and a density of 10.5 g/cm³. It has the highest thermal and electrical conductivity of any metal. Like gold, silver is used as a store of value. Pure silver is too soft for most uses; therefore, it is alloyed to increase its hardness and strength. For example, sterling silver contains 92.5% Ag and 7.5% Cu.

   Due to its photosensitive properties, silver finds applications in the photographic industry, which in the United States alone consumes 53 million ounces from the total industrial consumption of 123 million ounces yearly. Other applications of silver include tableware, coins, and electroplating. It is also used in certain cases such as transvenous leads in heart pacemakers.

4.5 NEWLY BORN ENGINEERING MATERIALS

The recent developments in engineering technology necessitates the innovation of various materials, such as nanomaterials, amorphous alloys, metal foams, and shape-memory alloys. Accordingly, it is important to discuss briefly these new engineering materials in terms of their possible impact on industrial applications.

4.5.1 NANOMATERIALS

This is a new field that takes a material science-based approach to nanotechnology. Nanotechnology will be considered later on in Chapter 22. *Nanomaterials* were first investigated in the early 1980s. They have some properties that are often superior to those of traditionally and commercially available engineering materials. These properties stem from the nanoscale dimension. Nanoscale is usually defined as a one tenth of micrometer (100 nm), in at least one dimension, although this term is sometimes also used for materials smaller than 1 μm. Nanolayers such as thin films, surface coatings, and computer chips have one dimension in the nanoscale and are extended in the other two dimensions. Materials having a nanoscale in two dimensions and extended into the third one include nanowires and nanotubes (Figures 4.6 and 4.7). Those of a nanoscale in three dimensions are called nanoparticles (Figure 4.8), such as precipitates, colloids, quantum dots, and tiny particles of semiconducting materials. Table 4.10 shows typical nanomaterials together with their corresponding size. The outstanding characteristics of nanomaterials also include strength, hardness, ductility, wear resistance, and corrosion resistance, which are basically required for structural and nonstructural applications. Moreover, nanomaterials should possess unique electrical, magnetic, and optical characteristics.

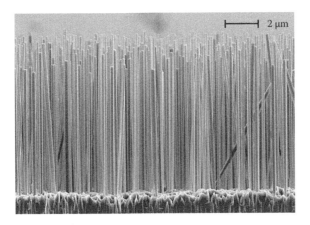

FIGURE 4.6 Nanowires. (From Austin Tech Happy Hour, Texas. With permission.)

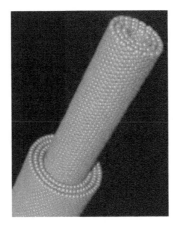

FIGURE 4.7 Nanotubes. (From Quanteq. With permission.)

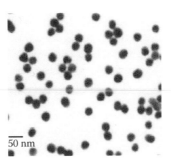

FIGURE 4.8 TEM image of 12-nm colloidal Au.

TABLE 4.10
Typical Nanomaterials

Nanotype	Size	Material
Nanoparticles	1–100 nm, diameter	Ceramic oxide
Nanocrystals and quantum dots	1–10 nm, diameter	Metals, semiconductors, magnetic materials
Nanowires	1–100 nm, diameter	Metals, semiconductors, oxides, sulfides, nitrides
Nanotubes	1–10 nm, diameter	Carbon, metal
2-D arrays of nanoparticles	Several nm^2 to μm^2	Metals, semiconductors, magnetic materials
Surface and thin films	1–1000 nm thick	Various materials
3D structures (superlattices)	Several nm	Metals, semiconductors, magnetic materials

Materials reduced to nanoscale can suddenly show very different properties, enabling specific applications. For instance, opaque substances such as copper become transparent, inert materials such as platinum and gold attain catalyst properties, stable materials such as aluminum turn into combustibles, solids such as gold turn into liquids at room temperature, and insulators such as silicon become conductors. The composition of nonmaterials can be any combination of the following elements or compounds: oxides, carbides, nitrides, metals and alloys, polymers, and composites. The preparation follows different techniques, such as inert-gas condensation, electrodeposition, plasma synthesis, sol-gel synthesis, and mechanical alloying or ball milling. Nanomaterials fall in different categories or shapes, such as

1. Nanoparticles: These are of great scientific interest, as they effectively bridge the bulk material and atomic or molecular structures. The high surface area to volume ratio provides a tremendous driving force for diffusion, especially at elevated temperatures. Sintering is possible at lower temperatures for shorter durations than for larger particles. The surface effect of nanoparticles also reduces the incipient melting temperature.

   Traditional polymers can be reinforced by nanoparticles, resulting in novel materials which can be used as lightweight replacements for metals. Nanoparticles are used in many fields, as shown in Table 4.11.

2. Nanowires: A nanowire is a nanostructure with the diameter of the order of a nanometer. Carbon nanowires are strong and have useful electrical properties. They are solid and therefore are stronger than nanotubes and are therefore used for nanoelectronics and ultrastrong fibers, friction-free bearings, and in space shuttle nose cones. The diameter of the nanowires is restricted to a few nanometers; however, there is no limitation in length,

TABLE 4.11
Nanoparticles Applications

Particle Type	Application
Palladium	Chemical vapor sensors to detect hydrogen gas
Quantum dots	To identify the location of cancer cells in the body
Iron	Clean up carbon tetrachloride pollution in ground water
Silicate	Provide a barrier to gasses (oxygen), or moisture in a plastic film used for packaging.
Zinc oxide	Industrial coatings to protect wood, plastic, and textiles from exposure to UV rays
Silicon dioxide	Filling gaps between carbon fibers, strengthening tennis racquets
Silver	To kill bacteria, making clothing odor-resistant
Titanium oxide	To remove germs and other pollutants from air
Manganese oxide	For removal of volatile organic compounds in industrial air emissions
Zinc oxide	Sensors capable of detecting a range of chemical vapors

with a typical aspect ratio of more than 1000. Nickel and platinum are typical metallic nanowires, Si and GaN are semiconducting nanowires, and SiO_2 and TiO_2 are insulating nanowires. These nanowires are used for electronic, optoelectronic, nanoelectrochemical, and high performance devices.

3. Nanotubes: Carbon nanotubes have the highest strength-to-weight ratio of any known material. NASA is combining nanotubes with other materials into composites to build lightweight spacecraft. Nanotubes can easily penetrate membranes such as cell walls, and function like a needle at the cellular level to deliver drugs directly to the diseased cells. Their electrical resistance changes significantly when other molecules attach themselves to the carbon atoms. They can therefore be used to develop sensors to detect chemical vapors such as carbon monoxide or other biological molecules.

4. Fullerenes: These are a class of allotropes of graphene sheets rolled into tubes or spheres, characterized by their mechanical strength and electrical properties.

Typical Applications: Current applications of nonmaterial include cutting tools, metal powders, computer chips, sensors, surgical and diagnostic minirobots, and various electrical and magnetic components (Chapter 22).

Safety: Nanomaterials are expected to cause health hazards such as cancer due to the increased rate of absorption associated with manufactured nanoparticles. The greater specific surface area (surface area/unit weight) of the nanoparticles may lead to increased rate of absorption through the skin, lungs, or digestive tract, causing health hazards. The Swedish Korolinska Institute conducted a study in which various types of nanoparticles were introduced to a human lung. The results released in 2008 showed that iron oxide nanoparticles caused little DNA damage and were nontoxic. Titanium dioxide caused only DNA damage. Copper oxide was found to be the worst; it causes a clear health risk.

4.5.2 Amorphous Alloys

An *amorphous alloy* is a metallic material with a disordered atomic structure. In contrast to most metals which are *crystalline* and therefore have a highly ordered arrangement of atoms, amorphous alloys are *noncrystalline*. These alloys have no grain boundaries, and their atoms are randomly and tightly packed, and because their structure resembles that of glasses, these alloys are referred to as *metallic glasses* or *glassy metals*.

The first reported metallic glass was an alloy (75 Au, 25 Si) produced at Caltech by Klemet, Wilens, and Duwez in 1960. The alloy had to be cooled extremely rapid to avoid crystallization. This was achieved by melt spinning (Figure 4.9), in which the alloy is melted by induction in a ceramic crucible and propelled under high gas pressure at very high speed against a rotating copper disc and chilled rapidly (splat cooling). Since the realized rate of cooling is on the order of

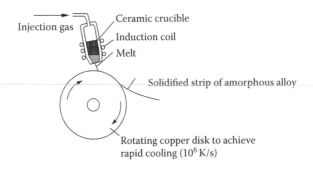

FIGURE 4.9 Production of amorphous alloy by melt spinning.

1 megakelvin per second (10^6 K/s), the molten alloy does not have sufficient time to crystallize. The metallic glass could only be produced as very thin ribbons, foils, or wires, so that heat could be extracted quickly enough to achieve the necessary cooling rate. The rapid solidification results in a significant extension of solid solubility, grain refinement, and reduced microsegregation.

More recently, a number of alloys with critical cooling rates low enough to allow formation of amorphous structure in thicknesses up to 10 mms has been produced. In 1992, the first aerospace commercial amorphous Vitreloy (41.2 Zr, 13.8 Ti, 12.5 Cu, 10 Ni, 22.5 Be) was developed by Caltech. Vitreloy has a tensile strength that is almost twice that of high-grade titanium.

Generally, metallic glasses at room temperature are not ductile, and tend to fail suddenly when stressed in tension, which limits the material applicability in critical applications. Therefore, there is a considerable interest in producing metal matrix composite materials consisting of a metallic glass matrix containing dendritic structure fibers of a ductile crystalline metal.

Amorphous metal is usually an alloy rather than a pure metal. The alloy contains atoms of significantly different sizes, leading to higher viscosity, which prevents the atoms from moving enough to form an ordered lattice. Moreover, rapid cooling results in low shrinkage of the material, thus promoting resistance to wear, plastic deformation and corrosion. Amorphous alloys, while technically glasses, are much tougher and less brittle than oxide glasses and ceramics. The thermal and electrical conductivity of amorphous materials are lower than those of crystals.

Typical Applications:

1. Amorphous alloys of boron, silicon, phosphorus, and other glass formers with magnetic metals (Fe, Co, Ni) are magnetic, with low coercivity and high electrical resistance. Such characteristics lead to low eddy current losses when subjected to alternating magnetic fields, a property useful in magnetic cores of transformers, generators, motors, and amplifiers.
2. Since amorphous alloys are true glasses, they soften and flow upon heating. This allows for easy processing, such as injection molding, the same way as polymers. As a result, such alloys have been commercialized for use in sports equipment, medical devices, and electronic equipment.
3. The amorphous 40 Ti, 36 Cu, 14 Pd, 10 Zr is believed to be noncarcinogenic. It is about three times stronger than Ti, and its elastic modulus nearly matches bones. It has a high wear resistance and does not produce abrasion powder. Biologically, it allows better joining with bones. The amorphous 60 Mg, 35 Zn, 5 Ca is used for implantation into bones as screws, pins, or plates to fix fractures. Unlike traditional steel or Ti, this material dissolves in organisms in a rate of roughly 1 mm/month, and is replaced with bone tissues. This rate may be adjusted by varying the Zn content.

4.5.3 Metal Foams

A metal foam (or cellular metal) (Figure 4.10) is a very intriguing material with extraordinary properties, where cavities cover almost 80%–95% of the total volume of foam. The very low density (0.03–0.2 g/cm³) shows some of the metal properties of which they are made up, together with other characteristics due to its particular structure. Thus, metallic foams are materials with a combination of very special physiochemical and mechanical properties, to such an extent that the package of properties they show is not currently covered by other materials. At the same time, they improve the variability of their applications.

Therefore, metal foams, as the name indicates, are metallic materials with porous structure (Figure 4.10). They can take either the form of an *open structure* of interconnected pores, similar to a bath sponge, or a *closed structure* of pores not connected to each other. For a certain foam material, the mechanical and physical properties depend on the porosity and the size of pores. The

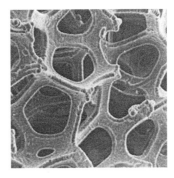

FIGURE 4.10   Porous structure of metal foam ($\rho = 0.4–0.65$ g/cm³).

porous structure results in a very low weight compared with full material, unbelievable rigidity, unexpected high strength and wear resistance, along with unprecedented surface area.

Metal Foam Preparation: There are a number of manufacturing methods to produce metal foams, one of which is by blasting air into a molten metal and tapping the froth that forms at the surface of the bath, which solidifies into foams. The foam can be machined, cut, drilled, rolled to form, and brazed or welded with care to offer flexible material with controlled porosity. Metal foams may be pure metals (Al, Ni, Cu, Pb, Co, Fe, Cd, Ti, and Ta) or alloys (Fe–Cr, Co–Cr, Ni–Cr, Ni–Cu, Ni–Cr–Al). Closed-pore metal foams are almost exclusively centered on Al foams.

Three important materials have been produced so far are

1. Ni foam (densities 0.4–0.45 g/cm³), for batteries and electrolysis applications, due to its good conductance.
2. Alloy version Ni–Cr foam (densities 0.6–0.65 g/cm³) has a good performance in structures, where strength, high temperature resistance, and corrosion resistance (comparable with Inconel) are required.
3. Alloy version Ni–Cr–Al foam to improve oxidation resistance at high temperatures, such as found in premixed gas burners.

Fields of Application: Metallic foam is a versatile and an attractive material applied in many engineering domains, including aviation, marine, chemical, and biomedical applications. Due to their cellular structure and extreme lightness, metal foams have enormous potential for use in a never-ending (multifunctioning) list of applications in diverse industrial sectors, such as

- Noise and vibration absorbers; e.g., reducing noise on roads and bridges
- Catalyst carriers
- Battery and electrolysis electrodes
- Filters
- Flame arresters and spark catchers for diesel engines
- Fire protection in buildings
- Lightweight beams
- Improving passive safety of ultralight structures and automobiles by providing sandwich-type foams
- Premixed gas burners
- Orthopedic implants
- Architectural maquette work (Figure 4.11) and decorative purposes

FIGURE 4.11 Architectural maquette: grayish foam simulating concrete and Silverish stainless steel plate. (From Mrs. Ineke Segers, Hogeschool voor de Kunst, Utrricht, Holland. With permission.)

In conclusion, metal foams can be seen as a future technology with high growth prospects in industrial use in the short term for their outstanding characteristics such as reduced weight, safety in transport, noise reduction, and so on. All of these, at a competitive cost, contribute substantially to business efficiency and overall economy.

4.5.4 Shape Memory Alloys

Shape memory alloys (SMAs) are alloys that can be deformed and then return to their original shape by heating (Figure 4.12). Since they *remember* their shape, they are also called *memory alloys*. This unusual property was first observed by Oelender (1932), but serious research advances in the field were not made until the 1960s. These materials, however, do not in themselves have what is considered intelligence; in fact, they are pretty stupid. The most effective and widely used SMAs include Ni–Ti, Cu–Zn–Al, and Cu–Al–Ni. The alloy 55Ni–45Ti is commercially known as nitinol (an acronym for nickel–titanium, Naval Ordnance Laboratories) and has many diverse applications.

Shape Memory Effect: A piece of straight wire of SMA can be wound into a helical spring. When heated, the spring uncoils, returning to its original shape. This behavior is reversible, such that the shape can change back and forth repeatedly upon application and removal of heat. In most SMAs, a temperature change of only about 10°C is enough to initiate a martensitic/austentic phase transformation on heating (and vice versa on cooling). Austenite is stronger as compared to martensite.

Typical Applications: SMAs find many applications in aircraft, satellites, automobiles, thermostats, robotics, orthopedic surgery, dentistry, optometry, and many others. Some examples are provided.

1. Joining critical piping in jet aircraft, nitinol is used. It is austenitic at room temperature, where the joint is machined, with the inner diameter of the external tube smaller than

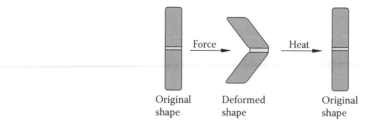

FIGURE 4.12 Behavior of shape memory alloy.

the outside diameter of the internal tube. The external tube is then cooled in liquid nitrogen to transform into the martensitic phase. It is stored in this stage in liquid nitrogen until installation. Once placed in position on the tube joint, it heats to room temperature, returns to its original machined diameter (shape memory), and establishes a leak-proof shrink fit.

2. Orthopedic surgery, in which bone plates made of nitinol represent surgical tools, which are used to assist in holding broken and fractured bones:

   a. The nitinol plate is first cooled well below its transformation temperature. It is then placed on the broken bones, after compressing bones together, and being held under some slight pressure, which helps to speed up healing the joint (Figure 4.13). When the nitinol plate heats up to body temperature, the plate attempts to contract (shape memory), applying sustained pressure for far longer than stainless steel or titanium.

   b. Figure 4.14 shows how even a badly fractured face can be reconstructed using nitinol bone plates.

Advantages and Drawbacks of SMAs: Advantages

- Excellent biocompatibility and cytocompatibility
- Diverse fields of application
- Good mechanical properties
- High corrosion resistance

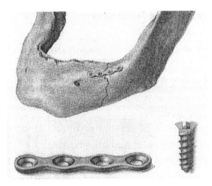

FIGURE 4.13   Nitinol bone plate used to heal broken bones.

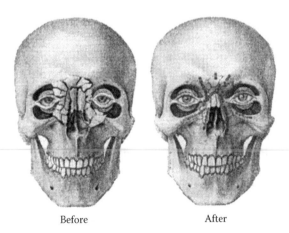

Before   After

FIGURE 4.14   Badly fractured face reconstructed using thin nitinol plates.

Drawbacks

There are some difficulties that must be overcome before SMAs can be employed to their full potential.

- These alloys are still expensive to manufacture and machine compared to conventional metals.
- Most are not resistant to high load.
- They have poor fatigue properties, such that a steel component may survive for more than 100 times more cycles than an SMA.
- Memory shape recovery is not accurate.
- With slight overheating, memory can be erased.

4.6　REVIEW QUESTIONS

1. Why are HSLA steels preferred to be used in automobile, railway cars, bridges, and building structural members?
2. What is the role of Cu as an alloying element in HSLA steels?
3. What is the function of Al used as an alloying element in Ti-based alloys?
4. State some typical applications of pure Ti and Ti alloys.
5. Differentiate between the three different structures of Ti alloys.
6. List three applications for each of the following metals: aluminum, copper, lead, and platinum.
7. Carbon appears in different forms in various types of CIs. What are the various forms and how can they be produced as desired?
8. What characteristic properties of nonferrous metals dictate and justify their use in spite of their high cost?
9. What are various types of brasses? Which brasses are best adapted to cold working? What is a cartridge brass?
10. What do you understand about steel?
11. How are plain carbon steels classified depending upon carbon content?
12. Distinguish between plain carbon steel and alloyed steel.
13. How are the mechanical properties of an untreated plain carbon steel influenced by carbon content?
14. Differentiate between wear-, corrosion-, and heat-resisting steels.
15. Distinguish among gray, white, malleable, and ductile CIs. How is each produced? Enumerate their characteristics and applications.
16. What is the composition of the following alloys? (Permalloy–Invar–Kovar) What are their special characteristics and applications?
17. What is a free-cutting steel? What elements are usually added to make a steel free cutting?
18. Which two elements are alloyed to form brass? What are the differences in composition and uses of α- and β-brass?
19. What is the main difference between brass and bronze?
20. State the composition of different bronzes used in engineering.
21. How does Al compare to Cu in terms of electrical conductivity?
22. What are some of attractive properties of Ti and Ti alloys?
23. What property of Monels accounts for a majority of their applications?
24. What is the primary application of pure Zn?
25. Which metals are classified as refractory metals?
26. What is the significance of the last two digits in a typical four-digit AISI-SAE steel designation?

27. Why should ferritic stainless steels be given first consideration when selecting a stainless steel?
28. What is a duplex stainless steel?
29. What are some of the unique properties of austenitic stainless steels?
30. What is a tool steel?
31. What alloying elements are used to produce hot-worked tool steels?
32. Define nanomaterials. What are their fields of application? What is their dimensional range? Give some material used in production of nanomaterials.
33. What are some differences between properties of nanomaterials as compared to traditional ones?
34. Explain why an amorphous alloy has a noncrystalline structure.
35. Why can an amorphous metal only be produced as thin wires and strips?
36. Amorphous steels are considered to be an ideal solution for transformer cores. Comment.
37. Enumerate some applications of amorphous alloys.
38. Describe the method of manufacturing metallic foam. Mention some fields of application.
39. Describe the shape memory effect.
40. What is nitinol? Explain two application examples of this material.
41. What is the typical analysis of Duralumin? Mention its characteristics and applications.
42. What is the lightest engineering metal?
43. What alloying element promotes the castability of Mg?
44. What are the two types of babbits?
45. Mention two reasons for applying Zn as a galvanizing coating for steel wires and plates.
46. What is the role of Cu as an alloying element for Zn-based die casting alloys?
47. What is the optimum tin content of a soft solder?
48. What is a self-lubricating bronze?
49. Write down typical compositions of German silver, Sterling silver, Monels, and Constantan. What are their typical applications?
50. Mention four commercial names of nichromes, and what are their compositions, characteristics, and applications?
51. What are the outstanding characteristics of a nanomaterial?
52. What are the compositions, characteristics, and applications of precipitation-hardened bronzes?

BIBLIOGRAPHY

Doyle, L.E., Keyser, C.A., Leach, J.L., Schrader, G.F., and Singer, M.B. *Manufacturing Processes and Materials for Engineers*. Upper Saddle, NJ: Prentice-Hall.

Klemet, W., Willens, R.H., and Duwez, P. 1960. Non-crystalline structure in solidified gold–silicon alloys, *Nature* 187, pp. 869–870.

Kumar, G., Tang, H., and Sckroeys, J. 2009. Nanotechnology with amorphous metals, *Nature* 457, pp. 868–872.

Schey, J.A. 2000. *Introduction to Manufacturing Processes*, 3rd ed. Singapore: McGraw-Hill.

Thornton, P.A., and Colagelo, V.J. 1985. *Fundamentals of Engineering Materials*. Upper Saddle, NJ: Prentice-Hall.

Van Vlack, L.H. 1973. *A Textbook of Materials Technology*, 1st ed. Massachusetts: Addison-Wesley.

Van Vlack, L.H. 1973. *Materials Science for Engineering*, 1st ed. Massachusetts: Addison-Wesley.

Van Vlack, L.H. 1975. *Elements of Materials Science and Engineering*, 1st ed. Massachusetts: Addison-Wesley.

http://en.wikipedia.org/wiki/Nanomaterials.

http://techon.nikkeibp.co.jp/english/NEWS_EN/20090610/171551/?pd. Maruyama, M. (Jun 2009) Japanese Universities Develop Ti-based Metallic Glass for Artificial Finger Joint.

http://physscsworld.com/cws/article/news/40573. Fixing Bones with Dissolvable Glass. Physics World (Oct. 2009).

5 Heat Treatment of Metals and Alloys

5.1 INTRODUCTION

Heat treatment (HT) embraces many processes, employing combinations of heating and cooling operations, so applied to solid state metal or alloy to induce desired properties. It can alter the mechanical properties of the metal by changing the size and shape of grains of which it is composed. The various HT processes may be broadly classified into

- Softening processes such as annealing, stress relieving, normalizing, and spheroidizing
- Tempering processes used for reheating quench-hardened or normalized steel to a temperature below the transformation range and then cooling at any desired rate
- Hardening processes such as full-depth (through) hardening, and anticrack hardening (martempering and austempering)
- Case hardening processes such as carburizing, nitriding, cyaniding, and induction and flame hardening

Specifically, HT processes have the following objectives:

1. Refining grain size
2. Improving machinability
3. Relieving induced stresses due to cold working
4. Relieving induced stresses due to welding operations
5. Improving mechanical properties such as strength, toughness, hardness, shock resistance, fatigue strength, etc.
6. Improving magnetic and electrical properties
7. Increasing wear resistance, heat resistance, and resistance to corrosion
8. Producing extra-hard surface on a ductile interior

5.2 HEAT TREATMENT OF STEELS

5.2.1 SIMPLIFIED IRON–IRON CARBIDE EQUILIBRIUM DIAGRAM

Consider again the Fe–Fe$_3$C equilibrium diagram illustrated in Figure 3.23. It is the basic diagram for analyzing ferrous alloys. It is useful in studying the equilibrium and nonequilibrium changes that iron carbon alloys undergo during heating and cooling. Such usefulness stems from the fact that this diagram is relatively unaffected by the small concentrations of Mn, Si, S, and P ordinarily present in conventionally refined steels. Focusing on steels (2% C), the Fe–Fe$_3$C equilibrium diagram can be simplified considerably. The portion of the phase (peritectic), which is totally ignored for heat treatment purposes, and that greater than 2% C are of little significance and are deleted. The resulting diagram (Figure 5.1) focuses therefore on the eutectoid reaction, and is quite useful in understanding the properties of steels and their heat treatment.

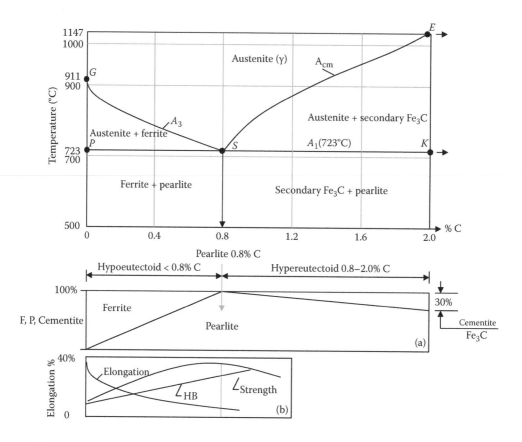

FIGURE 5.1   Part of the iron–iron carbide diagram.

The line GS indicates the beginning of decomposition by precipitation of ferrite (α) from the austenite (γ). The critical points along the GS line are designated as A_3 (A_{c3} in heating and A_{r3} in cooling). The line SE indicates the temperature at which the austenite begins to decompose with the precipitation of excess carbon as cementite. This latter is called secondary cementite to distinguish it from the primary cementite that precipitates from the liquid phase. Temperatures along the SE line are designated as A_{cm} points. Point S, corresponding to 0.8% C, shows the minimum temperature (723°C) at which austenite may exist in a state of equilibrium. At point S, austenite decomposes with simultaneous precipitation of ferrite and cementite, which form a eutectoid mixture known as pearlite. Pearlite consists of thin alternating plates or lamellae of cementite and ferrite called lamellar pearlite. As a result of a special heat treatment annealing process (spheroidizing), the so-called granular pearlite (rounded globules of cementite in a ferritic matrix) may be formed. The decomposition of austenite with the formation of pearlite corresponds to the line PSK (723°C) for all iron–carbon alloys. The critical temperature at which pearlite is formed is designated as A_1 (A_{c1} for heating and A_{r3}, for cooling). Therefore, on the basis of carbon content, carbon steels may be divided into three compositions (Figure 5.2).

1. Hypoeutectoid steels containing less than 0.8% C, which consist of ferrite + pearlite (ferrite + cementite) when they are completely cooled. The higher the carbon content of such steels, the more pearlite and the less ferrite they will contain. With increasing carbon content, the elongation is reduced, and both the hardness and strength increase (Figure 5.2a).
2. Eutectoid or pearlitic steel containing approximately 0.8% C. At a temperature of 723 (A_1), the austenite will decompose into a ferrite–cementite eutectoid mixture (Figure 5.2b).

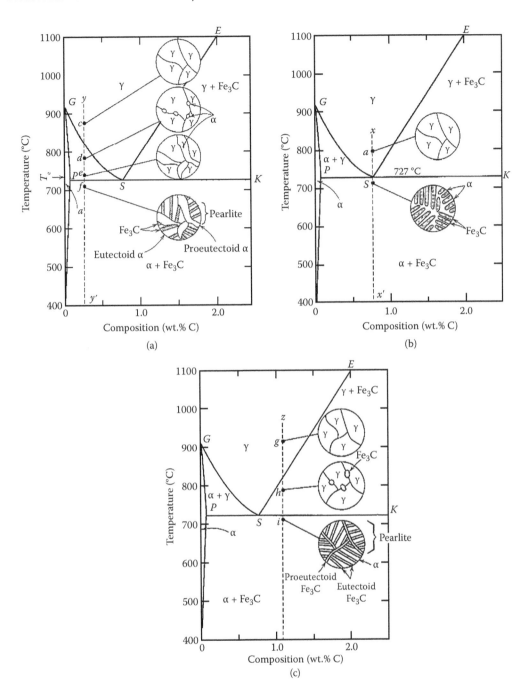

FIGURE 5.2 Schematics of microstructures of hypoeutectoid, eutectoid, and hypereutectoid steels. (From Callister, W. D., *Materials Science and Engineering—An Introduction*, 4th ed., John Wiley, New York, 1996. With permission.)

3. Hypereutectoid steels containing more than 0.8% C, up to 2.0% C. When the temperature drops to the line ES, the austenite precipitates cementite. Upon further cooling below ES, a two-phase state will exist, consisting of austenite (γ), and secondary cementite Fe_3C. The completely cooled hypereutectoid steels will have a structure consisting of pearlite and cementite (Figure 5.2c). Figure 5.2 shows schematics of microstructures above and below

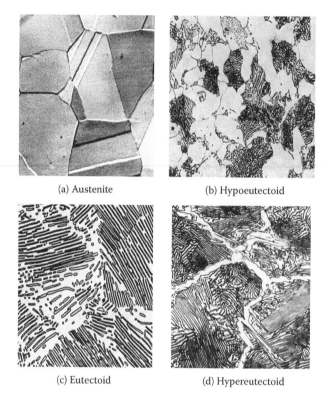

(a) Austenite   (b) Hypoeutectoid

(c) Eutectoid   (d) Hypereutectoid

FIGURE 5.3 Photomicrographs of (a) austenite and (b–d) corresponding microstructures after complete transformation. (From *Metals Handbook*, Vol. 9, Metallography and Microstructure, ASM International, Materials Park, OH, USA, 1985. With permission.)

eutectoid temperature of the three above compositions of steels. Figure 5.3 shows a photomicrograph of (a) austenite and (b–d) the corresponding microstructures, respectively, after complete transformation.

5.2.2 Effect of Alloying Elements

If alloying elements are added to the iron–carbon alloy, the position of A_1, A_3 and A_{cm} boundaries and the eutectoid composition are changed. Classical diagrams introduced by Bain (1939) show the variation of A_1 and the eutectoid carbon content with increasing amounts of a selected number of alloying elements (Figure 5.4). It needs to be mentioned that

1. All important alloying elements decrease the eutectoid carbon content.
2. The austenite-stabilizing elements Mn and Ni decrease A_1.
3. The ferrite-stabilizing elements Cr, Si, Mo, W, and Ti increase A_1.

Effect of Individual Alloying Elements: Various alloying elements are added to steels to improve hardenability, strength, toughness, wear resistance, workability, weldability, and machinability. The various alloying elements with their beneficial and detrimental effects are summarized.

- Carbon (C) improves hardenability, strength, and wear resistance while reducing ductility, weldability, and toughness.
- Nickel (Ni) improves strength, toughness, corrosion resistance, and hardenability.

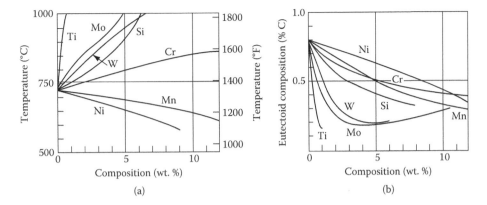

FIGURE 5.4 Influence of alloying element additions on (a) eutectoid temperature and (b) eutectoid carbon content. (Adapted from Bain 1939, *Functions of Alloying Elements in Steel*, ASM International, USA.)

- Chromium (Cr) forms a complex series of Cr-carbides, thus increasing the depth of hardness penetration. These carbides are very hard. Cr improves strength, hardness, wear resistance, toughness, and corrosion resistance. A combination of Ni and Cr is used to improve the mechanical properties of steels. Cr tends to promote coarse grain structure and increases the difficulty of heat treatment; this is counteracted by Ni, which refines the grain size.
- Molybdenum (Mo) improves hardenability, wear resistance, toughness, heat resistance, and creep resistance and minimizes temper embrittlement.
- Steels containing Cr and Ni suffer from temper brittleness, which makes these steels brittle if held at a temperature between 250°C and 500°C. Consequently they lose their shock resistance. This trouble can be overcome by adding about 0.25% Mo. Addition of Mo also hampers grain growth in steel at elevated temperatures, making the steel finer grained and tougher.
- Manganese (Mn) improves hardenability, strength, abrasion resistance, and machinability. It deoxidizes the molten steel and reduces hot shortness (hot shortness is the embrittlement of steel at elevated temperatures caused by a low melting constituent segregated at grain boundaries during solidification). It decreases weldability. Steels containing 1%–1.5% C and 11%–14% Mn are resistant to wear and work hardening, and are resistant to abrasion and shocks. It counteracts the effects of sulfur.
- Silicon (Si) is soluble in ferrite. It improves strength, hardness, corrosion resistance, and electrical conductivity. It decreases magnetic hysteresis loss, machinability, and cold formability when used up to 2.5%. It increases strength without sacrificing ductility.
- Cobalt (Co) and tungsten (W) improve strength and hardness at elevated temperature. W refines grains and raises the hot hardness of HSSs.
- Vanadium (V) is a strong carbide builder; it improves strength, toughness, abrasion resistance, and hardness at elevated temperatures; it inhibits grain growth during heat treatment.
- Selenium (Se) improves machinability.
- Boron (B) is added in small quantities (up to 0.001%) to improve hardness and mechanical properties without loss of, or even with some improvements in, machinability and formability.
- Niobium, or columbium, (Nb), and tantalum (Ta) impart fine grain size and improve strength and toughness, though they may reduce hardenability.
- Titanium (Ti) improves hardenability and deoxidizes steels.
- Copper (Cu) is added from 0.2% to 0.5% to increase atmospheric corrosion resistance. It adversely affects hot working characteristics and surface quality.

- Aluminum (Al) is an effective deoxidizer. Its addition controls grain growth. In nitriding steels, it is used from 0.9% to 1.5% for surface hardening due to the formation of a case of stable aluminum nitride.
- Lead (Pb) improves machinability. It causes liquid metal embrittlement.
- Phosphorus (P) improves the machinability of low carbon steels.
- Sulfur (S) improves machinability when combined with Mn. It lowers impact strength and ductility, and impairs surface quality and weldability.

Alloying Elements Rendering Desired Steel Characteristic:

- Mn, Cr, Mo, W, Ni, and Si are metals which delay the rate at which austenite is transformed to pearlite upon cooling. They allow hardening without drastic quenching to eliminate distortion and cracking.
- Mo, Cr, V, and W provide higher hot hardness.
- Mn, Ni, Si, C, Cr, and Mo provide higher strength at elevated temperatures.
- V and Al inhibit grain growth in austenite during heat treatment.
- Cr and Ni provide corrosion resistance.
- V, Mo, Cr, and W provide abrasion and wear resistance.
- Si, Al, and Ti combine with oxygen to prevent blowholes.
- Mn combines with sulfur, which otherwise causes brittleness.
- S, P, and Pb improve machining properties.

5.2.3 Time–Temperature–Transformation Diagram (T–T–T Diagram)

The T–T–T diagram (also called S-curve) is the most useful tool in presenting an overall picture of the transformation behavior of austenite, which enables the metallurgist to interpret the response of steel to any specified heat treatment, and to plan practical heat treatment operations to get desirable microstructure to control limited hardening or softening and soaking times. If the steel is heated to the austenitic region and held there until carbon is dissolved (austenite has an affinity to carbon), and is then cooled rapidly by quenching, the carbon is not given a chance to escape and is trapped as dispersed atoms in a strained low temperature ferritic lattice. That sets up a distorted structure (martensite), which is quite hard and strong, but brittle. The structure formed during continuous cooling of steel from above A_{c3} can be understood by studying the constant temperature (isothermal) transformation of austenite, thus separating the two variables—time and temperature.

Transformation at Constant Temperature (Isothermal Cooling): When carbon steel is quenched in baths held at constant temperature, the austenite transforms at a speed characteristic of each temperature. The times of beginning and completion of austenitic transformation are plotted against temperature to give the isothermal S-curve shown in Figure 5.5. The logarithmic time scale is used to condense results in small space. Austenite is completely stable above A_{c3}, partially unstable between A_{c3} and A_{c1}, and completely unstable below A_{c1} (Figure 5.1). Figure 5.5 shows schematically the isothermal transformation for the eutectoid (pearlitic) steel (0.8% C), cooled to below the critical temperature, where the temperature is held constant for a period (cooling curves 1 and 2) while transformation takes place. No change occurs in the area to the left of the S-curve. For the cooling regime 1, the structure begins to transform to coarse pearlite (a), and the transformation is completed at (b). On using the cooling regime 2, the structure begins to transform to bainite (c), and the transformation is completed at (d).

Transformation on Continuous Cooling: Since the most heat treatment operations involve continuous cooling, the previously discussed isothermal transformation may not be presentable. A continuous cooling transformation curve also for pearlitic steel (0.8% C) (Figure 5.6) is a modified

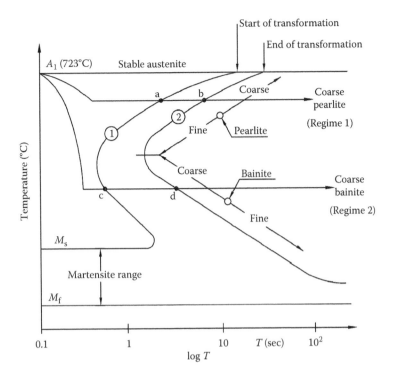

FIGURE 5.5 Isothermal S curve of pearlitic carbon steel.

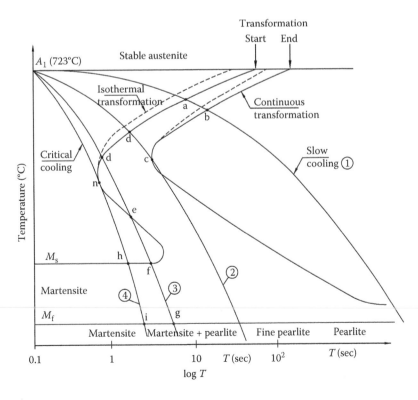

FIGURE 5.6 Continuous cooling S curve of pearlitic carbon steel.

form of the isothermal over a range of temperatures rather than at one temperature. The isothermal S-curve is represented as dotted for the purpose of comparison. In continuous cooling, the transformation begins later and at lower temperatures. The formation of bainite may be disregarded in continuous cooling in case of carbon and some in alloyed steels. Thus, the bainite region does not appear (Figure 5.6). This is not the case for many alloyed steels.

A number of cooling curves are shown superimposed on the S-curves (Figure 5.6), each representing a different rate of continuous cooling. Curve 1 indicates that a slow cool, such as in annealing, allows transformation to pearlite to start at (a) and to complete at (b). From lower temperatures than (b) the rate of cooling can be rapid without affecting the hardness and microstructure of steel. Curve 2 represents the fastest cooling rate at which austenite can transform entirely to pearlite (c), which however, is much finer than that developed by curve 1. With curve 3, partial transformation to fine pearlite will occur between (d) and (e), no change occurs between (e) and (f), and the remaining austenite is transformed then to martensite between (f) and (g). Any steel rapidly cooled along a line to the left of the nose (n) of the curve is kept as austenite until it reaches the M_s (start of martensite) temperature. The following empirical formula can be used for calculating M_s (°C) from the chemical analysis of the steel, provided the all carbides have been dissolved in the austenite:

$$M_s (C°) = 651–474 \ (\% \ C)–33 \ (\% \ Mn)–17 \ (\% \ Ni +\% \ Cr)–21 \ (\% \ Mo),$$

M_f (finish of martensite) is about 215°C below M_s.

Curve (4) represents the critical cooling rate, since it just touches the nose (n) of the S-curve, and transformation of austenite occurs between (h) and (i).

In most cases, the S-curve obtained under conditions of continuous cooling occupies a position downward and to the right, with respect to the isothermal S-curve (Figure 5.6). It should be understood that each analysis of steel has its own S-curve and many have been published by steel-making companies and professional institutions such as ASM. The S-curves of alloyed carbon steels are generally shifted to the right with respect to those of plain carbon steels. Figure 5.7 illustrates the relative position of S-curves for hypoeutectoid, hypereutectoid, and eutectoid carbon steels. The

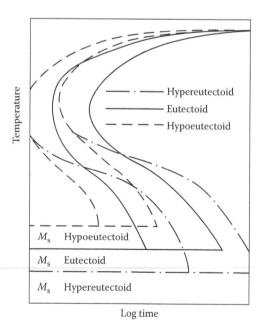

FIGURE 5.7   Relative positions of S curves for hypoeutectoid and hypereutectoid steels.

lower the temperature of transformation for a given steel, the harder and stronger the product. Thus, medium pearlite is harder than coarse pearlite and fine pearlite is harder than medium pearlite. The finer pearlite spacing offers more resistance to the flow of dislocations. For the same reason, bainite is harder than pearlite, and martensite is the hardest. Bainite and martensite contain both widely and minutely dispersed particles that provide even more resistance compared to pearlite.

5.3 BASIC HEAT TREATMENT OPERATIONS OF STEELS

The most commonly used heat treatment operations of steel are represented in the taxonomic chart in Figure 5.8.

5.3.1 ANNEALING PROCESSES

Annealing is the softening process in which steels are heated above their transformation range, held there for a proper time (soaking), then cooled slowly (at a rate of 30°C/h–150°C/h) below the transformation range, usually in the furnace itself. Several annealing procedures are employed to enhance the properties of steels and steel alloys. However, before they are discussed, some comment relative to the labeling of phase boundaries is necessary. Figure 5.9 shows the proportion of iron–iron carbide phase diagram in the vicinity of the eutectoid. The horizontal line at the eutectoid temperature, conventionally labeled A_1, is termed the lower critical temperature, below which, under equilibrium, all austenite is transformed into ferrite and cementite phases. The phase boundaries A_3 and A_{cm} represent the upper critical temperature lines for hypoeutectoid and hypereutectoid steels, respectively. For temperatures and compositions above these boundaries, only the austenite phase prevails. Other alloying elements shift the eutectoid and the positions of these phase boundary lines.

The objectives of annealing processes are

- Softening the steel so that it can be cold worked
- Reducing hardness and increasing ductility
- Improving machinability

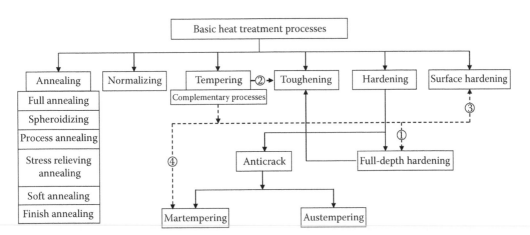

FIGURE 5.8   Classification scheme of the commonly used heat treatment of steels.

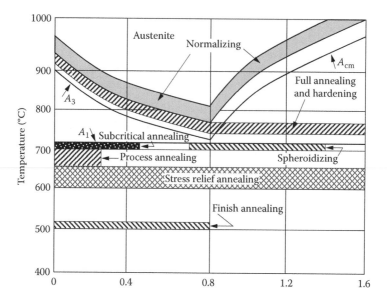

FIGURE 5.9 Iron–iron carbide diagram in the vicinity of the eutectoid, indicating heat treatment ranges of plain carbon steels.

- Refining grain size and realizing uniformity
- Preparing the steel for subsequent heat treatment
- Relieving internal stresses
- Producing desired mechanical, physical, electrical, and magnetic properties
- Realizing desired microstructures

Annealing processes are generally time consuming, and hence are costly. The recommended soaking times may be 1–2 h, depending on the article size. Cooling generally necessitates several hours. Annealing is frequently used for steels of carbon content ranging from 0.3 to 1.6% C. Carbides dissolve rapidly in hypoeutectoid steels, whereas they agglomerate in hypereutectoid steels.

Any annealing process consists of three stages

1. Heating to a desired temperature
2. Holding or soaking at that temperature
3. Cooling to room temperature

Both the temperature and the time are governing factors for the transformation reaction developed in various types of annealing processes. The selection of the annealing process often depends on the desired objectives; however, steel composition and its carbon content strongly influence the choice. Process annealing is restricted to low carbon steels (<0.25% C), and spheroidizing is a known treatment for high carbon steels. Full annealing (and normalizing) can be applied for all carbon contents. Since different cooling rates do not produce a wide variation of properties in low carbon steels, the air cooling of the normalizing treatment often produces acceptable uniformity, and hence is preferred over full annealing. For higher carbon contents (0.4%–0.6% C), different cooling rates can produce wide property variations, and the uniform furnace cooling of full annealing is often preferred.

Depending on the final resulting microstructure desired, the following annealing processes are commonly adopted

1. Full annealing: This is often utilized in low and medium carbon steels that will be machined or will experience plastic deformation. The alloy is austenitized by heating to 15°C–40°C above the A_3 or A_1 lines, as illustrated in Figure 5.9, until equilibrium is achieved. The alloy is then furnace cooled, which takes several hours. The microstructure of this annealing is coarse pearlite that is relatively soft and ductile. The full-anneal cooling procedure (also shown in Figure 5.5) is time consuming.

2. Spheroidizing: High carbon steels having a microstructure containing coarse pearlite, and even tool steels containing a large amount of free cementite may still be too hard to conventionally machine or plastically deform. These steels may be annealed to develop a spheroidite structure. Spheroidized steels have maximum softness and ductility to be easily machined and deformed. The spheroidizing heat treatment consists of heating the steel at a temperature of 700°C (20°C–40°C below the eutectoid line A_1) in the α +Fe$_3$C region of the phase diagram (Figure 5.9); the temperature is held constant for a considerable soaking time (e.g., 4 h for a 25-mm diameter piece). Due to this annealing, there is a coalescence of Fe$_3$C to form the spheroid particles (Figure 5.10). This treatment is carried out on medium- and high-carbon steels (0.6% C–1.4% C). Spheroidizing is seldom used on low-carbon steels because they become extremely soft. However, on low-carbon steels, it is used to permit severe deformation. As a conclusion, the objectives of spheroidizing are
   - To reduce strength
   - To increase ductility
   - To improve machinability
   - To give a basic tool structure for a subsequent hardening process

3. Process annealing: This is an annealing procedure for low-carbon steels that is used to negate the effects of strain-hardened metal. It is commonly utilized during fabrication procedures that require extensive plastic deformation, to allow a continuation of deformation without fracture or excessive energy consumption. Recovery and recrystallization processes are allowed to occur. Ordinarily, a fine-grained microstructure is desired, and therefore, the heat treatment is terminated before appreciable grain growth has occurred. The steel is heated to a temperature ranging from 550°C to 700°C, soaked, and then frequently cooled in still air (Figure 5.9). The material is not heated to as high a temperature as in full annealing (or normalizing), so process annealing is somewhat cheaper and of less tendency to produce scaling. Surface oxidation or scaling may be prevented or minimized by annealing at a relatively low temperature (but above recrystallization temperature) or by heating in a nonoxidizing atmosphere.

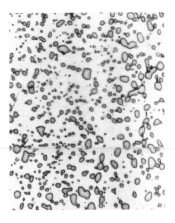

FIGURE 5.10 A photomicrograph of straight carbon steel having spheroidizing microstructure cementite embedded in α-matrix.

4. Stress-relief annealing: This is used with the purpose of reducing internal residual stresses developed due to
   - Plastic deformation associated with machining and grinding
   - Nonuniform cooling of a part that was processed at an elevated temperature, such as welding or casting
   - A phase transformation that is induced upon cooling

   Distortion and warping may result if these residual stresses are not removed. They are eliminated by stress-relief annealing in which large steel castings, welded assemblies, and cold-formed products are heated to a temperature below the A_1 (550°C–650°C), held there long enough to attain uniform temperature, and finally cooled to room temperature in air (Figure 5.9). The annealing temperature is usually a relatively low one, such that effects resulting from previous cold working and other heat treatment processes are not affected.

5. Subcritical (soft) annealing: In subcritical annealing, the cold worked steel is annealed at temperatures of 700°C–780°C to achieve nearly the ductility obtainable in full annealing but with less risk of distortion (Figure 5.9). It is used before moderate to severe cold-forming operations. Steels after this process are usually not suitable for general machining.

6. Finish annealing: This is a low-temperature annealing treatment (510°C) applied to cold-worked steels with low- or medium-carbon contents (Figure 5.9). As a compromise treatment, it lowers the level of residual stresses so as to reduce the risk of distortion in machining, and at the same time it also retains the machinability lost by cold working.

5.3.2 NORMALIZING

This is a heat treating process in which iron-based alloys are heated to 55°C above the upper transformation temperature A_3 for hypoeutectoid and A_{cm} for hypereutectoid steels, and held there for a specified period to ensure that a fully austenitized structure is produced. The austenitization process is then followed by cooling in still or slightly agitated air (Figure 5.9).

Figure 5.11 compares the time-temperature cycle of normalizing to full annealing, as represented on a continuous cooling S-curve. The normalized steel consists of ferrite + fine pearlite for the hypoeutectoid steel, and cementite + fine pearlite for the hypereutectoid steel. Fully annealed steel provides coarse pearlite instead of fine pearlite (Figure 5.11). Normalizing is a final treatment process for parts subjected to relatively high stresses. It is carried out to improve machining

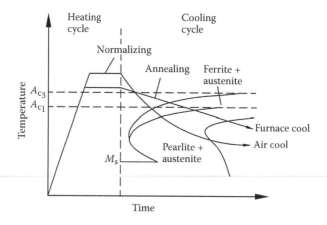

FIGURE 5.11 Comparison between full-annealed and normalized carbon steel as represented on a continuous cooling S curve.

characteristics, refine grain size, homogenize microstructure, improve cast dendritic structure, and provide the desired mechanical properties. The parts subjected to normalizing are characterized by higher yield and tensile and impact strength. Steel castings and forgings are normalized to modify their grain structure and to relieve internal stresses. Carbon steels containing less than 0.25% C are seldom normalized.

5.3.3 TEMPERING

This is the reheat process being carried out under subcritical temperatures after a hardening process. Therefore it is a complementary process which permits the trapped martensite to transform into troostite, or sorbite, depending on the tempering temperature. Troostite and sorbite consist of ferrite and finely divided cementite. Accordingly, tempering is a process used to modify the properties of hardened steel, striving at increasing its usefulness. The principal objectives of tempering are specifically

1. To achieve increased toughness and ductility at the expense of hardness and strength (Figure 5.12) and to improve the notch impact strength.
2. To relieve residual stresses. In this respect, tempering is similar to annealing. The stresses may be serious enough to cause distortion or even cracking.
3. To stabilize structure and dimensions by eliminating the retained austenite developed, depending upon several factors such as alloying elements in steel, austenitizing temperature, and the quenching rate of a previous hardening treatment. Retained austenite undergoes decomposition at room temperature to form martensite. Tempering achieves structural stability by hastening the decomposition of martensite and by causing either decomposition or stabilization of retained austenite. Dimensional changes are thus minimized or eliminated. It should be pointed out that stability and dimensional changes are of greatest importance in gauges and similar devices, where tolerances should be small.

The three most important considerations in tempering are

1. Tempering temperature
2. Soaking time
3. Effect of alloying elements

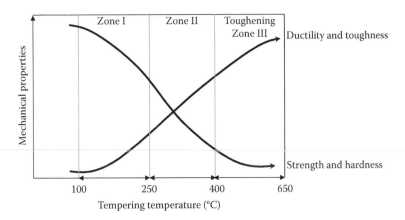

FIGURE 5.12 Schematic for the effect of tempering temperature on mechanical properties of carbon steels. The three tempering zones are visualized.

The cooling rate usually has no effect on the results, except that residual stresses are introduced by very rapid cooling. A noticeable effect of alloying elements is to increase the resistance to softening as a result of tempering. The elements exerting the greatest tendency in this respect are the strong carbide formers such as Ti, V, Mo, W, and Cr. Tempering is divided into three temperature ranges (Figure 5.12) depending on the application of steel.

I. Low-temperature tempering (100°C–250°C): This is used to retain hard martensitic micro-structure, but to give up brittleness and relieve internal stresses. This type is applied for cutting and measuring tools of carbon and low-alloy steels, and parts which are surface hardened and case carburized.

II. Medium-temperature tempering (250°C–400°C): This is employed for coil and leaf springs, as it provides the highest attainable elastic limit in conjunction with ample tough-ness. This treatment offers the steel a troostite structure. Plain carbon and low-alloy steels show decrease of toughness (known as blue brittleness) when tempered in the tempera-ture range near to 350°C. Precipitation of cementite in massive form around the troostite needles is believed to be the cause of blue brittleness.

III. High-temperature tempering (400°C–650°C): A treatment of quenching and tempering in this temperature range is frequently referred to as toughening, which provides the steel with the most favorable strength-to-toughness ratio (i.e., produces the maximum ratio of elastic limit to tensile strength). That is the miraculous feature of toughening. The steel develops a tough and ductile sorbitic structure, which is commonly found in heat-treated construc-tional steels, such as in axles, levers shafts, crankshafts, etc., subjected to dynamic stresses. Soaking at the tempering temperature depends on the thickness of the work piece. In order to avoid distortion, cooling is carried out in dry air. Figure 5.12 illustrates schematically the effect of tempering temperature on the mechanical properties of steel. It visualizes also the different temperature ranges of tempering.

5.3.4 ANTICRACK HARDENING TECHNIQUES

Two variations have been developed to produce strong structures and at the same time reduce the likelihood of distortion and cracking. These are the austempering and martempering techniques. In both techniques, the steel must be quenched from the austenitizing temperature to a temperature above the martensite start temperature (M_s).

5.3.4.1 Austempering (Isothermal Quenching)

In austempering, the part must be cooled rapidly from above A_3 (to prevent the formation of pearl-ite) in a vigorously agitated salt bath, maintained at a temperature within the bainite zone (350°C). The part is held at this temperature until the transformation of austenite is completed. The part is then cooled in air to room temperature (Figure 5.13a). The resulting bainite may be as hard as the martensite produced by conventional quench and temper hardness (Figure 5.14). Bainite usually has sufficient toughness, so that a temper is not required. Austempering is applied to small parts and shapes of small and delicate cross sections such as fine wires, and it is mainly used for aircraft engineering parts. It appears to be most beneficial in steels with carbon content ranging from 0.5% C to 1% C. The process is very costly and time consuming.

5.3.4.2 Martempering (Interrupted or Marquenching)

Martempering is a method for forming martensite with minimum distortion and residual stresses by reducing the difference in temperature between the inside and the outside of the martempered part. This technique is illustrated in Figure 5.13b. Martempering is applied to steel sections and irregular shapes which are likely to crack during conventional quenching. The initial quench is approximately the same as that for austempering, except that the part is not held at the quenching temperature long

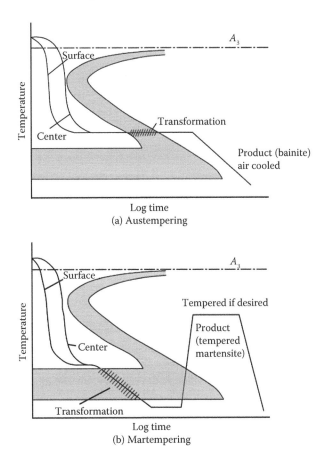

FIGURE 5.13 Anticrack hardening techniques.

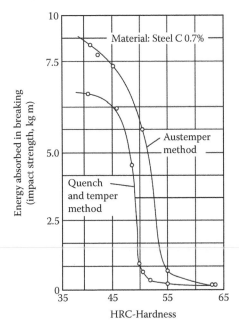

FIGURE 5.14 Toughness (impact strength) of asustempered and quench-hardened steel: a comparison.

enough to allow any bainite to transform. Instead, it is only held in a vigorously agitated salt bath (180°C–300°C) for a sufficient time to allow for equalization of temperature throughout the part (surface and center). When this is achieved, it is slowly cooled in air to room temperature, during which transformation occurs slowly and uniformly to martensite. Thus, the high stresses which accompany conventional quenching to martensite are avoided. The final product has some of the properties of tempered martensite, such that a tempering process is not always necessary. However, when further tempering is desirable, the part can be reheated to the suitable tempering temperature for this purpose.

5.3.5 Full-Depth Hardening (through Hardening)

This is the heat treating process in which steel is heated at a temperature directly above the transformation range, soaking at this temperature for a considerable period to ensure thorough penetration of the temperature inside the part, followed by quenching at critical cooling rate. Finally, the part should be tempered. Therefore, full-depth hardening comprises the following three steps.

1. Heating: It is carried at 30°C–40°C above the upper critical temperature A_3 in case of hypoeutectoid steel, and the same above the lower critical temperature A_1 in case of hypereutectoid (and eutectoid) steel (Figure 5.9).
2. Quenching: Upon cooling at critical cooling rate (3.33°C/s), austenite is transformed into α- iron, a needle-like microstructure known as martensite in case of hypoeutectoid. Martensite is a supersaturated solution of carbon in α-iron, and the hardness of steel is due to this hard microstructure. In hypereutectoid steel, a certain amount of cementite was not dissolved in austenite. Hence, on cooling, it is retained in the structure in addition to martensite. Such structure has a higher hardness and wear resistance than that which would be obtained upon quenching from a temperature above A_{cm}. This is because cementite is harder than martensite. Besides, heating to temperatures above A_{cm} will inevitably lead to coarsening of grains and warping of the part during quenching. A necessary condition when hardening hypereutectoid steels is the presence of excess cementite as separate small grains (Figure 5.15a). Excess cementite having the form of network (Figure 5.15b) will increase the brittleness of hardened steel and promote the formation of cracks.
3. Tempering: Fully hardened steels are usually tempered for relief of quenching stresses and for recovery of a limited degree of toughness and ductility. This is done by heating the steel to a temperature of 100°C–250°C. The achieved hardness depends on the carbon content of steel, as shown in Figure 5.16, from which it is noted that steel containing less

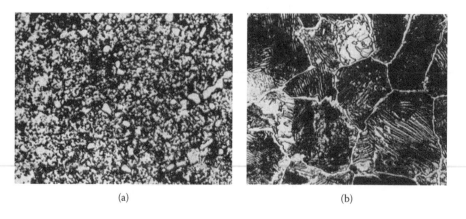

(a)   (b)

FIGURE 5.15 Microstructures of hardened hypereutectoid steel: (a) excess cementite as separate small grains ×500; (b) excess cementite in the form of network ×500, increases brittleness and promotes cracking.

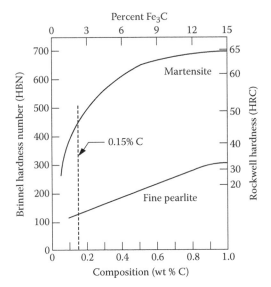

FIGURE 5.16 Effect of carbon content on hardenability of carbon steel (steels of C < 0.15% are not hardenable).

than 0.15% C does not respond to full-depth hardening. If the cooling rate is slightly less than the critical cooling rate, then austenite will be transformed into a fine form of peartlite called troositic pearlite. At a still slightly lower rate, austenite is transformed into another form of pearlite called sorbitic pearlite. Both forms are hard, strong, and brittle but not as hard and brittle as martensitic steel.

When straight medium carbon steels are quenched, the center hardness is lower than the surface hardness. The smaller the section, the higher the center hardness. If sections larger than a critical size must be dealt with, then proper addition of alloys such as Cr, W, Ni, and Mo will generally enable such larger sections to be hardened throughout, or nearly so, even when quenched in oil, and that is the main advantage of alloyed steels as compared to plain carbon steels. Effective heat treatment involves not only critical heating rates and temperatures, but also critical cooling rates and temperatures, both depending primarily upon section size, types and precipitates of alloying elements, and carbon content. For optimum results in quenching, it is necessary that the quench bath has adequate heat-extracting power, and that it extracts heat uniformly from all of the surfaces of the part being quenched, so that uniformity in hardening is achieved not only throughout a given part, but from one part to another. The rate of heat extraction in quenching varies widely, depending upon:

- The mass of the part and amount of surface area available
- The type of quenching medium and the degree of circulation or agitation

The first requirement of an appropriate quenching medium is that it must provide the desired speeds of cooling. Other factors influencing its selection are its low cost, general availability, ease of handling in regards to equipment and personnel, and its ability to remain fairly stable during extended use. The quenching media that most readily answer to the above demands are air, oil, sodium chloride (brine solutions), sodium hydroxide (caustic soda solutions), and water. Table 5.1 shows the comparative heat-extracting power of various quenching media. It indicates a considerable gap in the cooling rates of various coolants.

TABLE 5.1
Quenching Power of Various Media: Reclassified and Modified

Circulating or Agitating Condition	Severity of Quench H[a]			
	Air	Oil	Water	Brine
No circulation or agitation	0.02	0.25–0.30	0.9–1	2
Mild circulation or agitation	–	0.30–0.35	1.0–1.1	2–2.2
Good circulation	–	0.40–0.50	1.4–1.5	–
Violent circulation or agitation	–	0.80–1.10	4	5

Source: Tool Engineers Handbook, 2nd ed., 1959, ASTME, McGraw-Hill, New York.

[a] Severity of quench H is based on a value of 1.0 for still water.

Ultrasonic energy imparts vibratory waves directly to the quenching bath. These vibrations provide the mechanical breakdown of the vapor phase inherent in quenching. As a result, the part is continuously exposed to a fresh supply of quenching medium, thus providing a rapid and uniform quenching action. Due to the cavitation effect associated with ultrasonic vibrations, ultrasonically quenched parts have less heat-treat scale than those quenched in conventional baths, and there is some evidence that grain refinement is promoted.

Quenching, however, may lead to formation of cracks. For this reason, the quenching medium must be just severe enough to produce the desired hardness and strength. High flash point oils are the best for this application. Ordinary lubricating oils are not suitable. The articles to be hardened should be introduced into the quenching bath vertically to avoid distortion, and places where bubbles of vapor are likely to cling should be sealed. If possible, oil should be kept in turbulence by directing jets of oil on the part, or by steady flow of oil by pumps or rotating paddles.

The shape of parts for quenching must be considered. Sharp corners and asymmetrical or abruptly changing shapes are likely to develop cracks and should be avoided. Thin sections which will readily overheat in the furnace should also be avoided. Similarly, blind holes and long thin sections can also be problematic. Changes in cross sections must be made gradually to minimize stress-concentration. Holes should be correctly located. Holes in articles and tools to be quenched in water or brine may cause cracks. These are therefore blocked with wet asbestos. If the holes are threaded, these may be blocked by screw plugs.

Small parts are often heated in pans and are simply dropped in a quenching tank in which a netted basket is immersed. For thin articles and tools (such as screw taps, reamers, and so on), such a method is unsuitable, as distortions may occur and nonuniform hardness may develop. These parts must be immersed in the quenching tank in an exactly vertical position. Soft spots are formed if big articles are clamped by tongs. Special tongs with sharp bits must be used in hanging such articles. If the parts have holes, these may be held by wire passing through the holes. To prevent distortion of springs while quenching springs of large length and comparatively small diameter, these are tightly fitted into hollow mandrels.

5.3.6 Surface Hardening Techniques

Some parts of machinery must be very hard to resist surface wear, and yet must possess adequate ductility and toughness to resist impact loads. To meet these requirements, two basic techniques of surface heat treatment are used to provide a hard wear-resistant surface and a relatively soft and tough core. The first technique comprises the selective hardening methods which mainly deal with medium carbon and alloyed steels of carbon content 0.3%–0.7%, whereas the second technique comprises the diffusion methods which are generally applied to steels of relatively low carbon content (up to 0.25%); however, they are applicable for higher carbon contents if necessary.

5.3.6.1 Methods of Selective Hardening

Selective hardening is typically achieved by localized heating and quenching without any modification of the surface chemistry. It generally involves transformation hardening (heating and quenching). In this technique, the surface is heated to a point above the critical temperature A_3, then directly quenched. It is usually performed on steels with sufficient carbon content (above 0.3% C) that will respond to heating and quenching (Figure 5.16), without adding C or N_2 to the surface. Most surface hardening is done on 0.35% C–0.45% C steels; however, almost any medium of high carbon steel may be surface hardened if necessary. The range of hardness for this treatment is generally 50–65 HRC. Normally, distortion is very slight on surface-hardened parts; however, if parts are not treated as symmetrically as possible, distortion may not be avoided. Methods of selective hardening are described below.

1. Induction hardening: In induction hardening, the rotating part is placed inside a water-cooled copper conductor coil, but not touching it, or the conductor coil is placed in a stationary part (Figure 5.17). The conductor coils are then energized with a high-frequency ac current (f = 100–100,000 Hz). The changing magnetic field induces eddy currents in the surface, producing a very rapid rise in temperature due to the electric resistance of steel. In most cases, a rise of 722°C is produced in 2–3 s. When the proper temperature has been achieved to the desired depth, the power is shut off and the part is quickly cooled. Quenching is accomplished by dropping the part into a quenching bath or by using a quenching spray. The depth to which the current penetrates in induction hardening is determined from the formula

$$y = 50,000 \sqrt{\frac{\rho}{f}} \text{ mm}$$

where y is the depth of current penetration (mm), ρ is the electrical resistivity ($\mu\Omega$ cm), μ is the magnetic permeability of steel (gauss/oersted), and f is the current frequency (Hz).

y increases with temperature. It increases sharply (several times) at temperature above the Curie point (A_2 = 768°C) where steel is transformed from the ferromagnetic to the paramagnetic state.

Induction hardening is particularly suited for surface hardening processes, since the rate and depth of hardening can be easily controlled through the current and frequency provided by generators. Each job requires its own timing sequence and coils. Consequently, the process is better suited to mass production of small parts. Speed, cleanliness, and

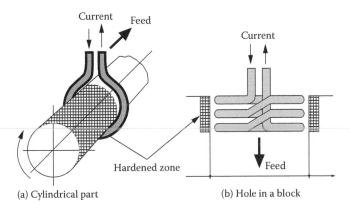

(a) Cylindrical part   (b) Hole in a block

FIGURE 5.17 Induction hardening process.

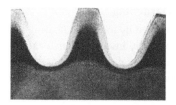

FIGURE 5.18 Tocco induction hardened gear. (From Tocco Incorporation, 1972. Cited in Degarmo, E. P., Black, J. T., and Kohser, R. A., *Materials and Processes in Manufacturing*, 8th ed., Prentice-Hall, Upper Saddle River, 1997, p. 33. With permission.)

freedom from scale and distortion are among the advantages of induction hardening; however, the process involves high relative cost as compared to flame hardening. Induction hardening is frequently used for shafts and cylindrical parts. It can be also adapted for more complex geometries when special fixtures and techniques are used. It can be used for surface hardening of crank shaft journals. Figure 5.18 shows the cross section of an induction-hardened gear (Tocco Induction Hardening), where hardening has been applied to those areas subjected to wear. Distortion during hardening is negligible, since the dark areas remain cool and rigid throughout the entire process.

2. Flame hardening: In this process, a high-intensity oxyacetylene flame is used to raise the surface temperature of the part high enough to form austenite. Heat input is quite rapid, and is concentrated on the surface. The part is then water quenched without the danger of surface cracking, then tempered to the desired level of toughness. Short heating times leave the interior at low temperature, and hence, free from distortion and any significant changes. In flame hardening, flexibility is provided through easy control of hardness and hardening depth. Similar to induction hardening, the hardness achieved is 50–65 HRC and the hardening depth can range from thin films of 0.7 mm to 6 mm. The process is applicable for large objects, since alternative methods are limited by both size and shape. Rapid quenching is achieved by water spray from a shower head which is connected as an integral part with the preceding heating torch. They move together with respect to the workpiece (Figure 5.19).

When flame progresses, it keeps on heating the surface to the critical temperature A_3, and simultaneously, the surface is quenched behind the flame. Thus, the operation becomes continuous. If the jet and flame remain fixed in position, and the work keeps on traversing at the calculated rate, this permits automated hardening of long work. Flame hardening equipment varies from simple hand-held torches to fully automated and even

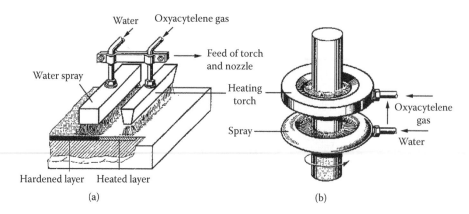

FIGURE 5.19 Flame hardening of (a) block and (b) shaft.

computerized units. In comparison to induction hardening, flame hardening is cheaper as initial investment is less. Moreover, it is more versatile, because it can be used for all sizes and shapes. On the other hand, in induction hardening, the hardening depth can be controlled more accurately by controlling the frequency, and also it is clean and fast, as previously mentioned.

3. Laser beam hardening (LBH): This is widely used to harden localized areas of steel and cast iron machine components. The heat generated by the absorption of the laser light is controlled to prevent melting and is therefore used in the selective austenitization of local surface regions which transform to martensite as a result of rapid cooling (self-quenching) by the conduction of heat into the bulk of the workpiece. This process is sometimes referred to as laser transformation hardening. There is no chemistry change produced by LBH. Other methods of laser surface treatments include surface melting and surface alloying. Laser surface melting refines the microstructure due to rapid quenching from the melt. In surface alloying, alloying elements are added to the melt pool to change the composition of the surface.

   LBH produces thin surface zones that are heated and cooled very rapidly, resulting in very fine martensitic microstructures, even in steels with relatively low hardenability. High hardness and good wear resistance with less distortion result from this process. Also, the laser light is reflected by mirrors to the focusing lens, which easily controls the width of the heated spot or track.

   Materials hardened by laser include plain carbon steels (1040, 1050, 1070), alloyed steels (4340, 52100), tool steels, and CIs (gray, malleable, and ductile). Typical case depths for steels are 0.25 to 0.75 mm, and for cast irons, about 1 mm. The flexibility of laser delivery systems and the low distortion and high surface hardness obtained have made lasers very effective in the selective hardening of wear-and fatigue-prone areas on irregularly shaped machine components such as camshafts and crankshafts. A disadvantage of LBH is that the work surface needs to have a dark coating to absorb light, and that adds to the cost. On the other hand, a laser does not need a vacuum like EBH, and the laser beam can travel to almost any lengths in air, and is not deflected by magnetic fields.

4. Electron beam hardening (EBH): Laser treatment is used to harden the surfaces of steels. The EB heat-treating process uses a concentrated beam of high-velocity electrons as an energy source to heat selected surface areas of ferrous parts. Electrons are accelerated and are formed into a directed beam by an electron beam gun. After exciting the gun, the beam passes through a focus coil, which precisely controls the beam density levels (spot size) at the workpiece surface, and then passes through a deflection coil. To produce an electron beam, a high vacuum of 10^{-5} torr (1 torr = 1 mm Hg) is needed in the region where the electrons are emitted and accelerated. This vacuum environment protects the emitter from oxidizing and avoids scattering of the electrons while they are still traveling at a relatively low velocity.

   Like laser beam hardening, the EBH process eliminates the need for quenching media but requires a sufficient workpiece mass to permit self-quenching. A mass of up to eight times that of the volume to be EB hardened is required around and beneath the heated surface. EBH does not require energy-absorbing coatings, as does LBH. The vacuum chamber of EB is small and subsequently the workpiece must be limited in size. The beam is controlled by a computer, impinging it as an array of dots on the surface to control temperature. Rocker arm pads are surface hardened over an area of about 5000 mm in 2.2 s at a cost of 0.365 cent.

5. Grind hardening (GH): During grinding, almost the whole mechanical energy induced at the surface of the workpiece dissipates into thermal energy. The generation of the imperfection of surface integrality (SI) due to the associated thermal energy is a well-known problem. However, this thermal energy may be utilized intentionally for the HT process, leading to grind-hardening or hardening by grinding. So GH is an integrated machining/

heat treating method aimed at strengthening and hardening of the surface layer of a component, thus realizing both economical and ecological advantages. It is well established that GH utilizes the heat flux mainly generated at abusive grinding conditions to induce controlled martensitic phase transformation in the workpiece shell of annealed (or tempered) medium- and high-carbon steels. GH is characterized by high grinding forces, excessive wear of grinding wheel, and high surface roughness. The application of coolants as chilling elements in GH is not essential, because the critical cooling rate is achieved by the self-quenching effect of the bulky workpiece.

GH can therefore be classified as a rapid thermomechanical surface hardening and strengthening process. Due to the mechanical effect exerted by the grinding wheel (GW), the metallurgical transformation does not need a long time to occur, and it can be performed only in one pass of the GW, which lasts only a fraction of a second. Consequently, GH is a short-time process, and has proven to be productive as compared to conventional surface hardening processes. The realized hardening depth by GH (up to 1 mm) is comparable with those achieved by other surface hardening methods.

Moreover, GH realizes economical benefits due to its increased integration level. By adopting GH as a new HT process, both hardening and grinding can be combined in one step, leading to shortened process sequence and less transportation. The process can specially be used for manufacturing components which are subjected to medium or low operating impact loads and just need a hardening depth of a few tenths of a millimeter to improve their wear resistance. Grooves for locking rings, guideways, and gearbox levers are some examples. Prehardened steels should not be considered for GH. It is a good practice after GH to adopt a finish cut to improve the surface integrity, thus enhancing the fatigue strength by removing the white etched area from the part surface. The condition of finish cut should be carefully selected; it may be a number of sparking out strokes, since only a small amount of heat flux is required to properly temper the martensitic structure developed by GH. Low-alloy, hypoeutectoid steels are recommended for GH rather than hypereutectoid steels.

6. Ion implantation: This is a surface modification process (not diffusion), in which ions with very high energy are selectively driven into a substrate. Ions of almost any atom species can be implanted, but nitrogen is widely used to improve corrosion resistance and tribological properties of steels and other alloys. Ion implantation machines accelerate ions, generated by specially designed sources at very high energies (10–500 keV). Ion implantation can be performed at room temperature, thereby minimizing the diffusion-controlled formation of precipitates and coarsening of the subsurface microstructure because the temperature of application is low and the process is carried out in accelerators with very good vacuums (10^{-5} torr). Clean surfaces are ensured and undesirable surface chemical reactions such as oxidation are reduced, and dimensional stability is excellent. For the coverage of areas larger than the beam, the specimen must either be translated or the ion beam must be faster over the specimen surface.

Because of the virtual absence of diffusion-controlled case formation during ion implantation, case depths are shallow (<0.25 μm). Very high strengths or hardness (70 HRC) of the nitrogen-implanted surface layers compensate for the shallow case depths of implantation. The properties of ion-implanted surfaces and shallow case depths make ion implantation suitable for very special applications. Examples include the surface hardening of razor blades and knives, a variety of tool steel applications and the implantation of 52100 bearing steels with Ti and/or N to improve rolling-contact fatigue resistance. In the latter applications, Ti was found to reduce the coefficient of friction and N to raise the hardness by intermetallic compound formation.

7. Electrolytic surface hardening: Electrolytic surface hardening is a special hardening method employing electrolysis in an aqueous solution under particular conditions, especially using

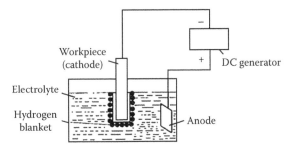

FIGURE 5.20 Schematic illustrating the principles of electrolytic surface hardening of medium carbon steel.

a pulsed dc current. The workpiece, which serves as cathode, is immersed in an electrolyte (5% concentration of NO_2CO_3) (Figure 5.20). The anode and the cathode are connected to a pulsed dc generator. When direct current at a high voltage (240–280 V) passes through the electrolyte, a hydrogen blanket is formed surrounding the cathode. Due to its high resistivity, the blanket is rapidly heated to about 2000°C, and a considerable portion of its heat is transferred to the workpiece. The workpiece is violently heated to the austenitizing temperature. Upon disconnecting the electric circuit, the hydrogen film disappears, and the workpiece surface is quenched immediately by the cold electrolyte in its vicinity so that the workpiece hardens due to the austenitic–martensitic transformation.

The advantages of electrolytic surface hardening are
- It has a short cycle time (2–6 s).
- It is easy to automate, and it is suitable for mass production of small parts.
- It is of relative low capital and running cost.
- The part surface will not be oxidized as it is protected by the hydrogen film.
- Less distortions of the work are developed.

The disadvantages are
- It is not applicable for large-size work or complicated shapes.
- Safety precautions are to be followed, since operating voltage is quite high.

5.3.6.2 Diffusion Methods (Thermochemical Hardening)

Surface hardening by diffusion (case hardening) involves chemical modification of the surface of steel by addition of C, N, or both. The basic process used is thermochemical because some heat is needed to enhance the diffusion of hardening species into the surface and subsurface of the part. The depth of diffusion exhibits time-temperature dependence, such that

$$\text{Case depth } \alpha \ K\sqrt{time}$$

where the diffusivity constant K depends on temperature, the chemical composition of steel, and the concentration gradient of a given hardening species. In terms of temperature, the diffusivity constant exponentially increases as a function of absolute temperature. Factors influencing the suitability of a particular diffusion method include the type of steel, desired case hardness, and the case depth.

The objectives of case hardening are

- To obtain a hard and wear-resistant surface on a part, with enrichment of the surface with carbon, nitrogen, or both
- To obtain a tough core

- To obtain close tolerances on the part
- To enhance the fatigue limit and mechanical properties of the part

Diffusion processes comprise the following techniques.

1. Carburizing: This involves adding carbon to the outer layer of steel through (a) solid, (b) gaseous, or (c) liquid media at a temperature and long enough time to obtain the desired penetration of carbon. After carburizing, the material is generally quenched to produce a hardened case. Anticipated distortion of the part and the amount of subsequent grinding along with service conditions will determine the depth of case required. The case depth is a function of the time and temperature.
   a. Pack carburizing: Pack carburizing is applicable for low carbon steels (<0.25% C). It consists of packing the roughly machined parts in some material rich in carbon, in boxes covered with lids and brazed to exclude air and escape of gases (Figure 5.21). The boxes are heated at 900°C–970°C for a period of 40–90 h, depending on the depth of the case desired. Any portions which are not required to be hard are protected by covering with some material which does not absorb carbon, such as asbestos or fireclay, or 0.1 mm thick Cu plating. The carburizing material usually consists of bone charcoal, charred leather, petroleum or coke, together with an energizer such as barium carbonate $BaCO_3$, which promotes rapid action. A specific composition of a carburizing medium is

   80%–85% charcoal of grain size 3–10 mm diameter.

   10%–12% $BaCO_3$, 1%–3% $CaCO_3$, 1% Na_2CO_3

   The carburizing of steel is due to atomic carbon, which is liberated by decomposition of CO. Air is always present in the carburizing box, even when filled with carburizer. At high temperature, O_2 in the air reacts with C in the carburizer to produce CO.

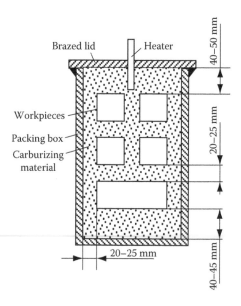

FIGURE 5.21 Pack carburizing of workpieces arranged in their packing box.

$$2C + O_2 \rightarrow 2\,CO$$

$$BaCO_3 \xrightarrow{Heat} BaO + CO_2$$

$$CO_2 + C \rightarrow 2\,CO$$

$$2CO \xrightarrow{Heat} CO_2 + C\ (atomic)$$

Low carbon austenite + nascent C (absorbed)

Low carbon austenite + nascent C (absorbed) $\xrightarrow{Heat}$ high carbon austenite, providing hard case after quenching.

The carbon is appreciably soluble in austenite. Therefore, the operation must be carried out above the A_3 temperature. Once carbon has entered the surface, it tends to diffuse inward at a rate depending upon the composition of steel, and increases with the operation temperature. This produces a carbon gradient from the outside to the inside. After casing, the parts are cooled down in the box, or quenched if the risk of distortion is not important. A disadvantage of pack carburizing is the time and cost of packing, and also, it is a rather messy process. A case depth up to 1.5 mm is a usual depth in common practice; however, a maximum depth of 10 mm could be achieved.

b. Gas carburizing: There is no difference in principle between gas and pack carburizing, since the processes are carried out by gases in both cases. However, gas carburizing is considerably simpler. It is accomplished by heating the part in a gaseous medium containing carbon such as methane CH_4 and other hydrocarbons. Accordingly, the chief reaction is achieved by dissociation of methane, producing a high potential of atomic carbon diffused to the surface.

$$CH_4 \xrightarrow{Heat} C_{at} + 2H_2$$

This method requires the use of a multicomponent atmosphere, whose composition must be very closely controlled to avoid deleterious side effects such as the formation of surface and grain boundary oxides. In addition, separate equipment is required to generate the gas and control its composition. Despite this complexity, gas carburizing has become the most effective and widely used variation (50%) of all carburizing techniques. Gas carburizing has many advantages over pack carburizing, and it is generally applied in mass production. Its advantages are

- Possibility of better regulation to obtain accurate case depth and hardness through controlling the gas composition.
- Less time is required, since there is no need to pack boxes filled with low-heat conductive substances.
- It is not a messy process.
- It needs only half as much floor space as required for pack carburizing.
- Many laborious, time-consuming operations and health hazards are excluded.
- The operation may be feasibly mechanized; it can be combined with subsequent hardening.

The depth of case vary from 0.25 mm on light duty articles, 0.5–1 mm for parts of automobile industries, and may be more for roller bearings and ball races where compressive stresses are high. The carbon content of the case influences the part service, and a maximum of 0.9% is aimed for, although this can be exceeded if grinding is carried out after casing. With such high carbon contents, free cementite networks are formed, which are extremely hard and brittle. Consequently, during grinding or

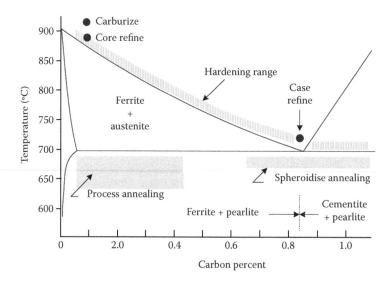

FIGURE 5.22   Core and case refinement of carburized case hardened steels.

service, cracks start at the cementite networks and cause layers to flake off, a phenomenon known as exfoliation.

Heat treatment after carburizing: As a result of prolonged heating at a high carburizing temperature, both core and case exhibit overheated coarse structures, which would be unsatisfactory. The articles should be heat treated, thereby, (1) refining the core, and (2) refining and hardening the case.

1. Core refining: This is accomplished by heating just above A_3 point (870°C) (Figure 5.22), at which point the coarse grained structure is replaced by fine grained austenite. After soaking, the part is quenched in water or oil to give a fine dispersion of ferrite in martensite. At the same time, all of the cementite in the case will go into solution at 870°C, and the rapid cooling prevents any coarse networks from forming again. The case is of high carbon content; however, it has coarsened because the temperature was well above its critical range. The coarse martensite of the case is brittle, which necessitates another heat treatment.

2. Refining and hardening of the case: This is produced by water or oil quenching from 760°C, which is just above A_1 point (Figure 5.22). Thereby, the fine grained austenite is formed in the case without appreciable growth occurring, since soaking is not prolonged, and on quenching, a fine martensite is formed which is hard but not extensively brittle. At the same time, the core undergoes tempering. Finally, it is advisable to temper at 150°C to relieve the quenching stresses. Table 5.2 illustrates the composition and applications of some case-hardened plain carbon steels.

TABLE 5.2
Case Hardened Plain Carbon Steels

Composition (%)		
C	Mn	Applications
0.06–0.15	0.6	Maximum core toughness, thin sections
0.10–0.18	0.8	For general purposes, i.e., gears, shafts
0.20–0.30	0.5	High load carrying capacity roller bearings

Manganese aids carburization and increases depth of hardening, but also increases liability to cracking of the hardened case. Silicon retards carburization, induces graphitization, and is usually below 0.3%. Since toughness of the core is one of the most important characteristics of case-hardened steels, the carbon content is usually below 0.2%, although it may be as high as 0.3% where greater support for the case is desired. For the latter, it is advisable to use an alloyed steel with high strength combined with toughness. Also, with alloyed steels, the desired properties may be obtained by oil quenching, thereby reducing the risk of distortion to which water-quenched carbon steels are liable.

c. Liquid carburizing: This carburizing is performed in baths of molten salts containing 75%–85% sodium carbonate Na_2CO_3, 10%–15% sodium chloride NaCl, and 6%–10% silicon carbide SiC (carborundum). The carburizing effect is due to carbon monoxide evolved according to

$$SiC + 2\,Na_2CO_3 \;\rightarrow\; \underbrace{Na_2SiO_3 + Na_2O}_{\text{Slag on the surface of the bath}} \quad + 2\,CO + C$$

The diffusion layer related to the holding time at a carburizing bath temperature of 850°C is given by

Holding time (h)	0.5	1.0	1.5	2.0	2.5	3.0
Depth of diffusion layer (mm)	0.2	0.25	0.35	0.4	0.45	0.5

The principal advantages of salt both carburizing
- Uniform heating
- Possibility for direct heating from the bath
- Less distortion of the part

The addition of 3%–5% ammonium chloride (NH_4Cl) enables the bath to be employed for saturating the part simultaneously with both C and N, thus accelerating the process and promoting the hardness to a certain extent, Table 5.3.

2. Carbonitriding: This consists of gas carbonitriding or liquid carbonitriding (cyaniding).

a. Gas carbonitriding: This is a modified form of gas carburizing, rather than a form of nitriding. The modification consists of introducing ammonia (20%–30%) into the gas carburizing atmosphere to add N to the carburized case as it is being produced. Nascent nitrogen forms at the work surface by the dissociation of ammonia in the furnace atmosphere. The nitrogen diffuses into the steel simultaneously with C. Typically, carbonitriding is carried out at a lower temperature and for a shorter time than is gas carburizing, producing a shallower case than is usual in carburizing, Table 5.3. In its effects on steel, carbonitriding is similar to liquid cyaniding. Because of problems in disposing of cyanide-bearing wastes, gas carbonitriding is often preferred over liquid cyaniding. In terms of case characteristics, carbonitriding differs from carburizing and nitriding in that carburized cases normally do not contain N, and nitrided cases contain N primarily, whereas carbonitrided cases contain both. Gas carbonitriding imparts a hard, wear-resistant case generally from 0.075 to 0.75 mm deep. A carbonitrided case has better hardenability than a carburized case (N increases the hardenability of steel). Consequently, by carbonitriding and quenching, a hardened case can be produced at less expense within the case depth indicated if either carbon or low-alloy steel is used. Hardness with low distortion can be achieved with oil quenching.

Steels commonly carbonitrided include those according to AISI, 1000, 1100, 1200, 1300, 1500, 4000, 4100, 5100, 6100, 8600 and 8700 series, with carbon contents up to

TABLE 5.3
Summary of Thermochemical Diffusion Hardening Processes, Reclassified, and Summarized

Process		Nature of Case Diffused	Process Temperature (°C)	Typical, Minimum, and Maximum Depth (mm)	Case Hardness HRC	Typical Substrate of Steel	Process Characteristics
Carburizing	Pack	C	815–1090	0.125–1.5	50–63[a]	Low carbon and low alloy steels	Low equipment cost, case depth difficult to control accurately.
	Gas	C	815–980	0.075–1.5	50–63[a]		Good control of case depth, could be dangerous.
	Liquid	C, N (possible)	815–980	0.050–1.5	50–65[a]		Rapid and clean, salt bath, requires frequent maintenance.
Carbonitriding	Gas	C, N	760–870	0.075–0.75	50–65[a]	Low carbon and low alloy steel	Lower temperature than carburizing, less distortion, slightly harder.
	Liquid (cyaniding)	C, N	760–870	0.005–0.125	50–65[a]	Stainless steels	Lower temperature than carburizing, less distortion, toxic, salt disposal is a problem.
Gas nitriding		N, N-compound	480–590	0.125–0.75	50–70	Alloy, stainless, and nitre-alloy steels	Quenching not required, slow (2–4 days), suitable for batch processes.

Source: Budinski, K. G., *Surface Engineering for Wear Resistance*, Prentice-Hall, Upper Saddle River, NJ, 1988. With permission.

[a] Requires quench from austenitizing temperature. Low-carbon alloy steels (oil quenching) are more desirable than plain carbon steels. Therefore less distortion and more toughness, which are important factors.

about 0.25%. Also, many steels in the same series with carbon contents 0.3%–0.5% are carbonitrided to case depths up to about 0.3 mm when a combination of reasonably tough, through-hardened core and a hard, long-wearing surface is required (shafts and transmission gears). Steels such as 4140, 5130, 5140, 8640, and 4340 for heavy duty gearing are treated by this method, at 840°C.

Often, carburizing followed by carbonitriding is used to achieve much deeper case depths and better engineering performance for parts than could be obtained by using only the carbonitriding process. This is applicable particularly with steels having low case hardenability; that is, the 1000, 1100, and 1200 series steels. The process generally consists of carburizing at 900°C–955°C to give the desired total case depth (2.5 mm), followed by carbonitriding for 2–6 h in the temperature range of 815°C–900°C to add the desired carbonitrided case depth. The parts can then be oil-quenched to obtain a deeper effective and thus harder case than would have resulted from the carburizing process alone. The addition of the carbonitrided surface increases the case residual compressive stress level and thus improves the fatigue resistance and increases the case strength gradient.

Advantages of gas carbonitriding as compared to gas carburizing
- Considerable acceleration of the process
- Higher wear resistance of case obtained
- Lower cost due to lower processing temperature, which reduces fuel consumption, and increases the service life of the furnace and its accessories

b. Liquid carbonitriding (cyaniding): Cyaniding is the process in which both C and N in the form of cyaniding salt are added to the surface of low (0.2% C) and medium carbon (0.3%–0.5% C) steel to increase the hardness and wear resistance. This method is also effective for increasing the fatigue limit of medium and small-sized parts such as gears, shafts, pins, and so on. The process involves the heating of parts in a molten cyanide salt bath maintained at a temperature of 780°C–870°C, then quenching in an oil or water bath. The bath contains sodium cyanide (NaCN). The cyanide salt reacts with the oxygen in the air and is oxidized. The outcome of the developed reactions is atomic nitrogen and carbon, which are diffused in the part surface.

$$2NaCN + O_2 \rightarrow 2NaCNO$$

$$2NaCNO + O_2 \rightarrow Na_2CO_3 + CO + 2N_{at}$$
$$\hookrightarrow 2CO \rightarrow CO_2 + C_{at}$$

The cyaniding time is determined by the case depth desired; it varies from 5 to 10 min. The resulting case contains about 0.6%–8.0% C, and about 0.4%–0.5% N. The hardness on the case varies between 50 and 60 HRC and the depth varies from 0.0025 to 0.125 mm, Table 5.4.

The principal advantages of cyaniding
- Bright finish can be obtained.
- Distortion can be easily avoided.
- Fatigue limit is improved.
- Less time is needed.

The limitations of cyaniding
- High cost
- Toxicity of cyanide salts

3. Nitriding: Nitriding is a diffusion process that introduces N into the surface of steel at a temperature range 500 to 550°C, while it is in the ferritic state. The finished machined and

TABLE 5.4
Composition, Previous Heat Treatment, and Hardness of Typical Nitrided Steels

Composition (%)					Previous HT (°C)		Hardness VH 30	
C	Al	Cr	Mo	Ni	Oil Quench	Tempered	Core	Case
0.4	1.1	1.6	0.2	–	900	650	270	1050
0.3	–	0.9	1.1	0.9	875	640	290	600
0.3	–	3.2	0.6	–	900	625	300	900
0.5	–	1.3	1.0	1.8	870	620	–	700

heat-treated parts are heated for a prolonged time in an atmosphere of ammonia NH_3. A case of 0.7mm takes about 10 h. About 30% of the ammonia dissociates ($NH_3 \leftrightarrows 3H+N_{at}$), and atomic (nascent) nitrogen is absorbed by the surface layer of steel forming iron nitrides Fe_4N and Fe_2N.

Steels responding to nitriding:
- Alloyed steels containing one or more of the alloying elements, Cr, Ni, Al, Mo, V.
- Best steel to be nitrided is the nitride-alloyed steel (about 1% Al, 1.0% Cr, 0.2% Mo).

Plain carbon steels are seldom nitrides. Table 5.3 illustrates some typical nitrided steels. Aluminum appears to form stable nitrides which do not diffuse readily, and a shallow but intensively hard case (50–70 HRC) is formed. Cr contributes to the hardness and also the flatness–hardness gradient below the surface, thus reducing the risk of spalling. Mo also increases hardenability and prevents embrittlement. The desire for better core properties led to the introduction of steels with lower Al content (0.4%). Al-free steels containing Cr and Mo are now being used when extreme hardening is not necessary, such as in the case of aeroengine crankshafts, where the case is relatively tough, with a hardness of 750 VH30.

The carbon contents of nitriding steels vary from 0.2 to 0.5%, the lower content being used for lightly stressed wearing parts such as spindles and gears, while medium carbon steels are used to withstand high local pressures, as with die blocks of metal forming and plastics. The stems of austenitic steel valves of aeroengines are also nitrided after previous deposition of Cu on the surface to be hardened by nitriding (N does not diffuse in austenite). By suitable initial hardening and tempering (oil quenching and tempering, Table 5.4), a wide range of tensile strengths can be obtained, even in large sections. Cast iron containing about 1.5% each of Cr and Al can be nitrided at 510°C to give a case of 0.35 mm thick and 900 VH30. It is principally used for previously machined cylinder liners.

Advantages of nitriding
- No heat treatment is needed after nitriding.
- It can be accomplished with minimum distortion and with excellent dimensional control.
- No machining is required after nitriding.
- Achieved hardness of 1000–1100 VH30 (70 HRC), which cannot be obtained by any other process.
- No grain growth is developed.
- Local surface areas can be kept soft by insulating.

Disadvantages of nitriding
- Only a few alloys respond to nitriding.
- Nitriding takes a very long time, and hence it is a costly process.
- If the correct alloy is not used, spalling is liable to occur.

4. Electrolytic diffusion surface hardening: The principle of this process is based on traditional thermochemical surface hardening (diffusion process), discussed before. One application of this process is electrolytic case hardening by carburization, where low carbon steel

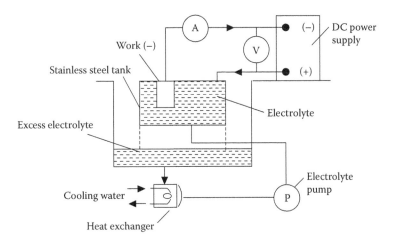

FIGURE 5.23 Schematic illustrating the principles of electrolytic diffusion hardening of low carbon steels.

containing 0.1% C–0.25% C is heated to elevated temperatures (900°C–1000°C) by the electrolyzing current, in the presence of carbon-rich electrolyte of a nominal composition of

$$\text{10 liter of water, 1 liter of glycerin } C_3H_8O_3, \text{ 1 kg } NH_4Cl.$$

The solution is prepared by solving NH_4Cl first in water, then glycerin is added to the solution. The conductivity is attained by NH_4Cl, whereas the glycerin acts as the carbon source. A dc power supply of about 200 V is applied to the part, connected as a cathode, while immersed slowly in the electrolyte by a stepper motor (Figure 5.23). A gas film on the part rapidly forms during the slow dipping process, and the part is heated to the austenitizing temperature. This temperature is attained for a period of time ranging from 1 to 60 min. On switching off the current, the part is quenched in the electrolyte to room temperature, thus developing the austenitic/martensitic transformation of the surface layer. The part is then cleaned by ethyl alcohol. Figure 5.24 illustrates the influence of the carburizing time on the case depth. It attains 25 and 80 μm for carburizing times of 1 and 60 min, respectively. A case hardness of 870 HV30 could be achieved using this process. Table 5.3

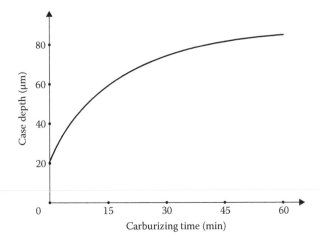

FIGURE 5.24 Influence of carburizing time on the case hardening depth of electrolytic diffusion hardening. (From Tarakci, M. et al., *Journal of Surface and Coatings Technology*, 199, 205–212, 2005. With permission.)

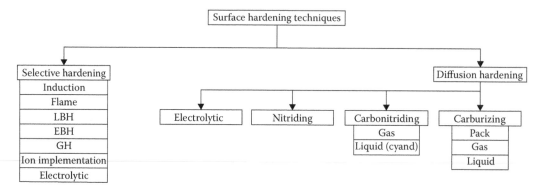

FIGURE 5.25 Classification of surface hardening techniques of steels.

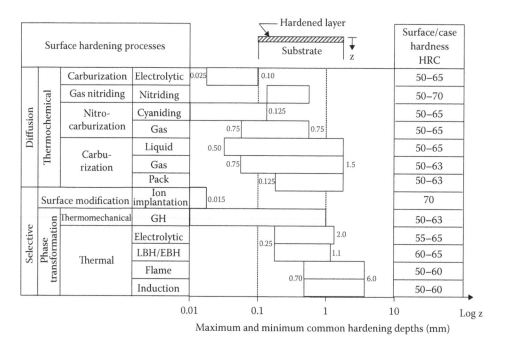

FIGURE 5.26 Maximum and minimum hardening depth of diffusion and selective hardening processes: a comparison.

summarizes the thermochemical diffusion hardening processes, whereas Figure 5.25 classifies all surface hardening techniques. As a conclusion, Figure 5.26 illustrates a comprehensive chart that summarizes typical maximal and minimal hardening depths that can be achieved by both diffusion and selective surface hardening processes, as well as the surface hardness.

5.4 HEAT TREATMENT OF CAST IRON

The main types of heat treatment processes performed on cast irons could be summarized as

1. Stress-relief annealing of complicated shape castings
   - Heating to 500°C–550°C at a slow rate of 75°C/h–100°C/h. Stress is completely removed at 650°C, but grain growth commences at 550°C, and is serious at 600°C.
   - Soaking for 2–5 h (1 h/inch section).

- Slow cooling in a furnace for 50°C/h–60°C/h to a temperature of 200°C. Alternatively, heavy castings can be naturally aged in stores for a long time (1 year), whereby the stresses are partially relieved.

2. Annealing to improve machinability by reducing the hardness
   - Heating to 850°C–900°C
   - Soaking for 1–2 h, to precipitate graphite from free cementite ($Fe_3C \rightarrow 3\ Fe + C$)
   - Slow cooling

3. Normalizing of gray CI castings
   - Heating to 850°C–870°C to dissolve a part of the free carbon in austenite
   - Cooling in air to develop a sorbitic structure

4. Hardening of gray CI castings
   - Heating to 820°C–900°C.
   - Isothermal hardening is performed, thus increasing the strength and improving the wear resistance several times over, or induction surface hardening of areas subjected to wear such as guideways.
   - Tempering after hardening by heating to 180°C–250°C for stress relieving, or by heating to 500°C–600°C to attain a tempered sorbitic structure.

5.5 HEAT TREATMENT OF NONFERROUS ALLOYS AND STAINLESS STEELS (PRECIPITATION HARDENING)

In several important classes of engineering alloys, including Al, Cu, Mg, and Be, as well as some stainless steels, the effect of decreasing solubility is utilized to produce a stronger material than can be achieved with solid solution strengthening alone. This process is referred to as *precipitation* or *age hardening*. It involves thermal treatment of the material to develop the final desired properties. In precipitation hardening, small particles (precipitates) of different phases are uniformly dispersed in the matrix of the original phase. The precipitates form because the solid solubility of one element in the other is exceeded.

The precipitation hardening process involves a composition which can be heated into a single-phase region (Figure 5.27). The example used here is an Al alloy containing 4% Cu heated to approximately 550°C. This portion of the process is called *solution treatment* and serves to dissolve Cu in Al, producing a homogeneous κ-phase (X). Then this solid solution is rapidly cooled to room temperature by quenching in water. This produces a supersaturated solid solution (A) (Figure 5.27), which contains considerably more Cu than it would under equilibrium conditions. Cu is uniformly distributed throughout κ. This solution-treated alloy may spontaneously begin to harden and strengthen after a time (1–10 h) at room temperature with no loss in ductility. This is called *natural aging*, which involves the precipitation of submicroscopic particles of the θ phase (a hard intermetallic compound of $CuAl_2$), throughout the κ-phase matrix (B) (Figure 5.27).

Since the precipitation reaction is diffusion controlled, the aging process can be accelerated by heating to an elevated temperature (not exceeding the solvus) for an appropriate period of time. Such acceleration of the precipitation process is referred to as *artificial aging*, which involves higher yield strength with some loss of ductility. Furthermore, *overaging*, which consists of heating for too long a period, results in coalescence of the very fine precipitate particles (C) (Figure 5.27). This process is accompanied by a loss of strength and hardness. Although weaker, an overaged part has better dimensional stability.

On the other hand, if the alloy after solution treatment (X) (Figure 5.27) cools slowly through the solvus, it provides a material that is soft, weak, and not especially ductile (much less than pure aluminum). This annealed alloy is generally unsatisfactory for engineering purposes; it contains large areas of soft κ, which is almost pure Al. The brittle, intermetallic $CuAl_2$ phase θ precipitates along grain boundaries, and consequently, it is the locus of crack propagation when fracture occurs, (D),

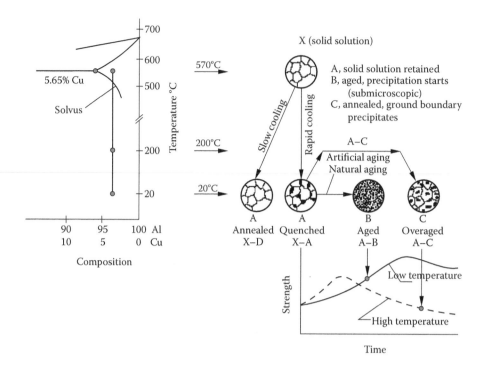

FIGURE 5.27   Precipitation hardening.

Figure 5.27. Table 5.5 illustrates the tensile and yield strengths, and the ductility of the age-hardenable alloy (96% Al–4% Cu) for the structures A, B, C, and D, shown in Figure 5.27. Such an alloy (96% Al–4% Cu) is extensively used in the heat-treated condition in aircraft construction and for other high strength-to-weight ratio applications. Figure 5.27 shows also the aging and overaging process for the Al–Cu alloy 2014 (0.9% Si, 4.4% Cu, 0.8% Mn, 0.5% Mg), from which it is depicted that both aging and overaging occur in less time at higher temperature, and the maximum strength is greatest with low aging temperatures. No benefit is derived if the precipitates are not coherent. For this reason, Al–Mg alloys are not hardenable.

Combined Hardening: Sometimes, it is desirable to combine two methods of hardening. The cold-working of an alloy which has previously been age-hardened increases the hardness still further. However, there are some practical difficulties encountered in this sequence. Age hardening increases resistance to slip and consequently increases the forces and energy required for cold-working, and also decreases ductility, so that rupture is liable to occur during cold working. A better alternative is to cold-work prior to precipitation hardening treatment. The temperature of the aging process, which follows cold-working, may relieve some of the strain hardening, and causes a slight decrease

TABLE 5.5
Mechanical Properties of the Age-Hardenable Alloy 96% Al–4% Cu

Treatment	TS (MPa)	YS (MPa)	Ductility (%) in 5 cm
Solution-treated then quenched (A)	240	105	40
Age-hardened (B)	415	310	20
Overaged (C)	170	70	~20
Annealed (D)	170	70	15

TABLE 5.6
Tensile Strength of Strain- and Age-Hardened Alloy (98% Cu and 2% Be)

Type of Treatment	Tensile Strength (MPa)
Annealed (870°C)	240
Solution-treated then quenched from 870°C	500
Age-hardened only	1200
Cold-worked only (37%)	740
Age-hardened, then cold-worked	1380 (liable to cracking)
Cold-worked, then age-hardened	1340

in hardness. Although it does not produce hardness as great as obtained from those of the first sequence, the final hardness is greater than that developed by using either method alone. Table 5.6 shows the tensile strengths of strain- and age-hardened Cu–Be alloy (98% Cu, 2% Be).

5.6   REVIEW QUESTIONS

1. Why are metals heat treated?
2. Draw the portion of the Fe–Fe$_3$C phase diagram that includes all steels ignoring the high temperature peritectic transformation. Give the approximate composition and temperature for the eutectoid transformation. Identify phase fields and the associated crystal structure, giving the name of each phase.
3. Steels AISI 1010 and 1045 are heated to 900°C. What phase of each crystal structure is present? The steels are then quenched in water. One of them becomes hard, and the other remains soft. Which one, and why? Refer to the metallographic structure of each steel.
4. Describe: annealing, tempering, hardening, and normalizing. What is the purpose of each? Define toughening. What are its main characteristics? State some of its applications.
5. What is the S curve (TTT diagram), and what does it show?
6. How are S curves developed? Explain the structural change indicated by a cooling curve on the S curve.
7. How many types of case hardening do you know? Which methods do you recommend for
   a. Crankshaft to be finish ground afterwards
   b. Small finished gear
   c. Absolutely true shaft
8. What two factors must be controlled if grain growth is to be limited when full annealing a piece of metal?
9. What is the major difference between full annealing and normalizing? Which process is preferred from the point of view of strength and grain refinement?
10. Which type of steel would be a candidate for
    a. Process annealing
    b. Spheroidizing
    c. Case hardening
    d. Nitriding
    e. Electrolytic diffusion hardening
11. What is the basic mechanism of the hardening of steel?
12. Describe three processes of diffusion surface hardening of steel. What are their applications, hardening depths, and merits?

13. Describe the phenomenon of exfoliation in gas and pack carburizing and spalling in nitriding. What are the measures to be utilized to avoid such phenomena?
14. Describe electrolytic hardening.
15. Is plain carbon steel treatable by nitriding or not?
16. Enumerate three important heat treatment processes of cast irons.
17. Mention how spheroidal CI can be obtained.
18. In what application do you recommend martempering?
19. Differentiate between induction and flame hardening, emphasizing cost, application, and control of hardening depth.
20. Define grind hardening. What is the hardening depth that would be realized by this process? What are the main advantages of GH?
21. Enumerate some manufacturing examples in which GH may be used.
22. Define ion implantation and mention some of its applications. What is the hardness and case depth that can be realized in such a process?
23. Enumerate advantages and disadvantages of electrolytic surface hardening of medium carbon steels.
24. What are the main objectives of the case hardening process?
25. What are the disadvantages of pack carburizing?
26. What are the advantages of gas carburizing over pack carburizing?
27. Enumerate advantages and limitations of cyaniding and nitriding.

BIBLIOGRAPHY

ASM International. 1985. *Metals Handbook, Vol. 9, Metallography and Microstructure*, Materials Park, OH.
ASM International. 1989. *Metals Handbook, Vol. 3, Heat Treatment*, Materials Park, OH.
ASTME. 1959. *Tool Engineers Handbook, 2nd Edition*, New York: McGraw-Hill.
Bain, E. C. 1939. *Functions of Alloying Elements in Steel*, ASM International, USA.
Budinski, K. G. 1988. *Surface Engineering for Wear Resistance*, Upper Saddle River, NJ: Prentice-Hall.
Callister, W. D. 1996. *Materials Science and Engineering—An Introduction*, 4th ed. New York: John Wiley.
Degarmo, E. P., Black, J. T., and Kohser, R. A. 1997. *Materials and Processes in Manufacturing*, 8th ed. Upper Saddle River, NJ: Prentice-Hall.
Jain, R. K. 1993. *Production Technology*, 13th ed. Delhi, India: Khanna Publishers.
Keyser, C. A. 1959. *Basic Engineering Metallurgy*, 2nd ed. Upper Saddle River, NJ: Prentice-Hall.
Lakhtin, Y. 1968. *Engineering Physical Metallurgy*, 3rd ed. Moscow: Mir Publisher.
Rollason, E. C. 1961. *Metallurgy for Engineers*, 3rd ed. London: Edward Arnold Publishers.
Seyfert, W. L. 1982. *Grundfachkunde Metall.*, 3. Auflage. Ernst Klett-Verlag, Stuttgart.
Tarakci, M. et al. 2005. Plasma electrolytic surface carburizing and hardening of pure iron. *Journal of Surface and Coatings Technology* 199: 205–212.
Van Vlack, L. H. 1973. *A Text of Material Technology*. Reading, MA: Addison-Wesley.
Van Vlack, L. H. 1975. *Elements of Materials Science and Engineering*. Reading, MA: Addison-Wesley.

6 Smelting of Metallic Materials

6.1 INTRODUCTION

Smelting is a major processing operation aimed at extracting metals from their original ores and refining them into the composition required by the users. The engineering and scientific accomplishments required to realize this are too complex to deal with here. However, we should cite the basic principles that underlie the production of metals by smelting because this information often relates to the selection and application of such metals. The ores of metals, except for gold and a small fraction of the Cu and Ag supply, are not metallic phases. These are most commonly oxides or sulfides. Copper, silver, lead, and zinc sulfides are significant sources of those metals. In either case, the metal must be extracted from its ore. Even when the ore is a sulfide MS, where M is the metal ion and S is the sulfide ion, it is common to first oxidize the ore to MO. The prime extraction step is one of reduction. The ease with which reduction occurs varies from metal to metal, depending on the reduction energy required. For example, lead oxide, PbO, and silver oxide, Ag_2O, reduce readily to metallic lead and silver, whereas it requires an input of considerable energy to affect the reduction of aluminum Al_2O_3. The reaction energy released (oxidation energy) as the metal burns to form oxide is exactly the same amount of energy which is required (reduction energy) to separate oxygen to form the metal. These energies considerably decrease as the melting temperature of the metal is approached.

The metallurgist has another practical choice to force oxide reduction, which involves the use of a reducing agent, such as carbon. In effect, carbon attracts the oxygen that is in the oxide to form CO_2, as illustrated in the reduction of lead oxide (PbO).

$$C + 2Pb\,O \xrightarrow{327°C} 2Pb + CO_2 \uparrow \tag{6.1}$$

According to this reaction, the energy released by forming CO_2 gas (94,400 cal/mol) is more than that required to separate oxygen from lead (38,150 cal/mol), at the melting temperature of lead (327°C). Furthermore, when the resulting CO_2 gas is continuously removed, there is no tendency for the reaction to reverse.

Another example is the reduction of Fe_2O_3 (hematite). It is affected after the initial formation of CO gas, according to the following reactions.

$$2\,C + O_2 \longrightarrow 2\,CO \,\big\downarrow$$
$$Fe_2O_3 + 3\,CO \xrightarrow{1600°C} 2\,Fe + 3\,CO_2 \uparrow \tag{6.2}$$

These reactions are most widely encountered in iron making in a blast furnace to produce several thousand tons of molten iron/day. Again, note that elevated temperatures (1600°C), favor this reaction, since the energy required for the reduction of ore decreases, while the energy released by the formation of CO_2 increases. Most metals, even when they are reduced to remove the oxygen, are not pure enough for commercial use; they must be refined. For example, an aluminum ore contains some iron. Since iron reduces more readily than aluminum, any iron in the ore will also appear in the metal. Likewise, a certain fraction of Si or P in an iron ore appears in extracted metallic iron,

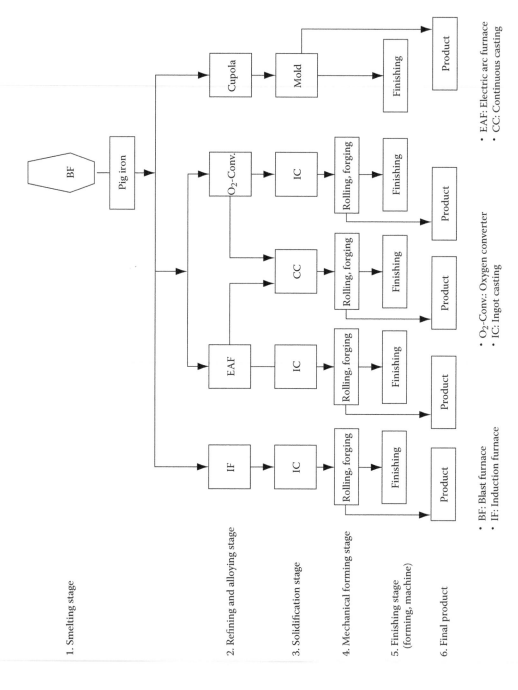

FIGURE 6.1   Production stages of a product from smelting to final stage.

because these elements are partially reduced along with the iron. As a rule, some of these impurities are not desired; they must be removed or eliminated by *refining processes*.

6.2 SMELTING OF FERROUS METALS

Blast furnaces are used for smelting ferrous metals, which are then refined in steel refining furnaces to produce steels or in cupolas to produce cast irons.

6.2.1 BLAST FURNACE

Pig iron, which contains impurities amounting to ca. 7% by weight, is made in a blast furnace by reduction of iron ore. Pig iron is then refined to produce CI in cupolas, or steels in steel refining furnaces such as oxygen converters, electric arc, and induction furnaces (Figure 6.1).

The availability of raw materials, particularly fuels and ores, is a major factor in determining the location of blast furnaces and steel mills. Raw materials for making 1 ton of pig iron are

2.0 tons of iron ore
1.0 ton of coke
0.5 ton of limestone
3.5 tons of air

Typical analysis of pig iron: 4% C, 1% Si, 1% Mn, 0.04% S, 0.4% P, and the balance Fe. However, the exact composition of pig iron depends upon

1. Composition of iron ore
2. Flux and coke
3. Operating conditions of the blast furnace

Table 6.1 illustrates the iron contents of different iron ores and their main location (origin). The charge of the blast furnace is composed of

1. Ore or agglomerates (pellets): These are the major sources. Minor sources are scrap, roll scale, and slag. The principal ores used are hematite and limonite, and less frequently used are magnetite and other ores. Pellets are agglomerates made from very fine particles of iron ore concentrate, fuel, binder (bentonite clay), and water. The agglomerates are placed

TABLE 6.1
Iron Ores: Composition, Iron Content, and Locations

Composition	Iron Content	Location
Fe_2O_4 (magnetite)	60%–70%	Sweden, Norway, Russia
Fe_2O_3 (hematite)	40%–60%	Spain, USA, Canada
$Fe_2O_3 \times H_2O$ (limonite)	30%–35%	Germany, France
$FeCO_3$ (siderite)	20%–40%	Austria, Hungary
FeS_2 (pyrite)	40%–50%	Germany, Spain, Russia

Source: Seyfert, W. L., Grungfachkunde Metall, Ernst-Verlag, Germany, 1982. With permission.

in a rotating drum, where they are formed into small rounded pellets of 15 mm diameter. Pellets are then fired to achieve strength to prevent them from crumbling when dripped in the blast furnace.

2. Flux: Fluxes are inorganic compounds that refine the molten metal by removing dissolved gases and various impurities. Limestone $CaCO_3$, and/or dolomite $(Ca\,Mg\,(CO_3)_2)$ are the principal blast furnace fluxes. The role of the flux is to react with the principal impurities, such as alumina Al_2O_3 and silica SiO_2, forming a low-melting slag. Since slag is lighter than the molten iron, it floats on the iron in the bottom of the furnace (Figure 6.2). At temperatures encountered in the furnace, calcining takes place according to the equation.

$$CaCO_3 \rightarrow CaO + CO_2 \uparrow \qquad (6.3)$$

The basic lime CaO then reacts with the acidic impurities Al_2O_3 and SiO_2 to form slag. The slag is a byproduct of the blast furnace, which is used in cement and bricks production as well as in thermal insulting materials.

3. Coke: The coke used in a blast furnace must be strong enough to prevent crumbling and blocking of the air passages. Since most of the sulfur of the iron originates in the coke, the sulfur content of the coke should be not more than 1.2%. Ash and phosphorus contents should also be low. The coke should have high calorific value, should be free from fines, and the pieces should not be too large, to keep optimum burning rates.

   The coke serves two functions
   - To provide heat for attaining desirable chemical equilibrium and adequate rates of reaction.
   - To provide reducing CO gas that is largely responsible for the reduction of iron oxide (Equation 6.2).

   Additional sources of heat include hot-air blast (up to 1100°C), and gaseous fuel injected with air. Table 6.2 demonstrates a typical energy balance for a blast furnace producing 2500 tons/day.

4. Air: In addition to solids charged into the furnace at the top, air is blown in at the bottom of the furnace. The air is at a temperature between 750 and 1100°C and is under pressure of 1.5–2 bar. It is enriched with gaseous fuels and moisture. These additions control the temperature and produce extra quantities of reducing gases. Coke alone does not produce enough reducing gases for maximum performance.

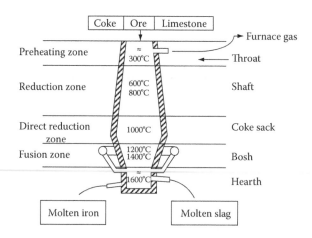

FIGURE 6.2 Temperature in different zones of a blast furnace.

TABLE 6.2
Energy Balance of a Blast Furnace

Type of Energy	MJ/kg of Pig Iron
1. Energy input:	
• Coke	14.9
• Gas fuel	1.1
• Hot air blast	2.0
Total input	18.0
2. Energy output:	
• Energy consumed in reduction of Fe and other oxides, heat lost in metal and clay	10.6
• Heat carried away by flue gases	7.4
Total output	18.0

Source: Mc Gannon, H. E., United States Steel Corp., Pittsburgh, 1971.

Reactions in the blast furnace:

$$2C + O_2 \rightarrow 2CO$$

$$C + H_2O \rightarrow H_2 + CO$$

$$Fe_2O_3 + 3CO \rightarrow 2Fe + 3CO_2$$

$$CO_2 + C \rightarrow 2CO$$

$$Fe_2O_3 + 3H_2 \rightarrow 2\,Fe + 3H_2O$$

$$H_2O + C \rightarrow CO + H_2 \tag{6.4}$$

Calcining reactions:

$$CaCO_3 \rightarrow CaC + CO_2$$

$$[CaC + Al_2O_3 + SiO_2] \rightarrow Slag$$

$$\downarrow \qquad \downarrow$$

Base   Acidic

lime   impurities
$$\tag{6.5}$$

Liquid slag and metal are tapped from the bottom of the furnace. Gases and dust escape from the top and are directed to stoves for heating the blast air. The temperature distribution in different zones of the furnace is illustrated in Figure 6.2. As the solids in the charge progress downward, they are heated. At the same time, gases become cooler as they approach the top of the furnace. In the bosh (Figure 6.3), iron and slag melt and drip from the still solid portion of the charge. The reduction in diameter of the bosh, along with a central pillar of the unburned coke, supports the charge. The assigned ratios of the solid charge are supplemented from the top in layers, alternatively.

Special refractory bricks are used as blast furnace lining. At the top, a super-duty, hard-fired brick is used, which is resistant to the abrasion of charge dumped into the furnace. High-temperature

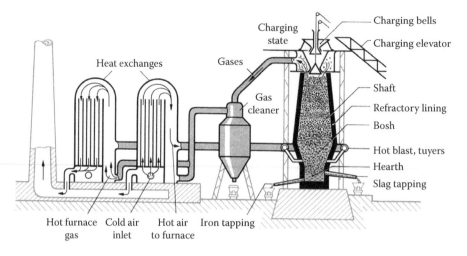

FIGURE 6.3 Blast furnace installation.

resistance is not a factor here. Brick in the in-wall zone (Figure 6.3), should withstand moderately high temperatures and moderate abrasion while the charge moves downward. The brick in the hearth and bosh must withstand very high temperatures as well as corrosion and slag attack. Carbon brick is preferred to be used in the hearth. The refractory brick in the in-wall and bosh are water cooled using hollow copper plates fitted between the bricks, as well as through the tuyere linings. A blast furnace producing about 2500 tons/day uses ca. 35,000 m³ of water for cooling purposes.

The gases leaving the top of the furnace are passed through a dust catcher, then burned in stoves used for preheating the air blast. The dust is agglomerated to be used again in the charge. Blast furnaces operate continuously for 5–7 years, then they must be rebuilt.

6.2.2 STEEL REFINING PROCESSES

Steel is made by refining either pig iron, steel scrap, or a combination of both materials. In steel produced from pig iron, the basic problem is to oxidize the impurities, which are then removed either as a gas (in case of carbon), or in slag (in case of Si, Mn, P, and S).

In the past three decades, there have been new concepts related to the technology of steelmaking. These are the development of steel converters, in which the air blast (Bessemer) is replaced by oxygen blast (L-D process), and the innovation of the continuous casting process since the early 1980s. There are different furnaces which are used for steelmaking. These are Bessemer converter, basic oxygen converter, electric arc furnace, induction furnace, crucible furnace, and continuous casting. The open hearth furnace has lost its importance, and therefore will not be discussed. The selection of a furnace depends on the type of operation; i.e., batch or continuous melting. In a batch, melting crucible or induction furnaces are used.

6.2.2.1 Basic Oxygen Converter (L-D Process)

The old Bessemer converter relied on blowing air from nozzles in the bottom of the converter through the molten pig iron charge (Figures 6.4 and 6.5). Oxidation of impurities supplied not only enough heat to keep the charge molten, but also enough to maintain favorable chemical equilibriums.

The basic oxygen converter is a new development of the old Bessemer process. It is the most widely used method today, and is known as the L-D process (derived from Linz Donawitz, Austria). In this process, the air of the Bessemer process is replaced by pure oxygen, which is introduced through a lance (Figure 6.6). The end of the water-cooled lance is suspended about 90 cm above the surface of the charge, and supplies oxygen at a pressure of 22 bar. The oxygen converter is a cylindrical vessel

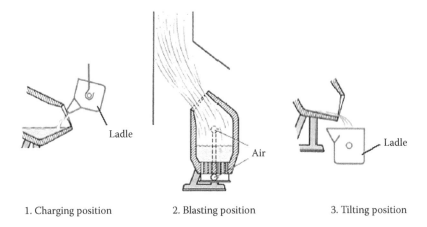

1. Charging position   2. Blasting position   3. Tilting position

FIGURE 6.4   Bessemer converter in its three operating positions.

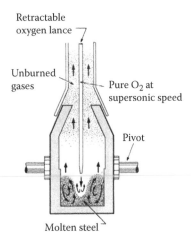

FIGURE 6.5   Seventy-ton Bessemer converter in its blasting position.

FIGURE 6.6   Basic oxygen converter. (From Guy, A. G., *Physical Metallurgy for Engineers*, Addison-Wesley, Reading: MA, 1962. With permission.)

of about 9 m high, 5 m inside diameter, and with a mouth of about 3 m in diameter. It is tilted to receive its charge (12%–13% scrap, and then molten pig iron), after which it is brought to a vertical position under a cooled-water hood (Figure 6.6). The lance is lowered to be just above the surface of the molten metal, and the blow starts. The oxygen quickly produces iron oxide in the melt,

$$2Fe + O_2 \rightarrow 2FeO \tag{6.6}$$

And this in turn oxidizes carbon, causing vigorous agitation of the melt to CO and dissolved silicon to SiO_2,

$$FeO + C \rightarrow CO + Fe \tag{6.7}$$

$$2FeO + Si \rightarrow SiO_2 + 2Fe \tag{6.8}$$

The CO leaves the vessel as a gas and completes combustion producing CO_2 in the air,

$$2CO + O_2 \rightarrow 2CO_2 \tag{6.9}$$

And the SiO_2 dissolves in the basic CaO-rich slag.

Fluxing agents such as lime and fluorspar are dropped from a hopper after oxygen blow has started. The lance is removed after the impurities have been oxidized. The converter is then tilted, first to one side to tap steel through a tap hole, and then to the other side to pour off the slag. Final adjustment of composition is made by ladle additions. The blowing time is typically 25 minutes. Basic oxygen converters vary from 50 to 350 tons capacity.

The same grades of steel are produced in the basic oxygen converter as in the open hearth furnace, and in addition, some grades of stainless and higher-alloy steels can also be produced which could not be made in an open hearth. The quality of products is even better than those of the open hearth. Nitrogen content is also low because pure O_2 rather than air is used for the blow. Sulfur is low due to the avoidance of sulfur-bearing fuels as a heat source. Manganese savings have also been realized because of better control of the process. About half of the basic oxygen steel produced is used for sheets, plates, and structural steel of welding quality, and the other half is used for rimmed steels of low carbon grade, from which deep-drawn automobile body parts are made.

6.2.2.2 Electric Arc Furnace

The features of electric arc furnaces are shown in Figure 6.7. The top is swung aside when the electrodes are raised so that the charge materials are dropped into the furnace. The top is then replaced, and electrodes are lowered to create an arc between the carbon electrodes and the metal charge.

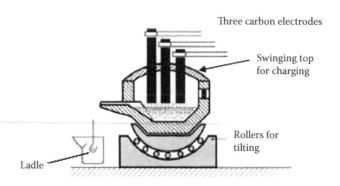

FIGURE 6.7   Electric arc furnace.

The arc is generated by applying high current (20,000 A), and low voltage (400V) electric energy. Commonly, three graphite electrodes are used in a vertical position, and thus, height in the furnace is adjusted depending on the charge size, and the electrodes are fed at a rate depending on their wear rate. The temperature achieved due to this arc reaches about 1900°C, which makes electric arc furnaces mainly used for smelting and casting of steels. The charge is usually steel scrap (pellets may be added), with additives to control and refine the chemical composition of the molten steel. Furnace capacities vary from 1 to 400 tons, but capacities ranging from 15 to 150 tons are most common. Since hydrocarbon fuels are not needed, the highest quality steels are generally produced in electric furnaces; moreover, they are less pollutant. They have the ability to hold the molten metal for any length of time for alloying purposes.

As the melting proceeds, slag forms from oxidized impurities and by reaction with lime or the furnace lining. After oxidization is complete, this slag is drawn off and replaced by a new slag cover in which the principal ingredients are lime, silica, magnesia, and calcium carbide. As soon as the final analysis of slag and metal have been adjusted to proper levels, necessary alloy additions are made to produce high quality steels of almost desired composition. When the furnace is tilted for tapping (Figure 6.8), the molten steel remains covered and protected by slag until the furnace is empty. The time elapsed from charging to tapping is dependent upon the size of the furnace and nature of the product, though typically it is about 4 hours.

Although these furnaces provide good mixing and homogeneity of the metal bath, the main disadvantage is that the noise and level of particle emissions is rather high, and the consumption of electrodes, refractory materials, and power results in high operating cost. However, the higher cost of the process is justified where small-scale, intermittent steel production requirements would

FIGURE 6.8 Electric arc furnace tilted for pouring. (Courtesy of Pittsburgh Lectromelt Furnace Corporation.)

not support a basic oxygen converter installation. The bottom-pouring electric arc furnace is a new development to avoid the time needed for tilting, which results in increased productivity.

6.2.2.3 Induction Furnaces

Induction furnaces are especially used in small foundries to produce composition-controlled smaller melts. Because of their high rapid melting rates and relative ease of controlling pollution, induction furnaces have become popular for melting virtually all common metals. They are especially useful for composition controlled melts, and have a range of capacities up to 65 tons. Figure 6.9 shows a coreless induction electric furnace, which consists of a crucible completely surrounded with a water-cooled coil, through which an HF current passes. There is a strong electromagnetic stirring action during induction heating; hence, this furnace has excellent mixing characteristics for alloying and for adding new charges to metal. For this reason, it is especially adapted for alloyed steel production. Because there is no contamination from the heat source, induction furnaces produce very pure metals. Since their operation is generally on a batch-size basis, they are not adaptable to continuous casting operations. Additionally, induction furnaces are used for the refining of nonferrous materials and cast iron. Figure 6.10 shows an induction furnace in pouring position.

6.2.2.4 Ingot Casting

During the oxidation process in the steel refining stage, a considerable amount of oxygen is dissolved in the molten metal. The molten metal solidifies when it is poured (teemed) from a ladle into an ingot mold. The oxygen and other gases are rejected as the gas solubility level decreases with decreasing temperature (Figure 6.11). The rejected oxygen combines with the atomic carbon in steel, forming CO gas and solid iron oxide as the temperature of the ingot drops. The CO bubbles are trapped in the solidifying mass. The porosity may take the form of small dispersed voids or large blowholes. The iron oxides form nonmetallic inclusions which are harmful to the mechanical properties of steel. The control of the amount of CO gas evolved is very important and leads to three different types of steel ingot. These are killed, semi-killed, and rimmed steels.

1. Killed steel (fully deoxidized) is steel in which oxygen gas has been removed, and thus porosity is eliminated. In the deoxidization process, the dissolved oxygen in the molten metal is made to react with elements such as aluminum (Al-killed steel), that are added to the melt. Other elements such as V, Ti, and Zr can also be added. All these elements have an affinity for oxygen, forming metallic oxides. The term *killed* means the steel lies quietly in the mold. The oxide inclusions in the molten metal float out and adhere to or dissolve in

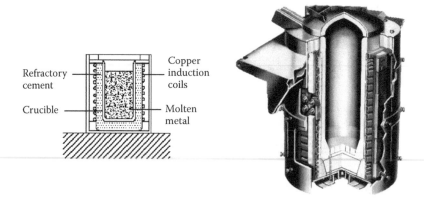

FIGURE 6.9   Coreless electric induction furnace, schematic cross section. (From Inductotherm Corp., USA. With permission.)

FIGURE 6.10   Induction furnace in pouring position. (From Inductotherm Corp., USA. With permission.)

slag. The killed steel ingots are the soundest, since they are free from gas bubbles, and have lower entrapped inclusions. The chemical and mechanical properties are relatively uniform throughout the ingot. However, because of material shrinkage during solidification, an ingot of this type develops a pipe (shrinkage cavity) of funnel-like appearance at the top. This pipe constitutes a substantial portion of the ingot and has to be cropped and scrapped (Figure 6.12a). These steels are of high carbon contents and of poor surface quality.
2. Semi-killed steel (partially deoxidized) contains some porosity in the upper portion of the ingot (Figure 6.12b). Enough deoxidant is added to partially suppress bubble evolution but not enough to completely eliminate the effect of oxygen. Appreciable additions of Al and ferrosilicon are dropped in the ladle. Semi-killed steels provide a compromise between good yield and quality. The quality is being suitable for structural steels and heavy plates. These steels are of intermediate carbon contents. They have poor surface quality.

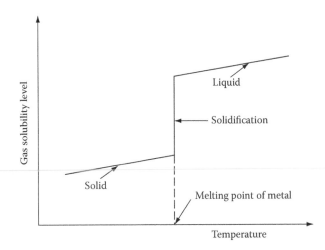

FIGURE 6.11   Effect of temperature on the solubility level of gas in a metal: a schematic.

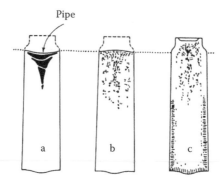

FIGURE 6.12   Types of steel ingots.

3. Rimmed steel (not deoxidized). A small amount of Al may be added to the steel to prevent excessive carbon monoxide gas evolution. The bubbles are trapped below the first layers to solidify, and near the top of the ingot (Figure 6.12c). The evolution of gas compensates for the shrinkage which accompanies solidification and eliminates pipe formation. This results in larger yields per ingot. The gas bubbles in rimmed steel are far enough below the rim or outer skin, and hence, the term *rimmed*. The surface of the rimmed steel is nearly pure iron. After hot and cold rolling, these ingots are well suited for deep drawing sheets and strips. Rimmed steels have a ductile skin with good surface finish. They have a generally low carbon content (less than 0.2%).

6.2.2.5   Continuous Casting

This is also called strand casting. It is a casting process which produces a continuous supply of metal of constant cross section from a bottomless water-cooled Cu mold. The process precedes hot rolling operation. About 90% of all steel production in highly industrialized countries is now made by the continuous casting method, rather than by the traditional ingot casting processes (Figure 6.1). This chapter would be incomplete if a brief description of the basic principles of this process is not included. Continuous casting has been developed since the early 1980s to overcome a number of ingot-related difficulties such as piping, entrapped slag, pores, and structural variations along the length of the product.

Like all casting processes, continuous casting begins with molten metal, but unlike other molding processes, it is not poured directly into the mold or ingot. Instead, the molten metal flows from a ladle to a settling tank called a tundish, which acts as a buffer store holding as much as three tons of liquid metal (Figure 6.13). With a residence time of 10 minutes in a tundish, any slag present is allowed to float to the surface and not be poured into the mold.

Out of the bottom of the tundish flows a steady stream of molten metal into the mold called a strand. This mold consists of a water-cooled Cu alloy box having the same cross section as the strand to be cast, but it has no bottom. Owing to the very rapid and efficient cooling action around the mold walls, the strand can be drawn out of the bottom of the mold as a continuous length of metal whose outside surface has solidified just sufficiently to retain its basic shape. The center of the strand is still molten at this stage, and therefore secondary cooling in the form of water sprays is essential to solidify the strand over its whole cross section (Figure 6.13).

If the primary cooling around the mold is inadequate, or if the time that the metal takes to pass through the mold is too short, the outer skin will not have solidified sufficiently to retain the still liquid center. Molten metal will then break out, spill, and solidify over adjacent parts of the casting machine, causing a serious and expensive breakdown. The molds should be internally coated

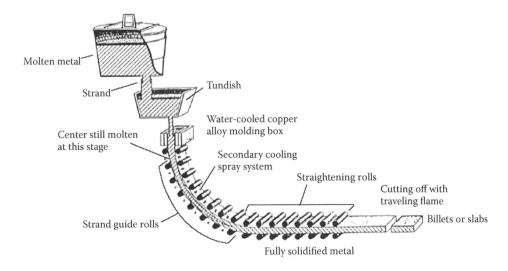

Molten metal

Strand

Tundish

Center still molten
at this stage

Water-cooled copper
alloy molding box

Secondary cooling
spray system

Straightening rolls

Cutting off with
traveling flame

Strand guide rolls

Billets or slabs

Fully solidified metal

FIGURE 6.13   A schematic of continuous casting. (Courtesy of AISI, Washington, DC.)

with solid lubricants such as graphite to reduce friction and adhesion at the mold/metal interfaces. Further protection is realized by oscillating or reciprocating the mold. The process is started with a dummy bar in the mold, upon which the first metal is poured and cooled. The continuously cast metal may go on for hundreds of heats at a rate of 2 m/min. The hot metal can be bent and unbent to reduce the height of building. Slab- or bar-strand may be cut into desired lengths during its movement by a flying saw or torch, or it may be fed directly into a rolling mill for further reductions in thickness and for shape rolling of products such as channel or I-beams. With the strand still being at a sufficiently high temperature to permit hot working, this avoids reheating cost when the strand is required later for rolling. Although the thickness of a steel strand is usually about 250 mm, recent developments have led to the reduction of this thickness to about 15 mm or less. The thinner strand reduces the number of rolling passes required, and improves the overall economy of the operation. After rolling, steel plates or shapes undergo one or more further processes, such as cold rolling, punching, stamping, deep drawing, cleaning, or pickling.

Nonferrous metals such as Al and Cu are also continuously cast, but by far the greatest tonnage poured involves the production of steels. Large diameter and long iron water pipes with flanged ends are cast in a semicontinuous manner with a separate charge of molten metal for each pipe. Finally, continuous casting is characterized, as compared to traditional ingot casting, by the following advantages and drawbacks.

Advantages of continuous casting
- Highly consistent dimensional quality, metallurgical properties, and uniform chemical composition
- Better surface quality
- Major energy saving
- Increased production rate
- No cropping and hence negligible waste of material
- With a simple adjustable mold design, section changes are easy to achieve, even without stopping the process
- Double-strand machines are now available
- Permits high degree of automation and low unit labor cost
- Cheaper than rolling from ingots

Drawbacks of continuous casting
- A shrinkage cavity sometimes forms, which must be eliminated during subsequent hot rolling operation.
- Molding equipment is complex and hence expensive.
- Metal breakouts are costly, and can cause considerable damage to the equipment.
- Thicknesses of less than 15 mm are currently difficult to produce consistently.
- Not suitable for batch or foundry production. It is more adapted for large tonnage production.

6.2.3 CAST IRON REFINING—CUPOLA FURNACE

A large amount of gray, nodular, and white CI is still melted in cupolas, although many foundries have now converted to the less pollutant-inducing furnaces. Cupolas will be dealt with in a detailed manner in Chapter 7.

6.3 SMELTING AND EXTRACTION OF NONFERROUS METALS

This section will discuss in brief the smelting, extraction, and refining of some important nonferrous metals such as aluminum, copper, titanium, tungsten, zinc, lead, and tin.

6.3.1 ALUMINUM

Aluminum is primarily used to produce pistons, engine and body parts for cars, and beverage cans. It may also be used as sheet metal, aluminum plate and foil, rods, bars and wire, aircraft components, windows, and door frames. Aluminum can either be produced from bauxite ore or from aluminum scrap. Refinement of aluminum ore is expensive, so that the production of aluminum from aluminum scrap covers about 40% of aluminum in the U.S. Primary aluminum refining is performed using the Bayer process, which refines the bauxite ore to obtain aluminum oxide (Al_2O_3). The ore used as the source of aluminum is the bauxite, a hydrated aluminum oxide $Al_2O_3.3H_2O$ that contains silica SiO_2, titana, TiO_2, and iron oxide Fe_2O_3. Impurities are first removed from crushed bauxite by treating with hot sodium hydroxide. High temperatures and pressures cause reactions in the ore/sodium hydroxide mixture. The result is dissolved aluminum oxide and ore residue. The residues, which include silicon, lead, titanium, and calcium oxides, form insoluble sludge in the bottom of the digester. The aluminum oxide is evaporated off and condensed.

$$2NaOH + Al_2O_3.3H_2O \leftrightarrow 2NaAlO_2 + 4H_2O \qquad (6.10)$$

The solution is then filtered while still hot to remove the TiO_2 and Fe_2O_3 and most of SiO_2. The concentrated sodium aluminate $NaAl_2.3H_2O$ is cooled, seeded with $Al_2O_3.3H_2O$ crystals, and precipitation of $NaAl_2.3H_2O$ occurs. The pure hydrated aluminum oxide is washed and heated to form pure alumina (Al_2O_3).

$$Al_2O_3.H_2O \xrightarrow{1000°C} Al_2O_3 + 3H_2O \qquad (6.11)$$

Aluminum oxide from the Bayer process is then reduced to aluminum using the Hall–Heroult process. In this process, aluminum is produced by extracting it from the aluminum oxide (Al_2O_3) through an electrolysis process driven by electrical current. The process uses molten salts called Cryolite (Na_3AlF_6) as an electrolyte that is capable of dissolving the alumina. Carbon anodes are

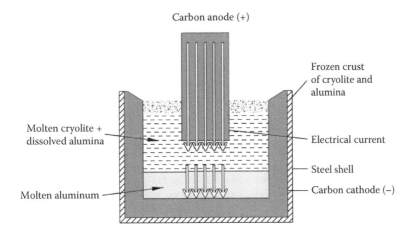

FIGURE 6.14 Aluminum smelter.

immersed into the electrolyte carrying electrical current, which then flows into the molten cryolite containing dissolved alumina. As a result, the aluminum is deposited in the bottom of the cell, where a molten aluminum deposit is found, while the oxygen reacts with the carbon of the anodes, producing carbon dioxide (CO_2) bubbles. The alumina reduction process is described by the following reaction:

$$2Al_2O_3 + 3C \rightarrow 4Al + 3CO_2 \tag{6.12}$$

During the smelting operation, some of the carbon is consumed as it combines with oxygen to form carbon dioxide. In fact, about half a pound (2 N) of carbon is used for every pound (4 N) of aluminum produced. Some of the carbon used in aluminum smelting is a byproduct of oil refining, and additional carbon is obtained from coal.

Alumina is extracted from the ore bauxite by means of the Bayer process at an alumina refinery. An aluminum smelter consists of a large number of pots and steel containers lined with carbon, in which the electrolysis takes place as shown in Figure 6.14. Smelting is run as a batch process, with the aluminum metal deposited at the bottom of the pots and periodically drained off. Because aluminum smelting involves passing an electric current through a molten electrolyte, it requires large amounts of electrical energy. On average, production of 2 lb (10 N) of aluminum requires 15 kWh of energy. The cost of electricity represents about one-third of the cost of smelting aluminum.

Due to high energy requirements, the major primary aluminum producers are located very close to large power stations and near ports, since almost all of them use imported alumina. Aluminum producers refine and melt the aluminum and pour it into ingots that are shipped to metal casting plants or other shaping plants for molding or rolling.

6.3.2 COPPER

Copper is commonly used to produce a wide variety of products, including electrical wire, pipes and tubes, automobile radiators, and many others. It is combined with zinc to produce brass, and with tin to produce bronze. Pure copper is usually combined with other chemicals in the form of copper ores. Sulfide ores, in which the copper is chemically bonded with sulfur, are the most common. Other ores include oxide ores, carbonate ores, or mixed ores, depending on the chemicals present. Many copper ores also contain significant quantities of gold, silver, nickel, and other valuable metals, as well as large commercially useless materials. Most of the copper ores mined in the United

States contain only about 1.2%–1.6% copper by weight. The most common sulfide ores are chalcopyrite, $CuFeS_2$, and chalcocite, Cu_2S.

The process of extracting copper from copper ore varies according to the type of ore and the desired purity of the final product. Each process consists of several steps in which unwanted materials are physically or chemically removed, and the concentration of copper is progressively increased. The steps used to process the sulfide ores are as follows (Figure 6.15):

1. Concentrating: This is normally used to remove the waste materials by the flotation method. The ore is crushed in a series of cone crushers, screened, and the crushed ore is then ground smaller (0.25 mm diameter) by a series of rod mills mixed with water. The slurry is mixed with various chemical reagents, which coat the copper particles. Pine oil or long-chain alcohol are often used as a frothier. This mixture is pumped into the flotation cells, where air is injected into the slurry through the bottom of the tanks. The chemical reagents make the copper particles cling to the bubbles as they rise to the surface, leaving the gangue, which contains SiO_2, Al_2O_3, Ca_2CO_3, etc. The frothier forms a thick layer of bubbles, which overflows the tanks and is collected in troughs. The bubbles are allowed to condense and the water is drained off. The resulting mixture, called a copper concentrate, contains about 25%–35% copper along with various sulfides of copper and iron, plus smaller concentrations of gold, silver, and other materials. The remaining materials in the tank (gangue) are pumped into settling ponds and allowed to dry.

2. Smelting: The copper concentrate is fed into a flash furnace, along with a silica material as a flux. Most copper smelters utilize oxygen-enriched flash furnaces, in which preheated, oxygen-enriched air is forced into the furnace to combust with fuel oil. The copper concentrate and flux melt and collect in the bottom of the furnace. Much of the iron in the concentrate chemically combines with the flux to form a slag, which is skimmed off the surface of the molten material. Sulfur in the concentrate combines with the oxygen to form sulfur dioxide, which is treated in an acid plant to produce sulfuric acid. The remaining molten material in the bottom of the furnace (matte) is a mixture of copper sulfides and iron sulfides that contain about 60% copper by weight.

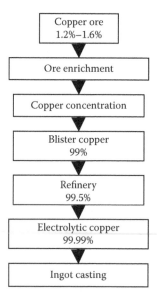

FIGURE 6.15   Smelting of copper.

The molten matte is drawn from the furnace and poured into a second furnace called a converter, where additional silica flux is added and oxygen is blown through the molten material. The chemical reactions in the converter are similar to those in the flash furnace. The silica flux reacts with the remaining iron to form a slag, and the oxygen reacts with the remaining sulfur to form sulfur dioxide. After the slag is removed, a final injection of oxygen removes all but a trace of sulfur. The resulting molten material is called the blister, which contains about 99% copper by weight.

3. Refining: The blister copper is heated in a refining furnace, which is similar to the converter described above. Air is blown into the molten blister to oxidize some impurities. A sodium carbonate flux may be added to remove traces of arsenic and antimony. The molten copper, which is about 99.5% pure, is then poured into molds to form large electrical anodes, which act as the positive terminals in an electrolytic cell. Each copper anode is in an electrolytic cell where a sheet of copper is placed on the opposite end of the tank to act as the cathode, or negative terminal. The cell is filled with an acidic copper sulfate solution, which acts as an electrolyte, that conducts electrical current between the anode and cathode. When an electrical current is passed through the cell, the copper is dissolved off the anode and is deposited on the cathode. Most of the remaining impurities fall out of the copper sulfate solution and form slime at the bottom of the tank. After the current is turned off and the cathodes are 99.95%–99.99% pure, copper is removed. The slime that collects at the bottom of the tank contains gold, silver, selenium, and tellurium. It is collected and processed to recover these precious metals. After refining, the copper cathodes are melted and cast into ingots, cakes, billets, or rods, depending on the final application.

6.3.3 TITANIUM

The aerospace industry is the largest user of titanium products because of the high strength-to-weight ratio and high temperature properties. It is typically used for airplane parts and fasteners, gas turbine engines, compressor blades, casings, engine cowlings, and heat shields. Because of having good compatibility with the human body, titanium is used in the production of human implants such as pacemakers, defibrillators, and elbow and hip joints. The primary ores used for titanium production include ilmenite, leucoxene, rutile, anatase, perovskite, and sphene. Ilmenite and leucoxene are titaniferous ores. Ilmenite ($FeTiO_3$) contains approximately 53% titanium dioxide. Leucoxene has a similar composition but has about 90% titanium dioxide. Rutile is relatively pure titanium dioxide (TiO_2). Anatase is another form of crystalline titanium dioxide and has just recently become a significant commercial source of titanium.

Titanium is produced using the Kroll process, which involves extraction, purification, sponge production, and alloy creation for forming and shaping (Figure 6.16). These materials are put in a fluidized-bed reactor along with chlorine gas and carbon. The material is heated to 900°C and the subsequent chemical reaction results in the creation of impure titanium tetrachloride ($TiCl_4$) and carbon monoxide. The reacted metal is put into large distillation tanks and heated to remove metal chlorides, including those of iron, vanadium, zirconium, silicon, and magnesium.

$$TiO_2 + 2Cl_2 \rightarrow TiCl_4 + CO_2 \qquad (6.13)$$

The purified titanium tetrachloride is transferred as a liquid to a stainless steel reactor vessel. Magnesium is then added and the container is heated to about 1100°C. Argon is pumped into the container so that contamination with oxygen or nitrogen is prevented. The magnesium reacts with the chlorine, producing liquid magnesium chloride and leaving pure titanium solid, since the melting point of titanium is higher than that of the reaction (1725°C). The titanium solid is removed from

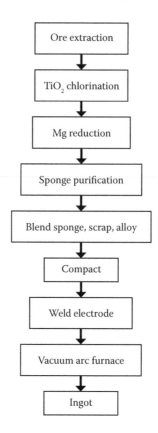

FIGURE 6.16   Titanium smelting process.

the reactor by pouring, and then treated with water and hydrochloric acid to remove excess magnesium and magnesium chloride. The resulting solid is a porous metal called a sponge.

$$TiCl_4 + 2Mg \rightarrow Ti \ (solid) + 2MgCl_2 \tag{6.14}$$

The pure titanium sponge is converted into a usable alloy via a consumable electrode arc furnace. The sponge is mixed with various alloy additions and scrap metal. The sponge electrode is then placed in a vacuum arc furnace (VAC) to melt the sponge electrode and form an ingot. After an ingot is made, it is removed from the furnace and inspected for defects. Figure 6.16 shows the main steps of titanium smelting.

6.3.4 MISCELLANEOUS METALS

Due to the great similarity between the extractive metallurgy of iron, aluminum, copper, and titanium, and the extractive metallurgy of other metals, the remaining paragraphs outline the general scheme followed for the production of magnesium, zinc, lead, tin, and tungsten.

Magnesium: The raw materials include magnesium chloride ($MgCl_2$) which contains 21% Mg and carnallite (9% Mg) that are derived from salt water or sea deposits. Other raw materials include brucite (42% Mg), magnesite (29% Mg), and dolomite (14% Mg). The Dow chemical electrolytic process is the most important method for production of magnesium. Accordingly, hydrated $MgCl_2$ is obtained from natural brines by recrystallization. It is then dehydrated and mixed with NaCl and

KCl to form a mixture that melts at about 700°C. Current is passed through the melt in an electrolytic cell. Magnesium (99.9 percent pure) is formed at the cathode while chlorine forms at the anode and is collected as a valuable byproduct.

$$MgCl_2 \rightarrow Mg \ (99.9\% \ pure) + Cl_2 \ (gas) \qquad (6.15)$$

Zinc: Zinc is refined from the ore known as zincblende (ZnS). Concentrated sulfide ores are converted to oxide or sulfate by heating in air. Zinc oxide ores, carbonate ores, and the converted zinc sulfate or zinc oxide can be heated with coal. The latter acts not only as a source of heat but provides the reducing medium for reduction to metallic zinc. Since the retorts operate above the boiling point of zinc, the metal passes off as a vapor and is condensed as a liquid near 480°C.

$$ZnO + CO \rightarrow Zn \ (vapor) + CO_2 \qquad (6.16)$$

A small amount of zinc is produced by electrolysis of zinc sulfate solutions. In this method, after roasting the ore and the reduction of silicates and carbonates, the ore is leached with sulfuric acid that creates a liquid solution of zinc sulfate ($ZnSO_4$).

$$ZnO + SO_2 \Rightarrow ZnSO_4 \qquad (6.17)$$

The liquid solution of zinc sulfate is put in an electrolytic cell where zinc is deposited on the negative electrode. The deposit is then stripped, melted, and recast for shipment.

$$ZnSO_4 \rightarrow Zn \ (pure) + SO_2 \qquad (6.18)$$

Lead: The principal ore is galena or PbS, which is concentrated by a flotation process and is then roasted to convert it to PbO.

$$2PbS + 3O_2 \rightarrow 2PbO + 2SO_2 \qquad (6.19)$$

The lead oxide is charged into a blast furnace with coke and with a flux of lime and iron oxide. Molten lead collects in the bottom of the furnace, from which it is tapped at intervals. Impurities are also tapped periodically and treated to recover valuable quantities of other metals.

$$PbO + C \rightarrow 2Pb + CO_2 \qquad (6.20)$$

Further refining of the lead is often made by heating the lead in a reverberatory furnace in the presence of air. Most of the impurities form oxides which pass into the exhaust gases, or float into a skin which can be skimmed or settle out so that they can be removed.

Tin: The principal ore contains the mineral cassiterite (SnO_2) that contains difficult to remove impurities such as copper, iron, lead, zinc, bismuth, and antimony. Washing and concentrating are first done, followed by roasting to remove sulfur and to form oxides and chlorides that may be removed with a dilute acid. Smelting is then done in a reverberatory furnace that produces high purity tin and high-tin-content slag, which can be further refined. Very high purity tin (99.9% pure) is obtained by electrolytic refining. The deposit is stripped, melted, and recast for commercial use.

Tungsten: Tungsten has such a high melting point (about 3400°C) that it is not easily converted to the liquid state. Tungsten is mined as an ore containing the minerals wolframite, $FeWO_4$, or scheelite, $CaWO_4$. This is converted to tungstic acid, H_2WO_4, which is reduced to tungsten powder by hot hydrogen. The powder is compressed to form a briquette and sintered in a hydrogen atmosphere at about 1100°C using the techniques of powder metallurgy.

6.4   REVIEW QUESTIONS

1. What raw materials are required for the production of pig iron?
2. Distinguish between iron ore and pellets.
3. What functions are performed in the blast furnace by limestone, ore, and air?
4. What are the principal impurities in pig iron, which are largely removed when it is converted to steel?
5. What is the heat source in a basic oxygen converter?
6. What are the advantages of a basic oxygen converter?
7. Differentiate between rimmed, killed, and semi-killed steels.
8. Enumerate important iron ores and list their iron contents.
9. Describe the smelting process in a blast furnace.
10. Define the calcining reaction involved in a blast furnace.
11. Rolling equipment is increasingly being added down stream in continuous casting. Why is this?
12. What is continuous casting, and what advantages does it offer?
13. What are the drawbacks of continuous casting?
14. How are the principal impurities removed from bauxite to make pure alumina?
15. How is aluminum obtained from pure alumina?
16. Describe (a) concentration, (b) roasting, (c) smelting, (d) converting, and (e) refining as applied to the extraction of copper from its ores.
17. Describe in brief how zinc, tungsten, and tin are extracted from their ores.

BIBLIOGRAPHY

Cottrell, A. 1982. *An Introduction to Metallurgy*, 2nd ed. London: Edward Arnold.

Degarmo, E. P., Black, J. T., and Kosher, R. A. 1997. *Materials and Processes in Manufacturing*, 8th ed. Upper Saddle River, NJ: Prentice-Hall.

Doyle, L. E., Keyser, C. A., Leach, J. L., Schrader, G. E., and Singer, M. B. 1985. *Manufacturing Processes and Materials for Engineers*, 3rd ed. Upper Saddle River, NJ: Prentice-Hall.

Guy, A. G. 1962. *Physical Metallurgy for Engineers*. Reading, MA: Addison-Wesley.

Kalpakjian, S., and Schmid, S. R. 2003. *Manufacturing Processes for Engineering Materials*, 4th ed. Upper Saddle River, NJ: Prentice-Hall.

Pollack, H. W. 1981. *Materials Science and Metallurgy*, 3rd ed. Upper Saddle River, NJ: Reston Publishing, Prentice-Hall.

Schey, J. A. 2000. *Introduction to Manufacturing Processes*, 3rd ed. New York: McGraw-Hill.

Seyfert, W. L. 1982. *Grundfachkunde Metall*. Germany: Ernst Klett-Verlag.

Van Vlack, L. H. 1973. *A Textbook of Materials Technology*. Reading, MA: Addison-Wesley.

7 Casting of Metallic Materials

7.1 INTRODUCTION AND CLASSIFICATION

Metal casting is the process through which the metal is molten, and allowed to flow by gravity or under pressure into a mold, where it solidifies in the shape of the mold cavity. The product is also termed *casting*. Casting includes both the casting of ingots or slabs for the primary metal industries, and the casting of shapes or near final shape products. Traditionally, the workshop where the casting process is carried out is named a foundry. Thus, a foundry is a factory equipped for making molds, melting metals in furnaces, transferring molten metal to the molds, performing the casting process, and the cleaning and finishing of castings. Typical casting products include engine blocks, carburetors, car wheel rims, machine bases, pipes, taps, valves, . . . etc.

The process is unique among metal forming processes for a variety of reasons. The most obvious is the wide variety of the process techniques, which provide the possibility of producing complicated shapes ranging in weight from a few grams to several hundred tons. The process is applicable and economically viable for a single product, or for a small number of products, while some casting methods are quite suited to mass production. Virtually any metal that can be molten is produced by casting. Metal casting ranks second only to steel rolling in the metal producing industry. It is estimated that castings are used in at least 90% of all products and in all machinery used in manufacturing. These advantages are contrasted by some disadvantages. The most serious is the safety hazards and environmental problems associated with the processing of hot molten metal. Other disadvantages for some casting methods include porosity, limitations on mechanical properties, and poor dimensional tolerances and surface finish.

The heart of the casting process is the mold, which contains the cavity where the molten metal is poured. Molds are made of a variety of materials, including sand, plaster, ceramic, and metal. The various casting processes are often classified according to the type of the mold, whether used once (expendable) or for producing many products (permanent) as shown in Figure 7.1.

7.2 HISTORICAL DEVELOPMENT OF CASTING

The casting of metal is a prehistoric technology. The melting of copper for casting of axes was practiced in the period between 5000 and 3000 BC, preceding the Bronze Age. The melting furnaces were elementary, but they demonstrated control of the fire developed from charcoal to achieve the melting temperature. Melting was carried out in a crucible. The molds were made of carved stone, which allowed the casting followed by deformation and assembly of elaborate jewelry in the form of bracelets and neckpieces.

The first bronze alloy was made of arsenical copper containing from 4% up to 12% As, which announced the beginning of the bronze age in the near east before 300 BC. The use of 5 to 10% tin as an alloying element with copper offered the advantage of lowering the melting point, deoxidizing the melt, improving the strength, and producing a shiny, polished cast surface. Tin and silver were added to the molten copper to produce tin bronze alloys. The elements of bronze were found in the eastern deserts of Egypt and Taurus Mountains of south central Anatolia. The open mold was later developed to permanent two-part molds for axially symmetric products such as axes and swords. Lost wax casting of small bronze and silver objects was also developed in the same period.

The Bronze Age began in the Far East (China and Thailand) in about 2000 BC, a millennium later than its start in the Near East, where casting was the prime forming method. The Chinese mold was

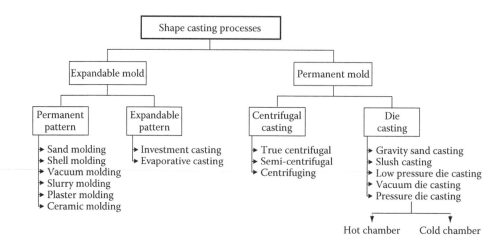

FIGURE 7.1   Classification of shape casting processes.

made of ceramic, typically of many separate parts. The poured alloy was mainly leaded tin bronze to improve fluidity of the molten metal, and accordingly, the surface details on cast parts. Ceramic cores were used to cast hollow sections. Excavated vessels also showed cast decorations which were incorporated by impressing decorative elements into the mold, or by casting on metal elements.

Cast iron was known in China in about 600 BC, and has appeared in excavated statues. Smelting was based on the use of coal, and therefore the percentage of phosphorus and sulfur were high, resulting in lower melting point (similar to that of bronze). Cast iron was not introduced in Europe before the 15th century, and the mass production of the alloy as a structural material began in the United Kingdom in the 18th century, where coal was first coked. The produced iron was used in erecting the iron bridge, and led to many other architectural applications.

The casting of brass started in Africa, with the brass having high enough zinc content to take on a golden hue. Recent discoveries of zinc furnaces in India, and the very long trade routes that were opened in the 17th century, suggest that the metal may have been brought from India. The most commonly applied casting technique was the lost wax technique. It was used in Ghana (The Gold Coast) for casting gold weights and decorative objects.

Large castings were made in bolted or welded sections, or were cast sequentially. However, traditionally large bells and guns were cast in one piece. The great bronze bell in the Kremlin was cast in 1735, and weighs 193 tons. After the development of cast steel in 1740, bells of cast steel became a specialty of Sheffield in the United Kingdom. Gun barrels were made by casting "gun metal" made of bronze containing 10% Sn. Cannons were cast around a core to form the bore.

The centrifugal casting process, which involves pouring of the molten metal into a rotating cylindrical mold, was developed in the United Kingdom in 1809. The method was soon adopted by the pipe foundries, and was adopted in the United States in 1848. The centrifugal casting of steel was first introduced in 1889, and railroad car wheels were spin-cast in 1901.

Following the early development of the centrifugal method, a permanent mold method known as slush casting was introduced. The process involved pouring the molten metal into a usually bronze split metal mold, until the mold was filled. The mold was then inverted and the metal that was still liquid was allowed to run out after a thin shell was formed on the mold wall. The process was limited to the production of hollow parts made of lower melting point metals, such as lead and zinc alloys. Die casting was developed in the early 1900s.

Aluminum is the most plentiful metal on the earth's crust. Aluminum casting, was developed late in 1855, and aluminum castings have found wide applications since World War II. The development of magnesium as a casting metal parallels the history of aluminum.

7.3   EXPENDABLE MOLD CASTING PROCESSES

Expendable molds are prepared by consolidating a refractory material such as sand, plaster, or ceramics around a pattern that defines the shape of the cavity, and also includes the sprue, runner, and riser required to fill the mold. The used refractory material should withstand the temperature of the molten metal. When the casting solidifies, the mold is broken up to remove the casting. The pattern can be made of a material that can be used many times (permanent pattern), or it can be made of a material that melts or evaporates with temperature (expendable pattern) as shown in Figure 7.1.

7.3.1   Sand Casting

Sand molds are the most conventional and most applicable types of expendable molds. Therefore, they will be used to describe the basic features of a casting mold. Figure 7.2 shows a cross-sectional view of a typical sand casting mold. The mold usually consists of two halves; the upper is known as the cope, and the lower is the drag. These two mold parts are contained in a halved box known as a flask. The two parts of the mold are separated at the parting plane. This allows opening of the mold to remove the pattern after packing the sand mold to provide the required mold cavity.

7.3.1.1   Patterns

Patterns are usually made of wood, but can be made of metal or strong plastic for greater durability and dimensional stability. Wooden patterns are the most widely used, for their ease of manufacturing, low density, and repairability. However, due to their limited wear resistance and the possibility of dimensional changes when used many times, plastic and metallic (aluminum, cast iron, and steel) are superior for large batches of products. Simple patterns are made in one piece, but for symmetric shapes, they are usually split into two halves along the parting plane, as shown in the pattern in Figure 7.2. Patterns are usually coated for ease of removal from the mold.

A pattern is made larger than the finished product dimensions to account for

a. Shrinkage allowance: All dimensions are increased to allow for the contraction of the casting during solidification and during cooling from the melting point to room temperature. This allowance is a function of the thermal characteristics of the product material, and is normally taken as a percentage of the dimension. Typical shrinkage allowances for most commonly cast materials are given in Table 7.1. Foundries were supplied by pattern-maker's rules that are longer by the shrinkage allowance, but recently CAD/CAM programs can account for all allowances.

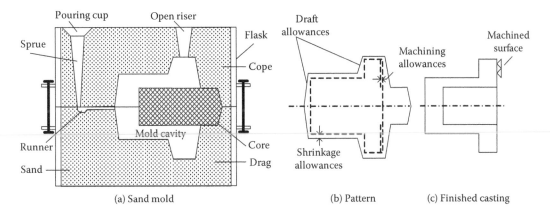

(a) Sand mold   (b) Pattern   (c) Finished casting

FIGURE 7.2   Characteristic elements of a sand mold.

TABLE 7.1
Shrinkage Allowances for Different Metallic Materials

Alloy	% Contraction Allowance
Aluminum and aluminum alloys	1.3
Copper	1.6
Brass	1.56
Bronze, aluminum	2.32
Bronze, phosphorus	1.0–1.6
Gray cast iron	0.9 to 1.04
Ductile cast iron	0.6 to 0.8
Lead	2.6
Nickel alloys	2
Carbon steels	1.6 to 2.0
Tin	2.0
Zinc alloys	1.18

b. Machining allowance: If some surfaces on the casting are to be machined, the thickness to be removed during machining should be added to the pattern. Machining allowance is usually in the range of few millimeters, depending on the size of the part to be machined and the type of the machining process.

c. Draft allowance: To facilitate withdrawal of the pattern from the mold without damaging the surrounding sand, surfaces perpendicular to the parting plane should be given a draft (tilting angle). This draft is usually in the range from 0.5° to 2°.

7.3.1.2 Cores

Patterns cannot have cavities or internal holes; otherwise, the sand filling these cavities will be demolished during pattern withdrawal from the mold. Therefore, any cavities or holes in the cast shape could be filled by the insertion of cores. The pattern is made solid and provided with an extra part to create a nesting support for the core called the "core print," as shown in Figure 7.2. Cores greatly increase the variety of shapes that can be formed by casting.

Cores are made of refractory material with higher bond strength than the sand mold in order to withstand the flow of molten metal without being collapsed. However, they must still be removable after solidification and cooling of the casting. If the weight of the core cannot be supported by only the core print, the free end can be additionally guided by "chaplets," often made of perforated metal parts that will melt into the casting alloy during solidification. Core prints and chaplets should also resist buoyancy of the core when the liquid metal fills the mold cavity. Cores are molded into core boxes made of wood or metal. Cores are usually made in one part, but they may be made in halves or several parts and pasted together using suitable bonding materials.

7.3.1.3 Gating System

The feeding system consisting of the sprue, runner, and risers should be added around the pattern before molding. The sprue is the downward passage through which the molten metal flows. It starts at the mold top with the pouring cup. This should allow smooth flow of the molten metal to the sprue, and restrain scale or any solid contaminants. The runner is the horizontal channel that carries the molten metal from the sprue to the mold cavity. The riser (known as feeder) is a reservoir that supplies supplementary metal to the casting during solidification to prevent formation of shrinkage cavities in the casting. The riser is usually added at the highest point of the mold cavity, as shown in Figure 7.2, so that molten metal goes through it after complete filling of the mold cavity. This is known as open riser, or top riser. Risers are not needed if the molten metal fills the mold under pressure. Sometimes a blind riser that is not open to the atmosphere is used. A blind riser is usually bigger than an open riser.

7.3.1.4 Types of Sand Molds

The widely used sand in foundries is silica (SiO_2) in the form of quartz, or silica mixed with other minerals. Other refractory sands suitable for special applications include: zirconium silicate ($ZrSiO_4$), olivine (Mg_2SiO_4), chromite ($FeCr_2O_4$), and aluminum silicates. Sand is the product of rocks over extremely long periods of time, and therefore it is widely available and inexpensive. The sand is suitable for casting because of its refractoriness, withstanding the high temperature of the commonly used molten metals in casting. The sand suitable for foundries should satisfy several requirements concerning grain size and shape. Small grain size provides close packing and smooth surface casting, while large grain size offers higher permeability of the mold (to allow escape of the gases and vapor arising during pouring). Irregular grain size permits better cohesion of the mold sand, yet it restricts permeability. Advanced foundries usually have sand testing laboratories for proper selection of the sand grain size, shape, and distribution of the mixtures, as well as the strength and permeability. Several indicators are used to determine the quality of the sand mold. These include

1. *Strength* determines the mold ability to maintain its geometric shape and resist the stresses and erosion caused by the flow of the molten metal.
2. *Permeability* is the ability of the mold to permit the escape of air, gases, or vapor during the casting process.
3. *Thermal stability* is the ability of the sand mixture at the surface of the mold cavity to resist cracking upon contact with the molten metal.
4. *Flowability* is the ability of the sand mixture to flow over and fill the mold. More flowability is useful for a more detailed casting.
5. *Collapsibility* is the ability of the sand mixture to collapse under force. It allows the casting to shrink freely during the solidification phase of the process. Otherwise, hot tearing or cracking will develop in the casting.
6. *Reusability* (reclaiming) is the ability of the mold sand mixture to be reused to produce other castings in subsequent manufacturing operations.

According to the types of additives to be mixed with the sand to form the mold, sand molds can be classified into six types; namely, green sand mold, dry sand mold, no-bake mold, silica gel molds, cement molds, and oil sand.

1. Green sand mold: To prepare the mold sand, the sand grains are bonded by a mixture of clay (bentonite; a hydrated aluminosilicate with layered structure) and water. A typical mixture consists of 90% sand, 7% clay, and 3% water. The mixture is made uniform by using mulling machines. This mixture is known as green molding sand, with the term *green* referring to the moist or damp mixture. This is the cheapest molding sand, with good permeability and sufficient strength for most applications. It is also reusable after breaking the mold to extract the casting.
2. Dry sand mold: Dry sand molds can be made using organic or inorganic binders instead of clay, and the mold is baked in an oven at a temperature range of 200°C–320°C. Baking increases the strength of the mold and provides better dimensional accuracy and surface finish of castings. However, baked molds may suffer distortion and lower collapsibility, which may lead to hot tearing and cracking of castings. The process is also expensive and the production rate is lower; therefore, it is commonly limited to large castings in low to medium production rates. In order to get the partial advantages of dry sand molds, the green sand mold surfaces surrounding the cavity are dried to a depth of 10 to 25 mm. This

is known as the *skin-dried mold*. Drying is done either with a torch, a heating lamp, or by hot air flow.

3. No-bake mold: Chemically bonded molds that cure at room temperature have been developed recently to replace the conventional clay-water binder and to avoid baking. These are known as no-bake molds. Some of these binding materials include furan resins (consisting of furfuryl alcohol, urea, and formaldehyde), and thermosetting resins such as phenolics. These no-bake molds offer good dimensional control in high production rate applications.

4. Silica gel mold (CO_2 process): Silica gel is used as a bonding agent. The sand is mixed with 3%–5% water glass in the form of a liquid, and the mold is formed. CO_2 gas is bubbled through the mold to react with the water glass and form a very firm bonding gel. This mold is applicable for large castings with high dimensional accuracy.

5. Cement mold: The hydration of cement results in the formation of a high strength binder. Cement is added to the sand at a ratio of 10 to 15%. This mold is used occasionally for large steel castings molded in a bit. The main drawback of this mold is that it is hard to break away from the finished casting, and cannot be reused.

6. Oil sand: Oil sand consists of sand mixed with a vegetable oil such as linseed oil and some serial flour. When heated to a temperature of about 230°C, the oil forms a polymer, and the sand is bonded with what could be regarded as a flour-filled polymer, achieving high enough strength for cores.

7.3.1.5 Sand Molding Techniques

Sand mixtures are compacted by different techniques selected according to the required product quality, the number of castings, and the production rate. These techniques include

1. Manual ramming for a limited number of simple castings, the pattern is located on a pattern plate, surrounded by the half flask, and the sand mixture is shoveled into the flask and rammed manually using hand tools (hand rammers and trowels). It requires highly skilled foundrymen to produce uniformly packed molds.

2. Sand slinger ramming where the consolidation and ramming is obtained by the impact of the sand mixture at a high velocity from an overhead impeller. The sand is conveyed to the molding station and dropped or slung onto the pattern surrounded by a flask at a rate of 500 to 2000 kg/min. The speed of the impeller shaft controls the mold hardness and strength. Slingers are most often used for ramming medium to large size molds at low rates of production.

3. Molding machines are cost effective for large production batches of high quality castings. The main two functions of molding machines are to pack the sand mixtures tightly and uniformly, and to draw the pattern from the mold. To achieve the first function, one of the following techniques is followed.

   a. Squeezer technique where the molding sand is squeezed between the machine table and an overhead squeeze board, pneumatically or hydraulically, until the mold attains the required hardness. Squeezing pressure could be provided by the top board (*top squeezer machine*), or by the table side (*bottom squeezer machine*). The main limitation of the squeezer technique is that the sand is packed more densely on the side where the pressure is applied, i.e., packing is not uniform along the depth, as shown in Figure 7.3a. Alternately, squeezing pressure can be applied on a number of equalizing pistons to control the pressure distribution according to the depth of the sand at different locations across the mold (Figure 7.3b).

   b. Jolting technique where the flask is first filled with the molding sand, and then the table supporting the flask is mechanically raised and dropped in succession. The sudden change in inertia at the end of each fall helps to pack the sand evenly around the

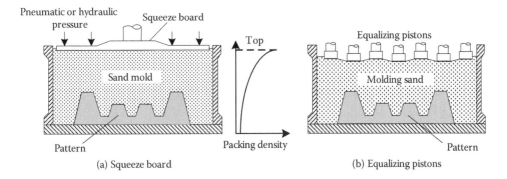

FIGURE 7.3 Squeezer sand molding machine.

pattern. This action is known as jolting. A pneumatic system is usually implemented to control the amplitude of the vertical movement. The drawback of this technique is that the sand is packed more densely at the parting plane. A schematic of a jolting machine is shown in Figure 7.4.

c. Jolt-squeeze machine: This machine overcomes the drawbacks of either squeezing or jolting machines. A jolting action is used to consolidate the sand on the pattern side, and is followed by a squeezing action to impart the required density throughout the mold.

To control the second function of the molding machine according to the method of drawing the pattern from the mold, one of the following techniques is used.

- Straight draw molding machine; where the pattern is fixed on the pattern plate, which is fixed to the table. After filling and squeezing, the flask is lifted from the pattern by stripping pins, as shown in Figure 7.5a.
- Turnover molding machine; where the flask together with the work table is rotated 180°, and the pins lift the table with the pattern out of the mold (Figure 7.5b). This is used for large size molds.

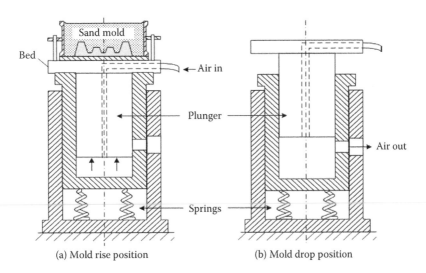

FIGURE 7.4 Jolting sand molding machine. (From Choudhry, S. K., *Elements of Workshop Technology*, 2nd ed., Media Promoters Publishers (MPP), Bombay, India, 1985. With permission.)

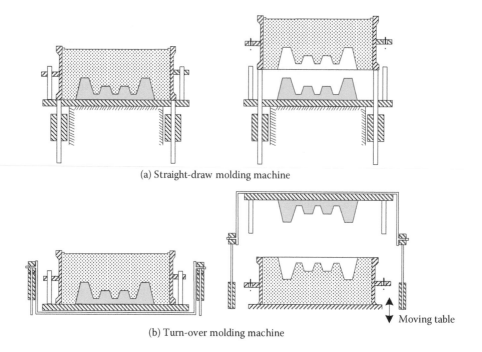

(a) Straight-draw molding machine

(b) Turn-over molding machine

▲▼ Moving table

FIGURE 7.5　Sand molding machine. (From Choudhry, S. K., *Elements of Workshop Technology*, 2nd ed., Media Promoters Publishers (MPP), Bombay, India, 1985. With permission.)

　4. Flaskless molding: When the compacting pressure is high enough, the sand mold acquires enough strength to maintain its integrity without a supporting flask. The mold halves are packed sequentially in one master flask, and then pushed together horizontally in a single production line on a conveyor with their parting planes and pouring cups vertical, as shown in Figure 7.6. Production rates can be as high as 250–275 molds/hour. The high strength of the molds minimizes sand movement during solidification and provides higher accuracy of castings. Highly automated production lines can be specially designed using this flaskless molding.
　5. Dynamic (impact) molding: Where the sand mold is compacted by the pressure wave resulting from a controlled explosive charge or sudden release of compressed gas. This ensures uniform mold strength and superior permeability.

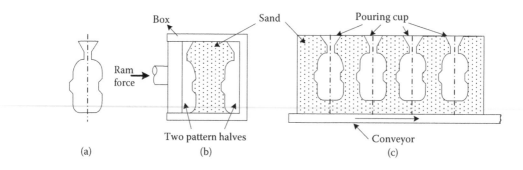

FIGURE 7.6　Flaskless molding process.

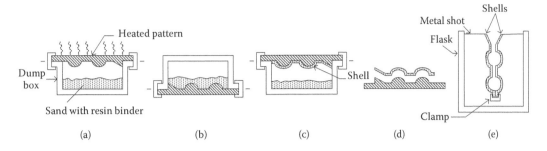

FIGURE 7.7 Shell molding process. (From Groover, M. P., *Fundamentals of Modern Manufacturing,* 3rd ed., John Wiley & Sons, London, 2007. With permission.)

7.3.2 SHELL MOLDING

Shell molding is a sand casting process in which the mold is a thin shell, of a few millimeters thickness, made of a mixture of sand and a 2.5%–4% heat-curing thermosetting resin binder such as phenol formaldehyde. This process was developed in Germany in the 1940s. The process is limited to small- to medium-size castings, but offers higher dimensional accuracy and surface finish compared to the conventional green sand mold. A dimensional tolerance of 0.5%, and surface finish of 1.25–3.75 μm, rms can be obtained, which reduces the need for further machining. Collapsibility of the mold is quite sufficient to avoid tearing and cracking of castings. The pattern is made of metal because it should be heated to 200°C–260°C before covering with sand. After coating with a parting agent, the pattern plate is located on top of a box containing the sand mixture, and the assembly is inverted so that the mixture settles on the pattern and a thin shell cures and hardens in situ, accurately reproducing the pattern details, as shown in Figure 7.7. When the shell is thick enough, the shell is turned back to get rid of the unbonded sand. The shell is stripped, assembled with the other half, placed in a flask, and supported on the outside by steel shots, ready for pouring the molten metal. The process can be highly automated for mass production and is very economical for large production rates. Typical shell molded products include gears, valve bodies, bushings, and camshafts.

7.3.3 VACUUM MOLDING

Vacuum forming, also known as the V process, was developed in Japan around 1970. The process is based on using a sand mold held together by vacuum pressure rather than by a binder. The pattern is covered tightly by a thin sheet of plastic, and the flask is placed in position and filled with sand. A second sheet of plastic is placed on top of the mold. Through a side hole in the flask, air is evacuated from the sand, leaving it hard and tightly packed so that the pattern can be drawn away. The second flask half is prepared and both are assembled as shown in Figure 7.8. During pouring, the mold remains under vacuum. After solidification of the casting, the vacuum is turned off, and the sand falls away. The sand can be fully reused without reconditioning. Large size castings can be obtained by vacuum molding, but the process is relatively slow, and not readily adaptable to mechanization. Therefore, it is most suitable for small batches of production.

7.3.4 SLURRY MOLDING

Where finer grained refractory may be made into slurry with water and poured in the flask around the pattern without ramming. The slurry is more heat resistant than the bonded sand, and therefore the shrinkage of mold and casting can be closely controlled. Higher dimensional accuracy and better surface finish can be obtained, and the process is considered a precision casting process.

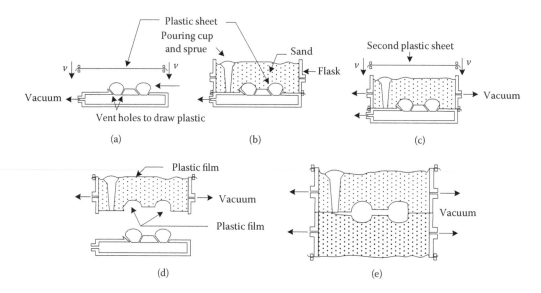

FIGURE 7.8 Vacuum molding. (From Groover, M. P., *Fundamentals of Modern Manufacturing*, 3rd ed., John Wiley & Sons, London, 2007. With permission.)

7.3.5 PLASTER MOLDING

This is due to the high flowability of plaster of Paris slurry (gypsum, $CaSO_4.2H_2O$). It is used as a mold for nonferrous castings with good surface finish and tight tolerances. The mold needs baking for several hours to remove the moisture and reach the required strength. Various inorganic fillers such as talc and silica flour may be added to improve the mold strength and permeability, as well as controlling contraction and setting time. Patterns should be made of plastic or metal since wooden patterns are affected by contact with water in the plaster slurry. Since gypsum is damaged at 1200°C, only metals or alloys with lower melting point such as aluminum, magnesium, and some copper-based alloys can be produced. Typical plaster-molded castings include metal molds for plastic and rubber molding, pumps, and turbine impellers. Casting sizes are small, with a maximum 100 kg, but most products weigh less than 10 kg.

7.3.6 CERAMIC MOLDING

Ceramic molding is similar to plaster molding except that the mold is made of refractory ceramic materials that can withstand higher temperatures than plaster, and thus can be applied for casting all metallic materials. Typical refractory powders for the process are zircon ($ZrSiO_4$), alumina (Al_2O_3), or fused silica (SiO_2), with various patented bonding agents. The fine-grained ceramic slurry is applied as a thin facing to the pattern and is packed with lower cost fire clay. The mold is fired at about 1000°C, and the molten metal is poured while the mold is still hot. The process is suitable for producing large constructional castings, as well as forging and casting dies, with high dimensional accuracy that does not require further finishing.

7.3.7 EXPENDABLE MOLD, EXPENDABLE PATTERN CASTING

In this category of casting processes, an expendable pattern is made of a material that can be either melted out before pouring, or burnt up during casting. Accordingly, patterns can be made in one piece and there is no need for parting planes in the mold, draft angles, or even cores for internal cavities. Expendable patterns are made by injecting the pattern material into the cavity of a pattern mold, and complex shapes can be made in several segments that can be easily assembled by heat or by adhesives. Shrinkage allowances are higher than those of metal, and therefore must be considered during pattern design. Casting processes based on expendable patterns are commonly named according to the type of the pattern material. These include investment casting (lost wax process), and evaporative casting (lost foam process).

7.3.7.1 Investment Casting (Lost Wax Process)

The term *investment* here means outer layer or covering, referring to the coating of the refractory material around the wax pattern. As the name "lost-wax" implies, the pattern is made of wax, which is the oldest thermoplastic material known to man. Originally, beeswax was used, but today the name "wax" applies to any material with wax-like properties. Modern blends for investment casting wax are compounds of hydrocarbon wax, natural ester wax, synthetic wax, natural and synthetic resins, organic filler materials, and water. All of these compounds are either straight chained carbon atoms (aliphatic compounds), or ring structured carbon atoms (aromatic compounds). The process was used in ancient Egypt and China more than 4000 years ago to manufacture ornamental products. However, it has found widespread industrial applications since the Second World War in the 1940s to manufacture precision, intricately shaped components for jet engines.

The sequence of the process can be presented in the following steps, as shown in Figure 7.9:

1. Wax patterns can be easily made in large quantities by injection molding in metallic dies.
2. These patterns are assembled in groups with wax sprues and runners in the form of tree branches, simply by local joining of mating parts using a hot plate or knife.
3. The pattern tree is precoated with a thin layer of fine (almost powder) refractory material to assure a smooth surface finish and capture the intricate details of the pattern. The

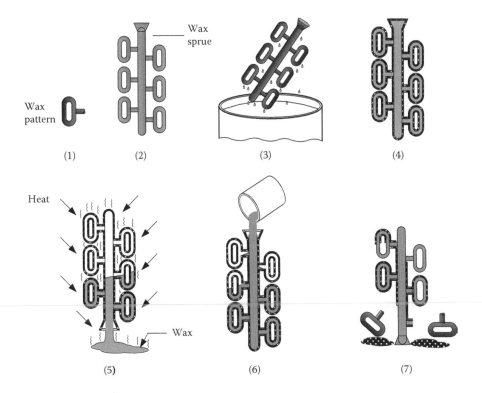

FIGURE 7.9 Steps of investment casting.

precoated pattern is dipped in the refractory slurry several times to form a sufficiently thick layer that completely covers the pattern.

4. The mold is allowed to dry in air for several hours to allow hardening of the binder.
5. The mold is held in an inverted position in an oven at temperatures of 90°C–175°C to allow the wax pattern to run out of the cavity, and the heat is increased to 700 to 1000°C to ensure high strength of the mold, and eliminate the danger of gas formation during casting.
6. The molten metal is poured and left to solidify and cool down.
7. The mold is broken up and castings are separated from the sprue.

Investment castings are usually small in size (less than 35 kg), but with net intricate geometries. All types of metals including steels, stainless steels, and other high temperature alloys are suitable for investment casting. Typical products include turbine blades and other components of turbine engines, complex machinery parts, jewelry, gears, cams, valves, and dental fixtures. The wax can be recycled for further use.

7.3.7.2 Evaporative Casting (Lost-Foam Process)

In this casting process, the pattern is made of expanded polystyrene foam, which is left in the mold to evaporate and burn up during casting. The foam can be easily formed to the required shape in one part, or in segments that can be joined together, or to the gating system by rubber cement or hot melt adhesives. The plastic foam is firm, but can be damaged by high compacting forces. The pattern is first coated with a fine water-based slurry, and then dried, placed in a flask, and surrounded with sand mixtures or steel shots. The molten metal is poured without removing the pattern. The heat of the molten metal evaporates the polystyrene, and the resulting gases are vented into the surrounding mold. The process has been applied for mass production of cylinder heads, crankshafts, brake components, and manifolds in the automotive industry. Recent developments of the process are

1. The use of polymethyl-methacrylate (PMMA), and polyalkylene carbonate as alternative pattern materials for ferrous castings.
2. Embedding the polymer pattern with fibers or particles, which do not evaporate during pouring, but mix with the molten metal, forming metal matrix composites (MMC). Similarly, grain refiners and modifier master alloys can be added to the polymeric pattern to mix with the molten metal for improving the casting structure.

7.4 PERMANENT MOLD CASTINGS

The main disadvantage of any of the expendable mold casting processes explained before is that the mold is only used once, and a new mold is required for every casting. In permanent mold casting, the same mold is used repeatedly for many castings. The mold is made of metal. This category of casting processes includes die casting and centrifugal casting.

7.4.1 Die Casting Processes

Die casting processes use a metal mold called a die set or simply a die. Solidification rates in die casting are much higher than in refractory molds. Therefore, production rates are much higher, and the grain size of castings is finer. The die metal must have a sufficiently high melting point to withstand erosion by the liquid metal at the pouring temperatures, a high strength to resist deformation under repeated use, and high thermal fatigue resistance to resist cracking. Common die materials are alloyed steels and cast iron. Recently, refractory metal alloys are used, such as precipitation-hardenable molybdenum alloys, to retain strength at high temperatures. Metals that are commonly cast in permanent molds include aluminum, magnesium, and copper alloys.

The die is usually constructed of two halves that can be tightly closed to pour the molten metal, and easily opened to eject the casting after solidification. Sometimes the die is made of several parts to cope with shape complexity. The cavity and the gating system are usually engraved into the two halves by machining, to provide accurate dimensions and high surface finish of the product. Metal cores can be used to provide internal cavities in castings. They should be designed with shapes that facilitate removal from the casting. Collapsible sand cores can also be compulsory in specific configurations. Vents must be provided to avoid entrapped gases. Ejector pins are necessary to eject solidified products, particularly if the process is mechanized. Coatings made of refractory powder in suspending medium are applied to the die surface for protection and to reduce heat transfer. Graphite or silicon films are used to eliminate adhesion and facilitate ejection of the casting. Die casting processes can be subdivided to gravity die casting, slush casting, low-pressure die casting, vacuum die casting, and pressure die casting.

7.4.1.1 Gravity Die Casting

This is the simplest type of die casting, where the molten metal fills the mold cavity under gravity. The machine is basically made of stationary and movable mold halves supported on a bed. The movable half may be manually operated using a suitable mechanical clamping system as shown in Figure 7.10, or hydraulically actuated. The process is widely used for casting of aluminum alloys, as well as magnesium and copper alloys.

7.4.1.2 Slush Casting

Slush casting is a variant of the gravity die casting process for the production of hollow sections. It is based on inverting the mold after it has been filled so that the metal that is still liquid is allowed to run out after a thin shell is formed on the mold wall. The thickness of the cast wall depends on the time interval between filling and inverting of the mold, the chemical and physical properties of the alloy, and the mold temperature. The outer surface of the product is highly controlled by the finishing of the die cavity, but the inner surface is rough. The process is limited to castings made of lower melting point metals such as lead, tin, and zinc alloys. Typical products are lamp bases (pedestals), candle sticks, and small statues.

7.4.1.3 Low Pressure Die Casting

In low pressure casting, the mold is placed upside-down right above the melting or holding furnace, and metal is fed under air pressure through the bottom gate into the mold cavity, as shown in Figure 7.11. This facilitates smooth filling of the die, and solidification is directed from the top downward, with no porosity. Another advantage is that the metal filling the cavity is drawn from the center of the ladle, while in gravity casting the surface layer of the molten metal is subject to oxidation, which affects the product quality. Air pressure is released as soon as the cavity is filled with solid metal, ensuring minimal material losses. The holding pressure of the die halves is larger than in gravity

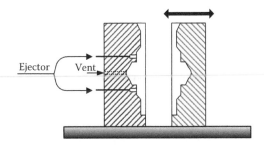

FIGURE 7.10 Gravity die casting process. (From Schey, J. A., *Introduction to Manufacturing Processes*, McGraw-Hill, New York, 1987. With permission.)

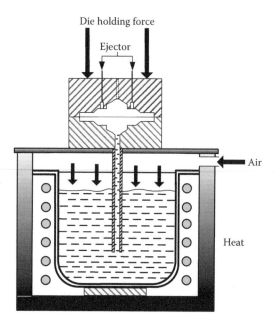

FIGURE 7.11 Low-pressure die casting. (From Schey, J. A., *Introduction to Manufacturing Processes*, McGraw-Hill, New York, 1987. With permission.)

die casting to resist the forces developed by the liquid metal pressure in the cavity. The process is widely applied for aluminum alloys.

7.4.1.4 Vacuum Die Casting

This is a variation of low pressure casting, where vacuum is used to draw the molten metal into the mold cavity instead of forcing it under low pressure. Vacuum reduces the pressure in the cavity to about two thirds of the atmospheric pressure. This leads to elimination of the air porosity, and imparts greater strength to the casting. This process should not be confused with the vacuum molding process. It is mostly applicable for thin-walled, complex shapes. Carbon steels and stainless steels weighing up to 70 kg have been vacuum cast.

7.4.1.5 Pressure Die Casting

In pressure die casting, the molten metal is forced into the die cavity under a moderate to high pressure range from 0.7 to 150 MPa, to fill intricate details. Therefore, small wall thicknesses in the range of 0.5 mm can be cast to net shapes. Clamping and holding forces for the die halves are higher due to these high pressures, and thus high capacity hydraulic cylinders up to 3000 tons are used for this purpose. The process is faster than gravity or low pressure die casting, yielding production rates up to 1000 (shots)/hour with small machine sizes. There are two different types of pressure die casting machines; namely, hot chamber and cold chamber.

- Hot-chamber process: The molten metal is forced to the mold cavity directly from the holding pot by an immersed cylinder and plunger system, as shown in Figure 7.12a. The system is designed to force specific batches of metal with each stroke under pressure ranges up to 40 MPa. The metal is held under pressure until solidification is complete. Dies are commonly cooled by a circulation of water to allow for rapid cooling and improving the die life (elimination of thermal fatigue). A tree branch type gating system to fill a group of cavities, similar to that of the lost wax method, can be used for high rate production of small products. This process is applicable for low melting point metals such as zinc, lead, and tin alloys.

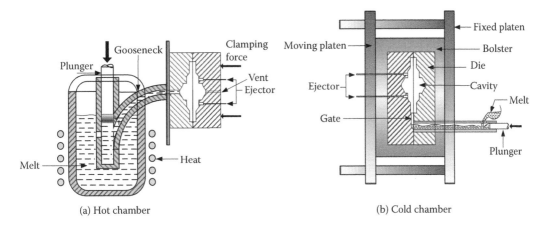

(a) Hot chamber   (b) Cold chamber

FIGURE 7.12 Pressure die casting machines. (From Schey, J. A., 1987, *Introduction to Manufacturing Processes*, McGraw-Hill, New York, 1987. With permission.)

- Cold-chamber process: Where a quantity of molten metal just sufficient for one shot is individually transferred by a ladle to the cylinder, and forced by the plunger to the die cavity, as shown in Figure 7.12b. Higher pressures up to 150 MPa can thus be applied. The process has long been established for zinc, magnesium, and aluminum alloys, as well as brasses.

For both types of machines, the plunger velocity is vital for the die life and product quality. Excessive velocity results in mold erosion and gas entrapment in the casting, whereas too slow filling results in incomplete products (misruns), and cold shuts. Plunger velocities can reach 40 m/s in casting zinc alloys. Lubricants should be added to the surfaces of the cavities between successive fillings. Dies are usually made of die steels. The most commonly used lubricants are graphite or sodium sulfide, suspended in a water base. The evaporation of water from lubricants after spraying is effective in keeping the die cool. Recently developed pressure casting machines are highly automated, and all process variables are numerically controlled.

7.4.2 Centrifugal Casting

When pouring molten metal in a rotating mold, the metal is pushed out by centrifugal force to the outer walls, where solidification starts and progresses inward. This force is high enough to eliminate porosity. Since inclusions in the molten metal tend to have lower density, they segregate toward the center and can be easily machined if required. Therefore, centrifugally cast products have sound fine grains which are free from porosity, and have a chemically clean structure. Further, the use of gates, cores, and risers is eliminated, reducing production cost. There are three types of centrifugal casting; namely, true centrifugal casting, semicentrifugal casting, and centrifuging.

7.4.2.1 True Centrifugal Casting

Used for casting of hollow cylindrical parts in cylindrical rotating molds, as shown in Figure 7.13. Long products such as tubes, cast iron water supply and sewage pipes, and gun barrels are usually cast in molds rotating around a horizontal axis, while shorter lengths such as bearing rings or bushings can be made in molds rotating vertically. The mold is made of steel and lined with graphite, a refractory sprayed layer, or even with green or dry sand lining. The outside of the mold can be water cooled to accelerate solidification. The outer contour of the casting can be made rectangular, hexagonal, or any shape, but the inner surface remains cylindrical. High surface quality is achieved on the outside surface, whereas the inner surface cannot have the same quality.

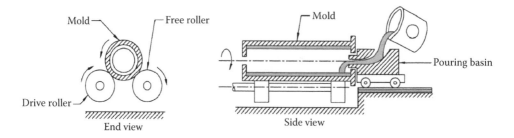

FIGURE 7.13   True horizontal centrifugal casting.

The process has recently been applied for casting of metal matrix composites (MMC), where particulate or fiber materials are added to the molten metal during pouring, such as graphite, aluminum oxide, silicon carbide, boron, beryllium, or tungsten. The process is also used for producing functionally graded materials (FGM), where the centrifugal force distributes the additives in MMC according to their densities, such that the mechanical properties vary across sections to achieve functional requirements. Typical applications are rocket nozzles and turbine components, which work under aggressive environments of steep thermal gradient. These are made of metal ceramic composites with gradual variation from near pure metal on one side to near pure composite at the other side of the section. The process is also used currently for producing bulk metallic glass (BMG), which represents a novel class of engineering materials with amorphous structure, having unique mechanical, thermal, magnetic, and corrosion properties. These properties are attractive compared with conventional crystalline alloys, and are very useful in a wide range of engineering applications.

7.4.2.2   Semicentrifugal Casting

Semicentrifugal casting is used to form axial-symmetric shapes by pouring metal in the central sprue of a rotating mold, usually around a vertical axis. The metal flows out to the outer areas by centrifugal force, as shown in Figure 7.14. The central sprue acts as a riser to eliminate central cavities. Rotating speed is relatively low. A dry sand core can be used if a central bore is required. Typical applications are large axial-symmetric castings such as gears, disc wheels, propellers, and pulleys.

7.4.2.3   Centrifuging (Centrifuged Casting)

In this process, a number of identical molds are arranged circumferentially on a rotating table, and the molten metal is poured in a central sprue to flow in radial runners to these molds under

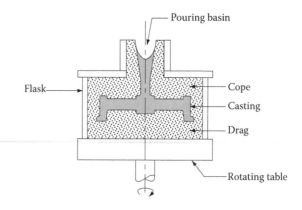

FIGURE 7.14   Semicentrifugal casting.

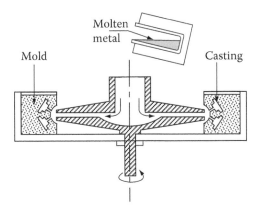

FIGURE 7.15 Centrifuging.

centrifugal force, as shown in Figure 7.15. The process is applicable for large production rates of small-sized, intricately shaped products (similar to investment casting).

7.5   MELTING FURNACES

In all casting processes, the metal or alloy must be heated to the molten state in a furnace and poured, or forced, into the mold. The types of furnaces most commonly used in foundries can be classified to cupolas, crucible furnaces, electric arc furnaces, and induction furnaces. Cupolas and crucible furnaces will be introduced in this section, while electric arc furnaces and induction furnaces are presented in Chapter 2.

Selection of the right furnace for a certain casting process depends on the type, melting point, and amount of the metal to be molten, as well as economic and environmental parameters. The economic aspect is vital, since the metal casting industry is one of the most energy-intensive manufacturing sectors, with the melting process accounting for 55% of its energy consumption. Melting and pouring involves a series of steps that encounters material and energy losses. These losses can be attributed to several factors, including undesired conduction, radiation and convection, stack loss (flue gases), and metal loss. The extent of the losses depends on the furnace design, the fuel used, and the method of imparting heat to the metals.

7.5.1   Cupolas

A cupola is a vertical cylindrical furnace made of a steel tubular construction, lined with refractory bricks and equipped with a tapping spout near its base. The furnace is mounted on steel columns, as shown in Figure 7.16. It is used only for melting cast iron, but it has the advantage of being operated continuously, yielding large production rates and large amounts of molten metal. About 60% of iron casting tonnage is melted in cupolas. The energy efficiency of cupola melting ranges from 40 to 70%.

The furnace is charged sequentially with layers of metal, coke, and flux through a door located near the middle of the height. The metal is commonly a mixture of pig iron and scrap (including sprues, runners, and risers cut from previous castings), as well as alloying elements if required. Coke is the solid fuel used to heat the furnace. Air is blown through radial openings (tuyeres) near the bottom of the cylinder for combustion of the coke. The flux is basically limestone which reacts with coke ash and other impurities to form slag. The slag floats on top of the melt, protecting it from reaction with the gases inside the cupola, and reducing heat loss. The molten metal is periodically tapped through the bottom tapping spout, and slag is tapped through a slag spout at a higher level, as shown in the figure. In principle, the furnace is similar to the "blast furnace" used in the production

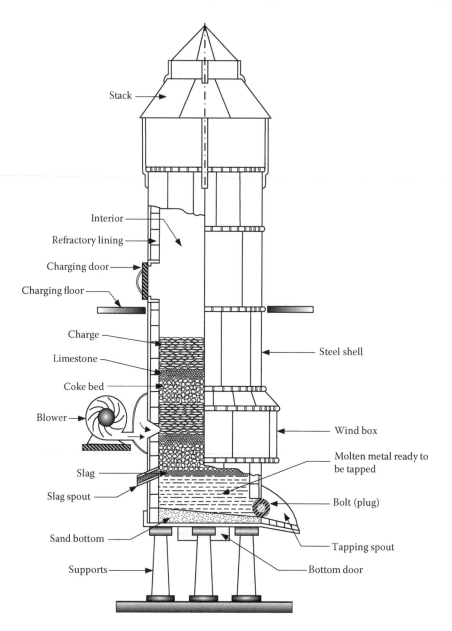

Stack

Interior

Refractory lining

Charging door

Charging floor

Charge

Limestone

Coke bed

Blower

Slag

Slag spout

Sand bottom

Supports

Steel shell

Wind box

Molten metal ready to be tapped

Bolt (plug)

Tapping spout

Bottom door

FIGURE 7.16 Cupola furnace.

of pig iron, but at a smaller scale. The efficiency of modern cupolas is improved by water cooling of the furnace, using new cement lining, hot air or oxygen-enriched air blast, or by using a plasma-fired blast. This last technique has been developed by Westinghouse Electric Corporation, USA. A plasma torch is installed in the tuyere zone of the cupola where blast air is supplied and mixed with the plasma torch flow in a special mixing chamber. This arrangement is reported to increase productivity (up to 60%).

A *cokeless cupola* was developed about 30 years ago using fuels such as natural gas, propane, diesel oil, or pulverized coal instead of coke. This type has been adopted in some large foundries in Europe and India. Air openings in a conventional cupola are replaced by burners, as shown in

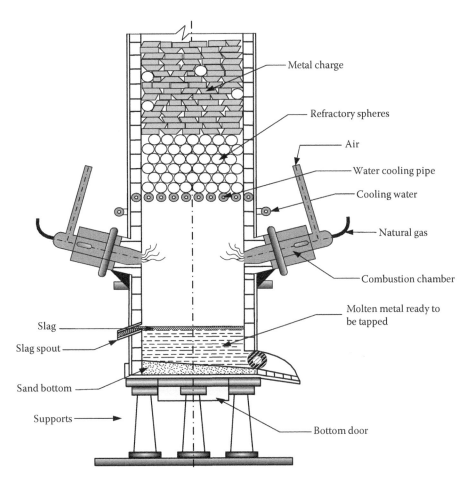

FIGURE 7.17   Cokeless cupola.

Figure 7.17, which blow hot gases that develop partially reducing conditions inside the cupola to decrease oxidation losses. Above the burners, a grid of cooling water supports specially developed refractory spheres which act as heat exchangers. The hot gases keep the spheres at high temperature to melt the scrap. The molten metal is collected in the well, and continually injected by a carburizer to control the carbon content. This technique eliminates CO emission considerably, and results in lower sulfur content in the metal and reduced slag production.

7.5.2   CRUCIBLE FURNACE

Crucible furnaces are used to melt the metal in a closed container (crucible) without direct contact with fuel or fuel products; therefore, they offer clean molten metal with minimum contamination. They are used for melting nonferrous metals such as aluminum, zinc, and copper alloys in small batches. Crucible furnaces offer the least expensive melting method for small volumes. However, their efficiency is very limited (7% to 19%), with more than 60% of the heat content lost in radiation.

The crucible is made of a refractory material, usually a clay–graphite mixture, or of a high temperature steel alloy. Means of heating may be natural gas, fuel oil, fossil fuel, or powdered coal. Some furnaces for low melting temperature alloys are heated by electrical resistance (immersion

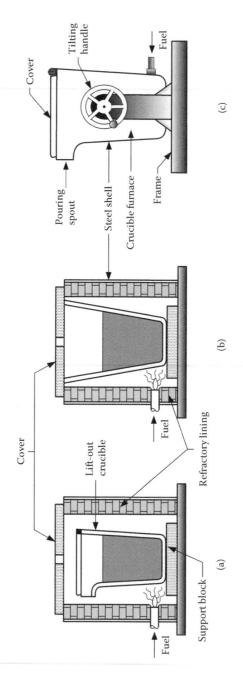

FIGURE 7.18 Types of crucible furnaces: (a) lift-out type, (b) stationary pot type, and (c) tilting pot type. (From Groover, M. P., *Fundamentals of Modern Manufacturing*, 3rd ed., John Wiley & Sons, London, 2007. With permission.)

heaters). According to the method of handling the molten metal, there are two basic types of crucible furnaces.

    a. Lift-out type: Where the crucible including the metal charge is located in the furnace for melting the metal charge, and then lifted out of the furnace and used as a pouring ladle (Figure 7.18a).

    b. Pot furnaces: Where the heating furnace and crucible are one integral unit. This type may be a *stationary pot furnace*, where the whole assembly is stationary, and the molten metal is ladled out of the furnace (Figure 7.18b), or a *tilting pot furnace*, where the entire assembly can be tilted after melting for pouring in a ladle (Figure 7.18c).

7.6 CLEANING AND FINISHING OF CASTINGS

After the casting is cooled down in expendable molds, the first step is to release it from the mold and remove the cores. For green and dry sand mold castings, this is usually done by shaking. Most of the cores fall off of the casting easily, as the binding material deteriorates due to the high temperature during casting. Otherwise, they are removed by shaking or by chemical dissolving of the bonding agent. Solid cores (in die casting) are removed by hammering or pressing. Residual sand is removed by shot blasting. The gating system (sprues, runners, and risers) as well as the flash that may form due to flow of the molten metal into the gaps between the mold halves must be trimmed by sawing, shearing, or abrasive wheel cutting. In the case of brittle casting alloys, and when the casting is relatively small, these extra parts can be easily broken off. The complete surface of the casting may be further cleaned by different operations including shot blasting, tumbling in a medium of refractory material, wire brushing, or chemical pickling. This step is usually not required in permanent mold casting.

The finished casting may be subjected to annealing to eliminate residual stresses and improve mechanical properties. Some surfaces may functionally require extra machining or grinding (machining allowance). Figure 7.19 shows a flow diagram for the steps of the sand casting process until a sound product is obtained.

Hot isostatic pressing (HIP) could also be used to eliminate porosity and improve some properties (ductility, stress rupture, and fatigue resistance) without substantially changing the shape or dimensions of castings. This is carried out by placing castings into a specially designed pressure vessel, loaded by an inert gas such as argon, at a temperature up to 1250°C, and pressurized at up to 100 MPa. This process is typically applied for special castings made of superalloys, and the titanium alloys used in jet engines.

7.7 QUALITY OF CASTINGS

Quality control and inspection at all production steps are of great significance to ensure sound castings. There are several chances of defects in castings which affect their quality. Some of the most frequently encountered defects as well as their causes are summarized in Table 7.2.

7.7.1 METHODS OF INSPECTION OF DEFECTS

    1. Visual inspection is the most widely used method for detecting obvious casting defects.
    2. Dimensional measurements are also made to ensure conformance of the casting to tolerances.
    3. Liquid penetrant inspection could be carried out to detect some surface defects such as pin holes or cracks.
    4. Surface and subsurface defects can be investigated using magnetic particle inspection.
    5. Internal soundness of high quality castings is examined by expensive ultrasonography, radiography, or eddy current inspection techniques (aircraft parts, crankshafts, and connecting rods for the automotive industry).

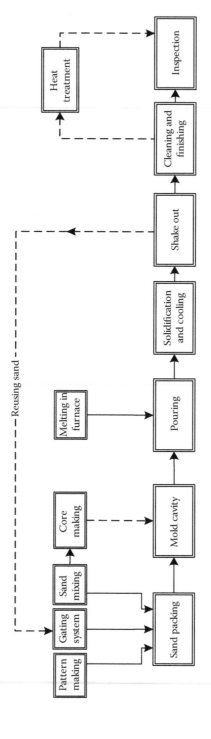

FIGURE 7.19 Flow diagram for the steps of sand casting.

TABLE 7.2
Common Defects in Castings and Their Causes

	Defect	Definition	Typical Causes	Shape
1	Misrun (short run)	Solidification before complete filling of the mold cavity	1. Insufficient fluidity 2. Pouring temperature too low 3. Too slow pouring 4. Section too thin	
2	Cold shuts (cold laps)	Lack of fusion between two streams of metal due to premature freezing	1. Insufficient fluidity 2. Pouring temperature too low 3. Too slow pouring 4. Section too thin	
3	Shrinkage cavity	Depression in the surface or internal void near the top of the casting (pipe)	Solidification shrinkage, with lack of sufficient feed from the riser	
4	Microporosity	Network of small voids distributed throughout the casting	Localized shrinkage of the final molten metal in dendritic structure (Alloys)	
5	Hot tearing (hot cracking)	External cracks having rugged edges, taking place immediately after solidification	1. Abrupt change of section 2. Sharp edged corners 3. High tensile stresses	
6	Sand wash in sand casting only	Irregularity in the surface due to sand erosion during pouring	1. Poor ramming of the mold sand 2. High flow rate of molten metal (turbulence)	
7	Mold shift	A step in the casting at the parting line	Shift between the two mold halves during assembly (mismatch)	
8	Core shift	Displacement of the core in the vertical position	Buoyancy of the molten metal	

(continued)

TABLE 7.2 (Continued)
Common Defects in Castings and Their Causes

	Defect	Definition	Typical Causes	Shape
9	Pinholes and blow holes (in sand casting only)	A cluster of small gas cavities formed at (pin holes) or slightly below (blow holes) the casting surface	1. Release of gases during pouring 2. Lack of permeability of the mold sand 3. Excessive moisture of the sand mold	Pinholes / Mold
10	Scabs (sand skin)— (in sand casting only)	Rough areas on the casting surface	Sand flakes separated and imbedded in the metal surface	Scab
11	Fins	An unintended thin projection of metal at the parting plane	1. Incorrect assembly of mold halves 2. Insufficient clamping force	Pins
12	Penetration (in sand casting only)	Liquid metal penetrates into sand mold or sand core, forming a mixture	1. High fluidity of the metal 2. Mold not hard enough	Penetration
13	Warpage	Unintentional and undesirable deformation during or after solidification	Different rates of solidification in different sections	Warped section
14	Swell	Localized or overall enlargement of the casting	Enlargement of the mold cavity by the pressure of molten metal	Swell

6. Metallurgical study of the crystal structure and chemical analysis for the material constituents could be necessary.
7. Mechanical testing of properties such as hardness, tensile strength, and fatigue can also be required to give a complete assessment of quality.

7.8 MODELING OF CASTING

A successful casting process depends on proper design of the gating system to ensure smooth fluid flow of the molten metal to fill all of the mold cavity, to avoid premature cooling, turbulence, and gas entrapment, as well as to control the solidification process to produce a sound casting. Precautions must be taken to avoid oxidation of the molten metal's surface, and impurities and gases in the mold cavity.

7.8.1 FLUIDITY

During mold filling, the heat content of the molten metal is reduced, and solidification begins while flow is taking place. Therefore, the temperature of the molten metal should be higher than the melting point to allow for the heat loss. The difference between the pouring temperature T_p and the melting point of the metal T_m is known as the *superheat*. Increasing superheat leads to larger grain size of the material, whereas lower superheat leads to premature solidification.

In order to characterize materials under complex mold filling conditions, technological tests were developed to provide quantitative comparison measures of the flowability of different metallic materials. These are known as fluidity tests. Therefore, fluidity is the mold-filling ability of a metal. The most famous fluidity tests are based on how far the molten metal spreads in a spiral path or in a longitudinal path, as shown in Figure 7.20. The length of the filled spiral or thin plate is known as fluidity index. The gating system for these tests is identical to that used in the casting mold. Fluidity increases with increasing superheat, which lowers viscosity and delays solidification. Increasing the mold temperature increases fluidity, but slows down the solidification process and the rate of production. Fluidity depends also on the freezing range of solid solutions, being lower with longer ranges. Further, fluidity is affected by the surface tension of the molten metal, as well as the mold material, size and geometrical configurations of the sprue.

7.8.2 FLUID FLOW THROUGH THE GATING SYSTEM

The gating system of a casting mold should be designed in accordance with the principles of fluid flow. Flow through the system should be laminar, since turbulent flow leads to erosion of the mold and entrapment of slag, mold material, and gases. Positive pressure should be maintained during filling so that no gas is sucked in anywhere. The pouring cup or basin should accommodate the stream of metal, and is often shaped to ensure smooth flow of the melt. It may include a skimmer to hold back inclusions and oxides. As the fluid goes down the sprue, its potential energy is converted to kinetic energy, leading to increasing velocity. To keep a constant mass flow and maintain a positive pressure differential, the area of the sprue should be decreased downward, i.e., the sprue should be tapered. Therefore, the fluid flow along the gating system is governed by two relations, namely, the continuity equation and Bernoulli's equation.

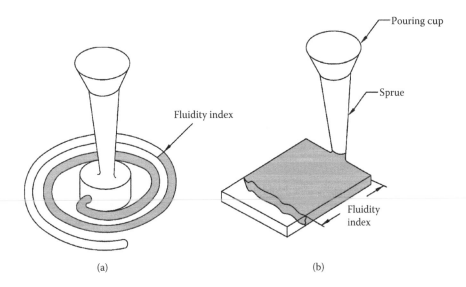

(a) (b)

FIGURE 7.20 Fluidity tests: (a) spiral mold and (b) thin plate mold.

7.8.2.1 Continuity Equation

For incompressible flow, the flow rate in any part of the system should be kept constant

$$Q = A_0 v_0 = A_1 v_1 \tag{7.1}$$

where Q is the volume passing through any cross-section per unit time, A is the cross-sectional area of a section, and v is the velocity at that point.

7.8.2.2 Bernoulli's Equation

Under steady flow conditions, the total energy of a unit volume of the fluid must be kept constant at every part of the system. The total energy has four components: Pressure energy due to the pressure p, kinetic energy due to velocity v, potential energy due to height h above a reference plane, and friction energy loss f of the melt against the system walls. The equation is written in the form

$$p_0 + \frac{\rho v_0^2}{2} + \rho g h_0 = p_1 + \frac{\rho v_1^2}{2} + \rho g h_1 + f \tag{7.2}$$

where ρ is the density of the fluid, and g is the gravitational acceleration ($g = 9.8$ m/s^2). The scripts 1 and 2 represent two different elevations. This relation can be used to calculate the velocity of the molten metal at the base of the sprue (lowest point) as follows.

Assuming that point 0 represents the top of the sprue, then the initial velocity of metal is zero ($v_0 = 0$), and there is no head ($h_0 = 0$). Similarly, the base of the sprue is represented by the point 1, then the head is the length of the sprue ($h = h_1$), and the metal velocity is $v = v_1$. Ignoring frictional losses ($f = 0$), and assuming that the system remains at atmospheric pressure throughout ($p_0 = p_1$), then substitution in Eq. 7.2 gives

$$v = \sqrt{2gh} \tag{7.3}$$

7.8.3 Mold Filling Time

The runner that transmits the metal from the sprue to the mold cavity is most commonly positioned horizontally on the separation line between the mold halves (Figure 7.2). Therefore, the head h is the same as the sprue base, and the rate of metal flow through the runner to the mold cavity remains equal to vA at the base (Eq. 7.1). Then the time required to fill the mold cavity of volume V can be calculated as

$$T_{mf} = V/Q = V/vA \tag{7.4}$$

where T_{mf} is the mold filling time, and V is the volume of the mold cavity. It should be noted that this calculated time is actually the minimum time, since it overlooks frictional losses and possible constriction of flow in the runner or the gate.

7.8.4 Solidification Time

As soon as the molten metal enters the mold, a thin solidified skin begins to form at the cold walls of the mold, and grows progressively inward. The total solidification time is the time required for the casting to solidify after pouring. This time t_s is directly proportional to the volume of the mold cavity V which governs the heat content, and inversely proportional to the surface area of the cavity through which heat is extracted. The solidification time can be calculated according to *Chvorinov's rule* in the form

$$t_s = C(V/A)^2 \tag{7.5}$$

where C is a constant that depends on the mold material as well as molten metal's thermal properties and temperature. The rule indicates that a casting with a higher volume-to-surface area ratio needs more time to solidify than one with a lower ratio. This principle is applied in the design of the riser. To perform its function of feeding molten metal to the mold cavity to compensate for shrinkage, the metal in the riser should remain in the liquid phase longer than the casting, i.e., t_s for the riser must be longer than t_s for the casting. This can be achieved by designing the riser to have a larger volume-to-surface area ratio. A blind riser losses less heat than an open riser, since the latter is open to the atmosphere, and thus subject to more heat extraction.

7.9 REVIEW QUESTIONS

1. Define the terms *casting* and *foundry*, and what are the advantages and limitations of the process?
2. What are the allowances that should be considered when designing a pattern for a sand mold?
3. Why are cores used in sand casting, and what are the factors to be considered in designing these cores?
4. What are the types of sand used in casting, and define the quality indicators for the sand characteristics.
5. What are the types of bonding materials which are added to mold sand?
6. Discuss the different techniques for sand molding, emphasizing the advantages and limitations of each technique.
7. Compare the shell molding and vacuum molding casting processes.
8. What are the limitations of plaster molding as a casting process?
9. Compare investment casting and evaporative casting as expendable pattern casting processes.
10. What are the objectives of permanent mold casting processes and what are their classes?
11. What are the differences between gravity die casting and low pressure casting processes?
12. What is the difference between the hot chamber and cold chamber machines in pressure die casting?
13. Compare vertical and horizontal true casting processes.
14. For small, intricately shaped cast products, differentiate between investment casting and centrifuging.
15. What are the types of furnaces used in foundries, and what are the main features of each?
16. Describe the procedure to be followed to produce molten cast iron in a cupola.
17. What are the advantages of a cokeless cupola compared to the conventional cupola?
18. Discuss the stages of cleaning and finishing of casting, emphasizing the role of heat treatment in controlling the characteristics of a cast product.
19. What are the parameters affecting the fluidity of the molten metal during pouring, and how are these parameters controlled?

7.10 PROBLEMS

1. The requirement is to produce a flat CI slab 250 × 200 × 100 mm by sand casting. The sprue is made 400 mm high with a conical shape (as shown in Figure 7.2) with a diameter a 40 mm at mid height. Calculate the following:
   a. The sprue diameter and the velocity of metal flow at the parting line.
   b. The rate of metal flow.
   c. The time for filling the mold cavity.
2. If you have three cylindrical castings of identical volume, with height/diameter ratios of 1, 2, and 4, respectively, calculate which solidifies faster, and comment on the results.

BIBLIOGRAPHY

ASM International. 1988. *ASM Handbook, vol. 15: Casting*, Materials Park, OH.

Brown, J. R. 2000. *Foseco, Ferrous Foundryman's Handbook.* Foseco International, Butterworth-Heinemann.

Brown, J. R. 2000. *Foseco Nonferrous Foundryman's Handbook.* Foseco International, Butterworth-Heinemann.

Campbell, J. 2003. *Castings*, Butterworth-Heinemann.

Choudhry, S. K. 1985. *Elements of Workshop Technology*, 2nd ed. Bombay, India: Media Promoters Publishers (MPP).

Genick, B.-M. 2009. *Fundamentals of Die Casting Design*, www.potto.org/DC/dieCasting.pdf.

Groover, M. P. 2007. *Fundamentals of Modern Manufacturing*, 3rd ed. London: John Wiley & Sons.

Nowosielski, R., and Babilas, R. 2007. Fabrication of bulk metallic glasses by centrifugal casting method. *Journal of Achievements in Materials and Manufacturing Engineering*, 20(1–2).

Schey, J. A. 1987. *Introduction to Manufacturing Processes*, 2nd ed. New York: McGraw-Hill.

U.S. DOE. 2005. Advanced Melting Technology: Energy Saving Concepts and Opportunities in the Metal Casting Industry. BCS Incorporated, U.S. Department of Energy.

8 Fundamentals of Metal Forming

8.1 INTRODUCTION

Metal forming processes involve applying mechanical force through a tooling system to a workpiece to change its shape permanently (deform it plastically) to the required shape which is predetermined by the die geometry. Additionally, these processes lead to higher mechanical properties of the product through control of the crystal structure of the material to be formed. Therefore, comprehensive understanding the fundamentals of metal forming processes should comprise both the mechanical aspects and metallurgical aspects. Mechanical aspects include the analyses of forces, stresses, strains, strain rates, plastic deformation, work hardening, and limiting stress and strain conditions for the flow of metal during the process. Metallurgical aspects, on the other hand, include the effect of plastic deformation under different temperatures on the crystal structure of the material, and accordingly, on the resulting mechanical properties. Details of both aspects are presented in this chapter.

8.2 SIMPLE STRESSES AND STRAINS

Stresses and strains are essential in studying forming processes, as they describe the nature and limits of metal flow during the process. They are also used to design the tooling required to perform deformation, and to select the capacity of the necessary equipment. To achieve this objective, one has to appreciate first the principles of stresses under simple loading conditions before moving to the general three-dimensional conditions which may arise in forming processes.

8.2.1 ELASTIC STRESSES AND STRAINS

The forces applied in forming processes may be tensile, compressive, or shear forces. These forces produce stresses in the material, which could be tensile, compressive, or shear stresses, as shown in Figure 8.1. Tensile and compressive stresses are perpendicular to the plane on which they act, and thus they are called normal stresses. They are commonly given the notation σ, with a positive sign for tension and negative sign for compressive stress. Shear stresses, on the other hand, act parallel or tangential to the area (tangential stress), as shown in the figure. They are given the notation τ and are considered positive when they act in the clockwise direction or negative when they act in the counterclockwise direction, as shown in the figure. Stresses can be presented by the following relations:

$$\sigma = F/A_o \text{ MPa} \quad \text{and} \quad \tau = F/A_s \text{ MPa} \tag{8.1}$$

where F is the applied force, A_o is the initial area on which the stresses act, and the subscript s refers to shear. The unit MPa or MN/m^2 represents the stress values commonly developed in forming processes.

These stresses cause deformation of the material, which is expressed in terms of strain (the change in length per unit length). Strains are similarly either longitudinal strains (tensile or compressive), which are expressed by ($\varepsilon = \Delta L/L_o$), or shear strains, expressed by the angle (γ), as indicated in Figure 8.1.

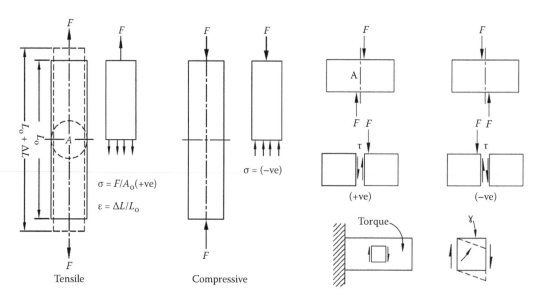

FIGURE 8.1   Simple normal and shear stresses and strains.

Detailed simple stress–strain relations are usually represented by the engineering stress–strain curve drawn from the standard tension test of a mild steel specimen, as shown in Figure 8.2. The first part of the curve shows an elastic linear behavior of the material. The term *elastic* means that the material returns to its original dimensions when the stress is removed. This linear relation is represented by the well-known Hook's law

$$\sigma = E\varepsilon \qquad\qquad (8.2)$$

where E is the modulus of elasticity of the material. This modulus is a measure of the material's stiffness, i.e., its resistance to elastic deformation under loading. This linear relation is valid up to a stress limit known as the proportional limit, or approximately up to the yield strength (σ_y).

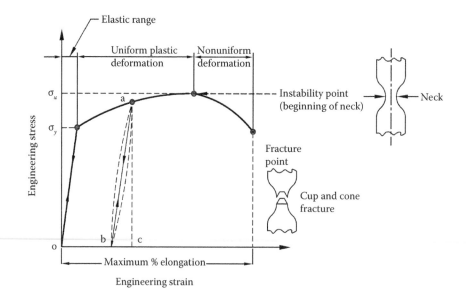

FIGURE 8.2   Simple tensile stress–strain curve for mild steel.

8.2.2 Plastic Stresses and Strains

Beyond the yield stress point, the material behaves plastically, or exhibits permanent, nonrecoverable deformation. In this range of the stress–strain relation, there is rapid increase of plastic strain without correspondingly high increase of the stress, as shown in Figure 8.2. The plastic strain remains uniform along the gage length of the tensile specimen up to the highest loading point representing the ultimate strength (σ_u). Beyond this point the stresses are concentrated in a limited length, causing severe reduction in the neck area up to the fracture point. The slope of the stress–strain curve in the uniform plastic deformation range represents the strain hardening (work hardening) of the material, which is a measure of the increase of the strength and hardness of the material due to plastic deformation. This strengthening occurs because of the movement of dislocations along the slip planes within the crystal structure. Therefore, the main parameters required from the stress–strain curve for plastic forming are the yield stress and ultimate tensile stress as the initial and final limiting stresses, as well as the slope of the curve within these limits to determine the strengthening of the material during forming.

If the load is removed in the plastic range at a certain stress level a, the unloading follows a line ab, parallel to the elastic line, resulting in a permanent plastic strain ob in the specimen. The strain cb represents the recoverable elastic strain (elastic recovery) when the load is removed. A second load cycle commencing with the permanent initial elongation ob in the specimen would follow the line ba, and continue along the original curve up to the fracture point. Thus, the repeated load cycle is associated with increase of the yield stress from σ_y to σ_a, without changing σ_u. Practically, the unloading from a to b and reloading from b to a will not follow precisely straight lines, but will form a small hysteresis loop, as shown in Figure 8.2.

The capacity of the material to allow large tensile plastic strain is the ductility, which is commonly measured by the maximum percent elongation strain to fracture. Another measure is the percentage reduction in area (of the necked portion). Materials with limited or no ductility are brittle materials that do not exhibit plastic deformation. A property closely related to ductility is malleability, which defines the ability of the material to be permanently deformed under compression, allowing lateral extension in lateral directions. A typical example of a malleable material is lead.

The capability of different metallic materials for plastic deformation depends to a large extent on the type of the crystal structure (Chapter 4). Metals having FCC structure possess four closely packed planes in their unit cell, with three slip directions in each plane, giving a total of 12 slip systems (combinations of slip planes and directions). Metals such as Pb, Al, Cu, Ni, Ag, and γ-Fe are highly deformable at essentially all temperatures. In metals with BCC structure, there are 12 slip systems, but packing in closely packed planes is not as close as in FCC metals. These metals such as α-Fe, W, and Mo, possess high strength and moderate ductility. In HCP metals such as Zn, Mg, Ti, and Be, the number of slip systems is limited to three, and therefore they are difficult to deform (they are rather brittle).

8.2.2.1 Poisson's Ratio

In the tensile specimen shown in Figure 8.1, the longitudinal strain ε was identified as the increase in length per unit length. The specimen will also exhibit reduction in the cross-sectional area, or lateral contraction as a result of elongation. If the specimen is cylindrical with initial diameter D_o, and the change in diameter is ΔD, then the lateral strain is ($\varepsilon_{lat} = -\Delta D/D_o$). The negative sign indicates that the lateral strain is opposite in sense to the longitudinal strain. The ratio of the lateral and longitudinal strains will always be constant, and is known as the Poisson's ratio v, where,

$$\text{Poisson's ratio } (v) = \frac{\text{lateral strain}}{\text{longitudinal strain}} = \frac{\Delta L/L_o}{\Delta D/D_o} \qquad (8.3)$$

The Poisson's ratio is constant for each material in the elastic range of stresses, and differs from one material to another, ranging from 0.25 to 0.33 for most engineering materials. The maximum limit of v in the elastic range should be less than 0.5. In the plastic range of stresses, the Poisson's ratio is the same for all ductile metals, and is equal to 0.5 due to volume constancy.

Substituting the longitudinal strain in Equation 8.3, in terms of stress from Equation 8.2, the lateral strain can be expressed as,

$$\text{Lateral strain } \varepsilon_{lat} = -v\frac{\sigma}{E} \tag{8.4}$$

8.2.3 TRUE STRESS AND TRUE STRAIN

The engineering stress is based on dividing the applied force on the initial cross-sectional area (A_o), as stated in the previous sections. However, it is due to the large reductions in the area supporting the load in the plastic range as the specimen elongates, then the instantaneous area should be used to represent the true stress in the plastic deformation range. Then, the true stress is expressed as,

$$\text{True stress } \sigma = P/A_i = P/A = \frac{P}{A_o}\frac{A_o}{A} = \sigma_{eng}\frac{l}{l_o} = \sigma_{eng}(1 + \varepsilon_{eng}) \tag{8.5}$$

where A_i is the instantaneous area, which is commonly referred to as A without suffix.

Similarly, due to the large elongations in the plastic range, the true strain is the sum of increments of strains, and thus if the length increases from L_0 to the instantaneous length L_i through increments $L_1, L_2, \ldots$, then the true strain is calculated as

$$\varepsilon = \frac{L_0 - L_1}{L_0} + \frac{L_1 - L_2}{L_1} + \ldots = \sum_{i=0}^{i=i}\frac{\Delta L}{L_0} = \ln\frac{L_i}{L_0} \tag{8.6}$$

Then,

$$\varepsilon_{true} = \ln\frac{L_0 + \Delta L}{L_0} = \ln\left(1 + \varepsilon_{eng}\right) \tag{8.7}$$

Being expressed in terms of the natural logarithm of the instantaneous length to the initial length ratio, the true strain is commonly known as the natural strain, or logarithmic strain. The true stresses and true strains are almost the same as the engineering stresses and engineering strains in the elastic range, but deviations increase with the increase of the plastic strain, as shown in Figure 8.3a. Therefore, in plastic deformation calculations, true stresses and true strains should be used.

Example 1

A bar of length L_o is elongated to a final length of 1.5 L_o, and then compressed again to its original length L_o. Calculate the engineering and true strains in both cases, and comment on the results. When the bar is elongated

$$\varepsilon_{eng} = \frac{1.5L_0 - L_o}{L_o} = 0.5$$

$$\varepsilon_{true} = \ln\frac{1.5L_o}{L_o} = 0.41$$

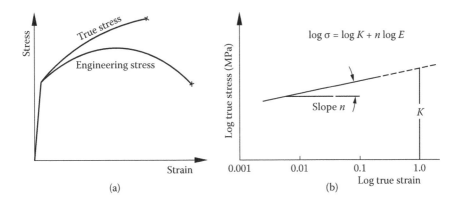

FIGURE 8.3 True stress–strain curve as plotted on the (a) engineering stress–strain curve and (b) as plotted on a log–log scale to determine K and n.

When the bar is compressed from $1.5\,L_0$ to L_0

$$\varepsilon_{eng} = \frac{L_0 - 1.5L_0}{1.5L_0} = -0.33$$

$$\varepsilon_{true} = \ln \frac{L_0}{1.5L_0} = -0.41$$

The results show that the engineering strain magnitude in compression is different from that in tension despite the fact that the physical strain in both cases should be the same with opposite signs. The results of true strain in both cases give the same magnitude with opposite signs, which validates the physical conditions. True strains express real deformation under large strain. The true strain is less in magnitude than engineering strain under elongation, and the deviation error is 0.22%. This error increases with the increase of the strain.

8.2.4 Empirical Relations for the Stress–Strain Curve

The true stress–strain curve shown in Figure 8.3a is thus more rational for metal forming applications. The elastic part of the curve is represented by Hook's law (Equation 8.2), while the plastic part is commonly represented by the empirical power equation

$$\sigma = K\,\varepsilon^n \tag{8.8a}$$

where K is known as the strength coefficient and n is the strain hardening exponent. These are material parameters (constants) that depend on the temperature, strain rate, microstructure, and heat treatment. They should be determined for a specimen that contains no effect of previous cold work before the tensile test (usually annealed specimens should be used in the tension test). The power law is reliable for plastic strains which are appreciably larger than elastic strains. A generalized power law, suggested by Swift, to account for the prestrain or previous cold work (ε_0) that may have been induced in the material before the plastic tensile strain (ε^p) during the test is given as

$$\sigma = K(\varepsilon_0 + \varepsilon^p)^n \tag{8.8b}$$

Typical values of K and n for different materials are given in Table 8.1.

TABLE 8.1

Typical Values of K and n of the Power Equation at Room Temperature

Material	K (MPa)	n
Pure lead	25	0.0
Aluminum-1100	180	0.2
Aluminum-2024	690	0.16
Copper, annealed	315	0.54
Brass (70/30), annealed	895	0.49
Phosphor bronze, annealed	720	0.46
Low-carbon steel, annealed	530	0.26
Low-carbon steel, cold rolled	760	0.08
Medium-carbon steel, cold drawn	640	0.15
Stainless steel 410, annealed	960	0.10
Stainless steel 304, annealed	1275	0.45

It should be noticed in the table that n has the minimum limit (zero) for pure lead because it is a perfectly plastic (nonstrain-hardening) material. Lead is deformed under hot working conditions at room temperature. The upper limit of n is 1 where the material is elastic. Practical limits of n for most metallic materials range from 0.1 to 0.5.

Plotting the plastic part of the true stress–strain curve, which is represented by the power Equation 8.8 on a log–log graph, is represented by a straight line, as shown in Figure 8.3b. The slope of the line determines the strain hardening exponent n, and the stress at a true strain equal to 1 is the strength coefficient K.

It has been found in the simple tension test that the maximum load is reached at the ultimate stress σ_u beyond which necking starts, leading to nonuniform deformation. The true strain at the onset of necking is found to be equal to the strain hardening exponent n. Therefore, materials of higher n exhibit more uniform strain before necking, and thus they should be selected for forming processes involving stretching. The true stress at the maximum load can be expressed as

$$\text{True } \sigma_u = K\, n^n \tag{8.9}$$

Since the power relation can be extended to large plastic strains beyond those that could be obtained in a simple tensile stress–strain curve, the power curve is better applied in the analysis of plastic forming processes. The curve is then called the flow curve, and any point on it represents the instantaneous yield stress (usually called the flow stress (σ_f), corresponding to any instantaneous plastic strain.

Example 2

During a tensile test with a metal that obeys the power curve, the tensile strength is found to be 340 MPa. Reaching the maximum load required an elongation of 30%. Calculate K and n for this metal.

From Equation 8.7: $\varepsilon_{true} = \ln(1 + \varepsilon_{eng})$

Then at elongation 30%: $\varepsilon_{true} = n = \ln(1 + 0.3) = \ln 1.3 = 0.26$

The ultimate strength is usually given in terms of engineering stress (340 MPa).

To calculate the true ultimate stress, Equation 8.5 is applied

$$\sigma_{true} = \sigma_{eng}(1 + \varepsilon_{eng}) = 340(1 + 0.3) = 442 \text{ MPa}$$

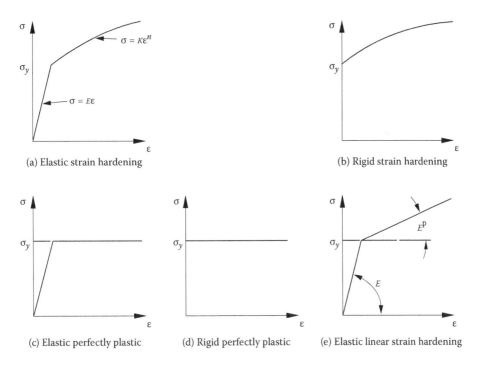

(a) Elastic strain hardening

(b) Rigid strain hardening

(c) Elastic perfectly plastic   (d) Rigid perfectly plastic   (e) Elastic linear strain hardening

FIGURE 8.4 Idealized stress–strain curves. (From Ragab, A., and Bayoumi, S., *Engineering Solid Mechanics*, CRC Press, Boca Raton, FL, 1999. With permission.)

From Equation 8.9: $(\sigma_u)_{\text{true}} = K\, n^n$

$$442 = K\,(0.26)^{0.26}$$

$$K = 627.4 \text{ MPa}$$

8.2.4.1 Idealized Stress–Strain Curves

Adopting the power relation to describe the plastic range in the true stress–strain curve modifies the simple tensile stress–strain curve to an idealized curve for an elastic–strain hardening material, as shown in Figure 8.4a. The elastic and plastic parts of the curve are represented by Equations 8.2 and 8.8a, respectively. Often, further simplifications are introduced to facilitate the analytical analysis of forming processes. For instance, in forming processes involving large plastic deformation, the elastic part could be completely neglected, leading to the rigid–strain hardening curve shown in Figure 8.4b. For materials exhibiting no or slight strain hardening, such as in hot forming processes, the behavior may be represented by the elastic–perfectly plastic model shown in Figure 8.4c ($n = 0$). Again, if the elastic range is neglected due to large plastic strains, the behavior turns to rigid–perfectly plastic, as shown in Figure 8.4d. Another significant idealization that has proved to be useful in the analysis of problems concerning small plastic strains is the elastic–linear strain hardening model demonstrated in Figure 8.4e.

8.3 TWO- AND THREE-DIMENSIONAL STRESSES AND STRAINS

Consider the simple tensile specimen shown in Figure 8.1a and assume a plane inclined to the direction of the applied force. The stress acting on this plane should be in the direction of the applied force as shown in Figure 8.5, to keep equilibrium of the specimen. Assuming the axis of

FIGURE 8.5   Stresses on an inclined plane in a simple tensile loaded bar.

the specimen to be x-axis, this stress should be ($\sigma_x = F/A$), as described earlier. This stress can be resolved into a normal component σ perpendicular to the inclined plane, and a tangential or shear component τ. Again, the shear stress can be resolved to any two perpendicular components in the plane, say τ_a and τ_b, as indicated on the side view of Figure 8.5. Therefore, the stresses on any plane in a loaded body can be resolved to a normal stress component perpendicular to the plane, and two perpendicular shear components in the plane.

Generally, in order to study the stresses at any point O in a three-dimensional body under general loading conditions, as shown in Figure 8.6a, an infinitesimal cube is assumed surrounding this point in the directions of a Cartesian coordinate system xyz. The stresses on each surface of this cube are resolved into one normal component and two shear components as stated earlier, leading to the three-dimensional stress system demonstrated in Figure 8.6b. Thus, the total stress components on this infinitesimal cube are three normal components σ_x, σ_y, σ_z perpendicular to the planes x, y, and z respectively, and six shear components τ_{xy}, τ_{yx}, τ_{yz}, τ_{zy}, τ_{xz}, τ_{zx} tangential to these planes in the specified directions. These stress components can be presented in a stress tensor form:

$$\sigma = \begin{vmatrix} \sigma_x & \tau_{xy} & \tau_{xz} \\ \tau_{yx} & \sigma_y & \tau_{yz} \\ \tau_{zx} & \tau_{zy} & \sigma_z \end{vmatrix} \tag{8.10}$$

Considering that the sum of moments of these components about each of the coordinate axes x, y, and z should be zero, the shear component $\tau_{xy} = \tau_{yx}$, $\tau_{yz} = \tau_{zy}$, $\tau_{xz} = \tau_{zx}$, which reduces the total stress components to six; only three normal components and three shear components. If the stresses

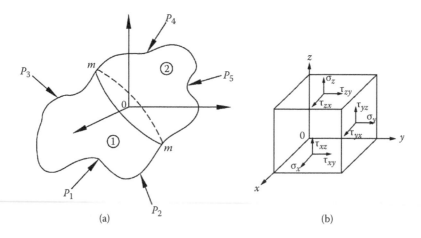

(a)   (b)

FIGURE 8.6   (a) Three-dimensional body in equilibrium under general loading conditions. (b) Stress components on an infinitesimal cube in the Cartesian directions xyz at point O in the body. (Adapted from Chakrabarty, J., *Theory of Plasticity*, Elsevier, 2006.)

are reduced to two dimensions xy, the stress components are reduced to only three; two normal components σ_x and σ_y and one shear component τ_{xy}.

Three-dimensional stresses on a body generally develop three-dimensional strains. Similar to the stress components, the developed strain components include three normal strain components ε_x, ε_y, ε_z, and three shear strain components γ_{xy}, γ_{yz}, γ_{xz}. These strain components can also be arranged in a strain tensor:

$$e \equiv \begin{vmatrix} e_x & \dfrac{1}{2}\gamma_{xy} & \dfrac{1}{2}\gamma_{xz} \\[2ex] \dfrac{1}{2}\gamma_{yx} & e_y & \dfrac{1}{2}\gamma_{yz} \\[2ex] \dfrac{1}{2}\gamma_{zx} & \dfrac{1}{2}\gamma_{zy} & e_z \end{vmatrix} \tag{8.11}$$

8.3.1 PRINCIPAL STRESSES AND STRAINS

If the infinitesimal cube is rotated to different coordinate directions x_1, y_1, and z_1, the values of the stress components will change. It is possible to find a set of axes along which the shear stress components are zero. In this case the normal stresses are called the principal stresses σ_1, σ_2, and σ_3, and the planes on which they act are called the principal planes, as shown in Figure 8.7. The magnitudes of the principal stresses are the roots of a third-order equation of the stress

$$\sigma^3 - I_1\sigma^2 - I_2\sigma - I_3 = 0 \tag{8.12}$$

where

$$\begin{aligned} I_1 &= \sigma_x + \sigma_y + \sigma_z \\[1ex] I_2 &= \sigma_x\sigma_y + \sigma_y\sigma_z + \sigma_z\sigma_x - \tau_{xy}^2 - \tau_{yz}^2 - \tau_{zx}^2 \\[1ex] I_3 &= \sigma_x\sigma_y\sigma_z + 2\tau_{xy}\tau_{yz}\tau_{zx} - \sigma_x\tau_{yz}^2 - \sigma_y\tau_{xz}^2 - \sigma_z\tau_{xy}^2 \end{aligned} \tag{8.13}$$

I_1, I_2, and I_3 are called the invariants of the stress tensor and are independent of the orientation of the stress axes. The roots are arranged such that ($\sigma_1 > \sigma_2 > \sigma_3$). The shear stresses have maximum

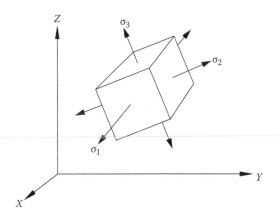

FIGURE 8.7   Principal stresses in three dimensions.

values on planes tilting 45° on the principal planes. There are three pairs of these maximum shear stress planes, as shown in Figure 8.8, and the maximum shear stresses are

$$\tau_{12} = \frac{\sigma_1 - \sigma_2}{2}, \quad \tau_{23} = \frac{\sigma_2 - \sigma_3}{2}, \quad \tau_{13} = \tau_{max} = \frac{\sigma_1 - \sigma_3}{2} \tag{8.14}$$

Derivations of Equations 8.12, 8.13, and 8.14 are beyond the scope of this book, but it is essential to know the principal stresses and maximum shear stresses in forming processes.

When $I_3 = 0$ in Equation 8.12, Equation 8.12 reduces to a second-order equation, and stresses are reduced to two dimensions xy. The roots of the equation are the principal stresses σ_1 and σ_2, and the maximum shear stress is τ_{max} as given in the following equations

$$\sigma^2 - I_1\sigma + I_2 = 0 \tag{8.15}$$

$$\sigma_{1,2} = \frac{\sigma_x + \sigma_y}{2} \pm \sqrt{\left(\frac{\sigma_x - \sigma_y}{2}\right)^2 + \tau_{xy}^2} \tag{8.16}$$

and

$$\tau_{max} = \sqrt{\left(\frac{\sigma_x - \sigma_y}{2}\right)^2 + \tau_{xy}^2} \tag{8.17}$$

Figure 8.9 illustrates a two-dimensional stress system (σ_x, σ_y, and τ_{xy}) and the equivalent principal stresses σ_1, σ_2, and τ_{max}.

8.3.1.1 Mean (Hydrostatic) Stress and Stress Deviators

In a three-dimensional stress system, the mean stress is given by the equation

$$\sigma_m = \frac{\sigma_1 + \sigma_2 + \sigma_3}{3} = \frac{\sigma_x + \sigma_y + \sigma_z}{3} = \frac{I_1}{3} \tag{8.18}$$

This stress is called the hydrostatic stress as it represents the pressure in a fluid system. The hydrostatic stress is responsible for change of volume of the material in the elastic range of stresses. Under

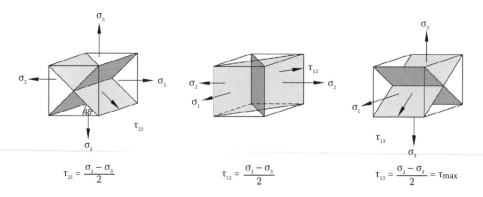

FIGURE 8.8 Planes of maximum shear stresses in three dimensions. (From Dieter, G. E., *Mechanical Metallurgy*, McGraw-Hill, 1988. With permission.)

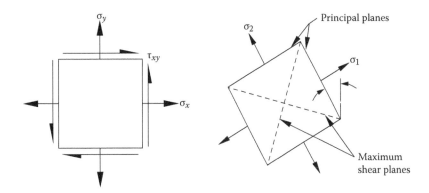

FIGURE 8.9 Two-dimensional stresses in the xy directions and in the principal directions. (Adapted from Dieter, G. E., *Mechanical Metallurgy*, McGraw-Hill, 1988.)

plastic conditions, the volume of the material is constant and the hydrostatic component of the stress has no effect on deformation. The differences between the stress components and the mean stress are known as the stress deviators, i.e.,

$$\sigma'_x = \sigma_x - \sigma_m, \quad \sigma'_y = \sigma_y - \sigma_m, \quad \sigma'_z = \sigma_z - \sigma_m$$

or

$$\sigma'_1 = \sigma_1 - \sigma_m, \quad \sigma'_2 = \sigma_2 - \sigma_m, \quad \sigma'_3 = \sigma_3 - \sigma_m \qquad (8.19)$$

The stress deviator components along with the shear components are responsible for the change of shape of the material under loading, while the mean or hydrostatic stress is responsible for change in volume in the elastic range only, as stated earlier.

8.3.1.2 Principal Strains

The principal strains $\varepsilon_1 > \varepsilon_2 > \varepsilon_3$ are calculated by solving a third-order equation similar to that given in Equation 8.12, after replacing σ by ε, and τ by $\frac{1}{2}\gamma$. Maximum shear strains are also found on planes tilting 45° on the principal planes as

$$\frac{1}{2}\gamma_{12} = \frac{\varepsilon_1 - \varepsilon_2}{2} \quad \text{or} \quad \gamma_{12} = \varepsilon_1 - \varepsilon_2,$$

and similarly,

$$\gamma_{23} = \varepsilon_2 - \varepsilon_3 \quad \text{and} \quad \gamma_{13} = \gamma_{max} = \varepsilon_1 - \varepsilon_3 \qquad (8.20)$$

Example 3

A machine shaft is subjected to a bending stress 30 MPa and a shear stress 40 MPa. Calculate the principal stresses and maximum shear stress on the shaft. Determine also the stress responsible for volume change in the elastic range of stress.

The bending stress acts in the shaft axis direction (say x-axis), then

$$\sigma_x = 30 \text{ MPa}, \ \sigma_y = 0, \text{ and } \tau_{xy} = 40 \text{ MPa}$$

Using Equation 8.16

$$\sigma_{1,2} = \frac{\sigma_x + \sigma_y}{2} \pm \sqrt{\left(\frac{\sigma_x - \sigma_y}{2}\right)^2 + \tau_{xy}^2} \,,$$

$$\sigma_{1,2} = \frac{30+0}{2} \pm \sqrt{\left(\frac{30-0}{2}\right)^2 + (40)^2} = 15 \pm 42.72$$

$$\sigma_1 = 15 + 42.72 = 57.72 \text{ MPa}$$

$$\sigma_2 = 15 - 42.72 = -27.72 \text{ MPa}$$

To calculate the maximum shear stress, use Equation 8.14

$$\tau_{max} = \frac{\sigma_1 - \sigma_2}{2} = \frac{57.72 - (-27.72)}{2} = 42.72 \text{ MPa}$$

The stress responsible for volume change is the mean or hydrostatic stress (Equation 8.18)

$$\sigma_m = \frac{\sigma_1 + \sigma_2 + \sigma_3}{3} = \frac{57.72 + 0 - 27.72}{3} = 10 \text{ MPa}$$

8.3.2 General Stress–Strain Relations in the Elastic Range

It can be found from Equation 8.2 that the direct stress σ_x is linearly related to the direct strain in the same direction ε_x in the form $\sigma_x = E\,\varepsilon_x$. According to Equation 8.4, the same stress generates lateral strains in the directions y and z as

$$\varepsilon_y = \varepsilon_z = -v\varepsilon_x = -v\frac{\sigma_x}{E}$$

Similarly, the direct stresses σ_y and σ_z develop comparable strains in the three directions x, y, and z. Superposition of the strain components in each direction leads to the general three-dimensional elastic strain–stress relations

$$\varepsilon_x = \frac{\sigma_x}{E} - v\frac{\sigma_y}{E} - v\frac{\sigma_z}{E} \quad \text{or} \quad \varepsilon_x = \frac{1}{E}[\sigma_x - v(\sigma_y + \sigma_z)]$$

Similarly,

$$\varepsilon_y = \frac{1}{E}[\sigma_y - v(\sigma_x + \sigma_z)] \tag{8.21}$$

$$\varepsilon_z = \frac{1}{E}[\sigma_z - v(\sigma_x + \sigma_y)]$$

The shear stress–strain relations are given in the form

$$\tau_{xy} = G\gamma_{xy}, \quad \tau_{xy} = G\gamma_{xy}, \quad \tau_{xy} = G\gamma_{xy} \tag{8.22}$$

where G is known as the modulus of rigidity, which is related to the modulus of elasticity by the relation

$$G = \frac{E}{2(1+v)} \tag{8.23}$$

The stress–strain relations in the elastic range of stresses in Equations 8.20 and 8.21 involve three terms, E, G, and v, which are known as the material constants. An essential parameter in the elastic range is the volumetric strain θ, where

$$\theta = \varepsilon_x + \varepsilon_y + \varepsilon_z = (\sigma_x + \sigma_y + \sigma_z)\frac{(1-2v)}{E} = \frac{3\sigma_m(1-2v)}{E} \tag{8.24}$$

Equations 8.20, 8.21, and 8.23 can be derived in terms of the principal stresses and strains as follows:

$$\varepsilon_1 = \frac{1}{E}[\sigma_1 - v(\sigma_2 + \sigma_3)]$$

$$\varepsilon_2 = \frac{1}{E}[\sigma_2 - v(\sigma_1 + \sigma_3)] \tag{8.25}$$

$$\varepsilon_3 = \frac{1}{E}[\sigma_3 - v(\sigma_1 + \sigma_2)]$$

$$\tau_{12} = G\gamma_{12}, \quad \tau_{23} = G\gamma_{23}, \quad \tau_{13} = G\gamma_{13}$$

$$\theta = \varepsilon_1 + \varepsilon_2 + \varepsilon_3 = (\sigma_1 + \sigma_2 + \sigma_3)\frac{(1-2v)}{E} = \frac{3\sigma_m(1-2v)}{E}$$

Since v in the plastic range of strains is 0.5 as stated earlier, substituting in Equation 23 shows that the volumetric strain = zero during plastic deformation, i.e.,

$$\varepsilon_x + \varepsilon_y + \varepsilon_z = \varepsilon_1 + \varepsilon_2 + \varepsilon_3 = 0 \quad \text{in the plastic range of strains.}$$

8.3.3 SOME SPECIAL CONDITIONS OF STRESS AND STRAIN

There are certain special stress and strain conditions that often find engineering applications, especially in metal forming. Some of these conditions are

1. Plane strain: where a body is constrained against lateral deformation in a direction normal to the plane of loading. If loading is in the x–y plane, then the material is restricted in the z direction. Then $\varepsilon_z = \gamma_{xz} = \gamma_{yz} = 0$.
   Substituting in Equation 8.21

$$\varepsilon_z = \frac{1}{E}[\sigma_z - v(\sigma_x + \sigma_y)] = 0$$

Then

$$\sigma_z = \nu(\sigma_x - \sigma_y) \quad \text{or} \quad \sigma_3 = \nu(\sigma_1 - \sigma_2) \tag{8.26}$$

During plastic deformation $\nu = 0.5$, and thus the stress in the direction where there is no strain is the average of other normal stress components. This is the condition of flat plate rolling, where there is no strain in the plate width direction.

2. Plane stress: Where the stress components are in one plane, say x–y, then $\sigma_z = \tau_{zx} = \tau_{zy} = 0$. This is a two-dimensional stress condition. However, there is a strain component in the z direction. Substituting $\sigma_z = 0$ in Equation 8.21 gives

$$\varepsilon_z = \frac{-\nu}{E}(\sigma_x + \sigma_y) \tag{8.27}$$

A typical example of a plane stress condition is a thin sheet loaded in the sheet surface direction. This may lead to strain in the direction perpendicular to that surface.

Example 4

In Example 3, it has been found that there are two principal stresses only 57.72 and –27.72 (plane stress problem). If the machine shaft material is mild steel, with a modulus of elasticity 210 GPa and Poisson's ratio 0.3, calculate the principal strains, volumetric strain, and the modulus of rigidity of the shaft material.

The principal stresses are $\sigma_1 = 57.72$ MPa, $\sigma_2 = 0$ MPa, and $\sigma_3 = -27.72$ MPa.
Applying Equations 8.25

$$\varepsilon_1 = \frac{1}{E}[\sigma_1 - \nu(\sigma_2 + \sigma_3)]$$

$$\varepsilon_1 = \frac{1}{210 \times 10^3}[57.72 - 0.3(0 - 27.72)] = 3.145 \times 10^{-4}$$

Similarly,

$$\varepsilon_2 = -0.429 \times 10^{-4},$$

$$\varepsilon_3 = -2.145 \times 10^{-4}$$

$$\theta = \varepsilon_1 + \varepsilon_2 + \varepsilon_3 = (3.145 - 0.429 - 2.145) \times 10^{-4} = 0.571 \times 10^{-4}$$

From Equation 8.23

$$G = \frac{E}{2(1+\nu)} = \frac{210}{2(1+0.3)} = 80.77 \text{ GPa}$$

3. Axial symmetry: If a geometrically axisymmetric body is under axial load, the state of stress is simplified. Employing polar cylindrical coordinates (r, θ, z), where symmetry is around the z-axis (principal direction), then $\tau_{r\theta} = \tau_{z\theta}$. This condition is prevalent in many forming processes, such as extrusion of symmetrical sections, wire drawing, deep drawing, and forging of symmetrical sections.

4. Pure shear: The case of pure shear results due to torsion. The only stress component is τ_{xy}. The stresses on planes tilting 45° on the shear planes are principal stresses, with magnitude

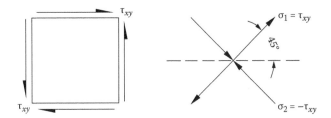

FIGURE 8.10 Pure shear stresses and associated normal stresses. (From Hearn, E. J., *Mechanics of Materials*, 3rd ed., Butterworth-Heinemann, 1997. With permission.)

$\sigma_1 = \tau_{xy}$, tensile in one direction and compressive ($\sigma_2 = -\tau_{xy}$) in the perpendicular direction, as shown in Figure 8.10.

8.4 YIELD CRITERIA

Under simple tension loading condition, the yield stress σ_y is the onset of plastic deformation in the material. However, under general three-dimensional loading conditions, a question arises as to whether or not the loading system causes yielding (plastic deformation of the part). Different criteria have been developed to recommend the onset of yielding, among which two criteria are the most applicable in metal forming, as follows.

1. Tresca yield criterion (maximum shear stress theory): Tresca proposed that initial yielding occurs when the highest of the maximum shear stresses attains a critical value determined from a simple tensile (or pure shear) test. The criterion can be written in the form

$$\sigma_1 - \sigma_3 = \sigma_f \tag{8.28}$$

where σ_1 and σ_3 are the maximum and minimum principal stresses, and σ_f is the yield or flow stress. The maximum shear stress at yielding (τ_f) can be determined from Equation 8.14, as $\tau_f = 0.5\,\sigma_f$. Note that this criterion is independent of the intermediate principal stress σ_2, which inserts a limitation to its versatility to all three-dimensional stress conditions.

2. Von Mises yield criterion (maximum distortion energy theory): Von Mises assumed that initial yielding occurs when the strain energy due to distortion attains a critical value, as determined from a simple tensile test. The criterion is commonly written in the form

$$\frac{1}{\sqrt{2}}\sqrt{(\sigma_1 - \sigma_2)^2 + (\sigma_2 - \sigma_3)^2 + (\sigma_3 - \sigma_1)^2} = \sigma_f \tag{8.29}$$

The criterion takes all three principal stresses into account, and therefore it is more realistic under all stress conditions, in spite of its relatively more complicated form. According to this criterion, the flow shear stress τ_f can be determined by substituting the pure shear condition into Equation 8.29, which gives,

$$\tau_f = \sigma_f / \sqrt{3} = 0.577\sigma_f\,.$$

The two criteria are plotted under two-dimensional stresses, as shown in Figure 8.11. The figure indicates that they coincide in specific stress cases, such as simple tensile or compressive stresses, as well as equal biaxial tensile or compressive stresses. The maximum

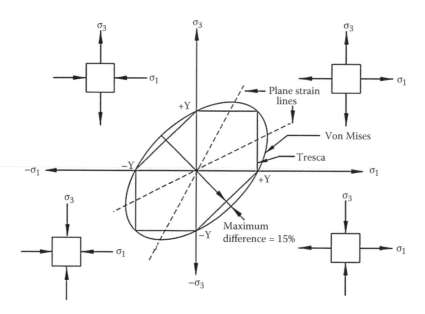

FIGURE 8.11 Comparison of Tresca and Von Mises yield criteria in plane stress. (Adapted from Ragab, A., and Bayoumi, S., *Engineering Solid Mechanics*, CRC Press, Boca Raton, FL, 1999.)

difference in predicting yielding according to both criteria is under pure shear condition, or plane strain condition, as shown in the figure. The yield strength under these conditions, according to Von Mises criterion, is equal to $(2/\sqrt{3} = 1.15$ that predicted by Tresca), i.e., the maximum difference is 15%. Therefore, it is easier to apply Tresca yield criterion in modeling forming processes, and taking the difference (15%) into consideration when doing the calculations in these special conditions.

8.5 GENERAL PLASTIC STRESS–STRAIN RELATIONS (THEORY OF PLASTICITY)

The stress strain relation in the plastic range is nonlinear, as shown in a tensile stress–strain curve presented in Figure 8.2. A more complicated situation in the plastic range is that the strains are not uniquely determined by the stresses, but depend on the history of loading or how the stress state was reached. It is therefore necessary to calculate the differentials or increments of the plastic strain dε throughout the loading history and then obtain the total strains by integration or summation along the stress–strain curve. Notice that the slope of the stress–strain curve or the power curve at each strain increment is different.

In the plastic range, the plastic strain increment components are proportional to the stress deviator components (Equation 8.19), according to the Levy–Mises relation, which is the basis of the theory of plasticity

$$\frac{d\varepsilon_1}{\sigma_1} = \frac{d\varepsilon_2}{\sigma_2} = \frac{d\varepsilon_3}{\sigma_3} = d\lambda \qquad (8.30)$$

where $d\lambda$ is a nonnegative constant which may vary throughout the history of loading (according to the slope of the stress–strain relation at each strain increment). Elastic strains are ignored in these equations. Thus, they can be applied to problems of large plastic flow such as metal forming processes.

Strain increment relations are derived from the Levy–Mises relation, similar to the elastic strain–stress relations in the form

$$d\varepsilon_1 = \frac{d\bar{\varepsilon}}{\bar{\sigma}}\left[\sigma_1 - \frac{1}{2}(\sigma_2 + \sigma_3)\right]$$

$$d\varepsilon_2 = \frac{d\bar{\varepsilon}}{\bar{\sigma}}\left[\sigma_2 - \frac{1}{2}(\sigma_1 + \sigma_3)\right] \qquad (8.31)$$

$$d\varepsilon_3 = \frac{d\bar{\varepsilon}}{\bar{\sigma}}\left[\sigma_3 - \frac{1}{2}(\sigma_1 + \sigma_2)\right]$$

where $\bar{\sigma}$ is the effective stress which is equivalent to the flow stress in Equation 8.29, or the flow stress on the power curve. Similarly, $d\bar{\varepsilon}$ is the effective plastic strain increment. These relations are analogous to the elastic strain–stress equations 8.21 and 8.25. The term $d\bar{\varepsilon}/\bar{\sigma}$ is the reciprocal of the slope of the flow curve, analogous to $1/E$ in the elastic range. It should also be noted that the Poisson's ratio in the elastic relations is replaced here by 0.5, which is the value of Poisson's ratio for all metals in the plastic range.

8.5.1 PLASTIC WORK

Referring to Figure 8.1, the work done to deform a bar of initial length L_0 and area A_0 subjected to a force F, is $F\,dL$.

Then the work per unit volume $= F\,dL/A_0\,L_0 = (F/A_0)\,(dL/L_0) = \sigma\,d\varepsilon$.

Under three-dimensional stresses

$$dw = \sigma*d\varepsilon = \sigma_1*d\varepsilon_1 + \sigma_2*d\varepsilon_2 + \sigma_3*d\varepsilon_3 \qquad (8.32)$$

8.6 EFFECT OF TEMPERATURE ON PLASTIC DEFORMATION

It has been described in Section 8.2.2 that for most metallic materials, plastic deformation is associated with strain hardening, i.e., with increase of the strength (and hardness) of the material. This is often utilized in strengthening metals by cold working processes such as rolling or drawing. Prior cold work causes the yield stress and ultimate tensile strength to rise in subsequent tensile testing, while ductility is reduced considerably. These trends are shown in Figure 8.12. Cold forming also distorts the crystals of the material by elongation in the direction of the major deformation, as shown in the figure. The increase in strength, drop in ductility, and distortion of crystals all put a limit on the amount of cold forming that could be performed in one operation.

Further cold forming of the part, or elimination of the effects of previous cold work, requires an appropriate heat treatment to soften the material. The most commonly known process is annealing, which involves heating and holding the metallic material at some elevated temperature for a certain time, as presented in a previous chapter. There are basically two different approaches for this process; namely, recovery and recrystallization, and grain growth may take place at elevated temperatures.

1. Recovery: This is through which the previously cold worked material is heated to a range of less than 0.3 of the melting point (T_m) on the absolute scale. Given enough time, such recovery increases atomic mobility, allows rearrangement of dislocations, and thus restores part of the original ductility and relieves the residual stresses without changing the distorted crystal structure, as shown in Figure 8.13. The strength of some metals is not affected by

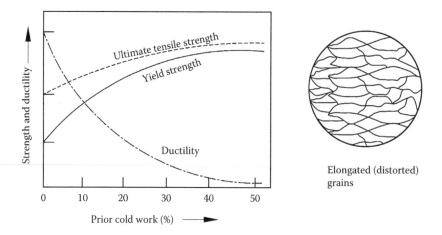

FIGURE 8.12   Effect of prior cold work on the strength, ductility, and microstructure of metallic materials. (Adapted from Schey, J. A., *Introduction to Manufacturing Processes*, McGraw-Hill, 1987.)

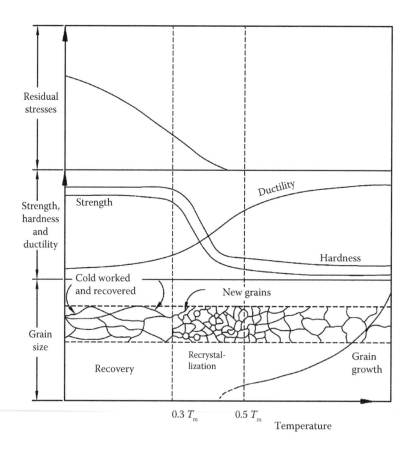

FIGURE 8.13   Effects of recovery, recrystallization, and grain growth on the mechanical properties and crystal structure of metallic materials. (Adapted from Kalpakjian, S., *Manufacturing Engineering Technology*, 4th ed., Addison-Wesley, 2003.)

TABLE 8.2
Recrystallization and Melting Temperatures for Various Metals and Alloys

Metal	Pb	Sn	Zn	Al (pure)	Co	Brass (60–40)	Ni	Fe	W
Recrystallization temperature (°C)	−4	−4	10	80	120	475	370	450	1200
Melting point (°C)	327	232	420	660	1085	900	1455	1538	3410

recovery. Therefore, it is considered a useful method for producing a material of higher strength yet reasonable ductility. Additionally, physical properties such as thermal and electrical conductivities are recovered to their pre–cold-worked states.

2. Recrystallization: In which the metal is heated to a temperature range between approximately 0.3 and 0.5 T_m, where atoms can move and diffuse across grain boundaries to form new equiaxed and strain-free crystals, replacing the distorted grains as shown in Figure 8.13. Initial ductility of the material before cold forming is restored, and similarly, strength and hardness are decreased. In addition to depending on temperature, the process requires sufficient time for atom movement and diffusion, depending on the prior cold work. Increasing the percentage of cold work enhances the rate of recrystallization and lowers the required temperature. There is a critical amount of cold work below which recrystallization is impractical. This critical limit is normally between 2% and 20% cold work. Recrystallization is known as annealing, or process annealing, when the purpose is softening of the work piece for further cold working. Alloying raises the recrystallization temperature up to 0.7 T_m. Some low melting point metals such as Pb, Sn, and Zn recrystallize at about room temperature, as shown in Table 8.2.

3. Grain growth: After recrystallization is complete, the equiaxed grains tend to grow if the metal is held at or above the recrystallization temperature for a prolonged time, as indicated in Figure 8.13. This phenomenon is called *grain growth*, which is undesirable, as it reduces the strength and toughness of the material. Therefore, it is necessary to have tight control of the annealing temperature and time to avoid grain growth.

8.7 COLD, WARM, AND HOT FORMING

To take advantages of the effects of temperature on plastic deformation as described in the previous section, forming processes can be carried under cold, warm, and hot working conditions.

8.7.1 COLD FORMING

This is performed at room temperature. It achieves several advantages, including

1. Better accuracy, involving closer dimensional tolerances and higher surface finish. This minimizes the need for finishing processes and allows the production of net shape or near net shape processes.
2. Higher product strength and hardness due to strain hardening.
3. Elongated crystals give the opportunity for desirable directional properties in the product.

Owing to these advantages, many cold forming processes are applied as mass production operations in industry, including cold rolling, wire and tube drawing, and most sheet forming processes. The main limitations of cold forming are

1. Higher forces and power are required.
2. The amount of cold work to be performed in one operation is limited. For a larger amount of deformation, intermediate annealing is necessary.

8.7.2 WARM FORMING

This is performed by heating the material below the recrystallization temperature (usually in the range of 0.3 T_m). This reduces the force and power required for deformation, and allows more intricate geometrical configurations with no need for annealing. This process provides a compromise between cold forming and hot forming.

8.7.3 HOT FORMING

This involves plastic deformation at temperatures above the recrystallization temperature. In practice, hot forming is conducted at higher temperatures where the softening process is fast. However, it should be noted that the plastic deformation process itself generates heat, which increases work temperatures in localized regions of high deformation. This may lead to local melting, which is highly undesirable. Therefore, hot working temperatures are usually maintained within the range of 0.5 T_m to 0.75 T_m. The most significant gain of hot forming is the ability to produce substantial plastic deformation far beyond what could be achieved in cold or warm forming. The flow curve of the metal during hot forming has a much lower strength coefficient (K), and strain hardening exponent (n) approaching zero, and substantially higher ductility (almost perfectly plastic condition, Figure 8.4d). All of these effects offer the following advantages to hot forming:

1. Much lower forces and power required.
2. Severe changes of shape and sections can be achieved.
3. The material structure is isotropic (same properties in all directions).
4. No intermediate annealing is required.

Unavoidable disadvantages of hot forming include poor dimensional tolerances, poor surface finish, work surface oxidation (scale formation), higher total energy required due to the high thermal energy, and shorter tool life. In practical forming processes that require high amounts of plastic deformation, such as in rolling, several passes are performed under hot forming conditions, and the final passes are performed by cold forming.

8.8 EFFECT OF STRAIN RATE ON PLASTIC DEFORMATION

Since the softening process associated with hot forming requires movement and diffusion of atoms, the time available for this process is critical. Then hot forming is substantially sensitive to the strain rate ($\dot{\varepsilon}$), where

$$\dot{\varepsilon} = \frac{d\varepsilon}{dt} = \frac{d}{dt} \ln \frac{l}{l_o} = \frac{1}{l}\frac{dl}{dt} = \frac{v}{l} \tag{8.33}$$

where $d\varepsilon$ is the strain increment, dt is the time increment, l is the instantaneous length of the part being deformed, and v is the deformation velocity (v), or the velocity of the press ram, or the moving element of the equipment. The strain rate $\dot{\varepsilon}$ is conventionally expressed in units of s^{-1}.

The flow stress–strain curves plotted at constant elevated temperatures and constant strain rate are shown in Figure 8.14a, at different strain rates. The figure indicates that at higher strain rates, the stress increases with strain, expressing that the rate of softening could not keep pace with strain hardening. Plotting the flow stress against the strain rate at a constant strain on a log–log scale

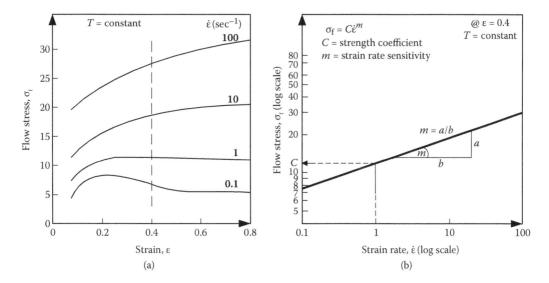

FIGURE 8.14 (a) Typical flow stress–strain curves at different strain rates. (b) Plot of the flow stress versus strain rate on a log–log scale. (From Schey, J. A., *Introduction to Manufacturing Processes*, McGraw-Hill, 1987. With permission.)

gives a straight line relation as shown in Figure 8.14b. This is indicative of an exponential relation between the flow stress (σ_f) and the strain rate $\dot{\varepsilon}$ in the form

$$\sigma_f = C\,\dot{\varepsilon}^m \tag{8.34}$$

where C is known as the strength coefficient and m is the strain rate sensitivity exponent. The figure shows that the value of C is found at a strain rate = 1 and m is the slope of the line. Eventually, both C and m change with temperature and with strain. Increasing temperature usually increases the strain rate sensitivity m and reduces the strength coefficient C. The strain rate sensitivity increases

TABLE 8.3
Typical Values for *C* and *m* for Different Materials

Material	Temperature (°C)	C (MPa)	m
Aluminum and alloys	200–500	310–14	0–0.2
Copper and alloys	200–900	415–14	0.02–0.3
Lead	100–300	11–2	0.1–0.2
Steel, low carbon	900–1200	165–48	0.08–0.22
Steel, medium carbon	900–1200	160–48	0.07–0.24
Stainless steel	600–1200	415–35	0.02–0.4
Titanium	200–1000	930–14	0.04–0.3
Superplastic alloys			
Lead–tin (eutectic) ($\dot{\varepsilon}$ = 0.01–0.2 min^{-1})	22	8.8	0.5
Zn–Al (eutectoid) ($\dot{\varepsilon}$ = 0.1–10 min^{-1})	250	6.68	0.5
Titanium–6Al–4V ($\dot{\varepsilon}$ = 0.01–0.1 min^{-1})	950	46.6	0.8
Brass (60/40) ($\dot{\varepsilon}$ = 0.01–0.1 min^{-1})	600	21	0.48
Al–33 Co ($\dot{\varepsilon}$ = 0.01–0.1 min^{-1})	500	20	0.75
Thermoplastics (polystyrene)	150	11	0.85
Molten glass	>1000	0.001	1 (Newtonian fluid)

TABLE 8.4
Typical Ranges of the Strain Rate Sensitivity m for Different Forming Conditions

m Range	$0 < m < 0.05$	$0.05 < m < 0.4$	$0.3 < m < 0.85$
Forming conditions	Cold forming	Hot forming	Superplastic forming

with increasing the recrystallization temperature. Typical values for C and m are given in Table 8.3 for different materials.

Typical ranges of m for different forming conditions are shown in Table 8.4.

At room temperature, the effect of the strain rate is almost negligible, indicating that the power relation (Equation 8.8) is quite satisfactory for cold forming processes. As temperature is increased, the strain rate plays a more significant role in determining the flow stress, indicating that Equation 8.32 should be adopted for modeling hot forming processes.

8.9 SUPERPLASTICITY

Strain rate sensitivities of 0.5 or greater promote extremely large elongations in some metallic alloys under hot tensile stresses by preventing localization of the deformation in a neck form, thus delaying fracture. This effect is known as superplasticity, and it can be achieved under the following conditions:

1. Extremely fine grain size (a few microns or less), with uniform and equiaxed grain structure
2. High temperatures ($T > 0.4\ T_m$, recrystallization temperature or higher)
3. Low strain rates ($10^{-2}\ \text{s}^{-1}$, or lower)
4. Stable microstructure without grain growth during deformation

Under these conditions, extremely high elongations at very low flow stresses are observed (200% up to as much as 2000%). There are two useful aspects of superplasticity.

1. The accompanying low flow stresses at useful working temperatures permits creep forging of intricately shaped parts and reproduction of fine details
2. The extremely high tensile deformations permit forming of sheets with large thickness using simple tooling

Common examples of superplastic materials are zinc–aluminum, bismuth–tin, titanium alloys, aluminum alloys, aluminum–lithium, and some stainless steels which have very fine grains, typically less than 10 to 15 μm. The behavior of the material in superplastic forming is similar to that of bubble gum or hot glass, which when blown expands to many times its original diameter before it bursts. Particular alloys such as Zn–22Al and Ti–6Al–4V can also be formed by bulk deformation processing, including forging, coining, and extrusion. Superplastic sheet forming processes will be presented in Chapter 10. The forming temperatures and the C and m values for some of these alloys are given in Table 8.3. Notice in the table that some thermoplastics and molten glass behave similar to superplastic materials.

The main advantages of superplastic forming include

- Complex shapes can be formed out of one piece, with fine details and close tolerances.
- Weight and material savings due to high formability.
- Little or no residual stresses develop in the formed part.
- Tooling cost is relatively low due to the low stresses required for forming (low-alloy steels are commonly used).

8.10   EFFECT OF FRICTION AND LUBRICATION IN METAL FORMING

In most forming processes, deformation is usually associated with sliding of the material against the harder surfaces of the tool or die parts. Consequently, friction between sliding parts is inevitable. It is well known that friction is commonly described by the coefficient of friction μ, as shown in Figure 8.15a, which can be represented as

$$= \frac{\text{Sliding force } F}{\text{Normal force } P} = \frac{\text{Shear stress}}{\text{Interface pressure}} = \frac{F/A}{P/A} = \frac{\tau_i}{p} \qquad (8.35)$$

where F is the force required to move the body parallel to the surface, P is the normal force, A is the contact area, τ_i is the shear stress at the interface, and p is the interface pressure. Equation 8.33 indicates that at constant interface pressure p, the shear stress τ_i increases linearly with the coefficient of friction. Practically, the interface shear stress τ_i cannot rise beyond a maximum value given by the shear flow stress τ_f of the work piece material, where the material refuses to slide on the tool, and instead deforms by shearing inside the body. Since the flow stress in shear $\tau_f = 0.5\ \sigma_f$, according to Tresca yield criterion, and $= 0.577\ \sigma_f$, according to Von Mises, then the maximum value of μ lies within the range 0.5 and 0.577, as shown in Figure 8.15a, when $\sigma_f = p$. At higher interface pressures, the maximum value of μ is lower, as shown in Figure 8.15b. Generally, it is much more true to state that the coefficient of friction becomes meaningless when $\tau_i = \tau_f$, since there will be no relative sliding at the interface, which is known as sticking friction. The material will not actually stick to the die surface but will suffer internal shearing flow, which generates severe problems in some forming processes.

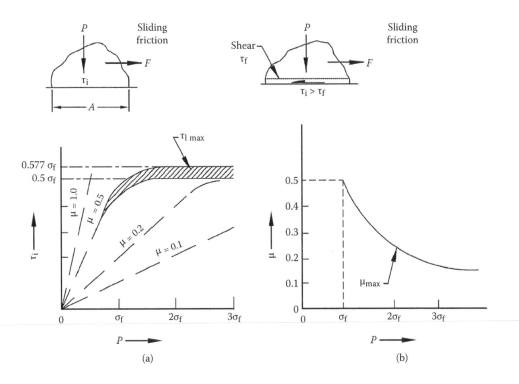

FIGURE 8.15   (a) Limit of frictional stresses by the interface shear between the material and die. (b) Effect of the die pressure on the coefficient of friction. (Adapted from Schey, J. A., *Introduction to Manufacturing Processes*, McGraw-Hill, 1987.)

Friction increases the pressures, forces, and power required to perform the process, and thus limits the amount of deformation that can be achieved. Suitable lubricants are used to improve frictional conditions by reducing μ, reducing forces, eliminating the possibility of sticking, reducing die wear (which affects dimensional accuracy), cooling the system in cold forming, preventing heat loss in hot forming, and controlling the surface finish of the formed part. The lubricant must be nontoxic, nonallergic, inflammable, noncorrosive, and easy to apply and remove.

Lubricants used for cold forming processes include mineral oils, fats and fatty acids, water-based emulsions, soaps, and phosphate coatings. Lubricants for hot forming include: Mineral oils, graphite, and molten glass. Graphite contained in water or mineral oil is a common lubricant for hot forging of various materials. Molten glass is effective for hot extrusion of steels.

8.11 REVIEW QUESTIONS

1. What are the objectives of plastic forming processes, and what are the main aspects affecting them?
2. What is the effect of the material crystal structure on its ability for plastic deformation?
3. Which of the idealized stress–strain curves is applicable for each of the following processes:
    a. Hot forging of copper product
    b. Cold rolling of steel sheets
    c. Lead sheathing of a cable
    d. Press fitting of a bearing onto a shaft
    e. Shrink fitting of a bush onto a shaft
4. What is the effect of temperature on the mechanical properties of metallic materials?
5. How does temperature affect plastic forming process?
6. Classify plastic forming processes according to the applied temperature.
7. How does the strain rate affect plastic forming processes?
8. Define superplasticity, and what are the conditions to reach superplastic behavior?
9. Give examples of superplastic materials and typical examples of their applications in industry.
10. How can the effect of friction in plastic forming processes be reduced?
11. What are the most common lubricants used in forming processes?
12. What are the limitations of the coefficient of friction in plastic forming processes?

8.12 PROBLEMS

1. A bar of 200 mm initial length is elongated by drawing to a length of 380 mm in three steps. The lengths after each step are 260, 320, and 380 mm, respectively.
    a. Calculate the engineering strain for each step separately and compare the sum with the overall engineering strain.
    b. Repeat calculations using the true strain, and comment on the results.
2. A mild steel specimen of diameter 12.5 mm and length 50 mm is subjected to a tensile test at room temperature. The loads and lengths are recorded as follows:

Load (kN)	31	35.2	36.4	37.4	40	44.5	49	51.9 Necking	41.8 Fracture
Length (mm)	50.1	50.4	50.7	52	52.3	53	55	64.2	70.3

    a. Plot the true stress–strain power relation on a log–log graph and determine K and n for the material.

    b. Determine the stress required to deform the material to a plastic strain of 0.4.

    c. On the same graph, plot the relation $\sigma = 315\ \varepsilon^{0.54}$ representing annealed copper and the relation $\sigma = 960\ \varepsilon^{0.1}$ representing stainless steel 410.

3. During a tensile test using a metal that obeys the power law, the tensile strength is found to be 340 MPa. Reaching the maximum load required an elongation of 30%. Find K and n and calculate the percent elongation when the stress is 300 MPa.

4. Consider a stress state where $\sigma_x = 70$ MPa, $\sigma_y = 35$, $\tau_{xy} = 20$ MPa, and $\sigma_z = \tau_{yz} = \tau_{xz} = 0$. Calculate the principal stresses.

5. At a point in a piece of elastic material, direct stresses of 90 MN/m^2 tensile and 50 MN/m^2 compressive are applied on mutually perpendicular planes. The planes are also subjected to unknown shear stress. If the maximum principal stress is limited to 100 MN/m^2 (tensile), calculate the value of the shear stress, the minimum principal stress, and maximum shear stress.

6. A copper sheet 500 mm long, 100 mm wide, and 2 mm thick is pulled in the length direction under tensile force of 200 N, and pressure P across the width on the thickness. The width dimension will not be changed. What value of P is needed to cause only yielding, according to both Tresca and Von Mises criteria, if the yield stress of the copper is 150 MPa?

7. In sheet metal stretching of a steel strip, a grid of circles 5 mm diameter is printed on the sheet surface. One of these circles changed after stretching into an ellipse whose major and minor diameters are 6.5 and 5.5 mm, respectively. The stresses in the sheet plane are such that $\sigma_1 > \sigma_2$, and the final value of σ_2 is 300 MPa. Determine the strain components, the maximum principal stress and the flow stress.

8. During the tension test of a medium carbon steel specimen, the strain rate is suddenly increased by a factor of 8. This has led to a corresponding rise of the stress level by 1.8%. What is the strain rate sensitivity exponent for this material?

BIBLIOGRAPHY

Banabic, D. 2000. *Formability of Metallic Materials*. Berlin: Springer-Verlag.

Cahn, R. W. 2007. *Thermomechanical Processing of Metallic Materials*. UK: Elsevier.

Chakrabarty, J. 2006. *Theory of Plasticity*, 3rd ed. UK: Elsevier Butterworth-Heinemann.

Hearn, E. J. 1987. *Mechanics of Materials I and II*. UK: Butterworth-Heinemann.

Kalpakjian, S. 2001. *Manufacturing Engineering Technology*, 4th ed. Addison-Wesley.

Ragab, A., and Bayoumi, S. E. 1999. *Engineering Solid Mechanics*, 3rd ed. Boca Raton, FL: CRC Press.

Rees, D. W. A. 2006. *Basic Engineering Plasticity, An Introduction with Engineering and Manufacturing Applications*. UK: Elsevier Butterworth-Heinemann.

Schey, J. 1987. *Introduction to Manufacturing Processes*, 2nd ed. McGraw-Hill.

Singh, U., and Dwivedi, M. 2009. *Manufacturing Processes*, 2nd ed. UK: New Age International Publications.

9 Bulk Forming of Metallic Materials

9.1 INTRODUCTION

Most metallic materials are subjected to plastic deformation at one stage or another during processing, to reach final or near to the final shape. This is carried out by applying forces to the material through a system of tooling to reach the plastic (permanent) deformation condition, i.e., to change its shape without volume change, and without bringing it to fracture or to the molten state. Metal forming also provides better control of the mechanical properties of the product. The term *bulk forming* implies that the product has large thickness compared to other dimensions, and that thickness is substantially changed during forming. The material to be deformed may be heated before processing to cause large deformation or to form complicated shapes. Plastic forming processes start from a cast block, and cover a wide range of processes performed sequentially to reach a final product, as shown in Figure 9.1.

The design and control of such processes depend on an understanding of the characteristics of the workpiece material, the conditions at the tool/workpiece interface, the mechanics of plastic deformation (metal flow), the equipment used, and the finished product requirements. These factors influence the selection of tool geometry and material as well as processing conditions. Because of the complexity of many metalworking operations, models of various types, such as analytical, physical, or numerical models are often relied upon to design such processes.

Metal forming is one of the oldest and most applied technologies used to produce metallic products. The earliest records of metalworking describe the simple hammering of gold and copper in various regions of the Middle East around 8000 BC. The forming of these metals was crude because the art of refining by smelting was unknown and because the ability to deform the material was limited by impurities that remained after the metal had been separated from the ore. The beginning of copper smelting around 4000 BC led to the ease of purifying metals through chemical reactions in the liquid state. Later, in the Copper Age, it was found that the hammering of metal brought about desirable increases in strength (strain hardening). The search for higher strength materials led to the utilization of alloys of copper and tin (the Bronze Age), and some 1300 years later, to iron and carbon (the Iron Age, which can be dated back to around 1200 BC). Most metalworking was done by hand until the 13th century, where the tilt hammer using water power to raise a lever arm and let it fall under gravity was developed and used primarily for forging bars and plates. This relatively simple device remained in service for some centuries.

The development of rolling mills followed that of forging equipment. Leonardo da Vinci's notebook included a sketch of a machine designed in 1480 for the rolling of lead for stained glass windows. In 1495, da Vinci is reported to have rolled flat sheets of precious metal on a hand-operated two-roll mill for coin-making purposes. In the following years, several designs for rolling mills were utilized in Europe. However, the development of large mills capable of hot rolling ferrous materials took almost 200 years. Early mills, driven by water wheels, employed flat rolls for making plates and sheets until the middle of the 18th century.

During the Industrial Revolution at the end of the 18th century, processes were devised for making iron and steel in large quantities to satisfy the demand for metal products. The need for forging equipment with larger capacity led to the invention of the high-speed steam hammer and the hydraulic press. From such equipment came products ranging from firearms to locomotive parts. Similarly, the

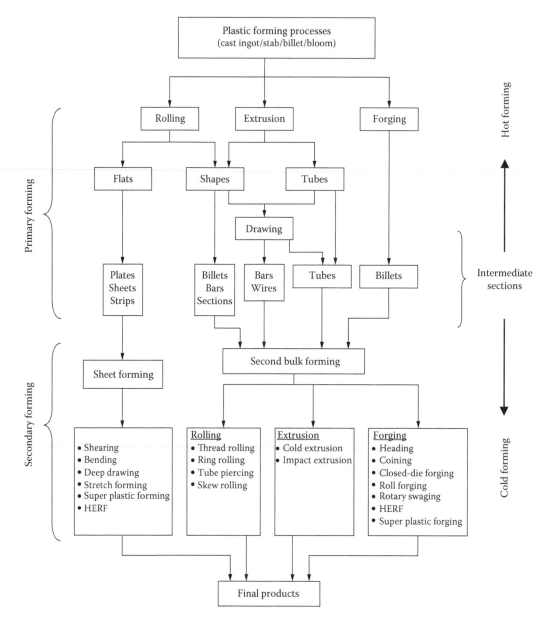

FIGURE 9.1 Sequence of forming processes leading to final products.

steam engine spurred developments in rolling, and in the 19th century, a variety of steel products were rolled in significant quantities. The past 100 years have seen the development of new types of metal forming equipment, including mechanical and screw presses and high-speed tandem rolling mills.

9.2 CLASSIFICATION OF FORMING PROCESSES

Forming processes can be classified from different points of view as follows.

    1. According to the purpose of deformation
       • Primary processes, where the aim is to eliminate the defects of the cast structure (porosity, blow holes, dendritic structure, etc.) through successive deformation steps,

to control the mechanical properties (improve ductility and toughness) and to produce intermediate shapes suitable for further processing, such as plates, sheets, strips, rods, tubes, standard sections, etc.

- Secondary processes, using the products of primary processes (intermediate sections) and shaping them into finished parts for final products.

2. According to the extent of the deformation zone
    - Bulk deformation processes, where the thickness is substantially changed. These processes can be further divided to
        - Steady state processes where all product parts are subjected to the same mode of deformation (except for the process beginning and end), as in rolling, extrusion, etc.
        - Non–steady-state processes, where the geometry changes continually during deformation, as in impression die forging.
    - Sheet metal forming, where the change in the sheet metal thickness is fairly limited, as in bending, deep drawing, etc.

3. According to the working temperature
    - Hot forming processes, where metal is deformed at temperature that exceeds its recrystallization temperature, which is about 50% of its absolute melting temperature. This helps in performing more extensive deformation, and higher shape complexity (usually required for primary or bulk forming).
    - Cold forming processes, when the finished part is to be used in the strain-hardened condition, or when small dimensions, tight tolerances, or higher surface finish is required.
    - Warm forming processes, heating below recrystallization temperature to reduce forces without allowing change of the crystal structure.

4. According to the type of applied force
    - Direct compression processes, where compressive force is applied to the work surface, and the metal flows at right angles to the direction of compression (forging and rolling).
    - Indirect compression processes, where the applied forces are tensile (drawing), or compressive (extrusion), but the indirect compressive forces developed on the die walls are very effective in metal flow.
    - Tension type forming, as in stretch forming, where the metal sheet is wrapped to the contour of a die under the application of tensile forces.
    - Bending type processes, as in sheet bending.
    - Shearing processes, where excessive shearing force is applied to rupture the metal (blanking, cropping, etc.).

Figure 9.2 shows the types of forming processes representing that classification.

9.3 FORGING PROCESSES

Forging is the process of plastic forming of a metallic material under direct compression using a press or a hammer, between two die halves. These die halves may be just the flat platens of a press or a hammer (open-die forging), or they may form the product's geometrical configuration when closed under pressure to the final position (closed-die forging).

9.3.1 OPEN-DIE FORGING

Forgings are made by this process when

- The forging is too large to be produced in closed dies, such as ship propeller shafts weighing up to 600 tons (6 MN).

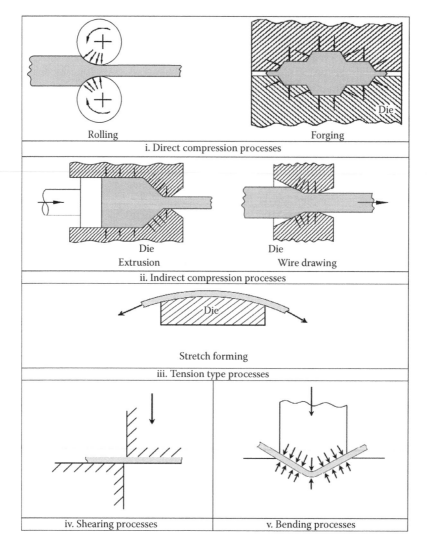

FIGURE 9.2 Classification of forming processes according to the type of the applied force. (Adapted from Dieter, G. E., *Mechanical Metallurgy*, McGraw-Hill, New York, 1988.)

- The aim is to eliminate the defects of the ingot cast structure, and generate a new equiaxed structure with more controlled mechanical properties.
- The quantity required is too small to justify the cost of closed dies.

Most open-die forgings are produced in a pair of flat dies; one attached to the hammer head or to the press ram, and the other to the anvil or press base. Swage dies (curved), V-dies, and V-die and flat-die combinations may be used. Typical open-die forging operations are shown in Figure 9.3.

The process requires highly skilled operators using various auxiliary tools (such as mandrels, sizing blocks or spacers, fullers, punches, etc.). Advanced material handling equipment such as overhead cranes and manipulators are also necessary. Complex shapes can be produced in open dies, but require extensive effort and a long time. They may need several reheatings of the ingot to reach the final shapes. Typical products are shafts (with round, square, rectangular, hexagonal or octagonal sections), billets with high mechanical properties (for rolling, further forging, or for

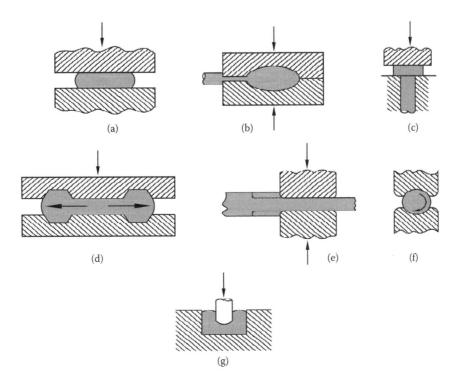

FIGURE 9.3 Typical open-die forging operations: (a) upsetting, (b) edging, (c) fullering, (d) drawing, (e) swaging, (f) piercing, and (g) punching. (From Dieter, G. E., *Mechanical Metallurgy*, McGraw-Hill, New York, 1988. With permission.)

die manufacturing), simple pancake shapes produced by upsetting (for making gears, wheels, etc.), mandrel forgings (long and hollow parts), and contour shapes as turbine wheels. Typical open-die forging products are presented in Figure 9.4, which shows the steps to reach the final shape.

Descaling of the part before and during open-die forging is done by busting and blow-off. The best practice includes the use of compressed air to blow away the scale as it breaks off. High-pressure water is also sometimes used to loosen scale, especially in hard-to-reach locations, such as the inside diameter of a mandrel forging. Failure to remove the scale causes it to be forged in, resulting in pits and pockets on the forged surfaces. The total amount of scale formed in open-die forging is usually greater than in closed-die forging because the hot metal is exposed to the atmosphere for a longer time; that is, open-die forgings usually require more forging strokes and sometimes require reheating. Metal loss in open-die forging through scaling usually ranges from 3 to 5%.

Cold heading is a cold forging process in which the force developed by one or more strokes (blows) of a heading tool is used to upset the metal in a portion of a wire or rod blank in order to form a larger section than the original. The process is applied to produce bolts and rivets without loss of material. Cold working also increases the tensile strength of the product and controls grain size. Typical cold heading operations are shown in Figure 9.5.

In the first scheme of head forming between punch and die, if the ratio of the free length of the bar to the initial bar diameter (known as upsetting ratio) exceeds 3, buckling would take place instead of heading.

Cold heading: This is most commonly performed on low-carbon steels, copper, aluminum, stainless steels, and some nickel alloys. Other nonferrous metals and alloys such as titanium, beryllium, magnesium, and the refractory metals and alloys, are less formable at room temperature and may crack when cold headed. These metals and alloys are sometimes warm headed.

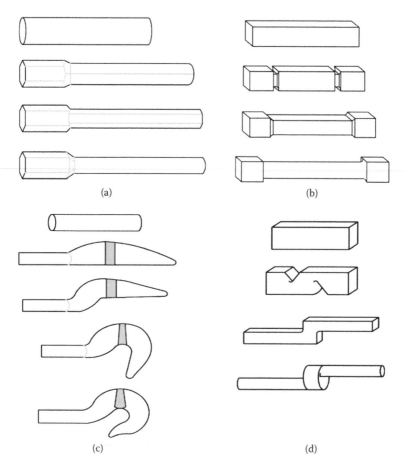

FIGURE 9.4 Typical steps to produce common shapes in open dies. (Adapted from ASM International, *ASM Handbook, vol. 14: Forming and Forging,* Materials Park, OH, 1988.)

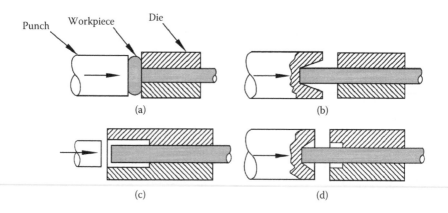

FIGURE 9.5 Cold heading of the end of a bar: (a) head formed between punch and die, (b) head formed in punch, (c) head formed in die, and (d) head formed in punch and die. (From ASM International, *ASM Handbook, vol. 14: Forming and Forging*, Materials Park, OH, 1988. With permission.)

9.3.2 Close-Die (Impression) Forging

Close-die forging is the forming of hot metallic material completely within the walls or cavities of two die halves that come together to enclose the part on all sides. The process permits the production of complex shapes at high reductions within closer dimensional tolerances than those achieved in open-die forging. The impression for the forging can be either entirely in a die half or can be divided between the top and bottom dies. The forging stock is commonly cut to length from a round or square bar to provide the volume of metal needed to fill the die cavities, in addition to an allowance for flash. The flash allowance covers the positive tolerance added to the stock to ensure complete filling of the die cavity before outward flow of the flash. It also acts as a relief valve for the extreme pressure produced in closed dies. Close-die forgings can weigh from hundreds of grams up to 25 tons. However, 70% of forgings weigh less than 1 kg.

The materials most suitable for hot die forging are those having high forgeability to allow a high amount of deformation without failure. The most commonly used materials are alloys of aluminum, magnesium, copper, nickel, and titanium, in addition to carbon steels, low-alloy steels, and stainless steels. In most hot-forging operations, the temperature of the stock material is higher than that of the dies. Metal flow and die filling are largely determined by the resistance and the ability of the forging material to flow (flow stress) and forgeability, by the friction and cooling effects at the die/material interface, and by the complexity of the forging shape. Lubricants prevent sticking of the part to the die walls, and reduce the pressures required for plastic deformation. Various types of lubricants are used, and they can be applied by swabbing or spraying. The simplest is a high flash point oil swabbed onto the dies. Colloidal graphite suspensions in either oil or water are frequently used. Synthetic lubricants can be employed for light forging operations. The water-based and synthetic lubricants are extensively used primarily because of cleanliness.

Considering shape complexity of the product, spherical and block-like shapes are the easiest to forge in impression or closed dies. Parts with long, thin sections or projections (webs and ribs) are more difficult to forge because they have more surface area per unit volume. Such variations in shape maximize the effects of friction and temperature changes and therefore influence the final pressure required to fill the die cavities. There is a direct relationship between the surface-to-volume ratio of a forging and the difficulty in producing that forging.

Flash design: Forging pressure increases with decreasing flash thickness and with increasing flash land width, because of combinations of increasing restriction, increasing frictional forces, and decreasing metal temperatures at the flash gap. Figure 9.6 shows the stages of filling the die cavity until complete filling at pressure P_2. Extra metal bulges in the flash with the pressure increase from P_2 to P_3. Therefore, the dimensions of the flash determine the final load required for closing the dies. The flash can be removed by sawing, grinding, or trimming in a die when the number of products is large.

Isothermal and hot-die forging: These are special categories of close-die forging in which the die temperatures are significantly higher than those used in conventional hot-forging processes. In isothermal forging, the die is maintained at the same temperature as the forging stock. This eliminates the die chill completely and maintains the stock at a constant temperature throughout the forging cycle. The process permits the use of extremely slow strain rates, thus taking advantage of the strain rate sensitivity of flow stress for certain alloys. The process is capable of producing net shape forgings that are ready to use without machining, or near-net shape forgings that require minimal secondary machining. The hot-die forging process is characterized by die temperatures higher than those in conventional forging, but lower than those in isothermal forging. Typical die temperatures in hot-die forging are 110 to 225°C lower than the temperature of the stock. When compared with isothermal forging, the lowering of die temperature allows wider selection of die materials, but the ability to produce very thin and complex geometries is compromised. Both techniques lead to the saving of material by increasing the ability to produce thinner sections, and reducing machining cost, as shown in Figure 9.7. Therefore, they are mainly used for expensive and difficult-to-machine

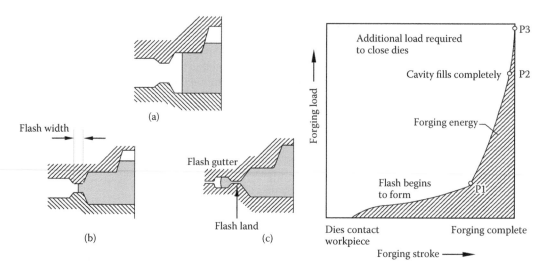

FIGURE 9.6 (a–c) Metal flow and load-stroke curve (d) in closed-die forging: (a) upsetting, (b) filling, and (c) flash filling. (From Kalpakjian, S., and Schmid, S. R., *Manufacturing Processes for Engineering Materials*, Prentice-Hall, Upper Saddle River, NJ, 2003. With permission.)

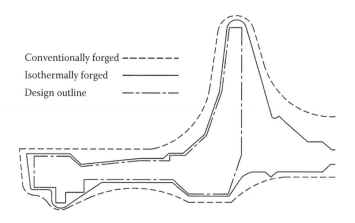

FIGURE 9.7 Weight reduction obtained in the production of a nickel-based disk by isothermal forging instead of conventional forging. Sonic outline is the outline determined through design. (From ASM International, *ASM Handbook, vol. 14: Forming and Forging*, Materials Park, OH, 1988. With permission.)

alloys such as titanium and nickel-based alloys. They are also ideal for forging superplastic materials, which are discussed in the previous chapter.

These processing techniques are primarily used for manufacturing airframe structures and jet-engine components made of titanium and nickel-based alloys, but they have also been used in steel transmission gears and other components.

9.3.3 SPECIAL FORGING PROCESSES

9.3.3.1 Rotary Swaging

This is a process for reducing the cross-sectional area of bars, tubes, or wires by repeated radial blows with two to four die segments. The die segments move simultaneously in the radial and circumferential directions relative to the work piece. The work is elongated as the cross-sectional area

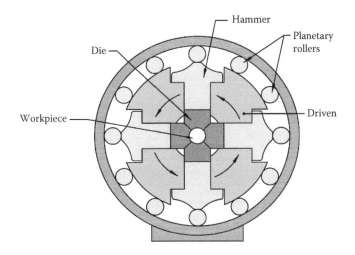

FIGURE 9.8 Principle of rotary swaging.

is reduced. The starting stock is usually round, square, or generally symmetrical in cross section. The maximum reduction in area is usually limited; however, it may reach as high as 70% under hot working conditions. A schematic sketch of the process is shown in Figure 9.8.

The process offers many advantages, including increase in thickness of tubes, different wall thicknesses to suit different strength requirements, increased strength (when cold formed), achieving net shape tolerances, and high surface quality.

9.3.3.2 Radial Forging

The process is similar, in effect, to the rotary swaging process, with the exception that the work piece is the rotating part between two die halves on a hammer or a press, as shown in Figure 9.9. The workpiece is fed between the dies, which are given a rapid periodic motion as the work piece rotates. In this manner, the forging force acts on only a small portion of the work piece at any time.

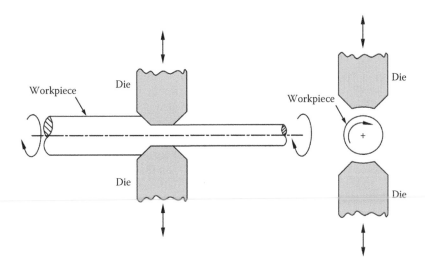

FIGURE 9.9 Radial forging process. (Adapted from ASM International, *ASM Handbook, vol. 14: Forming and Forging*, Materials Park, OH, 1988.)

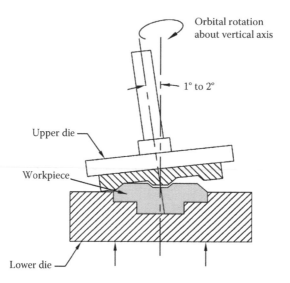

FIGURE 9.10 Orbital forging process. (Adapted from Doyle, L. E. et al., *Manufacturing Processes and Materials for Engineers*, 3rd ed., Prentice-Hall, Upper Saddle River, NJ, 1985.)

9.3.3.3 Orbital Forging

Orbital forging is a two-die forging process that deforms only a small portion of the work piece at a time in a continuous manner. As shown in Figure 9.10, the upper die, tilted with respect to the lower die, rotates around the workpiece. The tilt angle and shape of the upper die result in only a small area of contact (footprint) between the workpiece and the upper die at any given time. Because the footprint is only a small part of the workpiece surface area, rotary forging requires considerably less force than conventional forging. The tilt angle between the two dies plays a major role in determining the amount of forging force that is applied to the workpiece. A larger tilt angle results in a smaller footprint; consequently, a smaller amount of force is required to complete the same amount of deformation as compared to a larger contact area. Tilt angles are commonly about 1 to 2°. The larger the tilt angle, however, the more difficult are the machine design problems, because the drive and bearing system for the tilted die is subjected to large lateral loads and is more difficult to maintain. This process is used for forming bevel gears, wheels, and bearing rings.

9.3.3.4 Coining

Coining is a closed-die forging operation, usually performed cold, in which all surfaces of the workpiece are confined or restrained, resulting in a well-defined imprint of the die on the product. It is applied in sizing or sharpening or changing a radius or profile. Coining is often the final operation in a progressive-die sequence. The prepared blank is loaded above the compressive yield strength and is held in this condition during coining. Dwell time under load is important for the development of dimensions in sizing and embossing. It is also necessary for the reproduction of fine details, as in engraving. The process develops extremely high forces to ensure small but permanent deformation on the product surface area, and therefore it is limited to small-sized products. Typical applications include the production of coins, patterned tableware, medallions, and metal buttons.

9.3.4 Forging Equipment

The equipment required to perform forging processes can be classified according to their principle of operation to presses and hammers. Presses apply load gradually or statically, and are classified into mechanical and hydraulic presses. The ability of mechanical presses to deform the work material is determined by the length of the press stroke and the available force at various stroke positions.

Mechanical presses are therefore classified as stroke-restricted machines. Hydraulic presses are termed force-restricted machines because their ability to deform the material depends on the maximum force rating of the press. Hammers deform the workpiece by the kinetic energy of the hammer ram, and they are therefore classed as energy-restricted machines.

9.3.4.1 Mechanical Presses

Mechanical presses are based on mechanical mechanisms that transfer the rotation of an electric motor to oscillatory strokes on the press ram. The ram carries the top, or moving die, while the bottom, or stationary die, is clamped to the press frame. Energy stored in a flywheel, is controlled by a clutch, and is used to deform the workpiece at the end of the stroke. The most common types of mechanical presses are crank presses, eccentric presses, knuckle joint presses, and screw presses, as seen in Figure 9.11. The ram stroke is shorter than in hydraulic presses or hammers. Forging presses deliver their maximum force within about 3.2 mm from the end of the stroke, where the maximum force is associated with flash formation. The knuckle joint mechanism allows higher forces near the end of the stroke and is therefore suitable for coining processes. The maximum force rating of mechanical presses ranges from 2.2 to 142 MN.

9.3.4.2 Hydraulic Presses

The ram of a hydraulic press is driven by a hydraulic piston and cylinder. Control valves allow rapid approach speed of the ram followed by a slow squeezing speed of the work after contact. Therefore, hydraulic presses are suitable for both open- and close-die forging. The maximum press force can be limited to protect tooling.

Hydraulic presses are rated by the maximum force. Open-die presses are built with capacities ranging from 1.8 to 125 MN, and closed-die presses range in size from 4.5 to 640 MN. Ram speeds during normal forging conditions vary from 635 to 7620 mm/min.

Figure 9.11 gives the types of presses used commonly in forging processes, including crank, knuckle joint, and screw mechanical presses, as well as the hydraulic press.

9.3.4.3 Hammers

Hammers are the most widely used equipment for forging. They can be classified according to the method used to derive the ram, as shown in Table 9.1. The main components of a hammer are the ram, anvil, and the frame assembly. The simplest type is the falling mass or gravity drop hammer, where the ram is lifted to the highest position by a board, by steam, or by air pressure, and then released to fall under its own weight. The impact energy on the workpiece is the kinetic energy of

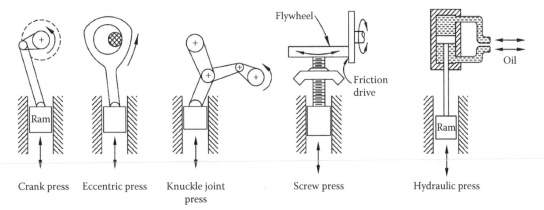

FIGURE 9.11   Schematic illustration of the types of presses used in forging processes. (Adapted from Doyle, L. E. et al., *Manufacturing Processes and Materials for Engineers*, 3rd ed., Prentice-Hall, Upper Saddle River, NJ, 1985.)

TABLE 9.1
Types and Main Specification Ranges of Hammers

Type of Hammer	Ram Mass (kg)	Maximum Energy (kJ)	Impact Speed (m/s)	Number of Blows per Minute
Gravity drop	40–3400	0.5–40	3–5	45–60
Power drop	680–31,750	20–600	4–8	60–100
Electrohydraulic	450–9980	Up to 100	3–4.5	50–75
Counterblow	—	Up to 200	3–6	10–60

the falling mass. In the power drop hammer, steam or pneumatic pressure is used to accelerate the ram during the downward stroke and to increase the hammer impact energy. Electrohydraulic hammers have been recently introduced, where the ram is lifted with oil pressure against an air cushion. The compressed air slows the upstroke of the ram and contributes to its acceleration during the down stroke blow. Electronic blow-energy control is added to allow the user to program the drop height of the ram for each individual blow. Thus, the operator can set automatically the number of blows desired in forging in each die cavity, and the intensity of each individual blow. This increases the efficiency of the hammer operations and decreases the noise and vibration associated with unnecessarily strong hammer blows. The counterblow hammer is another variation of the power-drop hammer. These hammers develop striking force by the movement of two rams simultaneously from opposite directions, and meeting at the work material, which absorbs almost the whole energy at higher (combined) speed. In this type, the vibration of impact is reduced, losses to the anvil or foundations are eliminated, and large inertia blocks and foundations are not needed.

9.3.5 FORGING DEFECTS

Defects that may arise in a forged product could be due to faults in the material, due to the prior heating process, or due to poor forging techniques or badly designed dies. Overheating of the metal or soaking at high temperature for a long time leads to excessive oxidation of the surface. This develops scale pieces, which are driven into the work surface during forging. Further cleaning of the product surface may leave shallow surface depressions (scale pockets) in the forging.

The most serious defects which may arise during the forging process are

a. *Laps, folds, or cold shuts:* These are usually caused by poor die design, or by incorrect positioning of the billet in the die cavity. Sharp corner changes in the die cause the metal to flow across the corner rather than to follow the die contour, and as the die closes, the metal folds back over itself, giving rise to a cold shut, as shown in Figure 9.12a.

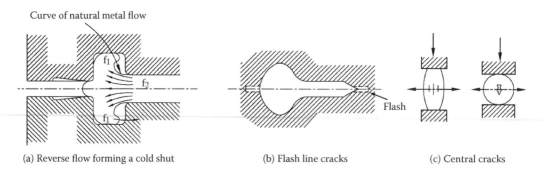

(a) Reverse flow forming a cold shut   (b) Flash line cracks   (c) Central cracks

FIGURE 9.12   Schematic presentation of forging defects.

b. *Flash line cracks:* These cracks may develop after removal of the flash, or even after subsequent heat treatment. The flash of a forging which undergoes severe reduction in thickness develops a fiber structure which is weak in the normal direction. Any undue strain in this direction may therefore cause fracture (see Figure 9.12b).

c. *Mismatched forgings:* These are produced when the upper and lower die blocks are misaligned during hammering or pressing.

d. *Incomplete forging:* This may be due to small billet size, low forging temperature, or poor die design so that the metal is unable to flow sufficiently.

e. *Central cracks:* Due to excessive secondary tensile stresses in open die forging (Figure 9.12c). The possibility of this defect is eliminated in close-die forging due to the lateral compressive stresses arising by the side walls of the die cavity.

9.3.6 MODELING OF FORGING PROCESSES

Ideal deformation of a billet with a circular or rectangular cross section between flat platens leads to a reduction in height and uniform increase in area. Considering volume constancy, the area corresponding to any height is calculated as

$$V = A_0 h_0 = A_1 h_1 \tag{9.1}$$

However, lateral friction at the billet–platen interfaces leads to restriction of material flow on these surfaces, causing a barrel shape of the lateral sides, and a consequent increase in the pressure required to carry out the upsetting process, as shown in Figure 9.13. The pressure ascends symmetrically from the edges of the billet, reaching maximum value at the center, forming what is known as the friction hill. The average pressure, and accordingly, the force required to carry out the upsetting process, will therefore increase due to friction.

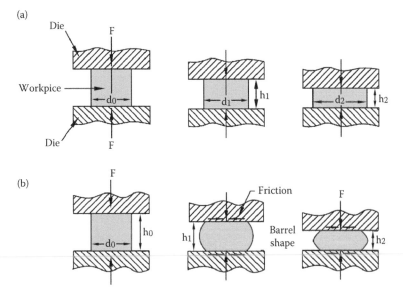

FIGURE 9.13 (a) Ideal deformation of a solid cylindrical specimen compressed between flat frictionless dies. (b) Deformation in upsetting with friction at the die–billet interfaces. (Adapted from Kalpakjian, S., and Schmid, S. R., *Manufacturing Processes for Engineering Materials*, Prentice-Hall, Upper Saddle River, NJ, 2003.)

9.3.6.1   Direct Compression (Upsetting) in Plane Strain

Figure 9.14 represents a billet with a rectangular section under compression through platens. The billet width is a, and height is h. The figure also shows the stresses on a vertical element in that section. As discussed in Chapter 8, plane strain condition assumes that the thickness of the section in the third direction (b) is much larger than the side (a), such that there is no strain in that thickness, i.e., ($h < a \ll b$). This condition resembles actual open-die forging of long shafts or blooms between flat platens.

Considering a force balance on the element (slab) in the x direction

$$\sigma_x h - 2\mu(-p)dx - (\sigma_x + d\sigma_x)h = 0$$

$$2\mu p dx = h d\sigma_x$$

σ_x and σ_y (taken as $-p$) are principal stresses. For plane strain deformation, the flow stress (σ_f) is calculated from Equation 8.28, and considering the maximum difference between the two yield criteria as stated in Section 8.4, then

$$\sigma_x - (-p) = \sigma_f' = 1.15\,\sigma_f,$$

where σ_f' is the flow yield stress under plane-strain = 1.15 σ_f.

Considering the flow stress to be constant (perfect plastic material), differentiation of the above equation gives: $d\sigma_x = -dp$

$$2\ pdx = -hdp \quad \text{or} \quad \frac{dp}{p} = \frac{-2}{h}\,dx$$

At $x = a/2$, $\sigma_x = 0$, and $p = \sigma_f'$, so the solution is

$$\frac{p}{\sigma_f'} = \exp\left[\frac{2}{h}\left(\frac{a}{2} - x\right)\right] \tag{9.2}$$

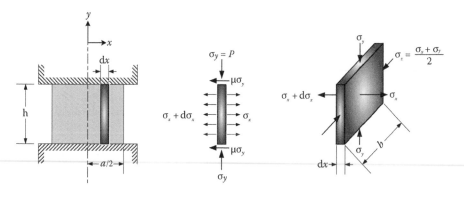

FIGURE 9.14 Stresses on an element in plane-strain upsetting between flat dies. (From Kalpakjian, S., and Schmid, S. R., *Manufacturing Processes for Engineering Materials*, Prentice-Hall, Upper Saddle River, NJ, 2003. With permission.)

The maximum value of P occurs at the center line, with a maximum value

$$\left.\frac{p}{\sigma_f}\right|_{\max} = \exp\left(\frac{a}{h}\right) \tag{9.3}$$

A plot of p/σ_f versus x gives the friction hill, as shown in Figure 9.15a.
To calculate σ_x

$$\sigma_x = p - \sigma_f = \sigma_f \left(e^{\frac{2}{h}\left(\frac{a}{2}-x\right)} - 1 \right) \tag{9.4}$$

Average pressure: The mean, or average, pressure is of great interest, as it can be used to calculate the applied force on the contact area.

$$p_a = \frac{2}{a}\int_0^{a/2} p\,dx = \frac{2}{a}\int_0^{a/2} \sigma_f\, e^{\frac{2}{h}\left(\frac{a}{2}-x\right)}\,dx$$

$$p_a \approx \sigma_f\left(1+\frac{a}{2h}\right) \tag{9.5}$$

Then the upsetting force is given by: $F = p_a\, ab$.

Sticking friction at the interface: To avoid shearing of the workpiece at the billet–platen inter-face, $\mu p \leq \tau_f$, as explained in Section 8.10, where τ_f is the shear yield strength of the billet material (equal to half the flow stress). Since, $p/2\tau_f \geq 1$, thus $\mu \leq 0.5$ if sliding friction is to take place. If the limit at which sliding friction is exceeded, sticking takes place. To calculate the pressure distribution under sticking, the frictional forces indicated previously as μp are replaced by the shear yield strength $\tau_f = 0.5\,\sigma_f$ in the previous analysis, which leads to

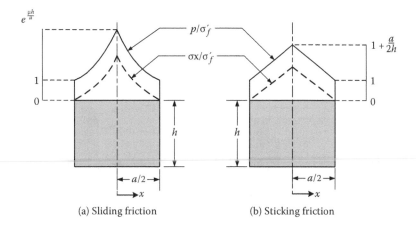

(a) Sliding friction   (b) Sticking friction

FIGURE 9.15 Pressure distribution and horizontal stress distribution in plane-strain upsetting with (a) sliding and (b) sticking friction.

$$\frac{p}{\sigma_f} = 1 + \frac{a/2 - x}{h} \tag{9.6}$$

This equation represents a linear variation of p with x, i.e., linear friction hill. The maximum value which occurs at the centerline is

$$p_{max} = \sigma_f \left(1 + \frac{a}{2h}\right) \tag{9.7}$$

A plot of the pressure distribution under sticking friction is shown in Figure 9.15b, for comparison with the case of sliding friction. The average pressure with sticking friction is

$$P_a = \sigma_f \left(1 + \frac{a}{h}\right) \tag{9.8}$$

Since pressure is maximum at the center, then sticking starts primarily at the center of the interface area and spreads outward. However, there could be a case where the outer parts may still be under sliding friction. The point of intersection x_1 between the exponential relation represented by Equation 9.2, and the linear relation represented by Equation 9.6, is the point separating the sticking and the sliding area, where

$$x_1 = \frac{a}{2} - \frac{h}{2} \ln \frac{1}{2} \tag{9.9}$$

9.3.6.2 Upsetting a Solid Cylindrical Specimen

The upsetting of a cylindrical workpiece is modeled using the same analysis followed with a rectangular specimen in the previous section. However, cylindrical coordinates are applied as shown in Figure 9.16, using stress components σ_r in the radial direction, σ_θ in the tangential direction, and σ_z in the axial direction.

Following a similar approach for the case of plane-strain, similar relations can be derived. Equilibrium leads to the relation

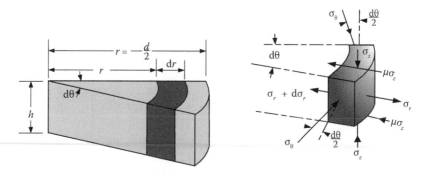

FIGURE 9.16 Stresses on an element during upsetting of a cylindrical specimen. (From Kalpakjian, S., and Schmid, S. R., *Manufacturing Processes for Engineering Materials*, Prentice-Hall, Upper Saddle River, NJ, 2003. With permission.)

$$\sigma_r hrd\theta + 2 \quad Pr\,d\theta dr + \frac{2\sigma_\theta hdrd\theta}{2} - (\sigma_r + d\sigma_r)h(r + dr)d\theta = 0$$

$$2\mu Prdr + h\sigma_\theta dr - h\sigma_r dr - hrd\sigma_r = 0$$

For axisymmetric flow, $\varepsilon_r = \varepsilon_\theta$, then; $\sigma_r = \sigma_\theta$.

For yielding, $\sigma_z - \sigma_r = \sigma_f$, back substitution gives the pressure distribution

$$\frac{p}{\sigma_f} = e^{\frac{2}{h}\frac{d}{2}-x} \tag{9.10}$$

Notice the similarity with Equation 9.2, in that the flow stress is used in Equation 9.10 directly without multiplying by 1.15. This is indicative that in the case of axial symmetry, the two yield criteria coincide; Section 8.4.

The average pressure $\hspace{4cm} p_a \approx \sigma_f \left(1 + \frac{2}{3}\frac{r}{h}\right) \tag{9.11}$

The upsetting force is $\hspace{4cm} F = p_a(\pi d^2/4)$

The value of the coefficient of friction μ can be estimated to be 0.05 to 0.1 for cold upsetting, and 0.1 to 0.2 for hot forging.

Under sticking friction, the stress distribution is linear

$$p = \sigma_f \left(1 + \frac{(d/2) - x}{h}\right) \tag{9.12}$$

Notice the similarity of equations for both plane-strain and cylindrical billets, and same trends of the results representing change of the upsetting pressure with the width/height (aspect) ratio at both conditions, as shown in Figure 9.17. It should be observed that higher aspect ratios lead to higher pressures at the same frictional conditions. Notice also that the pressures are higher for a

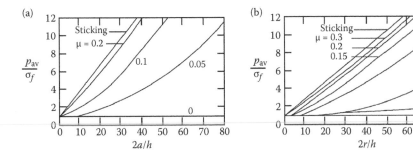

FIGURE 9.17 Ratio of the average pressure to yield stress as a function of friction and aspect ratio of the specimen: (a) plane-strain upsetting and (b) upsetting of a cylindrical specimen. (From Kalpakjian, S., and Schmid, S. R., *Manufacturing Processes for Engineering Materials*, Prentice-Hall, Upper Saddle River, NJ, 2003. With permission.)

plane-strain specimen compared to a cylindrical specimen with the same aspect ratio, and the same frictional conditions.

9.3.6.3 Modeling of Impression-Die Forging

Accurate calculation of forces in impression-die forging is difficult. To simplify force calculation, a pressure-multiplying factor K_p is recommended

$$F = K_p \sigma_f A \tag{9.13}$$

where F is the forging load, A is the projected area of the forging (including the flash), and σ_f is the flow stress of the material. Typical K_p ranges are shown in Table 9.2.

The capacity of the press to be used should have a rated maximum force well above the estimated load. During forging under hot working conditions, the strain rate affects the stresses required to carry out the process. The strain rate is calculated as given in Equation 8.31

$$\varepsilon = \frac{d\varepsilon}{dt} = \frac{d}{dt} \ln \frac{h_0}{h_1} = \frac{1}{h_1} \frac{dh_0}{dt} = \frac{v}{h_1}$$

where v is the relative velocity between the platens. To obtain an approximate estimate of the strain rate in impression die forging

$$\varepsilon_m = \frac{v}{h_m} = v \frac{A}{V} \tag{9.14}$$

where V is the volume of the metal and A is the projected area.

Similarly, an average strain may be estimated as

$$\varepsilon_m = \ln \frac{h_0}{h_m} = \ln \frac{h_0 A}{V} \tag{9.15}$$

As hammers are rated according to their energy, the energy required for forging is calculated approximately from the equation

$$E = \sigma_f \varepsilon_m V K_h \tag{9.16}$$

where E is the forging energy and K_h is the multiplying factor for hammers obtained from Table 9.2. The rated hammer capacity should be greater than the estimated energy. In hot forging, forces are much reduced with heated dies, but these preheated dies require very slow forging speeds.

TABLE 9.2
Pressure-Multiplying Factors for Closed-Die Forging Processes

Forging Shape	K_p	K_h
Simple shapes, without flash	3–5	2–2.5
Simple shapes, with flash	5–8	3
Complex shapes, with flash	8–12	4

Example 1

A 150 × 450 × 100 mm height specimen, made of annealed 4135 steel ($K = 1015$ MPa, $n = 0.17$), is upset between flat dies to a height of 50 mm at room temperature. Assuming that the coefficient of friction is 0.1, calculate the maximum pressure and the force required at the end of the stroke. Check for sticking.

Solution
Given

$a_0 = 150$ mm, $h_0 = 100$ mm, $h_f = 50$ mm, $b_0 = b_f = 450$ mm, $K = 1015$ MPa, $n = 0.17$, $\mu = 0.1$

$$\text{Volume} = \text{constant} = a_0 b_0 h_0 = a_f b_0 h_f \text{ (no change in } b)$$

$$a_f = a_0 h_0/h_f = 150 \times 100/50 = 300 \text{ mm}$$

$$\varepsilon_h = \ln(h_0/h_f) = \ln(100/50) = 0.693$$

The flow stress: $\sigma_f = K\varepsilon^n = 1015 \times (0.693)^{0.17} = 953.65$ MPa
For plane-strain: $\sigma_f' = 1.15\,\sigma_f = 1.15 \times 953.65 = 1096.7$ MPa
From Equation 9.3: $p_{max} = \sigma_f' \exp(\mu a_f/h_f) = 1096.7 e^{0.1 \times 300/50} = 1998.3$ MPa

From Equation 9.5: $p_a \approx \sigma_f' \left(1 + \dfrac{a}{2h}\right) = 1096.7 \left(1 + \dfrac{0.1 \times 300}{2 \times 50}\right) = 1425.71$ MPa

Force: $F = p_a(a_f\,b) = 1425.71\,(300 \times 450) \times 10^{-6} = 192.47$ MN
Check for sticking.

From Equation 9.9: $x_1 = \dfrac{a}{2} - \dfrac{h}{2}\ln\dfrac{1}{2} = \dfrac{300}{2} - \dfrac{50}{2 \times 0.1}\ln\dfrac{1}{2 \times 0.1} = -252.36$

x_1 is beyond the size of the specimen, then there is no sticking.

9.4 ROLLING PROCESSES

The primary objectives of the rolling process are to reduce the cross section of the incoming material while improving its properties, and to obtain the desired section at the exit from the rolls. The process can be carried out hot, warm, or cold, depending on the application and the material involved.

During rolling, the desired shape of metal is obtained by plastic deformation between two work rolls rotating in opposite directions at the same speed. Rolling of metallic materials is perhaps the most widely used metal forming process. More than 90% of all of the steel, aluminum, and copper produced worldwide goes through the rolling process at least one time. The principal advantage of rolling lies in its ability to produce desired shapes from relatively large pieces of metals at very high speeds in a somewhat continuous manner. Because other methods of metalworking, such as forging, are relatively slow, most ingots and large blooms are rolled into billets, bars, structural shapes, rods (for further drawing into wire), and rounds for making seamless tubing. Steel slabs are rolled into plates, sheets, strips, and foils. Accordingly, the sequence of rolling processes for producing different sections is demonstrated in Figure 9.18.

9.4.1 FLAT ROLLING

It is the process of reducing the thickness of a slab of metal between cylindrical (flat) rolls to get thinner and much longer products. Typical flat rolled products are

a. *Plates:* Usually hot rolled to thicknesses over 6 mm and widths of 1800–5000 mm. These plates are used in shipbuilding, boiler bodies, storage tanks, large structures, etc.

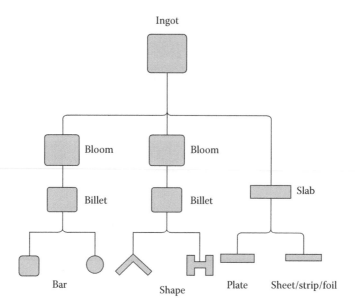

FIGURE 9.18   Sequence of rolling processes for production of bars, shapes, and flat products from blooms and slabs. (From ASM International, *ASM Handbook, vol. 14: Forming and Forging*, Materials Park, OH, 1988. With permission.)

b. *Sheets and strips:* Hot rolled sheets with thicknesses of 0.8 mm up to 6 mm are required for cold pressing of structural parts of vehicles and heavy machinery, as well as for making welded tubes. Thinner gauges, better surface finish, and closer tolerances are obtained by cold rolling in coiled up form. Coils are further slit to strips (narrower width), or cut into shorter lengths, or both.

c. *Foils:* It is possible to roll steel sheets to foils with a thickness of down to 3 µm. Aluminum foils are produced by rolling to a thickness of 8 µm, and are widely used as disposable trays and for heating food.

Flat rolling mills: The principle of a flat rolling mill is shown in Figure 9.19, where two rolls of equal diameter rotate at the same speed in opposite directions. Each roll includes a cylindrical part called a barrel, two necks of smaller diameter to be mounted on the mill frame bearings, and tenon ends (or wobblers) for coupling to the spindles for power and speed transmission. Rolls are usually made of cast iron or forged steel. The mill frame should have very rigid construction to withstand the high rolling forces. The gap between rolls is precisely controlled through axial movement to allow setting up of the required reduction.

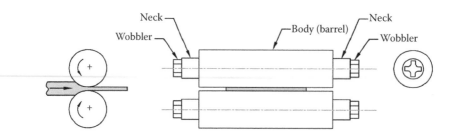

FIGURE 9.19   Flat rolling mills.

Flat rolling mills can be conventionally classified with respect to the number or the arrangement of rolls, as shown in Figure 9.20.

    a. *Two-high mill* is the simplest and most common type of rolling mill for hot rolling. The rolls are rotated only in one direction. When rolling to a specific reduction is finished, the slab is returned to the mill entrance, and the gap is reduced before the next reduction.

    b. *Two-high reversible mill*, where the work can be passed forward and backward through the rolls by reversing their direction of rotation. This type is called *slabbing mill*, for rolling of slabs, or *cogging mill*, for blooms and billets, and is suitable for hot rolling of *steel ingots.*

    c. *Three-high mill*, which consists of upper and lower driven rolls and a middle roll rotating by friction. This type is introduced to speed up the *production rate* and reduce the power and *transmission mechanisms*. It has also been adopted for hot rolling.

    d. *Four-high mill.* To roll thin gauges to close dimensional tolerances, small diameter rolls are needed. However, these may deflect under high forces *during rolling*, leading

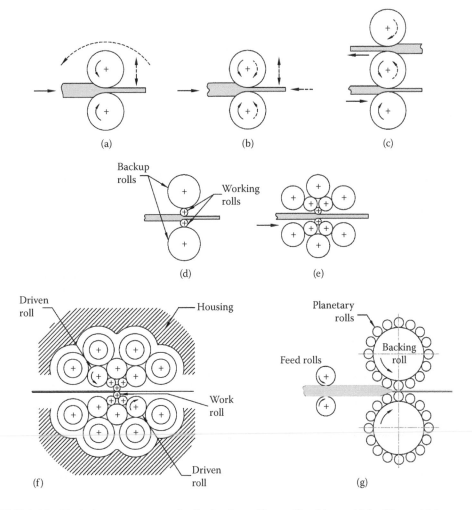

FIGURE 9.20 Typical arrangements of rolls for flat rolling mills: (a) two-high, (b) two-high reversing, (c) three-high, (d) four-high, (e) cluster, (f) Sendzimir, and (g) planetary.

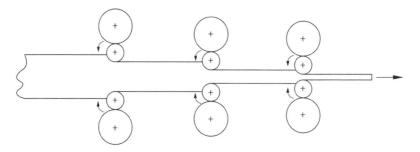

FIGURE 9.21   Schematic presentation of a three-stand tandem mill for slab rolling to coiled sheets.

to a section with *higher thickness* in the middle. This deflection can be eliminated by supporting the working rolls with two larger diameter backup rolls.

e. *Cluster mill.* For very thin sheets, small diameter work rolls need to be supported by more backup rolls. The cluster mill may have 6 or 12 rolls.

f. *Sendzimir mill* is a modification of the cluster mill, with 20 rolls. It is adopted for cold rolling of hard materials such as stainless steel.

g. *Planetary mill* was introduced by Krupp for the production of thin strips up to 40 mm wide. Backing rolls carry the work rolls on cages. Each working roll adds very little reduction, and thus the mill can yield high reductions of 80%–90% in a single pass.

Continuous Rolling Mill (Tandem Mill): For high production rates, a series of rolling mills are installed in tandem form, as shown in Figure 9.21. Each set of rolls is called a stand. Different reductions are carried out at different stands, and speeds are synchronized such that the delivery speed from each stand is the same input speed of the next one. This technique was recently adopted for thermomechanical rolling, which involves control of both thermal and mechanical treatments through the rolling stand for the purpose of producing thin-gauge steel sheets with high strength and high toughness/ductility (HSLA steels).

9.4.2 SECTION ROLLING

For the manufacturing of bars and standard structural sections (refer to Figure 9.18), section rolling is applied by using grooved rolls. The process starts with hot rolling in the cogging mill (usually the two-high reversible mill), to break down the ingot structure and to reach the bloom size (usually 500 × 500 mm square section). A further reduction by hot rolling leads to square billets with a minimum cross section of 40 × 40 mm. Blooms and billets represent the starting material for section rolling operations. The finished shape is reached through a number of passes that deform the material uniformly, avoiding excessive secondary tensile stresses which may lead to crack formation. This can be achieved by rotating the material sidewise, especially in the early passes. A typical sequence of passes used in the production of 12.5 diameter wire rod from a 100 mm square billet is shown in Figure 9.22. Notice the alternate sidewise rotation of the material from oval to diamond-shaped grooves to achieve uniform flow. Fins bulging through the roll grooves during deformation should be avoided.

Roll pass design for the production of complicated shapes such as standard structural sections is based on experience and differs from one manufacturer to another, even for the same final rolled section geometry. Relatively few quantitative data on roll pass design is available in literature. Recently, some manufacturers have developed their own computer program packages for that purpose. Progressive profile rolling of a channel section is shown in Figure 9.23.

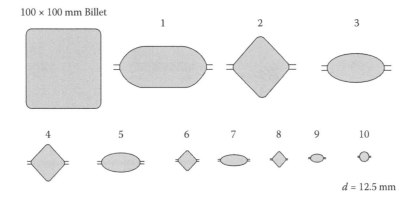

FIGURE 9.22 Typical sequence of roll passes for the production of round bar. (From Amstead, B. H. et al., *Manufacturing Processes*, 7th, John Wiley & Sons, New York, 1979. With permission.)

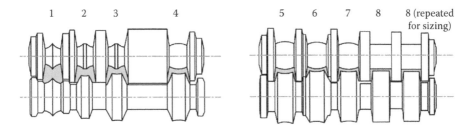

FIGURE 9.23 Progressive profile rolling of a channel section. (From Waters, F., *Fundamentals of Manufacturing for Engineers*, Taylor & Francis, Boca Raton, FL, 1996. With permission.)

9.4.3 TUBE ROLLING

The rolling techniques that are adopted for producing tubes and pipes are classified into seamless and seamed tubes. The first category is based on rotary piercing of bars and on subsequent size adjusting processes such as the Pilger process. The second type is based on bending plate or sheet strips and welding the contact edges.

9.4.3.1 Rotary Piercing (Mannesmann Process)

In this process, the heated bar is rotated between two slightly tapered barrel-shaped rolls, as shown in Figure 9.24. The axes of rotation of the rolls are skewed at a small angle (α) on either side of the rod axis of rotation to guarantee natural feeding of the bar between rolls. Both rolls rotate in the same direction. The spinning action of the rolls on the bar creates lateral tensile stresses at the center, as shown on Part c of the figure, leading to the formation of a cavity. The mandrel ensures the axial formation of the bore, and controls its internal surface.

9.4.3.2 Three-Roll Rotary Piercing

In the Mannesmann process, internal cracks may form in an irregular manner which may lead to tearing apart of the bar. The recently established three-roll piercer provides a uniform state of lateral compressive stresses at the bar center ahead of the mandrel, which reduces or eliminates internal defects. However, the pressure exerted by the mandrel is supplied externally to form the bore. The three rolls are oriented at 120° around the center, as shown in Figure 9.25, and their axes are inclined to the horizontal pass line by skew angle less than 15°. This process allows the production

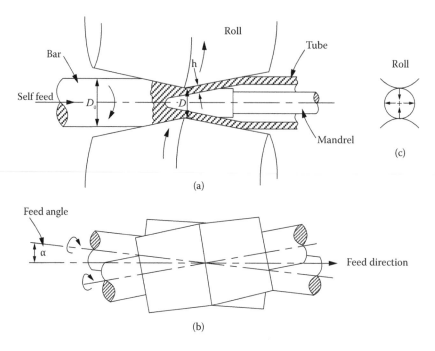

FIGURE 9.24 Schematics of the (a) rotary piercing Manessmann process, (b) tools orientation of the rolls with respect to the bar, and (c) stresses in the center of the bar. (Adapted from Koshal, D., *Manufacturing Engineer's Reference Book*, Butterworth-Heinemann, 1993.)

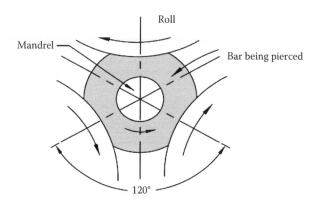

FIGURE 9.25 Schematic presentation of three-roll rotary piercing. (Adapted from Koshal, D., *Manufacturing Engineer's Reference Book*, Butterworth-Heinemann, 1993.)

of a seamless tube directly from continuously cast steel billet. Three-roll piercers can handle billets of up to 200 mm diameter and 3.8 m long.

9.4.3.3 The Pilger Process

The Pilger process is a special rolling process for reducing the size of tubes manufactured by rotary piercing. The Pilger mill is a two-high mill fitted with semicircular grooves that taper from a large to a small diameter over a small angle, and remains constant for the rest of the circumference. The tube to be sized is connected to a mandrel, which overhangs into the roll cavity, and the mandrel

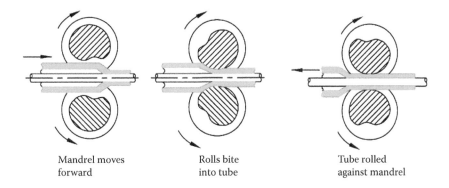

| Mandrel moves forward | Rolls bite into tube | Tube rolled against mandrel |

FIGURE 9.26   Pilger rolls and principle of the Pilger process. (Adapted from Higgins, R. A., *Engineering Metallurgy, 2: Metallurgical Process Technology*, Hodder and Stoughton, 1974.)

itself is connected to the piston of a pneumatic cylinder through a mechanism that rotates the tube 90° at each process step, in addition to moving it forward and backward; Figure 9.26.

When the rolls rotate to the open position (Figure 9.26a), the piston is pushed forward, so that the tube and mandrel are pushed into the roll gap. Continued rotation of the rolls brings the narrow section into contact with the tube so that they bite into it, isolating a small collar of metal, as shown in Figure 9.26b, and deforms it onto the mandrel. The whole mandrel assembly is forced backward against the pneumatic pressure (Figure 9.26c), which irons out that portion of the tube to the smaller outside diameter. The rolls finally reach the open position, and a new cycle starts. As the deformed tube becomes longer, it moves off the end of the mandrel to a bed of rollers. Finally, a small rim remains at the end of the tube, which is cut off, and the tube passes several times through sizing rolls (Figure 9.27), being turned through 90° between passes. This ensures uniformity of the outside diameter of the tube.

9.4.3.4   Rolling of Seamed Tubes

As stated earlier, seamed or welded tubes are produced by bending plates or sheets either in a three-roll bending machine or in a series of roll bending stands, and then welded along the line of

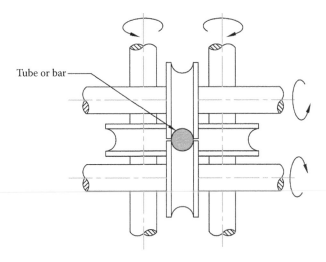

Tube or bar

FIGURE 9.27   Sizing rolls. (Adapted from Higgins, R. A., *Engineering Metallurgy, 2: Metallurgical Process Technology*, Hodder and Stoughton, 1974.)

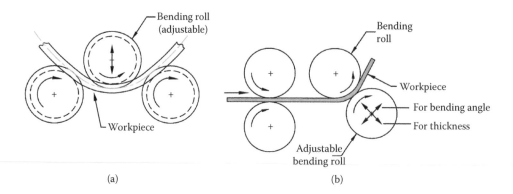

FIGURE 9.28   (a) Three- and (b) four-roll bending arrangement. (Adapted from ASM International, ASM International, *ASM Handbook, vol. 14: Forming and Forging*, Materials Park, OH, 1988.)

contact. In a three-roll bending machine (Figure 9.28a), the axes of the two bottom rolls are fixed in a horizontal plane, while the top roll is vertically adjustable to control the diameter of the produced tube. The three rolls are power driven. A roll arrangement for four-roll bending is shown in Figure 9.28b. The plate or sheet is driven between the two powered rolls on the left, and the lower bending roll is adjusted in two directions according to the thickness and bending angle.

9.4.4   Special Rolling Processes

These are based on special purpose rolling equipment developed for the production of special product shapes. Typical examples are

- a. *Rolling of rail wheels*, where the wheel is first forged and then rolled to the required final shape, as shown in Figure 9.29.
- b. *Ring rolling*, for controlling the sizes of annular forgings that are accurately dimensioned, and have circumferential grain flow. As the radial cross-section decreases, the ring diameter increases. The work rolls may be plain, producing uniformly rectangular ring cross-sections, as shown in Figure 9.30(a), or may have grooves or flanges to produce contoured ring cross-sections. Radial rolls can be used to control the ring thickness, as shown in Figure 9.30b.
- c. *Rolling of steel balls*, from bar for the ball bearing industry (Figure 9.31).

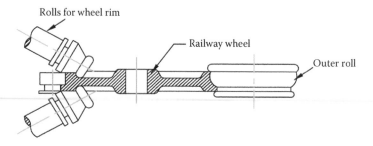

FIGURE 9.29   Rolling of rail wheel. (Adapted from Polukhin, P., *Metal Process Engineering*, Peace Publishers, Moscow, 1973.)

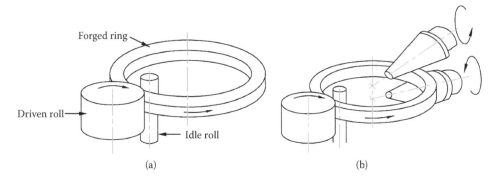

FIGURE 9.30 Schematic presentation of ring rolling: (a) single-pass (radial) rolling and (b) two-pass (radial–axial) rolling.

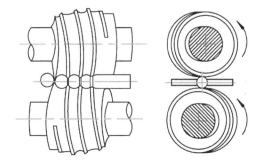

FIGURE 9.31 Rolling of steel balls. (Adapted from Lange, K., *Handbook of Metal Forming*, McGraw-Hill, New York, 1985.)

  d. *Cross-rolling of multidiameter shafts* (such as dumbbell shapes), using aligned roll axes, and the shaft is rotating in the same plane, as shown in Figure 9.32. The process is applicable for limited bar diameters of 10–130 mm, and lengths of up to 700 mm.

  e. *Roll forging* (hot forge rolling), for reducing the cross-sectional area of heated bars or billets by passing them between two driven rolls that rotate in opposite directions and have one or more matching grooves in each roll (Figure 9.33).

  f. *Thread rolling*, for the production of screw threads, which offers higher productivity (up to 500 threads/min), and higher fatigue resistance compared to machined threads. Thread rolling could be (a) between flat dies or (b) between two rolls (Figure 9.34).

9.4.5 ROLLING DEFECTS

There are many defects that may appear on a rolled product as a result of inaccuracies in the rolling mill or on the work material. These defects include

  a. *Bowing of the sheet:* Due to nonparallelism of the roll gap. If one edge of the sheet is smaller in thickness than the other, the length at the smaller edge will be longer.

  b. *Out-of-flatness:* Minor differences in elongation between different parts of the sheet during rolling, due to bending of the rolls, affect the flatness of thin sheets, and give rise to waviness, which may result in wavy edges or central cracks (zipper breaks), as shown in Figure 9.35. The obvious solution to roll bending is to use a cambered roll (barrel

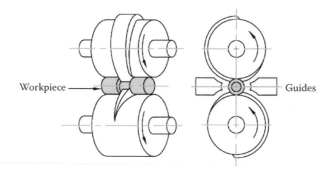

FIGURE 9.32   Cross-rolling of a dumbbell-shaped product.

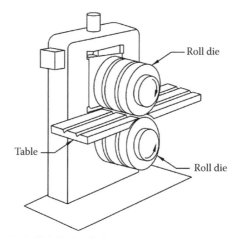

FIGURE 9.33   A roll forging machine.

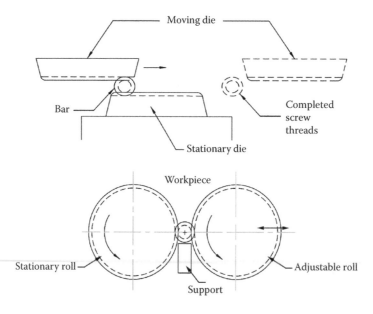

FIGURE 9.34   Thread rolling processes: (a) with flat die and (b) between two rolls. (Adapted from Lange, K., *Handbook of Metal Forming*, McGraw-Hill, New York, 1985.)

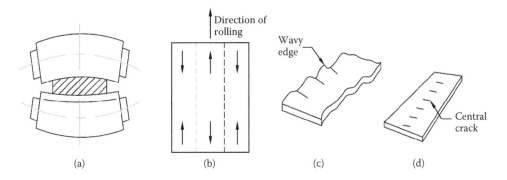

(a)   (b)   (c)   (d)

FIGURE 9.35 Consequences of roll bending (a), producing tensile stresses in the center, and compression on the edges (b), which results in wavy edge (c), or central cracks (zipper break) at the center (d). (From Dieter, G. E., *Mechanical Metallurgy*, McGraw-Hill, New York, 1988. With permission.)

shaped) so that, when deflected, it will present a parallel gap. The camber is generally less than 0.25 mm. A better solution is to use a rolling mill with hydraulic jacks, which permits elastic distortion on the rolls to correct deflection. Alternatively, a large number of backup rolls is applied, as shown in Figure 9.20d–f. A recently developed innovation is applied in thin sheet rolling, called "continuously variable crown," or CVC, where rolls are ground to an S shape and are allowed axial shift to control the thickness, as shown in Figure 9.36.

c. *Center buckles*, which result when the edges of the sheet are restricted relative to the central part. This effect is opposite to the out-of-flatness defect.

d. *Defects due to lateral spread*, despite plain strain conditions which restrict deformation in the sheet width, there is a tendency for expansion, which is opposed by frictional force. The result is a friction hill with maximum pressure at the sheet center, which suggests higher tendency of the edge parts to spread laterally, with less elongation in the length direction. This may develop a slight rounding at the ends, as shown in Figure 9.37(a). Alternatively, edge cracking may arise; Figure 9.37b. Under severe conditions, the center of the sheet may split, as indicated in Figure 9.37c. Edge cracking may also arise due to inhomogeneous deformation in the thickness direction. These effects can be eliminated by employing two vertical rolls at the edges during rolling to stop lateral expansion.

e. *Alligatoring* may arise due to excessive barreling of the side surfaces of the sheet during rolling, which creates secondary tensile stresses. These stresses lead to lateral tension on the surface and compression at the center. With the help of metallurgical weakness

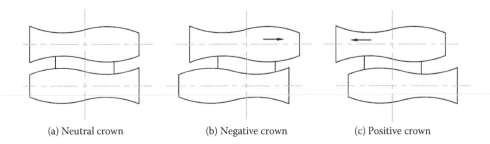

(a) Neutral crown   (b) Negative crown   (c) Positive crown

FIGURE 9.36 The continuously variable crown (CVC) principle: (a) neutral crown control, (b) negative crown control, (c) positive crown control. (From Nandan, R. et al., *Materials and Manufacturing Processes*, Taylor & Francis, Boca Raton, FL, 2005. With permission.)

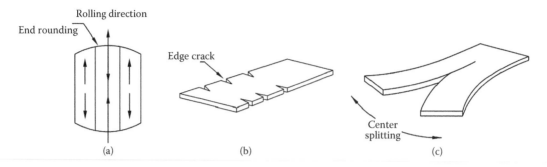

FIGURE 9.37 Defects due to lateral spread: (a) end rounding, (b) edge cracking, or (c) central split. (From Dieter, G. E., *Mechanical Metallurgy*, McGraw-Hill, New York, 1988. With permission.)

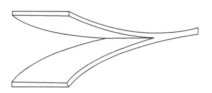

FIGURE 9.38 Alligatoring of the slab central plane.

along the central plane of the slab, the end of the sheet is split on the central plane, as shown in Figure 9.38.

f. *Residual stresses:* Small diameter rolls or small reductions tend to work the material plastically at the surface, which generates compressive residual stresses at the surface and tensile compressive stresses at the center, as shown in Figure 9.39a. On the other hand, large diameter rolls and high reductions tend to deform the central part of the thickness more than the surface, leading to opposite residual stresses; Figure 9.39b.

g. *Fissures* are internal defects due to incomplete welding of pipe or blowholes in the ingot before rolling.

h. *Longitudinal stringers* of nonmetallic inclusions. In severe cases, these and fissures may lead to lamination, which reduces the strength drastically in the thickness direction.

9.4.6 MODELING OF FLAT ROLLING

The flat rolling of plates and sheets is a typical plane-strain operation, since the reduction in the thickness from h_0 to h_f leads to increase of the length, with little or no widening of the width ($w_0 = w_f = w$) as shown in Figure 9.40.

Due to volume constancy of the strip, $h_0 w L_0 = h_f w L_f$.

Dividing the length L_0 and L_f by time introduces the strip speed before deformation V_0 and the final speed after deformation V_f, respectively. Therefore

$$h_0 V_0 = h_f V_f \tag{9.17}$$

A principal definition in flat rolling is the percentage reduction in thickness, where

$$\% \text{ reduction in thickness} = \frac{h_0 - h_f}{h_0} \times 100 = \frac{\Delta h}{h_0} \times 100 \tag{9.18}$$

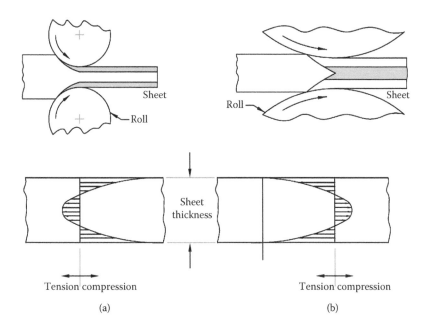

FIGURE 9.39   Residual stresses in flat rolling (a) small rolls, or small reduction in thickness, and (b) large rolls, or large reduction in thickness. (From Kalpakjian, S., and Schmid, S. R., *Manufacturing Processes for Engineering Materials*, Prentice-Hall, Upper Saddle River, NJ, 2003. With permission.)

The angle α is the contact angle between the roll and strip, or the angle of bite, and the roll arc in contact with the strip is the arc of contact. The contact length l is the projection of the arc of contact.

To calculate the contact angle in terms of the roll diameter D, and the reduction in thickness Δh, using simple trigonometry, the following equation is used:

$$\cos \alpha = 1 - \frac{\Delta h}{D} \tag{9.19}$$

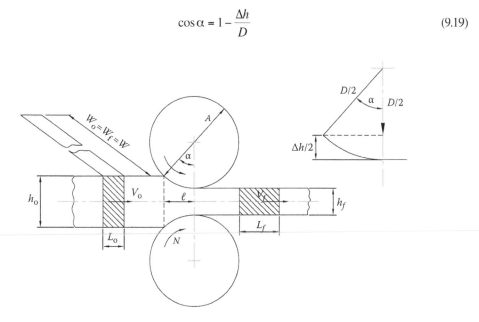

FIGURE 9.40   Schematic presentation of the flat rolling model.

The contact length l is calculated from the equation

$$l = \frac{D}{2}\sin\alpha = \sqrt{\frac{D}{2}\Delta h - \frac{(\Delta h)^2}{4}} \approx \sqrt{\Delta h \frac{D}{2}} \tag{9.20}$$

Condition of biting: At the beginning of contact between the roll and strip, a normal force N, and a tangential friction force ($T = \mu N$) arise, as shown in Figure 9.41. If the strip is to be pushed to deform between the rolls, the resultant of the horizontal components of these forces must be in the positive direction of rolling, i.e.,

$$T\cos\alpha \geq N\sin\alpha, \text{ or } \mu \geq \tan\alpha \tag{9.21}$$

Then the rolls bite the strip if the coefficient of friction is at least equal to tan α. If this condition is not fulfilled, a pushing force is given to the strip, or it should be accelerated before biting to develop an inertia force in the rolling direction. In hot rolling of ingots or slabs with large thickness, the roll surface is ragged to increase the coefficient of friction. For cold rolling, α ranges from 3° to 8°, and μ ranges between 0.02 and 0.3, while for hot rolling, α ranges between 18° and 24°, and μ ranges from 0.2 with good lubrication, to as high as 0.7, approaching sticking conditions. The maximum possible draft Δh in flat rolling is a function of friction and roll diameter

$$\Delta h_{max} = {}^2\frac{D}{2} \tag{9.22}$$

Relative velocities in the deformation zone: Referring to Equation 9.17, the strip velocity at the entrance V_0 increases through the deformation zone to the exit velocity V_f due to the progressive decrease in thickness. The horizontal component of surface velocity of the roll (V_r) is $V_r\cos\theta$ where θ is any angle between α at entry and zero at the exit. When plotting both strip velocity relation and horizontal roll velocity component relation, they intersect at a point n, as shown in Figure 9.42. At

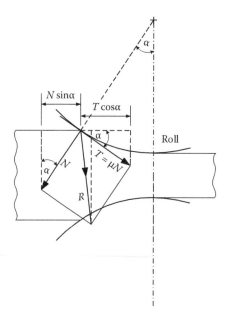

FIGURE 9.41   Condition of biting in flat rolling.

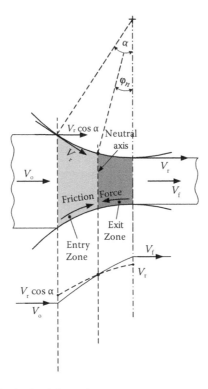

FIGURE 9.42 Relative velocities in the deformation zone.

this point, there is no relative velocity or slip between the roll and the strip, which is known as the neutral point, and the vertical axis cutting this point is the neutral axis. At the exit side of the neutral axis, the strip velocity is higher than the horizontal component of the roll velocity, which indicates forward slip of the strip relative to the roll, and frictional force toward the neutral axis. To the right of the neutral axis, on the other hand, the strip velocity is lagging, and frictional forces are in the direction of the neutral axis. Frictional forces are therefore shown on the figure in opposing directions, similar to upsetting (Figure 9.13). The relative movements and friction forces in the deformation zone in rolling are then analogous to the plane strain upsetting problem, with one exception that the flat platens in upsetting are replaced by circular platens in rolling. The friction hill phenomenon is thus valid for rolling, with the maximum pressure at the neutral axis.

Forward slip: This is a measure of the relative velocities involved at the exit

$$\text{Forward slip} = \frac{V_f - V_r}{V_r} \tag{9.23}$$

Roll pressure distribution: Considering the equilibrium of two elements on both sides of the neutral axis of the deformation zone, as shown in Figure 9.43, analysis of forces in the x direction, gives

$$(\sigma_x + d\sigma_x)(h + dh) - 2p \sin\phi \, Rd\phi - \sigma_x h \pm 2\mu p \cos\phi \, R \, d\phi = 0$$

Separation of variables and considering p and σ_x as principal stress, then

$$p - \sigma_x = \sigma_f = 1.15 \, \sigma_f$$

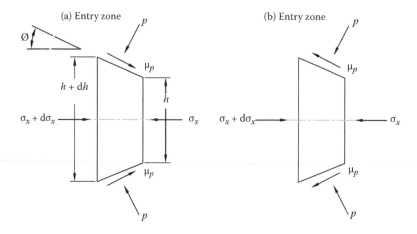

FIGURE 9.43 Stresses on two elements on both sides of the neutral axis. (From Kalpakjian, S., and Schmid, S. R., *Manufacturing Processes for Engineering Materials*, Prentice-Hall, Upper Saddle River, NJ, 2003. With permission.)

An exponential equation is reached in the form

$$\frac{p}{\sigma_f} = C \frac{h}{R} e^{\pm H}$$

where $H = 2\sqrt{\frac{R}{h_f}} \tan^{-1} \sqrt{\frac{R}{h_f}}\phi$, ϕ is the angle at any element.

At entry, $\phi = \alpha$; hence, $H = H_0$, with ϕ replaced by α. At exit, $\phi = 0$; hence $H = H_f = 0$.

Also, at entry and exit, $p = \sigma_f$. Therefore, the rolling pressure distribution is presented by the equations

Entry zone side $\qquad\qquad\qquad\qquad \dfrac{p}{\sigma_f} = \dfrac{h}{h_0} e^{(H_0 - H)}$ $\qquad\qquad\qquad$ (9. 24)

Exit zone side $\qquad\qquad\qquad\qquad \dfrac{p}{\sigma_f} = \dfrac{h}{h_f} e^{H}$ $\qquad\qquad\qquad\qquad$ (9.25)

Figure 9.44a shows the pressure distribution at different coefficients of friction, and Figure 9.44b shows the pressure distribution in the roll gap as a function of reduction in thickness. The pressure distribution indicates a friction hill similar to the upsetting model. Note that as friction increases, the maximum pressure increases, and the neutral point shifts toward the entry side. Without friction, the rolls slip instead of pulling the strip, and the neutral point shifts completely to the exit. Note also the increase in the area under the curves with increasing reduction in thickness, thus increasing the roll force.

9.4.7 Determination of the Neutral Point

The neutral point can be determined by equating the pressure distribution equations at entry and exit:

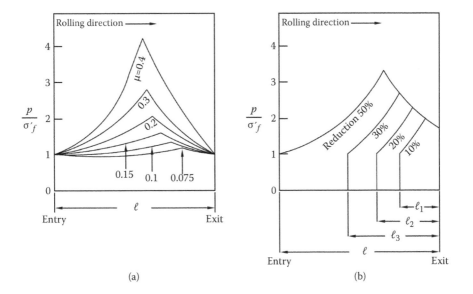

FIGURE 9.44 Pressure distribution in the roll gap as a function of the (a) coefficient of friction and (b) reduction in thickness. (From Kalpakjian, S., and Schmid, S. R., *Manufacturing Processes for Engineering Materials*, Prentice-Hall, Upper Saddle River, NJ, 2003. With permission.)

$$\frac{h_o}{h_f} = \frac{e^{H_0}}{e^{2\,H_n}} = e^{\,(H_0 - 2H_n)}$$

Or

$$H_n = \frac{1}{2}\,H_0 - \frac{1}{-}\ln\frac{h_0}{h_f}$$

Substitution into the *H* relation gives the angle of the neutral point in the form

$$\phi_n = \sqrt{\frac{h_f}{R}}\,\tan\,\sqrt{\frac{h_f}{R}}\,\frac{H_n}{2} \tag{9.26}$$

Effect of front and back tension: An effective method to reduce the roll pressure is to apply longitudinal tension on the workpiece in the axial direction. Tensions in rolling can be applied either at the entry (back tension, P_b) or at the exit (front tension, P_f) of the strip, or at both sides.

Entry zone $$p = (\sigma_f - P_b)\frac{h}{h_0}e^{\,(H_0 - H)} \tag{9.27}$$

And at exit zone $$p = (\sigma_f - P_f)\frac{h}{h_f}e^{\,H} \tag{9.28}$$

Front tension is controlled by the torque on the coiler (delivery reel). Back tension is controlled by a breaking system in the uncoiler (payoff reel). Tensions are particularly important in rolling thin, high-strength materials, where high forces are required. Figure 9.45 shows the effect of back tension (a) and front tension (b) on the pressure distribution. Note that applying front or back tension reduces the peak pressure on the rolls, and the pressure at the side where tension is applied. It also leads to the shifting of the neutral point and the reduction in the area under the curves.

Roll forces: The area under the pressure versus contact-length curve multiplied by the strip width w, is the roll force F

$$F = \int_0^{\phi_n} wpR\,d\phi + \int_{\phi_n}^{\alpha} wpR\,d\phi$$

With a rough approximation with low frictional conditions, the force is given as

$$F = lw\bar{\sigma}_f \tag{9.29}$$

where l is the contact length, and $\bar{\sigma}_f$ is the average flow stress in plane strain.

For higher frictional conditions, we may use the relation

$$F = lw\bar{\sigma}_f \left(1 + \frac{l}{2h_{av}}\right) \tag{9.30}$$

Note that

$$\bar{\sigma}_f = 1.15\bar{\sigma}_f = 1.15\frac{K\int_0^{\varepsilon_1}\varepsilon^n\,d\varepsilon}{\varepsilon_1} = 1.15\frac{K\varepsilon_1^n}{n+1} \tag{9.31}$$

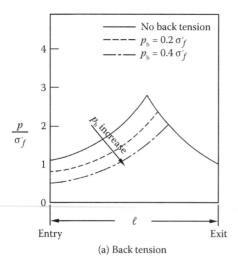

(a) Back tension

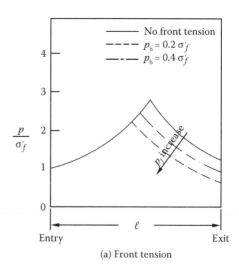

(a) Front tension

FIGURE 9.45 Pressure distribution in the presence of (a) back or (b) front tension.

Roll torque and power: The torque for each roll can be calculated as

$$T = \int_{\phi_n}^{\alpha} w \; pR^2 \, d\phi - \int_{0}^{\phi_n} w \; pR^2 \, d\phi$$

$$\text{entry zone} \qquad \text{exit zone}$$

Note that if the frictional forces are equal, the torque is zero.

A simpler relation is based on the assumption that the roll force F acts in the middle of the arc of contact, so, the torque per roll is

$$T = \frac{Fl}{2} \tag{9.32}$$

The power required per roll is

$$\text{Power} = \frac{\pi F l N}{60,000} \text{ kW} \tag{9.33}$$

where F is in newtons, l is in meters, and N is the rpm of the roll.

Forces in hot rolling: The average strain rate in flat rolling is obtained by dividing the strain by the time, which can be approximated as (l/V_r), so

$$\varepsilon = \frac{V_r}{l} \ln \frac{h_0}{h_f} \tag{9.34}$$

The flow stress σ_f corresponding to this strain rate can be obtained from the strain rate sensitivity equation.

Example 2

In a flat rolling process, it is required to reduce the thickness of a mild steel strip from 10 mm to 7 mm using a roll diameter of 400 mm at a rotational speed of 100 rpm. If the coefficient of friction is 0.13, calculate the following:

a. The biting angle and the length of the deformation zone
b. The coefficient of friction that ensures biting
c. The exit speed of the strip and the forward slip, given that the speed at entry is 100 m/min
d. The maximum pressure/flow stress ratio

Solution

a. Using Equation 9.19

$$\cos\alpha = 1 - \frac{\Delta h}{D} = 1 - \frac{10 - 7}{400} = 0.9925$$

Then the biting angle $\alpha = 7°$
Using Equation 9.20

$$l = \frac{D}{2}\sin\alpha = \frac{400}{2}\sin 7 = 24.37 \, mm$$

b. Using Equation 9.21: $\mu \geq \tan \alpha = \tan 7 \geq 0.12$.
   Since the coefficient of friction is 0.13, then biting takes place.
c. Using Equation 9.17: $h_0 V_0 = h_f V_f$

$$10 \times V_0 = 7 \times V_f \text{ then } V_f = 10 \times 100/7 = 142.9 \, m/min$$

Using Equation 9.23: Forward slip $= (V_f - V_r)/V_r$

$$V_r = \pi D N/1000 = \pi \times 400 \times 100/1000 = 125.7 \, m/min$$

$$\text{Forward slip} = (142.9 - 125.7)/125.7 = 13.7\%$$

d. Using the equation

$$H = 2\sqrt{\frac{R}{h_f}} \tan^{-1} \sqrt{\frac{R}{h_f}}$$

$H_0 = 6.19$
$H_n = 1.314$
$\Phi_n = 0.023 \text{ rad.} = 1.325°$
$h_i = 9.107 \, mm$
Using Equation 9.24

$$\frac{p}{\sigma_f} = \frac{h}{h_0} e^{(H_0 - H)}$$

$$\frac{p_{max}}{\sigma_f} = 1.158$$

9.5 EXTRUSION

Extrusion is the process of forcing a metallic material billet in a closed compartment to flow through a shaped die opening profile of the desired cross section. The process offers two main advantages over other forming processes, namely, its ability to produce very complex cross sections in large reductions, and deformation of materials that are relatively difficult to form, such as stainless steels and nickel-based alloys. Both advantages are due to the fact that during the process, the material only encounters compressive and shear stresses. The extrusion process can be done with the material in the hot or cold condition. Hot extrusion is used for mass production of long, straight metal products of constant cross section, such as bars, solid and hollow sections, tubes, and strips. On the other hand, cold extrusion is applied to produce axial symmetric final products of limited volume.

The first extrusion process was patented in 1797 for making lead pipes. The process was completed manually until 1820, when the first hydraulically powered press was used to produce long sections. In the year 1894, the process was expanded to the extrusion of copper and brass alloys.

9.5.1 CLASSIFICATION OF EXTRUSION PROCESSES

There are six basic types of extrusion processes, which will be presented in the following sections.

9.5.1.1 Direct and Indirect Extrusion

The two basic types of extrusion are the direct (forward) extrusion, and indirect (backward or inverted) extrusion, as shown in Figure 9.46. In forward extrusion, which is the most commonly used process (Figure 9.46a), the ram moves in the same direction of the extruded section and there is relative movement between the billet and the container, leading to high frictional forces. There is a reusable dummy block between the ram and the billet to keep them separated. In backward extrusion (Figure 9.46b), the billet does not move relative to the container, and a die fixed on a hollow ram is pushed against the billet, leading to flow of the extruded section in opposite direction to the ram movement. Frictional force on the billet/container interface is thus eliminated during indirect extrusion. Alternatively, the closed container end in backward extrusion can be forced to move against a fixed die and ram assembly.

The load distribution curves along the ram displacement for both forward and backward extrusion are demonstrated in Figure 9.47, which shows that the load rises sharply in forward extrusion as the billet is upset to fill the container and start extrusion through the die, exhibiting maximum load. Further movement of the ram is associated with fall of the load as the billet length decreases, and accordingly, frictional force decreases until a minimum is reached. Due to smooth flow of the metal

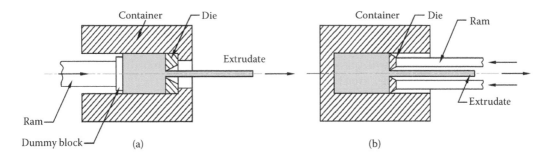

FIGURE 9.46 Basic types of extrusion: (a) direct (forward), (b) indirect (backward).

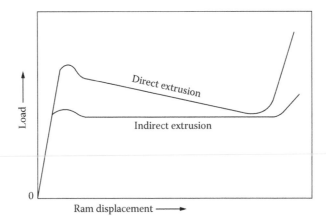

FIGURE 9.47 Typical load–ram displacement curves for forward (direct) and backward (indirect) extrusion.

through the die, a conical shape is formed in front of the die, leaving a dead metal zone inside the container. Forcing the ram further through the dead zone develops severe radial forces and raises the load curve sharply, as shown. Therefore, the process should be stopped when reaching the dead zone (about 10%–15% of the billet length), and the remaining disc (called the butt end) should be discarded. To reduce the effect of the dead zone, a conical die shape is used, which improves the flow pattern of the metal, as will be seen later. The load–displacement curve for indirect extrusion demonstrates a horizontal advance due to elimination of friction. Practically, the maximum load in backward extrusion is reduced by 25 to 30% compared to forward extrusion, which allows extrusion of larger billets, smaller cross sections, or increased speed. The dead metal zone is also much reduced. The advantages offered by indirect extrusion are counterbalanced by the design complications when using a hollow ram, and limitation of the cross-sectional area of the extrusion by the size of the gap inside the ram, and by the maximum force that it can take. The maximum length of the extrusion is ultimately dictated by the column strength of the hollow ram. Therefore, indirect extrusion is not as versatile as direct extrusion.

Both direct and indirect extrusion processes are carried out under hot working conditions to reduce the flow stress and to keep the material from strain hardening. They can also be carried out without lubrication to produce complex sections, with mirror surface finishes and close dimensional tolerances. However, lubrication is usually applied in direct extrusion to reduce the high frictional forces. Lubricating materials could be glass powder, grease, or the billet may be film coated by organic materials that reduce friction. To apply glass lubrication, the billet is heated to the required temperature, and then rolled on glass powder. The glass melts and forms a thin film of 0.5 to 0.75 mm, which separates the billet from the chamber walls and reduces the coefficient of friction considerably. Another advantage of this glass film is that it insulates the heat of the billet from the die. This technique is basically applied for steel extrusion, but it can be used for metals with higher melting point than steel. Another breakthrough in lubrication is the use of phosphate coatings. With this process, in conjunction with glass lubrication, steel can be cold extruded. The phosphate coat absorbs the liquid glass to offer even better lubricating properties.

9.5.1.2 Hydrostatic Extrusion

In hydrostatic extrusion, the billet in the container is extruded through the die by the action of a liquid pressure medium rather than by direct application of the load with a ram. The billet is completely surrounded by a fluid, which is sealed off and is pressurized sufficiently to extrude the billet through the die, as illustrated in Figure 9.48. This process can be done hot, warm, or cold; however, the temperature is limited by the stability of the fluid used. The process can be used to extrude brittle materials that cannot be processed by conventional extrusion, since ductility of the material is improved by applying hydrostatic pressure. The process also allows greater reductions in area (higher extrusion ratios) than conventional hot extrusion, and no billet residue is left on the container walls (no butt). The billets must be prepared by tapering one end to match the die entry angle. This is needed to form a seal at the beginning of extrusion. Usually, the entire billet needs to be machined to remove any surface defects. The main limitation of the process is the difficulty of maintaining fluid at high pressure, and the associated sealing problems. Hydrostatic extrusion presses usually use castor oil at pressure up to 1380 MPa. Castor oil is used because it has good lubricity and high pressure properties.

9.5.1.3 Tube Extrusion

Long tubes can by produced by extrusion by attaching a mandrel to the end of the ram so that it passes through the die. The wall thickness of the tube is controlled by the clearance between the die and the mandrel diameter, as shown in Figure 9.49. A billet used for tube extrusion could be hollow to allow for passing the mandrel through, or a solid billet can be pierced first by the mandrel before the start of tube extrusion. The pushing force of the mandrel to pierce the hot billet is supplied by a separate hydraulic system in the press (a double-action hydraulic press).

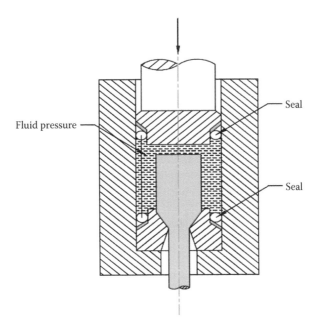

FIGURE 9.48 Hydrostatic extrusion.

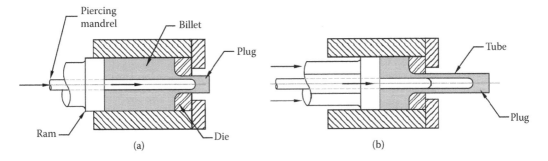

FIGURE 9.49 Tube extrusion using a double-action press for (a) piercing of the billet and (b) extrusion. (From Dieter, G. E., *Mechanical Metallurgy*, McGraw-Hill, New York, 1988. With permission.)

9.5.1.4 Porthole Die Extrusion

This process offers another extrusion technique for manufacturing long tubes of light weight alloys (such as aluminum and magnesium alloys) with a hollow section, as well as complicated hollow sections. The billet material is forced to split through several portholes around a central bridge, which supports a short mandrel. Then the separate streams are gathered within a welding chamber, surrounding the mandrel and leaving the die as a tube. An illustration of the porthole die is given in Figure 9.50. Due to the complicated die assembly and the complexity of metal flow, this porthole die extrusion has been conducted based on the experience of specialists. The extrusion ratio, speed, die shape, and bearing affect the welding pressure and consequently, the product quality. Typical applications are lightweight aerospace and automobile parts, such as aluminum tubes with high strength needed for door impact beams, seat side rails, and hood support.

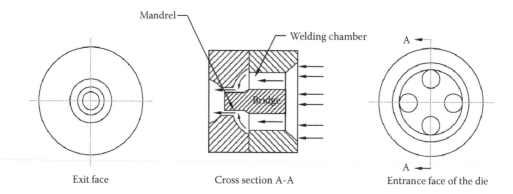

Exit face　　　　　　　Cross section A-A　　　　　Entrance face of the die

FIGURE 9.50 Porthole extrusion die. (From Dieter, G. E., *Mechanical Metallurgy*, McGraw-Hill, New York, 1988. With permission.)

9.5.1.5　Cold Extrusion and Impact Extrusion

Cold extrusion is applied to produce small volume, axial symmetric products at room temperature or near room temperature. The advantages of this process over hot extrusion are the elimination of oxidation, higher strength due to cold working, closer tolerances, good surface finish, and fast extrusion speeds or higher production rates. Material flow can be either in the ram movement direction (direct), or in the opposite direction (indirect), or a combination of both directions, as shown in Figure 9.51. The stresses on the tooling in cold extrusion are high, especially for high yield strength materials. Materials that are commonly cold extruded include: Lead, tin, aluminum, copper, zirconium, titanium, molybdenum, beryllium, vanadium, niobium, and steel. Lubrication is essential for this process, and the most effective technique is the phosphate coating. Examples of cold extrusion products include: Fire extinguisher cases, shock absorber cylinders, automotive pistons, and gear blanks.

Impact extrusion is a cold extrusion process carried out at high speed. The moving ram with the punch strikes the blank to extrude it either in the forward, backward, or both directions, according to the allowed clearances in the tooling. The process is usually applied to produce thin walled products with thickness/diameter ratios as small as 0.005. Concentricity between the punch and the blank is essential for uniform wall thickness. Typical products are dry battery cell tubes and collapsible tubes. Mechanical presses are used for impact extrusion at production rates up to 120 parts/minute.

9.5.1.6　Conform Extrusion

Conform extrusion was developed as a continuous extrusion technique in the United Kingdom by the Atomic Energy Authority in 1971 (less than 40 years ago). It has been used for the production of

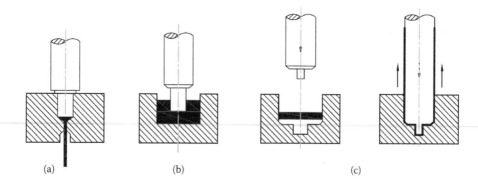

(a)　　　　　　　　　(b)　　　　　　　　　(c)

FIGURE 9.51 Cold and impact extrusion processes: (a) forward, (b) backward, and (c) combined.

aluminum coverings on steel wire. Further development has led to the application of the process for near net extrusion of irregularly shaped sections of copper products. Conventionally, direct extrusion uses billets of about 250 mm diameter as the input material, whereas conform extrusion uses 10–20 mm diameter wire. This wire is usually produced by rolling, which makes it cheaper.

In conform extrusion, a wire inserted in the groove of a wheel is drawn by friction against the wheel. The drawn material is dammed by an abutment inserted in the groove, and the extrusion pressure is generated by a shoe covering the groove. The material flows into the die chamber and is extruded to any form, as shown in Figure 9.52. The process develops frictional heat and heat due to conforming the wire to shear, which may raise the temperature up to 500°C without using heaters. The material is in a high plastic flow state so that near net shapes of irregular sections are possible. One more advantage of this temperature effect is the production of extrusions in a tempered state, and accordingly, the possibility to extrude continuously without having to stop the machine to join pieces of material together. Therefore, large coils of product can be obtained.

9.5.2 Extrusion Equipment

Most of the extrusion processes are carried out using hydraulic presses. These presses may be vertical or horizontal. Vertical presses require less floor space, but large vertical head. They allow uniform billet cooling inside the container, and consequently, uniform wall thickness and concentricity of tubes. Therefore, these presses are preferred for extrusion of thin wall tubes. Vertical hydraulic presses are available at capacities from 3 to 20 MN. Horizontal presses, on the other hand, require large floor space. The lower part of the billet in contact with the container cools faster, leading to warping of the product and nonuniform wall thickness. Therefore, these presses are applied for bars and shapes with relatively large cross sections. Horizontal press capacities range from 15 to 140 MN. Ram speeds are higher for high temperature alloys. Ram speeds of 0.4 to 0.6 m/s are used for refractory metals. Therefore, a hydraulic accumulator system is necessary for the press. Aluminum

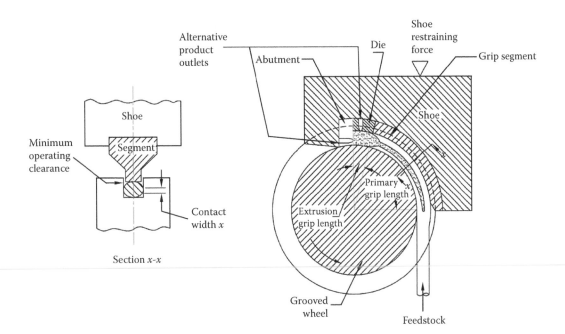

FIGURE 9.52 Conform extrusion arrangement. (From Koshal, D., *Manufacturing Engineer's Reference Book*, Butterworth-Heinemann, 1993. With permission.)

and copper, on the other hand, are prone to hot shortness, and ram speeds are restricted to few mm/s. A direct drive pumping system is used.

Extrusion tooling and specialty dies are essential for controlling product quality. They have to take extremely high stresses, and thermal shocks, with high resistance to fatigue, wear, and oxidation. Dies are made from high-alloyed tool steel. Whether they are flat or conical, they are supported on a die holder, bolster, and die head. Being subjected to extremely high internal pressure during extrusion, the container should have a liner shrink-fitted into a larger container or a compound cylinder to increase its pressure bearing capacity. The ram, which is loaded in compression, is protected from the hot surface of the billet by a dummy block, as stated earlier. Both the container liner and the dummy block should be replaced periodically after many cycles of thermal shocks.

In addition to the main tooling, other facilities should be provided such as: A billet heating furnace, automatic transfer of the billet to the container, hot saw to cut off the extrusion at the end of the cycle, a runout table with rollers to receive and hold products, and a straightener to correct minor warping of the extruded bars.

9.5.3 Extrusion Defects

Most common defects associated with extrusion processes include

- *Surface cracking*, which is often caused by the extrusion temperature, friction, or speed being too high. This defect ranges in severity from a badly roughened surface to repetitive cracking, called fir-tree cracking. These cracks are intergranular, and usually occur with aluminum, magnesium, zinc, and molybdenum alloys due to hot shortness. They can be eliminated by controlling the causing factors. It can also take place at low temperature when there is periodic sticking of the extruded bar with the die land.
- *Pipe*, which results when the flow pattern of the metal draws the surface oxides and impurities to the center of the product in the form of a funnel, or pipe. Such a pattern is often caused by high friction or faster cooling of the outer regions of the billet, which lead to a large dead zone. This defect can be reduced or eliminated by controlling friction and minimizing the temperature gradient. Alternatively, the billet surface can be machined prior to extrusion to eliminate oxides, or a dummy block with smaller diameter, thus leaving a thin shell along the container.
- *Center cracking*, which results due to excessive secondary tensile stresses at the center line of the deformation zone in the die. These cracks take the shape of repeated center bursts, arrowheads, or chevron cracks. These tensile stresses can be reduced by controlling the die angle and frictional conditions.
- *Variation of structure and properties* from the front to back end of extrusion in both the longitudinal and transverse directions. Regions of grain growth may be found after hot extrusion.

9.5.4 Modeling of Direct Extrusion

The ram force in direct extrusion can be calculated in different ways

a. Ideal deformation: The extrusion ratio R is defined in terms of the billet cross-sectional area A_0, and the final area of the extruded bar A_f as

$$R = \frac{A_0}{A_f} \tag{9.35}$$

Therefore

$$\varepsilon_1 = \ln \frac{A_0}{A_f} = \ln \frac{L_f}{L_0} = \ln R \qquad (9.36)$$

where L_0 and L_f are the lengths of the billet and extruded bar, respectively.
The work supplied by the ram force F, which travels a distance L_0, is written as

$$\text{Work} = FL_0 = pA_0L_0 = \text{total strain energy} = (\overline{\sigma}_f \, \varepsilon_1)A_0L_0$$

where p is the extrusion pressure and $\overline{\sigma}_f$ is the average flow stress of the material.

$$p = \overline{\sigma}_f \varepsilon_1 = \overline{\sigma}_f \ln \frac{A_0}{A_f} = \overline{\sigma}_f \ln R \qquad (9.37)$$

b. Ideal deformation and friction (slab analysis): Friction between the billet and the container increases the pressure required to carry out the process. This effect can be studied by considering an element as shown in Figure 9.53. Equilibrium of the forces in the axial direction

$$(\sigma_x + d\sigma_x)\frac{\pi D_0^2}{4} + \sigma_r \pi D_0 dx - \sigma_x \frac{\pi D_0^2}{4} = 0$$

Reducing this equation and applying Tresca yield criterion: $\sigma_1 - \sigma_3 = \sigma_f \therefore \sigma_x - \sigma_r = \sigma_f$

$$\frac{d\sigma_x}{\sigma_f - \sigma_x} = \frac{4}{D_0}dx$$

Integration of this last equation along the billet length, and considering the pressure at the end of the stroke to be equal to the ideal pressure, according to Equation 9.37, yields the following relation to calculate the pressure p at any billet length L.

$$p = \sigma_f \, (\ln R - 1)e^{\frac{4}{D_0}L} + 1 \qquad (9.38)$$

Forward movement of the ram during extrusion reduces the remaining length L, and the pressure is reduced, as explained before in Figure 9.47 for direct extrusion.

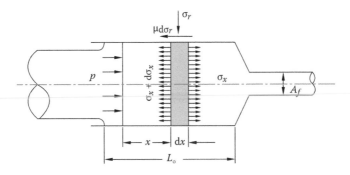

FIGURE 9.53 Equilibrium of an element in direct extrusion.

For strain hardening materials, σ_f should be replaced by $\bar{\sigma}_f$. To calculate the extrusion force, the pressure p is multiplied by the billet area A_0.

c. Optimum die angle: In the previous model, the die face is assumed to be perpendicular to the container wall (flat die). Smooth flow of the metal to the die orifice follows a conical shape near the die face, leading to a dead metal zone, as shown in Figure 9.54. To eliminate or avoid the dead metal zone, a conical die can be used with a die angle identified by the semicone angle α, which is equivalent to half the cone angle. This angle has an important effect on the extrusion forces. Increasing the die angle leads to shorter conical contact length, and thus reduces frictional forces along this length. However, deformation gets more inhomogeneous, and the redundant work increases. Therefore, an optimum die angle can be estimated corresponding to minimum total force, as shown in Figure 9.55. For most extrusion operations, the die optimum semicone angle is in the range 45°–60°.

Empirical relation for extrusion: Due to the difficulties in estimating the frictional work, redundant work and coefficient of friction, a convenient empirical formula has been developed in the form

$$p = \sigma_f (a + b \ln R) \tag{9.39}$$

where a and b are experimentally determined constants. An approximate value of a is 0.8 and b ranges from 1.2 to 1.5. Note that for the strain-hardening material, σ_f should be replaced by $\bar{\sigma}_f$.

Forces in hot extrusion: An approximate equation to calculate the average strain rate in hot extrusion for a semidie angle of 45° (as may be the case with a square die developing a dead zone), is given in the form

$$\varepsilon = \frac{6V_0}{D_0} \ln R \tag{9.40}$$

where V_0 is the ram speed.

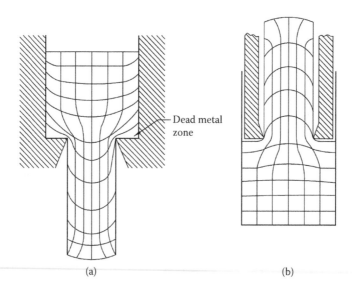

(a)   (b)

FIGURE 9.54 Patterns of metal flow in (a) direct and (b) indirect extrusion showing the dead metal zone in the first pattern. (Adapted from Deiter, G. E., *Mechanical Metallurgy*, McGraw-Hill, 1988.)

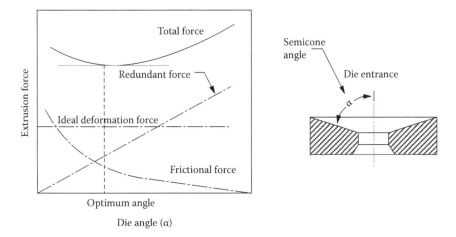

FIGURE 9.55 The extrusion force as a function of die angle α. The optimum die angle corresponds to minimum total extrusion force. (Adapted from Kalpakjian, S., and Schmid, S. R., *Manufacturing Processes for Engineering Materials*, Prentice-Hall, Upper Saddle River, NJ, 2003.)

This relation is used to calculate the flow stress under hot conditions using the strain rate sensitivity equation: $\sigma_f = C\dot{\varepsilon}^m$.

Increasing the extrusion speed increases the extrusion pressure, especially at higher temperatures. This does not allow the generated heat to be dissipated, and may lead to

a. Incipient melting of the workpiece material, causing defects
b. Circumferential surface cracks on the extruded bar caused by hot shortness, as described in 9.54

These problems can be eliminated by reducing the extrusion speed.

Example 3

In a forward extrusion process, a mild steel billet of 250 mm diameter and 400 mm length is extruded to 10 mm diameter rods at a speed of 0.2 m/s. The billet is heated to 950°C where the strain rate sensitivity parameters are $C = 125$ MPa and $m = 0.12$, and the coefficient of friction at the billet/container interface is 0.1. You are asked to calculate

a. The extrusion ratio and percent reduction
b. The number of 6 m rods to be produced, if 16% of the billet length is to be discarded due to the dead zone effect
c. The ideal pressure and the pressure at the start of plastic deformation
d. The required extrusion force

Solution

a. Using Equation 9.32: The extrusion ratio $= A_0/A_f = (D_0/D_f)^2 = 250^2/10^2 = 625$

$$\% \text{ reduction} = (A_0 - A_f)/A_0 = (250^2 - 10^2)/250^2 = 99.84\%$$

b. Constant volume: $A_0 L_0 = A_f L_f$, considering that 16% of the billet length is discarded

$$250^2 \times (400 \times (1 - 0.16)) = 10^2 \times L_f$$

$$L_f = 210 \text{ m}$$

Number of 6 m rods = 210/6 = 35 rods

c. Using Equation 9.34, the ideal pressure $p_i = \sigma_f \ln (R) = \sigma_f \varepsilon$
For hot extrusion, $\sigma_f = C\dot{\varepsilon}^m$
Using Equation 9.37

$$\varepsilon = \frac{6V_0}{D_0}\ln(R) = \frac{6 \times 0.2}{0.250}\ln(625) = 30.9 \text{ s}^{-1}$$

$$\sigma_f = 125 \times 30.9^{0.12} = 188.67 \text{ MPa}$$

$$p_i = 188.67 \ln(625) = 1.2 \text{ GPa}$$

Using Equation 9.35,

$$p = \bar{\sigma}_f \ (\ln R - 1)e^{\frac{4\,l}{D_0}} + 1 \ = 188.67 \ (\ln 625 - 1)(2.718)^{\frac{4 \times 0.1 \times 0.4}{0.250}} + 1$$

$$p = 2.134 \text{ GPa}$$

Note that when friction is considered, the pressure at the start of plastic deformation is higher than the ideal pressure, which is the equivalent to the pressure at the end of the process (just before the dead zone) by about 78%; refer to Figure 9.47(1).

d. Extrusion force = $p_{av} \times A_o = ((2.134 + 1.2)/2) \times 10^3 \times \pi \times (0.25)^2/4 = 81.83 \text{ MN}$

9.6   ROD, WIRE, AND TUBE DRAWING

In the drawing process, the cross-sectional area of the rod, tube, or wire is reduced by pulling through a die. The die geometry determines the cross-sectional area of the drawn product, and a reduction in area. Virtually any axial–symmetric cross section or square or rectangular sections can be drawn; however, most of the drawn sections are circular in cross section. Since the applied force is tensile, the maximum allowed reduction in area should be limited to avoid cracking. Drawing is usually carried out at room temperature using a number of passes or reductions through successive dies. Therefore, drawing permits excellent surface finishes, closely controlled dimensions along the product, and higher strength material due to cold working. Annealing may occasionally be necessary after a number of drawing passes to eliminate the strain hardening resulting from cold work before the drawing operation is continued. The deformation is accomplished by a combination of direct tensile stresses created by the pulling force at the die exit, and indirect compressive stresses developed by the die walls.

9.6.1   CLASSIFICATION OF DRAWING PROCESSES

The principle of rod, wire, and tube drawing is the same regarding the die and the pulling force; however, the equipment used is different according to the size and diameter of the products. Therefore, rod, wire, and tube drawing are presented separately in the following sections.

9.6.1.1 Rod Drawing

Rod (bar) drawing is usually carried out using a draw-bench to accommodate the straight length of the product. The rod end near the die is reduced or pointed by hammering or by rotary swaging so that it can be inserted through the die. A schematic sketch of the drawing process in a chain type draw bench is presented in Figure 9.56. In most modern drawing machines, equipment for tagging the rod end is included in the draw bench in the form of a push pointer, as shown in Part A of the figure. The push pointer jaws grip the rod and force it through the die so that the tagged portion passes between the jaws of the draw head clamps. The draw head clamps grip the rod end and are pulled either by chain drive or by a hydraulic system. Draw benches with up to 1 MN pulling force and 30 m long are commercially available. Draw speeds range from 0.15 to 1.5 m/s.

9.6.1.2 Wire Drawing

Wire drawing is carried out using a coiling spool or cylinder, where the end of the wire is connected to a fixed point on the cylinder after passing through the die, and continuous rotation of the wheel develops the pulling force. Floor space for wire drawing is then much smaller than that needed for draw benches, as shown in Figure 9.57. The process usually starts with a coiled small diameter, hot rolled rod. The rod is first cleaned by pickling to remove surface scale, which can lead to defects in the produced wire or excessive die wear. Then the rod is prepared for effective lubrication by coating with copper or tin sulfates. This coating helps maintain a film of lubricant on the rod's surface. The lubricant may be simply soap or mineral oil. Alternatively, the whole die is immersed in oil with an excessive pressure additive.

Since the reduction in area is limited in drawing by about 30%, to achieve large reductions, a number of passes is required using a multiple-die wire drawing machine. Multiple-die machines with one die and one spool for each drawing stage are commonly used, as shown in Figure 9.57a. Since the wire diameter is decreased after each pass, the length and speed are increased proportionally. Thus, the peripheral speed for each draw step must be increased to prevent slippage. One way to achieve this is to equip each step with a separate motor with variable speed control. Alternatively, a more economic setup is to use a single motor to drive a series of stepped cones, as shown schematically in Figure 9.57b. The diameter of each cone should be designed carefully to provide the required peripheral speed for each step, as shown in the figure. The drawing speeds in multiple-die machines may reach 10 m/s for steel wire, but with nonferrous wire drawing, speeds up to 30 m/s are common.

9.6.1.3 Tube Drawing

Tubes produced by extrusion or roll-piercing are often cold finished by drawing to get closer dimensional tolerance, better surface finish and higher strength, or to produce tubes with smaller diameter

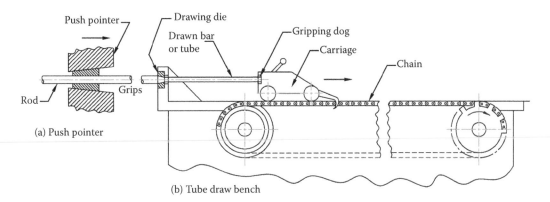

FIGURE 9.56 (a) Tagging of the rod by push pointer. (b) Schematic presentation of a chain-type rod or tube draw bench. (Adapted from Timings, R. L., *Manufacturing Technology*, vol. 1, 3rd ed., Longman, 1998.)

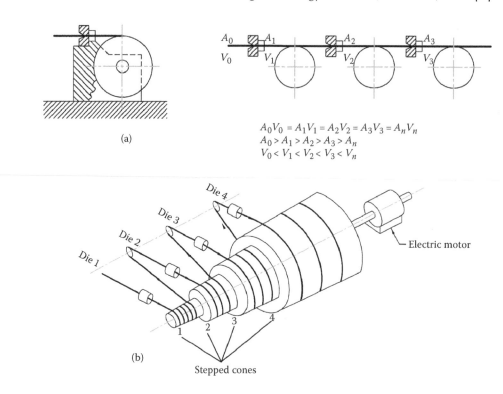

$$A_0 V_0 = A_1 V_1 = A_2 V_2 = A_3 V_3 = A_n V_n$$
$$A_0 > A_1 > A_2 > A_3 > A_n$$
$$V_0 < V_1 < V_2 < V_3 < V_n$$

FIGURE 9.57   Multiple die wire drawing machines: (a) single-step type and (b) stepped cone type.

or thinner wall thickness that can be obtained by other forming processes. Cold drawing of tubes can also be applied to produce tubes with irregular shapes. Most of the machines used for tube drawing are of the bench type, used for rod drawing; however, tubes can also be drawn in coil form when diameter is small and material is ductile, such as with copper or aluminum tubes. There are four basic types of tube drawing; namely, tube sinking, tube drawing with a fixed mandrel, with a floating mandrel, and with a moving mandrel, as shown in Figure 9.58.

a. Tube sinking, is the process of tube drawing without a mandrel. It is adopted for tubes with higher thickness-to-diameter ratios, or when the main objective of the process is to control outside diameter irrespective of the wall thickness. The tube end is first pointed to pass through the die and then fixed to the jaws of the draw head. Tube sinking leads to increase of the tube length and the wall thickness. The percentage of thickness increases and tube elongation depends on the flow stress of the drawn part, die geometry, and interface friction. There is no limit to the size of the tubes which can be produced using this method. Reductions of up to 35% can be obtained.

b. Tube drawing with a fixed mandrel, where a short mandrel or plug is held in position in the mouth of the die by means of a tie rod fixed to the end of the draw bench. Neither long tubes nor tubes with internal diameter of less than 6 mm can be produced using this method. This method is widely used for drawing large- to medium-diameter straight tubes with limited length. The resulting tubes are characterized by close dimensional tolerance on the internal diameter and uniformity of wall thickness. Frictional forces induced between the tube walls, the die, and the mandrel are high; however, this can be compensated for by using larger die angle and/or efficient lubrication.

c. Tube drawing with a floating mandrel, where the contour of the mandrel or the plug for this method is so designed that it adjusts itself to the correct position during drawing;

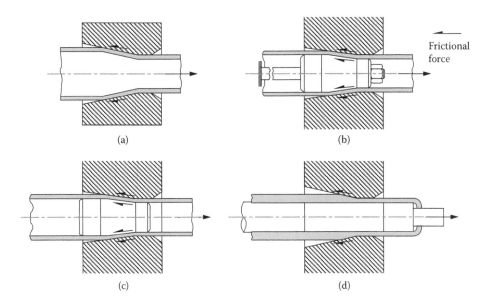

Frictional force

(a)   (b)

(c)   (d)

FIGURE 9.58 Tube drawing techniques: (a) tube sinking, (b) tube drawing with a fixed mandrel, (c) tube drawing with a floating mandrel, and (d) tube drawing with a moving mandrel.

      Figure 9.58c. There is virtually no limit to the length of tube which can be drawn, and therefore the method is suitable for drawing of coiled tubes at speeds as high as 10 m/s.

d. Drawing with a moving mandrel, where the mandrel is made of a heat-treated alloy steel rod equal in length and in the bore diameter to the finished tube. The rod moves through the die along with the drawn tube (Figure 9.58d). Movement of the mandrel at the same drawing speed of the tube prevents friction on the internal tube wall. A disadvantage of this method is that the drawn tube has to be stripped from the mandrel, which may require a further reeling process. Small diameter stainless steel tubing used in the manufacture of hypodermic needles is drawn using this method.

9.6.2 THE DRAWING DIE

The effective part of a drawing die is a smooth conical hole sunk in a material of considerable strength, fatigue resistance, and wear resistance, as shown in Figure 9.59. The approach angle length is the actual working surface where reduction in area takes place. The semicone angle of this part is a main design parameter that depends upon both the material of the die and the metal to be drawn as well as the percentage reduction in area. The entrance or bell part of the die has a larger angle than α but is still tapered to serve as a reservoir for the lubricant to be self-forced through the die during deformation. This part ends with a smooth curve at the open end of the die. The bearing is a cylindrical part that forms the die land, which controls the diameter of the drawn part, and must be of suitable length to increase the service life of the die by permitting several refinishing operations of the approach angle. The back relief clears the die from the drawn shape, while providing reinforcement for the working section of the die. It ends with a smooth curve at the free end. Most drawing dies are made of cemented carbide or industrial diamond (for wire drawing dies). They could be made of a die insert encased in a steel casing for protection.

9.6.3 MODELING OF THE DRAWING PROCESS

In drawing operations, the cross-sectional area of a rod, wire, or tube is reduced by pulling through a converging die. The major variables in drawing are the reduction in cross-sectional area, die angle

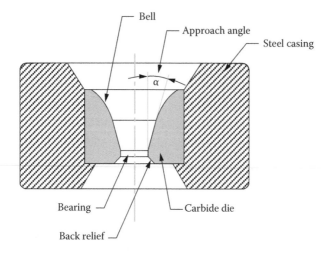

FIGURE 9.59 The main parts of a drawing die section.

α, and friction. Analysis of the mechanics of the process is similar to that followed for extrusion. A simple model of the process is shown in Figure 9.60.

a. Ideal deformation

For a round rod or wire, the drawing pressure p for the simplest case of ideal deformation with no friction can be obtained by the same approach used in extrusion.

$$p = \sigma_f \ln \frac{A_0}{A_f} = \sigma_f \ln R = \sigma_f \varepsilon_1 = \sigma_f \ln \frac{1}{1-r} \qquad (9.41)$$

where σ_f is the flow stress of the drawn material and r is the percentage reduction in area. For strain-hardening materials, σ_f is replaced by an average flow stress $\bar{\sigma}_f$ due to a total longitudinal strain $\varepsilon = \ln \dfrac{A_0}{A_f}$, as shown in Figure 9.60.

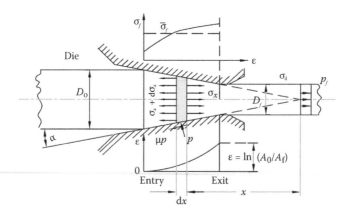

FIGURE 9.60 Equilibrium of an element in bar drawing and variation in the flow stress and strain along the deformation zone.

The drawing force is then given as

$$F = \bar{\sigma}_f A_f \ln \frac{A_0}{A_f} \tag{9.42}$$

b. Ideal deformation and friction

Friction increases the drawing force. Using the slab method by considering the equilibrium of horizontal forces on an element in the deformation zone as shown in Figure 9.60, the following expression is obtained:

$$p = \sigma_f \ln R(1 + \mu\cot \alpha) \tag{9.43}$$

where α is the semidie angle, and μ is the coefficient of friction along the bar–die interface. Notice that when $\mu = 0$, the equation is reduced to Equation 9.41 for ideal deformation.

For strain-hardening materials, σ_f is replaced by an average flow stress $\bar{\sigma}_f$.

c. Redundant work of deformation

Inhomogeneous (redundant) deformation adds an extra part to the drawing pressure, as shown in the following relation:

$$p = \sigma_f \{\ln R\left(1 + \cot\alpha\right) + \frac{2}{3}\tan\alpha\} \tag{9.44}$$

The equation indicates that the extra pressure due to redundant deformation increases with the increase of the die angle.

d. Maximum reduction in area per pass

The reduction in area in drawing

$$r = \frac{A_0 - A_f}{A_0},$$

Referring to Equation 9.41, the limit for deformation is set when $p = \sigma_f$, that is, when $\ln \dfrac{1}{1-r} = 1$, or $r = 63.2\%$, which is theoretically the maximum reduction allowed per pass. Practically, the maximum reduction per pass is limited to 20 to 30%.

e. Optimum die angle

Similar to extrusion, there is an optimum die angle in rod and wire drawing. The same figure for the optimum die angle in extrusion (Figure 9.55) is applied. Practically, the optimum die angle is taken between 8° and 10°, with a maximum of 12°. Figure 9.61 shows the effect of reduction in cross-sectional area on the optimum die angle in drawing. It is indicated in the figure that the optimum die angle increases with the increase of the reduction in area.

f. Die pressure

The die pressure p_d at any diameter along the die contact length can be obtained from the Tresca yield criterion: $p_d = \sigma_f - p$.

Note that the drawing pressure increases toward the exit, whereas the die pressure drops toward the exit. If back tension is applied similar to back tension in flat rolling, the drawing pressure is increased, but the die pressure decreases. Therefore, the multidie wire drawing presented in Figure 9.57 is associated with a reduced pressure on the die walls.

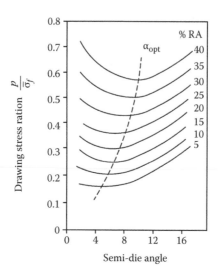

FIGURE 9.61 Effect of reduction in cross-sectional area on the optimum die angle in drawing. (From Altan, T., *Metal Forming Fundamentals and Applications*, ASM International, Materials Park, OH, 1983. With permission.)

Example 4

It is required to draw a 6 mm diameter wire at a reduction 20%, at a speed 3 m/s. The wire is made of annealed copper with strain hardening parameters $K = 315$ MPa, $n = 0.54$. The semidie angle is 6°, and the coefficient of friction is 0.1. Calculate the following:

a) The ideal pressure
b) The pressure when friction is considered
c) The total pressure
d) The drawing force and power

Solution

a. Using Equation 9.38

$$p = \sigma_f \varepsilon_1 = \sigma_f \ln \frac{1}{1-r}$$

$$\varepsilon_1 = \ln \frac{1}{1-0.2} = 0.223$$

$$\sigma_f = \bar{\sigma}_f = \frac{K\varepsilon^n}{n+1} = \frac{315 \times 0.223^{0.54}}{1.54} = 90.965 \text{ MPa}$$

$$p_i = 90.965 \times 0.223 = 20.28 \text{ MPa}$$

b. Using Equation 9.40

$$p_f = \sigma_f \ln R (1 + \mu \cot \alpha) = 90.965 \times 0.223 \times (1 + 0.1 \times \cot 6)$$

$p_f = 39.59$ MPa   (95% higher than ideal pressure)

c. Using Equation 9.41

$$p_t = \sigma_f \{\ln R(1 + \cot\alpha) + \frac{2}{3}\tan\alpha\}$$

$p_t = 90.965 \{0.223(1 + 0.1 \cot 6) + 0.667 \tan 6\} = 45.96$ MPa

The total pressure is 16% higher than the pressure due to friction.

d.

$$\varepsilon_1 = 0.223 = \ln\frac{A_0}{A_f} = \ln\frac{\pi 6^2/4}{A_f}$$

$$\therefore A_f = 22.6 \text{ mm}^2, \ d_f = 5.367 \text{ mm}$$

Using Equation 9.39

$$F = p_t A_f = 45.96 \times 22.6 \times 10^{-3} = 1.04 \text{ kN}$$

$$\text{Power} = FV = 1.04 \times 3 = 3.12 \text{ kW}$$

9.7 REVIEW QUESTIONS

1. Why do we have to perform plastic forming processes?
2. Discuss briefly the development of metal forming processes and technology throughout history.
3. Classify forming processes according to the working temperatures and according to the type of the applied forces.
4. What are the reasons for using open-die forging processes, and what are the necessary precautions and accompanying operations?
5. Suggest two alternative techniques to form hexagonal heads for bolts, and represent them schematically.
6. Why do we have to use lubricants in closed-die forging, and what are the suitable types of lubricants?
7. How does the surface area-to-volume ratio of a forging affect the design of the die?
8. Why is flash necessary in closed-die forging, and how is it designed?
9. What are the benefits of hot die forging and isothermal forging, and what are the differences between them?
10. Differentiate between rotary swaging, radial forging, and orbital forging as special forging techniques.
11. Compare mechanical presses, hydraulic presses, and hammers according to their maximum capacity, constriction, and applications in forging processes.
12. What are the major types of forging defects and how can we eliminate them?
13. Classify rolling processes according to the produced shape.
14. Describe the different arrangements of rolling mills to perform rolling processes for flat shapes with different thicknesses.
15. Why is a tandem mill used, and what precautions should be considered when arranging the roll stands?
16. Describe briefly the sequence of roll passes from a billet to reach a final round or channel section.

17. Discuss the possible rolling processes for manufacturing tubes, showing the limitations of each.
18. Describe some special rolling processes, considering the directions of rotation of rolls, alignment of roll axes, and the shape and direction of movement of the workpiece.
19. What are the possible defects in rolling and how can we eliminate them?
20. What are the advantages of extrusion compared to other metal forming processes?
21. Compare forward and backward extrusion, considering friction, die design, and the required loads.
22. What are the advantages and limitations of hydrostatic extrusion?
23. How is porthole extrusion different from tube extrusion, and what are the applications of each?
24. Why was the conform extrusion process developed in industry?
25. What are the common types of extrusion defects?
26. How do we determine the optimum die angles for both extrusion and bar drawing processes, and what are the typical ranges for each process?
27. Why is drawing of rods, wires, and tubes performed under cold forming conditions?
28. Compare rod drawing and wire drawing processes.
29. What are the types of multidie wire drawing machines, and what are the advantages of each?
30. What are the possible arrangements for tube drawing processes, and when do we apply each?
31. How is the die pressure in the wire drawing processes reduced?

9.8 PROBLEMS

1. Plane strain compression is performed on a slab of metal having a yield stress 280 MPa. The width of the slab is 200 mm, while its height is 25 mm. Assuming an average coefficient of friction 0.10, calculate the following:
   a. The maximum pressure at the onset of plastic flow.
   b. The average pressure at the onset of plastic flow.
   c. Repeat calculation if sticking friction prevailed at each interface, and compare results.
2. A cylindrical specimen made of annealed 4135 steel ($K = 1015$ MPa, $n = 0.17$) has a diameter of 150 mm, and is 100 mm high. It is upset between flat dies to a height of 50 mm at room temperature. Assuming that the coefficient of friction is 0.2, calculate the force required at the end of the stroke.
3. It is expected that in strip rolling, the rolls will begin to slip if the back tension P_b is too high. Derive an expression for the magnitude of back tension required to make the rolls begin to slip.
4. A 300 mm wide mild steel strip is hot rolled at a temperature 900°C ($C = 165$ MPa, $m = 0.08$) from a thickness of 12 mm to 8 mm using 400 diameter rolls rotating at 200 rpm. Assuming a coefficient of friction 0.2, calculate
   a. The biting angle and contact length
   b. The forward slip and maximum possible reduction
   c. The approximate roll force and torque
   d. The horsepower for the mill
5. A copper billet 125 mm diameter and 250 mm long is extruded to a diameter 50 mm at 800°C at a speed of 0.25 m/s. Using square dies and assuming poor lubrication (dead metal zone at 45°), estimate the force required in this operation for copper at the given temperature, $C = 130$ and $m = 0.06$.

6. A round rod of annealed 302 stainless steel (K = 1300 MPa and n = 0.3) is being drawn from a diameter of 10 mm to a diameter of 8 mm at a speed 0.5 m/s, using a semidie angle of 8°. Calculate the percentage reduction, the applied force due to ideal deformation, friction, and inhomogeneous deformation. Assume coefficient of friction of 0.1. Calculate the required power, process efficiency, and the die pressure at the exit.

BIBLIOGRAPHY

Altan, T., Oh, S., and Gegel, H. 1983. *Metal Forming Fundamentals and Applications*. Materials Park, OH: ASM International.

Altan, T., Ngaile, G., and Shen, G. 2005. *Cold and Hot Forging: Fundamentals and Applications*. Materials Park, OH: ASM International.

Amstead, B. H., Ostwald, P. F., and Begeman, M. 1997. *Manufacturing Processes*, 7th ed. New York: John Wiley & Sons.

ASM International, 1988. *ASM Handbook, vol. 14: Forming and Forging*. Materials Park, OH.

Dieter, G. E. 1988. *Mechanical Metallurgy*. New York: McGraw-Hill.

Heinz, T. 2006. *Metal Forming Practice: Processes–Machines–Tools*. The Netherlands: Springer.

Hosford, W. F., and Caddell, R. M. 2007. *Metal Forming Mechanics and Metallurgy*, 3rd ed. Cambridge University Press.

Kobayashi, S., Oh, S., and Altan, T. 1989. *Metal Forming and the Finite Element Analysis*. Oxford University Press.

Koshal, D. 1993. *Manufacturing Engineer's Reference Book*. Elsevier.

Lange, K. 1985. *Handbook of Metal Forming*. New York: McGraw-Hill, 1985 (published by SME, 2006).

Marciniak, Z., and Duncan, J. 2002. *Mechanics of Sheet Metal Forming*. Butterworth-Heinemann.

Nandan, R., Rai, R., Jayakanth, R., Moitra, S., and Chakraborti, N. 2005. *Regulating Crown and Flatness During Hot Rolling, Materials and Manufacturing Processes*, pp. 20, 459–478.

Ragab, A., and Bayoumi, S. 1999. *Engineering Solid Mechanics Fundamentals and Applications*. Boca Raton, FL: CRC Press.

Rowe, G. W. 1977. *An Introduction to the Principles of Metalworking*. London: Edward Arnold.

Schuler, L. 1998. *Metal Forming Handbook*. The Netherlands: Springer.

Waters, F. 2002. *Fundamentals of Manufacturing for Engineers*. UCL Press, Taylor & Francis.

10 Sheet Metal Forming Processes

10.1 INTRODUCTION AND CLASSIFICATION

Sheet metals are characterized by very large area-to-thickness ratio. They are the output of flat rolling with thicknesses up to about 6 mm, as described in the previous chapter. Sheet forming processes have been developed tremendously during the last century to cope with the vast improvement of many industries, such as automotive, aerospace, electronics, household appliances and utensils, etc. Sheet metal products have high strength, good dimensional accuracy and surface finish, and economical mass production for large quantities.

In sheet forming processes, the stresses applied in the plane of the sheet are generally tensile, since compressive stresses in this plane lead to buckling or wrinkling. This is different from bulk forming stresses, where the objective is mainly to reduce the thickness through direct or indirect compressive stresses. Most sheet forming processes are performed under cold working conditions.

Sheet metal forming processes may be broadly classified into shearing processes, bending processing, deep drawing, stretch forming, and superplastic forming, according to the product shape, as shown in Figure 10.1.

10.2 SHEARING PROCESSES

The first step in any sheet forming process is to cut the correctly shaped and sized blank from a larger sheet through a shearing process. Thus, shearing is the separation of a part of the sheet into two parts. It generally leads to complete separation into two pieces, but some shearing operations are designed to get partial separation, which could be followed by bending of the partially cut part.

10.2.1 CLASSIFICATION OF SHEARING PROCESSES

Shearing operations are subclassified into open contour shearing and closed contour shearing, as shown in Figure 10.2.

10.2.1.1 Open Contour Shearing

This separates the sheet into two parts along a straight or curved line, as shown in the left side of Figure 10.2. Such operations are usually carried out using shears with parallel (guillotine) or inclined blades (alligator shears), but circular shears with one or more pairs of circular blades are sometimes employed. Guillotine shears are usually applied for straight cuts of small thickness sheets, but their main drawback is the high force required to induce simultaneous shearing of the sheet along the entire length. Shears with inclined blades act as scissors. They are used to shear the strip gradually to reduce the shearing force, and thus they are suitable for higher thicknesses, up to 20 mm, with a cutting length of 3 m or more. The angle of inclination is normally 2° to 5°, and should not exceed a certain limit; otherwise, the sheet will be pushed forward between the blades instead of being sheared. Alligator shears and circular shears are applied for cutting curved lines.

Hydraulic, mechanical, and pneumatic shears are available. Hydraulic shears are operated by hydraulic cylinders. They permit longer strokes for larger thicknesses, but the number of strokes per minute is lower than a mechanical shear. Mechanical shears are driven by eccentric links with a flywheel. The flywheel stores energy and thus allows using a small motor for intermittent cuts. Mechanical shears offer a higher number of strokes per min (up to 100 spm). For both hydraulic and

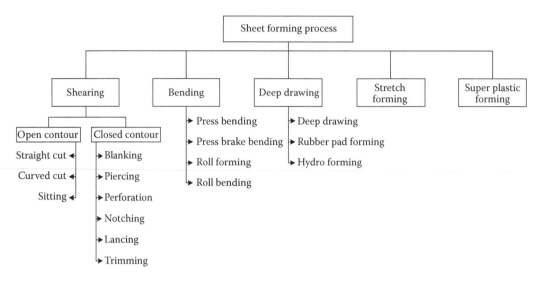

FIGURE 10.1 Classification of sheet metal forming processes.

mechanical shears, the sheet must be held rigid by hold-down devices. Hold-down pressure must be greater than the forces generated during shearing of the sheet material. These forces depend on the blade clearance, blade rake angle, and sheet thickness. Auxiliary devices may be added, such as back or front gauges, to control the length between cuts for repetitive shearing, and squaring arms to control perpendicularity of cuts to one sheet side. Pneumatic shears are exclusively designed for small thicknesses of less than 1.5 mm and for short pieces.

Shearing blades are made in one piece from tool steel (commonly, AISI D2 steel). They should be as hard as possible, especially for long ones, to reduce wear but not to the limit leading to breaking of the blade. Blades made of D2 tool steels are usually hardened to 58 to 62 HRC. Clearance should be allowed between the moving and fixed blades. Usually, clearance is fixed in shears, and is set for thinnest thickness to be sheared. Excessive clearance leads to wiping of the metal between blades, resulting in burrs, or even flanging of the sheet. On the other hand, too small clearance causes double shearing.

Circular shears are commonly applied for longitudinal *slitting* of the sheet to smaller widths (strips). A slitting line for cutting wide-coiled stock into narrower widths consists essentially of an uncoiler for holding the coil, a slitter consisting of a number of pairs of circular blades on rotating arbors, and a recoiler for recoiling the slit strips, as shown in Figure 10.3.

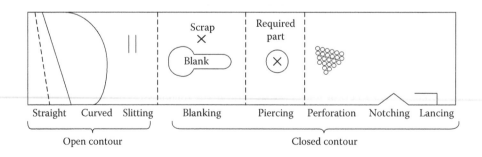

FIGURE 10.2 Shearing operations.

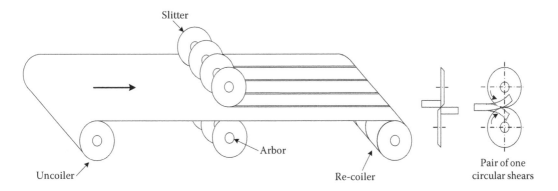

FIGURE 10.3 Circular shearing (slitting) of sheets.

10.2.1.2 Closed Contour Shearing

This includes different operations according to the required job, as shown in Figure 10.2.

a. Blanking is the shearing along a closed contour to produce a blank with the required outline. The surrounding remainder left from the sheet strip is scrapped.
b. Piercing is the same shearing operation as in blanking, but it produces a hole with the required shape in the sheet. The material punched out is removed as waste. When piercing a circular hole, the operation may also be called *punching*.
c. Perforation is the operation of piercing or punching a number of holes in a sheet blank for a certain purpose (e.g., to perform as a strainer).
d. Notching is the operation of removing part of the desired shape from the side or edge of the sheet strip or blank.
e. Lancing is the partial shearing of a part without removal. The lanced part could be further bent or formed.
f. Slitting is similar to lancing, where part of the sheet is partially sheared without removal, but shearing does not open to the edge. Lancing is usually accompanied by bending to form a part protruding from the sheet surface.
g. Trimming is the operation of cutting away excess metal in a flange or flash from a formed sheet.
h. Shaving is the operation of shear removal of burrs or extra parts resulting due to blanking or piercing.

All closed contour shearing processes are performed using die sets, with the main parts being the punch and die. These die sets are operated by a mechanical, hydraulic, or pneumatic press. Turret punch presses are special machines in which punches and dies are mounted in synchronized order to perform many shearing operations on a blank secured on a free floating table. These turret punch presses are produced as an enclosed rigid structure to allow minimum deflection during operation. The drive is a hydraulic system with CNC control. Turret presses usually incorporate 4 controlled axes. Two of them are for the movement of the sheet strip on the press table in the X and Y directions to bring the required position before the press ram. The third is for turret rotation, and the fourth for the rotation of the indexing tools. The turret usually has 32 stations, which help to finish all the work, even on very complicated sheet products. The indexing station provides punching at any angle.

10.2.2 SHEARING MECHANISM

As performed in all shearing operations, the sheet strip, or blank, is located between the stationary die (or lower fixed blade in open shearing contours), and a moving punch (or upper blade), with a clearance c between edges, as shown in Figure 10.4. The progressive relative positions between the travelling punch and stationary die leads to the following distinguished areas on the cut surfaces:

a. Indentation zone, where the punch indents the upper surface of the sheet and the die indents the lower surface, as shown in Figure 10.4a. This indentation is sometimes called the roll-over. The depth of the indentation zone is larger at higher thicknesses and in more ductile materials. It ranges from 6% to 10% of the thickness.

b. Penetration zone: When the stresses at the cutting edges reach the shear yield stress of the material, the punch penetrates into the upper side, and the die penetrates into the lower side, developing smooth burnished shear areas as shown in Figure 10.4b. The percentage length of the penetration zone depends on the ductility of the material being sheared, with longer percentages for materials of higher ductility. This percentage ranges between 30% and 60% of the thickness.

c. Crack propagation zone: As the punch approaches the end of the stroke, the remaining area of the sheet thickness between the punch and die becomes too small to resist fracture. Cracks develop from both ends and propagate, as shown in Figure 10.4c, leading to separation of the parts with fracture surfaces. Proper clearance between the punch and die results in meeting of the crack fronts, as shown in Figure 10.4d. The angle of the fractured surface is identified as the breakout angle, and it ranges from 7° to 11° for normal types of shear surfaces.

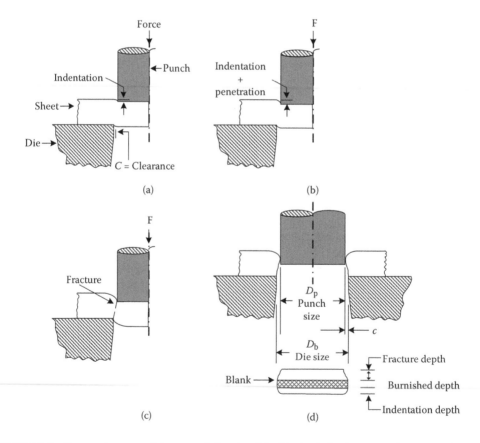

FIGURE 10.4 Progressive steps of sheet metal shearing (blanking and piercing): (a) indentation, (b) penetration, (c) fracture initiation and propagation, and (d) separation of the sheared part.

Figure 10.4 indicates that the sequence of the different parts of the shear surface on both the resulting hole and the separated part are reversed. The penetration zone of the hole is identical to the punch diameter, and the penetration zone of the separated part is identical to the die diameter. Therefore, the punch controls the diameter of the hole (piercing operation), and the die controls the diameter of the produced blank in a blanking operation. Consequently, in a piercing operation to produce a hole with diameter D, the punch diameter is chosen as $D_p = D$, and the die diameter is $(D + 2c)$, whereas, in a blanking operation, to produce a blank with diameter D, the die diameter is $D_d = D$, and the punch diameter is $(D - 2c)$.

Excessive clearance between the punch and die prolongs the tool life, but it leads to the formation of burrs along the top of the blank edge contour and the bottom of the edge of the produced hole, as shown in Figure 10.5a. On the other hand, too small clearance leads to offsetting of the two propagating cracks, and the punch has to complete the cut through a secondary torn zone between both cracks, as shown in Figure 10.5b. The optimum clearance is proportional to the sheet material thickness and strength and is inversely proportional to ductility. Average clearance values range from 8% to 10% for low-carbon steel and from 14% to 16% for high-carbon steel. Ranges of 5%–8% are applied for soft aluminum and copper alloys. Application of the right clearance leads to a maximum force based on the ultimate shear strength of the sheet material, but the force drops down sharply to zero value at a depth of about 0.4 of the sheet thickness. This results in minimum work or energy consumption, as shown in Figure 10.6. Excessive clearance reduces the peak force due to bending, but the work is higher due to delayed decay of the force to zero value at a depth almost equal to the sheet thickness. Insufficient clearance does not increase the force but results in higher work consumption in the tear zone, as indicated in the figure. The maximum shearing force (F_{max}) can be calculated from

$$F_{max} = t(\text{perimeter of profile to be cut}) \, \sigma_s \qquad (10.1)$$

where t is the sheet thickness and σ_s is the ultimate shear strength of the sheet material. It is possible to reduce this maximum force, and hence the press size required by chamfering either the punch or the die, which results in progressive shearing effect. This is similar to the inclined blade (alligator shearing) in open contour shearing. When blanking, the punch should be flat so that the part being pushed through the die remains flat; thus, chamfering is applied to the die. On the other hand, when piercing, the part pushed through the die is scrap, and it does not matter if it is bent. Hence, chamfering is normally applied to the punch.

Fine blanking eliminates indentation and fracture from blanks and produces a fine shear surface perpendicular to the blank surface. This can be carried out by using a pressure pad surrounding the

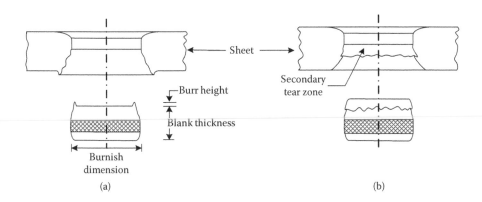

FIGURE 10.5 Effect of (a) excessive and (b) small clearance on the sheared areas.

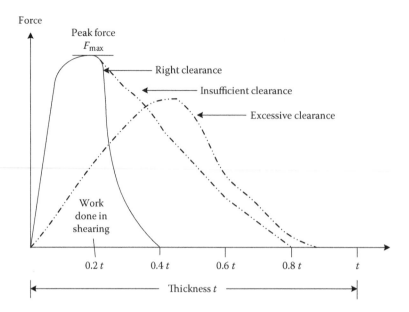

FIGURE 10.6 Effect of clearance on the shearing force and work. (From Waters, F., *Fundamentals of Manufacturing for Engineers*, Taylor & Francis, Boca Raton, FL, 2001. With permission.)

punch on the upper side of the sheet, reducing the clearance to about 1% of the sheet thickness, and using a lower pressure cushion to support the blank during shearing.

10.3 BENDING PROCESSES

Bending is the process of changing a flat blank into a curved product. It is one of the most commonly applied sheet forming processes in the majority of industries. Products may have one, two, or more bends with small radii, or bends with relatively large radii such as loops, tubes, or corrugated parts, as shown in Figure 10.7.

10.3.1 BENDING PARAMETERS

During bending, the inner fibers of the bend are compressed, and the outer ones are tensioned. The transition from compressive to tensile stresses passes through a neutral plane in (or near) the mid-section with no change in length, as shown in Figure 10.8. The length of the neutral axis in the bend zone is named the "bend allowance" (L_b), which determines the blank length required to produce the bend. Other parameters controlling the bend shape are the bend radius R, which is the radius of curvature of the inner side of the bend, and the bend angle α, which is the angle of the bent zone. Therefore, the bend allowance is determined from Equation 10.2:

$$L_b = \alpha(R - kt) \tag{10.2}$$

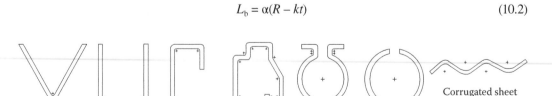

Corrugated sheet

FIGURE 10.7 Typical examples of bending products.

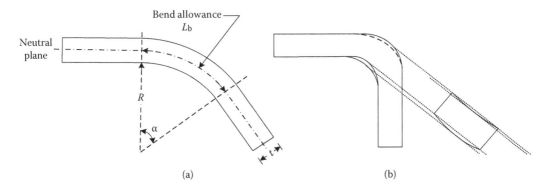

FIGURE 10.8 Dimensional parameters in (a) bending and (b) deformation of the cross section due to bending.

where t is the blank thickness, and k is a constant that equals 0.5 for ideal deformation with large radius, but may be reduced down to 0.33 for small radius ($R < 2t$). For values of k less than 0.5, the neutral axis is shifted toward the inner surface of the bend, and the radial blank thickness in the bend zone is reduced to keep the volume constant. The smaller the radius of curvature, the higher the reduction in thickness. The minimum bend radius is a forming limit which depends on the blank material. Highly ductile metals could have a bend radius of zero, allowing the blank to be bent flat upon itself ($\alpha = 180°$), whereas for high strength materials, minimum R should not be less than 5t. The minimum bend radius can be predicted from the percentage reduction in area r in a simple tension test, which is a measure of ductility. For r of less than 0.2, the shift in the neutral axis can be neglected, and R_{min} can be calculated from the equation:

$$\frac{R_{min}}{t} = \frac{1}{2r} - 1, \quad \text{for } r < 0.2 \tag{10.3}$$

When r is greater than 0.2, the shift of the neutral axis cannot be ignored, and the minimum radius is calculated from the equation:

$$\frac{R_{min}}{t} = \frac{(1 - r^z)}{2r - r^z}, \quad \text{for } r > 0.2 \tag{10.4}$$

The blank width W also affects the limit of bending. In bending narrow sheets, cracking or failure will occur at the blank edges, where tensile stresses in the width direction are maximum. The section at the width is distorted, as shown in Figure 10.8b. Wider blanks do not show this distortion, and plane strain conditions prevail at (W/t) ratios greater than 10.

10.3.2 SPRINGBACK IN BENDING

Springback is the dimensional change of the shape of the bent blank when the load on the bending tool is removed as a result of elastic recovery of the material, as shown in Figure 10.9. A quantity characterizing springback is the spring back factor, K_s, which is the ratio between the bend angle after and before springback. Since the bend allowance is the same before and after bending, then Equation 10.2 can be written in the form:

$$\text{Bend allowance} = (R_i + 0.5t)\alpha_i = (R_f + 0.5t)\alpha_f$$

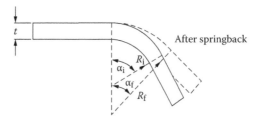

FIGURE 10.9 Springback in bending. (Adapted from Kalpakjian, S., and S. Schmid, *Manufacturing Processes for Engineering Materials*, 4th ed., Prentice-Hall, 2003.)

$$K_s = \frac{\alpha_f}{\alpha_i} = \frac{(2R_i/t + 1)}{(2R_f/t + 1)} \tag{10.5}$$

where R_i and R_f are the initial and final bend radii, respectively, and t is the sheet thickness. A factor $K_s = 1$ indicates that there is no springback and $K_s = 0$ indicates complete elastic recovery, as in a leaf spring. Springback increases with higher yield strength and/or smaller modulus of elasticity of the material. It also increases with higher plastic strain. Therefore, given the initial radius of curvature of the bend R_i, the yield stress σ_y, the modulus of elasticity E, and the thickness t, the final radius of curvature R_f can be obtained from the approximate relations.

For plane stress conditions ($W < 10t$)

$$\frac{R_i}{R_f} = 4 \left(\frac{R_i\sigma_y}{Et}\right)^3 - 3 \left(\frac{R_i\sigma_y}{Et}\right) + 1 \tag{10.6a}$$

For plane strain conditions ($W \geq 10t$)

$$\frac{R_i}{R_f} = 1.69 \left(\frac{R_i\sigma_y}{Et}\right)^3 - 2.25 \left(\frac{R_i\sigma_y}{Et}\right) + 1 \tag{10.6b}$$

To reduce or eliminate springback, one of the following techniques can be applied, as shown in Figure 10.10.

1. Overbending is the most common technique based on increasing or decreasing the angle of the punch or the die to compensate for the springback angle (Figure 10.10a). Alternatively, correction is based on the radius of curvature obtained from Equation 10.6.
2. Using material with higher modulus of elasticity or lower yield strength, or using larger radius of curvature and smaller bending angle.
3. Thining or ironing of the side walls of the bend (Figure 10.10b).
4. Coining (or bottoming) the bend by applying high localized compressive stresses through a nose on the punch (Figure 10.9c).
5. Stretch bending, where the strip is stretched during bending. Combined tension and bending stresses reduces stress nonuniformity and springback. Stretch bending will be introduced in a later section in this chapter.
6. Bending at high temperature to reduce the yield stress, and accordingly, springback.

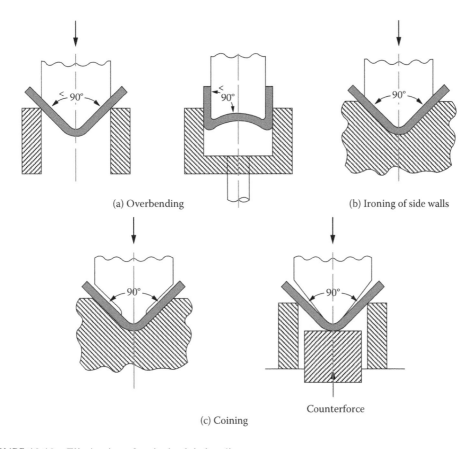

(a) Overbending   (b) Ironing of side walls

Counterforce

(c) Coining

FIGURE 10.10   Elimination of springback in bending.

10.3.3   RESIDUAL STRESSES IN BENDING

Another problem associated with bending is the residual stresses resulting due to nonuniform deformation across the section. When the bending stresses exceed yielding, plastic deformation starts at
the outer surfaces of the bend, and extends gradually toward the center, as shown in Figure 10.11.
Due to the change of the stresses from tensile to compressive around the neutral axis, there remains
an elastic portion around this axis and the stress distribution is nonuniform, as shown in Part b of

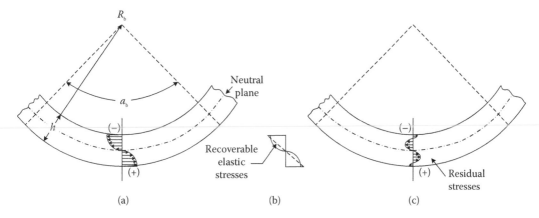

FIGURE 10.11   Residual stresses in bending.

the figure, which also represents the elastic recoverable stresses when the load is removed. The net residual stresses are indicated in Figure 10.11c, where compressive residual stresses exist at the outer surface of the bent part, which change to tensile till upon reaching the neutral axis. On the other hand, the inner part of the bend demonstrates tensile residual stress at the surface, which changes to compressive till upon reaching the neutral point. These residual stresses may lead to distortion or failure of the part when subjected to certain loading conditions during service, which may increase the total tensile stresses up to the limit of failure. Stress relieving is required to remove these internal residual stresses.

10.3.4 Bending Equipment

Bending processes can be carried out using conventional presses, press brakes, roll bending machines, and contour roll forming. On conventional presses, the tooling is a die set similar to that used in shearing or deep drawing processes. It consists mainly of a punch connected to the press ram, a die fixed on the press table, and accessories for location, fixation, alignment, etc. Other bending equipment are presented in this section.

10.3.4.1 Press Brake

Press brake is a special type of press with a long, narrow bed, used to bend, form, or punch relatively long sheet metal strips that are difficult to form using a conventional press. It is used for applications where production quantities are too small to justify the tooling cost for contour roll forming. Press brakes are adopted for simple v-bends or more intricate shapes, as well as other processes such as shearing, straightening, embossing, beading, and corrugating. The drive system is either mechanical, such as with an eccentric mechanism, or hydraulic. Press brakes are specified by the size (length of bed and length of stroke), and capacity.

Modern press brakes use *air bending*, where an advanced control determines the depth to which the punch or upper die extends into the lower die. This requires much less tonnage than bottom bending or coining. Typical dies used on press brakes, including air bending, are shown in Figure 10.12. More sophisticated press brakes interface with a *backgauge* through a CNC control system to allow the operator more accurate positioning of the strips for forming. Backgauges are typically have one up to seven axes of movement, each controlled by a separate electric motor.

10.3.4.2 Roll Bending Machines

Many curved sheet metal parts, such as rings, cylinders, cones, or segments of these parts are uneconomical to produce using press forming. Such parts are best produced using roll bending machines, where flat blanks are curved by passing them through a set of rolls. Most roll bending machines use three- or four-roll arrangements, as described in Section 9.4.3.4 (Figure 9.28). Another type of roll bending uses a two-roll bending machine. The upper roll is made of steel, and the lower roll is made of steel covered with urethane, as shown in Figure 10.13. The urethane deforms under the pressure of the top roll, which leads to uniform pressure distribution on the rolled material that forces it to conform to the shape of the top roll. In a way, the top roll serves as a rotary punch, whereas the urethane roll serves as a rotary female die. The material is completely contained during the bending process. This type is more expensive than a comparable three-roll bending machine, and is applied only for high rates (100 to 350 parts per hour) in large production runs. The soft urethane does not scratch the bent surface.

10.3.4.3 Contour Roll Forming

Contour roll forming is a continuous bending operation of a metal strip at room temperature using tandem sets of contour rolls (sometimes called roller dies) that shape the metal in a series of progressive stages until the desired cross-sectional configuration is obtained. Practically any contour section consisting of several bends, however complicated, can be roll formed, as shown in Fig-

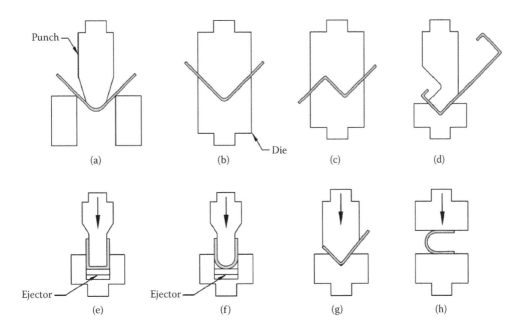

FIGURE 10.12 Typical dies used by press brakes: (a) air bending, (b) V-bending, (c) offset bending, (d) goose-neck punch for multiple bending, (e) channel bend dies, (f) U-bend dies, (g) acute angles, (h) flatting dies. (Adapted from *ASM Handbook, Volume 14: Forming and Forging*, ASM International, Materials Park, OH, 1988.)

ure 10.14. As a continuous process, roll forming is ideal for producing parts with long lengths or in large quantities. The process typically produces sections with thickness up to 20 mm, lengths up to 30 m, and runs at speeds from 10 to 180 m/min, depending on the required configuration, the material, and the desired tolerances. The process offers high-quality long products with close tolerances, uniformity of shape, and high surface finish. Typical applications in the aircraft industry are air frame stringers and longerons, trim and window frames, stiffeners, and honeycomb seals. In the industry of appliances, roll formed products include panels for refrigerators, stoves, microwave ovens, laundry and vending machines, shell fronts, ladder supports, etc. Figure 10.15 illustrates examples of the stages in roll forming of a U-shaped section. CAD/CAM roll design systems are available to predict the progressive tooling stages to produce almost any profile. Resulting data is

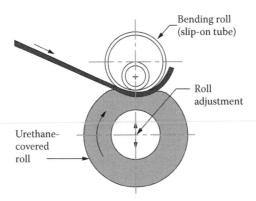

FIGURE 10.13 Principles of two-roll bending using urethane-covered roll.

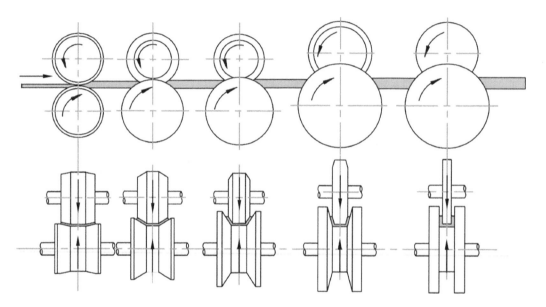

FIGURE 10.14 Different shapes produced by contour roll forming. (Adapted from Schuler, L., *Metal Forming Handbook*, Springer, 1998.)

FIGURE 10.15 Stages of contour roll forming of a U-section. (From Schuler, L., *Metal Forming Handbook*, Springer, 1998. With permission.)

directly fed to an NC lathe to cut the rolls to the required profiles. Programmable logic processors (PLCs) and microprocessors are used to control processing on roll forming lines.

10.4   STRETCH FORMING

In stretch forming, the metal strip is clamped at the edges and stretched over a form block to the required shape. This process is used to form aerospace parts from nickel, steel, aluminum, and titanium alloys, as well as other heat-resistant and refractory metals which are difficult to form using other processes. Typical products are automotive door panels and airplane wing skin panels with a large radius of curvature. Stretch forming cannot be used to produce sharp-cornered shapes.

The force required for the process is about 70% less than that required for any equivalent press forming process. The stretch forming machine and tooling are much cheaper than the cost of a press and die set; therefore, the process is more economical. The tensile strength of the product is increased by about 10% because stretching and hardness is accordingly increased (by about 2%). The uniform deformation leads to lower tendency for buckles or wrinkles. Springback and residual stresses are significantly reduced or eliminated due to uniform stress gradient. However, the process is suitable only for low-volume production.

10.4.1 STRETCH FORMING MACHINES AND ACCESSORIES

There are two major types of stretch forming machines

1. Stretch draw forming is usually carried out with a form block or sometimes with mating dies.
   a. Form block method, where a sheet metal strip is clamped at both ends, and the form block is moved by a hydraulic piston to stretch it, as shown in Figure 10.16a. Alternatively, the form block can be fixed to the machine bed, and stretching over the block is carried out by two hydraulic cylinders connected to the grippers.
   b. Mating-die method, using a two-piece die mounted in a single-action hydraulic press and two grippers that hold the strip and apply tensile force to it, as shown in Figure 10.16b. This method combines the advantages of stretch forming, which generates a moderately formed part, and press forming, which sets sharp contours as beads, feature lines, or even coining or stamping. Stretch draw press tooling for large parts, such as automobile roof panels, weighs only one-third of that for a conventional double-action press, and consumes only one-third of the force required for press forming.
2. Stretch wrapping (rotary table stretching), where the strip is fixed at one end to a stationary gripper, while the other end is fixed to a gripper connected to a hydraulic cylinder unit to apply enough tension to exceed the yield stress of the strip material. The form block, mounted on a turntable, revolves into the strip to reach the required contour, as shown in Figure 10.17. The stationary gripper may be mechanical or hydraulic. The hydraulic cylinder unit is free to swivel with table rotation such that the tensile force on the strip is

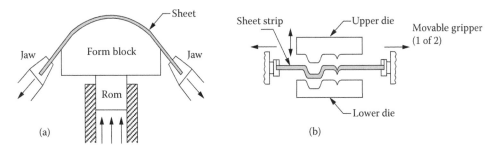

FIGURE 10.16   Stretch draw forming: (a) form block method and (b) mating dies method.

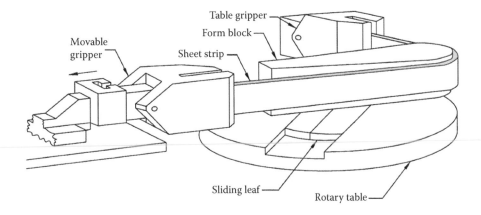

FIGURE 10.17   Stretch forming machine with rotary table. (From *ASM Handbook, Volume 14: Forming and Forging*, ASM International, Materials Park, OH, 1988. With permission.)

tangential to the contact point with the form block. This eliminates friction between the strip and form block during wrapping, and leads to accurate product size without spring-back. No lubrication is required for the process. Rotary table machines are available with capacities up to 10 MN.

The form block contour can vary throughout the bend, and the strip follows it accurately. Stretch forming blocks are made of low-carbon steel, cast iron, hard plastics, or hardwood. The jaws for grippers can be made segmented or contoured to allow equal stretch to all parts of the strip. Blanks for stretch forming are made longer than the required length to allow for trimming of the ends that may be damaged by the gripper jaws.

10.5 DEEP DRAWING

Deep drawing is the process of changing flat sheet blanks into cup-shaped products, such as cartridge cases, cans, bathtubs, shell cases, and automotive panels. This is carried out by locating a blank of calculated size over a shaped die, and pressing the metal into the die with a punch, as shown in Figure 10.18. To avoid wrinkling of the outside edges of the blank during drawing, blank holding pressure is applied as shown in the figure.

In deep drawing of a circular cup with a flat bottom, three types of deformation zones can be recognized, as shown in Figure 10.19.

a. The base (central segment): It is in contact with the punch head, and is wrapped around the punch. This base is under equal biaxial tension, resulting in thinning of the base.
b. The flange: Metal in the outer segment of the blank is drawn radially inward toward the die throat. Thus, the outer circumference is reduced gradually under hoop compressive stress and radial tensile stress. This results in increase of the thickness. The flange may be wrinkled under the effect of hoop compressive stresses unless blank holding pressure is applied.
c. The cup wall: As the metal passes over the die radius, it is bent and then straightened under tensile stress. The plastic bending under tensile stress leads to thinning, which modifies the thickening due to circumferential contraction. The cylindrical region (cup wall) is formed under tensile stresses generated due to the punch movement downward. Being constrained

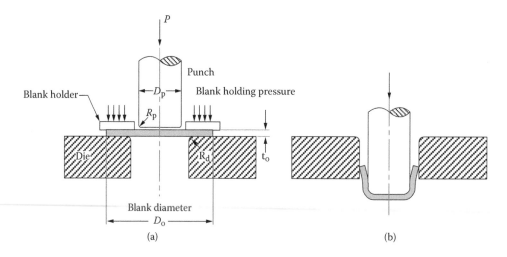

FIGURE 10.18 Deep drawing of a cylindrical cup: (a) before and (b) after drawing. (Adapted from Dieter, G., *Mechanical Metallurgy*, McGraw-Hill, 1988.)

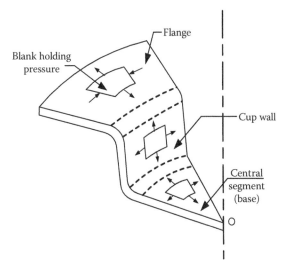

FIGURE 10.19 Stresses and deformation in a section from a drawn cup.

by the rigid punch, the cup wall is not allowed to contract in the circumferential direction (plane strain tension). The wall thickness is thinnest near the base just above the punch radius, and if the material strength does not support the applied tensile stress, local necking may occur in this region, leading to tearing. Measurements show that the average wall thickness, t_m, is in the range

$$1.04\, t_0 > t_m > 0.94\, t_0$$

It is often approximated that the average wall thickness is equal to the initial thickness.

10.5.1 BLANK HOLDING PRESSURE

The blank holding pressure plays an essential role in determining whether the sheet blank is purely drawn or stretched. Too high pressure increases the conditions for stretching, which leads to necking and tearing. On the other hand, too small pressure leads to smooth flow of the blank into the die cavity under pure drawing conditions, which leads to thickening of the flange and wrinkling of the outer circumference of the blank. The blank holding force should be just sufficient to prevent wrinkling. Blank holding pressure can be applied through one of three methods.

a. Incorporating a spring-loaded pressure pad surrounding the punch on a single action mechanical or hydraulic press, as shown in Figure 10.20. As the pressure pad touches the blank during the downward stroke of the press, the spring is compressed, creating pressure on the flange. The disadvantage of this method is that the clamping force increases as the punch moves further for deeper drawing.
b. Using a flat controlled pressure blank holder on a single-action press is preferred because the pressure can be controlled by supplementary pneumatic or hydraulic die cushion systems, as in Figure 10.18a.
c. Using a double-action hydraulic press that allows two concentric rams controlled separately. The inner ram is connected to the punch, whereas the outer ram applies the required blank holding pressure.

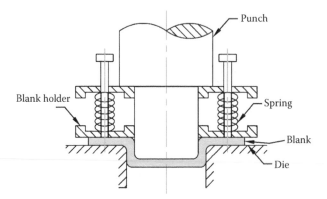

FIGURE 10.20　Pressure pad for blank holding pressure. (Adapted from Choudhury, S. K. et al., *Elements of Workshop Technology*, Vol. 2, Media Promoters and Publishers, 1984.)

When the relative thickness of the blank to its diameter is larger than 2%, a blank holder is not required.

10.5.2　IRONING

Clearance between the punch and the die walls should be larger than the blank thickness by 7% to 15% for deep drawing. It could be increased to 15% to 20% for minimal force requirement. If this clearance is less than the thickness produced by free thickening of the flange, the metal in the wall zone will be squeezed or ironed between the punch and die to produce a uniform thickness. The smaller the clearance, the greater the amount of ironing.

10.5.3　DEEP DRAWING FORCE

The force required on the punch for deep drawing is the sum of three parts: The ideal force of deformation, frictional force, and the force for ironing, if included. The trends of variation for these three components with the stroke length are shown in Figure 10.21. The ideal force of deformation

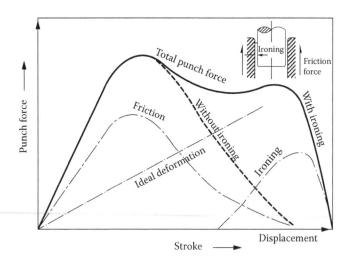

FIGURE 10.21　Force components for deep drawing. (From Dieter, G., *Mechanical Metallurgy*, McGraw-Hill, 1988. With permission.)

increases continuously with the stroke length due to the increase of the strain with punch movement, and due to the increase of the flow stress due to strain hardening of the material. Frictional force rises to the peak early, and then decreases with the continual decrease of the flange length in contact with the die and the pressure pad. The force required to produce ironing, if included, takes place at a later stage of the process. An additional factor is the force required for bending and unbending around the die radius. The following relation represents an approximate estimation of the force required for deep drawing, including these three components.

$$F = \pi D_p t \left(1.1 \bar{\sigma}_f \right) \ln \frac{D_o}{D_p} + 2F_p \frac{D_p}{D_o} \ e^{(\ \pi/2)} + B1 \tag{10.7}$$

where F is the total punch force, D_p is the punch diameter, D_0 is the blank diameter, $\bar{\sigma}_f$ is the average flow stress of the blank material, t is the wall thickness (or blank thickness), μ is the coefficient of friction, F_p is the force due to blank holding pressure, $B1$ is force due to bending and unbending (straightening) around the die radius.

The first term in the equation represents the ideal force, the second term represents frictional force under the blank holder, and the exponential term represents friction at the die radius.

10.5.4 LIMITING DRAWING RATIO

The drawability of a metal is measured by the ratio of the initial blank diameter to the diameter of the cup drawn from the blank, represented by the punch diameter. The limiting drawing ratio (LDR) for a material is the maximum of this ratio that can be carried out without tearing (failure). This ratio is limited by the condition: $\ln (D_0/D_p) \le 1$.

Then the maximum LDR is given as

$$(LDR)_{max} = \frac{D_o}{D_p} \bigg|_{max} = 2.72 \tag{10.8}$$

Because of friction and the effects of bending and unbending, the LDR is significantly lower than 2.72. Experiments show that for typical values of friction and tooling geometry, LDR is in the range of 1.9 to 2.2. Typical practical conditions that affect drawability are

- Die radius (R_d, Figure 10.18) should be about 10 times the initial blank thickness (t_0).
- Punch radius (R_p) should be 1 to 5 times t_0. Too small radius leads to local thinning and tearing.
- Clearance between the punch and die should be 1.15 to 1.2 t_0 to avoid ironing.
- Blank holding pressure is about 2% of the average flow stress of the material.
- Press speed should be in the range of 6–17 m/min for a single-action press and 11–15 m/min for a double-action press. When the operation includes ironing, the drawing speed should be reduced. In mechanical presses, the speed is considered in the midstroke (the speed changes from maximum to zero at the stroke end).
- Lubricate the die side to reduce drawing friction. Lubricants range from machine oil to pigmented compounds.

10.5.5 EFFECT OF ANISOTROPY

Sheet metals inherit anisotropic characteristics during rolling, that is, the change of properties with direction. There are two types of anisotropy that affect drawability: normal anisotropy, where the

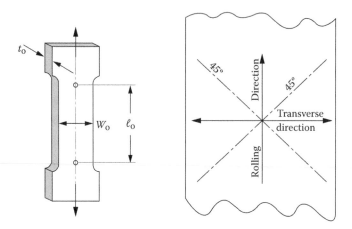

FIGURE 10.22   Tensile test specimen from a rolled sheet for measuring normal anisotropy ratio, R, in terms of width and thickness strains.

properties in the thickness direction differ from those in the plane of the sheet, and planer anisotropy, where properties vary with direction in the plane of the sheet.

Normal anisotropy is defined by the ratio R of the width strain ε_w to thickness strain ε_t in a tensile test specimen (Figure 10.22) cut from the sheet

$$R = \frac{\varepsilon_w}{\varepsilon_t} = \frac{\ln(w_0/w_f)}{\ln(t_0/t_f)} = \frac{\ln(w_0/w_f)}{\ln(w_f l_f/w_0 l_0)} \qquad (10.9)$$

Normal anisotropy is also called the strain ratio. The subscripts 0 and f refer to the original and final dimensions of the specimen, respectively. The final length and width in a test specimen are measured at elongations below necking.

Due to planer anisotropy of the sheet material, the R value of the tensile specimen depends on the orientation with respect to the rolling direction of the sheet (Figure 10.22). Therefore, an average value $\bar{R}$ of the normal anisotropy, representing all practical orientations (longitudinal direction L, 45° to the rolling direction and transverse direction T) is calculated as

$$\bar{R} = \frac{R_L + 2R_{45} + R_T}{4} \qquad (10.10)$$

Typical values of average normal anisotropy $\bar{R}$ range from 0.4 to 1.8 for most metallic materials, as shown in Table 10.1. Notice that the normal anisotropy of titanium alloys is exceptionally high (3.0–5.0).

TABLE 10.1
Typical Average Normal Anisotropy Ratios $\bar{R}$ for Various Sheet Metals

Metallic material	$\bar{R}$	Metallic material	$\bar{R}$
Zinc alloys	0.4–0.6	Aluminum alloys	0.6–0.8
Hot-rolled steel	0.8–1.0	Copper and brass	0.6–0.9
Cold-rolled rimmed steel	1.0–1.4	Stainless steel and HSLA steel	0.9–1.2
Cold-rolled aluminum killed steel	1.4–1.8	Titanium alloys (α)	3.0–5.0

Source: Kalpakjian, S., and S. R. Schmid, *Manufacturing Processes for Engineering Materials*, 4th ed., Prentice-Hall, 2003.

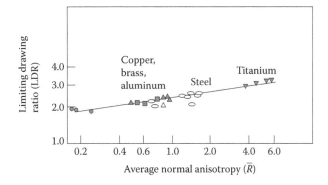

FIGURE 10.23 Effect of average normal anisotropy on the LDR. (From Atkinson, M., 1967, *Sheet Metal Industries*, vol. 44, pp. 167–178. With permission.)

When $\bar{R}$ equals unity, the material is isotropic in the normal direction. Higher values of $\bar{R}$ lead to better resistance of the material to thinning and thus deeper drawn heights or higher drawability can be achieved. Figure 10.23 demonstrates the linear relation between LDR and $\bar{R}$.

Planer anisotropy depends on the orientation in the sheet plane, and is measured by the variation in strain ratio in different directions in the plane of the sheet, ΔR, which can be expressed as

$$\Delta R = \frac{R_L - 2R_{45} + R_T}{2} \tag{10.11}$$

A completely isotropic material would have $\bar{R} = 1$ and $\Delta R = 0$. These two measures define plastic anisotropy.

Planer anisotropy causes undesirable earing of the material during drawing, as shown in Figure 10.24. The area containing ears can be trimmed, but it should be taken into consideration when calculating the original blank diameter.

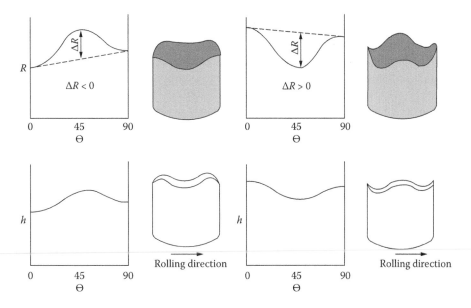

FIGURE 10.24 Relation between earing of deep drawn cups and angular variation of R. (From Hosford, W. F., and Caddell, R. M., *Metal Forming Mechanics and Metallurgy*, 3rd ed., Cambridge University Press, 2007. With permission.)

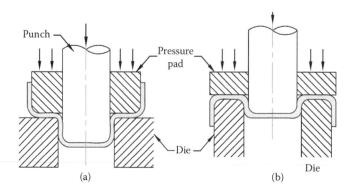

FIGURE 10.25 Types of redrawing: (a) direct and (b) reversed redrawing. (Adapted from Dieter, G., *Mechanical Metallurgy*, McGraw-Hill, 1988.)

10.5.6 REDRAWING

It has been shown that the LDR limits the diameter of the cup that can be drawn from a blank with a certain diameter. To make long, slender cups (such as cartridge cases, and long, closed-end tubes), successive drawing operations are essential. Reducing the diameter of a drawn product and increasing its height is known as redrawing. There are two basic types of redrawing, as shown in Figure 10.25.

a. Direct (regular) redrawing, where the outside surface of the first drawn part remains the same outside surface of the redrawn product (Figure 10.25a). In this case, the metal must go through a second bending and unbending around the die radius, leading to high strain hardening, which puts further limit to the LDR. This type of redrawing may be associated with buckling for some metals. The force required for redrawing could be reduced, and metal flow improved, by providing the first drawn cup with a taper (conical) cup.

b. Reverse (indirect) redrawing, where the cup is turned inside out so that the outside surface of the original cup becomes the inside surface of the redrawn product, as in Figure 10.25b. Using reverse drawing ensures better control of wrinkling.

The reduction obtained in redrawing is always less than that obtained in the first draw because of the higher strain hardening and the higher friction involved in redrawing. For successive redrawing processes, annealing is necessary after 50% to 80% total reduction.

10.6 RUBBER PAD FORMING (FLEXIBLE-DIE FORMING)

This process employs a flexible half tool made of rubber pad or polyurethane, requiring only one solid tool half (called form block) to form a blank to final shape through bending or drawing. The form block is usually similar to the punch in a conventional die, but it can take the form of the die cavity. The rubber acts somewhat like hydraulic fluid in exerting nearly equal pressure on all work-piece surfaces as it is pressed around the form block. The process is applied to produce moderately shallow and recessed parts with relatively simple configurations and flanges. Typical products are deeply recessed taillight reflectors for automobiles and drawn toaster shells.

Form block height is usually less than 100 mm to avoid buckling. It can be made of relatively cheap materials such as epoxy resin, zinc alloys, hardwood, aluminum, cast iron, or steel. Sheet thickness is up to 6 mm. The production rates are relatively high, with cycle times in the range of a few seconds up to less than a minute. Therefore, hydraulic presses are used with cycling rates as high as 1500 per hour. Required pressures are typically in the order of 10 MPa.

Advantages of the rubber-pad forming compared to conventional forming processes are

- Only a single rigid tool half is required to form a part.
- One rubber pad, or diaphragm, takes the place of many different die shapes, returning to its original shape when the pressure is released.
- Tools can be made of low cost, easy-to-machine materials due to the hydrostatic pressure exerted on the tools.
- The forming radius decreases progressively during the forming stroke, unlike the fixed radius on conventional dies.
- Thinning of the work metal is reduced considerably.
- Different metals and thicknesses, as well as laminated sheets with nonmetallic coatings, can be formed in the same tool.
- Parts with excellent surface finish can be formed, as no tool marks are created.
- Setup time is considerably shorter, as no lining up of tools is necessary.

Disadvantages of rubber pad forming are

- The pad has a limited lifetime, depending on the severity of forming and the pressure level.
- Lack of sufficient forming pressure results in parts with less sharpness or with wrinkles, which may require subsequent processing.
- The production rate is relatively slow, making the process suitable primarily for low-volume production.

10.6.1 RUBBER PAD FORMING TECHNIQUES

The oldest and most simple rubber pad forming technique is the Guerin process, as shown in Figure 10.26a. The rubber pad is relatively soft, with height in the range of 3 times the height of the product, and larger than the height of the form block. The pad is made of one piece or cemented slabs and is kept in a steel or cast iron retainer.

The Marform process is a more sophisticated technique developed to produce wrinkle-free shrink flanges through adoption of blank holding pressure controlled hydraulically, as shown in Figure 10.26b. Rubber pressure with such a technique is in the range of 35 to 70 MPa.

10.7 HYDROFORMING (FLUID-FORMING PROCESSES)

Fluid-cell forming is a further development of the rubber pad forming process to increase its capacity, and allow production of all flange configurations and intricately drawn shapes. The pressure is increased through a flexible hydraulic fluid cell to force the pressure pad to follow the contour of the form block under uniform pressure. This technique is known as the Verson-Wheelon process, which is shown in Figure 10.27a. It allows the addition of beads and ribs to the formed part.

A special type of hydraulic press developed for such a process is called the *hydroform press*. A lower hydraulic ram drives the punch upward, and the upper ram acts as a positioning part and is provided with a pump that delivers fluid under pressure to a pressure dome, as shown in Figure 10.27b. The blank folder is supported by a solid bolster, and is stationary during the operation. Deeper draws can be performed using this technique than in conventional drawing dies. The dome pressure may reach up to 100 MPa, and punch force capacities vary from 3.2 to 20 MN.

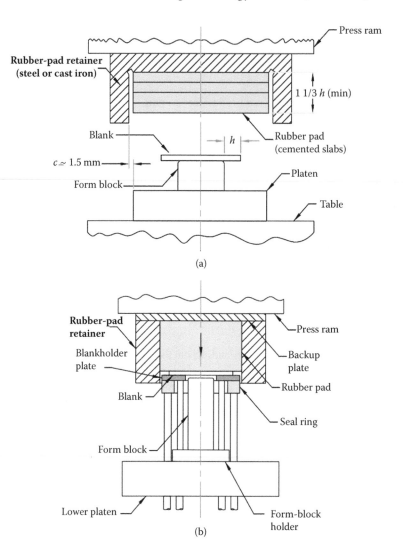

FIGURE 10.26 Rubber pad forming techniques: (a) Guerin process and (b) Marform process. (Adapted from *ASM Handbook, Volume 14: Forming and Forging*, ASM International, Materials Park, OH, 1988.)

10.8 SPINNING

Spinning is the forming of axial symmetric products over a rotating mandrel using rigid tools or rollers. The process can be classified into three categories, namely, conventional spinning, flow turning, and tube spinning.

10.8.1 Conventional Spinning

Conventional spinning is a process in which a circular blank is held against a form block which is rotated at high speed in a lathe-like machine spindle. The blank is progressively formed (in a sequence of passes) over the block using a wooden tool or small-diameter work rolls, as shown in Figure 10.28a. The tool may be manually operated, or by using mechanical power through a lever or hydraulic mechanism. The blank thickness does not change appreciably during this process. The upper limit of blank thickness increases with the increase of the material ductility. The blank

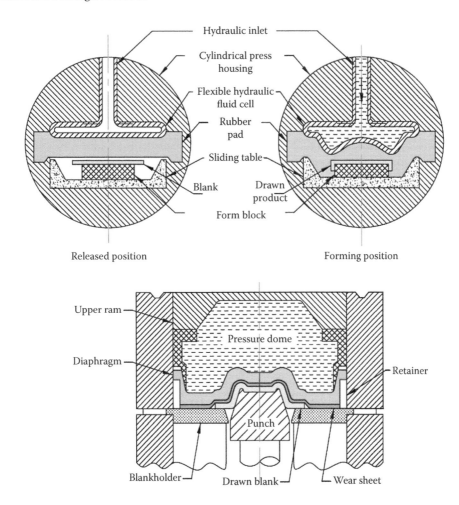

FIGURE 10.27 Hydroforming (fluid-forming) processes: (a) fluid cell forming (the Verson-Wheelon process), (b) hydroform press. (Adapted from *ASM Handbook, Volume 14: Forming and Forging*, ASM International, Materials Park, OH, 1988.)

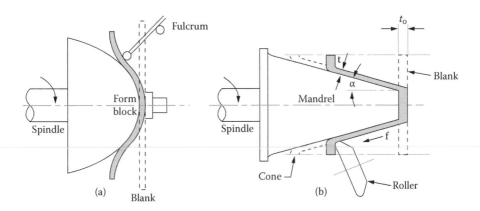

FIGURE 10.28 (a) Conventional spinning and (b) flow turning. (Adapted from Dieter, G. E., *Mechanical Metallurgy*, McGraw-Hill, 1988.)

is sometimes preheated to improve ductility, or to increase the thickness to be spun. The machine cost and tooling cost are relatively cheap, but production time is high, tolerances are high, and the process is adopted only for small production runs. The process is applied to produce flanges, rims, large cups, cones, bells, light reflectors, tank ends, and many aircraft components.

10.8.2　FLOW TURNING (SHEAR SPINNING)

This is a variation of conventional turning, where the blank diameter is kept the same but the thickness of the spun part is reduced over a conical mandrel, generating the form shown in Figure 10.28b. The severe cold working of the material during flow turning increases the product strength. The process is applied for manufacturing large conical or curvilinear shapes such as rocket motor casings and missile nose cones. Contradictory to conventional spinning, the production time is shorter, tolerances are smaller, and the process is suitable for mass production.

The thickness in flow turning to a conical shape is reduced according to the relation

$$t = t_0 \sin \alpha \qquad\qquad (10.12)$$

where α is the semicone angle. When small angles are required ($\alpha < 18°$), several passes can be used to decrease the angle gradually to avoid severe shearing of the metal.

A range of horizontal spinning lathe sizes is available for blank diameters from 6 mm up to 1.8 m. Vertical lathes permit blank diameters from 1.8 up to 4.9 m. Spinning lathes use a variable speed drive to allow for quick change of speed during operation as required. Surface speeds of 300 to 600 m/min are most commonly used. The feed rate in flow turning controls the surface finish of the product and the fit of the work on the cone surface. The feed rate increase holds the work piece tighter on the mandrel, while decreasing the surface finish. On the contrary, smaller feeds will cause a loose fit, while improving surface finish. Therefore, a compromise should be sought. Feeds are in the range of 0.25 to 2 mm/rev. These machines may be included in highly sophisticated automation facilities. Mandrels or form blocks for conventional spinning are made of steel, cast iron, aluminum, or coated hard wood. Tools and rollers are made of tool steels or aluminum bronze. Mandrels for flow turning are made of tool steels for high production rates. Lubricants in the form of grease, wax, or soap are used to reduce friction and the heat generated during forming.

10.8.3　TUBE SPINNING

In tube spinning, tubular shaped metallic material (preform) is spun over a cylindrical mandrel using rollers to reduce the thickness and increase length and strength. The maximum reduction in thickness per pass is limited by the limit of deformation that the material can withstand without failure. Tube spinning is also limited by the smallest percentage reduction in thickness that will ensure complete flow of the metal. This minimum reduction is usually 15% to 25%, depending on the metal and on the thickness of the original tube. The minimum size of the tube that can be spun is more likely to be the limiting factor rather than the maximum size, since smaller tubes require more slender mandrels and a slower speed that may not be available on the machine. The process is widely applied in the aircraft and automotive industries, and for making pressure vessels, as it offers improvement of the mechanical properties, high dimensional accuracy, and surface finish.

Tube spinning could be classified according to the direction of metal flow relative to the direction of roller movement to forward and backward spinning, as shown in Figure 10.29. In both methods, the preform is fixed in one position at one end, and the remaining length is free to slide along the mandrel.

a. Forward spinning, where the roller moves away from the fixed end, and the metal flows in the same direction of the roller. This method is used for bottomed or internally flanged performs, which can be fixed at the side of the tailstock. It offers closer control of the final

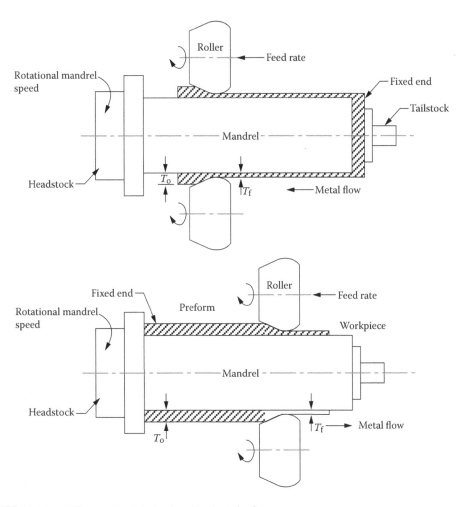

FIGURE 10.29 (a) Forward and (b) backward tube spinning.

tube length. A main drawback is the slower rate of production, where the roller has to traverse the whole finished length.
b. Backward spinning, where the preform should be an open-ended tube, which is fixed against a fixture in the headstock side. The roller advances toward the fixed end, while the metal flows in the opposite direction. The main advantage is that the roller traverses only a smaller length equal to the length of the preform, while in forward spinning it has to move along the final length of the product. The roller traverses only 25% of the final length for a 75% reduction in thickness. Therefore, production time is faster in backward spinning. A second advantage is that the compressive force of the roller against the fixture helps in the spinning of metals with relatively low ductility. The major disadvantage of backward spinning is that the spun length is free to move a long distance, which renders it susceptible to distortion.

Tube spinning machines are similar to conventional and flow turning lathe machines. However, tube spinning machines provide higher power capacities, due to the higher compressive forces required for tube drawing (similar to extrusion). Tube spinning machines are provided with two opposed rollers to minimize deflection of the mandrel due to high length-to-diameter ratios. For higher accuracy of product cylindricity, three or more rollers could be used to have the centering effect of a steady rest (for flow turning, one or two rollers only may be used). Modern tube spinning machines are numerically controlled by computers, which offer higher flexibility in controlling the length of traverse, number of passes, and the reduction per pass.

10.9 SUPERPLASTIC FORMING OF SHEETS

Superplasticity has been defined in Section 8.9, and the conditions for reaching superplastic behavior for some specific metals and alloys were presented. This phenomenon has been utilized in industry to apply extremely high elongations to form sheet blanks with large thickness at very low flow stresses using simple tooling. At present, there are a number of superplastic forming processes for sheet metals, which include blow forming, vacuum forming, thermoforming, superplastic forming/diffusion bonding (SPF/DB), and dieless drawing.

10.10 BLOW FORMING AND VACUUM FORMING

Blow forming and vacuum forming are basically the same process (sometimes called superplastic stretch forming) in that a gas pressure differential is imposed on the superplastic diaphragm, causing the material to form into the die configuration. In vacuum forming, the applied pressure is limited to atmospheric pressure (100 kPa), and the forming rate and capability are therefore limited. With blow forming, additional pressure is applied from a gas pressure reservoir, and the only limitations are related to the pressure rating of the system and the pressure of the gas source. Maximum pressures of 690 to 3400 kPa are typically used in this process.

The blow forming method is illustrated in Figure 10.30, which shows a cross section of the dies and forming diaphragm (blank). In this process, the dies and sheet material are normally maintained at the forming temperature, and the gas pressure is imposed over the sheet, causing the sheet to form into the lower die. The gas within the lower die chamber is simply vented to atmosphere. The lower die chamber can also be held under vacuum, or a back pressure can be imposed to suppress cavitation if necessary. This pressure is generally applied slowly rather than abruptly in order to prevent too rapid a strain rate and consequent rupturing of the part. The periphery of the sheet is held in a fixed position and does not draw in, as would be the case in typical deep-drawing processes.

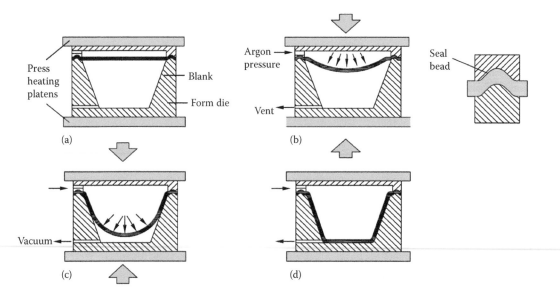

FIGURE 10.30 Schematic of the blow forming technique for superplastic sheet forming: (a) blank loading and (b) heating, (c) forming under pressure, and (d) final form.

It is common to use a raised land (seal bead) machined into the tooling around the periphery to secure the sheet from slippage and draw-in and to form an airtight seal in order to prevent leakage of the forming gas, as shown in the figure. This results in considerable thinning of the sheet for complex and deep-drawn parts, and it can also result in significant gradients in the thickness in the finished part. This process is being increasingly used to fabricate structural and ornamental parts from titanium and aluminum.

10.11   THERMOFORMING METHODS

These methods have been adopted from plastics technology for the forming of superplastic metals, and they sometimes use a moving or adjustable die member in conjunction with gas pressure or a vacuum. Figure 10.31 shows two examples of thermoforming methods. In Figure 10.31a, a female die is used to stretch form the superplastic sheet by application of gas pressure to force the sheet material against the die cavity. In Figure 10.31b, the first step involves blowing a bubble in the sheet away from the tool. The male tool is then moved into the bubble, and the pressure is reversed to cause the bubble to conform to the shape of the tool.

10.11.1   SUPER PLASTIC FORMING/DIFFUSION BONDING PROCESS

SPF is applied in combination with diffusion bonding (SPF/DB) to form aircraft titanium panels, as shown in Figure 10.32. Sheets (usually three) are put down together and diffusion bonded at selected locations by pressing them at high temperature. The resulting molecular bond offers a fully homogeneous strong joint at these locations. The unbounded regions are expanded into a mold under air pressure to form the V-shaped stiffening ribs shown in the figure. The SPF/DB process produces more complex, stiffer, stronger, and lighter structures in one piece, thus giving improved component integrity, accuracy, and the elimination of unnecessary joints.

10.12   SHEET METAL FORMABILITY

Sheet metal formability is defined as the ability of the sheet metal to undergo the desired permanent shape changes without failure (necking, tearing, or splitting).
   Major factors affecting formability

- Properties of the sheet metal
- Equipment and tooling characteristics (speed of deformation, tooling geometrical configurations, and surface finish)
- Lubrication at interfaces between sheet and tooling

10.12.1   TESTING FOR FORMABILITY

There is a wide variety of methods for testing formability, the most common of which include

a. Tension test: It is the basic and common test for formability. It determines the yield stress, ultimate tensile strength, the percentage elongation, strain hardening exponent n, and anisotropy parameters $\bar{R}$ and ΔR. Tensile samples are cut from the sheet in the rolling direction, transverse direction, and 45° directions, as shown in Figure 10.22, to determine anisotropy parameters. However, simple mechanical property measurements made from this test do not give all required conditions of formability.

b. Cupping tests: The Swift flat-bottom cupping test shown in Figure 10.33a is a standardized test for deep drawing to determine the LDR referred to earlier. In the Olsen and Erichsen tests, a sheet metal specimen is clamped between two ring dies, and a steel hemispherical

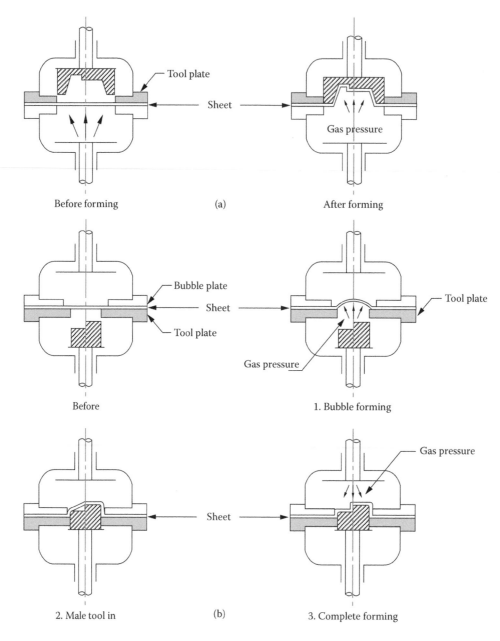

FIGURE 10.31 Thermoforming methods that use gas pressure and movable tools to produce parts from superplastic alloys using (a) female and (b) male dies.

punch (Figure 10.33b) is pushed hydraulically to stretch the sheet until a crack appears on the surface. The depth of the bulge until fracture is measured. These tests subject the sheet mainly to stretching, while the Swift test subjects the material to pure deep-drawing conditions. Virtually all practical sheet forming processes subject the sheet metal to both stretching and deep-drawing conditions. Another common cupping test is the hole expansion test, where a circular blank with a central hole is drawn using a flat die, as shown in Figure 10.33c. The hole expansion is observed during the test until the appearance of cracks on its perimeter.

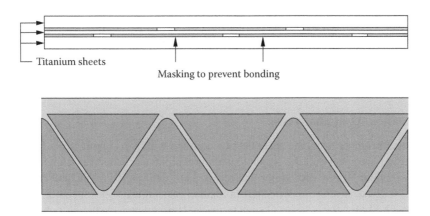

FIGURE 10.32   Superplastic forming/diffusion bonding (SPF/DB) process.

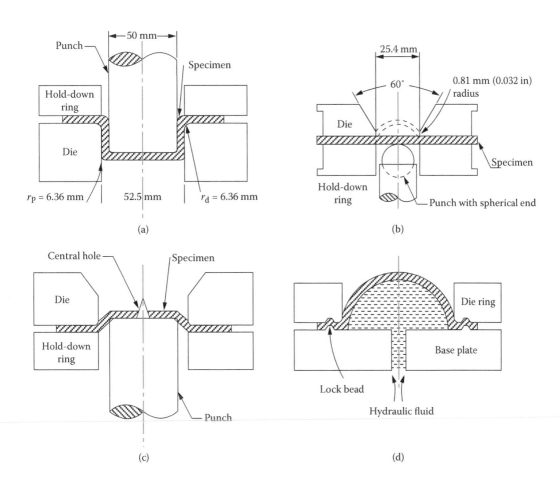

FIGURE 10.33   Common cupping sheet formability tests: (a) Swift deep drawing test, (b) Olsen and Erichsen tests, (c) hole expansion test, (d) hydraulic bulge test. (Adapted from *ASM Handbook, Volume 14: Forming and Forging*, ASM International, Materials Park, OH, 1988.)

c. Bulge test: Where a circular blank is clamped at its periphery and is bulged by hydraulic pressure, as shown in Figure 10.33d, thus replacing the hemispherical punch, as shown in Figure 10.33b. The process is one of the pure stretch forming tests, and no friction is involved. It can be applied to obtain the effective stress-effective strain curves for biaxial loading under frictionless conditions.

10.12.2 Forming Limit Diagrams

In the Olsen and Erichsen tests, the sheet blank is marked with a grid pattern of circles 2.5–5 mm in diameter using chemical etching or photoprinting techniques. After stretching, the circles are deformed to elliptical shapes with major and minor axes, defining the principal strains, as shown in Figure 10.34. These strains are measured in regions near failure zones (necking or tearing) and are plotted on the FLD. After a series of such tests is performed on a particular type of sheet metal, the boundaries between safe and failed regions are plotted on the FLD, as shown in Figure 10.35. The major strain on the deformed sheet is mainly tensile, and is plotted on the vertical axis of the

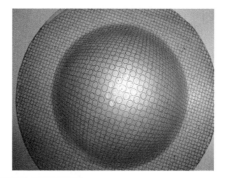

FIGURE 10.34 Circular grid marks on deformed specimen, giving the magnitude and direction of principal strains.

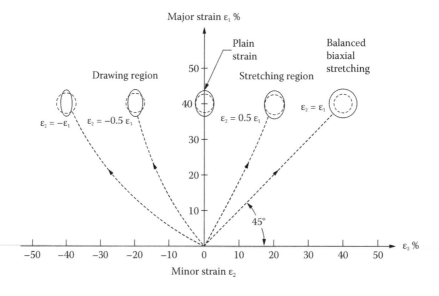

FIGURE 10.35 Illustration of several major strain/minor strain combinations. (Adapted from *ASM Handbook, Volume 14: Forming and Forging*, ASM International, Materials Park, OH, 1988.)

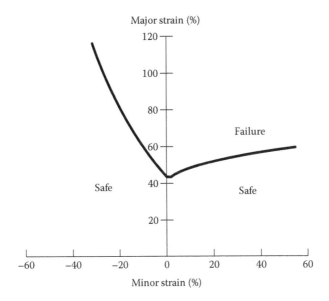

FIGURE 10.36 Typical forming limit diagram for steel. (From *ASM Handbook, Volume 14: Forming and Forging*, ASM International, Materials Park, OH, 1988. With permission.)

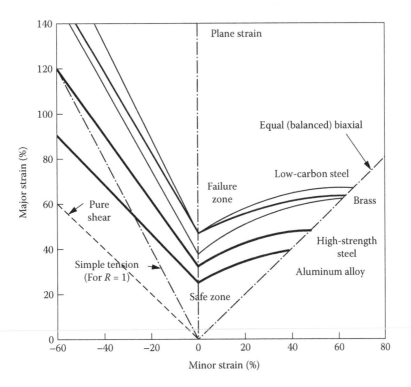

FIGURE 10.37 Forming limit diagram (FLD) for various sheet metals. (From Kalpakjian, S., and Schmid, S. R., *Manufacturing Processes for Engineering Materials*, Prentice-Hall, 2003. After Hecker, S. S., and A. K. Gosh. With permission.)

diagram. The minimum, or minor strain, can be positive or negative according to whether deformation is stretching or drawing, and is plotted on the horizontal axis. There are five possible different strain paths illustrated in the figure, each leading to the same major strain of 40%, but with minor strains ranging from −40% to 40%. The ellipses shown were originally circles (shown dashed) in the undeformed sheet. On the right side of the figure, the circles have changed to ellipses that are larger in all directions. This is the region of *biaxial stretching*, and the limit is the diagonal (45°) direction, representing *balanced equal biaxial stretching*, where the circles have expanded without changing shape. On the left side of the figure, the circles have transformed into ellipses which are larger in one direction but smaller in the perpendicular direction than the original circles. This is the region of *drawing*, including the strain state developed in *the tensile test* for an isotropic material, and the limit of pure shear (where transverse strain is equal to the longitudinal strain, but in the compression side). On the vertical axis, the ellipses are larger in one direction, but unchanged dimensionally from the original circles in the perpendicular direction, which represents the *plane strain* condition. A typical FLD curve for a steel sheet is shown in Figure 10.36, with the areas above the curve representing failure due to deformation, and the areas below the curve representing safe forming conditions. The FLD for various sheet metals is shown in Figure 10.37. The figure indicates that the higher the curve, the better the formability of the material. It should be noticed also that it is desirable for the minor strain to be negative to have larger safer major strain. Thicker sheets lead to higher forming limit curves, and accordingly, are more formable. However, a thick sheet develops cracks during bending around a small radius.

10.13 REVIEW QUESTIONS

1. Classify sheet forming processes according to the shape of the product.
2. What are the types of sheet metal shearing machines? And what are their blades made of?
3. Classify closed-contour shearing processes, giving examples of the applications of each.
4. Describe the zones shown on the sheared edges of a blank and on the pierced part on a sheet, and show how they are correlated.
5. What is the effect of increasing or decreasing the punch-die clearance in shearing processes on both the quality of the blank and on the consumed energy during the process?
6. How to reduce the force required for blanking or piercing without affecting the product quality.
7. How to eliminate the defects of the sheared surface.
8. What is the effect of the radius of curvature in bending on the quality of the bent surface?
9. What are the factors affecting springback in bending and how can it be eliminated?
10. Why are there residual stresses after bending, and how can they be eliminated?
11. Why is a press brake used in sheet forming? Compare it with a conventional press.
12. What are the applications of roll bending machines? Compare two- and three-roll bending machines, regarding their applications and economic considerations.
13. What are the applications of stretch forming, and why is it different compared to bending and deep drawing?
14. What are the types of stretch forming machines, and the functions of each?
15. What are the types of stresses on different parts of a metallic cup during deep drawing, and what are the effects of these stresses on their thicknesses?
16. What is the role of a blank holder in a deep-drawing process? How and when should this pressure be applied?
17. Discuss the effect of punch/die clearance during deep drawing.
18. What is the maximum value of the limiting drawing ratio (LDR)? Why are typical LDRs less than this maximum value?
19. Define anisotropy and explain how it affects deep drawing.

20. When and how should we use redrawing?
21. What are the advantages and limitations of rubber pad forming compared to conventional deep drawing?
22. Discuss hydroforming processes and when should we apply them?
23. What are the types of spinning processes? What are the applications of each?
24. Compare forward and backward tube spinning processes.
25. What are the types of superplastic sheet forming processes and the applications of each?
26. Describe how to test sheet formability. Why not use ductility alternatively?

10.14 PROBLEMS

1. It is required to cut washers from 2-mm-thickness mild steel strips using two punch/die sets on a mechanical press. The outside diameter of the washer is 30 mm and the inside diameter is 18 mm. If the ultimate tensile strength of this steel is 450 MPa, the shear strength is 70% of the ultimate strength, and the optimum clearance is 8%. Calculate the required forces and the diameters of the punches and dies.
2. A mild steel strip with a thickness of 0.9 mm, width of 10 mm, and length of 20 mm is bent to a radius of 12.5 mm. Assuming that the yield stress is 280 MPa and the modulus of elasticity is 210 GPa, calculate the radius of the part after bending, the springback factor, and the initial bending angle.
3. A 1000 × 200 × 1.2 mm annealed mild steel strip is stretched on a tool with circular shape to form an arc with a radius 500 mm and an angle of 120°. The material has a strain hardening relation of $\sigma = 530\ \varepsilon^{0.26}$. Calculate the final length, the percent reduction in the cross-sectional area and the maximum permissible stretching length before necking.
4. If the material in the previous problem suffers a decrease in thickness of 10% when fully stretched to the maximum length, calculate the normal anisotropy.
5. The table below lists the properties of several sheet materials to be applied in different forming processes:

Material	E (GPa)	σ_y (MPa)	R_0	R_{45}	R_{90}	n
A	207	220	1.9	1.2	2.0	0.25
B	207	241	1.2	1.0	1.2	0.22
C	72.4	172	0.7	0.6	0.7	0.22
D	114	138	0.6	0.9	0.6	0.5
E	69	7	1.0	1.0	1.0	0.00

You are asked to select the right material for each of the following applications and state the reason for your selection.
   a. Which of these materials would have the highest LDR in cupping?
   b. Which material would show the greatest amount of earing during cupping?
   c. Which material would have the greatest uniform elongation in a tension test?
   d. Excluding material E (because of its low yield strength), which material could be formed into the deepest cup by a hemispherical punch acting on a clamped sheet (no "drawing")?
6. For the materials presented in the FLD of Figure 10.36, if the major strain in a forming process is equal to 40%, find:
   a. The maximum positive minor strain for the aluminum alloy.
   b. The state of stress at the point of maximum positive minor strain for the alloy.
   c. The maximum negative minor strain for the same material.
   d. The maximum positive and negative minor strains for high-strength steel.

BIBLIOGRAPHY

Altan, T., Oh, S., and Gegel, H. 1983. *Metal Forming Fundamentals and Applications*. Materials Park, OH: ASM International.

Amstead, B. H., Ostwald, P. F., and Begemann, M. 1979. *Manufacturing Processes*, 7th ed. USA: J. Wiley & Sons.

ASM Handbook, Volume 14: Forming and Forging, 1993. Materials Park, OH: ASM International.

Atkinson, M. 1967. *Sheet Metal Industries*, vol. 44, pp. 167–178.

Degarmo, E. P., Black, J. T., and Kosher, R. A. 1997. *Materials and Processes in Manufacturing*, 8th ed. USA: Prentice-Hall.

Dieter, G. E. 1988. *Mechanical Metallurgy*. UK: McGraw-Hill.

Groover, M. P. 2007. *Fundamentals of Modern Manufacturing*, 3rd ed. USA: John Wiley & Sons.

Hosford, W. F., and Caddell, R. M. 2007. *Metal Forming Mechanics and Metallurgy*, 3rd ed. USA: Cambridge University Press.

Kalpakjian, S., and Schmid, S. 2003. *Manufacturing Processes for Engineering Materials*, 4th ed. USA: Prentice-Hall.

Kobayashi, S., Oh, S., and Altan, T. 1989. *Metal Forming and the Finite Element Analysis*. NY: Oxford University Press.

Koshal, D. 1993. *Manufacturing Engineer's Reference Book*, 13th ed. UK: Butterworth-Heinemann.

Lange, K. 1985. *Handbook of Metal Forming*. USA: McGraw-Hill, published by SME (2006).

Marciniak, Z., and Duncan, J. 2002. *Mechanics of Sheet Metal Forming*. Oxford: Butterworth-Heinemann.

Ragab, A., and Bayoumi, S. 1999. *Engineering Solid Mechanics Fundamentals and Applications*. Boca Raton, FL: CRC Press.

Rowe, G. W. 1971. *An Introduction to the Principles of Metalworking*. London: Edward Arnold.

Schey, J. A. 1987. *Introduction to Manufacturing Processes*, 2nd ed. Singapore: McGraw-Hill.

Schuler, L. 1998. *Metal Forming Handbook*. Germany: Springer.

Tschaetsch, H. 2006. *Metal Forming Practice: Processes–Machines–Tools*. Berlin: Springer.

Waters, F. 2002. *Fundamentals of Manufacturing for Engineers*. Glasgow: UCL Press.

11 High-Velocity Forming and High-Energy-Rate Forming

11.1 INTRODUCTION AND CLASSIFICATION

High-velocity forming (HVF) and high-energy-rate forming (HERF) processes are distinguished from conventional forming processes by their higher forming speeds. The range of speeds for conventional forming is 0.3 to 5 m/s. The upper limit of these forming speeds is attained when power drop hammers are used in forging processes, as shown in Section 9.3.4.3. The range of forming speeds in HVF and HERF is 5 to 300 m/s. These nonconventional forming processes began in 1960 and grew fast to cope with the Space Age requirements of tougher and more highly heat resistant materials, as well as larger product sizes.

HVF processes emerged from the principle of the proportionality of the kinetic energy of hammers to the square of the velocity. Therefore, high kinetic (mechanical) energy can be transported to the metal to deform it using a relatively small weight ram or die at high velocity. This has led to the development of a new family of high-speed hammers with smaller size (reduced cost) and shorter stroke (higher rate of production). The high power acceleration of the ram for these hammers is supplied by the burning of a mixture of air and a hydrocarbon (gasoline, kerosene, or diesel fuel) or a gaseous fuel (propane or natural gas), which is the same as the principle of an internal combustion engine. The maximum velocity for these machines is limited by the inertia of the moving parts. High-speed hammers with velocities in the range of 5 to 60 m/s are available. These hammers are used in performing HVF processes such as forging, extrusion, bar cropping, and blanking.

The need for forming metals at velocities higher than those achieved by HVF presses called for adopting systems that generate high energy at very short times (high energy rate), such as the pressure developed when detonating an explosive material (chemical energy), interaction of magnetic fields, or sudden electric discharge in a fluid. The energy is transferred directly to the metal to be deformed. This has led to the emerging of HERF processes, which include

1. Explosive forming
2. Electromagnetic forming
3. Electrohydraulic forming

These HERF processes are mainly applied for sheet and plate products, often with large size, to procure the advantages of increased formability and reduced springback. However, they can also be applied for specific forging processes. Deformation velocities in the range of 30 to 300 m/s are attained when using HERF equipment. These nonconventional forming processes can therefore be classified as shown in Figure 11.1.

11.2 CHARACTERISTICS OF HVF AND HERF PROCESSES

This family of nonconventional forming processes can be characterized by

a. Improved formability: In high-velocity forming, ductility far beyond the usual limits prescribed by the forming limit diagram for sheet metal forming can be achieved. This improved ductility is manifested between two critical velocity limits. Below the first

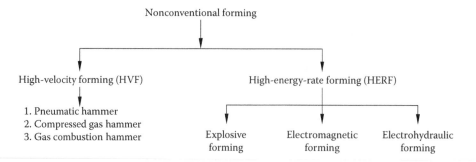

FIGURE 11.1   Classification of nonconventional forming processes.

critical velocity, ductility is not strongly affected by the strain rate. The second critical velocity (known as von Karman velocity) is related to the plastic wave speed in the material. This second critical velocity is between 30 to 150 m/s for most important engineering metals and alloys. Between these two critical velocities, enhanced formability is noticeable. Figure 11.2 demonstrates the increase in ductility for aluminum and copper rings expanded radially in an electromagnetic forming experiment.

b. More uniform strain distribution in one operation: In conventional forming processes, friction between the die parts and the work material causes the strain distribution in the material to become nonuniform. In most HERF operations, no contact with a die is required. This leads to elimination of frictional forces, and thus produces more uniform strain on the surface, as can be shown in Figure 11.3. In type 2 strain distribution, which is related to shock wave loading, the flat blank tends to deform vertically, producing a conical shape with little work being done in the circumferential direction. As a result, the maximum strain occurs at the apex of the cone. Type 3 strain distribution, producing more uniform pressure, can be achieved when using low explosive, explosive gas, or explosive cord in the form of a ring, as will be shown later.

c. Reduced wrinkling: The increase of inertia forces at high forming velocities suppresses the formation of wrinkling in sheet forming. Consider a simple example of a ring being compressed radially inward. If this is done at low velocity, the ring will collapse like a drinking straw under suction. At high velocities, inertia forces will restrict each material point on the ring to move radially, keeping a circular section during deformation. This is shown in

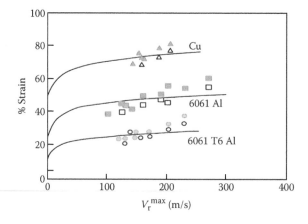

FIGURE 11.2   Improved ductility with the forming velocity in a ring expansion electromagnetic forming experiment. (From *ASM Handbook, vol. 14.B, Metalworking, Sheet Forming*, Appendix D, ASM International, Materials Park, OH, 2006. With permission.)

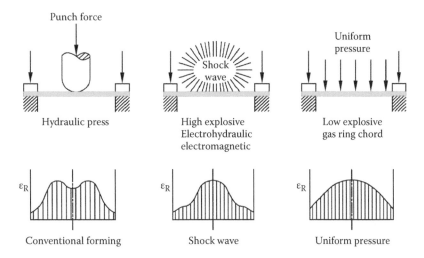

FIGURE 11.3 Three basic types of radial strain distribution for deep pressing of spherical shapes (dome ends) from flat blanks. Type 1: Deep recessing with a punch (conventional forming); type 2: shock wave (HERF); type 3: uniform pressure (low explosive or explosive gas).

Figure 11.4, representing ring compression under electromagnetic force at different energy levels (velocities). The figure indicates that at higher energies, wrinkling is eliminated.

d. Embossing and coining: The high pressure generated in HVF and HERF is large enough to produce significant plastic deformation on the workpiece surface, which enables embossing and coining-like operations on large areas which cannot be produced by similar conventional forming operations.

e. Light weight tooling and equipment: In conventional forming, huge press or hammer capacities are required to form large areas. Most of the HERF processes are based on shock wave propagation to apply the pressure required for forming in a very short time,

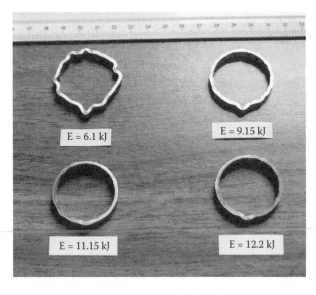

FIGURE 11.4 Electromagnetic high-velocity crimping of a thin aluminum ring onto a mandrel. (From *ASM Handbook, vol. 14.B, Metalworking, Sheet Forming*, Appendix D, ASM International, Materials Park, OH, 2006. With permission.)

without the need for hard tools to withstand these pressures. Very light framework can easily contain and dissipate developed kinetic energy.

f. Reduced springback: The large impact pressure generated by HERF causes the sheet metal to conform closely to the die shape. The plastic wave front propagating through the thickness causes the residual elastic strains to be minimum, and therefore reduces springback even if the contour is too shallow.

11.3 HIGH-VELOCITY FORMING MACHINES

In HVF processes, the energy required to deform the workpiece is delivered at high speed, using a high-speed forming hammer. This high-speed deformation is increasingly vital for industrial production due to several reasons

- The development of new high-strength alloys that are difficult to deform conventionally
- The requirement of large size parts for aerospace, marine, and storage facilities industries that require economically unfeasible huge press or hammer capacities
- The need to produce parts of more complicated forms that cannot be formed through conventional forming processes
- The appeal of manufacturers to eliminate the wasteful cutting processes or reduce the number of forming steps, for economical considerations

The use of high-speed hammers began in the mid 1950s. They are based on the process of accelerating the ram of the hammer to high speeds by the sudden release of compressed gas, or by the instantaneous conversion of the energy stored in a hydrocarbon into kinetic energy through ignition (analogous to a combustion chamber). These hammers are mainly applied for hot or cold bulk forming processes (forging or impact extrusion), for shearing processes, as well as for powder compaction. Available high-speed hammers include pneumatic, compressed gas, and gas combustion hammers.

11.3.1 PNEUMATIC (COMPRESSED AIR) HAMMER

The simplest high-speed hammer is based on a differential piston mechanism that releases the energy stored in the form of compressed air at a high rate. The principal cycle of the hammer is shown schematically in Figure 11.5. The system consists of a cylinder divided into upper and lower

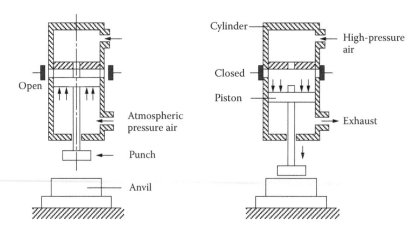

FIGURE 11.5 Pneumatic high-velocity hammer. (From Nagbal, G. R., *Metal Forming*, Khanna Publishers, New Delhi, India, 1998. With permission.)

compartments, separated by a disc with a small orifice at the center. A piston with an upper plunger of small cross-sectional area reciprocates inside the lower compartment so that the plunger fits into the orifice at the uppermost position. Atmospheric pressure is forced at the lower compartment of the cylinder to raise the piston upward to the highest position that closes the upper compartment, as shown in part a of the figure. Forcing compressed air through the upper compartment generates a very large force on the small area of the upper plunger of the piston, producing a high-speed stroke downward as soon as the side valve is closed. During this stroke, the air at atmospheric pressure in the lower compartment is exhausted out, as shown in part b of the figure. Such hammers can deliver up to 55 kJ of energy at speeds of 60 m/s.

11.3.2 COMPRESSED GAS FORMING HAMMER

A schematic presentation of the compressed gas hammer is shown in Figure 11.6. In the starting position, the ram of the hammer is held in the upper position by compressed gas (nitrogen or air) at pressure of about 20 MPa acting on the ram shoulders, and fitting an upper plunger of the ram in a sealed orifice. Pushing gas from the upper valve to act on the small plunger area breaks the equilibrium of the system and pushes the ram downward, where the pressure in the chamber acts on the whole area of the ram, causing quick acceleration to strike the workpiece (position c). When the forming operation is complete, two auxiliary hydraulic pistons return the ram back to the starting position. An air cushion absorbs the frame reaction against the ram acceleration. The first hammer built on this principle was in the USA in the mid 1950s, and it was called the DynaPak. The work capacities for these hammers are up to 400 kN.m, at impact speeds of about 20 m/s, and a stroke index of 8 strokes/min.

11.3.3 GAS COMBUSTION HIGH SPEED HAMMERS

In these hammers, volatile hydrocarbons such as gasoline or kerosene are mixed with air and burned in a closed chamber (analogous to a combustion engine). Air fills a combustion chamber through the inlet valve at a pressure of 3–15 MPa, as shown in Figure 11.7a. Fuel is injected through a nozzle, and a spark plug ignites the mixture, as in Part b of the figure, raising the pressure in the combustion chamber by 5 to 7 times. The resulting expansion accelerates the ram of the hammer in the downward stroke to impact the workpiece as indicated in (c). Pressurizing air at 2–5 MPa in the pressure chamber returns the ram in the reverse stroke to the initial upper position (d). The exhaust valve opens during the reverse stroke. The first machine built on this principle was in the UK, and

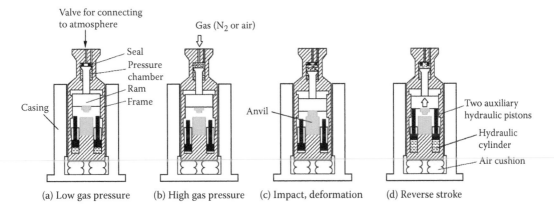

| (a) Low gas pressure | (b) High gas pressure | (c) Impact, deformation | (d) Reverse stroke |

FIGURE 11.6 Principle of operation of a high-speed hammer based on compressed gas: (a) starting position, (b) beginning of working stroke, (c) starting deformation, and (d) reverse stroke (From Davies, R., and Austin, E. R., *Developments in High-Speed Metal Forming*, Machinery Publishing Corporation, 1985. With permission.)

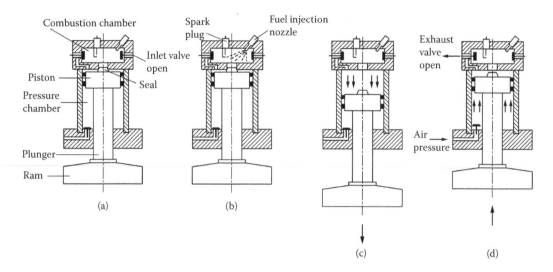

FIGURE 11.7 Principle of operation of gas-combustion hammer: (a) charging position, (b) ignition of fuel, (c) working stroke, and (d) reverse stroke.

was called the Petro-Forge, and was applied in cold forging and bar cropping operations. The work capacity for this hammer is 20 kN.m, at impact speeds of about 18 m/s, and a stroke index of 60 strokes/min.

An alternative of a high-speed forming hammer is a *pneumatic accelerator* (energy converter), that can be adapted easily onto a conventional mechanical press to convert the low-speed operation (0.06 to 1.5 m/s) into the high-speed operation of a hammer (up to 10 m/s). This system is presently applied in high-speed blanking of thick sheet metals for the purpose of reducing distortion, and improving the surface quality and dimensional precision of obtained blanks. A typical design of the accelerator is shown in Figure 11.8. The pneumatic accelerator is fitted on a conventional mechanical press having 400 kN punching capacity, to cut mild steel blanks of 30 mm diameter and 10 mm thickness. The accelerator contains two chambers; the piston chamber V_p, and the hammer chamber V_h, as shown in part a of the figure. Compressed air is charged at a preset pressure into the piston chamber through a check valve, leading to upward movement of the piston, and at the same time, lifting the hammer through the lifting rod. When the mechanical press is triggered on, the press ram pushes the piston downward, and this will compress the air in the piston chamber, as shown in part b of the figure. At the same time, the lifting rod is dislodged from the hammer, but the hammer is held in position due to the pressure action on the under side of the flange of the hammer (Figure 11.8c). When the piston comes close to the end of the stroke, as shown in part d of the figure, the highly compressed air in the piston chamber rushes into the hammer chamber through the neck of the lifting rod, which acts as a nozzle. This sudden exposure of the large area of the hammer to the high-pressure air causes the hammer to move downward. Once the top of the hammer clears the port holes of the hammer cover, the high pressure air rushes into the hammer chamber through the port holes, thereby forcing the hammer downward at very high speed (in the range of 10 m/s), which is more than enough for high-speed forming. On the return stroke of the press ram, the piston forces the press hammer upward, since the pressure in the piston chamber is still very high. At the same time, the lifting rod lifts the hammer upward, causing the air in the hammer chamber to flow into the piston chamber through the check valve. The cycle repeats itself and the number of strokes per minute of the energy converter will be the same as the number of strokes per minute of the press. The products manufactured using this pneumatic accelerator have higher surface quality than the surface of the products of the conventional mechanical presses, such that they do not need secondary finishing.

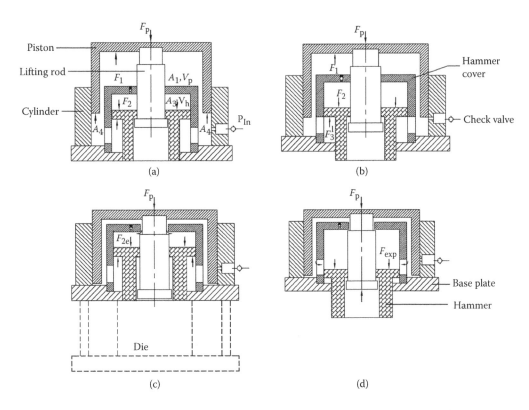

FIGURE 11.8 Pneumatic accelerator operated by a press to apply high-impact velocity on the workpiece to be formed. (From Yaldiz, S. et al., *Materials and Design*, 28: 889–896, 2007. With permission.)

11.4 HIGH-ENERGY-RATE FORMING PROCESSES

As described earlier, HERF processes are based on systems that generate high energy at very short times (high energy rate), such as explosive forming, electromagnetic forming, and electrohydraulic forming (EHF). A detailed study of these processes will be presented in this section.

11.4.1 EXPLOSIVE FORMING

Explosive forming uses the instantaneous high pressure resulting from the detonation of an explosive charge to deform a metal blank or tube plastically. This process was discovered in the late 1800s, but practical applications on thick plate forming began in the 1930s. Between 1950 and 1970, the U.S. Government invested huge funds on research in development and application of HVF and HERF processes, especially on explosive forming technology for complex and large structural parts for the aerospace industry.

11.4.1.1 Principles and Types of Explosives

Explosives can be generally categorized as high and low explosives, depending on the speed of reaction. Low explosives like gunpowder burn by generating heat and hot gas. The speed of reaction front is less than the sound speed. They are used for propelling bullets. High explosives have detonation velocities in the order of 3500 to 8000 m/s, and energies in the order of 1 MJ/kg. They are available in the form of granules, powder, castable compounds, and detonation cords (the cord consists of a filament of explosive material covered by a protective water-repellent coating), which are most applied in high-speed forming, as they allow fairly precise handling of the required explosive quantities. Detonation cords usually range from 2 to 10 grams/meter. Table 11.1 shows some of the high explosives used in explosive forming.

Explosive forming is inexpensive, as it needs only a few cents worth of explosives to generate the energy required to form very large products (tubes up to 1.4 m diameter, and 9 m length, or dome diameters up to 12 m). However, restricted safety regulations regarding the transportation, storing, and use of explosives put severe constraints on their application.

11.4.1.2 Classification of Explosive Forming Methods

Explosive forming methods may be classified from two different points of view, namely

a. According to the relative position between the explosive and work material: These methods may be classified into

Contact operation: Where the explosive is placed directly (or with a thin protective rubber layer) onto the work surface. This operation is applied when high speed or high pressure are required, such as in cladding.

Standoff operations: These are more common in sheet forming, where the explosive charge is separated from the work by a calculated standoff distance, and all are contained in a suitable transfer medium such as water (other transfer systems such as sand may be used in some applications).

b. According to the way the explosion is contained: These operations are classified to

Open (unconfined) operations, where the work, a single element die with a blank holder, and the explosive charge are immersed in a water tank, as shown in Figure 11.9a. This is the most commonly used system. The standoff distance should be precisely controlled according to the span of the blank over the die cavity. This system is inherently inefficient, where only part of the energy developed from explosion is used to deform the blank, and significant energy is lost in kinetic agitation of the water bath. Larger tank diameters are preferred to reduce the shock waves reaching the walls from centrally located charges. To reduce the stresses on the walls of the tank, one of the following approaches is applied:

TABLE 11.1
Properties of Some of the High Explosives Used in Explosive Forming

Explosive	Relative Power (% TNT)	Form of Charge	Detonation Velocity (m/s)	Energy (kJ/kg)	Storage Life	Maximum Pressure (GPa)
Trinitrotoluene (TNT)	100	Cast	7010	780	Moderate	16.5
Cyclonite (RDX)	170	Pressed granules	8380	1270	Very good	23.4
Pentrite (PETN)	170	Pressed granules	8290	1300	Excellent	22.1
Pentolite (50/50)	140	Cast	7620	950	Good	111.3
Tetryl	129	Pressed granules	7835	—	Excellent	—
60% straight dynamite	109	Cartridge granules	3810	715	Fair	4.3
Primacord, 8.5 g/m	—	Plastic or cotton cord	6340	—	Excellent	—

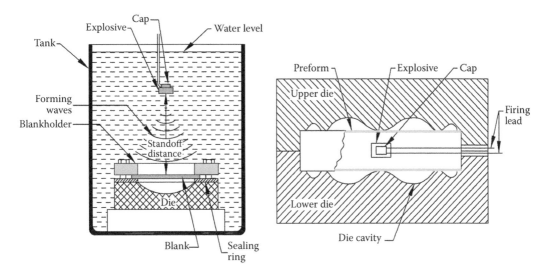

FIGURE 11.9 (a) Open (unconfined) and (b) confined systems for explosive forming. (From *ASM Handbook, vol. 14, Forming and Forging*, 9th ed., ASM International, Materials Park, OH, 1988. With permission.)

- An inflated rubber tube is used on the inside wall of the tank to act as a cushion. It can reduce up to 80% of the stresses on the wall. However, it is difficult to maintain the tube position, and to prevent its damage during explosion.
- A curtain of air bubbles is generated on the inside wall of the tank through an air line with holes at the tank base. The bubble curtain is controlled by the size of holes in the line and the air flow and pressure in the line. This is the best approach for economic water tank construction.

Confined system operations: These operations use a two-piece die that completely confines the workpiece in a strong and rigid enclosure, as shown in Figure 11.9b, for tubular shape forming. Higher efficiency and dimensional tolerances as well as less noise are attained. However, the system is limited to relatively small-sized products, for economic feasibility and ample die life.

Sheet and tube forming: Explosive forming is most commonly applied to form thin sheets and tubular forms using standoff operations. Typical applications are shown in Figure 11.10. Part a of the figure shows a typical die-filling operation, where a vacuum is necessary to prevent entrapped air between the die and the workpiece, which could affect the product accuracy. Figure 11.10b represents the formation of a corrugated sheet. Part c shows a detonating cord shaped in a loop form, located close to the outer periphery of large blanks for better distribution of pressure, to force the sheet to take the form of the die cavity without the need for vacuum. In part d of the figure, a tube is formed in a closed die using a detonation cord to form a tubular product.

It should be noticed that for all produced shapes, a one-sided die is used, which reveals the advantage of cheaper tooling in explosive forming compared to other conventional forming techniques, especially for large products.

Another attractive feature is the versatility in die making procedures using castable die materials, such as glass reinforced plastic (GRP), urethane, epoxy resin, ductile cast iron, kirksite (a zinc-based casting alloy), or concrete. A simple wooden or styrofoam pattern can be used to manufacture these dies. For higher pressures and longer tool life, hardened steel dies are used. The tanks and dies should have circular shapes, as sharp corners are susceptible to failure due to the shock waves.

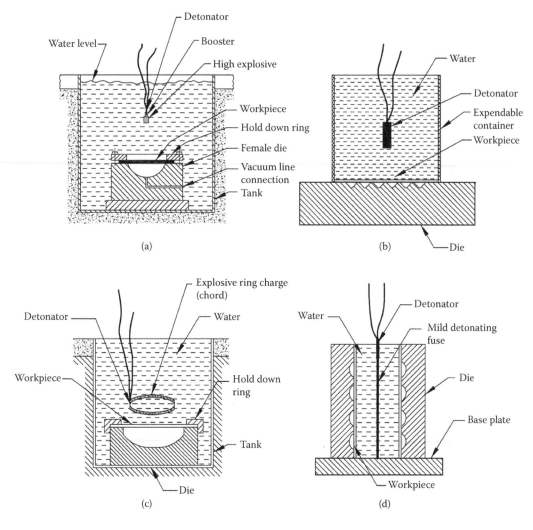

FIGURE 11.10 Explosive sheet forming operations: (a) vacuum die forming, (b) forming a flat panel, (c) using a detonation cord to prescribe the pressure distribution in an open forming system, (d) using a detonation cord to form a cylinder (open- or closed-die system). (From *ASM Handbook, vol. 14.B, Metalworking, Sheet Forming*, Appendix D, ASM International, Materials Park, OH, 2006. With permission.)

For safe detonation in explosive forming, a firing box for the electric blasting caps is used. It is operated by a 6-V battery or directly from a single-phase current line.

Plate forming: Explosive forming represents a unique economic method for manufacturing large sections from thick plates. Thus, it has often been used to make bulkheads for large pressure vessels, sections of ships, and large vehicles, and several large nuclear reactor components. Contact forming is often used in these operations. Concrete dies are often used for such large products. As stated earlier, springback is eliminated and accordingly, high dimensional tolerances are achieved. Commonly, tolerances in the order of 0.025 mm are obtained in sheet and plate forming of fairly large structures.

11.4.1.3 Process Variables

The main variables affecting the explosive forming process are the type and amount of explosive, the standoff distance, the medium through which the shock wave is transmitted, and the work material and geometry.

1. Type and amount of the charge: The type of charge has been discussed earlier. The amount of the charge required for developing the pressure that deforms the workpiece is determined by equating the energy required for deformation and the fraction of the charge energy which falls onto the workpiece. For a circular blank formed to a hemispherical dome, this equation takes the form

$$\sigma_y t\pi h^2 = \frac{Ec\eta\theta^2}{4} \tag{11.1}$$

where

σ_y = yield strength of the blank material
t = blank thickness
h = central height of the formed part
Ec = chemical energy released in the detonation = me_s
m = weight of the explosive charge
e_s = specific energy of the explosive charge
η = efficiency of energy transfer = 0.35 to 0.4 (typical value)
2θ = solid angle subtended by the blank at the charge

In general, a small amount of explosive forming charge is required to produce large energy. Typically, one gram of explosive can produce 4 to 5 kJ of energy.

2. Standoff distance: This is the distance between the workpiece and the explosive. It should be as minimum as possible. However, too small standoff results in excessive local deformation and thinning of the workpiece. Therefore, smaller standoff distances are used for deep drawing, and larger ones are used for shallow drawing. Generally, the standoff distance S is taken as a function of the blank diameter D as follows:

$$S = D \text{ when } D \text{ is up to 0.7 m}$$

$$S = 0.5D \text{ when } D \text{ is more than 0.7 m.}$$

3. Hydrostatic head: The hydrostatic head in explosive forming is defined as the vertical distance that the charge is immersed under the water surface. This head should be large enough to reduce the lost energy of the shock wave. Common practice in explosive forming is to use a hydrostatic head $H = 2S$ (where S is the standoff distance).

4. Transmission media: As stated earlier, water is mostly used as a transmission medium to confine and increase the efficiency of the shock wave resulting when detonating the explosive charge. The size of the charge required for forming a specific product in water is approximately 80% smaller than that needed when forming it in air. Figure 11.11 shows the pressure difference between water and air media at variant standoff distances. Other liquids such as oil are used successfully as transmission media, but water is cheaper, more available, and easier to handle. Solid media such as rubber sheets, cast plastics, and metals may be used in explosive forming. They generally offer better pressure distribution, which eliminates the possibility of damage of the product surface by solid particles from the blasting cap striking it when a liquid medium is used. Rubber and plastics improve the formability of some metals. Plastics media are successful for close dimensional tolerances of tubular shapes. Sand or small glass beads are successfully used as transmission media for explosive forming of hot metals (such as tungsten), to reduce the transfer of heat to the explosive charge.

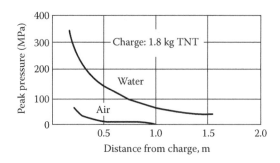

FIGURE 11.11 Peak pressure versus standoff distance for explosive charges fired in air and water. (From *ASM Handbook, vol. 14, Forming and Forging,* 9th ed., ASM International, Materials Park, OH, 1988. With permission.)

5. Work material: In general, all types of materials that can be conventionally formed can also be explosively formed. Some materials which are difficult to form on presses, such as Nimonics, titanium alloys, and high-strength steels, can be successfully formed explosively. Further, materials which are too brittle, such as tungsten domes, have been explosively formed from a flat, hot blank. Heating of the material can be carried out by direct resistance heating, radiation heating, induction heating, contact with a heated solid or liquid medium, or by means of the chemical reaction in the medium contacting the work surface (contact operation).

11.4.1.4 Failure of Explosive Formed Products

In standoff explosive forming, distributed pressure is applied over the workpiece surface so that different elements of the material tend to move in different directions and with velocity gradient. This results in metal thinning. If the pressure is too high, the product may fail due to one of the following reasons:

- Excessive thinning of the central part under maximum pressure in the centerline direction
- Excessive plastic deformation
- Work hardening of the product material

11.4.1.5 Advantages and Limitations

The main advantages of explosive forming can be summarized as follows.

- The process can form sheets and tubes in large sizes that cannot be formed by any other process.
- Metallic materials can be formed into shapes which cannot be produced economically by other conventional processes.
- Only half dies are required.
- Capital investment in tanks and dies is cheaper.
- Products can be formed to close dimensional tolerances.
- Tough materials as well as ductile materials can be explosively formed.
- One die can be used to form products in different thicknesses.

The only limitations of the process are

- The process should be carried out at safe and remote places, and standard precautions should be taken to deal with explosive materials and to handle explosion hazards.
- The process is recommended only for limited production rate.

- The process does not ensure high impact strength or high stress corrosion resistance of products.

11.4.2 Electromagnetic Forming

Electromagnetic forming (EMF) is a noncontact technique where large forces can be conveyed to any electrically conductive material (such as copper, aluminum, low-carbon steel, and molybdenum) by a pure electromagnetic interaction.

11.4.2.1 Principles of the Process

This system is based on storing large energy (in the range of 5 to 200 kJ) in a large capacitor or a bank of capacitors by charging to a high voltage (usually between 3 and 20 kV). The charge is switched over low inductance through a coil or actuator. Large currents run through the coil with peak values typically between 10 to 1000 kA, and periodic time in the order of tens of microseconds. This develops an extremely strong transient magnetic field in the vicinity of the coil, which induces eddy current in the nearby conductive workpiece material in opposite direction to the current in the primary coil. The opposed fields in the coil and the workpiece set up a repulsive electromagnetic force that produces high stresses in the workpiece, causing it to deform plastically at velocities exceeding 100 m/s. Being purely electromagnetic, the process is not limited to repetition rate by the mechanical inertia of moving parts. Since the timing of magnetic impulse can be synchronized precisely to microseconds, the EMF machine can be controlled to function at repetition rates of hundreds of operations per minute. With manual feed of work material, the production rate is limited only by the rate at which the operator can load and unload parts. When using suitably designed holding and clamping devices that secure the part during forming, production rates of 600 to 1200 per hour can be achieved. When fully automated equipment is incorporated, much higher production rates of up to 12,000 products per hour can be achieved.

The basic circuit used for electromagnetic compression of a metallic tube consists of a power supply, an energy storage bank of capacitors, switches, a forming coil, and a field shaper (if required), as shown in Figure 11.12. Figure 11.12a indicates that the magnetic field surrounding the forming coil does not penetrate the conductive metallic tubular workpiece. Instead, it is intensified by confinement in the limited annular space between them. The field shaper, shown in part b of the figure, is a massive current-carrying conductor, indirectly coupled to the forming coil. It is used to concentrate the magnetic field at the points at which higher forming pressure is desired. Accordingly, the standard forming coil can be used to produce a variety of shapes by changing the geometric configurations of the field shaper.

11.4.2.2 Basic Methods of Electromagnetic Forming

Depending on the geometrical configuration of the forming coil and field shaper and their relative position to the work material, EMF can be carried out by three different methods, as shown in Figure 11.13.

a. Compression forming, where the forming coil surrounds a tubular work metal, which develops compressive forces on the tube walls. Usually a tube is compressed against a grooved or suitably contoured insert, plug, or fitting inside the workpiece, as shown in Figure 11.13a.
b. Expansion forming, where the coil is inside the tube, as shown in Figure 11.13b. The die (or field shaper) surrounding the work tube depicts the final form of the product.
c. Contour forming, where the forming coil takes a pancake form, and the die shape determines the final shape of the product, as shown in Figure 11.13c.

Any deep cuts or slots on the workpiece tube deters the continuous electrical path and restricts the magnetic field, and thus eliminates the forming force. However, minor irregularities such as small perforations do not seriously interfere with the field flow, and thus do not affect formability.

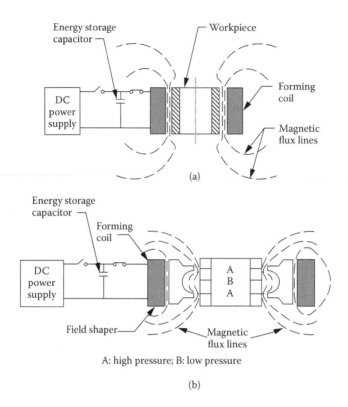

FIGURE 11.12 Basic circuit and magnetic field pattern for electromagnetic compression forming of a tubu-lar workpiece (a) without field shaper and (b) using a field shaper. (From *ASM Handbook, vol. 14, Forming and Forging*, 9th ed., ASM International, Materials Park, OH, 1988. With permission.)

This process has wide applications in the automotive and aircraft industries, in the manufac-turing of electrical equipment and nuclear reactors, and in assembly work. Typical applications include: Assembly of air conditioner accumulators, high-pressure hoses, steering wheels, gasoline fill tubes, universal joint yokes, drive linkages, cooling system ducts, and the sizing of tubing. Typical products are shown in Figure 11.14. This process is also applicable in the manufacturing of appliances, consumer goods, and computers.

Electromagnetic hammers and riveters: In addition to contour forming of tubular parts, flat coils are used in two unique aerospace applications: as an electromagnetic hammer and an electromag-netic riveter. The hammer is used to flatten surfaces on the aluminum skin of the central fuel tank of a space shuttle, while the riveter is used in the assembly of aircraft wings. The hammer can gen-erate a pressure up to 35 MPa to deform the aluminum surface in a very short time of 100 μs. The repulsive force on the coil causes it to jerk back to position. In both applications, high pressure can be exerted on a limited area with no contact, which eliminates the effect of mechanical strokes on the change of the material structure. Electromagnetic riveting is recommended for installing rivets to precision interference profiles for fluid tight and fatigue sensitive applications without massive stationary equipment. The riveter consists of a power supply and two opposed guns with special power transmission cables. The electromagnetic energy is converted into mechanical kinetic energy to drive the rivets. The process produces higher quality rivets at lower cost, in addition to being less noisy, more flexible, and smaller in size compared to conventional large automatic hydraulic rivet-ing machines. The electromagnetic riveter is usually mounted on a robot for automated assembly of components.

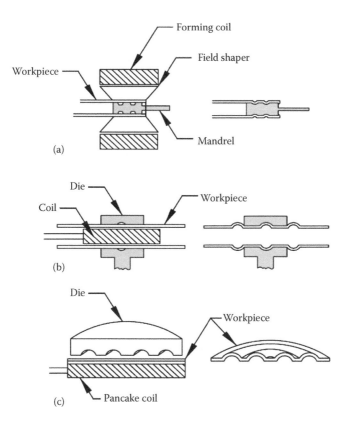

FIGURE 11.13 Three basic methods of EMF. (a) Compression, (b) expansion, and (c) contour forming. (From *ASM Handbook, vol. 14, Forming and Forging*, 9th ed., ASM International, Materials Park, OH, 1988. With permission.)

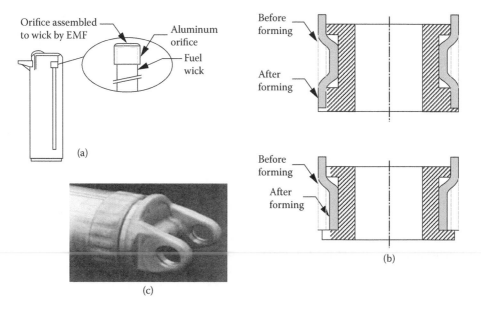

FIGURE 11.14 Typical EMF products: (a) disposable cigarette lighter component assembled by EMF, (b) axially loaded joints, (c) aluminum drive shafts (torque tubes) used in vans and trucks. (From *ASM Handbook, vol. 14, Forming and Forging*, 9th ed., ASM International, Materials Park, OH, 1988. With permission.)

11.4.2.3 Pressure Required for EMF

The static pressure (p) that acts on a tube with an outside diameter D_o, and a thickness of t, to reach a hoop stress equal to the yield strength σ_y; i.e., to reach plastic deformation, can be obtained from the relation

$$\frac{pD_o}{2t} = \sigma_y \text{ or } p = \frac{2t\sigma_y}{D_o} \tag{11.2}$$

Due to the inertia effect when deformation takes place at high strain rates, a multiplying factor N in the range of 2 to 10 is used to calculate the magnetic pressure p_m required to deform the tube. Therefore,

$$p_m = \frac{2tN\sigma_y}{D_o} \tag{11.3}$$

Example 1

What is the magnetic pressure required to form a tube with an external diameter of 100 mm, and a thickness of 2 mm, made of mild steel AISI 1015 (yield strength σ_y = 284.4 MPa)?
Solution
Assuming a strain rate factor N = 10, the magnetic pressure can be calculated from Equation 11.3, as:

$$p_m = \frac{2 \times 2 \times 10 \times 284.4}{100} = 113.76 \text{ MPa}$$

Practically, the EMF process is capable of applying up to 340 MPa in compression when using standard forming coils.

11.4.2.4 Advantages and Limitations of EMF

Electromagnetic forming has many advantages over other HVF and HERF processes.

- The forming time is in the range of a few microseconds, and the production rate is limitlessly high.
- No contact with a tool or the form coil. Therefore, no lubrication is required, and no tool marks affect the work surface. Previous anodizing or plating of the work metal is not affected by EMF.
- Precision: Due to the high control of the electromagnetic pulse, the forming pressure is highly precise, and exceptional repeatability of products can be achieved.
- Higher formability of the material due to the high strain rates and absence of friction with tooling. No intermediate annealing is required.
- Similar to other HERF processes, springback is eliminated.
- Tooling is relatively inexpensive. When dies are required, they are made of nonconductive materials to avoid any potential of inducing current.
- When using EMF to assemble two parts, the conductive material can be formed on any nonmetallic part, such as plastics, ceramics, composites, or rubber.
- Joints made by EMF are typically stronger than the parent metal.

The main limitations of the process include

- Only electrically conductive materials can be formed (materials with an electrical resistivity of 0.15 μΩ m or less).

- The speed of forming limits the amount of deformation. The process is not applicable for deep drawing.
- The maximum pressure applied in compression forming is limited to 340 MPa. The process is therefore limited to thin wall tubes or thin sheets.

11.4.2.5 Safety Considerations

The potential hazards of EMF operations can be divided into three categories, namely

a. Product forced ejection hazards: In compression forming of tubes, the forming coil should completely surround the part of the tube to be deformed, where the electromagnetic pulse develops radial forces on the tube. If the tube is accidentally only partially surrounded by the coil, the force acting on the end should be sufficient to throw the tube out (eject it) at high speed. In the same way, improperly positioned parts could be ejected from an expansion coil. Therefore, an appropriate coil shield should be provided to protect personnel, and carefully designed work positioning and holding locators must be used to ensure safe handling.

b. Noise hazards: The sound produced during the sudden flow of the deformed part has a moderate level. However, if the area to be deformed is large, or the distance moved by the tube is large, higher sound levels are produced, and additional sound conditioning is required.

c. Electrical hazards: The voltages used and the current developed in EMF are extremely high. They should be completely contained in a well-grounded heavy-gauge metal cabinet. Doors and panels are electrically interlocked to avoid unauthorized access to high-voltage components. The workpiece should never be held by hand, where there is a possibility of arc between it and the coil. Coils should be encased in metal shells that act as eddy current shields.

11.4.3 ELECTROHYDRAULIC FORMING

EHF can be considered as a hybrid between explosive forming and electromagnetic forming. Here, the explosive shock wave is replaced by an intense shock wave produced in the form of a spark in the gap between two electrodes immersed in liquid. The resulting liquid-based shock wave is very similar to what would be produced by a point or line explosive charge. The current pulse is produced using the same capacitance circuit used in EMF equipment. Therefore, the energy efficiency in EHF is typically similar to that of EMF. The process is better controlled than explosive forming, and accordingly, it is preferable in regular factory conditions. However, it is difficult to reach the same quantities of forming energy that explosive forming can achieve. Therefore, EHF has not been widely adopted in many industries.

11.4.3.1 Principles of the Process

The principle of the EHF circuit for forming a tubular part is shown in Figure 11.15. Two electrodes are immersed in water in the centerline of the tube, and connected via a switch to a bank of capacitors. The capacitors are charged from a power unit through a charging resistor to a predetermined voltage and energy level. When the switch is closed, the stored energy is discharged across the gap between the electrodes, resulting in a shock wave in the water, which forces the tube to be deformed according to the die configuration. The die is made of two halves to allow for product ejection after forming. Vacuum lines should be connected to the die to eliminate air entrapment, which may distort the formed shape.

Other possible arrangements of electrodes are shown in Figure 11.16. Arrangements a and b are used for cupping or deep drawing of flat blanks. Figure 11.16b shows an arrangement of two concentric electrodes with insulation in between. Parts c, d, and e demonstrate potential arrangements for forming tubular products.

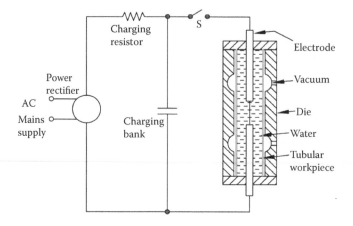

FIGURE 11.15   Basic elements of an EHF circuit.

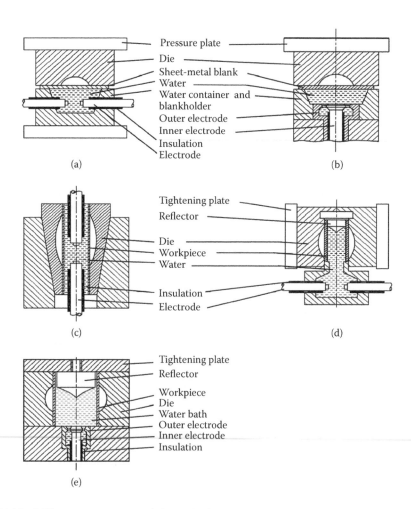

FIGURE 11.16   Different arrangements of electrodes in EHF.

11.4.3.2 Energy Requirements

In EHF, the energy stored in a charged capacitor is given by

$$E = 0.5 \; Cap. \; (Vc)^2 \tag{11.4}$$

where E is energy (J), *Cap.* is capacitance (farads), and *Vc* is charge voltage (V).

The efficiency of converting the stored electrical energy in the capacitors to plastic deformation of the workpiece varies considerably. Due to the high energy losses of the resulting shock wave in water, the efficiency is typically between 10% and 20%. The storage capacity in EHF machines is limited to 100 kJ for practical physical size and cost of the capacitor. Capacitors are normally charged from a standard single-phase main supply line (220 V) through a rectifier and a step-up transformer to get a charging voltage of up to 40 kV.

When the stored energy in the capacitors is discharged across the electrodes, very high current passes in the gap or in the initiating wire extending between them. The water in the gap is ionized very rapidly and heated to vaporization. The water surrounding the ionized channel offers inertial resistance to the expanding gas. The pressure in the gas pocket rises to a very high value in a very short time. Pressure release is controlled by the dynamic behavior of water and takes the form of a steep fronted pressure wave. The pressure of the shock wave on the workpiece accelerates it to deform plastically in the surrounding die cavity. The mode of deformation is similar to that of explosive forming.

11.4.3.3 Process Variables

The variables that affect the efficiency of the EHF process are

1. Electrode gap width: The gap between electrodes depends mainly on the charge voltage. The higher the charge voltage, the longer the gap. Typical gap lengths are 12 mm at 10 kV and 50 mm at 50 kV. Too small gap restricts discharge, while too large gaps will not vaporize the initiating wire completely, thus causing the efficiency to fall. The optimum gap when using distilled water is only about half that of when tap water is used.
2. Standoff distance: This is the distance from the discharge gap to the workpiece. Higher deformation can be achieved at lower standoff distance. Small standoff distances are used to produce conically shaped, free-formed components because of the concentration of energy at the center of the blank. The forming of large diameter tubes involves large standoff distances, and therefore peak pressures and the amount of deformation are proportionally reduced.
3. Charge energy: The workpiece central deformation (δ) increases with the discharge energy E, according to the power relation:

$$\delta = E^K \tag{11.5}$$

   where k is an empirical constant = 0.36 to 0.65. The value of this constant depends mainly on the time during which the discharge takes place. A low inductance circuit provides faster discharge time and is a far more efficient generator of shock waves.
4. Hydrostatic head: This is defined as the vertical head of water above the discharge gap. Since the majority of EHF operations take place in completely closed compartments, hydrostatic head is not significant. However, in arrangements where the hydrostatic head is a design factor, as in Figure 11.16a,b, the higher the head, the higher the amount of deformation that can be produced. The hydrostatic head should be at least equal to the standoff distance in order to obtain the maximum amount of deformation.
5. Initiating wire between electrodes: The initiating wire between electrodes is mainly used for spark ignition. It should be made of a good conducting material, such as copper or

aluminum. The wire diameter should be as small as possible so that it is vaporized due to the extremely high current passing through, and thus developing the discharge. Common diameters for the wire vary between 0.125 and 1.25 mm.

6. Electrodes: In EHF, the electrodes are usually made of mild steel, stainless steel, copper, or brass. The electrode diameter should be selected to withstand the extremely high current developed when the capacitance is discharged.
7. Speed of deformation: The range of forming speeds is similar to those obtained during explosive forming. Typical speed values for free forming are 30 m/s to 60 m/s for copper and aluminum, and about 100 m/s for stainless steel.
8. Die material: Die materials are similar to those used in explosive forming. Commonly used die materials are kirksite (zinc–aluminum alloy), epoxy resin, or steel.

11.4.3.4 Advantages and Limitations

The main advantages of EHF are

- The process can be used for both tubes and flat blanks to form tubular or dished shapes, respectively.
- Conductive or nonconductive materials can be formed.
- Higher production rates than explosive forming can be achieved.
- The requirement of only a half die made of a relatively cheaper material is highly economical.
- The process is inherently suitable for higher automation compared to explosive forming.

The main limitations of the process are

- The process is applicable only to form relatively smaller components, compared to explosive forming.
- Materials with critical impact velocities of less than 30 m/s are not appropriate for this process.
- Capital cost is higher than that of an equivalent explosive forming process.

11.5 FUTURE OF HVF AND HERF

It has been shown through this chapter that materials can develop much higher plastic deformation at higher velocities compared to what can be achieved in conventional quasistatic forming processes. This extended formability under high velocity is recently defined as hyperplasticity, and this topic is presently under extensive investigation to fully understand the mechanisms of plastic deformation under dynamic loading conditions.

HERF processes have a long and significant history, and have been successfully applied for the production of parts which are impossible or difficult to produce using conventional forming techniques. However, these processes have never achieved the place that they ought to have among mainstream manufacturing methods. This might be attributed to the fact that there is little connection between the communities that understand capacitor bank discharge or explosive engineering and the metal forming engineers. As a result, neither toolmakers nor manufacturing engineers have become well acquainted with or trained in these methods, and hence they are not fully utilized. There remain significant opportunities in their adoption, development, and application in industry. The future trends for such development can have effect in the following directions.

1. Continuous increase of the velocity ranges and capacity limits of HVF hammers, as well as further improvement of their efficiency for handling larger amount of deformation or more operations.
2. Further development of the explosive forming process through better control of the detonation systems, and better isolation of the shock waves for safer operation. The development

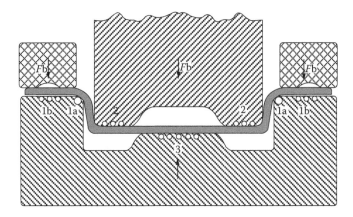

FIGURE 11.17 Hybrid process of press tooling with a large number of embedded electromagnetic forming coils. (From *ASM Handbook, vol. 14.B, Metalworking, Sheet Forming*, Appendix D, ASM International, Materials Park, OH, 2006. With permission.)

concentrates on the confined system operations for higher efficiency, and lower noise levels.

3. Higher emphasis will be given for the development of electromagnetic coils configuration and the field shapers for the purpose of handling complicated product geometries. Recent research is also going toward development of hybrid processes that combine the hard tooling of conventional forming processes with electromagnetic techniques only in the regions where the special HERF advantages of better pressure distribution, improved formability, and reduced springback are required. This can be carried out by embedding electromagnetic forming coils into specific areas of traditional stamping tools, as shown in Figure 11.17. The main idea here is that the capacitor bank is optimized for quickly producing a large number of small electromagnetic pulses coupled to coils in regions of the tool where increased strains are required in order to make a product.

4. EHF is undergoing extensive research to control the pressure distribution for higher efficiency and additional product geometrical configurations.

11.6 REVIEW QUESTIONS

1. Differentiate between the principles HVF and HERF as nonconventional forming processes.
2. What are the main characteristics that distinguish HVF and HERF processes from conventional forming?
3. Why are HVF and HERF processes needed in industry?
4. Classify high-velocity hammers according to their drive systems.
5. Compare the hammers based on compressed gas with those based on gas combustion.
6. How can a pneumatic accelerator develop high velocity for improving the quality of formed parts?
7. Why are explosives used in forming processes?
8. Classify the types of explosives and recommend the types suitable for metal forming.
9. Describe the differences between the contact type and standoff distance of explosive forming methods.
10. What are the differences between confined and unconfined explosive forming systems?
11. What are the aspects considered in designing water tanks for explosive forming?

12. What are the different transmission media used in explosive forming? Give the advantages and limitations of each.
13. What are the reasons of failure of explosively formed parts?
14. What is the principle of EMF and what are the main parts of the circuit required to form the process?
15. What is the effect of the field shaper in EMF?
16. What are the methods of EMF, and how do they affect the product shape?
17. What is the effect of slots or undercuts in the tube to be formed on the EMF process?
18. What is the principle of electromagnetic hammers and riveters and what are their applications?
19. Discuss the hazards of EMF and what are the safety precautions to avoid these hazards?
20. What are the main advantages and limitations of EMF over other HERF processes?
21. Compare between EMF and EHF, considering the principle of operation, maximum capacity, and applications.
22. What are the variables affecting EHF processes?
23. Discuss the sources of losses that limit the efficiency of EHF machines.
24. What are the future trends for HERF processes?

11.7 PROBLEMS

1. A 300 gm RDX explosive charge (1270 kJ/ kg) is used to form a dished end for a storage tank from a mild steel plate 1.2 m diameter, and 10 mm thickness. You are asked to
   a. Suggest a suitable standoff distance.
   b. Calculate the hydrostatic pressure.
   c. Calculate the solid angle subtended by the blank at the charge.
   d. Calculate the central height of the formed part if the efficiency is 40% and the yield strength of mild steel is 350 MPa.
   e. Suggest an alternative explosive for this process.
2. The charging voltage of an EHF machine is 16 kV and the bank of capacitors is 120 μF.
   a. Calculate the machine capacity, given that the process efficiency is 20%.
   b. Suggest the electrode gap width for this machine.
3. A brass tube (σ_y, = 300 MPa) of 150 mm outer diameter, and 2.5 mm thickness is to be formed using EMF. Calculate the magnetic pressure required to deform it, using a multiplying factor of 8 for the high strain rates. What is the hydrostatic pressure inside the tube?

BIBLIOGRAPHY

ASM International. 1988. *ASM Handbook, vol. 14: Forming and Forging, Sheet Forming*, 9th ed., Appendix D. Materials Park, OH.

ASM International. 2006. *ASM Handbook, vol. 14B: Metalworking, Sheet Forming*, Appendix D: High-Velocity Metal Forming, pp. 405–418. Materials Park, OH.

Davies, R., and Austin, E. R. 1985. *Developments in High-Speed Metal Forming*, Machinery Publishing Corporation.

High Velocity Forming of Metals, 1968. ASTME.

Kalpakjian, S., and Schmid, S. R. 2003. *Manufacturing Processes for Engineering Materials*, 4th ed. Upper Saddle River, NJ: Prentice-Hall.

Nagbal, G. R. 1998. *Metal Forming*. New Delhi, India: Khanna Publishers.

Yaldız, S., Saglam, H., Unsacar, F., and Isik, H. 2007. Design and application of a pneumatic accelerator for high speed punching, *Materials and Design*, 28, Elsevier, pp. 889–896.

12 Powder Metallurgy and Processing of Ceramic Materials

12.1 INTRODUCTION

Powder metallurgy (PM) is a processing technology of metallic materials which involves producing them first in the form of powder, and then consolidating them into a solid form by the application of pressure (compaction), and heat (sintering) at a temperature below the melting point of the main constituent. For a growing number of manufacturers, PM is one of the preferred technologies in the manufacture of precision (net shape) metal components. The technique offers a cost-effective alternative to other metal forming processes. The sequence of operations for PM that will be presented in this chapter can be summarized as shown in Figure 12.1. Powder metallurgy is also adopted for manufacturing a wide variety of ceramic materials. Therefore, these ceramic materials as well as their processing techniques are introduced in this chapter.

Powder metallurgy offers numerous advantages, which include

- Diversity and flexibility of material properties to meet product performance requirements.
- PM is applied to form components from refractory metals such as molybdenum, tantalum, and tungsten with high melting temperatures that exclude conventional casting processes.
- Alloys with immiscible phases, which will not form solutions under normal melting conditions, can be made to form alloy systems by the intimate mixture of particles prior to compaction and sintering.
- Production of conventionally impossible material combinations such as graded structures or composite metals.
- The production of net shape or near net shape products, thus reducing the need for further machining or finishing processes.
- Higher utilization of raw material (up to 100%).
- Higher strength and corrosion resistance.
- Material saving (up to 40%) compared to casting or forming followed by machining processes.
- Higher productivity and fast mass production at lower cost.
- Less energy consumption compared to casting and forming processes.
- Unmatched design flexibility.
- High dimensional accuracy and excellent surface finish.
- Materials can be heat treated for higher strength or increased wear resistance.
- Controlled porosity and permeability for self-lubrication or filtration.
- Environmentally friendly technology. It allows recycling, conserving raw materials and manufacturing processes, and yielding low emissions.

The main limitations of PM are

- The process is capital intensive, requiring expensive compaction equipment and sintering furnaces. Therefore, only long production runs are appropriate.

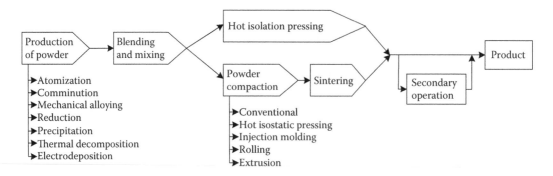

FIGURE 12.1　Sequence of processes for powder metallurgy.

- The cost of powdered metals is relatively high; however, this is offset by economical usage.
- The restricted size of the product due to the required high compaction pressure. The technique is not suitable for producing too large dimensions, or for thicknesses of less than 1.5 mm.

12.2　HISTORICAL DEVELOPMENT OF POWDER METALLURGY

The production of powders of some metals such as gold, copper, and bronze, and some oxides, such as iron oxide, was developed for decorative and ornamental purposes on other materials, such as bases for inks and paints, and in cosmetics, since the beginnings of recorded history. In about 3000 BC, the ancient Egyptians heated iron oxide in charcoal and crushed shell fire intensified by air blasts from blowers to produce sponge iron. The resulting hot sponge iron was then hammered (forged) to weld the particles together for making tools. Later, this sponge iron was broken into powder particles, washed and sorted to remove slag and large pieces, and either compacted or sintered into a porous material which was subsequently forged. These crude forms of PM eventually led to the development of one of the presently viable techniques for producing iron powder, by grinding sponge iron to fine powder, heating in hydrogen to remove oxides and annealing or softening the powder.

During the 18th and 19th centuries, powdered platinum was developed in Europe, leading to the expansion of modern powder metallurgy. In 1781, Achard produced a fusible platinum–arsenic eutectic alloy (containing 87% Pt) that melts at 600°C. Then he formed solid platinum by hot hammering the spongy alloy after volatilization of the arsenic. In about 1790, platinum chemical vessels were commercially produced in Paris. Other metals were later produced using the same technique, such as palladium (using sulfur instead of arsenic), and iridium (using phosphorus). Solid malleable platinum was also produced in almost the same period without using arsenic, by heating chemically precipitated platinum powder in a clay crucible to soften, and forging it to shape. In 1813, it was reported that malleable platinum sheets were formed by drying out successive layers of powder suspended in turpentine and heating the resulting films at high temperature without pressure. In 1826, the high-temperature sintering operation was applied commercially for the first time in Russia, to previously compacted platinum powder. The final product was highly workable. Wollaston devised the foundation of modern powder metallurgy in 1829, by concentrating on the preparation of powder compaction at high pressure in a horizontal toggle press that he specially developed for the purpose. He succeeded in making rolled thin platinum sheets that formed superior crucibles. This technique only became obsolete in 1859, with the advent of the platinum fusion procedure, which was less expensive and faster.

In 1830, copper powder reduced from precipitated copper carbonate could be sintered into a compact to form copper coins with impressions. It was then realized that lower reduction temperatures led to fine powders for better sintering. The powder was directly used or stored in glass bottles to eliminate atmospheric contamination. Similarly, medals of silver, lead, and copper were produced by PM. In the second half of the 19th century, the PM industry developed bearing materials and self-lubricated bearings, which allowed running shafts at higher speeds without the need for continuous lubrication.

Before the end of the 19th century, PM was used to produce carbon filament for incandescent lamps by extrusion and sintering. Early in the 20th century, these filaments were replaced by osmium wire filaments produced by chemical precipitation of the powder. The osmium lamp was succeeded by tantalum filament lights, followed by zirconium, vanadium, and finally tungsten filaments. To improve the ductility of tungsten wire filaments, the powder was mixed with 2 to 3% Ni, pressed into a compact and sintered in hydrogen to form bars that were drawn to the required size. Nickel was removed from the final filament by a vacuum heat treatment at high temperatures. This process was an important step toward the industrial development of cemented carbides and composite materials. Tungsten was soon recognized as the best material for lamp filament, and is still in use by reducing tungsten oxide powder in hydrogen, pressing it into a compact, presintering at 1200°C, and final sintering at 3000°C. The same PM procedure is now used for producing other refractory metals such as molybdenum, niobium, thorium, and titanium.

Cemented carbides, which have become one of the greatest inventions in the 20th century, were developed as die materials for drawing of tungsten wire and filaments, as well as for machining of harder materials. They started by sintering a mixture of tungsten granules and carbon at high temperature to give an extremely hard compound. Later on, the Germans (1914–1925) bonded finely divided hard particles of tungsten carbide with metallic cementing agents such as cobalt for close chemical affinity of carbide particles. Krupp perfected the process in 1927, by producing and marketing "Widia" which means "like diamond." In 1928, the material was introduced in the United States (by General Electric Co.), which developed a carefully controlled PM procedure for manufacturing it as a tool material.

Metal matrix composites (MMC) represent the next development in PM to be used for heavy-duty contacts, electrodes, counterweights, and radium containers. All of these composite materials contain refractory metal particles, usually tungsten, and a cementing material with different proportions of lower melting point metals such as copper, copper alloys, silver, iron, cobalt, and nickel. Some combinations also contain graphite. The first attempt started shortly after 1900, and new MMC materials are still being developed for various industry applications.

Porous metal bearings and filters signify another crucial area of PM that gained attention during the 1900s, by developing "self-lubricating" bearings. Modern types are made of copper, tin, and graphite powders, and impregnated with oil. Self-lubricating bearings have been introduced in the automotive industry since 1927, and later in the home appliance industry, such as in refrigerator compressor bearings. Metallic filters were the next stage in the development of these porous metals. The most spectacular development in this area was the paraffin-impregnated sintered iron driving bands for military projectiles, which were extensively used during the World War II.

Modern PM has found the most intensive applications in the automotive industry. Iron and iron-based PM components such as gears, cams, and other structural shapes have become dominant. In the second half of the 20th century, the aerospace and nuclear industries flourished by using refractory and reactive metals such as tungsten, molybdenum, niobium, titanium, and tantalum, and nuclear metals such as beryllium, uranium, zirconium, and thorium. These are all recovered from their ores, processed and formed by PM techniques. Wrought PM products emerged in the 1960s by developing hot isostatically pressed superalloys, PM forgings, PM tool steels, roll compacted strip, and dispersion-strengthened copper. In the 1980s, injection molding and ultrarapid solidification introduced further development in PM techniques.

12.3 METAL POWDER PRODUCTION

The production of metal powder represents a preliminary step for PM technology, and could be looked at as a separate industry. Virtually all metals can be made into powder form. The methods for metal powder production are numerous, but they can be generally divided into three main categories: thermal/mechanical, chemical, and electrolytic methods. The thermal/mechanical methods include atomizing of the powder from the molten metal, comminution, and mechanical alloying. Chemical methods include reduction of metal oxides, precipitation from solution, and decomposition of metal salts such as carbonyls. The electrolytic method is based on electrochemical deposition of metallic powders. Chemical and electrolytic methods generally produce finer sizes and higher purity of metals. Figure 12.2 represents these classes of the methods of metal powder production.

12.3.1 THERMAL/MECHANICAL METHODS

12.3.1.1 Atomization

This method is based on injecting molten metal through an orifice to form a stream which is broken down (atomized) into droplets by a jet of inert gas, air, or water, or through centrifugal force. These droplets solidify into powder. There are several techniques for streaming and atomization, as shown in Figure 12.3. Part (a) of the figure indicates siphoning of the molten metal by a high-velocity inert gas flow through a nozzle. The gas sprays the molten metal into droplets that solidify and gather as powder in a container. Figure 12.3b shows the same process, but the molten metal flows under gravity and is atomized by air jets. Air may be replaced by a high-speed water jet, as shown in Figure 12.3c. This is known as water atomization, and is the most commonly used process for metals that melt below 1600°C, where the cooling rate is higher and the powder shape is irregular. The drawback of this process is the oxidation of the particle surface, which can be offset by using synthetic oil instead of water. In both air and water atomization, the particle size is controlled mainly by the velocity of the fluid jet. The particle size is inversely related to velocity. Streaming the molten metal on a rapidly rotating disk leads to spraying the metal in all directions to produce powder, which is known as centrifugal atomization, as shown in Figure 12.3d.

Atomization is the most versatile method of production for metal powders at a wide range of production rates (from 1 to 10^5 tons/year), and a wide variety of particle sizes, from 10 μm to 1 mm.

12.3.1.2 Comminution

Mechanical comminution includes crushing, milling in a ball mill, or grinding of brittle or less ductile metals and oxide powders into small particles with irregular shape. It can be applied for getting finer sizes from powders produced by atomization. This technique is not applicable for ductile metals, as it produces particles in the form of flakes, which has no direct applications in PM.

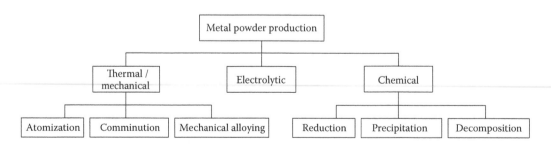

FIGURE 12.2 Methods of metal powder production.

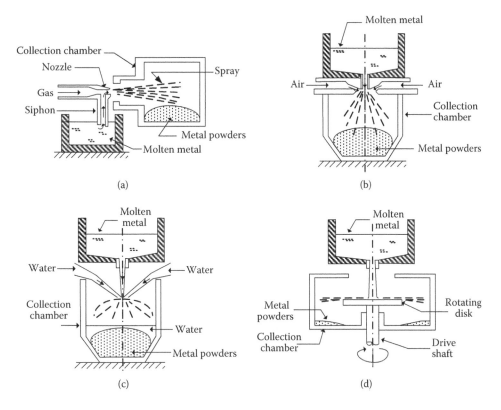

FIGURE 12.3 Atomization methods for producing metal powder: (a) gas atomization, (b) air atomization, (c) water atomization, and (d) centrifugal atomization. (From Groover, M. P., *Fundamentals of Modern Manufacturing,* 2007, John Wiley & Sons, New York. With permission.)

12.3.1.3 Mechanical Alloying

This is a relatively recent technique, developed in the 1960s by crushing mixed powders of two or more pure metals in a ball mill. Under the impact of the hard balls of the mill, the metal particles are fragmented, and the generated heat agglomerates fine particles into alloy powders.

12.3.2 Chemical Methods

12.3.2.1 Reduction

Chemical reduction includes a variety of chemical reactions through which metallic compounds are reduced to elemental metal powders. A common process involves reducing oxides by using reducing gas agents such as hydrogen or carbon monoxide. Produced powders are spongy and porous, and have uniformly sized spherical or angular shapes. Typical examples are the production of iron, tungsten, and copper powders.

12.3.2.2 Precipitation from Solution

The production of metal powder could be based on leaching an ore or ore concentrate, followed by precipitating the metal from the leach solution. Metal precipitation can be accomplished by electrolysis, cementation, or chemical reduction. The process is applicable for lower-grade ores. Typical examples include the separation and precipitation of copper, nickel, and cobalt from salt solutions by reduction with hydrogen.

12.3.2.3 Thermal Decomposition

Both iron and nickel are produced by thermal decomposition of the respective carbonyls. These carbonyls are obtained by passing carbon monoxide over spongy metal at specific pressure and temperature to obtain iron pentacarbonyl $Fe(CO)_5$, or nickel tetracarbonyl $Ni(CO)_4$, which are both liquid at room temperature. When the liquid temperature is raised to boiling, the carbonyl decomposes to reform the metal and carbon monoxide (reversible reaction). The chemical purity of the obtained powder can be very high (over 99.5%), and the particle size can be closely controlled (to less than 10 µm for iron).

12.3.3 ELECTRODEPOSITION

An electrolytic cell is set up, in which the source of the required metal is the anode, which is slowly dissolved under the applied current, transferred through the electrolyte, and deposited on the cathode. The deposit takes the form of a loosely adhering powdery or spongy layer that can be easily disintegrated into fine powder (such as copper or silver), or a dense, smooth, brittle layer is deposited, which should be removed by crushing, then washed and dried, giving a very high-purity metallic powder (such as iron or manganese). This process is commonly applied for producing powders of beryllium, copper, iron, silver, tantalum, and titanium.

12.4 POWDER METAL CHARACTERIZATION

The various powder production methods presented in the previous section provide a wide variety of metal powders with specific control limits of the chemical composition and physical characteristics to meet the requirements of a large variety of PM applications. Metal powders (as well as ceramic powders) are characterized by their geometric features, friction between particles, density and porosity (packing), as well as chemistry and surface films.

12.4.1 PARTICLE GEOMETRIC FEATURES

The geometry of particles can be defined by the particle size and distribution, particle shape, and surface area. The most common method to define the particle size is to screen the powder through a series of sieves with different mesh sizes. The higher mesh size indicates a smaller size of particles (mesh size 200 represents a screen with 200 openings in an inch). The screening method has a practical upper limit of 500 (corresponding to particle size of 25 µm) due to the difficulty of making finer screen and because of tendency for agglomeration of finer powders. Table 12.1 represents the standard U.S. sieve series for different particle sizes.

Another method for examining finer sizes is microscopy. Optical microscopes can be used for measuring particles of a small diameter of 0.5 µm, and electron microscopes can measure particle diameters of 0.01 µm.

Metal powder shapes can be categorized into various types, as shown in Figure 12.4. These shapes are assessed through microscopy. Logically, spherical and rounded types can be represented by the diameter. Other shapes require at least two measures to assess their size. Therefore, the

---

TABLE 12.1
Standard Sieve Series for Different Particle Sizes

Designation mesh	18	20	25	30	35	40	45	50	60	70	80
Sieve openings (µm)	1000	850	710	600	500	425	355	300	250	212	180
Designation mesh	100	120	140	170	200	230	270	325	400	450	500
Sieve openings (µm)	150	125	106	90	75	63	53	45	38	32	25

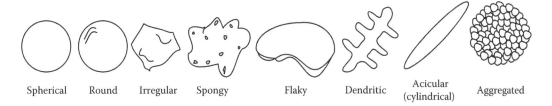

Spherical Round Irregular Spongy Flaky Dendritic Acicular (cylindrical) Aggregated

FIGURE 12.4 Possible particle shapes in PM. (Adapted from Groover, M. P., *Fundamentals of Modern Manufacturing*, 2007, John Wiley & Sons, New York, and Kalpakjian, S., and S. R. Schmid, *Manufacturing Processes for Engineering Material*, 4th ed., Pearson Education, 2003.)

shape is usually described in terms of the aspect ratio, or shape factor (index). Aspect ratio is the ratio of the largest to the smallest dimensions of the particle, with a minimum ratio 1 for spherical particles to a maximum of 10 for flaky or acicular particles. The shape factor (index), K_{sf}, is a measure of the surface area to the volume of the particle, with reference to a spherical particle of equivalent diameter. For a sphere with surface area $As = \pi D^2$ and volume $V = \pi D^3/6$, where D is the diameter of the particle,

$$K_{sf} = \frac{Asp}{vp} D \qquad (12.1)$$

For a spherical specimen, $K_{sf} = 6$, and for other particle shapes, K_{sf} is larger than 6. Higher shape factor means higher surface area for the same total volume or weight of metal powders, and correspondingly greater area for surface oxidation to occur. Small powder size also leads to more agglomeration of particles, which is a disadvantage in automatic feeding of the powder.

The size distribution of particles is an important aspect, as it affects the processing of powders. It is usually given in terms of a frequency distribution plot.

12.4.2 INTERPARTICLE FRICTION AND FLOW CHARACTERISTICS

Friction between particles restricts the ability of powder to flow willingly, thus leading to a higher pile when the powder is poured freely from a narrow pile. A practical measure of interparticle friction is the *angle of repose* (side-to-base angle) for this pile. Smaller particle size indicates greater friction and thus larger angle of repose. Spherical shapes lead to the lowest friction and smallest angle. Higher shape factors lead to higher friction.

Flow characteristics are significant in die filling and compaction. During pressing of the powder in the compaction die, resistance to flow increases density variations in the compacted part, which is unfavorable. A common measure of flow is the time required for a certain weight of powder to flow through a standard-sized funnel. Larger flow times demonstrate more difficult flow and higher interparticle friction. Lubricants are usually added to reduce interparticle friction and facilitate flow during compaction.

12.4.3 DENSITY AND PACKING

There are two measures of density in powder metallurgy. *The bulk density* is the density of the powders in the loose state. This includes the effect of the pores between particles. *The true density*, on the other hand, is the density of the true volume of the material after being solidified from the molten state. The true density is larger than the bulk density.

The ratio of the bulk density to the true density is known as the *packing factor*, and thus,

Packing Factor = Bulk Density / True Density (12.2)

Typical values of the bulk density for loose powders range between 0.5 and 0.7. This packing factor depends on the particle shape and the distribution of particle sizes. The packing factor can be increased by one of the following actions:

a. Large variation of powder sizes in a sample, which is associated with fitting of the smaller particles in the interstices of the larger ones.
b. Vibrations which lead to settling of the particles more tightly.
c. Application of external pressure during compaction which leads to rearrangement and deformation of particles.

Porosity represents an alternative means of assessment of the packing factor of a powder. It is defined as the ratio of the volume of pores (empty spaces) in the powder to the bulk volume. Therefore

$$\text{Porosity} = 1 - \text{Packing Factor} \tag{12.3}$$

12.4.4 CHEMISTRY AND SURFACE FILMS

Metallic powders are classified as either elemental, consisting of pure metal, or prealloyed, where each particle is an alloy. Surface films represent an obstacle in PM technology because of the large area of powders per unit weight of metals. Possible surface films include oxides, silica, absorbed organic materials, and moisture. Such films necessitate extra processes for removal prior to shape processing.

12.5 BLENDING AND MIXING OF POWDERS

Blending and mixing represent the first step in powder metallurgy processes. In order to obtain sound compaction and sintering, metallic powders must be thoroughly homogenized beforehand. The term *blending* is commonly used when powders of the same chemical composition but different particle sizes are intermingled. Different particle sizes are often blended to reduce porosity. The term *mixing* refers to powders of different chemistries being combined to form alloys that are difficult or impossible to produce using other methods. This offers a great advantage of the PM technology. The distinction between the two terms is not always precise in industrial practice.

Blending and mixing are performed using mechanical drums and mixers. Rotating drums could be cylindrical, rotating horizontally around their axis, or double cone shaped, rotating vertically around a horizontal axis. These are usually designed with internal baffles to prevent free-fall during blending of powder with different sizes. Best results are obtained when the container is 20 to 40% full. Mixers could be of the screw type for agitation, or rotating blade type for stirring. Schematic presentation of mixing and blending equipment is shown in Figure 12.5. Blending and mixing equipment should not permit contamination or deterioration of the powder. Contamination is avoided by mixing in an inert atmosphere to avoid oxidation, or in a liquid. Excessive mixing may change the shape or size of particles, or may strain-harden them, leading to deterioration of their characteristics.

Other ingredients are commonly added to the metallic powders during blending and mixing. These additives include

a. Lubricants such as stearic acid or zinc stearate are added to reduce friction between particles (improve their flow characteristics) and at the die wall during compaction. Typical weight percentage of a lubricant ranges between 0.5% and 5%.
b. Binders are added to reach a particular required strength level during compaction.
c. Deflocculants, which inhibit agglomeration of fine powder to maintain better flow during compaction.

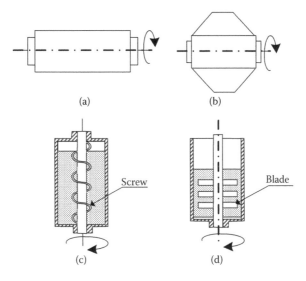

FIGURE 12.5 Schematic illustration of blending and mixing equipment: (a) cylindrical drum, (b) double-cone drum, (c) screw mixer, and (d) blade mixer. (Adapted from Groover, M. P., *Fundamentals of Modern Manufacturing*, 2007, John Wiley & Sons, New York.)

12.6 POWDER COMPACTION

Compaction involves applying pressure to the powder to accomplish the required shape. This can be carried out either by directly applying uniaxial pressure (conventional compaction), or hydrostatically from all directions (isostatic compaction). The latter technique ensures more uniform density distribution within the part after compaction.

12.6.1 Conventional Compaction

Conventionally, compaction is carried out by pressing the powder contained in a die between opposing punches to form the *green compact*, as shown in Figure 12.6, which indicates that two punches

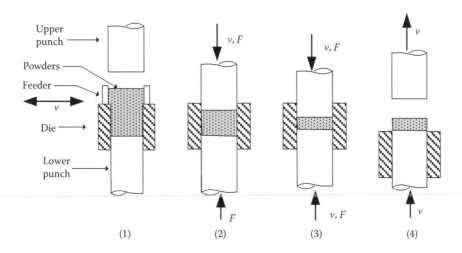

FIGURE 12.6 Powder compaction by pressing, (1) filling the initial die cavity with powder, (2) & (3) compacting force, (4) ejection of green product. (From Groover, M. P., *Fundamentals of Modern Manufacturing*, 2007, John Wiley & Sons, New York. With permission.)

TABLE 12.2
Compacting Pressures for Various Powders

Material	Compaction Pressure (MPa)
Aluminum	70–275
Brass	400–700
Bronze	200–275
Iron	350–800
Tungsten and tantalum	70–140
Aluminum oxide	110–140
Carbon	140–165
Cemented carbides	140–400
Ferrites	110–165

Source: Kalpakjian, S., and Schmid, S. R., *Manufacturing Processes for Engineering Material*, 4th ed., Pearson Education, 2003.

should apply uniaxially opposite forces in order to attain uniform density distribution on the product. Compaction pressure repacks the powder into a more efficient arrangement by reducing the pore spacing, increasing the number of contacting points between particles, and plastically deforming them, resulting in interparticle contact areas. Accordingly, the density is increased from the bulk density to the *green density*, which is as close as possible to the true density depending on the compaction pressure.

Compaction pressures are applied by either mechanical or hydraulic presses. For smaller size products, crank or eccentric mechanical presses are used, while relatively larger sizes are compacted in toggle or knuckle joint presses. Large products are compacted in hydraulic presses with capacities as high as 45 MN. Generally, double-action presses are preferred to provide force and velocity for both punches. Smaller press velocities are preferred to eliminate the possibility of air entrapment in the die cavity. Typical compaction pressures range from 70 MPa for aluminum powders to up to 800 MPa for iron and steel powders, as shown in Table 12.2.

12.6.2 Isostatic Compaction

Isostatic pressing involves applying pressure from all directions for achieving greater uniformity of compaction and density distribution compared to uniaxial compaction. This technique is advantageous for complicated part geometries, and for large diameter-to-height ratios. The powders are contained in a flexible mold, and hydraulic pressure is used to achieve compaction. Isostatic compaction can be carried out under cold or hot conditions.

a. Cold isostatic pressing (CIP) involves carrying out the process at room temperature. The mold is made of rubber, neoprene, urethane, or any other elastomer material, filled with the powder mix, and placed inside a closed compartment, as shown in Figure 12.7. Water or oil is pressurized around the mold, leading to the required isostatic compaction. The initial size of the mold should be larger to allow for shrinkage during compaction. Solid cores can be used with hollow parts, as indicated in the figure. The pressure ranges from 400 MPa to up to 1000 MPa. The tooling cost in CIP is less expensive for a relatively smaller-sized product, and for a limited number of products.

b. Hot isostatic pressing (HIP) is carried out at high temperatures and pressures. Molds are made in the form of sheet metal capsules to withstand higher temperatures, and the pressurizing medium is inert gas (most commonly argon) to avoid chemical reaction with the

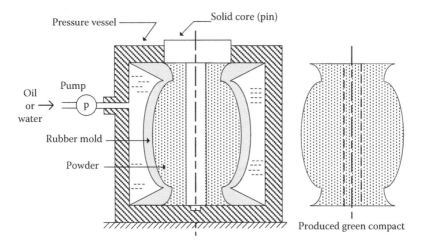

FIGURE 12.7 Cold isostatic pressing (CIP): (1) powder filling flexible mold and applying hydrostatic pressure, (2) green part after mold removal. (Adapted from Groover, M. P., *Fundamentals of Modern Manufacturing*, 2007, John Wiley & Sons, New York.)

material. After densification, the capsule has to be stripped off. The main advantage of HIP is the ability to reach the full density of the product, thus accomplishing compaction and sintering in one step. However, the process is relatively expensive, and limited for relatively simple shapes. Therefore, it is mainly used for producing superalloy parts for the aerospace industry.

12.6.3 Powder Injection Molding

Very fine metal powders (<10 μm) and ceramics can be compacted by injection molding similar to plastics. Powders are blended with suitable binders and injected into a mold cavity after heating to the molding temperature. When the part is cooled, it is removed from the mold, the binder is removed using thermal or solvent techniques, and the part is ready for sintering. The binder acts as a carrier for the particles to provide adequate flow during molding. Suitable binding materials are phenolics (thermosetting polymers), polyethylene (thermoplastic polymers), water, gels, or waxes. Polymers are the most frequently used binders. PIM is not cost effective compared to conventional compaction for simple axial symmetric shapes. It is most economical for small thicknesses (in the range of 5 mm), and complex configurations at high rates of production. Typical products are components of watches, guns, surgical knives, and small automotive parts.

12.6.4 Powder Rolling

Powders can be compacted between two rolls in a rolling mill operation to form long metal green strips that can be directly fed to a sintering furnace, as shown in Figure 12.8. The process is also named "roll compaction." Rolling speeds of up to 0.5 m/s can be applied. Typical products are sheet metal strips for electrical and electronic components and coins.

12.6.5 Powder Extrusion

Powders can also be compacted by extrusion to produce long parts with different profile areas at high densification pressures. Rotating screw extruders with heaters, similar to those used in the extrusion of plastics, can be adopted. Binders such as water or organic materials are added to provide adequate plasticity for forming.

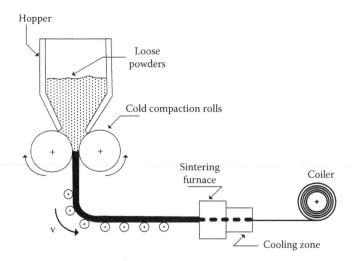

FIGURE 12.8 Powder rolling, sintering, and coiling. (Adapted from ASM International, 1998. *Metals Handbook, vol. 7: Powder Metal Technologies and Applications*, Materials Park, OH.)

12.7 SINTERING

The green compact lacks strength and hardness, which can be improved by a special heat treatment process known as sintering. The process involves heating the green compact in a furnace with controlled atmosphere to a temperature below its melting point, but sufficiently high to allow bonding of the individual particles. The sintering temperature is usually in the range of 0.7 to 0.9 of the melting temperature of the metal or alloy. The time for sintering varies significantly depending on the type of powder and the particle size. It ranges from a minimum of 8 minutes for iron and iron-graphite, to up to 10 hours for ferrites, as shown in Table 12.3.

Most sintering furnaces are continuous for medium-to-high rates of production. They include three chambers, as shown in Figure 12.9, to achieve the following stages of the sintering cycle:

1. Preheat chamber, where the lubricants and binders are burned off and volatized
2. Sintering chamber, where the whole volume of the product reaches the sintering temperature to gain the required strength and hardness
3. Cool down chamber, to obtain the finished part

TABLE 12.3
Sintering Temperature and Time Ranges for Different Metallic Materials

Material	Copper and its Alloys	Iron and Iron-Graphite	Nickel	Stainless Steels	Ferrites
Temperature (°C)	760–900	1000–1150	1000–1150	1000–1290	1200–1500
Time (min)	10–45	8–45	30–45	30–60	10–600

Material	Tungsten Carbide	Molybdenum	Tungsten	Tantalum
Temperature (°C)	1430–1500	2050	2350	2400
Time (min)	20–30	120	480	480

Source: Kalpakjian, S., and Schmid, S. R., *Manufacturing Processes for Engineering Material*, 4th ed., Pearson Education, 2003.

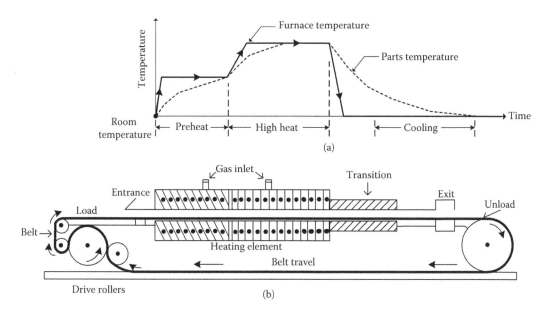

FIGURE 12.9 Continuous sintering furnace (mesh belt type), including three chambers, and temperature cycle for the furnace and the parts. (Adapted from German, R. M., *Powder Metallurgy of Iron and Steel*, John Wiley & Sons, New York, 1998.)

The furnace atmosphere should be highly controlled to protect the part particles from oxidation, or to control the carburization and decarburization of iron and iron-based compacts. Common sintering furnace atmospheres are inert gas, nitrogen-based, dissociated ammonia, hydrogen, and natural gas. Vacuum atmosphere is necessary for sintering specific materials such as stainless steels and tungsten.

The most dominant mechanism of sintering is based on bonding of particles by diffusion (solid-state bonding). These bonds between particles form necks and transform them into grain boundaries, as shown in Figure 12.10. The formation and growth of these bonds is associated with reduction in the surface energy of the particles and accordingly, the size of pores, which leads to shrinkage of the product during sintering. The density is accordingly increased to the true density and the strength rises distinctly to its maximum value. Ductile materials exhibit an increase in ductility, as shown in the figure. Further heating or longer sintering time after maximum strength is reached may lead to migration of the grain boundaries and increase of the grain size, with a consequent drop in strength. Remaining pores may be left behind (inside the new grains) and become stable, and thus impair the compact properties.

Another sintering mechanism takes place when the mix includes particles of different metals, with one metal of a lower melting temperature. Under such conditions, the particles of lower melting point melt and spread to envelop the solid particles due to surface tension. This mechanism is known as "liquid phase sintering." A typical example of this mechanism is cobalt in tungsten carbide tool tips, where the melting point of cobalt is 1495°C, while tungsten carbide melts at 2870°C.

12.8　SECONDARY OPERATIONS

Additional processes may be performed to sintered parts to reach the required properties and geometric configurations of PM products. These processes include

1. Densification and sizing, by applying a repressing operation in a closed die to increase the product density (densification), improve the mechanical properties, achieve close geometrical accuracy (sizing), or add details on the product surface (coining).

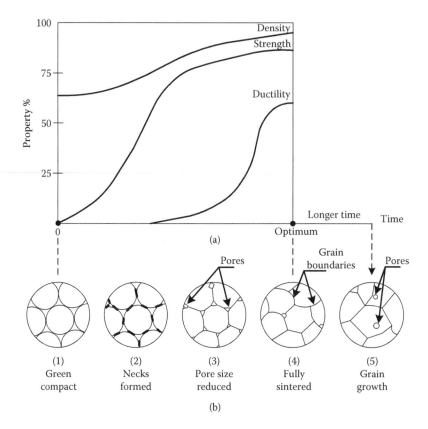

FIGURE 12.10 Mechanism of sintering to gain true density, and maximum strength. Particles are (1) bonded at the contact points, (2) necks are formed, (3) pore size is reduced, and (4) grain boundaries are developed. (5) Grain growth and pores are developed in the grains at longer time. (Adapted from Schey, J., *Introduction to Manufacturing Processes*, 2nd ed., McGraw-Hill, New York, 1987.)

2. Machining, to produce geometric features that cannot be obtained by pressing, such as threading, recessing, or side holes. Grinding may be carried out to control dimensional accuracy and surface finish of the product.

3. Impregnation and infiltration. Since porosity is a unique and inherent characteristic in PM, it can be utilized to develop special products by filling the pores with oils, grease, polymers, or metals with lower melting temperature than the powder metal. A typical application of impregnation is the production of bearings and bushings, with up to 30% of their volume internally lubricating oil. Universal joints are now manufactured using PM techniques and impregnated with grease to act as self-greasing units. A typical application of infiltration is allowing molten copper to infiltrate the pores of iron-based parts by capillary action to improve hardness, strength, and corrosion resistance.

4. Electroplating and coating to achieve higher wear and corrosion resistance. Common electroplating materials include copper, nickel, chromium, zinc, and cadmium.

5. Heat treatment to improve hardness and strength.

12.9 CERAMIC MATERIALS

Ceramics are introduced in this chapter, since a wide variety of them are manufactured by powder metallurgy. Ceramics are inorganic, nonmetallic materials that consist of metallic and nonmetallic elements bonded together primarily by ionic and/or covalent bonds. The word *ceramic* is derived

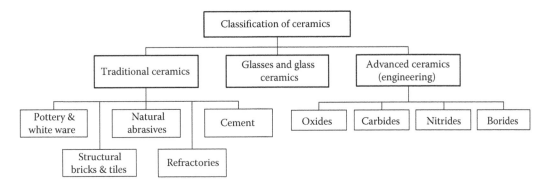

FIGURE 12.11 Classification of ceramic materials.

from the Greek word *Keramikos* which means "burnt clay." This suggests that the term included all products made from fired clay, such as bricks, tiles, fireclay refractories, basins, and other sanitary ware, ornaments, porcelains, and pottery tableware. Recently, the term *ceramics* is extended to cover almost all hard and brittle materials of mineral origin with refractory characteristics (high melting temperature), high chemical stability in hostile environments, and good electrical and thermal insulation. These include cement, glasses, metallic oxides such as alumina, magnesia, and zirconia, and metallic carbides such as tungsten carbide, titanium carbide, and boron carbide. The most recently developed ceramics, such as silicon nitride, cubic boron nitride, and sialons, are used in high technology applications. There are different approaches to classify ceramics, but they can be broadly classified as traditional ceramics, which are based on natural materials, and advanced or engineering ceramics, which are finely manufactured in almost pure form to suit specific applications. A third class is glasses and glass ceramics, which are based on natural silica, but are distinguished by particular characteristics. These classes are shown in Figure 12.11, and are briefly described in this section.

12.9.1 TRADITIONAL CERAMICS

The sources of traditional ceramics are natural raw materials such as clay minerals and quartz sand (silica), which are the most abundantly available materials in nature. These are solid crystalline compounds formed and mixed in the earth's crust over billions of years through complicated geological processes. Traditional ceramics can be classified into 5 categories according to the product families, as shown in Figure 12.11.

1. Pottery and whiteware ceramics: This category is one of the oldest, dating back thousands of years, yet it still has wide demand for its products. The row materials for this category are clays, silica sand, and feldspar. Clays are the most widely used raw materials in ceramics. They consist of fine particles of hydrous aluminum silicate. The most common clays are based on the mineral *kaolinite* ($Al_2Si_2O_5(OH)_4$). Other *clay minerals* vary in composition due to the addition of other elements such as Mg, Na, and K. *Silica* (SiO_2) is the principal constituent in making glass, and an important ingredient in other ceramic products, including abrasives, refractories, and whitewares. The most common source of silica in nature is quartz, and the main source of quartz is *sandstone*. The low cost of *sandstone* and relative ease of its processing accounts for the widespread use of silica in ceramic products. It is generally mixed in various proportions with clay and other minerals to achieve the required characteristics in the final products. *Feldspar* refers to a group of rock-forming silicate minerals (aluminum silicates, combined with other elements such as calcium, potassium, sodium, and barium), which make up as much as 60% of the earth's crust.

Mixtures of clay, silica, and feldspar, in various proportions, change to plastically form-able and moldable material when mixed with water (wetted), and harden to a ceramic prod-uct when dried or heated to a sufficiently high temperature, in a process known as *firing*. Through these mixing, wetting, forming, drying, and firing processes, different products are made, including pottery, whiteware (dishes, plates and pots), wall tiles, and sanitary ware. *Glazing* involves application of a surface coating, usually a mixture of oxides such as silica and alumina, to give pottery products an attractive appearance, and eliminate the effect of moisture. Stoneware has lower porosity due to closer control of ingredients and higher firing temperatures. China is fired at even higher temperatures, which produces the translucence in the finished products. The higher temperatures convert much of the ceramic material to a glassy (vitrified) phase, which is relatively transparent compared to the polycrystalline form. Porcelain is made of the same constituents, but fired at still higher temperatures (1200°C–1400°C) to achieve very hard, tough, dense, and glassy products. In addition to being used in tableware and cookware, porcelain has several other applications, including electrical insulation and bathtub coatings.

2. Structural bricks and tiles: Bricks are blocks of ceramic materials used in masonry con-struction, usually laid using various kinds of mortar. Bricks are made of clay, calcium sili-cate, concrete, or shaped from quarried stone. Clay is the most common material, mixed with up to 30% silica sand to reduce shrinkage. The mixture is mixed with water to the required consistency, formed by press molding or extrusion, and fired in a kiln at 900°C to 1000°C to reach the required strength. Lime, ash, and other organic materials may be added to improve quality. Calcium silicate bricks include lime mixed with quartz and min-eral colorants. The mixture is wetted, molded, and cured in an autoclave. Clay pipes, roof, floor, and drain tiles are also structural ceramics made of various low-cost clays containing silica and gritty matter widely available in natural deposits. The same procedure of mix-ing, wetting, molding, and firing is used in their manufacture.

3. Refractory ceramics: Refractory ceramics are designed to have acceptable mechanical and chemical properties at high temperatures. They may take the form of bricks for lining melting furnaces (as well as kilns, incinerators, and reactors), and shaped products such as crucibles, coatings, or insulating fibers. Most refractories are based on oxide compounds such as alu-mina, silica, magnesia (MgO), zirconia (ZrO_2), and lime (calcium oxide CaO). Most alumina is processed from the mineral *bauxite* (an impure mixture of hydrous aluminum oxide and aluminum hydroxide). Fire clays are also widely used in the manufacture of refractories. Silicon carbide and carbon (graphite) are two other refractory materials used in severe temperature conditions, but they cannot be used in contact with oxygen, as they oxidize and burn.

Based on chemical composition, refractories can be classified as acidic, basic, or neu-tral. Common acidic refractories are based on silica and alumina. Magnesia is the core material for most basic refractories. It is generally more expensive than the acidic materi-als, but is often required in metal processing applications to provide compatibility with the metal. Neutral refractories containing chromite (Cr_2O_3) are often used to separate the acidic and basic materials, since they tend to attack one another. The combination is some-times attractive when a basic refractory is necessary on the surface for chemical reasons, and the cheaper acidic material is used beneath to provide strength and insulation.

4. Natural abrasives: Because of their high hardness, ceramic materials are often used for abrasive applications, such as cutting, sanding, grinding, polishing, and lapping. The hard-est known natural abrasive is diamond. *Corundum* contains a high percentage of alumina (Al_2O_3), which is used in grinding wheels and sandpaper. Similarly is *carborundum* (sili-con carbide SiC), which is harder than alumina, although the latter gives better results when grinding steel, which is the most widely used metal. Softer minerals such as calcite (calcium carbonate $CaCO_3$) are used as polishing agents in toothpaste. The topic of abra-sive technology is dealt with in detail in the chapter about abrasive cutting.

5. Cement: Cement is a binding material to bring two parts together. It has been developed since ancient times for masonry. Cement used in construction is characterized as hydraulic or nonhydraulic. Hydraulic cements (e.g., Portland cement) harden because of chemical reactions that do not involve "drying out"; they can harden even under water or when constantly exposed to wet weather. The chemical reaction that results when the dry cement powder is mixed with water produces hydrates that are not water soluble. Nonhydraulic cements (e.g., lime and gypsum plaster) must be kept dry in order to gain strength.

Portland cement is made by heating limestone (calcium carbonate) with limited amount of clay at about 1450°C in a kiln, resulting in a hard material known as clinker. Clinker is then ground with a small percentage of gypsum into a cement powder. The most important use of cement is in the production of mortar and concrete; the bonding of natural or artificial aggregates to form a strong building material which is durable against common environmental effects.

12.9.2 GLASSES AND GLASS CERAMICS

As a type of ceramics, glass is an inorganic, nonmetallic compound (or a mixture of compounds), that cools from the melt to a rigid condition without crystallizing; i.e., forming a glassy (vitreous or amorphous), commonly transparent structure. The oldest known glass specimens, in the form of beads and other simple shapes, date back about 4500 years. They were found in Mesopotamia and ancient Egypt. The principal ingredient in all glasses is silica, most commonly found as quartz in sandstone and silica sand. Silica glass has a very low coefficient of thermal expansion and therefore it resists thermal shock. These properties are ideal for designing products for heating to elevated temperature, such as Pyrex and chemical glassware. Adding other oxides such as Na_2O, CaO, Al_2O_3, MgO, K_2O, PbO, and B_2O_3 reduce the melting point of glass for easier processing and improve control of properties. However, silica is the best glass former, as it remains the main component in any type of glass, in the range of 46 to 96%. Other constituents are contained in a solid solution with SiO_2 to achieve one or some of the following functions:

- Acting as a flux to promote fusion during heating
- Increasing fluidity of the molten glass for ease of processing (CaO and Na_2O)
- Retarding the tendency to crystallize from the glassy state (*devitrification*) (MgO)
- Increasing hardness and retarding thermal expansion in the final product (Al_2O_3)
- Improving the chemical resistance to acids, alkalis or water
- Adding color to the glass
- Reducing hardness and increasing the index of refraction for optical applications, e.g., lenses, (PbO)

The most widely available category of glass is window glass, which is made of quartz (0.1 to 0.6 mm particle size), limestone ($CaCO_3$), and soda ash (Na_2CO_3). The same glass can be used for making bottles and other containers, by cooling it faster from the melt. The glass used in light bulbs and other thin glass items is high in soda and low in lime, in addition to small amounts of magnesia and alumina. Addition of boron oxide in amounts up to 13% produces glass for laboratory purposes (a typical example is Pyrex glass). Optical glasses for the lenses of eyeglasses and optical instruments may have a low index of refraction (crown glass), or a high index of refraction (flint glass). Glass fibers are manufactured for composite materials applications, as discussed in Chapter 14.

Glass ceramics: Glass can be converted through heat treatment into a crystalline form. The crystalline phase in the final product typically ranges between 90% and 98%, with the remainder being unconverted vitreous material. The grain size is appreciably finer than the grain size of conventional ceramics (commonly between 0.1 and 1.0 μm). This fine crystal microstructure makes glass ceramics much stronger than the glasses from which they are converted. Crystallization leads to opacity of the glass ceramics, usually producing gray or white color. The heat treatment process involves reheating the previously shaped glass product for a prolonged period of time, at a temperature sufficient to form a dense network of crystal nuclei throughout the material. This high density of nucleation sites inhibits grain growth of individual crystals, leading to the ultimately fine grain size. The tendency for nucleation of crystal structure is based on nucleating agents such as TiO_2, P_2O_5, and ZrO_2. Once nucleation is initiated, keeping the part for a longer time at higher temperature causes growth of the crystalline phase. Glass ceramics have higher strength, high creep resistance at high temperature, absence of porosity, low coefficient of thermal expansion (near to zero), and high resistance to thermal shock. Typical applications include cooking ware (e.g., Pyroceram and Pyroflam tradenamed cooking utensils), heat exchangers, and housings of radar antennas.

12.9.3 ADVANCED (ENGINEERING) CERAMICS

The terms *advanced ceramics*, *engineering ceramics*, or *new ceramics* refer to ceramic materials that have been developed synthetically, with tightly controlled chemical composition, over the past few decades. The most commonly developed ceramics to cope with advanced functional applications in industry can be organized into chemical compound categories: Oxides, carbides, nitrides, and borides, as presented in Figure 12.11. These categories will be introduced in this section, and the new techniques specially developed for manufacturing them are introduced in the coming section.

1. Oxide ceramics: The most widely used new oxide ceramic is alumina. It has been discussed above as a traditional ceramic for use as a refractory and an abrasive. Synthetically produced alumina has a thorough control of purity, additives, and particle size of the product, which leads to much higher strength, toughness, hot hardness, and thermal conductivity, compared to natural alumina. Engineering alumina is used in a wide variety of applications, including abrasives, artificial bones and teeth, electrical insulators, electronic components, coatings, alloying ingredients, refractory bricks, cutting tool tips, and spark plug bodies. Zirconium oxide is also synthetically produced in the form of *cubic zirconia* in various colors for use as gemstones (best known as diamond stimulants).

2. Carbide ceramics: These include silicon carbide (SiC), tungsten carbide (WC), titanium carbide (TiC), tantalum carbide (TaC), and chromium carbide (Cr_3C_2). Silicon carbide is widely used as an abrasive, but there are plenty of other applications, including resistance heating elements, car brakes, ceramic plates in bullet-proof vests, light-emitting diodes, semiconductor electronics, and additives in steel making. WC, TiC, and TaC are extensively used in cutting tools due to their high hot hardness and wear resistance. Cr_3C_2 is an extremely hard refractory material, usually processed by sintering. It is used in the coatings of bearings, seals, and orifices. It is also used as an additive in corrosion- and wear-resistant materials. All described carbides, except SiC, are bonded in a metal framework, forming *cermets* (reduced from ceramics and metals), which are used as tool tips in metal cutting. Carbon black is the usual source of carbon in synthesizing these carbides.

3. Nitride ceramics: Recently developed nitride ceramics include silicon nitride (Si_3N_4), boron nitride (BN), and titanium nitride (TiN). Si_3N_4 has high thermal shock resistance, wear resistance, and ability to withstand high structural loads at high temperatures. Therefore, it is used in high temperature structural applications such as gas turbines, rocket engines, cutting tools, bearings, melting crucibles, and insulating films in electronics. Cubic boron nitride (CBN) has a hardness comparable to that of a diamond. It is widely used in cutting tools and abrasive wheels. Hexagonal boron nitride is similar to graphite, and is used as a lubricant and an additive in cosmetics. Titanium nitride has high hardness and all other mechanical properties of other nitrides, except that it is an electrical conductor. It is ideally used as a surface coating on cutting tools in a thickness of 6 μm.

A new ceramic material related to the nitride and oxide groups is the oxinitride ceramic called *sialon*. Its name is derived from its ingredients: Si–Al–O–N. Its chemical composition is variable, a typical composition being $Si_4,Al_2O_2N_6$. The properties of sialon are superior to those of silicon nitride, regarding oxidation resistance at high temperature. It is used in cutting tools and other high temperature applications (see Chapter 15).

4. Boride ceramics: Borides, including titanium boride (TiB_2), zirconium boride (ZrB_2), and chromium boride (CrB_2) have high melting points, strength, and oxidation resistance. They are used in turbine blades, rocket nozzles, and combustion chamber liners.

12.10 CERAMIC MANUFACTURING PROCESSES

Ceramics can be manufactured using different methods, some of which have their origins in the early civilizations, especially for traditional ceramics. The main concern in this section is to introduce the manufacturing processes for producing a solid ceramic form with the desired shape such as fiber, powder, film, or monolith (a massive body analogous to the green product). Most of these processes commonly end with a sintering process to get the required part. These are commonly used for the production of advanced ceramics with specifically designed structure, such as oxides, carbides, and nitrides. These processes can be grouped depending on the starting material, whether in a gaseous, a liquid, or a solid phase or combinations of different phases, as shown in Table 12.4.

12.10.1 CHEMICAL VAPOR DEPOSITION

Chemical vapor deposition (CVD) is a process by which reactive molecules in the gas phase are transported to the hot surface of a substrate (wafer), at which they chemically react and form a high purity solid film or coating. It is a highly developed technique that can be used to deposit all classes of materials, including metals, ceramics, and semiconductors. Large areas can be coated, and the process is applicable on a mass production scale. Thick films or even massive bodies (monoliths) can also be produced by prolonging the deposition process so that the desired thickness is achieved.

Details of the chemical reactions are discussed in Chapter 20. Generally, they are carried in reactor vessels usually made of quartz to be inert for the reactions and to withstand the high temperatures which range from 400°C up to 2400°C, depending on the type and rate of deposition. Substrate heating is required, and the temperature of the substrate influences the deposition rate and structure. Generally, high temperature produces crystalline deposits, whereas low temperature yields amorphous materials. The pressure in the reactor also affects the uniformity of the

---

TABLE 12.4
Common Ceramic Manufacturing Processes

Starting Materials	Method	Product
Gases	Chemical vapor deposition	Films, monoliths
Gas–liquid	Directed metal oxidation	Monoliths
Gas–solid	Reaction bonding	Monoliths
Liquid–solid	Reaction bonding	Monoliths
Liquids	Sol-gel process	Films, fibers
	Polymer pyrolysis	Fibers, films
Solids (powders)	Melt casting	Monoliths
	Sintering (powder metallurgy)	Monoliths, films

Source: Rahaman, M. N., *Ceramics Processing and Sintering*, 2nd ed., Marcel Dekker, 2003. With permission.

deposited layer, and pressure ranges of 1 to 15 kPa are applied. Frequently, volatile byproducts are also formed, which are exhausted out of the reactor. A schematic presentation of types of CVD reactors are shown in Chapter 20.

Typical applications include deposition of pyrolytic carbon or graphite, tungsten carbide (WC), or silicon carbide (SiC) coating, titanium oxide (TiO_2) or silicon (Si) or silicon oxide (SiO_2) films on electronic devices, silicon nitride (Si_3N_4) films on semiconductor devices or coatings on composites, and boron nitride (BN) blocks. A plasma-assisted CVD process has been recently applied to produce diamond films or coatings on a substrate at relatively low temperature (about 2000°C) and low pressure (less than atmospheric). The basic reaction involves pyrolysis of a carbon-containing precursor such as methane (CH_4).

CVD techniques provide a distinct advantage of fairly low fabrication temperatures for ceramics and composites with high melting points that are difficult to fabricate by other methods. The low reaction temperatures also increase the range of materials that can be coated by CVD, especially for the highly refractory coatings. However, a major limitation of the process is that the material deposition rate is very slow, typically in the range of 1–100 µm/h. The production of massive bodies can therefore be very time consuming and expensive. Another problem that is normally encountered in the fabrication of massive bodies by CVD is the development of a columnar microstructure of fairly large grains, which leads to fairly low intergranular strength. These limitations direct CVD primarily to the formation of thin films and coatings.

12.10.2 Directed Metal Oxidation

It is well known that direct reaction between an oxidizing gas and molten metal is impractical for the production of oxide ceramics, since the reaction usually forms a solid protective coating that separates the metal, and effectively stops the reaction. A novel technique has been developed for the production of porous and dense ceramics as well as composites by employing directed oxidation of a molten metal by a gas (e.g., air), adding alloying elements to the melt. As an example, aluminum oxide (Al_2O_3) can be manufactured by adding a few percent of Mg and an element from the carbon group in the periodic table (e.g., Si, Ge, Sn, Pb), and keeping the alloy in a temperature range 900°C–1350°C while passing the oxidizing air. The oxide coating becomes no longer protective, and small pores are formed, through which molten metal is drawn up to the top surface of the film, thus continuing the oxidation process, as shown in Figure 12.12a. The reaction is continued under the same conditions until the required thickness is reached. An amount of unreacted metal of 5 to 30% by volume is left, depending on the processing parameters.

For the production of composites, a filler material in the form of fibers, particles, or platelets is shaped into a preform with the required product form, and added on top of the molten metal, as shown in Figure 12.12b. The oxidation process continues from the molten metal surface upward into the preform, such that the oxidation product becomes the matrix of the composite. The growth of the matrix into the preform involves little or no change of dimensions. The technique is applied to produce composites with not only matrices of oxides, but also nitrides, borides, carbides, and titanates. A distinct advantage of the method is that the problems associated with shrinkage during densification in other fabrication techniques (such as powder processing) are avoided. A second advantage is the ability to produce large components with high dimensional control.

12.10.3 Reaction Bonding

Reaction bonding (or reaction forming) is a process in which a solid porous preform reacts with a gas (or a liquid) to produce the required chemical compound and bonding between grains. Similar to directed metal oxidation, the process is associated with little or no shrinkage of the preform so that close dimensional tolerances are achieved in the finished product. The process is widely applied for the production of silicon nitride (Si_3N_4) and silicon carbide (SiC).

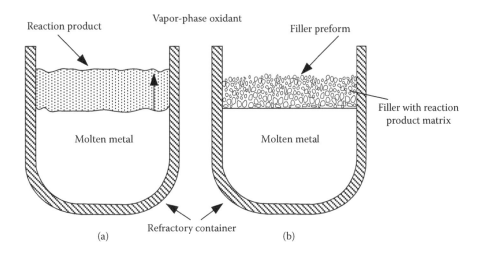

FIGURE 12.12   Schematic of the directed oxidation process of a molten metal to form an (a) oxide ceramic and (b) in the presence of a filler matrix to form a composite material. (From Rahaman, M. N., *Ceramics Processing and Sintering*, 2nd ed., Marcel Dekker, 2003. With permission.)

For the production of silicon nitride by reaction bonding (RBSN), silicon powder is first consolidated by pressing in a die, isostatic pressing, slip casting, or injection molding, to form a billet or a shaped part. The preform is then preheated in argon to about 1200°C, and machined to the required shape and dimensions. The part is then heated in nitrogen gas at atmospheric pressure to 1250°C to 1400°C, where reaction bonding takes place, producing Si_3N_4. The reaction is associated with increase of the density of Si (2.33 g/cm^3) to the density of Si_3N_4 (3.18 g/cm^3) which is higher by 22%. However, there is no size change, as stated earlier, since the expansion is compensated by reduction of the internal pores. The product has a final porosity of 15%–20%, and some residual unreacted Si. This porosity causes the strength of the RBSN to be inferior to that produced by hot pressing. However, the process is economically feasible where there is no need for expensive further machining.

Reaction bonded silicon carbide (RBSC) is a good example of the reaction between solid and liquid. Commonly, a mixture of SiC particles (5–10 μm), carbon, and polymeric binder is formed into a green body by pressing, extrusion, or injection molding. The mixture is infiltrated with liquid silicon at a temperature higher than the melting point of silicon (1410°C). The reaction product crystallizes on the original SiC grains and bonds them together. The infiltration and reaction processes are simultaneous. Capillary pressure provides the driving force for infiltration, and good wetting of the surfaces by liquid silicon is a key requirement. The grain size should be tightly controlled, since the strength of the material decreases with larger grain sizes. Commonly, the final product contains an interconnected network of unreacted silicon, but there is no unreacted carbon or pores. The optimum RBSC composition is about 90% volume SiC and 10% Si, but special products with 5%–25% volume Si are sometimes made. The presence of Si leads to a deterioration of the mechanical strength at temperatures above 1200°C.

12.10.4   SOL-GEL PROCESSING

The ceramic manufacturing techniques in which a solution of metal compounds is converted into a solid part are known as *liquid precursor processes* or *chemical solution deposition*. The sol-gel process is the most important of these processes for the production of oxides, and has been highly developed since the mid-1970s. The process is based on converting a solution of metal compounds (alkoxides in an appropriate alcohol) or a suspension of very fine particles in a liquid, referred to as sol, into a highly viscous mass, known as gel, through chemical reactions. The gelled material consists of colloidal particles joined together in the form of a network. Gels are then dried through slow evaporation, which is associated with considerable shrinkage. *Drying* can take a very long time, in the range of days, for a gel with a thickness of few centimeters. Dried gel is subjected to *densification* at temperature higher than 500°C (this temperature is much lower than what would be required to make the same material through sintering).

The sol-gel process offers several advantages over other processes. Because of the ease of purification of the starting liquid material (Sol), parts with high purity can be produced. Materials with exceptionally high chemical stability can be produced, since the mixing of constituents takes place at a molecular level during the chemical reactions. Another advantage is the relatively lower densification temperature. Disadvantages of the process include the high cost of the starting materials (Alkoxides are quite expensive), and shrinkage (and possibility of warping and cracking) during drying. Therefore, the process is limited to the fabrication of thin parts and films.

12.10.5 Polymer Pyrolysis

This process refers to the pyrolytic (transformation by the action of heat) decomposition of metal organic polymeric compounds to produce ceramics. The polymers used for this purpose are commonly known as *preceramic polymers*, since they form the *precursors* to ceramics. The chain backbone of preceramic polymers contains elements such as Si, B, and N in addition to or other than carbon (Chapter 13). Pyrolysis of the polymer results in a ceramic containing some of the elements present in the chain. The process is related to the sol-gel process described in the previous section, where metal-organic polymer gel is synthesized and converted to an oxide ceramic.

The characteristics of the ceramic product formed by polymer pyrolysis depend on the structure and composition of the polymer, and on the pyrolysis parameters. The effectiveness of the polymer is determined by its processing characteristics, the percentage of the ceramic yield (by weight) on pyrolysis, the microstructure and purity of the ceramic, and the cost of the process. Examples of the types of polymers and the resulting ceramics are given in Table 12.5.

The table shows that the process is applicable mostly for the production of nonoxide ceramics such as silicon carbide, silicon nitride, and boron carbides and nitrides. The main advantages of the process are the ease of processing into the desired shape, and the lower required temperature. However, pyrolysis is associated with a large change of volume (the decrease in volume may be up to 75%), which puts a limit to thin parts (in particular fibers) and coatings.

TABLE 12.5
Some Precursor Polymers and the Ceramics Produced by Their Pyrolysis

Polymer Precursor	Ceramic Yield (% wt.)	Atmosphere	Ceramic Product
Polycarbosilanes	55–60	N_2/vacuum	SiC–amorphous SiCxOy
Polymethylsilanes	~80	Ar	SiC
Polysilazanes	80–85	N_2	Amorphous SiCxNy
	60–65	N_2	Amorphous Si_3N_4–C
	~75	N_2	Amorphous SiCxNy
Polysiladiazanes	70–80	Ar	Amorphous Si_3N_4–SiCxNy
Polyborasilazanes	~90	Ar	Amorphous BSixCyNz
	~75	NH_3	Amorphous BSixNy

Source: Rahaman, M. N., *Ceramics Processing and Sintering*, 2nd ed., Marcel Dekker, 2003. With permission.

12.10.6 Melt Casting

Melt casting is one of two techniques for the production of a required ceramic part from an assemblage of fine powders by the action of heat. In its simplest form, the process involves melting the powder, followed by forming it into shape by one of several techniques including: Casting, rolling, pressing, blowing, and spinning. The problem with applying heat to ceramic powders is that they either have a high melting point (such as zirconium oxide, with a melting point about 2600°C), or they decompose before melting (such as silicon nitride). Therefore, melt casting is limited to glasses.

An important variation of glass processing is the *glass ceramic* process, which is used for producing the glass ceramics introduced in the last section. The glass in the amorphous (glassy) state is first obtained by melting and forming to shape by one of the processes outlined above. Then the glass is crystallized using a heat treatment process on two steps. The first step is to heat and hold the glass at a relatively lower temperature to induce nucleation of crystals. The temperature is then raised to one or more higher temperatures to promote growth of crystals throughout the glass.

12.10.7 Ceramic Processing Using Powder Metallurgy

All powder metallurgy processes described in the main part of this chapter are used to manufacture polycrystalline ceramics. These include blending and mixing, powder compaction, and sintering techniques. Alternatively, sintering is performed directly on masses of ceramic (monoliths) produced by one of the processes described in this section to sustain high purity or high homogeneity, or tight control of the structure characteristics. In the processing of ceramics, the term *firing* may be used as an alternative to sintering, since this is the term commonly used in producing traditional ceramics.

12.11 REVIEW QUESTIONS

1. What are the main advantages that distinguish powder metallurgy from other manufacturing processes?
2. What are the limitations of powder metallurgy?
3. Discuss the historical development of powder metallurgy as a means of development of new industrial applications that would have been impossible without this technique.
4. Classify the methods of production of metal powders.
5. Compare the different methods of atomization for metal powders, considering the type of material and the quality of particles.
6. What is the difference between comminution and mechanical alloying as a means of crushing of metal powders?
7. Compare reduction and precipitation as chemical techniques for metal powder production.
8. How can iron and nickel powders be produced using thermal decomposition?
9. How can electrolysis be utilized in the production of metal powders, and what are the forms of deposition of the powder on the cathode?
10. Differentiate between the aspect ratio and the shape factor as measures of the size of particles.
11. What are the possible types of surface films on the particles? What is the effect of this film and how can it be eliminated?
12. Distinguish between the terms "Blending" and "Mixing" in powder metallurgy.
13. What are the types of mixers used in blending powders, and what are the parameters that improve the mixing process?
14. What are the ingredients commonly added during mixing and blending of powders?

15. Compare cold isostatic pressing with hot isostatic pressing, considering the pressure medium, product type and shape, as well as applications.
16. Discuss the mechanisms of sintering, and how do they affect the properties of products?
17. What are the secondary operations that can be applied to control the geometric characteristics and properties of products?
18. What are the categories of traditional ceramics, and what are the applications of each?
19. What are the main constituents of glass, and what are the effects of additives?
20. Why are glass ceramics different from glass, and what are their applications?
21. What are the main characteristics and applications of each of the following advanced ceramics: Oxides–carbides–nitrides–borides?

BIBLIOGRAPHY

ASM International. 1998. *Metals Handbook, vol. 7: Powder Metal Technologies and Applications.* Materials Park, OH.

Degarmo, E. P., Black, J. T., and Kosher, R. A. 1997. *Materials and Processes in Manufacturing,* 8th ed. Upper Saddle River, NJ: Prentice Hall.

Fujiki, A. 2001. Present and future prospects of powder metallurgy parts for automotive industry. *Materials Chemistry and Physics,* 67: 298–306.

German, R. M. 1998. *Powder Metallurgy of Iron and Steel.* New York: John Wiley & Sons.

Groover, M. P. 2007. *Fundamentals of Modern Manufacturing.* New York: John Wiley & Sons.

Handbook of Non-Ferrous Metal Powders Technologies and Applications, 2008. Elsevier.

Höganäs, *Iron and Steel Powders for Sintered Components.* Sweden.

Kalpakjian, S., and Schmid, S. R. 2003. *Manufacturing Processes for Engineering Material,* 4th ed. Pearson Education.

Kubicki, B. 1995. *Sintered Machine Elements.* Ellis Horwood.

Narasimhan, K. S. 2001. Sintering of powder mixtures and the growth of ferrous powder metallurgy. *Materials Chemistry and Physics,* 67: 56–65.

Rahaman, M. N. 2003. *Ceramics Processing and Sintering,* 2nd ed. Marcel Dekker.

Schey, J. 1987. *Introduction to Manufacturing Processes,* 2nd ed. New York: McGraw-Hill.

Somiya, S. et al. 2003. *Handbook of Advanced Ceramics, Volume II: Processing and Applications.* Elsevier.

13 Polymeric Materials and Their Processing

13.1 INTRODUCTION

Polymers are molecular materials in which individual molecules or units are chemically bonded in a chain-like structure. The word *polymer* comes from the Greek roots *poly* (many) and *meros* (parts); these parts are known as monomers. Monomers are made up of organic materials in which atoms of carbon are joined in covalent bonds with other atoms of hydrogen, oxygen, nitrogen, silicon, chlorine, fluorine, or sulfur. Natural polymers such as silk, shellac, wood, rubber, and cellulose were available to the ancient man. However, synthetic or semisynthetic polymers, manufactured as industrial products, are recently developed materials. The term "plastics" is a general common name given to most synthetically developed polymers that may contain other constituents to improve performance and/or reduce product cost. Again, "plastic" is derived from the Greek *plastikos*, which means, *fit for molding*. It refers to the malleability or plasticity of these polymers during manufacturing that allows them to be cast, pressed, or extruded into a variety of shapes, such as films, fibers, plates, tubes, bottles, boxes, and much more.

Polymeric materials in the form of plastics, rubbers, and fibers have for many years played an essential role in everyday life as electrical insulators, tires, packaging facilities, food and beverage containers, textiles, etc. There is an ever increasing demand for replacing metallic materials in machinery and equipment with plastics for improving strength-to-weight ratio, safety and appearance, and for noise and cost reduction. This tremendous development of polymeric products has led to the presently accepted reality that the total volume of produced plastics now exceeds that of metals, and the rate of expansion is still growing faster.

Recently, most monomers are obtained by fractional distillation of petroleum, or natural gas (the petrochemical industry). These monomers are linked to form polymers through chemical reactions known as polymerization. The resulting polymers can be categorized into three main categories according to their properties and applications. These are known as thermoplastics, thermosets, and elastomers, as shown in Figure 13.1. Details of these categories will be presented in the following sections.

13.2 HISTORICAL DEVELOPMENT OF POLYMERIC MATERIALS

The development of polymeric materials and their manufacturing processes commenced only about 170 years ago. One of the earliest stages was the discovery of the vulcanizing effect of sulfur on natural rubber by Goodyear and Hancock in 1839. The first man-made plastic, a form of cellulose nitrate, was exhibited at the Great International Exhibition in London in 1862. It was described as a replacement for ivory and tortoise shell. Further research on cellulose nitrate by the American Hyatt led to the development of celluloid, which was patented in 1870 (he added camphor as a plasticizer under heat and pressure effects). This celluloid was later exploited in developing photographic films and windshields for carriages and early cars. Cellulose fibers were first manufactured around 1890, and it was named Rayon. Cellophane and cellulose acetate were developed around 1910. Cellophane was used in packaging, and cellulose acetate has been used as a base for photographic films.

Another important development in the early 1900s was Bakelite, made by polymerization of phenol and formaldehyde, by American Leo Baekeland. This further led to the development of

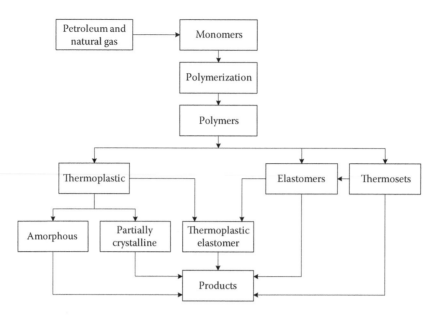

FIGURE 13.1 Classification of polymeric materials.

urea-formaldehyde in 1918, and melamine-formaldehyde in 1939, which was used in houseware products. Two great developments of the 1930s were nylon and polyethylene. Nylon was first used on a commercial level in 1939 in ladies' garments, but by 1941 it was used in self-lubricating bearings, wire insulation, surgical sutures, and catheters. Similarly, the British company, Imperial Chemical Industries (ICI), contributed to the synthesis of polyethylene, which was applied in insulating submarine and radar cables.

Several important developments in the 1940s included fluorocarbons (Teflon), silicones, polyurethanes, epoxy resins, and acrylonitrile-butadiene-styrene (ABS). In 1950, polyester fibers were developed, and later in the 1950s, polypropylene, polycarbonates, and high-density polyethylene followed. Up to the present moment, tremendous growth has been achieved in the field, regarding new types and applications of polymeric materials.

13.3 POLYMERIZATION

13.3.1 POLYMERIZATION REACTIONS

There are many different polymerization processes to change the basic chemicals (monomers) into polymeric (plastic) materials in the form of granules, pellets, powders, or liquids. Details of the chemical reactions in these polymerization processes are beyond the scope of this book, but a brief discussion of their basic procedures can be presented. Polymerization processes can be generally grouped into two techniques, namely chain polymerization and step reaction polymerization.

1. Chain (Addition) polymerization, in which the double bonds between carbon atoms in the monomers are induced to open so that they join with other monomer molecules to form a chain-like structure. Polymerization of ethylene molecules into polyethylene (PE) is a typical example of this process. Ethylene has a chemical composition C_2H_4, where molecules are arranged in the structural form shown in Figure 13.2a, representing the monomer. Adding a chemical catalyst (known as initiator) to ethylene during polymerization opens or breaks the double bond between the two carbon atoms, making the monomer highly reactive (Figure 13.2b), with unpaired electrons. These open ends capture other monomers

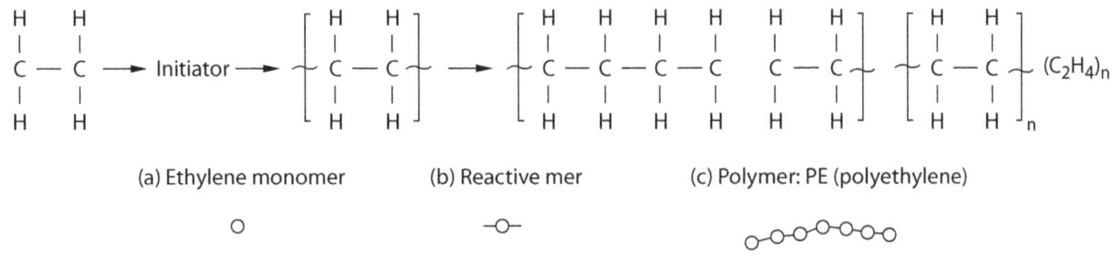

FIGURE 13.2　Chain polymerization of ethylene to polyethylene.

to form a chain that propagates by capturing further monomers successively until a large molecule (macromolecule) is formed and the reaction is terminated, as demonstrated in Figure 13.2c. The name "addition polymerization" represents the successive addition of monomers to the active ends in a chain reaction. Termination of the reaction occurs when the active ends of two rapidly growing chains collide, or by their collision with a terminator radical. The rate of growth of chain reaction is high (reaction rates of 10^4/sec have been recorded), but practically, it may take several minutes to complete the polymerization of a given batch, since all of the chain reactions in the mixture are not simultaneous. This method is applied for producing similar other chain-type polymers such as polypropylene (PP), polyvinyl chloride (PVC), polystyrene (PS), and polytetrafluoroethylene (PTFE, or Teflon), as shown in Figure 13.3. It should be noticed that the monomer for all of these

Polymer	Monomer	Polymer structure	Chemical formula
Polypropylene	H H | | C = C | | H CH3	H H | | C — C | | H CH3]n	(C_3H_6)
Polyvinyl chloride	H H | | C = C | | H Cl	H H | | C — C | | H Cl]n	$(C_2H_3Cl)_n$
Polyvinyl fluoride	H H | | C = C | | H F	H H | | C — C | | H F]n	$(C_2H_3F)_n$
Polystyrene	H H | | C = C | | H C6H5	H H | | C — C | | H C6H5]n	$(C_8H_8)_n$
Polytetrafluoroethylene (PTFE)	F F | | C = C | | F F	F F | | C — C | | F F]n	$(C_2F_4)_n$

FIGURE 13.3　Chain reaction polymers.

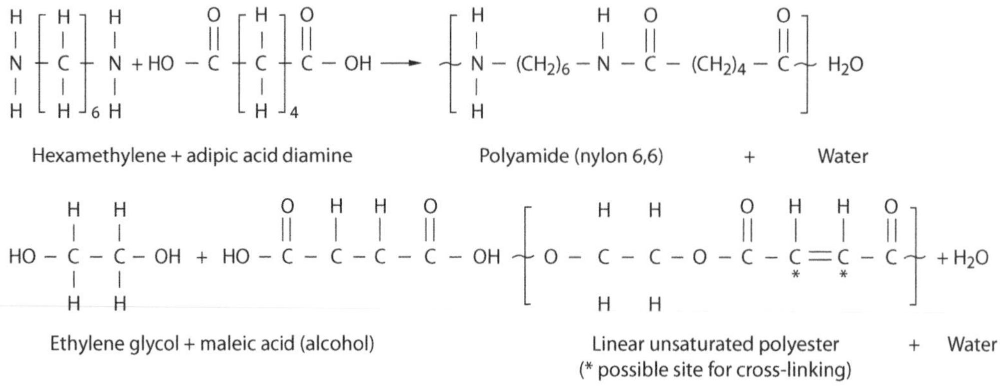

Hexamethylene + adipic acid diamine   Polyamide (nylon 6,6)   +   Water

Ethylene glycol + maleic acid (alcohol)   Linear unsaturated polyester   +   Water
                                                 (* possible site for cross-linking)

FIGURE 13.4 Examples of step reaction.

materials is similar to that of ethylene, with one or all of the hydrogen atoms replaced by another atom or molecule. Most of the polymers produced using chain polymerization are thermoplastics.

2. Step reaction (Condensation) polymerization, in which two dissimilar monomers are joined into a short group representing a new monomer. This new monomer reacts with similar ones, forming a chain or cross-linked structure in a stepwise procedure. A byproduct such as water is released and condensed. This is why the process is called condensation polymerization. Typical examples of a step reaction polymer is polyamide (Nylon-6,6), and polyester, as shown in Figure 13.4. Both thermoplastic and thermosetting polymers are produced by step reaction.

13.3.2 DEGREE OF POLYMERIZATION AND MOLECULAR WEIGHT

In both types of polymerization, the average length of the molecules can be controlled by terminating the reaction. Since the polymer chains produced in a structure are not all of equal length, the average length is determined statistically. The number of mers in the average molecule, which determines the length of the chain, is known as the *Degree of Polymerization* (DP). This DP represents the number n on the polymer structure, shown in Figures 13.2 and 13.3. Higher DP of a polymer increases its mechanical strength, but also increases the viscosity in the liquid state, which slows down subsequent forming processes. For low-density polyethylene (LDPE), DP = 700 units, while for ultra high-molecular weight-polyethylene (UHMWPE), DP = 170,000.

The *Molecular Weight (MW)* of a polymer is the sum of the molecular weights of the mers in the molecule (it represents the average weight in grams × Avogadro's number (6.02×10^{23} molecule)). For very low molecular weights, the strength is too low for the polymer material to have any useful commercial applications. On the other hand, at too high molecular weight, the strength reaches saturation level (to infinite molecular weight), beyond which there will be no further improvement of the property, as shown in Figure 13.5. Therefore, most commercial polymers have molecular weights between 10×10^3 and 10×10^6.

13.3.3 FORMS OF POLYMER CHAINS

a. Linear polymers: The structure of the polymers presented in Figures 13.2 and 13.3 is of the chain type. This is the simplest form of structure, known as linear polymer, which is the characteristic structure of thermoplastic polymers. The macromolecular chains do not have to be straight, but they can become randomly coiled or twisted, as shown in Figure 13.6a, and the polymer is amorphous.

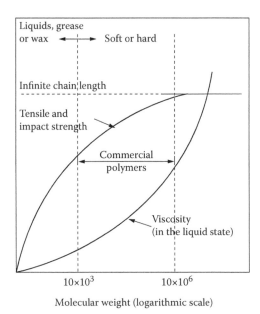

FIGURE 13.5   Effect of molecular weight on the strength and viscosity of polymers.

b. Branched polymers: Side chains are branched from the main chain during the polymerization reaction, as in Figure 13.6b. These branched chains restrict packing of the molecules,
   leading to lower density and higher deformation resistance compared to linear chains.
c. Cross-linked polymers: When two branched structures are linked by covalent bonds, they
   form a cross-linked structure, as indicated in Figure 13.6c. Lightly cross-linked structures
   are characteristic of vulcanized elastomers, which are elastic and resilient.
d. Network structure: When the polymer is highly cross-linked in three dimensions, it is
   called a network structure, as shown in Figure 13.6d. Thermosetting plastics have network
   structures with high cross-linking (network). They are commonly hard and brittle. Typical

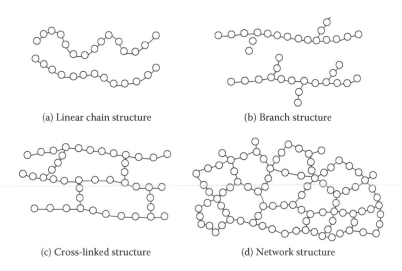

(a) Linear chain structure   (b) Branch structure

(c) Cross-linked structure   (d) Network structure

FIGURE 13.6   Types of polymer chains.

examples are epoxies, phenolics, and silicons. Cross-linking inhibits the reversibility of the polymerization reaction, which makes the polymer chemically set; i.e., when heated, it tends to degrade and burn, rather than to melt.

13.3.4 COPOLYMERS

Though the structure may vary, all previously discussed polymers are based on one kind of repeating unit, known as a homopolymer. For obtaining improved specific properties of polymeric materials, two different types of polymers can be polymerized to form a new copolymer (or binary copolymer), somewhat analogous to solid solution alloying. A typical example is the styrene-butadiene copolymer, which is widely used in automotive tires. Similarly, three component polymers can be made, which are known as *ternary polymers* or *terpolymers*. A good example is the ABS plastic (acrylonitrile-butadiene-styrene), which is used for helmets and refrigerator liners. A further possibility analogous to two-phase metal alloys is known as *polymer alloys* or *polymer blends*. These have two incompatible polymers (which do not enter into a joint chain) mixed together, with one of them acting as the matrix. This provides utilization of the favorable properties of different polymers. An example of polymer blends is to mix brittle polymers with small quantities of elastomers that scatter throughout the polymer, leading to higher toughness (such as rubber-modified polymers used in car bumpers).

There are four different arrangements of the constituting mers in binary copolymers, as demonstrated in Figure 13.7. If the constituents are given the general symbols A and B, these arrangements are

a. Alternating copolymers, where the mers are arranged in order (A-B-A-B-A-B). These are shown as alternate white and black spheres, as shown in Figure 13.7a.
b. Random copolymers, where the mers are in random order (A-A-A-B-A-B-B-A), according to the relative proportions of the mers, as in Figure 13.7b.
c. Block copolymers, in which mers of the same type group into blocks of random length alternately (A-A-A-A-B-B-B-B-B-A-A-A-A-B-B), as in Figure 13.7c. The frequency depends on the proportions of the constituents.
d. Graft copolymers, where a long chain is formed from one type (A-A-A-......), and short branches of the other type (B-B-B), emerge at different locations, as shown in Figure 13.7d.

13.3.5 CRYSTALLINITY IN POLYMERS

The chain-type molecules of polymeric material support an amorphous (glassy) structure. However, in some polymers, very long chains may fold in a repeated regular pattern, forming a crystalline region, as in Figure 13.8. These crystalline regions are called *crystallites*, and they change the material characteristics by increasing density, stiffness, strength, hardness, toughness, heat resistance,

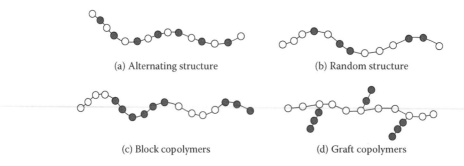

(a) Alternating structure   (b) Random structure

(c) Block copolymers   (d) Graft copolymers

FIGURE 13.7   Various structures of copolymers.

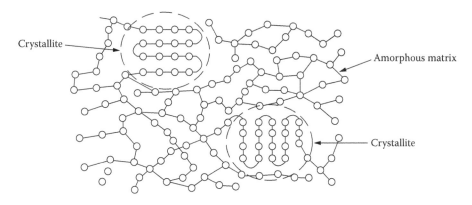

FIGURE 13.8   Partial crystallinity in thermoplastic polymers.

and solvent resistance, while reducing ductility. Polymers, which are transparent in the amorphous state, change to opaque when partially crystalline. The proportion of crystallites in the mass is known as the *Degree of Crystallinity*, which is always less than 100%; i.e., there is no perfect crystallinity in polymers, as in metals and ceramics. A partially crystalline polymer is considered as a two-phase system, with crystallites scattered in the amorphous matrix. Crystallinity may be imparted during polymerization or by deformation during subsequent processing.

Factors affecting the ability of a polymer to crystallize

a. Complexity of the chain: Crystallization is easier for simple linear polymers such as polyethylene, and difficult for complex chains with large side groups, branched, or cross-linked chains. Copolymers do not crystallize.
b. Cooling rate: Slow cooling allows more time for the chains to form crystallites, and allows their growth.
c. Annealing: Heating the polymer to just below the melting temperature can allow chains to align and form crystallites.
d. Deformation: Hot elongation of a thermoplastic tends to align the structure and form crystallites. Pronounced directionality of the structure results in anisotropy of properties, with higher strength in the length direction of molecules.
e. Degree of polymerization: It is more difficult to crystallize longer chains.
f. Plasticizers: These are chemicals added to the polymer for softening. They reduce the degree of crystallinity.

13.3.6   Glass-Transition Temperature and Melting Point of Polymers

At low temperatures, amorphous polymers are rigid, hard, and glassy. When heated, molecules are excited, exhibiting slow rate of increase in the specific volume up to a certain temperature known as the glass transition temperature T_g (similar to the behavior of glass when heated), as shown in Figure 13.9. At this temperature, the material changes to a rubbery state, where movement of the molecules is higher (unfolding and untangling linear chains), exhibiting a viscous as well as elastic (viscoelastic) behavior. Beyond T_g, the rate of increase of the specific volume is higher (steeper). With further heating, the material changes gradually to a liquid, without showing a clear melting point T_m, or a change of the rate of increase of the specific volume in the liquid phase, as shown in the figure. The figure shows that faster rates of cooling from the melt leads to higher glass-transition temperature.

Crystalline and partially crystalline polymers, on the other hand, exhibit relatively sharp decrease of the specific volume at a distinctive temperature (or a narrow temperature range when cooled from the melt to the viscoelastic condition). This temperature is indicative of a melting point similar to

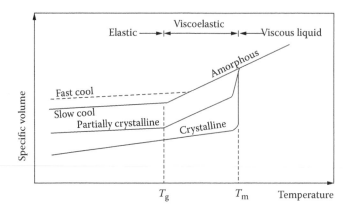

FIGURE 13.9 Change of characteristics of thermoplastic polymers with temperature.

metallic materials. It can be seen in the figure that partially crystalline materials still show a glass-transition temperature.

13.4 THERMOPLASTIC POLYMERS (THERMOPLASTICS TP)

Thermoplastics are solid materials at room temperature, but when heated above the glass-transition temperature T_g (or melting point T_m), they deform in a highly viscous manner, which allows them to be easily shaped in a mold or die into products. The process is reversible on subsequent heating and cooling; i.e., they can be reshaped without significant degradation. Therefore, it is customary to form these polymers into an intermediate shape, such as pellets, bars, tubes, sheets, or films, and ship them to other manufacturers to form them into final products.

13.4.1 THERMOMECHANICAL PROPERTIES OF THERMOPLASTICS

The defining property of the possibility of recycled heating and cooling of a thermoplastic is attributed to its structure, being amorphous with linear and/or branched macromolecules that do not cross-link when heated. However, some thermoplastics degrade gradually when subjected to elevated temperature several times. This long-term effect is known as *thermal aging*. The susceptibility of thermal aging differs from one TP to another, and the rate of aging depends on temperature. This is why high-quality products compel using new or virgin material in plastic molding, while others allow mixing a percentage of previously molded material (sprues or scrap) to the virgin material.

Mechanical properties of thermoplastics are highly affected by temperature. In the range from room temperature, up to the glass-transition temperature T_g, the material is rigid and glassy due to limited chain mobility. When deformed under load, it follows a purely elastic behavior; i.e., it returns to its original dimensions when the load is removed. The polymer behaves like a spring, similar to elastic deformation of metallic materials, as shown in Figure 13.10b, controlled by the relation:

$$\tau = G\gamma, \quad \text{where:} \quad G = \frac{E}{2(1+\nu)} \tag{13.1}$$

Where τ is the shear stress, γ is the shear strain, E is the modulus of elasticity, and ν is the Poisson's ratio, which ranges from 0.25 for stiff polymers to 0.5 for flexible ones. When the stress exceeds a critical value (the tensile stress), the thermoplastic breaks in a brittle behavior (glassy elastic behavior). Such types of polymers are used as engineering construction materials, but they lack toughness. In this temperature range, thermoplastics can only be processed by machining.

Above the melting point T_m, the material changes into a viscous fluid, exhibiting Newtonian viscous flow when loading. When the shear load is removed, the material remains in its deformed shape, which can be modeled by a damper or a dash pot, as shown in Figure 13.10a. Viscosity increases with increasing the molecular weight of the polymer.

In the temperature range between T_g and T_m, amorphous polymers exhibit rubbery (viscoelastic) flow. The increased mobility of molecules allows some deformation, in addition to relative sliding of molecules against each other. Therefore, when the load is removed, part of the strain is recovered as elastic deformation and part may remain as viscous flow. There are two models representing the viscoelastic flow. The most simple is the Maxwell element, based on a spring and a damper in series, where a linear flow pattern is assumed, as shown in Figure 13.10c, followed by regain of the elastic deformation when the load is removed. The alternative model is based on the Voigt element, represented by a spring and damper in parallel, where the elastic recovery is gradual, as shown in Figure 13.10d. Most polymers can be presented by either or both models. Thermoplastics are commonly processed in the viscoelastic region, between T_g and T_m. They should be cooled well below T_g prior to release from the mold, so that the new molecular arrangement is solid. This also implies that a thermoplastic product must never be allowed to heat above T_g during service. Otherwise it will be distorted when cooled due to elastic recovery.

Tougher thermoplastics, such as ABS and nylon, can exhibit large total elongation under tensile loading. Even though a neck may appear after limited uniform elongation, the neck resists fracture because of the high strain rate sensitivity of polymers and alignment of molecules, which strengthens the material, as shown in Figure 13.11. Total elongation of several hundreds or even thousands percent may be attained before fracture. This allows thermoforming of thermoplastics into complex shapes of thin sections. Alignment of the molecules in the elongation direction leads to material anisotropy. The material becomes stronger and stiffer in the stretching direction than in the transverse direction.

In thermoplastic polymers, toughness is significant only below T_g. It can be increased with moderate crystallinity, and in block copolymers. Many thermoplastics such as high-impact polyethylene show rise in toughness with temperature. Some thermoplastics such as polymethylmethacrylate (PMMA) and polystyrene are low impact-resistant, and they do not exhibit any rise in toughness with temperature. The great temperature sensitivity of polymers make creep deformation significant.

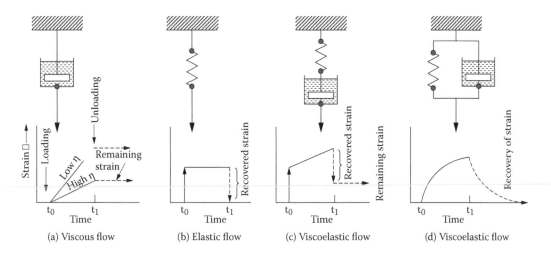

FIGURE 13.10   The different modes of deformation for polymers at different temperature ranges. (From Schey, J. A., *Introduction to Manufacturing Processes*, McGraw-Hill, New York, 1987. With permission.)

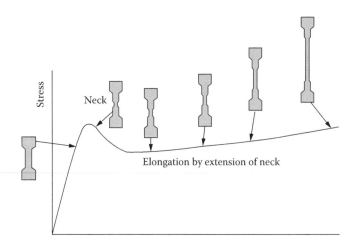

FIGURE 13.11 Tensile stress–strain curve for a tough thermoplastic material showing large elongation by extension of neck.

13.4.2 Major Families of Thermoplastics

The four major thermoplastics, polyethylene, polypropylene, polyvinylchloride, and polystyrene, together represent over 85% by volume of world plastics consumption. They are known in the field as the "big four" commodity thermoplastics. Because of their relatively lower prices, they dominate the market, and in any material selection design procedure, they are usually the first to be considered before turning to the more expensive thermoplastics. These four groups will be presented, followed by other families of thermoplastics that find vital engineering applications in descending order, to provide a comprehensive view of thermoplastics. Detailed properties of the presented TPs are given in Table 13.1.

1. Polyethylene (PE) was first synthesized in the 1930s, and it has now become the most widely produced and used plastic because of its low cost, easy processing, and being chemically inert. PE is classified into several categories based mostly on its density and chain branching. The most common grades are
   - Low-density polyethylene (LDPE): It has a high degree of short and long chain branching, with no or limited crystallinity and lower density. It is characterized by lower strength and higher ductility. LDPE is used for rigid containers, plastic bags, film wrap, and wire insulation.
   - High-density polyethylene (HDPE): has a more linear structure, with higher crystallinity and density. It has higher strength, stiffness, and higher melting point compared to LDPE. It is used to produce milk jugs, detergent bottles, garbage containers, and water pipes.
   - Ultra-high molecular weight polyethylene (UHMWPE): It has much higher molecular weight than other grades, as the name implies, which gives it an outstanding toughness and excellent wear and chemical resistance. It has the highest impact strength of all thermoplastics. However, it has limited crystallinity and low density. These characteristics allow using UHMWPE in a diverse range of applications, including moving parts on weaving machines, bearings, gears, chopping boards, and parts of implants for hip and knee replacement joints.
2. Polypropylene (PP) has the lowest specific gravity of plastics, but its strength-to-weight ratio is high. It has become a major plastic, especially for injection molding. It is compatible to HDPE because of many common properties and equal cost. However, its higher

TABLE 13.1

Structure and Properties of the Major Families of Thermoplastics

Thermoplastic	Abbreviation	Composition	Tensile Strength (MPa)	Modulus of Elasticity (MPa)	Elongation (%)	Specific Gravity	T_g (°C)	T_m (°C)
Polyethylene	PE							
*Low density	LDPE	$(C_2H_4)_n$	9.6–17.2	140	100%–500%	0.91–0.94	−100	105–115
*High density	HDPE		30	700	20%–100%	≥0.941	−115	135
*Ultrahigh molecular weight	UHMWPE		40		≥350	0.93–0.935	—	—
Polypropylene	PP		35	1400	10–500	0.9	−20	176
Polyvinylchloride	PVC	$(C_2H_3Cl)_n$	50–80	2900–3300	2 (rigid form)	1.39	82	212
Polystyrene	PS	$(C_8H_8)_n$	50	3000–3600	3–4	1.05	95	240
Acrylonitrile-butadiene-styrene	ABS	$[(C_3H_3N)_x.(C_4H_6)_y.(C_8H_8)_z]$	50	2100	10–30	1.06	105	200
Polymethyl-methacrylate	PMMA	$(C_5H_8O_2)_n$	55	2800	5	1.15–1.19	105	200
Polyamide: nylon-6,6	PA 6,6	$[(CH_2)_6(CONH)_2 (CH_2)_4]$	70	700	300	1.14	50	260
Polyethylene-terephthalate	PET	$(C_2H_4–C_8H_4O_4)_n$	55	2300	200	1.3	70	265
Polycarbonate	PC	$[C_3H_6(C_6H_4)_2CO_3]_n$	65	2500	110	1.2–1.22	150	267
Cellulose acetate	CA	$(C_6H_9O_5–COCH_3)_n$	30	2800	10–50	1.3	105	306
Polyoxymethylene	POM	$(CH_2O)_n$	70	2500	25–75	1.42	−80	175
Polytetrafluoroethylene	PTFE	$(C_2F_4)_n$	20	425	100–300	2.2	127	327

melting point makes it predominate in high temperature applications such as in bottles that require sterilization and in hydraulic heating pipes. It also has high fatigue and corrosion resistance. Applications include housings, automotive components, houseware parts, ropes, flexible hinges of flip-top bottles, textiles, and fibers for carpeting.

3. Polyvinylchloride (PVC) is the third most widely used plastic, after PE and PP. It can be made softer and more flexible by adding plasticizers to the polymer, which gives it broadly variable properties, and accordingly, a broad range of applications. In the rigid form (with no plasticizer), PVC pipes cover about 50% of the material consumption. These are used in construction, water and sewage systems, irrigation, fittings, window and door frames, etc. In the softened form, PVC is used in flexible hoses, electrical cable insulation, films, sheets, clothing, food packaging, flooring and roofing, inflatable waterbeds and structures, as well as toys. The vinyl chloride monomer has carcinogenic characteristics, and it should be handled with care during polymerization to produce PVC.

4. Polystyrene (PS) is a linear, solid, hard (glassy) thermoplastic at room temperature with limited flexibility. Pure PS is transparent, but can be made to take various colors. It degrades at elevated temperatures, and dissolves in different solvents. Solid PS is used in disposable cutlery, plastic models, CD and DVD casings, and smoke detector housings. It is usually manufactured by extrusion, stamping, and injection molding. Extruded PS is about as strong as unalloyed aluminum, but much more flexible and lighter. PS can also be produced in the form of expanded polystyrene foam (EPS) and extruded polystyrene foam (XPS) (which is given the trade name Styrofoam). These forms have much higher toughness, but transparency and tensile strength are reduced. These foams are used in packaging, thermal insulation in buildings, and architectural modeling. Copolymers based on PS are also available. A well known copolymer of PS with polybutadiene rubber is high-impact polystyrene (HIPS). Common applications of HIPS are in toys and product castings.

    Several other copolymers are also used with styrene, such as the well known engineering plastic *acrylonitrile butadiene styrene (ABS)*. This is a terpolymer with excellent mechanical properties, made of two phases. One is the hard copolymer styrene-acrylonitrile, and the other is the rubbery styrene-butadiene copolymer. The percentage of styrene varies from 40% to 60%. ABS is used in the production of electronic cases, musical instruments, golf club heads, automotive components, appliances, and sewage pipes, but ABS pipes can become brittle with time. For most appliances, it can be used between –25°C to 60°C. PS is also used in some polymer-bonded explosives (PBE), and in napalm.

5. Acrylics can be found in different forms, including chemical compounds, paint, resin, or fibers. They are all derivatives of acrylic acid with other chemicals. Acrylic fibers are made from polyacrylonitrile. They are soft, warm, and have a wool-like feel, which facilitates using them in clothing.

    The most significant thermoplastic in the group is polymethylmethacrylate (PMMA), which is commercially known as "acrylic glass" or "Plexiglas," with excellent transparency, as the name implies. It is competitive with glass in optical appearance, and therefore applied in construction of commercial aquariums, optical instruments, windows of aircrafts, and viewing ports of submarines, motorcycle helmet visors, as well as lenses for automotive exterior lights. PMMA is less than half the density of glass, and its toughness is much higher. The main limitation of PMMA, compared to glass, is its lower scratch resistance. However, scratching may easily be eliminated by polishing, heating, or by scratch resistance coating, which may also add other functions.

6. Polyamides (PA) are thermoplastics made by step-growth polymerization of amide monomers. There are many forms of PA, the most important of which are nylons and aramids. *Nylon* is a silky material, and is therefore used as a synthetic substitute of natural silk, as in parachutes, fabrics, carpets, musical strings, tire cords, and ropes. Solid nylon is used in mechanical parts such as screws and gears, as well as in hair combs. The most common

commercial grades are nylon-6, and nylon-6,6, according to the number of carbon atoms in the monomer. Properties of the latter grade are given in Table 13.1.

Aramids are aromatic polyamides. They are available in the form of strong, heat resistance fibers, which are used in aerospace and military applications, for fabrics, and as an asbestos substitute. *Kevlar* (developed by DuPont in 1965) is a well-known grade of aramids with a high strength equivalent to that of steel at much smaller weight. Typically, it is spun into ropes or fabric sheets. It is applied in bicycle tires, racing sails, and body armor.

7. Polyester is a category of polymers which contain the ester functional group in their main chain. Depending on the chemical structure and cross-linking, polyesters can be thermoplastics or thermosets. However, the most common polyesters are thermoplastics, with a well known example, "polyethylene terephthalate" (PET). It is used in synthetic fibers, photographic films, magnetic recording tapes, and blow-molded containers. Characteristics of PET are shown in Table 13.1. Some of the trade names of PET fabrics include Tergal, Dacron, Diolen, Terylene, and Trevira. Polyester fabrics may exhibit some advantages over natural ones, such as improved wrinkle resistance and low moisture absorption, which make them ideal for "wash and wear" garments, and home furnishings such as bed sheets, table sheets, and curtains. They are also used for tire reinforcement, ropes, conveyor belts, and safety belts. Liquid crystalline polyesters are used in LCDs, filters, dielectric film for capacitors, and film insulation for wires.

8. Polycarbonates (PC) represent a particular group of thermoplastics with high temperature resistance, high toughness, high creep resistance and transparency, which position them among commodity plastics and engineering plastics. Applications include compact discs, DVDs, Blu-ray discs, drinking bottles, lenses for sunglasses and eyeglasses, safety glasses, automotive headlamp lenses, advertisement signs, computer cases, and instrument panels.

9. Cellulose is the structural component of the primary cell for plants, and other living organisms, with the chemical structure $(C_6H_{10}O_5)_n$. Actually, it is the most common organic compound on earth. For industrial use, cellulose is obtained from wood pulp and cotton, and is used to make cardboard, paper, and textiles made of cotton, linen, or any plant fibers. It is also converted into a wide variety of products, such as cellophane and rayon. Cellulose is further used in manufacturing nitrocellulose (smokeless gunpowder), and water-soluble adhesives and binders. The first successful thermoplastic polymer made from cellulose was celluloid in 1872, but the compound was first synthesized without the use of any biological component in 1992. Celluloid is used in photographic and movie films.

   Cellulose itself cannot be used as a thermoplastic because it decomposes with temperature; however, it can form several plastics with various other compounds. A good example is *cellulose acetate (CA)*, the properties of which are given in Table 13.1. It is applied in film bases in photography, as a frame material for eyeglasses, and in cigarette filters.

10. Polyoxymethylene (POM) is also known as acetal, polyacetal or polyformaldehyde. It is an engineering semi-crystalline thermoplastic used in precision parts that require high stiffness, strength, wear resistance, and excellent dimensional stability. However, POM has a low impact strength and relatively low melting point. Applications in the solid form include low-friction wheel bearings, cams, impellers, flutes, and whistles. Main properties of POM are given in Table 13.1.

11. Fluoropolymers is a family of thermoplastic polymer in which atoms of fluorine replace the hydrogen atom in a hydrocarbon. It is characterized by high resistance to acids, solvents and bases, as well as low friction coefficient. It was discovered in 1938. The most commonly used member of the family is *polytetrafluoroethylene (PTFE)*, commercially developed by the manufacturer DuPont under the trade name *Teflon*. PTFE has the second lowest coefficient of friction against any solid. It is used as a non-stick coating for

cookware, as lubricants, as water seals between mating surfaces, and in containers and pipes carrying reactive and corrosive chemicals. PTFE also has excellent dielectric properties, and is therefore used as an insulator in cables and connector assemblies, and as a material for printed circuit boards used at microwave frequencies.

12. Polysulfones are known for their toughness and stability at high temperatures. They have the highest service temperature of thermoplastics. They were introduced in 1965 by Union Carbide. Due to their higher cost, they are used in specialty applications, and are often a superior replacement of polycarbonates. Typical applications are steam irons, coffeemakers and water heaters, medical parts that require sterilization, filtration media, flame retardants, and membranes for gas separation. Polysulfones can be reinforced with glass fibers, resulting in a composite material with double the tensile strength and three times the modulus of elasticity. They are also used as copolymers.

13.5   THERMOSETTING POLYMERS (THERMOSETS)

Thermosetting polymers, known as thermosets, are characterized by a highly cross-linked structure extending in a three-dimensional arrangement (network structure), as in Figure 13.6d. Effectively, the whole part or product is permanently set as one giant macromolecule with strong covalent bonds. This cross-linking (curing) reaction is irreversible; i.e., when reheated, the part tends to degrade and burn without melting. Therefore, thermosets do not exhibit a sharply defined glass-transition temperature, or a melting point. Some precursors, after mixing, will flow and cross-link at room temperature, with the heat required for curing resulting from the exothermic reaction.

13.5.1   General Characteristics of Thermosets

Due to the difference in structure, the general characteristics of thermosets are different from those of thermoplastics. Generally, thermosets are stiffer, harder, and more brittle. Their modulus of elasticity is two or three times higher than thermoplastics. Their strength and hardness are not affected by temperature changes; i.e., they are thermally stable, and can withstand higher service temperature. Further, thermosets are less soluble in common solvents.

13.5.2   Major Families of Thermosets

Owing to the brittleness of thermosets, and probably because of the complicated processing involved in their curing, they are not as widely used as thermoplastics. The most commonly used family of thermosets is that of phenolic resins, with an annual volume of about 6% of the total market of engineering polymers. Amino resins are less consumed, but represent the second family after phenolics. In this section, the major families of thermosets will be presented in descending order of consumption, when possible. Detailed properties of these thermosets are given in Table 13.2.

1. Phenolic resins (Phenolics) represent a class of thermosetting materials where phenol (carbolic acid) reacts with aldehydes (hydroxyl group). The most widely used phenolic resin is *phenolformaldehyde (PF)*, which was known in 1907 under the trade name *Bakelite*. The chemical structure and main properties of PF are given in Table 13.2. Phenolics are used to make molded products, including handles and knobs for kitchenware, telephone casings, panels, and boards. For these purposes, they can be combined with fillers such as wood flour, cellulose fibers, and minerals for strengthening and quality improvement. Phenolics are also used as coatings and adhesives for plywood, printed circuit boards, bonding material for brake linings, and abrasive wheels.

2. Amino resins are formed by polymerization of amines and aldehydes. They are generally used as adhesives and as coatings. They can be molded with fillers similar to phenolics.

TABLE 13.2
Structure and Properties of the Major Families of Thermosets

Thermoset	Abbreviation	Composition	Tensile Strength (MPa)	Modulus of Elasticity (MPa)	Elongation (%)	Specific Gravity
Phenolic resins: Phenol-formaldehyde		Phenol (C_6H_5OH)-Formaldehyde (CH_2O)	70	7000	<1	1.4
Amino resins:						
• Urea-formaldehyde	UF	$[CH_2\text{-}N\text{-}CO\text{-}N\text{-}CH_2]_n$				
• Melamine-formaldehyde	MF	$(C_3H_6N_6)\text{-}(CH_2O)$	50	9000	<1	1.5
Epoxy resins	—	—	70	7000	0	1.1
Polyester resins		$(C_4H_2O_3)+(C_2H_6O_2) + (C_8H_8)$	30	7000	0	1.1
Silicone resins		$(CH_3)_6\text{-}SiO)_n$	30	—	0	1.65

There are two well known thermosets in this family.

- *Urea-formaldehyde (UF)* is made from urea and formaldehyde heated in the presence of a mild base such as ammonia. It is competitive with phenolics as an adhesive for plywood and particle-board (medium-density fiberboard, MDF). UF is also used in molding solid products.
- *Melamine-formaldehyde (MF)* (can be shortened to melamine) is widely used for kitchen utensils (plates and bowls), and as a main constituent in high pressure laminated boards under the trade names Formica and Arborite. It can also be used in laminate flooring, wall panels, and whiteboard.

3. Epoxy resins are thermosetting polymers formed by reaction of an epoxide resin with polyamine (hardener or curing agent). Epoxies are known for their excellent adhesion, chemical and heat resistance, high mechanical properties, and electrical insulating properties. Epoxies have a wide range of applications, including general purpose adhesives, surface coatings, industrial flooring, molds and castings, as well as carbon and glass fiber-reinforced composites. As high performance adhesives, they are used in the construction of aircrafts, automotives, bicycles, boats, and snow boards, for almost all materials where high strength bonds are required. They can be made flexible or rigid, transparent or opaque/colored, fast curing or extremely slow curing. In the electronics industry, epoxies find wide applications as electrical insulators, as protectors against short circuiting, dust and moisture, and in making laminated printed circuit boards.

4. Polyester is a category of polymers which contain the ester functional group in their main chain. Although the most common polyesters are thermoplastics, as presented in Section 13.4, they can also be formed as thermosets, based on unsaturated polyester with relatively low molecular weight. Polyester thermosets are used in manufacturing casting products, and resins are used in making reinforced plastics (composites) to fabricate large pipes, tanks, yachts, and construction panels. One of the main classes of polyester resins is the *alkyd* resin, which is used as a base for paints, varnishes, and lacquers.

5. Silicone resins: Silicon (Si) is the most common metalloid on earth, which is naturally found in the form of silicon oxide (silica), or silicates. It has many industrial uses, in the form of a principal component of semiconductors, glasses, cements, and ceramics. It is also a constituent of silicones, a class of various synthetic polymeric materials made of (C, H, O, and Si), with common characteristics of impermeability to water, flexibility, and resistance

to chemical attack. They can be produced in the form of fluids, elastomers, and thermosetting resins with high cross-linking, which are relevant to this section. Silicon resins are used for paints, varnishes, and other coatings, pressure-sensitive adhesives, laminates of printed circuit boards, molding compounds, and mold-release agents. A typical silicone thermoset is given in Table 13.2.

13.6 ELASTOMERS

The term *elastomer* is derived from "elastic polymer," to represent a family of polymers that exhibit large elastic deformation, which may exceed 500%, under relatively low stresses. The term *rubber* is used as an alternative to elastomer, giving the same significance. Elastomers are amorphous polymers that exist at an ambient temperature above their glass transition temperature, so that they are soft and deformable when stretched, with a small modulus of elasticity, as shown in Figure 13.12. Compared with other families of polymers, the figure indicates that T_g for elastomers is below room temperature, and therefore, its large elastic deformation behavior at room temperature is analogous to viscoelastic flow of thermoplastics and thermosets beyond their glass transition temperatures, which are well above room temperature. Elastomers consist of long-chain molecules that are cross-linked, but the degree of cross-linking is significantly lower than in thermosets. Normally, these long molecules are tightly twisted and curled, but when stretched, they are straightened, as shown in Figure 13.6c. It should be noted that cross-linking implies that the material cannot be reshaped.

Natural rubber (NR) is derived from *latex*, which is a milky colloidal suspension found in the sap of a tropical tree (known as rubber or caoutchouc tree). The purified form of natural rubber is polyisoprene, which has no practical applications, being sticky in hot weather and stiff in cold weather. To form an elastomer with useful properties and applications, natural rubber must be cross-linked by vulcanization. The first vulcanized rubber was discovered by adding sulfur for cross-linking (curing) to make automotive tires. Later, other chemicals are added to sulfur to accelerate curing and achieve higher properties. At present, carbon black is an important additive that reinforces vulcanized natural rubber to improve tensile strength and resistance to tearing and abrasion. Other products of natural rubber include shoe soles, bushings, hoses, belts, seals, and shock-absorbing components. Distinctive additives, besides carbon black, used in natural rubber and synthetic elas-

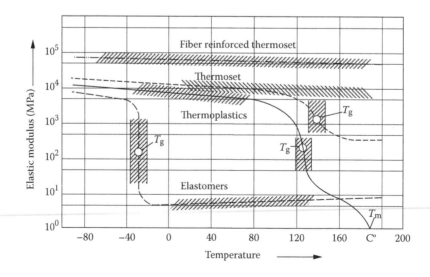

FIGURE 13.12 Modulus of elasticity–temperature relations for different classes of polymers.

tomers include clay, kaolin, silica, talc, and calcium carbonate. Typical characteristics of vulcanized natural rubber, along with those of other synthetic elastomers, are given in Table 13.3.

Synthetic rubbers were developed during the world wars of the last century when NR was difficult to obtain at reasonable cost, and better characteristics were sought. Petroleum was the predominant raw material for these synthetic rubbers. Now, about 80% of elastomers are synthetic. The major types of these synthetic elastomers will be presented in this section in descending order of consumption, when possible.

1. Styrene-butadiene rubber (SBR) is a copolymer consisting of styrene and butadiene at a ratio of about 1:3. Styrene is presented in Section 13.4 and in Table 13.1. Polybutadiene as a synthetic rubber is a very soft, almost liquid material, and is usually mixed with other polymers to form a tough and elastic rubber. SBR is now the most commonly produced elastomer (about 40%), and is mainly used in automotive tires (blended with NR). It has good abrasion resistance, and good aging stability (better than NR) when protected by additives. Other applications include shoe heels and soles, gaskets, chewing gum, sealing and binding agents in buildings, as well as wire and cable insulation.

2. Ethylene-propylene-diene monomer (M-class) rubber (EPDM) is a teropolymer of ethylene (45% to 75%), and propylene with small proportion (2.5%–12% by weight) of a diene monomer for cross-linking. The E, P, and D letters in the abbreviation refer to the first letter of the constituents, while the M refers to its classification in the ASTM standard D-1418. The M class includes rubbers having a saturated chain of polyethylene. EPDM rubbers have outstanding heat, steam, ozone, and weather resistance. They are used in vibrators, seals, roof water and weather proofing, radiators, washers, hoses, tubing, belts, electrical insulation, and speaker cone surrounds.

TABLE 13.3
Properties of Different Types of Elastomers

Elastomer	Abbreviation	Structure	Tensile Strength (MPa)	Modulus of Elasticity (MPa)	Elongation (%)	Specific Gravity
Natural rubber (polyisoprene)	NR	$(C_5H_8)_n$	25	18 (at 300% elongation)	700	0.93
Styrene-butadiene rubber	SBR	$(C_8H_8)_n +$ $(C_4H_6)_n$	20	17 (at 300% elongation)	700	0.94
Butadiene rubber	BR	$(C_4H_6)_n$	15	—	500 at failure	0.93
Ethylene-propylene-diene rubber	EPDM	$(C_2H_4)_n +$ $(C_3H_6)_n +$ Diene monomer for cross-linking	25	—	300 at failure	0.86
Polyurethanes	PU or PUR	Different structures	30	Varies widely	Depends on cross-linking	1.2
Polyisobutyl rubber	PIB	$(C_4H_8)_n +$ $(C_5H_8)_n$	20	7	700	0.92
Chloroprene rubber	CR	$(C_4H_5Cl)_n$	25	7 (at 300% elongation)	500 at failure	1.23
Isoprene rubber	IR	$(C_5H_8)_n$	25	17 (at 300% elongation)	500 at failure	0.93

3. Polyurethanes represent a large family of polymers including the urethane group (NHCOO) in their structure. According to variations in their processing and cross-linking, they can be thermoplastics, thermosets, or elastomers, the last two being the most commonly used. The largest application (about 75%) of polyurethane is in foams, which range from very soft and elastomeric to rigid with highest cross-linking. As flexible foams, polyurethanes are used in upholstery, bedding, and automotive and truck seating. Unfoamed polyurethane elastomers are molded into foot wear, gel pads, print rollers, and car bumpers. In the form of low-density rigid foam, it is used as filler material in hollow construction panels, and as thermal insulators in refrigerators and freezers. Polyurethanes are also used in paints, varnishes, and similar coating materials.

4. Butyl rubber is also known as polyisobutylene (PIB) which is produced by polymerization of about 98% of isobutylene with about 2% isoprene. It has excellent impermeability and good flex properties, which has led to its use in inflatable products such as inner tubes and liners in tubeless tires. Other applications include additives, fiberoptic compounds, sealants, and basketballs.

5. Chloroprene rubber (CR) is a colorless liquid monomer used for producing the elastomer poly-chloroprene, which is commonly known as *Neoprene* (a trade name given by DuPont). It has a good chemical stability and maintains flexibility over a wide temperature range. It has a variety of applications as orthopedic braces (wrist, knee, etc.), electrical insulation, membranes, and automotive fan belts. A foamed neoprene containing gas cells is used as an insulation material such as in wetsuits used by divers, as well as in shock-protection applications.

6. Isoprene rubber can be polymerized by synthesizing a chemical equivalent to natural rubber. Unvulcanized synthetic isoprene is softer and more easily molded than raw NR. Applications of the synthetic rubber are similar to those of natural rubber, but the cost is higher by about 35%.

7. Nitrile-butadiene rubber (NBR) is a copolymer of acrylonitrile (25%–50%) and butadiene (50%–75%). Its properties vary with the nitrile content, showing higher resistance to oils and lower flexibility with more nitrile. NBR is used in making disposable gloves (for cleaning, labs, and examination purposes), as well as fuel and oil handling hoses and seals in automotives. NBR can withstand a temperature range from −40°C to 108°C, which makes it ideal for extreme automotive applications. It is also applicable in molded products, footwear, additives, sealants, sponges, expanded foam, and floor mats.

8. Silicone rubber: Silicones can be thermosets or elastomers according to the degree of cross-linking. Silicone rubber is generally nonreactive, stable, and resistant to extreme environments and temperatures from −55°C to 300°C, while still maintaining its useful properties. It has wide applications in automotive components, food storage products, sportswear, footwear, electronics, prosthetic devices and implants, and in repair tools. One of the most common silicone rubbers is *polydimethylsiloxane (PDMS)*, the characters of which are given in Table 13.3. It is particularly known for its unusual rheological (flow) properties. Its applications range from contact lenses and medical devices to special purpose elastomers, in shampoos, caulking, lubricants, and heat-resistant tiles.

13.7 THERMOPLASTIC ELASTOMERS

This is a class of copolymers or a physical mix of thermoplastics and elastomers that aim to gain advantages typical of both rubbery materials and elastic materials. They derive their elastomeric properties not from chemical cross-linking, but from physical connections between soft and hard phases that make up the material. In order to qualify as a thermoplastic elastomer, a material must possess three essential characteristics:

a. The ability to be stretched to moderate, recoverable elongation
b. The ability to be processed as a melt at elevated temperature
c. Absence of significant creep

The most commonly used TPEs are

- Styrene-Butadiene-Styrene block copolymer (SBS)
- Thermoplastic polyurethanes (TPU)
- Thermoplastic copolyester
- Thermoplastic polyamides

13.8 PROCESSING OF POLYMERIC MATERIALS

It has been found in the previous sections of this chapter that polymeric materials possess wide variations in characteristics; however, their processing to reach a final product involves operations similar to those used to form and shape metals. They can be shaped by molding, casting, and forming, as well as machining and joining. Comparatively, since the temperatures required for melting or curing plastics are lower than metals, their processing is much easier and less energy consuming. The type of the process to be adopted for making plastic products depends on the characteristics and properties of the polymer, and the shape and form of the required product. As stated in previous sections, thermoplastic resins are commonly obtained as a linear polymer, which can be repeatedly molten or solidified by heating or cooling. Heat softens or melts the material so that it can be formed, and then subsequent cooling hardens or solidifies the material to the required shape. Thermosetting materials, on the other hand, are formed by cross-linking or curing, and then the material cannot be softened by heating. Therefore, thermosets are usually supplied as a partially polymerized molding compound or as a liquid monomer-copolymer mixture. They can then be shaped with or without pressure, and polymerized to the cured condition with chemicals or heat. Figure 13.13 provides a layout of the processing methods for polymers, as presented in the following sections.

13.8.1 EXTRUSION

Extrusion is one of the most common forming processes for thermoplastics and elastomers in large volumes. Similar to the extrusion of metals, it involves compression of the material to force it to flow through a die orifice, to get a long, continuous product with cross-sectional shape dictated by the shape of the orifice. Typical products are pipes, tubes, hoses, structural shapes such as windows and door frames, sheets, films, and continuous filaments. Extruders are also used for insulation sheathing of wires and cables, as well as coating of substrates such as cloth, paper, or foil. Production runs are continuous, and the products (*extrudates*) can be cut to the desired lengths.

Screw extruder: The feed supply in the form of pellets or powder is fed by gravity through a hopper to a cylindrical barrel, where it is heated to the molten state, and forced to flow through a die opening by means of a rotating screw, as shown in Figure 13.14. An oversimplified analogy of the extruder is the meat mincer, which is essentially a screw conveyor. The two main parts of the extruder are the cylindrical barrel and the screw.

The internal diameter of the barrel ranges from 38 to 200 mm, and gives a rough guide to the extruder capacity. The barrel length-to-diameter ratio ranges between 15 and 30, with the higher ratios applied for thermoplastics to provide a long route for heat transfer, and the lower ratios applied for elastomers where cross-linking does not take a long time. The barrel is surrounded by external heaters, which may be an oil, steam, electric resistance, or inductance type, with the last two being the most popular. The heat gained from the heater provides a fraction of the heat required to melt the polymer. The rest of the required heat is generated by friction due to the compression and shearing

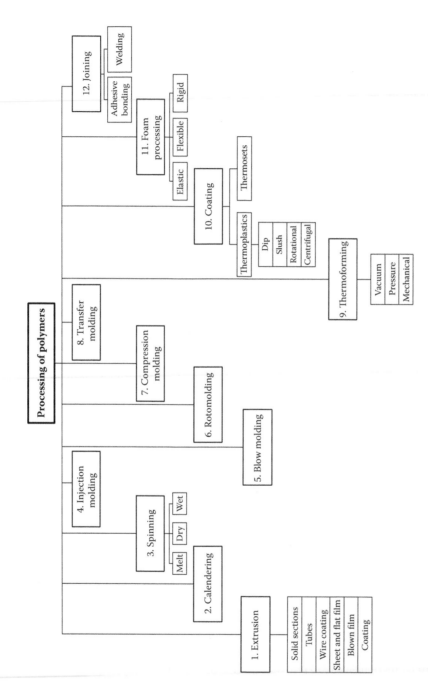

FIGURE 13.13　Layout of the processing methods for polymers.

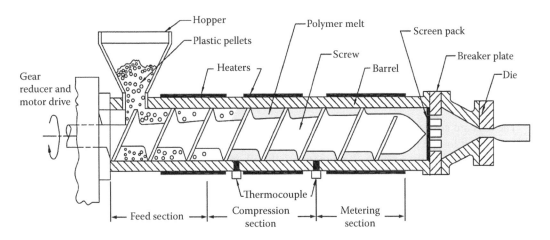

FIGURE 13.14 Main components of a single-screw extruder for polymers. (From Groover, M. P., *Fundamentals of Modern Manufacturing*, 3rd ed., John Wiley & Sons, London, 2007. With permission.)

action of the screw on the polymer. In modern high-speed screw extruders, most of the heat required for steady running is supplied by frictional heat.

The screw is the most essential part of the extruder. Through rotation at a speed of about 60 rpm, it conveys the material through the barrel toward the die orifice. The screw is made of an alloyed steel cylinder with a helical channel cut into it, as shown in Figure 13.14. The generated helical ridge is called the flight (equivalent to thread), which must be hardened to withstand high frictional conditions and resist wear. It has a constant pitch, with a helix angle chosen to optimize the feeding characteristics in the range between 12° to 20°. The flight depth at the hopper end is larger than the flight depth at the die end (the depth ratio is known as the compression ratio) to generate compressive force on the polymer as it moves toward the die. The compression ratios for single screw extruders range from 2 to 5. The flight diameter is smaller than the barrel diameter by a limited clearance of about 0.05 mm to prevent excessive buildup of the polymer on the inside wall, and thus minimize heat transfer. As shown in the figure, the screw is usually divided into three sections according to the function as follows.

a. The feed section picks up the pellets or powder from under the hopper mouth, and conveys them in the solid state to the next section. It is deep flighted to supply large amounts of material, to ensure complete filling of the forward sections.
b. The compression section through which the flight depth is gradually diminishing to generate large enough compression on the melting volume. This large compression forces the trapped air back through the feed section to ensure an extrudate free from porosity. Further, the relative motion of the helical ridges generate shearing action on the melt to produce good mixing, raise the temperature, and increase fluidity of the mix.
c. The metering section aims to force the molten polymer through the die at steady high pressure sufficient to eliminate pulsations.

The screw may be cored for steam heating, or more often, for water cooling, to avoid overheating of the polymer. The screw is motor driven at variable speeds through step-change gear boxes or variable-speed hydraulic systems.

Multiple screw extruders, using more than one screw, are developed for more uniformly blending plasticizers, fillers, pigments, and/or stabilizers into the polymer. In twin screw extruders with intermeshing screws, the material is conveyed forward with very low friction. Heat is therefore controlled through the outside heaters, and is not influenced by the screw speed. This is essential for extruding heat-sensitive polymers, such as PVC, and vinyls in general.

The die zone and its configurations: To prevent any unmolten polymer or entrapped dirt or lumps from entering the die, fine wire screens are placed into the stream just before entering the die. They also increase back pressure to improve mixing and homogenization. These screens are supported by a breaker plate with small axial holes to straighten the flow lines of the liquid polymer, as shown in Figure 13.14.

The dies are not part or component of the extruder. They are special tools that are designed and manufactured for the particular extrudate profile to be produced. They all start with a converging zone, which receives the molten polymer through the screen pack and the breaker plate, and delivers it at higher pressure and speed to the die orifice. When leaving the die, the polymer is still soft, and should be cooled by air, water spray, or by passing through a water bath. The soft polymer molecules partially recoil, leading to an increase of the dimensions of the extrudate, as shown in Figure 13.15a. This phenomenon is known as *die swelling*, which must be taken into account during die design. The longer the length of the die opening (die land), the smaller the die swelling. Generally, polymer extrusion profiles may be categorized to solid profile, hollow profile such as in tubes, wire and cable coating, sheet and film, and extrusion coating.

Solid sections include rounds, squares, polygons, and constructional shapes such as door and window frames, and automotive trims. For profiles other than round, the die orifice must be made slightly different from the desired profile so that the die swell compensates for the difference, producing the required profile. Typical examples for square and T-sections are given in Figure 13.15b. Since different polymers have different viscoelastic parameters, the extent of die swelling and the die shape correction depends on the material to be extruded.

Hollow sections (tubes) need a mandrel to form the hollow shape, similar to metal extrusion of tubes. There are several designs for such dies, one of which is demonstrated in Figure 13.16. The mandrel is held through three spider legs, as shown in Section A-A of the figure. The molten polymer flows in three separate tunnels, surrounding the legs, which reunite with forward movement into a complete tubular form. This process is analogous to the porthole die extrusion presented in Section 9.5.1.4. The mandrel is often provided with an air channel, through which an air stream is forced to keep control of the internal diameter of the tube during hardening of the thermoplastic polymer. Tube extrudates also require air or water cooling to the outside diameter, similar to solid sections. It should be noticed that tubes are also subject to die swelling, which should be accounted for to get the required dimensions. The extrusion of rubber tubing does not require external and

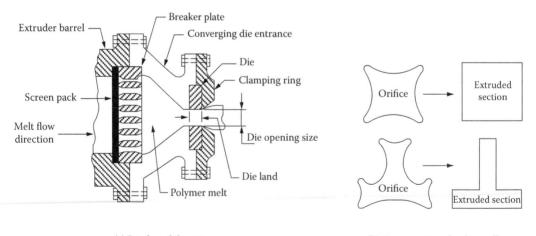

(a) Die for solid sections   (b) Compensation for die swelling

FIGURE 13.15   (a) Die swelling and (b) compensation for different configurations.

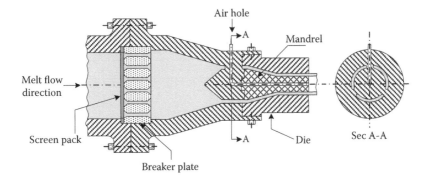

FIGURE 13.16   Cross-section of an extrusion die for producing tubes.

internal cooling precautions similar to thermoplastic tubes, since rubber gains dimensional stability due to the cross-linking reaction.

Wire and cable coating in continuous lengths with insulating plastics is a principal application, with large annual consumption. The core may be a single straight metal wire or strand, multiple wires, or a bundle of previously individually coated wires. A typical extrusion die for the purpose is shown in Figure 13.17, where the wire is pulled at high speed through the die. The pulling action is usually provided through winding onto a rotating spool. For coating by thermoplastics such as polyethylene, nylon, or plasticized PVC, the coating is hardened by water or air cooling. Rubber coatings are cross-linked by heating.

Sheet and flat film extrusion (slit-die). Most of the thermoplastic sheets and films are produced by extrusion, with specially designed dies and facilities to handle the product. The term *sheet* is given to flat parts with large surface area, and thickness ranges from over 0.25 mm up to about 12.5 mm. Sheets are used for flat products or for subsequent thermoforming applications. The term *film* is given to thicknesses from or less than 0.25 mm. Films may be flat or tubular. Flat films are extruded using the same sheet extrusion dies with additional cooling and coiling facilities, while tubular film processing will be studied in the following sections. Films are mainly used for packaging, covering, and lining. The most commonly used film material is LDPE. Others include PP, PVC, and cellophane.

The extrusion dies for sheets and flat films use a narrow slit (to control the thickness), and wide die openings, with widths up to 3 m. The die includes a manifold to spread the melt across the width

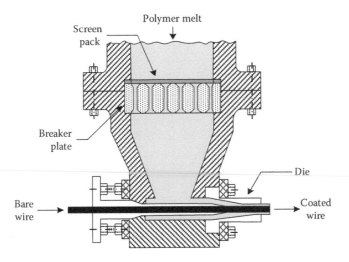

FIGURE 13.17   Extrusion coating of wires and cables.

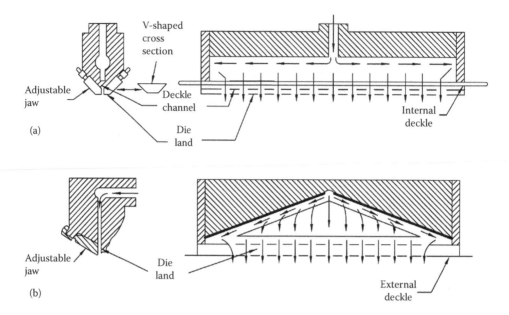

FIGURE 13.18 Extrusion dies for sheets and films: (a) T-type and (b) coat-hanger type. (From Chanda, M., and Ray, S., *Plastics Fabrication and Recycling*, CRC Press, Boca Raton, FL, 2007. With permission.)

to ensure uniform laminar flow, as shown in Figure 13.18. The deckle rods illustrated in the figure are used to adjust the sheet or film width. Practically, edges suffer some thickening, which must be trimmed. The sheet or film should be cooled below T_g or T_m, immediately after leaving the die orifice. This is achieved by passing the extrudate through a water bath or over two or more chrome-plated chill rolls, which are cored for water cooling. A schematic drawing of the chill roll operation for extruded films is shown in Figure 13.19. Several chill rolls are usually applied to maintain dimensional stability and close tolerances of the film before being pulled by the power carrier rolls

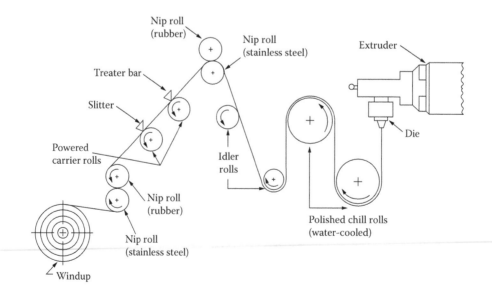

FIGURE 13.19 Schematic chill roll film extrusion. (From Chanda, M., and Ray, S., *Plastics Fabrication and Recycling*, CRC Press, Boca Raton, FL, 2007. With permission.)

and wound up. This process may be carried at very high production speeds of 5 m/s. The chrome-plated circumferences of the first rolls must be highly polished so that the film has high gloss and clarity.

Blown-film extrusion is widely used in the manufacturing of PE and other plastic films for packaging. It is a compound process involving extrusion, where the melt flows around a mandrel and emerges through a ring-shaped orifice in a tubular form, and blowing air under pressure through the mandrel to expand the tube into a bubble of the required diameter. The air contained in the bubble cannot escape, as it is sealed by the die at one end and the nip or pinch rolls at the other, so that it acts as a permanent mandrel, as shown in Figure 13.20. The air pressure must be kept constant to sustain uniform film thickness and tube diameter. The ratio of the bubble diameter to the tube diameter is known as the blow-up ratio, which may reach 4 or 5. Collapsing rolls (or plates) are used to squeeze the tube back to flat shape and direct it through pinch rolls for windup reel.

The film bubble is cooled below T_g by blowing air through a cooling ring around the die. With further cooling as the bubble moves upward, some thermoplastics such as PE exhibit crystallinity, and become cloudy compared with the clear amorphous structure. The transition line coinciding with this transformation is known as the frost line, as shown in the figure. The final blown film can be left in tubular form or cut at the edges.

The blown-film technique has many advantages compared to the split-die techniques.

a. The blown air results in isotropic strength properties to the film, which is superior (offers maximum toughness) to the directional properties resulting from stretching the film in one direction in split-die extrusion.
b. Easier control of the film width and gauge through control of the extrusion rate and air pressure. Thinner gauges of 0.1 to 0.2 mm, with blown films of diameters over 2 m,

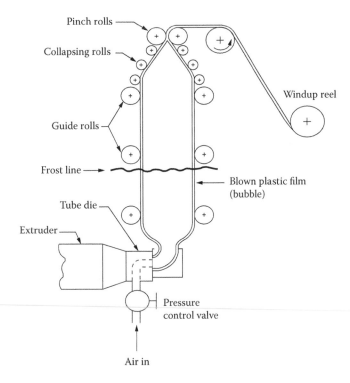

FIGURE 13.20 Blown-film extrusion. (From Groover, M. P., *Fundamentals of Modern Manufacturing*, 3rd ed., John Wiley & Sons, London, 2007. With permission.)

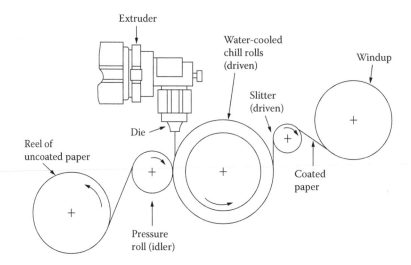

FIGURE 13.21 Extrusion coating of paper. (From Chanda, M., and Ray, S., *Plastics Fabrication and Recycling*, CRC Press, Boca Raton, FL, 2007. With permission.)

giving films of width 6.3 m can be produced, which have wide applications in agriculture, horticulture, and building.
c. Elimination of the end effects such as edge trim and nonuniform temperature that result in the flat film.

On the other hand, production rates are lower in blown-film extrusion.

Extrusion coating in which many substrates such as paper, paperboard, cellulose film, fiberboard, metal foils, or transparent films are coated with resins by direct extrusion. The resins most commonly used for coating are PE, PP, ethylene vinyl acetate copolymer, nylon, PVC, and polyester. Coatings are applied in thicknesses from .05 to 0.4 mm, and the substrate thicknesses range from 0.0125 to 0.6 mm. The die and equipment used for coating are similar to those used for the extrusion of flat film. Figure 13.21 shows a typical extrusion coating setup. The thin molten film from the extruder is pulled down into the nip between a chill roll and a pressure roll directly below the die. The pressure between the two rolls forces the film to be drawn to the required thickness, since the substrate moves faster than the extruded film. The coating is cooled by the water-cooled, chromium-plated chill roll. The pressure roll is covered by a pressure sleeve, usually neoprene or silicon rubber.

13.8.2 CALENDERING

Calendering was originally developed for processing rubbers, but is now widely adopted for producing thermoplastic films, sheets, and coating. Plasticized PVC represents a major portion of calendered film and sheet, ranging from 0.075 mm for film to 2.5 mm for vinyl tile for floor coatings. Other applications include shower curtains, vinyl table cloths, and pool liners, as well as inflatable boats and toys.

The process is based on feeding a heat-softened mass of polymer between two rolls, where it is squeezed into a sheet or film, which then passes around other rolls for cooling, controlling the thickness, and winding up. The last rolls in the calendaring unit control the final thickness as well as the surface quality, which may be glossy, matte, or embossed. Calenders are characterized by the number of rolls and their arrangement. A typical arrangement for calendering with an embossing unit is shown in Figure 13.22. Other arrangements may be adopted for double-plying, covering with

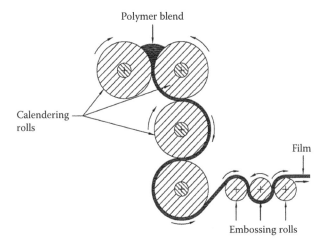

FIGURE 13.22 Typical arrangement for a calendering and embossing unit.

a rubber layer, or profiling with an engraving cylinder. Despite the simple appearance of the calender, it is expensive for the required close dimensional tolerances, temperature controls, and surface quality of the finishing rolls. The production rate is high and may reach 2.5 m/s.

13.8.3 SPINNING OF FIBERS

Fibers are extensively used in the textile industry and have recently been applied in reinforcing composite materials. A fiber is defined as a thin material shape having a length at least 100 times its diameter or cross-sectional area (aspect ratio). A filament is an individual strand of continuous length. Twisting together filaments into one strand gives a continuous filament yarn (thread), and a cord is formed by twisting together two or more yarns. Spinning is the principal manufacturing process for synthetic fibers, the most important of which is polyester, followed by nylon, acrylics, and rayon. These represent about 75% of the total fiber market, with the remaining 25% committed to natural fibers, including cotton, wool, silk, and jute. The process involves extruding the polymer melt or solution through a die with multiple holes, called a spinneret, to make filaments, which are then drawn and wound onto a bobbin. The number of holes in a 150 mm diameter platinum disc spinneret varies from about 10^4 for rayon spinning, to about 720 for tire core yarns, down to 10–120 holes for textile yarns. There are three major categories of synthetic fiber spinning processes. These are melt, dry, and wet spinning, as shown in Figure 13.23.

a. Melt spinning, where the polymer is heated in an inert atmosphere, and the viscous melt is pumped through the spinneret. Special pumps are used to operate in the temperature range of the melt (230°C–315°C). For a polymer with high melt viscosity, such as PP, a screw extruder is directly used to feed the spinneret. The filaments emerging from the spinneret are drawn and simultaneously cooled by air and conditioned by moisture, before being collected and spooled onto a bobbin, as shown in part (a) of the figure. The final diameter before winding may reach only one-tenth of the extruded diameter. Melt spinning is the most commonly used process, especially for polyesters and nylons.

b. Dry spinning, where the polymer is dissolved in a solvent at a concentration of about 20 to 40%. The solution is filtered and forced through the spinneret into a hot air chamber to reach the final dimensional stability of the filaments by evaporation of the solvent, as shown in Figure 13.23b. Hot air may be replaced by nitrogen or superheated steam for higher control of the fiber properties. The skin which forms first on the fiber by evaporation

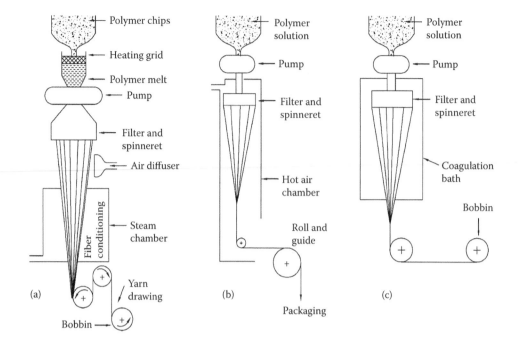

FIGURE 13.23 The principal types of synthetic fiber spinning: (a) melt spinning, (b) dry spinning, and (c) wet spinning. (From Chanda, M., and Ray, S., *Plastics Fabrication and Recycling*, CRC Press, Boca Raton, FL, 2007. With permission).

    gradually collapses and wrinkles as more solvent diffuses, leading to a thinner and irregularly lobed diameter. Dry-spun fibers have lower void concentrations than wet-spun fibers, which reflects the higher densities and lower dyeability of the first. Typical examples of the process applications are cellulose acetate dissolved in acetone, and polyacrylonitrile (PAN) dissolved in dimethylformamide.

  c. Wet spinning also involves pumping a solution of the polymer into the spinneret, and passing it in a nonsolvent bath for precipitating or coagulating the polymer into coherent strands that are collected onto bobins, as shown in part (c) of the figure. For example, PAN in dimethylformamide can be precipitated by passing a jet of the solution through a bath of water, which is miscible with the solvent but coagulates the polymer. The process is the most complicated of the three processes, as it may involve chemical reactions, and fibers having the highest void contents, which gives them increased dyeability. The surface is irregular, with longitudinal serrations.

Filaments produced by any of these three processes are subjected to a stretching operation to orient the structure into the axial direction. Drawing orients the crystallites in the axial direction so that the strength and modulus of elasticity in that direction is increased and elongation is decreased. Usually, stretching is carried out at a temperature above T_g of the material. Stretching is achieved by winding the filament onto a drum driven at a higher surface speed than the preceding one.

13.8.4 Injection Molding

Injection molding is the most widely used molding process for thermoplastics. Some thermosets and elastomers can be produced with modifications in equipment and process parameters to allow for cross-linking. The process involves feeding the polymer through a hopper to a cylinder (barrel) where it is heated to the viscoelastic state, and injecting it under the pressure of a plunger or screw to a mold cavity, where it hardens by cooling below the glass transition temperature. Then the mold is opened to eject the molded part.

Intricate shapes can be produced so long as the cavities can be formed in the mold to the required shape, and the part can be ejected. The mold may also contain more than one cavity for production of more than one product per cycle. Product sizes can range from as small as 50 gm, up to about 25 kg. Mold pressures range from 55 to 275 MPa, and cycle times of 10 to 30 seconds are common in injection molding machines.

The principles of conventional injection molding machines emerged from the die casting machine. The machine includes a plunger that pushes the granules into the heating zone, where the material is softened, and the molten polymer is further pushed forward through a nozzle to fill the mold cavity. Further development led to machines with two plunger units, one known as the plasticizer, which softens the polymer, and one known as the shooting plunger, that pushes the melt into the mold. Recent injection molding machines replaced the plunger with a screw similar to that of the extruder, as shown in Figure 13.24. The major difference is that the screw in the injection molding machine rotates and reciprocates. Rotation of the screw increases heating of the polymer by converting mechanical energy into heat and increasing heat transfer at the cylinder wall, as well as mixing and homogenizing the melt. Forward movement of the screw without rotation gives the action of a ram, which rapidly moves to inject the molten plastic into the mold. A non-return valve mounted near the tip of the screw prevents the melt from backward flow along the screw flights (threads). The product is left in the mold to cool under pressure to pack additional melt into the cavity to compensate for contraction. The screw is then rotated and withdrawn with the non-return valve open to allow fresh charge of polymer to flow into the forward direction of the barrel. Subsequently, the mold is opened, the product is ejected, and the mold is relocked under force, ready for a new cycle. The mold is held tightly closed by the clamping action of a hydraulic cylinder, as demonstrated in the figure. The clamping force should be high enough to resist injection force. This force may reach 15 meganewtons for large machines. Injection molding barrels are shorter than extruder barrels, with typical L/D ratios of 18 to 24, and these may reach 26 for fast-running machines.

Molds and mold design: An injection machine mold is specially designed according to the material type and geometrical configurations of the product. In addition to deciding the product shape, the mold performs several other functions. It conducts the hot plasticized material from the machine heating barrel to the cavity through a channel system, vents the entrapped air or gas, cools the part, and ejects it without damage. The mold design and manufacturing determines both the product quality and production cost.

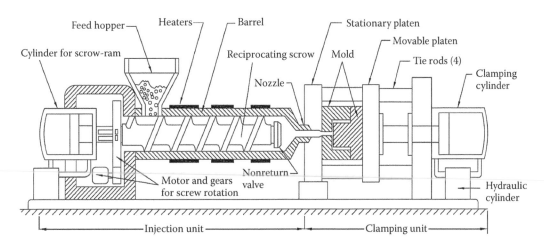

FIGURE 13.24 Cross-section of a screw-type injection molding machine. (From Groover, M. P., *Fundamentals of Modern Manufacturing*, 3rd ed., John Wiley & Sons, London, 2007. With permission.)

The design most commonly used is the two-plate mold design. One plate is fixed to the stationary platen at the side of the injection unit, and the other is attached to the moving platen which is connected to the clamping cylinder (refer to Figure 13.24). The cavity is formed by removing metal from the mating surfaces of the two halves, and the parting line (surface) determines where the mold opens to eject the product, as shown in Figure 13.25. The part demonstrated in this figure represents only one single cavity in a multiple-cavity mold for producing several products in a single shot. The stationary plate incorporates a distribution channel, to permit flowing of the polymer melt from the nozzle into the mold cavity through a sprue, a runner, and a gate, similar to casting processes. Ejector pins passing through the moving plate are used to eject the product and the sprue at the end of the cycle, when the two plates are opened. Usually, shrinkage of the product during cooling causes it to stick to the moving half. To compensate for shrinkage, which is large in thermoplastics, the dimensions of the mold cavity must be made larger than the specified part dimensions. Typical shrinkage values for molding of thermoplastics are given in Table 13.4.

Cooling is carried out by incorporating water channels in the mold halves, connected to an external pump for circulation. The cooling channels should be spaced evenly to prevent uneven temperatures on the mold surface. They should be set as closely as possible to the cavity surface without violating the mold strength. Most of the air in the mold cavity is evacuated through the small clearances between the ejector pins and the moving plate during flow of the polymer. Additionally, narrow air venting slots can be milled into the parting line, usually opposite the gate. The slots commonly range from 25 to 50 μm depth, and from 10 to 25 mm width. The small depth of the slot does not allow inflow of the viscous polymer. The mold material is a major factor in temperature control. Beryllium-copper has a high thermal conductivity, about twice that of steel and four times that of stainless steel. Therefore, a beryllium-copper mold should cool four times faster than SS molds.

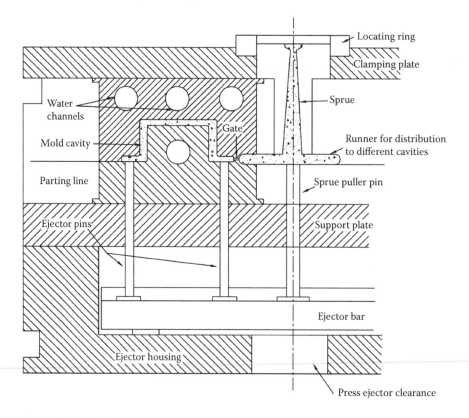

FIGURE 13.25 A typical two-plate injection mold design. (From Chanda, M., and Ray, S., *Plastics Fabrication and Recycling*, CRC Press, Boca Raton, FL, 2007. With permission.)

TABLE 13.4
Typical Shrinkage Values for Molding of Thermoplastics

Thermoplastic	ABS	Nylon-6,6	PC	PE	PS	PVC
Shrinkage (mm/mm)	0.006	0.020	0.007	0.025	0.004	0.005

Source: Groover, M. P., *Fundamentals of Modern Manufacturing*, 3rd ed., John Wiley & Sons, London, 2007.

Other possible designs of the mold may use a three-plate mold, with a third movable plate which allows central gating for each cavity to provide even distribution of the melt flow into the mold cavity. Figure 13.26 demonstrates a single-cavity, three-plate mold, which offers two advantages to a two-plate mold for the same product. First, the central flow of the melt from the part base ensures uniform flow around the sides, compared to the side flow in a two-plate mold. In side flow, the material moves in two streams around the core, and joins on the opposite side, which may lead to weak points at the weld line. The second advantage concerns enhanced automatic operation of the machine when the three plates open, leaving two spaces. The sprue and runner part drop from one opening under gravity or with blown air assistance to one container, while the product is ejected from the other and allowed to drop into another container.

Gas-assisted injection molding: This starts with partial injection of the polymer melt in the mold cavity, followed by injection of compressed gas into the core of the melt to pack the mold and obtain a hollow, rigid product, free of sink marks, as shown in Figure 13.27. This is known as the Asahi gas injection molding process (AGI). The hollowing out of thick-sectioned moldings results in reduction in weight, material saving, shorter cooling cycle, and reduced clamping force. Further, it eliminates uneven shrinkage, and supports molding complicated shapes in a single form, thus simplifying the mold design. Compressed gas may be introduced in the desired locations through gas channels to allow uniform shrinkage and to reduce warpage.

Injection molding of thermosets: Some thermosets in granular or pellet form are suitable for injection molding. The machines required for these thermosets have the same features of thermoplastic injection molding machines, with two differences. The barrel length for injection molding machines of thermosets is shorter to avoid premature curing and solidification of the polymer. The second difference is in the design of the feed screw. The screw used for thermosets has a zero

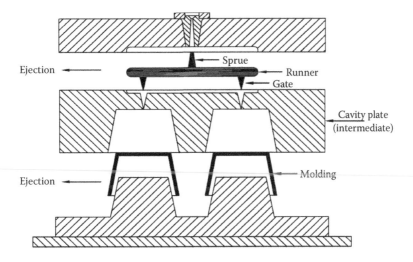

FIGURE 13.26 Three-plate mold for injection molding.

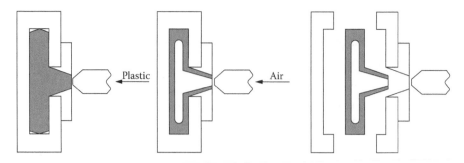

FIGURE 13.27 Schematic of the Asahi gas injection molding process (AGI).

compression ratio; i.e., the depths of the flights at the feed zone end and at the nozzle end are the same. By comparison, the compression ratios for the screws of thermoplastic molding machines ranges from 2 to 5, similar to those of extruders (refer to Section 13.8.1). These two differences help to keep temperatures in the barrel at relatively low levels, in the range from 50°C to 125°C, depending on the polymer.

The material is fed through the hopper to the barrel, and moves forward due to rotation of the screw, changing from solid to viscous fluid. When the required amount of fluid (charge), is accumulated at the nozzle end, it creates back pressure on the screw. The screw stops turning and moves forward under hydraulic pressure to force the material through the sprue, runner, and gate to the mold cavity. The mold is heated to 150°C to 230°C, where cross-linking occurs to harden the plastic (for thermoplastics, the mold is cooled). The mold is then opened and the product is ejected. Curing is the most time-consuming step of the cycle, especially for thicker sections. The average cycle time for injection molding of a 6.25 mm thickness thermosetting part is 30 seconds (compared to 45 seconds for thermoplastics).

The most commonly used injection-molded thermosets are phenolics. Other thermosets include melamine, urea-formaldehyde, unsaturated polyesters, alkalyds, and epoxies. Elastomers are also injection molded. Fillers such as glass fibers, clay, wood fibers, and carbon black are added in large proportions to thermosetting polymers to form composite materials that are injection molded to a wide variety of applications.

Another development of injection molding for a family of thermosets is *"reaction injection molding (RIM)."* This process involves mixing two highly reactive liquid ingredients and instantly injecting the mix into a mold cavity, where chemical reactions lead to solidification of the product. Typical examples of these catalyst-activated thermosets are urethanes, epoxies, and urea-formaldehyde.

13.8.5   BLOW MOLDING

Essentially, blow molding is used for manufacturing hollow, seamless thermoplastic products such as bottles and other containers; however, the process is also applied for producing toys, automotive gasoline tanks, and hulls for sail boards and small boats. Products range in size from small bottles of volume 5 ml, to large storage drums of 38,000 liters capacity. The process starts with a tube-like shape of the thermoplastic called a *parison*, manufactured by extrusion or injection molding. The parison is located in a two-half mold with a cavity having the required bottle or container shape. The two halves are closed, sealing both ends of the parison. Compressed air is blown into the parison, which is inflated to fill the contour of the mold cavity. When the polymer surface touches the cold walls of the metal mold, it is cooled rapidly below the glass transition temperature and hardens. The mold is then opened to eject the product, a new parison is introduced, and the cycle is repeated. Flash is trimmed if necessary. A schematic of the process sequence is shown in Figure 13.28.

Large production rates are usually achieved by adopting continuous extrusion blow molding. A molten parison is continuously produced from a screw extruder, and the molds are arranged on a

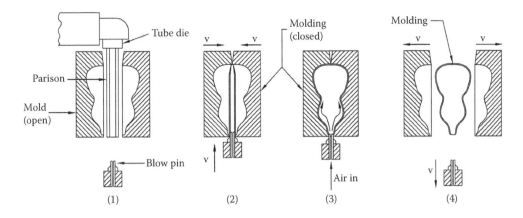

FIGURE 13.28 Extrusion blow molding.

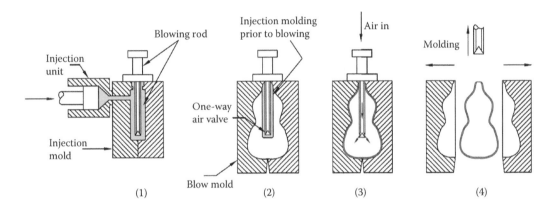

FIGURE 13.29 Injection blow molding.

rotating table, or on the periphery of rotating wheel stations. With continuous rotation of the table or wheel, the first mold is extruded, the second is air blown, the third is cooled, the fourth is ejected, and the cycle is repeated. Extrusion blow molding is more commonly used than injection blow molding. Almost any thermoplastic can be blow molded, but polyethylene products are the most predominant, particularly HDPE and UHMWPE. Polyethylene squeeze bottles represent a large percentage of blow-molded products. Other materials commonly used are PP and PVC.

In injection blow molding, the parison is formed by injection molding as a thick-walled tube around a blowing stick, which is transferred to the blow molding station to be inflated. A schematic demonstration of the process is shown in Figure 13.29. The process is relatively slow, but it allows good control of neck and wall thickness of the product.

Blow molds are made of machined or cast blocks, including the cavity, and provided with cooling channels, vents, pinchoffs, flash pockets, and mounting plate. The mold material is selected based on thermal conductivity, durability, cost, product quality, and the thermoplastic being blown. Commonly used mold materials are beryllium, copper, beryllium-copper alloys, aluminum, A-2 tool steel, and stainless steel 420.

13.8.6 ROTATIONAL MOLDING

Rotational molding (commercially known as "rotomolding"), is an alternative for blow molding for manufacturing large hollow products with complicated curves, uniform wall thickness, high surface

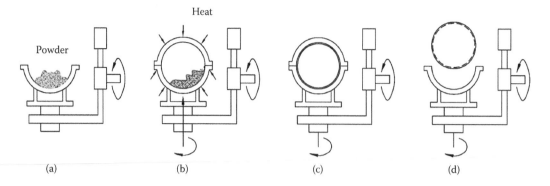

FIGURE 13.30 Rotational molding cycle: (a) Charging, (b) Heating, (c) Cooling, and (d) Unloading.

finish, and no residual stresses. It has been used for a variety of products such as car and truck body parts, industrial containers, storage tanks, garbage cans, ice buckets, light balls, large toys, and boat and canoe hulls. The process is principally applied for most thermoplastics, but applications for thermosets and elastomers are becoming more common. The most used thermoplastics are PE (especially HDPE), PP, ABS, and high-impact PS. The process includes four stages, as shown in Figure 13.30.

a. Loading a charge of predetermined weight of powdered plastic into a two-halved mold cavity with the outer required configurations of the product. The mold halves are then closed.

b. The mold is moved into a hot oven, where it is rotated simultaneously in two perpendicular planes. The oven is heated by hot air or by a liquid of high specific capacity 260°C up to 480°C, according to the polymer type. Rotation distributes the polymer on the walls of the cavity, forming a fused layer distributed evenly with uniform thickness. The particles are forced under gravity, since the rotational speed is relatively slow, such that centrifugal force is not effective.

c. Still rotating, the mold is moved from the oven to a cooling chamber, cooled by forced cold air or water spray, where the plastic skin solidifies.

d. Finally, rotation is stopped, and the mold is opened and the product is removed.

Molds in this process are simple and inexpensive compared to injection molding or blow molding, but the production cycle is much longer, lasting up to 10 minutes or more for large products.

13.8.7 COMPRESSION MOLDING

Compression molding is one of the oldest and most common molding processes for thermosets. It is equivalent to closed-die forging for metals. The process involves loading a charge of predetermined weight of the plastic, in the form of pellets or powder, into the bottom half of a heated mold. The top half of the mold is pressed, forcing the charge to flow and conform to the cavity shape. The mold is held at a temperature of 150°C to 200°C depending on the material, until curing is complete and the material is hard (solid). Then the mold halves are opened, and the molding is ejected. It has become common practice to preheat the charge before loading, to soften the polymer and reduce the cycle time. The mold must allow venting to provide for the escape of steam, gas, or air produced during the process. Special-purpose hydraulic presses can be used to provide application of initial pressure, followed by opening the mold slightly to release gases (known as breathing), and then applying the final pressure. The cure time depends on the molding material, preheating temperature, and part thickness. This time may range

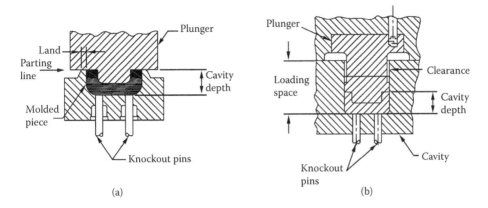

FIGURE 13.31 Compression molding using (a) simple flash mold and (b) positive mold. (From Chanda, M., and Ray, S., *Plastics Fabrication and Recycling*, CRC Press, Boca Raton, FL, 2007. With permission.)

from seconds to a few minutes. The pressure required depends on the polymer and usually ranges from 15 to 30 MPa. This pressure is commonly applied through a vertical hydraulic press.

Molds for compression molding are more simple than those for injection molding, since there is no need for a sprue or runner. The simplest design of the mold is the open flash mold, shown in Figure 13.31a, which allows any slight excess of the powder to escape in the form of a flash when the top and bottom parts of the mold are closed. Mold closing speed should be controlled, since fast closing might entrap gases or air, and may splash out finely powdered material. This mold is applicable for polymers with high melt viscosity. Since most rubbers possess this property, this mold is widely used for producing gaskets, shoe heels, door mats, and many similar products. In the positive mold, as in Figure 13.31b, an accurately measured charge must be loaded. Any excessive powder will not allow the mold halves to close, and insufficient powder results in a defective (incomplete) product. A small clearance of 10 to 30 μm between closing surfaces of the plunger (top part) and the cavity allows venting of entrapped gases. Brass and mild steel are used as mold materials, but for severe molding conditions, molds are made of special grades of tool steel.

A wide range of thermosets are compression molded. The most commonly used types are phenolics, urea-formaldehyde, melamine, epoxies, and alkyds. Elastomers can also be molded, as described in the previous paragraph. Thermoplastics can be compression molded, but the cycle time will be much slower, since the mold needs cooling before ejection, whereas for thermosets the mold is hot during the whole cycle. Therefore, applications for thermoplastics are limited. Typical thermosetting products include electric plugs and sockets, cookware handles, plates, buttons, buckles, and appliance housings. Recent applications include orthodontic retainers and pacemaker casings, as well as plenty of reinforced plastics moldings. For higher quality products, vacuum chambers are now added to the mold to eliminate the defects caused by entrapped air or water in the molding compound.

13.8.8 Transfer Molding

Transfer molding is a development of compression molding for intricate shapes with varying wall thicknesses. The uncured thermoset or elastomer powder is heated and pressurized by a plunger in a transfer chamber (or pot) outside the mold cavity until reaching the viscous condition. Then it is forced through a sprue to flow into the hot mold cavity, where it sets hard (cross-linked) to the cavity shape and cures with time. When the mold opens, the product is pushed out using ejector pins. The sprue comes out with the transfer plunger, where it is removed. These stages are shown in Figure 13.32. The pressures required for transfer molding are about three times larger than for

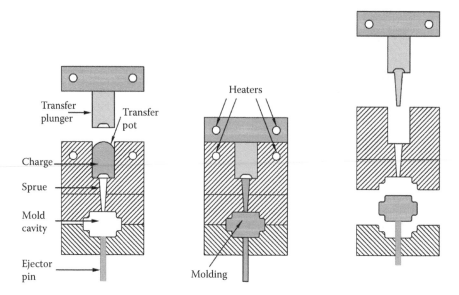

FIGURE 13.32   Transfer molding.

compression molding. Since the polymer enters the cavity in a molten state, the process enables production of moldings around thin pins and metal inserts (such as electrical plugs).

13.8.9 THERMOFORMING

When moderately heated, a thermoplastic sheet or film becomes as soft as a sheet of rubber, which can then be stretched to any required shape. The process is similar to stretching and deep drawing of metal sheets. According to the means applied in stretching the heat-softened sheet, the process can be subdivided into three categories: vacuum forming, pressure forming, and mechanical forming. Common thermoplastics processed by thermoforming include PS, cellulose acetate, PVC, ABS, PMMA, PE (low density and high density), and PP. These materials possess high strain rate sensitivity to withstand high uniform deformation during stretching. The majority of these processes are done with extruded sheets; however, cast, calendered, or laminated sheets can be used. The process is not applicable for thermosets or elastomers sheets, since they cannot be resoftened after cross-linking (curing) during sheet forming.

Being carried out on sheets, thermoforming processes are generally suited to products with large area-to-thickness ratios. Typical products include refrigerator and freezer door liners, washing machine covers, dishwasher housings, aircraft and automotive parts (instrument panels, interior panels, ceilings, and arm rests), large-patterned diffusers in lighting systems, advertising displays, bath tubs, serving trays, packaging, toys, etc.

    a. Vacuum thermoforming, in which the thermoplastic sheet is clamped and heated by a radiant heater until softening, as shown in Figure 13.33. The soft sheet is then sealed on the mold. Air is evacuated from the mold cavity using a suction pump, so that the sheet is forced to take the mold contour. The vacuum holes in the mold are very thin, in the range of 0.8 mm diameter, so that they do not affect the product surface. Vacuum is maintained until the product cools and becomes rigid. This was the earliest thermoforming process, which was developed in the 1950s.

    b. Pressure thermoforming, where positive pressure is applied to deform the heated sheet, opposite to vacuum forming, as shown in Figure 13.34. Vent holes in the mold allow

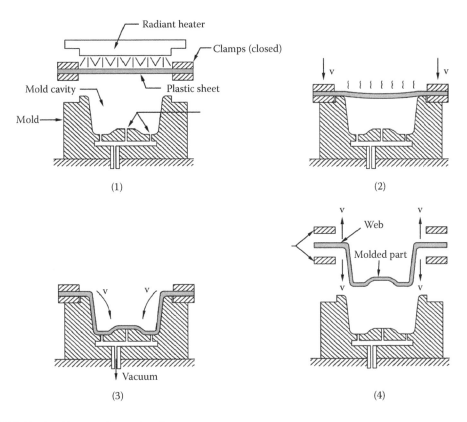

FIGURE 13.33 Vacuum thermoforming.

entrapped air to escape. Pressure is released after the product cools and hardens. As compared to vacuum forming, pressure forming offers a shorter production cycle, greater contour definition, and better dimensional control, since higher pressures can be adopted. Pressures of 3 to 4 atmospheres are common.

A variation of vacuum forming or pressure forming, known as free forming, is used with acrylic sheets to produce parts with superior optical quality (such as aircraft canopies). The sheet periphery is clamped, and deformation takes place only under pressure or vacuum without a mold cavity.

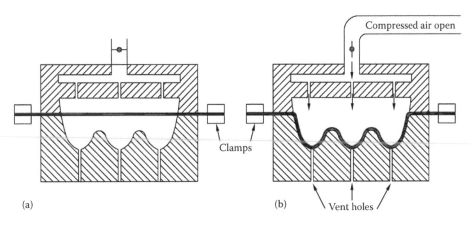

FIGURE 13.34 Pressure thermoforming.

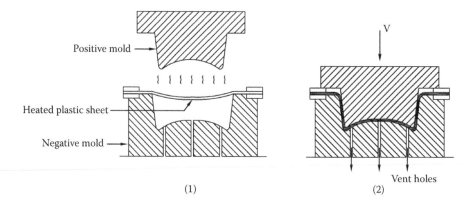

Positive mold

Heated plastic sheet

Negative mold

V

Vent holes

(1) (2)

FIGURE 13.35 Mechanical thermoforming.

c. Mechanical thermoforming, where two matched mold halves are used to form the heated sheet under the pressure of a press, as shown in Figure 13.35. The process offers excellent reproduction of mold details, and high dimensional accuracy. These advantages are offset by the higher cost of the mold, being made of two (positive and negative) halves.

13.8.10 Casting Processes

In polymer shaping, the term *casting* involves pouring a liquid resin into a mold to fill it under gravity and allowing the polymer to harden. Compared to other processes, casting requires a simple and less costly mold, and the cast product is free of residual stresses. The process is applied for low production quantities. Both thermoplastics and thermosets are cast.

13.8.10.1 Casting of Thermoplastics

The liquid resin can be a naturally liquid prepolymer or monomer, or a granular solid heated to a highly fluid condition. The first is polymerized in the mold to form a high-molecular-weight thermoplastic, while the second is left to cool and solidify in the mold. Typical cast thermoplastics include acrylics, polystyrene, polyamides (nylons), and PVC. For casting of simple shapes such as rods or tubes, a two-piece metal mold is usually used, with an entry gate for pouring. For making flat-cast acrylic Plexiglas or Lucite sheets, two pieces of polished plate glass separated by a gasket with the edge sealed and one corner open are usually used as a mold. Such sheets have superior flatness and clarity that cannot be achieved by flat sheet extrusion.

A third technique for casting of thermoplastics is known as *plastisol casting*. A plastisol is a suspension of PVC in a liquid plasticizer to produce a fluid mixture. When the fluid is poured into the mold and heated to about 175°C, the resin undergoes dissolution in the plasticizer, forming a gel that on cooling, solidifies as a flexible vinyl product having superior quality. Three different variations of plastisol casting are dip casting, slush casting, and rotational casting, as shown in Figure 13.36.

a. *Dip casting*, where a heated mold is dipped into liquid plastisol and then drawn slowly at a predetermined rate, as in Figure 13.36a. The built-up plastisol layer, with the mold, is then heated to about 175°C to 200°C. After cooling, the flexible plastic is stripped from the mold. Typical products include transparent overshoes and flexible gloves. The dipping process is also used for coating metal objects with vinyl plastic, such as wire dish drainers, coat hangers, and other industrial and household goods.

b. *Slush casting* is similar to metal slush casting (refer to Section 7.4.1.2). The plastisol liquid is poured into a preheated hollow metal mold, where a skin solidifies on immediate contact

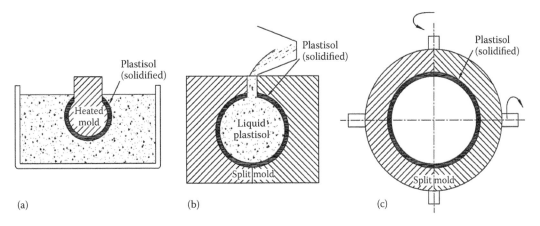

(a)   (b)   (c)

FIGURE 13.36 Plastisol casting processes: (a) dip casting, (b) slush casting, and (c) rotational casting.

with the mold wall, as in Figure 13.36b. The thickness of the cast is controlled by the time of stay in the mold. Excess liquid is poured out, and the solidified layer and mold are kept in the oven for a specific period, and then left to cool. The mold is then opened to remove the product. The process is suitable for hollow flexible products, such as squeezable dolls and boots. The process is also known as *shell casting*.

c. *Rotational casting*, where a predetermined amount of liquid plastisol is loaded in a heated, closed, two-piece mold. The liquid is uniformly distributed against the walls of the mold in a thin uniform layer by rotating the mold in two perpendicular planes, as shown in Figure 13.36c. The plastisol in the mold is solidified in an oven. The mold is then opened, and the part is removed. The method is used to make completely enclosed hollow objects such as dolls, plastic fruits, squeeze bulbs, and flushing system floats.

d. *Centrifugal casting*, which is similar to centrifugal casting of metals. This process involves rotating a heated tubular mold, which is charged with a powdered thermoplastic along its length. The powder is forced by centrifugal force toward the hot mold wall, and thus changes to a viscous fluid, building up a plastic layer. When the required thickness is reached, the heat source is removed, and the mold is cooled while still rotating to maintain uniform thickness. When cooling is finished, the plastic tube shrinks away from the mold, and is removed. Common tube diameters range from 150 to 750 mm, and lengths up to 2.5 m can be produced by this process.

13.8.10.2 Casting of Thermosets

This process involves pouring the liquid ingredients into the mold, and allowing them to cross-link (cure) with the aid of heat and/or catalysts, depending on the type of resin. The mold should be given enough time to cure before removing the cast. Typical cast thermosets include polyurethane, unsaturated polyester, phenolics, and epoxies.

13.8.11 FOAM PROCESSING

A foam polymer is a polymer and gas mixture having a porous or cellular structure, which offers the material significant inherent characteristics, including

a. High load bearing capacity per unit load (lower density than unfoamed polymers)

b. High thermal insulation due to the low conductivity of the contained gas

c. High energy absorption or energy dissipating capacity

Therefore, foamed plastics are widely used as buoyancy products (buoys or life jackets), as insulators (e.g., cups for hot drinks), as core panels for load-bearing structures, as packing materials for protection of products during shipping, and as cushioning materials for furniture and bedding. Other terms used for these foam materials are *cellular polymers*, *blown polymers*, and *expanded polymers*. The most common foam polymers are polystyrene (known commercially as Styrofoam), polyurethane, PVC, PP, ABS, epoxy, phenol-formaldehyde, urea-formaldehyde, silicones, and foamed rubber. It should be noticed that these foams belong to all families of polymers; i.e., thermoplastics, thermosets, and elastomers.

Foam plastics can be classified from different points of view. According to the foam elasticity, they can be classified as

a. *Elastic foam*, in which the polymer is rubber, capable of large elastic deformation.
b. *Flexible foam*, in which the polymer is a highly plasticized polymer, such as PVC.
c. *Rigid foam*, in which the polymer is a stiff thermoplastic such as PS, or a thermoset such as phenolic. Rigid foam is usually used for insulation, or in load-bearing structures.

Depending on chemical formulation and degree of cross-linking, polyurethanes can range over all three categories.

According to the nature of the foam cells, they can be classified into

a. *Closed-cell type*, where each individual pore cell, more or less spherical in shape, is completely closed in by the polymer matrix. These are usually formed under pressure, and are best used as buoys to prevent cells filling with water.
b. *Open-cell type*, where individual cells are interconnected, as in a sponge. These are typically produced under free expansion conditions, and are best used in cushioning and bedding applications, where compression allows air to flow from cell to cell and dissipate energy.

Most foaming processes produce both types.

According to the bulk density, foams are commonly classified into

a. Low-density foam, for a density range of 2–50 kg/m³
b. Medium-density foam, for a density range of 50–350 kg/m³
c. High-density foam, for a density range of 350–960 kg/m³

Foaming processes: Polymers are foamed using gases such as air, nitrogen, or CO_2. The percentage of gas in a foam bulk may reach 90% or more. Foaming processes include three different techniques for bringing the gas into the polymer.

a. Mixing air with the liquid resin by mechanical agitation, and then hardening the polymer through heat or chemical reaction.
b. Dissolving a physical blowing gas agent such as N_2, or pentane (C_5H_{12}), in the polymer melt under pressure. When pressure is released, the gas comes out of the solution and expands.
c. Mixing the polymer with chemical blowing agents that decompose at high temperature to release gases such as N_2, or CO_2 within the melt.

In all foaming processes, once the foam has been expanded or blown, the cellular structure must be stabilized rapidly; otherwise, it would collapse. For thermoplastic polymers, expansion is carried out above the softening point T_g, and the foam is then immediately cooled below this point, which is known as physical stabilization. For thermosets, the foam requires chemical stabilization by cross-linking immediately after expansion.

Foams are processed by shaping in the form of blocks, sheets, slabs, and boards, as well as molded and extruded shapes. They can also be sprayed to form coatings on substrates, or foamed in place between walls. Recently, conventional plastic processing machines such as extruders and injection molding machines have been adopted for processing foamed plastics. There are plenty of chemical reactions and chemical processing details involved in producing foamed products from each polymer group, which is beyond the scope of this book.

13.8.12 JOINING OF PLASTICS

Plastics can be joined or assembled by either adhesive bonding or by welding, as shown in Figure 13.37. Adhesives are most widely used by virtue of low cost and adaptability to high-speed production. Both adhesive bonding processes and welding processes will be introduced in this section.

A. Adhesive bonding processes

   Solvent cements: They function by attacking the surfaces of the joint so that they soften, and on evaporation of the solvent, will join together. The dope cement performs the same function in addition to containing a quantity (usually less than 15% resin) of the same plastic to be bonded in the solution to support the bonding action. Light pressure should be applied to the joint until it is fully hardened. Structural bonds of up to 100% of the parent polymer can be achieved with solvent cements. The solvent is applied by immersing the surfaces into it until softening, removing quickly, and bringing under light pressure for some time. Alternatively, when the joint is close-fit, a brush or eye dropper is used, allowing the solvent to spread by capillary action. Surrounding areas should be masked to prevent them from being etched.

   Monomeric or polymerizable cements: These consist of a reactive monomer compatible with the polymer to be bonded, together with a suitable catalyst and accelerator. The mixture polymerizes either at room temperature or at a temperature below the softening temperature of the thermoplastic to be bonded. This is generally known as adhesive bonding, since the adhesive mixture acts as an agent to hold two substrates together (as opposed to solvent cementing, where the parent materials become an integrated part of the bond, and the adhesion is chemical). A typical example is the mixture of epoxy resin and hardener, which provides an excellent adhesive to many materials.

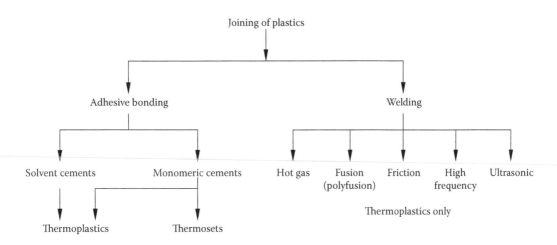

FIGURE 13.37 Classification of plastic joining processes.

B. Welding processes: Welding of polymeric materials is done by applying heat with or without pressure. Common heat welding processes include

a. *Hot gas welding*, which is comparable to gas welding of metals. It is particularly useful for joining thermoplastic sheets for producing tanks and ducts. The sheet edges are cleaned and beveled to form a V-shaped joint. A filler rod made of the same polymer type is used with the tip placed in the V-channel. All parts of the joint are heated by a hot gas stream directed from an electrically heated nozzle. The stream temperature ranges from 200°C to 400°C, depending on the type of the polymer, which is sufficient to fuse the polymer and join the parts. The gas used for heating is usually inert, such as N_2 to prevent oxidation, but hot air may be used for welding PVC.

b. *Fusion (hot-tool) welding* is where an electrically heated metallic plate or tool is used to heat the surfaces of the two parts to be joined. A typical example is the polyfusion process, to join tubes by means of injection molded couplings. Figure 13.38 represents a schematic of fusion welding of a socket on a tube. The tool heats the outside surface of the tube and inside surface of the coupling until softening. Then parts are firmly pressed and left to cool until reaching maximum strength. The tool is usually chrome-plated to prevent sticking to the polymer.

c. *Friction (spin) welding*, which is comparable to friction welding of metals. The process involves fixing one of the two parts to the spindle of a lathe-like machine, and fixing the other part to the tailstock. Applying pressure between both parts while one is rotating generates friction, which causes the polymer to fuse. The spindle is stopped, but the pressure is maintained until cooling. The process is limited to thermoplastic circular shapes.

d. *High-frequency welding*, which is used for thermoplastics with high dielectric-loss characteristics, such as ABS, PVC, and cellulose acetate. The parts to be joined are subjected to the electrodes of a radio transmitter, which generates high frequencies between 27 and 40 MHz. The high-frequency field generates vibrations in the plastic molecules that cause high enough heat to fuse the interfaces and weld the joint. Thermoplastics with low dielectric characteristics, such as PE, PP, and PS, cannot be welded by this technique.

e. *Ultrasonic welding*, where the parts to be joined are sufficiently disturbed by applying ultrasonic frequency that generates frictional heat, which causes melting and welding of the joint. The ultrasonic frequency is in the range of 20 to 50 kHz, and the transducer which converts the energy to mechanical vibrations is of either magnetostrictive or piezoelectric type. Energy is transmitted to the plastic through a titanium horn which vibrates at amplitudes 0.013 to 0.13 mm. Ultrasonic welding is generally used for welding thin sheets or films of thermoplastics with lower modulus of elasticity, such as PE, and plasticized PVC.

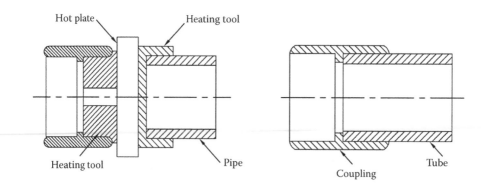

FIGURE 13.38 Schematic of fusion welding of a coupling on a tube.

13.9   REVIEW QUESTIONS

1. What are the main constituents of plastic products?
2. Classify polymeric materials, showing the main features of each class.
3. What are the different forms of polymer chains? Give examples of each form.
4. Compare different categories of copolymers with corresponding categories of metallic alloys, giving examples of each.
5. What is the effect of crystallites on the structure and properties of polymers?
6. Discuss the factors affecting crystallinity of polymeric materials.
7. Represent graphically the effect of temperature on the specific volume of both amorphous and crystalline polymers, and define the processing range on this graph.
8. Compare thermoplastics and thermosets, considering their structure, physical and mechanical properties.
9. Differentiate the extruders used for thermoplastics from those used for elastomers.
10. Compare the slit-die technique and the blown film technique for extrusion of thin plastic films.
11. What are the differences in the functioning and design of the extruder screw and the injection molding machine screw?
12. Compare the two-plate mold and three-plate mold in injection molding.
13. What are the differences between injection molding machines for thermoplastics and those for thermosets?
14. Compare the open flash mold and positive mold in compression molding.
15. List the different techniques of thermoforming.
16. Compare solvent cements and polymerizable cements as adhesive bonding materials.
17. Describe the effect of die swell in extrusion, and how to compensate for it.
18. Compare compression molding and transfer molding, emphasizing the differences in the molds.
19. Select five plastic products that we use daily. Suggest a suitable material and a manufacturing process for each.
20. Differentiate between high-frequency welding and ultrasonic welding for polymeric materials.

13.10   PROBLEMS

1. The molecular weights of UHMWPE and LDPE are 4×10^6 and 2×10^4, respectively, and the spacing between carbon atoms in the molecule is 1.26×10^{-4} µm. Calculate the degree of polymerization and length of stretched chain in both grades of polyethylene.
2. Given that the density of polystyrene is 1.05 g/cm^3, calculate the percentage of gas in expanded polystyrene foam having a density of 50 kg/m^3.
3. In the blow molding process presented in Figure 13.28, if an HDPE extruded parison is used with outside and inside diameters of 40 and 37 mm, respectively, calculate the thickness of the blown bottle if the mold diameter is 120 mm.

BIBLIOGRAPHY

Brent Strong, A. 2006. *Plastics Materials and Processes*, 3rd ed. Upper Saddle River, NJ: Pearson-Prentice Hall.

Brinson Hal, F., and Catherine Brinson, L. 2008. Polymer Engineering Science and Viscoelasticity: An Introduction. Springer.

Brydson, J. A. 1999. *Plastics Materials*, 7th ed. Butterworth-Heinemann.

Chanda, M., and Ray, S. 2009. *Plastics Fundamentals, Properties and Testing*. Boca Raton, FL: CRC, Taylor & Francis Group.

Chanda, M., and Ray, S. 2009. *Plastics Fabrication and Recycling*. Boca Raton, FL: CRC, Taylor & Francis Group.

Crawford, R. J. 1998. *Plastics Engineering*, 3rd ed. Butterworth-Heinemann.

Groover, M. P. 2007. *Fundamentals of Modern Manufacturing*, 3rd ed. London: John Wiley & Sons.

Kumar, A., and Gupta, R. K. 2003. *Fundamentals of Polymer Engineering*, 2nd ed. Marcel Dekker.

Kalpakjian, S., and Schmid, S. R. 2003. *Manufacturing Processes for Engineering Materials*, 4th ed. Upper Saddle River, NJ: Prentice-Hall.

McCrum, N. G., Buckley, C. P., and Bucknall, C. B. 1997. *Principles of Polymer Engineering*, 2nd ed. Oxford Science Publications.

Mills, N. 2005. *Plastics: Microstructure and Engineering Applications*, 3rd ed. Elsevier Science and Technology Books.

Schey, J. A. 1987. *Introduction to Manufacturing Processes*, 2nd ed. New York: McGraw-Hill.

Troughton, M. 2008. *Handbook of Plastics Joining*. William Andrew.

14 Composite Materials and Their Fabrication Processes

14.1 INTRODUCTION

The term "composite materials" signifies that two or more materials are combined on a macroscopic scale to form a new useful material. The combinations are among any two or more of the known material families, i.e., metals, ceramics, and polymers. The key is the macroscopic examination (by the naked eye or at low magnification) of the new material to identify its constituents. This differentiates composites from other combinations such as alloys, where different constituents come together on the microscopic scale, and the material is macroscopically homogeneous, i.e., components cannot be distinguished by the naked eye. The main objective of a composite material is that it utilizes the most distinguished qualities of its constituents to produce a material with the characteristics needed to perform design requirements. The most significant characteristics of composites are the strength-to-weight ratio and the stiffness-to-weight ratio, which can be several times greater than steel or aluminum. Other improved properties include fatigue life, toughness, wear resistance, corrosion resistance, appearance (aesthetic aspect), thermal conductivity or insulation, and acoustic insulation. The unique gain of the composites is the possibility of achieving combinations of properties that are not achievable with metals, polymers, or ceramics alone. Nevertheless, it should be noticed that the inherent structure of composites implies that they are heterogeneous and anisotropic.

Composite materials have a long history starting with man's existence on Earth, where he used and is still using wood for everyday life. Recorded history showed that Ancient Egyptians used straw to strengthen mud bricks. They also used plywood when they realized that wood layers could be rearranged to achieve superior strength, as well as resistance to thermal expansion, and to swelling due to moisture absorption. Medieval swords and armor were made of layers from different metals (Japanese swords or sabers were made of steel and soft iron for high flexural and impact resistance). Reinforced concrete (cement and gravel, reinforced with steel rods) is a well-known construction material. More recently, fiber-reinforced polymer matrix composites have become main constituents in aircraft and aerospace vehicles owing to their superior strength-to-weight and stiffness-to-weight ratios. A good example is shown in the list of composite material parts on the main structure of the Boeing 757-200 aircraft (Figure 14.1).

14.2 CLASSIFICATION AND CHARACTERISTICS OF COMPOSITES

According to the structure of the composite materials they can be classified to four commonly accepted types

1. Fiber-reinforced composites that consist of fibers or whiskers of one material, embedded in a matrix material
2. Particulate-reinforced composites that consist of particles of one material in a matrix material
3. Laminated composites that consist of layers of two or more materials
4. Combinations of some or all of the first three types, such as reinforced concrete

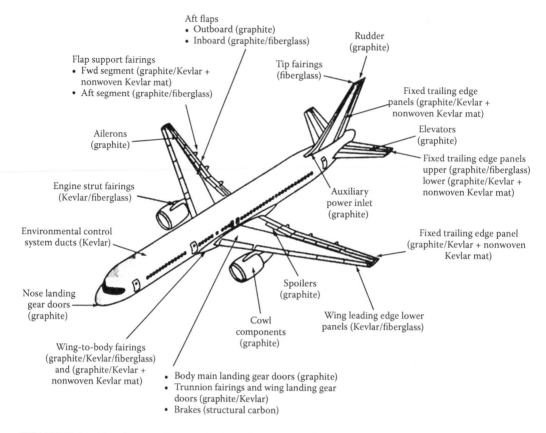

FIGURE 14.1   List of composite parts in the main structure of the Boeing 757-200 aircraft.

This classification is presented in Figure 14.2, and details of the definitions and structure of the constituents of these composite materials are explained in next sections.

14.3   FIBER-REINFORCED COMPOSITES

In order to appreciate the characteristics of fiber-reinforced composites, we should understand the types and characteristics of fibers, as well as the possible types of matrix materials.

14.3.1   FIBERS AND WHISKERS

Fibers consist of a large number of filaments, each filament having a diameter of 5–15 μm. A fiber is geometrically characterized by its high length-to-diameter ratio. Long fibers of various materials are inherently much stiffer and stronger in its axial direction than the same material in the bulk form. A typical example is glass, where the ultimate strength of a glass plate is only about 20 MPa; yet a glass fiber has strength in the range of 2800–4800 MPa in commercially available forms, and about 7000 MPa in a laboratory-prepared form. The reason for these differences is that in fibers, the crystals are more perfectly aligned around the fiber axis with minimum defects compared to the bulky structure, where defects such as dislocations, different orientations, and large grain boundary areas are inevitable. Fibers can be prepared in short or long lengths. Short fibers of lengths ranging from fractions of millimeters to a few centimeters are usually used as reinforcement in a matrix material, whereas long fibers are used as woven or in fabrics.

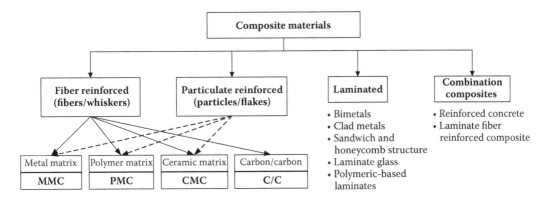

FIGURE 14.2 Classification of composite materials.

Table 14.1 represents the specific weight, strength, and modulus of elasticity (stiffness) for commonly known fibers in comparison with common structural metallic wire materials. The table shows also the strength- and stiffness-specific weight ratios, which are used as indicators of the effectiveness of fibers, especially in weight-sensitive applications such as aircrafts and space vehicles.

Notice in the table that Kevlar fibers offer the highest strength-specific weight ratio, and graphite provides the highest stiffness-specific weight ratio.

- The preparation methods and characteristics of the most important fibers used in composite materials are glass fibers: Obtained by pulling the glass (silicon + sodium and calcium carbonate) through the small orifices of a plate made of platinum alloy.
- Carbon fibers: Obtained by heating rayon, polyacrylonitrile (PAN) precursor fibers, or pitch (residues of the petroleum products) at high temperature of about 1700°C in an inert (nitrogen) atmosphere to carbonize the fibers. Then only the hexagonal carbon chains remain. To get graphite fibers the heating exceeds 1700°C, to graphitize the carbon fibers partially. Higher temperature increases the fiber modulus but decreases its strength. The fibers can be bent easily, as they are far thinner than the human hair. Thus, carbon or graphite fibers can be woven into fabric.

TABLE 14.1
Effective Properties of Commonly Known Fibers as Compared with Structural Metallic Materials

Fiber or Metal Wire	Diameter (μm)	Specific Weight (kN/m³)	Tensile Strength (GPa)	Modulus of Elasticity E (GPa)	Strength/Specific Weight (km)	E/Specific Weight (mm)
E-glass	16	25.0	3.4	72	136	2.9
S-glass	10	24.4	4.8	86	196.7	3.5
Carbon		13.8	1.7	190	123.2	13.8
Graphite	7	13.8	1.7	250	123.2	18.1
Kevlar	12	14.2	2.9	130	204	9.2
Beryllium	—	18.2	1.7	300	93.4	16.5
Boron	100	25.2	3.4	400	134.9	15.9
Silicon	14	21.6	—	95	—	4.4
Aluminum		26.3	0.62	73	23.6	2.8
Titanium		46.1	1.9	115	41.2	2.5
Steel		76.6	4.1	207	53.5	2.7

- Aramid (Kevlar) fibers: Aromatic polyamides obtained by synthesis at −10°C, and then drawn to get high modulus of elasticity; first commercially used in the early 1970s, as a replacement of steel wires in racing tires.
- Boron fibers: Made by chemical vapor deposition of boron chloride and hydrogen at about 1200°C. Boron fibers are deposited on a tungsten wire. The obtained boron fibers have a relatively large diameter of about 100 μm; thus, they cannot be woven or bent.
- Silicon fibers: Made by chemical vapor deposition of methyl trichlorosilane (a chemical compound containing silicon, hydrogen, and chlorine) and hydrogen at 1200°C, analogous to boron.
- Whiskers: Have basically the same perfectly aligned structure of a fiber but very short and stubby. The length is in the range of few hundred times the diameter of a whisker. Whiskers of iron have significantly higher strength than steel in the bulk form. The actual strength of iron whiskers is 13 GPa, while the strength of steels is in the range of 0.4–1.6 GPa.

14.3.2 MATRIX MATERIALS

In order to make use of the advantages of the fibers or whiskers, they have to be bonded together to take the form of a structural part that can carry loads. The binder material is called the matrix that supports the embedded fibers in place, as well as sharing the load with them. Generally, the matrix is of considerably lower strength and stiffness, and sometimes lower density than the fibers or whiskers. Nevertheless, the combinations of the fibers or whiskers with the matrix material leads to very high strength and stiffness yet still have low density.

The fibers reinforce the matrix along the line of their length. Therefore, reinforcement may be one- (1D), two- (2D), or three-dimensional (3D) according to the orientation of fibers, as shown in Figure 14.3. The 1D reinforcement gives maximum strength in one direction. In the 2D reinforcement, the fibers are distributed in two perpendicular directions, giving maximum strength in their orientations. In the 3D reinforcement, fibers are randomly distributed in the bulk of the matrix giving isotropic strength in all directions.

Matrix materials can be polymers, metals, ceramics, or carbon. The cost of a matrix increases in the same order as does the temperature resistance.

1. Polymer matrix: The composite materials having polymer matrix are called polymer matrix composites (PMC). Both thermoplastics and thermosets are used as matrix materials, with the latter being most widely used. Typical thermoplastic matrix materials include polypropylene, polyphenylene sulfone, polyamide, and polyether-ether-ketone (PEEK). Typical thermosetting matrix materials include epoxy, phenolic, polyester, polycarbonate, silicone, and polyurethane. Details of most of these materials are discussed in Chapter 11. A widely known example of PMC is fiberglass, which is a glass fiber-reinforced plastic (GFRP).
2. Metal matrix: The composite materials having metallic matrix are called metal matrix composites (MMC). Typical metal matrix materials include aluminum alloys, titanium alloys, and nickel–chromium alloys. MMC can be processed in the liquid state (casting) or in the solid state (powder metallurgy). The simplest method for processing continuously

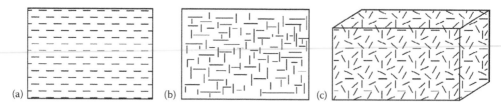

FIGURE 14.3   Fiber reinforcement orientations: (a) 1D, (b) 2D, and (c) 3D.

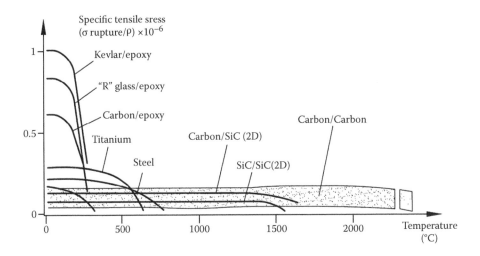

FIGURE 14.4   Effect of temperature on the specific strength of different composites. (From Gay, D. et al., *Composite Materials Design and Applications*, CRC Press, Boca Raton, FL, 2003. With permission.)

reinforced composites with Al- or Zn-based matrices is by liquid metal infiltration using squeeze casting or pressure die casting. Powder metallurgy is preferred for higher melting point alloys such as Cu-based alloys. In both techniques, interface adhesion can be modified by additions of alloying elements in the matrix.

3. Ceramic matrix: The composite materials having ceramic matrix are called ceramic matrix composites (CMC). These are the least common matrix materials. CMCs can be cast from molten slurry around stirred-in fibers with random orientation or with preferred flow direction. Alternatively, the ceramic matrix can be vapor deposited around already in-place fibers. Aluminum oxide and silicon carbide are typical matrix materials for CMCs, especially in high temperature applications.

4. Carbon matrix: The carbon matrix is commonly used with carbon fibers in what is known as carbon/carbon or C/C composite. These C/C composites retain their properties at temperatures over 2000°C. Figure 14.4 shows that the specific resistance of C/C composites is retained at much higher temperatures than other principal composites. A three-dimensional woven structure of carbon fibers is first prepared in the axial radial and circumferential directions. Then the preform is pressure impregnated with liquid pitch to fill the voids between the yarns. It is then heat treated at 2550°C to produce a rigid multidimensional reinforced graphite structure. The impregnation, densification, and graphitization cycles are repeated until the final desired density is obtained. Carbon–carbon is well suited to structural applications at high temperatures or where thermal shock resistance and/or a low coefficient of thermal expansion is needed. Typical applications include nose cones of ballistic missiles, wing leading edges of space shuttles, and brake discs and pads in racing cars.

14.4   PARTICULATE COMPOSITE MATERIALS

Particulate-reinforced composite materials consist of particles or flakes of one or more materials embedded in a matrix of another material. The particles can be metallic or nonmetallic, so is the matrix. Thus, there are four possible combinations of particulate composite materials.

1. Composites with nonmetallic particles in nonmetallic matrix: The most commonly known model of this composite is concrete. It consists of sand and gravel embedded in a matrix of a mixture of cement and water chemically reacted to form a hard bulk. Sand and gravel are

loose particles from rocks of sandstone, limestone, or basalt. Sand particles are usually less than 2 mm, while gravel particles range from 2 to 64 mm. Both are formed by natural abrasion and weathering of these rocks or by quarrying and crushing of hard-wearing rocks. Concrete has relatively high compressive strength but significantly lower tensile strength. Therefore, it is usually reinforced by materials that are strong in tension such as steel.

Flakes of mica (sheet like silicate mineral) or glass can form an effective composite material when embedded in a glass or plastic matrix. Flakes offer two-dimensional stiffness and strength in their plane, as opposed to the one axial dimension in fibers. Usually, flakes are packed in parallel order, resulting in higher density than fiber packing. The high overlap of flakes in a composite material makes it impermeable to fluids. Mica-in-glass composite materials are good insulators in electrical appliances. Glass flakes in plastic matrices offer stiffness and strength equivalent to or higher than GFRP because of the higher packing density.

2. Composites with metallic particles in a nonmetallic matrix: Aluminum paint is made of aluminum flakes suspended in paint (plastic matrix). Upon application, the aluminum flakes orient themselves parallel to the surface giving superior coverage and appearance. Similarly, silver flakes can be applied to give good electrical conductivity. Cold solder consists of a metal powder suspended in a thermosetting resin. When applied, the composite is strong and hard in addition to high thermal and electrical conductivity. Inclusion of copper in an epoxy resin greatly increases conductivity. Other metallic additives can improve conductivity, lower the coefficient of thermal expansion, and reduce wear.

Solid-rocket propellants are made of aluminum powder and perchlorate oxidizers (salts derived from perchloric acid $HClO_4$) in a flexible polymeric binder such as polyurethane or polysulfide rubber. The particles comprise as much as 75% of the propellant leaving only 25% for the binder. The composite material must be uniform to provide steady burning reaction and controlled thrust.

3. Composites with metallic particles in a metallic matrix: Metal particles are used to reinforce another metal matrix (without dissolving or alloying) in order to gain additional properties. Lead particles are embedded in copper alloys or steel to improve their machinability. Additionally, lead is used as a natural lubricant in bearings made of copper alloys. Another good example of these composites is the reinforcement of a ductile matrix metal with particles of tungsten, chromium, or molybdenum to gain the elevated temperature properties of the brittle constituents, yet having a ductile composite. The process used is liquid centering, which involves infiltration of the matrix material around the brittle particles.

4. Composites with nonmetallic particles in a metallic matrix: Ceramic particles are commonly embedded in metal matrix to form a composite material known as cermet (cer from ceramic and met from metal). A cermet is ideally designed to have the optimal properties of high temperature resistance, abrasion resistance, and hardness from ceramics and the ability of a metal to undergo plastic deformation. The ceramic particles may be oxides, borides, carbides, nitrides, or alumina. The metal matrix is usually nickel, molybdenum, or cobalt. Cermets could be classified as MMC if the metal matrix represents a considerable volume of the material, but when the metal volume percentage is limited, it is classified as CMC. Typical applications of cermets include cutting tool tips, wire drawing dies (tungsten carbides in a cobalt matrix), valves (chromium carbide in a cobalt matrix), turbine parts (titanium carbide in either cobalt or nickel), nuclear reactor fuel elements (uranium oxide in stainless steel), and control rods (boron carbide in stainless steel).

14.5 LAMINATED COMPOSITE MATERIALS

Laminated composite materials consist of layers of at least two different materials bonded together. Lamination is applied to merge the high characteristics of the constituent layers to gain a more useful material. Similar to other composite materials, the properties that can be emphasized by lamination include strength, stiffness, low density, corrosion resistance, wear resistance, thermal insulation, and acoustical insulation, as well as the attractive appearance. Typical examples of laminated composite materials are as follows.

14.5.1 BIMETALS

Bimetals are laminates made of two metals with significantly different coefficients of thermal expansion bonded together. Bonding is usually carried out by rolling strips of the two metals together under high pressure. Under temperature change, the two metals tend to expand to different lengths, but being bonded, they deflect in a pattern proportional to the temperature different. Therefore, they are used as temperature measuring and control devices known as bimetallic thermometers or thermostats. The simplest form of a bimetal thermometer is a cantilever strip usually made of bronze and invar (nickel-iron alloy) bonded together. When subjected to temperature change, the bronze tends to expand more than invar, which bends the strip as shown in Figure 14.5a. This is usually used for regulating temperature in ovens and irons. The same principle is applied by twisting the bimetal to a spiral and helical shape, so that the deflection due to temperature change causes a pointer to move. The helical type thermometer is shown in Figure 14.5b.

Bimetals have other applications such as tinned steel sheets that are used for making food and juice cans. The tin layer is corrosion resistant. Similarly, galvanized steel sheets are zinc coated for corrosion-resistant applications.

14.5.2 CLAD METALS

Cladding or sheathing is the covering of a metal with another one for gaining certain characteristics. In the 1960s, aluminum wire clad with about 10% copper was introduced as a replacement of the copper wire for electrical wiring. Aluminum wire by itself is economical and lightweight, but it overheats and is difficult to connect to terminals at wall switches and outlets. Aluminum wire connections break because of fatigue as a result of repeated thermal expansion and contraction when the current is turned on and off. This rupture causes short circuits and consequently potential fires. On the other hand, copper wire is relatively expensive and heavy but is much less sensitive to thermal expansion and is easier to connect to switches or outlets. The copper clad aluminum wire is lightweight and connectable, stays cool, and is less expensive than copper wire. Cladding is usually done by extruding both metals through a die or by rolling sheets together under high pressure. Many

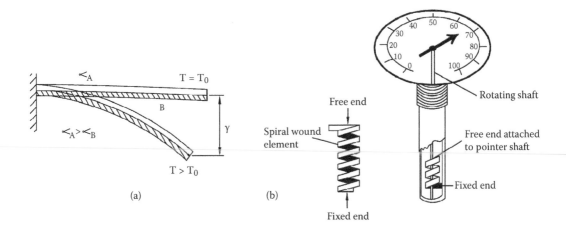

FIGURE 14.5 Bimetal thermometers: (a) cantilever strip type and (b) helical type.

countries currently use cladding to make coins from two metals, a cheaper metal as a filler such as zinc, and an impressive expensive clad.

14.5.3　Sandwich and Honeycomb Structures

A sandwich structural composite is another subclass of laminated composite materials. It is fabricated by attaching two thin but stiff skins to a lightweight but thick core by adhesive bonding or welding. The core material normally has low strength, but its higher thickness provides the sandwich composite with high flexural rigidity, with overall low density. The sandwich plate can also have excellent thermal insulation characteristics. However, sandwich materials are not damping (no acoustic insulation), and have the risk of buckling greater than classical structures.

The facing skin sheets can be made of metal (stainless steel or aluminum), laminated wood, thermoplastics, or asbestos. Core materials include expanded polymer (see section 13.8.11), wood plate, ribbed metal strips, or hexagonal shaped honeycomb cells, as shown in Figure 14.6. The table attached to the figure shows that for a certain design of a honeycomb structure, when the honeycomb layer thickness is 3 times the cover thickness, the stiffness of the structure increases 37 times, and the flexural strength increases 9.2 times, while the weight only increases by 6%. Honeycomb cores can be made of a wide variety of materials including aluminum, aramid, Kevlar, fiberglass, or carbon.

14.5.4　Laminate Glass

The concept of protection of a material layer by another, as described for metals, has been extended to manufacture safety glass by covering a layer of plastic with two glass layers. Ordinary window glass is durable enough to retain its transparency under extreme weather conditions. However, it is brittle and is hazardous when broken as it produces sharp edged pieces. On the other hand, polyvinyl butyral (PVB) is a plastic usually used for applications that require high optical clarity, toughness, and strong binding and adhesion to many surfaces. However, it is flexible and susceptible to scratching. Laminated safety glass is formed of a PVB layer, sandwiched by two layers of glass. The glass in the composite material protects the plastic from scratching and gives it stiffness. The plastic provides the toughness of the composite material and holds the glass from scattering when broken. In case of impact not enough to completely pierce the composite, it produces a characteristic "spider web" cracking pattern. Laminated glass is normally used when there is a possibility of human impact or where the glass could fall if shattered. Typical applications include skylight glazing, automotive windshields, exterior storefronts, and hurricane-resistant construction. The PVB

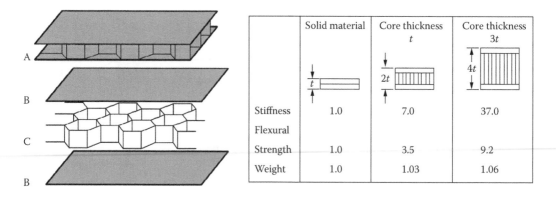

	Solid material	Core thickness t	Core thickness $3t$
Stiffness	1.0	7.0	37.0
Flexural Strength	1.0	3.5	9.2
Weight	1.0	1.03	1.06

FIGURE 14.6　Honeycomb structure: (A) the bonded structure and (B, C) the lamellae before bonding. The attached table shows the effect of thicknesses on the stiffness and the flexural strength of the structure.

interlayer also gives the glass a much higher sound insulation rating, owing to the damping effect, and also blocks 99% of transmitted ultraviolet light.

14.5.5 POLYMERIC-BASED LAMINATES

Many materials can be saturated with different types of plastic layers for a variety of applications. A typical example of polymer based laminate material is Formica, which is the trade name for heat-resistant, wipe-clean boards with a hard and durable surface. Formica is made of layers of heavy Kraft paper (made of chemical pulp from softwood, usually used for envelopes and packaging), impregnated with phenolic resin, overlaid by a plastic saturated decorative sheet and a cellulose mat. The layers are bonded together under heat and pressure. Similar laminate materials are further developed for the same purpose of durable surfaced boards, which are often referred to as high pressure decorative laminate (HPDL). Aluminum composite panel (ACP) is a useful variation of the same principle, describing flat panels that consist of a plastic core layer bonded between two aluminum sheets. Aluminum sheets can be coated with highly nonreactive polymer paints. ACP is frequently used for partitions, false ceilings, and external cladding of buildings, as it is very rigid and strong despite its light weight. Owing to the ability of painting the aluminum in any kind of color, ACPs are produced in a wide range of metallic and nonmetallic colors as well as patterns that imitate other materials, such as wood or marble.

Layers of glass or asbestos fabrics can be impregnated with silicones to provide a composite material with significant high temperature properties. Glass, Kevlar, or nylon fabrics can also be laminated with various resins to give an impact and penetration resistant composite that is exclusively applicable as lightweight personnel armor.

14.6 COMBINATIONS OF COMPOSITE MATERIALS

Numerous multiphase composite materials exhibit more than one characteristic of the different classes discussed before. A typical example is the reinforced concrete that combines particulate and fibrous composites. Also, laminated fiber-reinforced composite materials represent a hybrid class that obviously combines fibrous and laminated classes. It comprises layers of fiber-reinforced composite materials bonded together, with the fiber directions of each layer typically oriented in different directions to provide different strengths and stiffness of the laminate in various directions. Thus, the strengths and stiffness of the laminated fiber-reinforced composite material can be tailored to the specific design requirements of the structural element being built. Such materials are used in aircraft wing panels and body sections, rocket motor cases, boat hulls, and tennis rackets.

14.7 FABRICATION OF COMPOSITE MATERIALS

Composite materials are formed into shapes by many different processing technologies. The two or more phases of a composite are typically produced separately before being combined into the composite part geometry. The matrix materials are generally produced by the processes described in related chapters, i.e., a metal matrix is processed by casting for metals that can be molten or PM processes for refractory metals, a polymer matrix is processed by polymerization techniques, and a ceramic matrix is commonly formed by powder metallurgy techniques. Because particulate materials processing is discussed in the chapter on powder metallurgy, the emphasis in this chapter will be on the fabrication of PMCs.

Despite the diversity of fabrication processes of composite materials, they should ensure some general characteristics of the formed composites that are as follows.

1. Good bonding between matrix and fibers: To provide reinforcement by improving mechanical properties such as stiffness and strength. The bond is provided by the matrix material that transfers the tensile stress applied on the composite through fibers in the form of shear stress τ as shown in

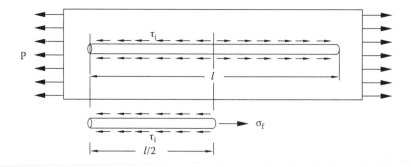

FIGURE 14.7 Shear stress transfer between fiber and matrix. (From Hoa, S. V., *Principles of the Manufacturing of Composite Materials*, DEStech Publishers, Lancaster, PA, 2009. With permission.)

Figure 14.7. (The fiber supports the load via its tensile strength σ_f, while the matrix provides the load transfer via shear strength τ_i where i refers to the interface between the matrix and the fiber.) The shear stress is usually maximum at the end of the fiber and minimum toward its center as the figure indicates. Assuming constant shear stress to simplify calculations, equilibrium of the half fiber shown in the figure gives

$$\frac{\pi}{4}d^2\sigma_f = \pi d \frac{l}{2}\tau_i \tag{14.1}$$

where d is the fiber diameter and l is its length. This equation provides the critical aspect ratio of the fiber l/d in the form

$$l/d = \sigma_f/2\tau_i \tag{14.2}$$

Equation 14.2 can be presented to give the critical length of the fiber l_c in the form

$$l_c = \frac{\sigma_f}{2\tau_i}d \tag{14.3}$$

Then the reinforcement effect depends on the aspect ratio l/d. If the aspect ratio is smaller than the critical value given by the equation, failure will occur owing to slipping between the reinforcement and the matrix, making the reinforcement ineffective. However, because of stress magnification at the end of the fibers, the smaller the number of ends of the fibers, the better the reinforcement effect. Accordingly, long continuous fibers give better reinforcement than short fibers. Spots of poor bonding are called dry spots, which cannot transfer the shear load between the fiber and matrix, and act as nuclei for cracks. However, these dry spot locations are useful in absorbing impact energy.

2. Proper orientation of fibers: Because properties such as stiffness and strength are very sensitive to this orientation. Deviation of the fibers by 10° from the required orientation leads to drop of stiffness by more than 30%. This deviation from the intended orientation may be due to improper position of the layer during the lay-up step or due to the flow of liquid resin that pushes the fibers during the filling period in liquid composite molding (LCM).

3. Good amount of volume fraction of fibers: Because fibers improve mechanical properties of the composite material; therefore, the greater the amount of fibers, the higher will be these properties. The amount of fibers is usually expressed in terms of volume fraction, v_f, which is defined as

$$v_f = \frac{V_f}{V_c} \tag{14.4}$$

where V_f is the volume of fiber and V_c is the volume of the composite material. The stiffness of a unidirectional composite along the axial direction is given by the rule of mixtures

$$E_c = E_f v_f + E_m v_m \tag{14.5}$$

where the subscript f refers to fibers and m refers to matrix.

The fiber volume fraction and matrix volume fraction are related by

$$v_f + v_m + v_v = 1 \tag{14.6}$$

where the last term refers to volume fraction of voids. However, for high-quality composite material, the amount of voids should be minimum (less than 1%), and Equation 14.6 can be reduced to

$$v_f + v_m = 1 \tag{14.7}$$

Substitution of Equation 14.4 into Equation 14.2, the modulus can be expressed as

$$E_c = (E_f - E_m)v_f + E_m \tag{14.8}$$

Equation 14.8 shows that the modulus is linearly proportional to the fiber volume fraction. Therefore, the larger the fiber volume fraction, the higher are the mechanical properties. However, it should be noticed that the volume fraction of fibers cannot reach a value of 1; otherwise, the composite will be a bundle of fibers with no matrix. Practically, fibers in a composite material should not touch one another to allow for the matrix bonding fibers together. This puts a limit to the maximum fiber volume fraction depending on the fiber type, diameter, length, and required arrangement of fibers in the composite. The theoretical maximum limit of the volume fraction of fibers is 0.785 for open arrangement packing of fibers and 0.907 for closed (hexagonal) packing. In practice, fiber volume fractions achieved are around 68% for hand lay-up using autoclave molding and may be 70% for pultrusion.

Example 1

Calculate the maximum theoretical volume fraction of fibers in the matrix, assuming square arrangement.

Assuming square arrangement of $n \times n$ fibers in the matrix with no gap between fibers as shown in Figure 14.8. On the basis of this arrangement, the volume fraction can be calculated as

$$v_f = \frac{\text{volume of fiber}}{\text{total volume}} = \frac{n^2 \times \pi \times d^2}{4 \times (nd)^2} = \frac{\pi}{4} = 0.785$$

Then the limiting maximum volume fraction under this condition is 0.785.

4. Uniform distribution of fibers within the matrix material: Nonuniform distribution of fibers in the matrix leads to a heterogeneous structure. The region where there is more matrix than fiber is known as a resin rich area, which has lower stiffness and strength. Under loading, such areas can act as locations for cracks to nucleate.

5. Proper curing of the resin in PMCs: The resin is usually added to fibers in the form of a low viscosity liquid so that it can wet the surface of fibers. After complete wetting, the resin needs to solidify and harden. For the thermoset resin, this is known as curing, and for the thermoplastic resin this is known as solidification. In either case, the resin needs to be hard and stiff to allow for the reinforcement effect to take place. In regions where the resin is not completely cured (not hard enough), crack nucleation areas can arise.

6. Limited amount of voids and defects: Voids and defects may develop during fabrication of composite materials. Voids can arise owing to lack of compaction between layers or to low pressure

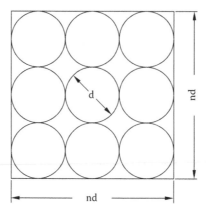

FIGURE 14.8 Square arrangement of the fibers in the matrix. (Adapted from Hoa, S. V., *Principles of the Manufacturing of Composite Materials*, DEStech Publishers, Lancaster, PA, 2009. With permission.)

of the resin during curing. Voids should be reduced to a commonly accepted limit of 1% as stated earlier. Other defects include delamination between layers, cracks, fiber misorientation, or nonuniform fiber distribution.

7. Good dimensional control of the final part: Polymeric resins shrink during curing or solidification by a percentage range of 5%–8%, depending on the material type. This shrinkage may lead to residual stresses, poor dimensional tolerances, or warping of the composite. For large structures such as wings of an aircraft, a few percentage of shrinkage of the composite part can lead to significant deformation of the structure. Further, the surface finish of parts such as automobile panels may be adversely affected by shrinkage. Low-profile additives are commonly added to resins to control shrinkage.

The fabrication processes can be generally classified into major groups which are molding processes, prepreg fabrication, filament winding, and pultrusion. These processes will be discussed in the following sections.

14.8 MOLDING PROCESSES

Molding processes are usually applied in fabricating PMCs. Forming of composites by molding varies depending on the nature of the part, the number of parts, and the cost. However, the steps used in all molding processes are almost the same as presented in Figure 14.9.

The fibers used in molding can be in the form of

- Single filaments with an average diameter of 10 μm.
- Tows: Untwisted bundles of a large number of continuous filaments. There are tows 3k, 6k, and 12k. Tows of 3k designate bundles of 3000 filaments.

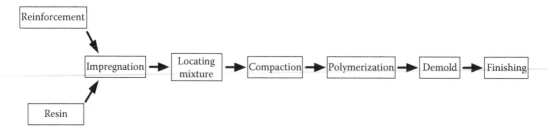

FIGURE 14.9 Steps of molding of composite materials.

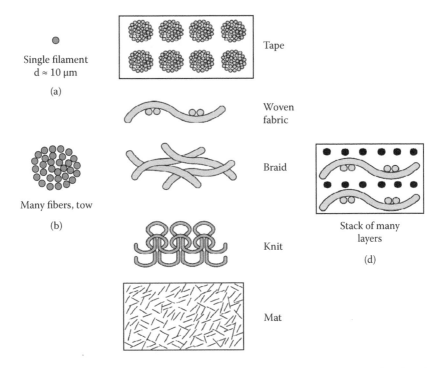

FIGURE 14.10 Forms of fibers at different scales. (From Hoa, S. V., *Principles of the Manufacturing of Composite Materials*, DEStech Publishers, Lancaster, PA, 2009. With permission.)

- Tapes: Combinations of tows in adhesives or
- Fabrics, in which tows are woven together to form mats, weaves, braids, or knits.
- Stack of many layers compacted together.

These different forms of fibers are shown in Figure 14.10.

The mold material can be made of metal polymer, wood, or plaster. Various molding processes are presented in this section.

14.8.1 CONTACT MOLDING

This is an open molding technique, where there is one mold, either male or female. It is also known as the hand lay-up technique. Dry tows or dry fabrics are laid on the mold, and liquid resin (with catalyst or accelerator) is poured and spread with a roller and left to cure. Instead, preimpregnated layers are placed on the mold with the fibers oriented alternately in opposite directions as shown in Figure 14.11. Compaction is performed using a roller to squeeze out air pockets. The setting (curing) of the polymer may take from a few minutes to a few hours depending on the accelerator and the part size; thus, the process may be relatively slow but necessary for tailored large products such as boats, as shown in Figure 14.11.

14.8.2 COMPRESSION MOLDING

In this technique, the impregnated layers are located in a mold half, and a counter mold is used for compression of the part as shown in Figure 14.12. Limited pressure of 1 to 2 bars is applied to the assembly through a press. Curing takes place either at room temperature or heating may be

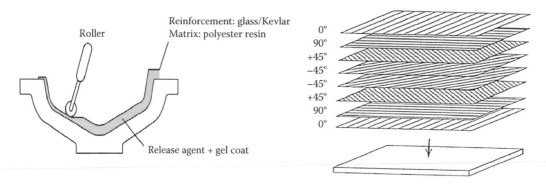

FIGURE 14.11 Contact (hand lay-up) molding of a boat from layers with different orientation.

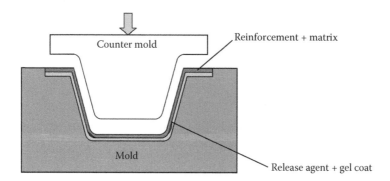

FIGURE 14.12 Compression molding. (From Gay, D. et al., *Composite Materials Design and Applications*, CRC Press, Boca Raton, FL, 2003. With permission.)

required. The process is applicable for average production rates of about 200 parts per day, and typical products are automotive and aircraft parts.

14.8.3 VACUUM MOLDING

A vacuum molding process uses a flexible molding skin (felt) having an imprint face arranged inside an airtight compartment. The process is also known as depression molding or bag molding. As in the case of contact molding, an open mold is used, and the impregnated reinforcements are placed with the felt on top. In the case of sandwich plates, the cores are also added. A sheet of soft nonadhesive plastic (polyvinyl or nylon) is used for sealing (adhesively bonded to the mold perimeter). Vacuum is applied under the plastic sheet as shown in Figure 14.13, which leads to compaction of the part and elimination of air bubbles. It also presses the flexible molding skin onto the plastic material to form a molded surface with sculptures and asperities giving a decorative or functional granular appearance. The molding skin is made of a flexible, anti-adhesive material, notably silicone, for ease of removal from the mold after the plastic material has hardened. The composite is polymerized in an oven or using an autoclave under pressure (7 bars in the case of carbon/epoxy to achieve better mechanical properties). Heating for polymerization can be applied through electron beams for total thickness up to 25 mm, or X-rays for total thickness up to 300 mm. This process has applications in aircraft industry with a rate of a few parts per day.

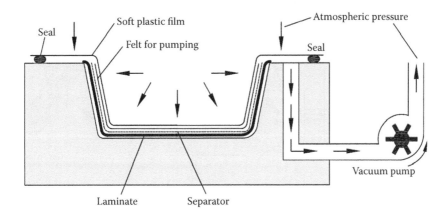

FIGURE 14.13 Vacuum molding/depression or bag molding. (From Gay, D. et al., *Composite Materials Design and Applications*, CRC Press, Boca Raton, FL, 2003. With permission.)

14.8.4 LIQUID COMPOSITE MOLDING

This process avoids the relatively lower quality of contact molding (due to lack of control of the compaction, entrapped air, and slow curing), and the higher cost of autoclave molding. The main steps of the process are shown in Figure 14.14 and discussed as follows.

1. Preforming: Where dry fibers are packed into a preform with the required configurations of the part to be produced. The starting materials can be tows, random mats, or woven fabrics. The finished preform is usually woven, compression molded, braided, or knitted together. Small amounts of adhesive or small stitches may be required to hold the preform in shape.
2. Positioning into the mold: Where the mold is made of two metallic halves. The finishing of the final product depends on the quality of the mold surfaces.
3. Resin infusion: The two halves of the mold are closed, and resin is infused under pressure to wet the fibers and fill up any cavity within the preform. Injection pressure is in the range of 6 to tens of MPa. The time duration of resin infusion depends on the size of the part and on the reactivity of the resin system. It may range from seconds for resins

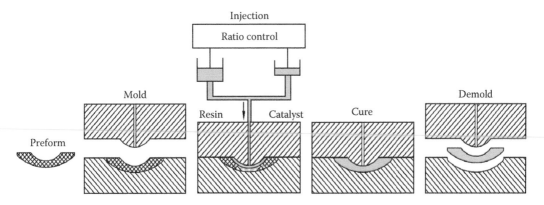

FIGURE 14.14 Resin injection molding of composite materials. (From Hoa, S. V., *Principles of the Manufacturing of Composite Materials*, DEStech Publishers, Lancaster, PA, 2009. With permission.)

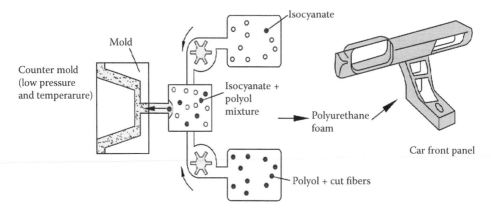

FIGURE 14.15 Foam injection molding of fiber reinforced polyurethane. (From Gay, D. et al., *Composite Materials Design and Applications*, CRC Press, Boca Raton, FL, 2003. With permission.)

with fast reactivity such as cyanate to the order of minutes or hours for resins with slower reactivity.

4. Curing: Takes place after complete infusion of the resin into all cavities of the preform. It is important that the resin does not gel during infusion; otherwise, short shots are obtained. The mold pressure is maintained during curing. Normally, curing agents and catalysts are injected with the resin as shown in the figure. Curing can be accelerated by heating.

5. Demolding: Where the part is removed from the mold after complete curing.

The cost-effective range for LCM is in the middle range in the production volume (in the order of 20,000 to 60,000 parts per year). Compared to compression molding, much less pressure is required. Therefore, the mold for LCM is less expensive. The process is applicable for fabrication of complex three-dimensional parts. It has found acceptance for automotive parts.

14.8.5 MOLDING BY FOAM INJECTION

The process is applicable for production of glass fiber–reinforced rigid polyurethane. Polyurethane is prepared by mixing polyisocyanate with polyol and cut fibers in addition to a catalyst and a chain extending agent. The mixture is injected into a mold as shown in Figure 14.15. Such products have good mechanical and thermal properties and an excellent finish. They remain stable for a long time. A typical application is the car front panel as demonstrated in the figure.

14.8.6 CENTRIFUGAL MOLDING OF TUBES

In this process, a suitable resin in the liquid form, mixed with cut fibers, are forced in a rotating mold. The centrifugal force provides homogeneous distribution of fibers in the resin during curing or solidification. The process is analogous to the centrifugal casting process (Chapter 5). The fabricated tube has good surface condition including the internal surface.

14.9 PREPREG FABRICATION

Prepregging involves the incorporation of the partially cured resin with the fibers in a sheet-like shape. A schematic of the prepregging machine is shown in Figure 14.16. The process involves feeding of dry fibers from reels through stations of combs where the fibers are spread out. The fibers

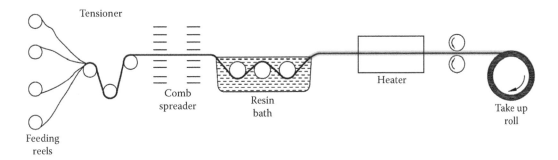

FIGURE 14.16 Schematic of a prepregging machine. (From Hoa, S. V., *Principles of the Manufacturing of Composite Materials*, DEStech Publishers, Lancaster, PA, 2009. With permission.)

then enter into a bath of liquid resin where they are wetted. Subsequently, the fiber/resin combination is heated to change the liquid resin into a partially cured state. During partial curing, about 30% of the cross links are formed. The partially cured resin is viscous enough to help keep the fibers in the configuration of flexible flat sheets. This fiber/viscous resin combination is called prepreg. The resin in the prepreg is usually thermoset such as epoxy. However, thermoplastic resins such as vinyl ester are recently utilized. The fibers are carbon, glass, or fiber at about 60% by volume. Prepregs are usually covered with backing paper on both sides for protection and handling purposes. Then the prepregs are rolled up for storing and shipping. They should be used to form a composite product within a few months; otherwise, complete curing may take place during storing for a long time leading to hardening of the resin and losing the sticky effect or the ability of bonding stacks of layers. To slow down curing during storing, prepreg rolls are stored in freezers at temperature about –5°C. During shipping, the prepregs are contained inside refrigerated bags.

Prepregs are available in the form of tapes with unidirectional fibers or woven fabrics. Widths of prepreg rolls vary from 25 to 300 mm. In order to use these sheets to form a composite product, they are taken out of the freezer and left at room temperature for a few hours to become pliable (to conform to the mold contour). Then they are cut to the desired configuration. Cutting of prepregs is similar to cutting cloth fabrics, except that the fibers need higher shearing force. It used to be done by hand, but recently it has become mechanized and computerized not only to speed up the process but to reduce waste, especially when cutting at an angle with the axial direction. Stacked prepregs can be molded using most of the molding processes described in the previous section.

14.10 FILAMENT WINDING

Filament winding is a fabrication technique for creating composite material structures with axial symmetry. The process involves winding rovings (narrow bundles of fibers) or monofilaments under varying amounts of tension over a male mold or mandrel. The mandrel rotates while a carriage carrying the filaments is traversed back and forth at a speed synchronized with the rotation speed, laying down fibers in the desired pattern. The most common filaments are carbon, Kevlar (Aramid), or glass fiber and are coated with resin just before winding by passing them through a resin tank as shown in Figure 14.17. Prepregs can also be used for winding. The simplest type of filament winding machine is the two-axis type for producing pipes. The availability of additional degrees of freedom can be useful in winding at the end of the part, such as spherical shapes or shapes with variations in cross section. Winding machines with multiple degrees of freedom robots are available for complicated shapes. Once the mandrel is completely covered to the desired thickness, the mandrel is placed in an oven or autoclave to cure (set) or solidify the resin. Then the mandrel is removed, leaving the hollow final product.

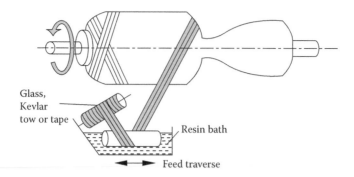

Glass,
Kevlar
tow or tape
Resin bath
Feed traverse

FIGURE 14.17 Filament winding on a complex mandrel.

Usually, all pressure vessels or pipes made by filament winding of composites have a liner. The function of this liner is to seal the liquid or gas inside the vessel or pipe. Generally, the fibers provide the strength and stiffness required for the structure. However, the structure is not seal tight against leak or sweep of the fluid out of the container. In case of inflammable fluid, this can be dangerous even if the fiber network is sufficient to withstand the pressure. Therefore, to seal the fluid, a flexible liner is used. It can be made of PVC, rubber, or thin aluminum container. If the liner is stiff enough to withstand the compression of the winding force, it can be used as a mandrel. If not, or in cases where it is necessary to take the mandrel out, an extractable, collapsible, breakable, or dissolvable mandrel can be used as follows.

- Extractable mandrel: Steel tubes can be used for winding fiberglass pipes. They are made longer than the required pipe length. After curing of the pipe, the mandrel is extracted by a winch. A release agent should be spread on the outside of the mandrel for ease of extraction. Alternatively, inflatable rubber bladder mandrels can be used. They should be inflated with enough air pressure to withstand the winding force. They may be used to supplement the stiffness of a thin liner.
- Collapsible mandrel: Consists of segmented surface made of several pieces that can be expanded to take the final shape of the mandrel by collapsible linkage (similar to the operation of the collapsible umbrella). When the wound part is cured, the mandrel is collapsed to be removed.
- Breakable mandrel: Made of plaster that can be molded to the required shape. When the product is cured, the plaster is broken and removed. For small tubular composite parts, glass mandrels can be used and broken to be removed when the product cures.
- Dissolvable mandrel: Made of a material that can be dissolved in solution after curing of the composite. It can be made of low melting alloys or eutectic salts that can be melted at moderate temperature after curing of the part. Another example is soluble plaster that can be easily washed out after product curing. Alternatively, a mixture of sand with polyvinyl alcohol is an excellent choice for the mandrels of relatively large diameters up to 1.5 m, at limited quantities. It dissolves readily in hot water but requires careful molding control.

Filament winding is well suited to automation, where the tension on the filaments can be carefully controlled. Filaments that are applied with high tension results in a final product with higher rigidity and strength, whereas lower tension results in more flexibility in the product. The orientation of the filaments can also be carefully controlled so that successive layers are plied or oriented differently from the previous layer. The angle at which the fiber is laid down will determine the properties of the final product. A high angle "hoop" will provide crush strength, while a lower angle pattern (known as a closed or helical) will provide greater tensile strength. Typical products include tubes and pipes, bicycle forks, power transmission shafts, yacht masts, pressure vessels, storage tanks, and missile casings. Applications of pressure vessels include oxygen tanks used in aircraft and by mountain climbers and compressed natural gas cylinders for vehicles. Pipes are used for conducting corrosive fluids.

14.11 PULTRUSION

Pultrusion is a continuous process of fabricating long parts of composite materials with constant cross section. The process is carried out on three steps as shown in Figure 14.18.

- Step a: Fiber tows are drawn from fiber racks, routed through a series of guides, and then traverse through a bath of low viscosity resin for impregnation.
- Step b: Upon exiting from the resin bath, fibers are collimated into an aligned bundle and entered into a heated die with the required shape. Through the die, the resin flows and wets the die, the assembly is compacted, and the resin cures such that the fiber/resin system becomes solid.
- Step c: Upon exiting from the die, the composite structural component is pulled by a puller and is cut to the required lengths.

The name pultrusion combines the terms "pull" and "extrusion," because the die is similar to that of an extrusion process for metallic materials. The process yields parts at relatively lower cost, because fibers in the tow form are less expensive than in the prepreg or woven form. However, there may be limitations on quality and unidirectionality of fibers. In order to keep low cost of the products, E glass fibers are mostly used, even though applications may include more expensive fibers such as S glass, carbon, and Kevlar. Resins are usually low cost polyester or vinyl ester, even though other resins such as epoxy, phenolic, and thermoplastic (such as PBT) have been used. The volume fraction of the fibers should be high for high strength and stiffness. Pultrusion offers high rates of production that vary between 0.5 and 3 m/min, depending on the configuration of the profile and the curing or polymerization rate. Table 14.2 represents the structural properties of pultruded glass fiber–reinforced polyester rods and bars, as compared to stainless steel and low-carbon steel rods. It should be noticed that the flexural strength and tensile strength of a pultruded bar are about 3 times larger than a stainless steel bar and about 3.5 times higher than a low-carbon steel bar, whereas the specific weight of the pultruded bar is only 0.27 that of steel. The impact

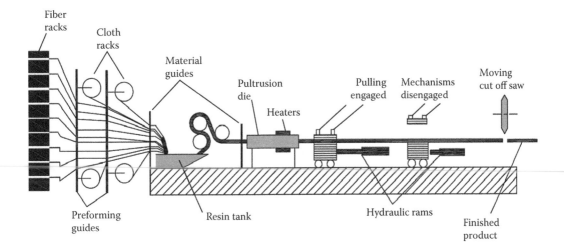

FIGURE 14.18 Schematic of the pultrusion process.

TABLE 14.2
Comparative Properties of Structural Materials Including Pultruded Materials

Structural Property	Pultruded Rod and Bar	Stainless Steel	Low-Carbon Steel
Glass content (% by weight)	70	—	—
Flexural strength (MPa)	690	207–241	193
Flexural modulus (GPa)	41	193	207
Tensile strength (MPa)	690	207–241	186–228
Tensile modulus (GPa)	41	193	207
Impact strength (kJ/m^2)	102.9	17.9–23.1	—
Thermal conductivity (W/m/K)	0.7	13.85–26.68	37.5–66.34
Specific heat (J/g/°C)	1.00	0.50	0.42–0.46
Rockwell hardness (scale)	80H	90B	72B
Specific gravity (25°C)	2.10	7.92	7.9
Thermal coefficient of expansion (mm/mm/°C × 10^6)	5.4	16.2–18	10.8–14.4

Source: PowerPoint presentation, Georgia Institute of Technology.

strength of the pultruded bar is about 5 times larger than that of stainless steel. Similarly, the thermal conductivity and coefficient of thermal expansion of pultruded bar are much lower than equivalent values for steel.

Low-cost commercial pultruded sections made of unidirectional glass rovings often include inexpensive forms of nonwoven broad layers called continuous strands and chopped strand mat. The random orientation of these materials provides some degree of off-axis strength and stiffness enhancement at very low cost. Woven materials used in the pultrusion process must be placed between more stable forms such as layers of unidirectional rovings, as shown in Figure 14.19.

Pultrusion is applied for the production of many lightweight, corrosion-resistant, and low electrical conductivity components. These include standard shapes such as rods, angles, I beams, panels, plates, and rebars for concrete reinforcement. Other components include side rails for ladders,

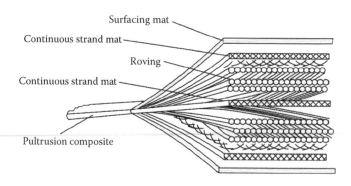

FIGURE 14.19 Exploded view of materials in pultrusion.

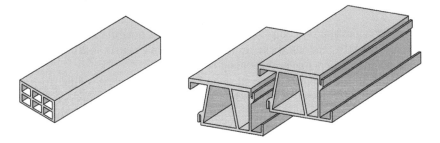

FIGURE 14.20 Typical composite sections fabricated by pultrusion.

fishing rods, electrical insulator rods, tool handles, buss components, sign posts, and sucker rods for oil drilling rigs. Pultruded sections are applied in large structures such as bridges, platforms, roof trim, building panel sections, highway delineator marks, etc. Typical pultruded sections are presented in Figure 14.20.

14.12 REVIEW QUESTIONS

1. How to differentiate a composite material from an alloy?
2. Define composite materials, and what are their general advantages and limitations?
3. Classify composite materials according to their structure and composition.
4. Why are material fibers much stronger than the bulk form? Explain using typical examples.
5. What are the major preparation methods for most used fibers and what are their main characteristics?
6. What are the functions of a matrix material, and what are the main types of matrix materials?
7. Why are carbon/carbon composites used, and what are their main applications?
8. Compare fibers and particles or flakes as reinforcing materials, giving examples.
9. What are the different combinations of particulate composite materials? Give examples of each.
10. Discuss the application of bimetals in thermometers and thermostats.
11. What are the functions and applications of laminated glass?
12. Why are sandwich structures and honeycomb structures used? Give examples of their advantages.
13. Describe two applications of the possible combinations of composite materials.
14. Discuss the general characteristics of the fabrication processes for composite materials?
15. What is the effect of volume fraction of fibers in PMCs, and what are the theoretical and practical limits for this fraction?
16. Compare contact molding with vacuum molding, emphasizing the differences in product quality.
17. What are the precautions to be followed when using prepregs to form a composite product?
18. What are the types of mandrel used in filament winding? Discuss the reasons for replacing the mandrel with a liner.
19. How does automation affect the filament winding process and control the quality of the product?
20. Why are pultruded sections cheaper than other composite sections?
21. Compare the properties of pultruded sections with those of similar metallic sections.

14.13 PROBLEMS

1. Given that for a glass/epoxy composite material the glass fiber diameter is 10 μm, and its strength is 3500 MPa, and the shear strength of the epoxy matrix is 20 MPa, what is the minimum length of fibers that provide good bonding?
2. For a carbon/epoxy composite material hand lay-up with a fiber volume fraction 0.68, calculate its stiffness, given that the modulus of elasticity of carbon fibers is 190 GPa and epoxy is 4.5 GPa. Compare this stiffness with the separate stiffness of each constituent.

BIBLIOGRAPHY

Campbell, F. C. 2004. *Manufacturing Processes for Advanced Composites.* New York: Elsevier.

Donaldson, S. L., and Miracle, D. B. (eds.). 2001. *ASM Handbook,* vol. 21, *Composites.* Materials Park, OH: ASM International.

Gay, D., Hoa, S. V., and Tsai, S. W. 2003. *Composite Materials Design and Applications.* Boca Raton, FL: CRC Press.

Hoa, S. V. 2009. *Principles of the Manufacturing of Composite Materials.* Lancaster, PA: DEStech Publishers.

Jones, R. M. 1999. *Mechanics of Composite Materials,* 2nd ed. Boca Raton, FL: CRC Press.

Kaw, A. K. 2006. *Mechanics of Composite Materials.* Boca Raton, FL: CRC Press.

Mazmudar, S. 2002. *Composites Manufacturing: Materials, Product, and Process Engineering.* Boca Raton, FL: CRC Press.

Vasiliev, V. V., and Marozov, E. V. 2001. *Mechanics and Analysis of Composite Materials.* New York: Elsevier.

15 Fundamentals of Traditional Machining Processes

15.1 INTRODUCTION

Traditional machining comprises two basic types, namely, the chipping (or cutting) processes and the abrasive processes. The chipping processes such as turning, boring, shaping, planing, milling, broaching, and so on, are those in which a single or multiedge cutting tools of definite tool geometry are used. The abrasive processes such as grinding, honing, lapping, and so on, are those in which cutting abrasives of nondefinite geometry, in a bonded form or as free powder, are used. In Chapter 16, the traditional general purpose machine tools and their operations are briefly dealt with. Details of metal cutting theory are outside the scope of this book; the reader is referred to original literature for such details. However, there are some facts based on observations, describing clearly the phenomenon of chipping and abrasion processes. Such facts could be understood, merely using imagination, supplemented by sound engineering judgment.

A comparison of abrasive machining with chipping using cutting tools in the present and in the future is shown in Figure 15.1. Chipping is preferred when there is a considerable stock to be removed. When high machining accuracy (MA) is required, abrasive machining would be the solution. In the past, these two machining technologies hardly competed for their application fields; however, this is not expected to continue. In the future, high accuracy can be obtained using ultraprecision machines equipped with conventional cutting tools. On the other hand, extremely high material removal rates (MRRs), sometimes higher than those obtained by chipping, can be attained with abrasive machining, for example, using cubic boron nitride (CBN) grinding wheels. To gain maximum benefit of abrasive machining, the process characteristics should be understood. The extremely small scale size of chips produced and the self-sharpening characteristic of the abrasives are key advantages. A simplified comparison of grinding, as representative of the abrasive machining, and metal cutting (chipping) is given in Table 15.1.

15.2 BASICS OF CHIPPING PROCESSES

Through close observation of a chipping process, the following facts are concluded.

1. Metal is cut by removal of chips, which may be in the form of continuous ribbon or discontinuous chips, depending on the properties of work material and machining conditions. The chip running on the tool face is always thicker than the undeformed chip thickness, meaning that it is upset (shortened) because of cutting (Figure 15.2). For simplicity, it is always assumed that there is no metal flow lateral to the direction of chip flow. Owing to the strain hardening accompanied by chip deformation, the chip hardness is usually much greater than the hardness of the parent metal. There is generally no crack extending ahead of the tool cutting edge, a misconception that found popular support many years ago.
2. There is a line AB (shear plane) that separates the deformed and undeformed zones of the work material. The chip (above AB) has been upset by a concentrated shearing action. The shear plane AB is a plane of maximum shear stress. The shear angle φ (Figure 15.2) depends on various cutting conditions.
3. There are three zones of interest in the chipping process (Figure 15.3). These are the shear plane 1, chip/tool interface 2, and work/tool interface 3.

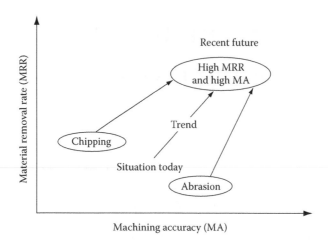

FIGURE 15.1 Comparison of chipping and abrasive machining in future and present, as related to MRR and MA.

TABLE 15.1
Comparison of Grinding and Chipping Operations

Item of Comparison	Chipping	Grinding
Specific energy, k_s (J/mm³)	1–2	2–50
Tolerance (μm)	20–500	5–100
Surface roughness, R_a (μm)	1–2	0.1–0.4
Chip thickness (mm)	0.1–2	0.01–0.1
Cutting speed (m/min)	30–300	1000–3600
Edge rake angle γ (deg)	(−5)–(30)	(−35)–(5)

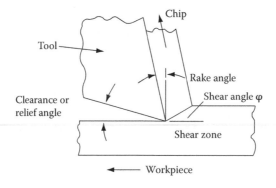

FIGURE 15.2 Chip upsetting.

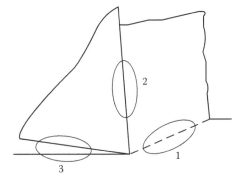

FIGURE 15.3   Three zones of interest in a chipping process.

4. Cutting is associated with a lot of heat, mainly generated because of plastic deformation (zone 1) and friction (zones 2 and 3). The heat generation should be kept to minimum by the following measures:
   - Having sharp edges of cutting tools
   - Better finish of tool face and flank
   - Increased sliding speed of chips on tool face by increasing cutting speed
   - Improved tool geometry
   - Application of cutting fluid
5. In metal cutting the built-up edge (BUE), a phenomenon described later on, is an undesirable phenomenon because it deteriorates the machined surface finish and induces vibrations and chattering. It significantly alters the cutting tool geometry.
6. The operational characteristics of a cutting process are generally described by a simple term, "machinability." The main aspects to estimate this term are the tool life, the surface finish, or the power consumed in cutting.

15.2.1   ELEMENTS AND KINEMATICS OF A CHIPPING PROCESS

Figure 15.4 illustrates the elements of a basic chipping process. These are the workpiece, the cutting tool, the chips, and the cutting fluid.

The cutting tool should have a basic wedge form. It is driven by a force to overcome friction and forces that hold material molecules of the work material together. Accordingly, the tool separates a layer of work material called a chip (unwanted material). This is, however, conditioned by the presence of relative movements between the tool and the workpiece. The work done because of the application of the cutting forces and relative movements is converted into heat energy, which is

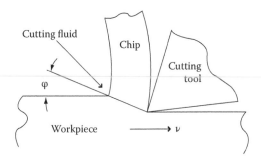

FIGURE 15.4   Elements of a chipping process.

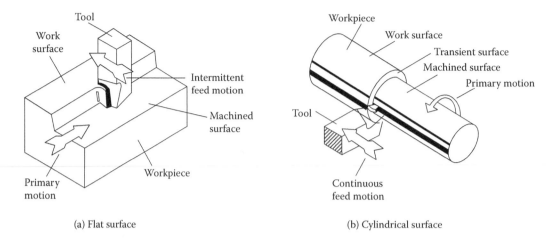

FIGURE 15.5 Generation of surfaces.

mainly conducted to the cutting tool, the chip, and the workpiece. In most cases, it is essential to cool these elements by appropriate cutting fluid (Figure 15.4).

In general, two kinds of relative motions must be provided to generate the desired machined surface. These are the main and feed motions, which are defined as

- Main or primary motion: Rapid relative motion between the tool and the workpiece. It is exerted by either the tool or the workpiece, which absorbs most of the machine power.
- Feed motion: Slow relative motion exerted by either the tool or the workpiece to bring a new material ahead of the tool to be cut. It may be actuated step by step (intermittent) or continuously. It usually absorbs a small proportion of the total machining power.

The most common forms of these movements are circular or linear. They are responsible for the creation of vast range of profiles and surfaces with the desired geometric characteristics. The simplest surfaces to create or generate are flat surfaces as well as external and internal cylindrical surfaces. If a workpiece is reciprocating forward and backward in a straight-line main motion, while the tool is incrementally fed over the reciprocating workpiece in a direction at right angles to the main motion of the workpiece, a flat surface will be generated on the workpiece (shaping or planing) (Figure 15.5a). Similarly, an external cylindrical surface can be generated by rotating the workpiece (main motion) while feeding the tool (feed motion) parallel to the axis of the workpiece rotation (turning) (Figure 15.5b).

15.2.2 GEOMETRY OF A SINGLE-EDGE TOOL

Most texts dealing with this topic refer to a cutting tool with single-cutting edge as a single-point tool. This is not altogether correct because all cutting tools cut with their edges rather than with their points. Therefore, the correct terminology, a tool with a single-cutting edge, is used here. This tool is applicable in many cutting operations such as turning, boring, shaping, slotting, and so on. Its geometry is concerned with the basic tool angles provided to the tool to give it the basic wedge form. Such wedge form enables the tool to penetrate in the work material; the work is subjected to rapidly increasing shear stress, so that it fractures along the shear plane if ultimate shear strength of work material is exceeded. Consequently, the chip is separated from the work leaving the machined surface in a torn, rough condition, depending on prevailing machining conditions. Obviously, the material of the cutting tool must be much harder and tougher than the material of the workpiece; this prevents the workpiece cutting the tool. Moreover, the cutting tool should be provided with sharp cutting edge to minimize rubbing and friction forces.

A typical single-edge tool and its terminology are illustrated in Figure 15.6. The single-edged tool may be shaped as a solid bar of cutting material (Figure 15.6a) or it may have inserts or tips that are either brazed (Figure 15.6b) or mechanically attached to a less costly alloy steel shank (Figure 15.6c). Referring to Figure 15.6a, the tool has a base supporting it and a shank of rectangular (or circular) cross section from which the tool can be clamped. The cutting edges are formed on the tool front (tool head), which is bounded by three surfaces, namely, the rake face $A\gamma$, on which chip is separated from the WP and flows over the face, the main (major) flank $A\alpha$, situated against the cutting surface, and the auxiliary (minor) flank $A_{\bar{\alpha}}$, situated against the machined surface. The line of intersection of the rake face $A\gamma$ and the main $A\alpha$ flank represent the main cutting edge S, which removes the bulk of the machined material. The line of intersection of the rake face A_γ and the auxiliary flank $A_{\bar{\alpha}}$ represents the minor cutting edge $\hat{S}$, which mainly controls the final surface finish of the machined surface. The geometry and nomenclature of cutting tools, even single-edged tools, are surprisingly complicated subjects. It is difficult to determine the appropriate planes in which various angles of single-edge tools should be measured. For example, it is so difficult to determine the slope of the rake face.

Different standards (ISO/DIN 3002 and BS 1296) suggested different reference planes to measure tool angles. In German standard (DIN), three planes are suggested (Figure 15.7). These are

- Basic plane 1, which contains the tool base and perpendicular to the direction of main motion.
- Cutting plane 2, which contains both the main cutting edge and the velocity vector. It is perpendicular to plane 1.
- Auxiliary plane 3, which is perpendicular to planes 1 and 2 and passing to the cutting point P.

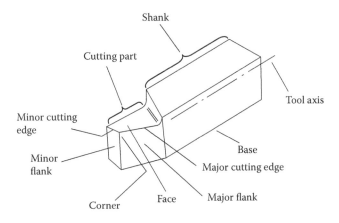

(a) Shaped on a solid bar

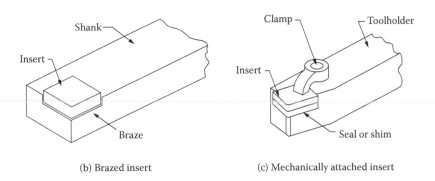

(b) Brazed insert

(c) Mechanically attached insert

FIGURE 15.6 Terminology of a single-edged tool.

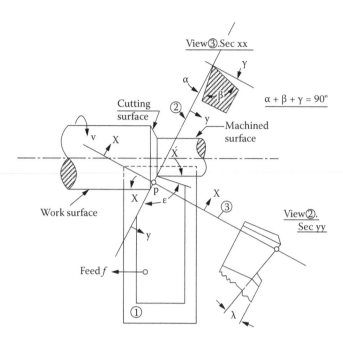

FIGURE 15.7   Angles of single-edged cutting tool.

15.2.2.1 Tool Angles

Figure 15.7 visualizes the main tool angles which specify the tool geometry according to the German standard DIN 3002, 1973. These are

Side relief (clearance) angle α: The angle between the main flank and the cutting plane 2 as measured in the auxiliary plane 3. This angle reduces friction between the tool and the work. It should be kept as small as possible to avoid weakening of the tool.

Wedge (tool) angle β: The angle between the tool face and the main flank as measured in the auxiliary plane 3.

Cutting angle δ: The sum of α and β,

$$\delta = \alpha + \beta$$

Side rake angle γ: The angle between the tool face and the plane passing through the cutting point P parallel to the basic plane 1 as measured in auxiliary plane 3.

Referring to Figure 15.7, then

$$\alpha + \beta + \gamma = 90°$$

The side rake γ is the most important angle of the cutting tool. It allows the chip to flow easily on the tool face and controls the chip formation. The cutting force and power consumed in cutting are reduced by increasing the side rake γ; however, the tool becomes more sensitive to shock loads. The rake angle γ may be positive when the tool face from the cutting edge slopes downward with respect to the tool basic plane, equals to zero when the tool face is parallel, and negative when the face slopes upward with respect to the tool basic plane. As indicated in Figure 15.8, the cutting edge with negative rake $(-\gamma)$ will be stressed mainly in compression, while if positive rake $(+\gamma)$ is used, the cutting force changes direction, lending higher moment leading to probable rupture of cutting tools, even at small loads.

Nose angle ε and nose radius r: ε is the angle included between the projections of the main and minor cutting edges S and Ś on the basic plane (Figure 15.7). The main and minor cutting edges

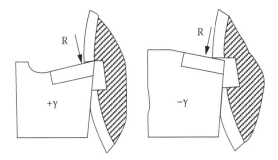

FIGURE 15.8 Forces acting on cutting tools with positive and negative rakes.

S and Ś intersect in an arc at tool nose or corner. The provision of tool nose radius r improves the tool life and surface finish. However, large nose radius may result in chattering that destroys the surface quality. Therefore, the nose radius should be selected properly. It ranges typically from 0.1 to 2 mm. The cross section of the tool shank should be correspondingly increased as the nose radius increases.

Angle of approach (setting angle) χ: The angle between the projection of the main cutting edge S on the basic plane and the direction of feed.

Secondary angle of approach χ': The angle between the projection of the minor cutting edge on the basic plane and the direction opposite to the feed f.

Referring to Figure 15.7, then

$$\chi + \varepsilon + \chi' = 180°$$

Angle of cutting edge inclination λ: The angle included between the main cutting edge S and a line passing through the basic plane as measured in the cutting plane (Figure 15.7). Edge inclination λ is considered to be negative if the tool nose is the highest point of the main cutting edge (Figure 15.9). Some of the functions of λ are to alter the chip flow direction and the chip thickness ratio. Representative chips using three angles of inclination λ are shown in Figure 15.10. All chips are produced from a length of cut of 100 mm; otherwise, machining conditions are the same. The change from flat spiral chip of orthogonal cut to the helical chip is a characteristic of a tool with inclination is evident.

Generally, tool angles are chosen with respect to material to be machined, tool material, and the machining process used. The harder and stronger the material to be machined, the greater must be the wedge angle β and the smaller must be the relief and rake α and γ, respectively. Positive rakes are used when machining ductile materials, while negative rakes are used when machining hard brittle material.

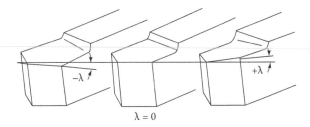

FIGURE 15.9 Tools with different angles of cutting edge inclination.

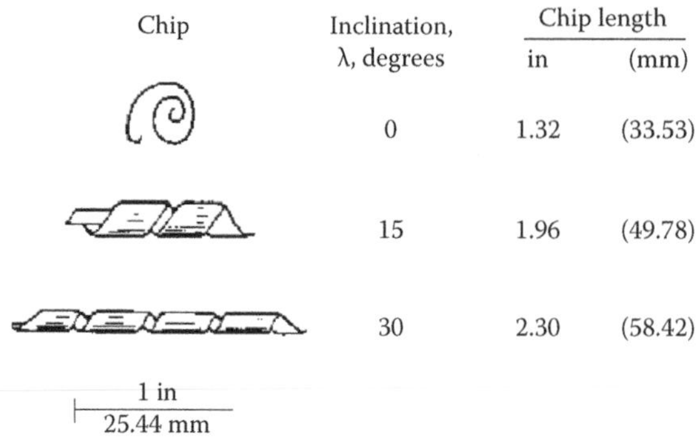

Chip	Inclination, λ, degrees	Chip length in	Chip length (mm)
(spiral chip)	0	1.32	(33.53)
(chip)	15	1.96	(49.78)
(chip)	30	2.30	(58.42)

$$\dfrac{1\ \text{in}}{25.44\ \text{mm}}$$

Length of cut = 100 mm, γ = 10, depth of cut = 0.13 mm, cutting speed v = 0.5 m/min, Coolant: carbon tetrachloride

FIGURE 15.10 Chips cut by tools with different angles of cutting edge inclination, at otherwise constant machining conditions. (From Shaw, M. C., *Metal Cutting Principles*, Clarendon Press, Oxford, 1984. With permission.)

15.2.2.2 Tool Signature According to American Standard Association

It is the numerical method of identification according to which seven elements comprising signature of single-edge tools are always stated in the following order in Table 15.2 (Figure 15.11). As shown in Figure 15.11, degrees for angles and units (inch) for nose radii are omitted. Only the numerical values are indicated. The most effective angles are the back and side rake angles (BRA and SRA). Table 15.3 visualizes their recommended values for different work and tool materials.

15.2.3 CHIP FORMATION–CHARACTERISTIC TYPES OF CHIPS–CHIP BREAKERS

These important topics of metal cutting are briefly discussed in the following sections.

15.2.3.1 Chip Formation

Upon inspection of the chip as produced from a chipping process, it is found that the chip under the surface is shiny or burnished because it has been rubbed along the tool face under conditions of high temperature and pressure. However, the upper surface of the chip, being formerly from original surface of the work, has a ragged appearance. As a result of strain hardening, the chip usually becomes harder, less ductile than the workpiece material. The increase in hardness and strength of the chip

TABLE 15.2
Tool Signature (ASA)

Angles, Nose Radius	Abbreviation	Tool Signature
1. Back rake angle	BRA	8
2. Side rake angle	SRA	14
3. End relief (clearance) angle	ERA	6
4. Side relief angle	SRFA	6
5. End cutting edge angle (χ' in German standard)	ECEA	20
6. Side cutting edge angle (90-χ' in German standard)	SCEA	15
7. Tool nose radius	NR	1/8

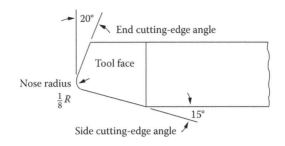

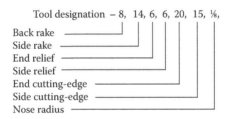

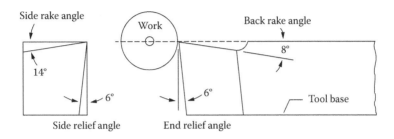

FIGURE 15.11   RH turning tool signature according to ASA nomenclature.

TABLE 15.3
Recommendations of Rake Angle (ASA)

				Tool Material			
				Cemented Carbide			
Work		**HSS**		**(Brazed)**		**(Throwaway)**	
Material	**B H N**	**BRA**	**SRA**	**BRA**	**SRA**	**BRA**	**SRA**
Al alloys	30–150	20	15	3	15	0	5
Cu alloys	40–200	5	10	0	8	0	5
CI	300–400	5	5	–5	–5	–5	–5
Steel	85–225	10	12	0	6	0	5
	325–425	0	10	0	6	–5	–5

Source: Machinability Data Handbook, Metcut, 1980. With permission.

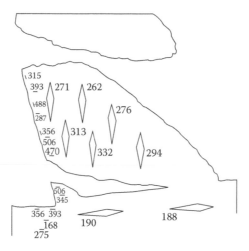

FIGURE 15.12 Knoob hardness distribution in cutting zone in case of a discontinuous chip.

depends on the shear strain (Figure 15.12). As the rake angle decreases, the shear strain increases and the chip becomes stronger and harder.

15.2.3.2 Characteristic Types of Chips

Three characteristic types of chips are distinguished. These are discontinuous, continuous, and continuous chips with BUE (Figure 15.13). The type of chips produced significantly influences surface roughness and integrity of the machined workpiece.

Factors influencing the type of chips produced include

- Properties of materials being machined
- Cutting speed, depth of cut, and feed rate
- Tool rake angle and tool surface finish
- Type of cutting fluid
- Cutting temperature

15.2.3.2.1 Discontinuous Chips

These are small segments that do not adhere loosely to each other, owing to fractures that occur perpendicular to the tool face. This type (Figure 15.13a) is always produced when machining hard

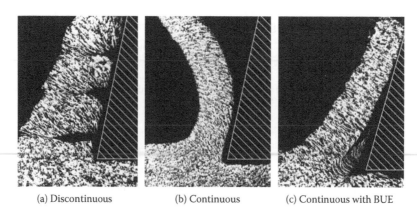

(a) Discontinuous   (b) Continuous   (c) Continuous with BUE

FIGURE 15.13 Characteristic types of chips.

and brittle materials like bronze, brass, and cast iron (CI) but also may be produced when machining ductile materials at very high speeds and feeds using tools of small or negative rakes. The formation of this type in brittle materials imparts reasonable finish, increases tool life, and consumes less power. With ductile materials, however, the surface finish is bad and the tool life is short, but smaller chips are easier to dispose of.

Factors leading to discontinuous chips include

- Machining of brittle materials
- Selecting low cutting speeds, large feeds and depths, small tool takes, and using no cutting fluids

15.2.3.2.2 Continuous Chips

These are formed by continuous plastic deformation of metal without fracture ahead of the cutting edge. This type (Figure 15.13b) is common when most ductile materials such as mild steel, Cu, and Al are machined at high cutting speeds. Continuous chips are desirable being associated with low power consumption, stable cutting without chattering, good surface finish, and long tool life. However, the disposal of such chips is problematic. This can be tackled by tools with chip breakers.

Factors leading to continuous chips include

- Machining of ductile materials
- Selecting high cutting speeds, small feeds and depths, tools having small rakes and smooth tool face, and using efficient cutting fluids

15.2.3.2.3 Continuous Chips with BUE

Such chips also appear in the form of long ribbons similar to the previous type with the difference that it has a BUE, adhering on the tool nose (Figure 15.13c). BUE forms, generally, by machining ductile materials with ferrous tools (TCS and HSS). It is less common in machining CI and materials that produce discontinuous chips. On closely observing the cutting edge, small fragments of the workpiece are found to be welded on it forming the BUE in the stagnant zone, in which elevated temperatures and high pressures are prevailing. As the cutting proceeds, the BUE grows with additional deposition of layers until it becomes large; then it breaks up. A part of it is carried away by the chip and the rest deposits on the workpiece, deteriorating the machined surface (Figure 15.14). The formation and destruction of BUE periodically occur in short intervals of 0.01–0.2 s, depending on cutting speed and other machining conditions. More deterioration of the machined surface

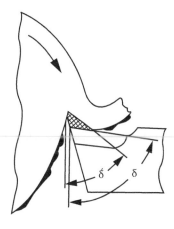

FIGURE 15.14 BUE phenomenon and its effect on surface quality and cutting-edge geometry.

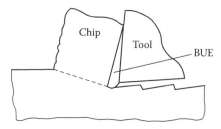

FIGURE 15.15   Surface roughness due to chattering induced by BUE.

quality occurs owing to vibration and chattering initiated because of the BUE phenomenon (Figure 15.15). The cutting edge remains active, even when totally covered by BUE. The BUE changes the tool geometry. It increases the effective rake angle of the tool (Figure 15.14). Owing to the strain hardening effect, the hardness of BUE may be 2 to 3 times that of the metal being machined.

Besides the type of tool and workpiece materials, other machining parameters that considerably affect the formation of BUE are cutting speed, tool geometry, and cutting fluids. At low cutting speed the use of cutting fluids prohibits the formation of BUE, whereas at high cutting speed, cooling promotes its formation. The tool rake angle and cutting speed have also a combined effect on its formation. Positive rake and low cutting speed prohibits the formation of BUE, while negative rake and high cutting speed promote its formation.

Factors leading to continuous chips with BUE include

- Machining of ductile materials producing continuous chips and those having the ability for strain hardening
- Using carbon tool steel and high-speed steel tools; carbide- and ceramic-tipped tools are less inclined to form BUEs
- Cutting tool geometry
- Machining parameters
- Using cutting fluids promotes or prohibits BUE formation, depending on the magnitude of cutting speed

15.2.3.3   Chip Breakers

Before the advent of carbide tools and the use of high cutting speeds, the chip produced in metal cutting was not a serious problem. At low speeds, the chips usually have a natural curl and tend to be brittle. However, cutting speeds have now increased to such an extent that chip control has become necessary, because the chip leaves the cutting edge at a very high velocity, which is dangerous to the operator as the chips are hot and sharp.

In turning, where the tool is continuously removing metal for a long period, a continuous chip can become entangled with the tool, the workpiece, or the machine parts. This type of chip can be hazardous and unless controlled properly can result in mechanical chipping of the tool edge. If coolant is applied, the chip may interfere with the flow. This causes alternating heating and quenching, resulting in thermal stresses, which can reduce the tool life, especially if carbide and ceramic tools are used. Moreover, the handling of continuous chips in bulk can represent a major economic problem. A handling characteristic of chips can be expressed by a factor called the bulk ratio, which is the total volume occupied by the chips divided by the volume of the solid material. Unbroken continuous chips have a bulk ratio of approximately 50, tightly wound chips a bulk ratio of about 15, and well-broken discontinuous chips have a bulk ratio of approximately 3.

The foregoing considerations have led to the development of chip breakers. These are located near the cutting edge, where they obstruct the flow of the chips, making them curl and thus attaining decreased radius of curvature. These then break into pieces (Figure 15.16). Basically, there are two

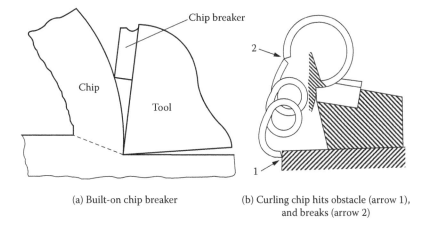

(a) Built-on chip breaker

(b) Curling chip hits obstacle (arrow 1), and breaks (arrow 2)

FIGURE 15.16 Action of built-on chip breakers.

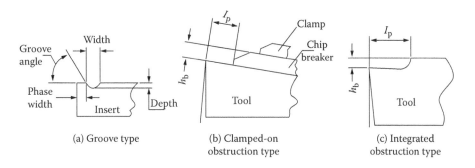

(a) Groove type

(b) Clamped-on obstruction type

(c) Integrated obstruction type

FIGURE 15.17 Types of chip breakers.

types of chip breakers: the groove and the obstruction. The obstruction types are of two categories, namely, clamped on and integrated (Figure 15.17). It must be considered that chip breakers may increase the tool cost, cutting forces, and power consumption by about 15%–25%.

15.2.4 Mechanics of the Chipping Process

Earlier investigations were done to understand the mechanism involved in a chipping process. The main objective of this aspect was to predict theoretically the important cutting parameters influencing that process, without the need of empirical testing. In all machining operations, two different models are generally adopted. These are the oblique and the orthogonal cutting models, in which a wedge shaped tool of single cutting edge is used to perform the cut (Figure 15.18). The majority of machining operations, in practice, are performed by oblique cutting (three-dimensional case), where the cutting edge is inclined by an angle λ to the line drawn at right angles to the direction of the main motion (Figure 15.18a). A special case of cutting, where the cutting edge is perpendicular to the direction of main motion is termed orthogonal cutting (Figure 15.18b). The orthogonal cutting (two-dimensional case) is simpler to investigate compared to three-dimensional oblique cutting. Hence, it is used in the following analysis to understand the basics of the chipping process.

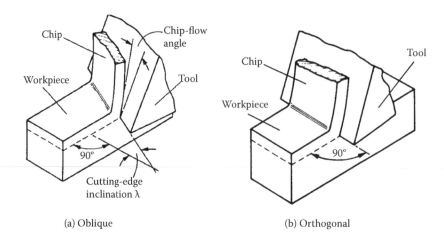

FIGURE 15.18 Oblique and orthogonal cutting.

15.2.4.1 Orthogonal Cutting

Figure 15.19 shows some cases of orthogonal cutting. These are

- Collar turning
- Tube turning
- Cutting on a planer (or shaper), where the cutting edge is in right angles to the cutting motion

15.2.4.1.1 Merchant Analysis

Merchant investigated the orthogonal cutting model. His analysis was based on the following assumptions:

- Using a tool with a perfectly sharp cutting edge
- Shear plane is a plane of maximum shear stress

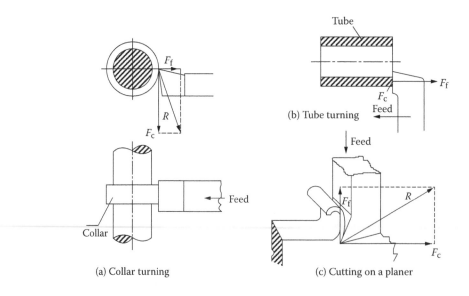

FIGURE 15.19 Three cases of orthogonal cutting.

- Small and uniform chip thickness of continuous configuration with no BUE
- Two-dimensional model, in which no lateral flow in either sides

Figure 15.20 represents a schematic of orthogonal operation. The chip is in equilibrium under the action of two forces, the cutting force R exerted by the tool and the reaction force R exerted by the workpiece, then

$$R = R \tag{15.1}$$

Referring to Figure 15.21, the resultant cutting force can be resolved into three sets of components, along and perpendicular to the direction of cutting speed (F_c, F_t), shear plane (F_s, F_p), and tool face (F_f, F_n) where

F_c = main cutting force component in the direction of cutting speed
F_t = thrust component, perpendicular to F_c
F_s = shear force component along the shear plane
F_p = component normal to the shear plane
F_f = friction force component along the tool face
F_n = component normal to the tool face

Therefore

$$R = \sqrt{F_c^2 + F_t^2} = \sqrt{F_s^2 + F_p^2} = \sqrt{F_f^2 + F_n^2} \tag{15.2}$$

The components F_s, F_p, F_f, and F_n can be expressed in terms of the components F_c, F_t, the shear angle φ, and the tool rake γ (Figure 15.21)

$$F_s = F_c \cos \varphi - F_t \sin \varphi \tag{15.3}$$

$$F_p = F_c \sin \varphi + F_t \cos \varphi \tag{15.4}$$

$$F_f = F_c \sin \gamma + F_t \cos \gamma \tag{15.5}$$

$$F_n = F_c \cos \gamma - F_t \sin \gamma \tag{15.6}$$

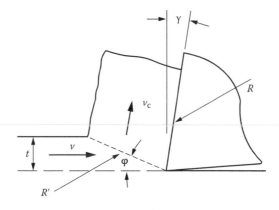

FIGURE 15.20 Chip in orthogonal cutting under equilibrium of R and R'.

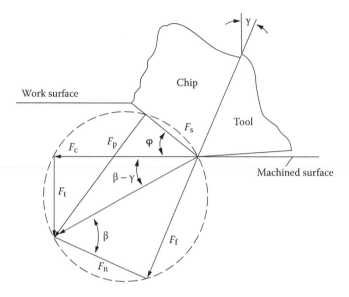

FIGURE 15.21   Resolution of the resultant force acting on the workpiece in three sets of force components.

From the same figure, F_s and F_c can be expressed as

$$F_s = R \cos(\varphi + \beta - \gamma) \qquad (15.7)$$

$$F_c = R \cos(\beta - \gamma) \qquad (15.8)$$

Then

$$F_c = \frac{F_s \cos(\beta - \gamma)}{\cos(\ + \beta - \gamma)} \qquad (15.9)$$

where β is the friction angle.

Shear angle relationship
Referring to Figure 15.22a,

$$AB = \frac{t_1}{\sin\ } = \frac{t_2}{\cos(\ - \gamma)}$$

$$r = \frac{t_1}{t_2} = \frac{\sin\ }{\cos(\ - \gamma)}$$

$$\tan\ = \frac{r \cos \gamma}{1 - r \sin \gamma} \qquad (15.10)$$

where t_1 is the depth of cut, t_2 is the chip thickness, and $r = t_1/t_2$ is the chip of thickness ratio ($r < 1$).

According to Equation 15.10, the shear angle φ increases with increasing chip thickness ratio r and increasing tool rake γ (Figure 15.23).

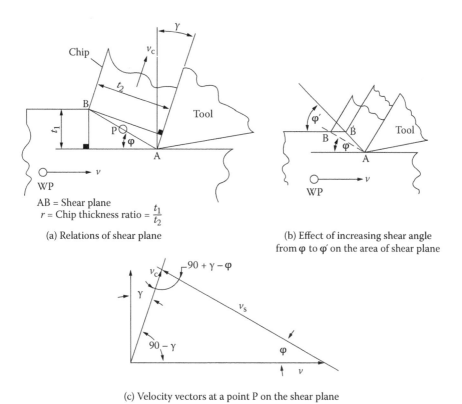

AB = Shear plane
r = Chip thickness ratio = $\dfrac{t_1}{t_2}$

(a) Relations of shear plane

(b) Effect of increasing shear angle from φ to φ′ on the area of shear plane

(c) Velocity vectors at a point P on the shear plane

FIGURE 15.22 Kinematics of orthogonal cutting.

Velocity vectors at shear plane
Figure 15.22c visualizes the velocity vectors at a point P on the shear plane. The chip velocity along the tool face v_c and shear velocity along the shear plane v_s can be expressed in terms of cutting velocity, shear angle φ, and chip thickness ratio r according to

$$v_c = \frac{\sin}{\cos(- \gamma)} \cdot v = r \cdot v \tag{15.11}$$

$$v_s = \frac{\cos \gamma}{\cos(- \gamma)} \cdot v \tag{15.12}$$

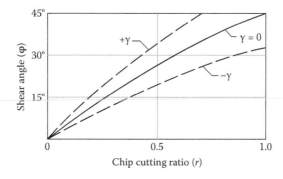

FIGURE 15.23 Relationship among shear angle φ, chip thickness ratio r, and tool rake γ.

According to Equation 15.11, the chip velocity v_c is always smaller than cutting speed, because the chip thickness ratio r is always less than unity.

Coefficient of friction on the tool face

The mean coefficient of friction μ_m on the tool face is given by

$$\mu_m = \tan\beta = \frac{F_c \sin\gamma + F_p \cos\gamma}{F_c \cos\gamma - F_p \sin\gamma} = \frac{F_p + F_c \tan\gamma}{F_c - F_p \tan\gamma} \tag{15.13}$$

15.2.4.1.2   *Other Shear Angle Relationships*

The shear angle φ is the most important parameter influencing the chip formation process. According to Equation 15.10, the increase of φ (by increasing γ and r) leads to decrease in the area of shear plan from AB to AB′ (Figure 15.22b) and consequently to decrease of cutting force and consumed power; that makes the cut easy. Figure 15.23 illustrates the relationship between the shear angle φ, the chip thickness ratio r, and the tool rake γ.

Besides the foregoing analysis of Merchant, two other models are considered in which the shear angle φ has been expressed in terms of the friction angle β and the tool rake γ. The first was developed by Ernst and Merchant. Their analysis was based on the assumption that the shear angle would take up such a value to reduce the main cutting force F_c (work done) to minimum $\dfrac{dF_c}{d\varphi} = 0$. The shear angle is accordingly expressed by

$$\varphi = 45° - \frac{1}{2}(\beta - \gamma) \tag{15.14}$$

The second analysis was developed by Lee and Shaffer. It was based on the Tresca theory of plasticity using the slip-field analysis,

$$\varphi = 45° - (\beta - \gamma) \tag{15.15}$$

Figure 15.24 shows both plots of Equations 15.14 and 15.15 together with experimentally determined values of φ from which the following can be concluded:

- Both relationships have the same trend, although they differ numerically except when $\beta = \gamma$, where $\varphi = 0$.
- For both theoretical relationships, φ decreases with increasing $\beta - \gamma$, thus making the cut more difficult (the workpiece behaves with less machinability).
- It appears that the model of Lee and Shaffer is the more realistic model yet proposed. That is clear from the closer agreement of this model and the experimental results shown in Figure 15.24.

15.2.4.2   Oblique Cutting

Figure 15.25 shows a case of three-dimensional oblique cutting as applied to tuning operation using a single-edge straight cutting tool.

15.2.4.2.1   *Cutting Forces*

Three basic cutting force components are involved, including

- Main cutting force F_c is a tangential force acting in the direction of velocity vector.
- Feed force F_f is an axial force acting in the direction of feed. It is perpendicular to F_c.

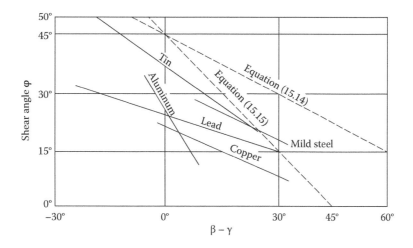

FIGURE 15.24 Comparison of experimental and theoretical shear angle relationships found by Ernst-Merchant and Lee-Shaffer.

- Radial force F_r is perpendicular to both F_c and F_f and is acting radially along the tool shank (perpendicular to the axis of workpiece rotation). It is also called the passive force component, because it exhibits no movement and hence no work done or power are exhibited in this direction.

The magnitude of the resultant cutting force R is given by

$$R = \sqrt{F_c^2 + F_f^2 + F_r^2} \tag{15.16}$$

Its direction depends on the magnitudes of the three components; however, experimental tests have indicated that the main cutting force F_c is far larger than either of the other two force components.

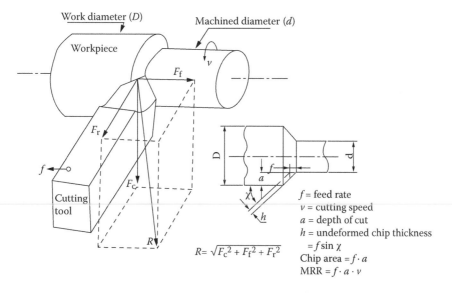

FIGURE 15.25 Force components in oblique cutting as applied on turning.

15.2.4.2.2 Power Requirement

The power P_c required to cut material is the product of main cutting force F_c (N), and the linear cutting speed v (m/min).

The linear cutting speed is given by

$$v = \frac{\pi D N}{1000} \text{ m/min} \tag{15.17}$$

where D is the diameter of the workpiece before cut (mm) (Figure 15.25) and N is the rotational speed of workpiece (rev/min).

Therefore, the power P_c is given by

$$P_c = \frac{F_c v}{60000} \text{ kW} \tag{15.18}$$

and the main motor power P_{mot} is given by

$$P_{mot} = \frac{F_c v}{60000 \eta_m} \text{ kW} \tag{15.19}$$

where η_m is the mechanical efficiency of the machine tool ($\eta_m = 0.7$ for lathes, drilling machines, and 0.6 for reciprocating machines like shapers, planers, and slotters). P_c is not the total power needed for cutting, because it is solely based on the main cutting force F_c. Additional feed power P_f of a magnitude far smaller than P_c should be exerted to feed the tool against the feed force F_f, where f is the feed in millimeters per revolution.

$$P_f = \frac{F_f f N}{60 \times 1000 \times 1000} \tag{15.20}$$

15.2.4.2.3 Specific Cutting Power and Specific Cutting Resistance

The volumetric material removal rate VRR (Z) is given by

$$Z = a \times f \times v \tag{15.21}$$

where a is the depth of cut and is equal to $\dfrac{D-d}{2}$ (mm) (Figure 15.25) and d is the diameter of the work after machining (mm).

If P_c is divided by VRR, the power required to cut a unit volume per unit time is termed the specific cutting power k_s.

$$k_s = \frac{P_c}{Z} = \frac{F_c v}{afv} = \frac{F_c}{af} \text{ N/mm}^2 \tag{15.22}$$

Because the specific cutting power k_s is expressed in N/mm², it is also often called the specific cutting resistance. This parameter depends on the type of material to be cut. However, it is not a material constant. For a given material, it depends on cutting conditions, especially the undeformed chip thickness h, which is proportional to the feed rate f (Figure 15.25). Kienzle and Victor (1956) expressed the relationship between k_s and h empirically by the following equation, which is valid for both orthogonal and oblique cutting.

$$k_s = C h^{-z} \tag{15.23}$$

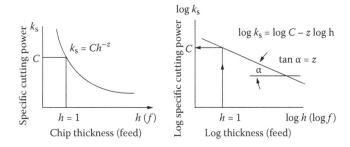

FIGURE 15.26 Schematic of the relationship between k_s and $h(f)$ as determined by Kienzle and Victor (1956, Werkstattstechnik und Maschinen, Vol. 46).

Therefore

$$\log k_s = \log C - z \log h \tag{15.24}$$

where C = const and $-z$ is a negative exponent.

Equations 15.23 and 15.24 are schematically shown in Figure 15.26. Figure 15.27 illustrates approximate values of k_s for various materials and operations at different undeformed chip thicknesses (feeds) as represented on double logarithmic ordinates. Figure 15.28 (Feldmuehle, Vogel-Verlag, August 1972) illustrates a nomogram for the determination of the motor power P_{mot} assuming a mechanical efficiency $\eta_m = 0.7$ of the lathe machine.

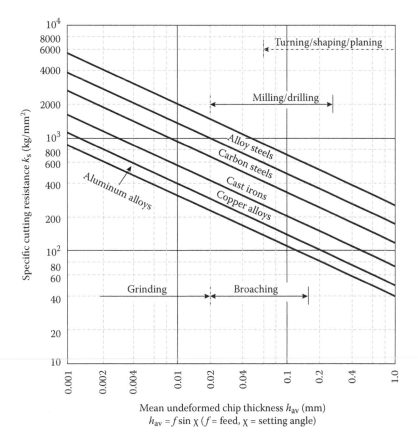

FIGURE 15.27 Approximate values of specific cutting resistance for different materials and operations.

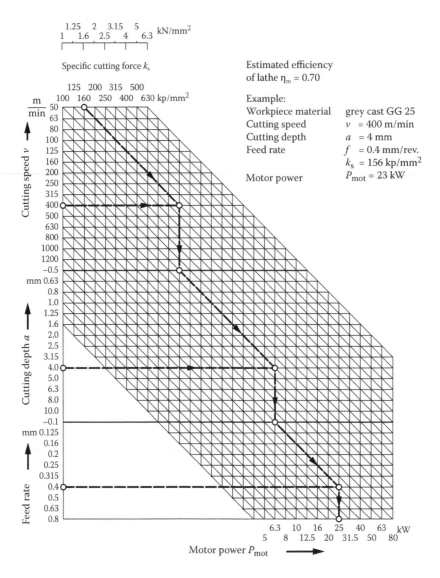

FIGURE 15.28 Nomogram for the determination of the motor power by turning. (From Werkzeugmaschinen International, Feldmuehle, Vogel-Verlag, Heft 4, 1972. With permission.)

Example

Calculate the motor power P_{mot}, required to machine gray CI, GG 25 (k_s = 1600 N/mm²) using the following cutting conditions: f = 0.4 mm/rev, a = 4 mm, and v = 400 m/min.

    Check your answer using the nomogram in Figure 15.28.

Solution

$$F_c = k_s a f = 1600 \times 4 \times 0.4 = 2560 \text{ N}$$

$$P_{mot} = \frac{F_c v}{60000 \eta_m} = \frac{2560 \times 400}{60000 \times 0.7} = 24.5 \text{ kW}$$

According to the nomogram P_{mot} = 25 kW.

15.2.5 Cutting Tool Materials

The steady development of civilization of the past thousands of years was certainly because of the great progress of tool materials, starting from hard wood, bone, rocks to copper, then to iron and steel up to the end of the nineteenth century. The most remarkable aspect is the increasing rate of introducing new tool materials in the past 100 years. By the beginning of the twentieth century, Taylor has introduced his high-speed steel (HSS) instead of tool carbon steel (TCS), thereby increasing the cutting speed to three times. In 1926, Krupp's developed a very hard sintered carbide material, used in cutting and forming, called Widia (German acronym: Wie Diamant, i.e., like diamond). Cutting speeds of 3 to 6 times of those of HSS have been realized. Ceramics were then introduced in 1950 as a new tool material, realizing cutting speeds of 6–8 times of those achieved by HSS. Latest developments are tool materials like cermets, coated carbides, and cubic baron nitride (CBN), which have hardness next to that of diamond. Now polycrystalline diamond tools are mainly used in finish turning, achieving cutting speeds of 2 km/min.

15.2.5.1 Characteristics of Cutting Tool Materials

The cutting tool is subjected to severe conditions, such as high temperatures of more than 1000°C, severe friction, and high dynamic stresses. Therefore, the tool should be strong enough to withstand the above mentioned conditions. Accordingly, the ideal tool material should be characterized by

- High hot hardness (red hardness or refractoriness), i.e., it would retain its hardness at high operating temperatures.
- High strength and toughness to withstand both static and impact loads. Unfortunately, extreme hardness poses limitation on toughness, so a compromise must be struck between adequate hot hardness and improved toughness, because the less brittle the tool, the less likely it is to break prematurely in service under impact loading and interrupted cuts (Figure 15.29).
- High fatigue strength (endurance limit) to withstand repeated loads.
- High wear resistance to resist mechanical and thermal abrasion, so that the tool resists edge chipping.

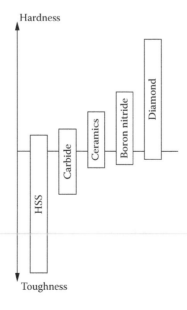

FIGURE 15.29 Hardness versus toughness of cutting tool materials. (From Sandwik Coromat, Halesowen. With permission.)

TABLE 15.4
Relative Costs of Tool Materials

Tool Material	Relative Cost
Tool carbon steel (TCS)	1
High-speed steel (HSS)	1–2
Cast nonferrous alloys	2–10
Cemented carbides	3–10
Sintered alumina oxides, cermets	20
Coated carbides	10–30
Borazon (CBN), diamond	100 and more

- Inertness and chemical stability, i.e., no metallurgical affinity to workpiece (WP) material.
- Resistance to thermal shocks, especially if coolant is suddenly turned on during cutting.
- High thermal conductivity to conduct away heat from the cutting edge efficiently.
- Low coefficient of friction at the tool–chip interfaces to guarantee minimum wear and hence reduced cutting force and consumed power.
- Low tool cost.
- Infinite tool life.
- Easy to form into the required shape.
- Easy to sharpen.

Self-evident, ideal tool material having all above mentioned characteristics does not exist. An increasingly wide range of tool materials are, however, available from which the production engineer must select the material that will best perform the task at the lowest unit cost. The relative costs of typical cutting tool materials in industrial use today are illustrated in Table 15.4. They may vary with market conditions, quantities, and tool shapes. The most important characteristic of any cutting tool material is its hot hardness. Figure 15.30 illustrates the hot hardness of some of the most common tool materials in current use.

15.2.5.2 Types of Tool Materials

The cutting tool materials are listed in ascending order of hot hardness and descending order of toughness. They may be ferrous (e.g., tool steels and HSSs) or nonferrous (e.g., cast nonferrous alloys, carbides, cermets, and ceramics).

15.2.5.2.1 Tool Steels

1. Carbon tool steel: It is the oldest kind, however, little used today. It contains 0.9%–1.3% C and softens at about 150°C–200°C. Therefore, it is limited to very low cutting speeds. It is easy to harden and to sharpen. Its main disadvantage is that it is liable to cracking and distortion because it must be quenched in water.

2. Low or medium alloy tool steels: They have alloying elements such as Mo and Cr, which improve hardenability, and W, Mo, and V, which improve wear resistance. These steels suffer less distortion and cracking because they are generally oil hardened. These tool materials lose their hardness rapidly (250°C–300°C). Consequently, they are used for comparatively inexpensive tools and for low cutting speed applications. Low alloy steels (0.8%–1.0% C, 1.0%–1.2% Cr, 1.2%–1.6% W) are used for manufacturing of drills, milling cutters, measuring gauges, and metal forming dies, whereas medium alloy steels (1.4%–1.7% C, 12% Cr, 0.7%–0.9% V) are used for knurling tools, broaches, and drawing and shearing dies.

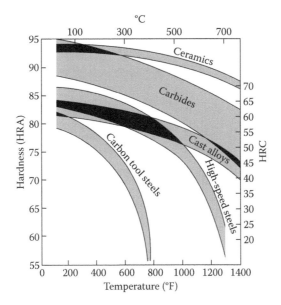

FIGURE 15.30 Hot hardness of some common tool materials in current use. (From Ingersoll Cutting Tool Co. With permission.)

15.2.5.2.2 High-Speed Steel

High-speed steel (HSS) is the material that was first introduced as lathe tool by Taylor and White in 1910. Its introduction at that time led to a significant increase in cutting speeds, which accounted for its name. However, today HSS is misnamed, because it is now considered the general purpose tool material for machining operations performed at low and moderate cutting speeds. HSSs contain mainly carbide forming alloying elements as W, Cr, V, and so on (10%–20% by volume), besides 0.7%–1.5% C. Therefore, their microstructures consist of a martensitic matrix, in which complex carbides (Fe_2W_2C, $Cr_{23}C_6$, and VC), and cementite Fe_3C are embedded. These carbides help in achieving HSS good hot hardness and sharp cutting edges at operating temperatures up to 600°C but rapidly soften at higher temperatures (Figure 15.30). Consequently, it could operate at cutting speeds 3 times which be achieved by those of tool steels.

HSS provides a good balance between hot hardness and toughness (Figure 15.29) to cope with intermittent cutting shocks. Moreover, it can be fully hardened with little danger of distortion and cracking, and the tool can be rapidly sharpened many times during its useful life. Because of its toughness, HSS is especially suitable for high positive rake angle tools and for low power machine tools with low stiffness that are subjected to vibration and chatter. The basic limitation of HSS is the relatively low cutting speeds. For high-production machining, it should be replaced almost completely by carbides and other nonferrous tool materials.

The first and most common 18-4-1 HSS is designated according to the percentage of its alloying elements W, Cr, and V, respectively. During World War II, when there was a shortage of tungsten, it had been largely replaced by the strategic available molybdenum. A smaller percentage of Mo (only 50%) behaves in the same manner of W regarding hot hardness. The American Iron and Steel Institute (AISI) has classified the commercially available HSSs into two groups, namely, the T group (tungsten HSSs), and the M group (molybdenum HSSs). The M group generally has higher abrasion resistance than the T group, which undergoes less distortion during heat treatment. Today, the M group constitutes about 95% of all HSSs produced in the states.

As illustrated in Table 15.5, the majority of HSSs contain 1%–2% V. Addition of more V (T-15) allows the formation of complex carbides, which inhibits grain growth at high temperature needed

TABLE 15.5
Some of Widely Used HSSs According to AISI

AISI Designation		Composition (wt.%)						Remarks
		C	W	Cr	V	Mo	Co	
T group	T-1	0.73	18	4	1	—	—	Taylor straight W grade
	T-4	0.75	18	4	1	0.6	5	W–Co grade
	T-15	1.55	12.5	4.5	5	0.6	5	W–Co–V grade
M group	M-1	0.80	1.75	3.75	1.15	8.75	—	Straight Mo grade
	M-30	0.80	1.8	4	1.2	8.25	5	W–Mo–Co grade
	M-36	0.85	6	4	2	5	8.25	Mo–Co grade

in heat treatment of HSSs; T-15 is commonly used for high strength materials such as Cr alloy die steels. The pronounced superiority of HSS containing Co (T-4, T-15, M30, and M36) is due to the tendency to increase hot hardness by reducing the amount of retained austenite γ after tempering to 5%, thus allowing these steels to be used at higher cutting speeds.

Improving the performance of HSS tools
The following techniques represent the recent developments toward improving the performance of HSS tools.

1. After heat treatment of HSS, it may be subjected to low temperature liquid carbonitriding (550°C–570°C).
2. Admitting stream of O_2 in the furnace during tempering of HSS leads to the formation of a blue oxide film Fe_3O_4 on the tool surface. This film is tightly adhered to the base metal. It has a high hardness. This technique is used for forming tools rather than cutting tools.
3. Physical vapor deposition (PVD) is used to apply very thin coatings (5–7 μm) metal carbides, nitrides, or oxides. The cost of HSS-coated tools is about twofold that of the normal HSS. TiC, TiN, and Al_2O_3 have been proposed as coating materials. Multilayer coatings of these are also possible. TiC coatings are effective for abrasive wear resistance. Al_2O_3 coating is a good thermal barrier due to its lower thermal conductivity. Hence it is used for high-speed and high-feed rate operations. TiN coating is of yellow-gold color, and it is the most preferred coating, especially effective in preventing adhesion at tool-chip interface; this reduces galling and greatly improves the tool life. However, the tools do not perform well at low cutting speeds, because this coating can be chipped off. Proper lubrication during cutting is a must. The wear of HSS-coated tools is generally reduced such that the tool life extends up to 3–4 times that of conventional grade. It should be marked that coated HSS tools must be hardened and tempered after coating. Special precautions must be taken to preserve the very thin coating layer. A limitation of coated HSS tools is that they cannot be resharpened.
4. HSS can be produced by powder metallurgy (PM). A stream of molten metal is broken into droplets (50–500 μm) by a gas or water jet. The droplets are rapidly solidified within fraction of a second, after which the sintering process is followed. PM is important for large-size tools. This technique is double the cost of regular HSS. PM-HSS tools exhibit better grindability, greater toughness, better wear resistance, higher hot hardness, and more consistent performance.

15.2.5.2.3 Cast Nonferrous Alloys
Cast nonferrous alloys are cutting tool materials that contain no iron except as impurity. They must be cast in shape then ground to size. Cast alloys are generally used for simple and large size tooling.

These materials serve best in speed range between HSS and sintered carbide. They are of two basic categories: Co-base (Stellites) and Cb-based (UCON) alloys.

1. Co-base cast alloy is of trade name Stellite; it was introduced in 1915 as a co-alloy of typical analysis

    38%–52% Co, 30%–32% Cr, 10%–12% W, 2% C

    Its structure is composed of Co-matrix, in which complex W–Cr carbides are embedded with volume ratio of 25%–30%. It is of a hardness ranging from 58 to 64 HRC.
    Stellite is characterized by
    - Weak in tension, fragile, resistant to corrosion, and not heat treatable
    - Not tough as HSS, hence only for jobs free from shocks, intermittent cuts, and vibrations
    - Maintains hardness at elevated temperatures (up to 750°C) so used at relatively higher speeds (25%) as compared to those of HSS
    - Available as bars of circular and square sections or as inserts, brazed or attached or lathe tools, and milling cutter bodies
    - Less tendency to form built-up edge (BUE)
    - Do not require cutting fluids except if special surface quality required
    Selective applications of Stellites include
    - Machining of plain carbon steel, CI, and hard bronze
    - Generally recommended for deep and continuous roughing operations at relatively higher feeds and speeds
    - Used in the production of slip and limit gauges
2. Cb-based cast alloy is nitrided refractory columbium-based alloy. UCON is the trade name of Union Carbide Company. Typical analysis of UCON is 50% Columbium (Cb), 30% Ti, 20% W, free from carbides. If the tool surface receives a nitriding treatment, UCON attains hardness up to 25% more than ceramics and 50% more than sintered carbides.
    UCON is characterized by
    - High hardness and reasonable toughness
    - Excellent thermal shock resistance
    - Excellent resistance to diffusion and chip welding
    - Relatively expensive as compared to sintered carbide
    - Edge life is 3–5 times as compared to sintered carbide
    - Difficult to machine
    - Available as throwaway insert

Applications and limitations of UCON include the following: UCON is not practical for machining many materials such CI, Ti, Ti alloys, stainless steels, and heat-resistant steels. However, it is used in machining of common steels at high-speed range of 300–500 m/min. It is not recommended for intermittent cuts.

15.2.5.2.4 Cemented (Sintered) Carbides: Widia

Cemented carbides (commercially known in Europe as Widia, the trade name of Krupps) are basically composed of finely ground carbides (WC + TiC + TaC). These carbides are extremely hard, but not tough enough to withstand impact loads during cutting. Therefore, they should be bonded by a soft metal, usually Co, just as bricks are cemented together by mortar, hence the name *cemented* carbides. These carbides are produced by sintering through powder metallurgy technique; hence they also acquired the name *sintered* carbides. Cemented carbides are composed of hard carbides (85%–95% by volume) bonded together in a Co-matrix.

Cemented carbides are classified into two fundamental groups, namely, the straight group WC + Co, and the multicarbide group WC + TiC + TaC + Co (Figure 15.31). The early tools of the straight group were only very effective when machining CI, nonferrous metals, and nonmetallic materials. However, they were not compatible with steels because crater wear is rapidly developed on the tool face. It was found that the crater wear considerably decreased, if TiC and TaC were added to produce the multicarbide group. Such addition promoted wear resistance but reduced the toughness. Therefore, the multicarbide group is best suited for cutting ferrous metals producing long chips like steels (Figure 15.31).

ISO Standard for Classification of Cemented Carbides: There are many systems for classification. However, it appears that the international system ISO is gaining ground, and it is now becoming a world standard. According to ISO 504-1975 Standard for Carbide Inserts, the cemented carbides for metal cutting applications are grouped into three grades, identified by the letters P, M, and K, coded by three different colors, blue, yellow, and red, respectively. Each grade is further subdivided into different types on scale of 01–40 to indicate hardness and toughness. Types 01 and 10 are used for finishing, while types 30 and 40 are suitable for rough cutting (Figure 15.31). Table 15.6 illustrates the physical, mechanical, and thermal properties, as well as the composition and typical materials being machined by each grade. Referring to this table, grade P is intended for machining of steels because it contains high percentage of TiC + TaC. Grade K is mainly composed of WC and therefore tougher than grade P; it is intended for machining materials of broken chips, producing less crater wear. Grade M is an intermediate grade, which is capable of machining materials producing both long and short chips. It is used in cases where it is required to cut several materials with the same tool upsetting.

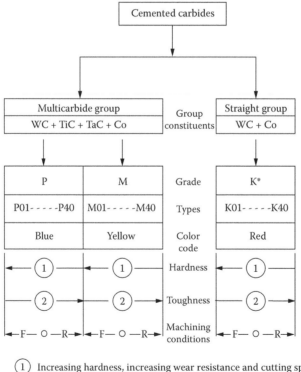

FIGURE 15.31 ISO classification of cemented carbides for metal cutting applications.

TABLE 15.6
ISO Classification (DIN-4990) of Cemented Carbide for Cutting Tools (Abstracted)

Grade Symbol (Color)	Type	Trend[a]	Composition			Specific Weight (g/cm³)	VH30 (kg/mm²)	σ_b	σ_c (kg/mm²)	E	$\mu_{l,exp}$ 10^{-6}/°C	k[b] (cal/cm°C s)	Materials to be Machined
			(TiC+TaC)	Co%	WC%								
P (Blue)	P01		64	6	30	7.2	1800	75	—	—	—	—	Ferrous metals with long chips
	P10	1	28	9	63	10.7	1600	130	490	53000	6.5	0.07	
	P20	1 2	14	10	76	11.9	1500	150	500	54000	6.0	0.08	
	P30	2	8	10	82	13.1	1450	175	500	55000	5.5	0.14	
	P40		12	13	75	12.7	1400	190	470	56000	5.5	0.14	
M (Yellow)	M10		10	6	84	13.1	1700	135	—	58000	5.5	0.12	Ferrous metals with long or short chips and nonferrous metals (multipurpose grade)
	M20	1 2	10	8	82	13.4	1550	160	500	57000	5.5	0.15	
	M30		10	9	81	14.4	1450	180	480	—	—	—	
	M40		6	15	79	13.6	1300	210	440	54000	—	—	
K (Red)	K01		4	4	92	15.0	1800	120	—	—	—	—	Ferrous metals with short chips (CI) Nonferrous metals and nonmetallic materials
	K10[c]	1 2	2	6	92	14.8	1650	150	570	63000	5.0	0.19	
	K20		2	6	92	14.8	1550	170	550	62000	5.0	0.19	
	K30		1	9	90	14.5	1400	190	480	58000	—	0.17	
	K40		0	12	88	14.3	1300	210	450	57000	5.5	0.16	

Note: $\mu_{l,exp}$ = coefficient of linear expansion, k = thermal conductivity.

[a] Trends: 1, increasing hardness, wear resistance, and cutting speed; 2, increasing toughness and feed.

[b] σ_b = bending strength, σ_c = compression strength, E = Young's modulus.

[c] Sometimes VC may be added.

Because of their high hardness over a wide range of temperatures (800°C–1100°C), high modulus of elasticity, high thermal conductivity, and low thermal expansion (Table 15.6), cemented carbides are the most popular cutting tool materials for production operations. They are capable of machining harder materials at high cutting speeds. Carbide tools can operate at cutting speeds of 3–5 times those allowed by HSS tools.

Precautions to be observed when using cemented carbide tools include

- The machine tool must be rigid enough and free from vibrations.
- The machine power and speeds must be adequate to allow the carbide tool to cut with high metal removal rates.
- The tool must not be allowed to rub after switching off the machine. It is a good practice to withdraw the tool while the feed is still engaged rather than disengaging the tool feed first.
- The tool and work piece must be rigidly clamped.
- Coolant, when used, must be effective and sufficient.
- Carbide tools should be sharpened dry, and should not be quenched after sharpening.

Inserts of Cemented Carbides: Cemented carbides are generally available as inserts or tips, which are provided with a number of cutting edges. These inserts are available in a great variety of shapes such as square, triangle, diamond, and round. A square insert, for example, has four cutting edges on each face, and it is stronger (strength depends on included angle only) than triangular or diamond inserts, while the round insert is stronger than all other shapes (Figure 15.32). Inserts are usually

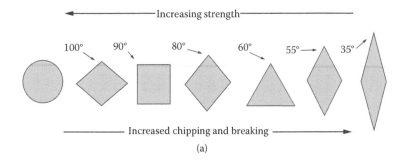

(a)

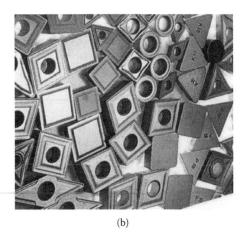

(b)

FIGURE 15.32 Carbide inserts: (a) typical shapes of carbide inserts (from Kennametal) and (b) collection of carbide inserts (from Microna, Gaelve-Sweden).

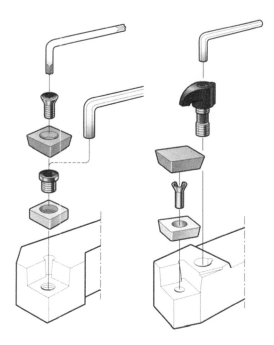

FIGURE 15.33 Typical designs of tool holders: Microna, Gaelve-Sweden.

clamped on tool shanks made of tool steel in turning, planing, boring, and slotting operations. Many ingenious designs have been developed to allow inserts to be held rigidly on their shank (Figure 15.33). Figure 15.34 shows a triangular threading insert when threading a right-hand (RH) saw screw. Figure 15.35 shows milling cutters equipped with mechanically clamped cemented carbide inserts. Less frequently used are the inserts brazed to the tool shanks. Considerable economic savings are achieved if mechanically clamped inserts (throwaway tips) are used instead of brazed types because

- High cost of insert sharpening is eliminated.
- When the cutting edge is worn out, the tip is rotated (indexed) to another cutting edge; therefore, the tool changing cost is greatly reduced.
- Brazing cost is avoided, and no elaborate brazing equipment is needed.

Carbide tips are usually provided with flat margins of negative rakes to protect them against chipping caused by mechanical impact. Tips are also provided with grooves formed in the rake face close to edges acting as chip breakers, which curl the chips and clear it from around the tool. This

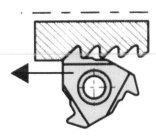

FIGURE 15.34 A triangular threading insert, while threading a RH saw screw.

FIGURE 15.35 Milling cutters equipped with mechanical clamped carbide inserts. (From Krupp, Widia, GmbH, Essen-Germany. With permission.)

is important for automated production. The ability to cut for long periods without stopping to clear chips has a considerable effect on machining cost.

Developments toward enhancing the durability of cemented carbides: Two recent developments are considered; these are

1. Cemented titanium carbides (TiC-based tools): More recently, cemented TiC with Mo and Ni as bonding material has been available in the form of disposable tips. This material is less tough but more wear resistant than conventional Co-cemented carbide grades. Developments of TiC tools have led to superior grades that allow cutting speeds approaching those of ceramic tools; these grades are used in semirough cutting of both CI and steels. With the relative shortage of tungsten supplies, TiC serves as a substitute for conventional carbide grades. Some commercial suppliers now include TiC-based grade in their catalogues. It is believed that in the near future, at least one half of all steel machining will be carried out using cemented TiC.
2. Coated carbide inserts: The basic requirement of efficient rough machining of steel is a tool material that exhibits the toughness of WC, while giving the superior wear resistance of TiC. Therefore, much interest is directed to coating of cemented WC substrate with a thin layer of harder materials such as TiC, TiN, and alumina Al_2O_3 using chemical vapor deposition (CVD) technique. These coatings are between 5 and 8 μm thick and are found to eliminate interdiffusion between the chip and the tool. When the coating has been worn away by abrasion, the wear rate becomes the same as that for uncoated tool. A large proportion of indexable tips are now coated.

Recently, multilayer coated carbides offer enhanced toughness, shock resistance, and longer edge life. These inserts are particularly effective in machining CI and steels. The principal materials used for coating are TiC, coated directly over to bond well with substrate, followed by Al_2O_3, and then the outer layer TiN to resist wear and has low thermal conductivity; the intermediate layer should bond well and be compatible with both layers. Such coatings preclude resharpening of inserts as the user cannot replace the coatings after sharpening.

The CVD technique is performed by heating the inserts in a sealed chamber in hydrogen gas at atmospheric or reduced pressure, to which volatile compounds are added to supply the constituents of coatings. The temperature range is 800°C–1050°C, and the heating cycle lasts for several hours.

Optimization of the CVD process leads to consistent deposition of uniform thickness strongly adhered to the WC substrate.

15.2.5.2.5 Ceramics (Alumina–Base Tools)

Ceramics are fine-grained pure aluminum trioxide Al_2O_3 particles (<3 µm). As cemented carbides, they are produced through powder metallurgy technique by pressing alumina powder, followed by sintering without binder at high temperature of 1700°C. Its basic material (alumina) is cheap and available; however, the processing is expensive, and therefore, ceramics are not cheap as compared to carbides. Ceramics usually are available as disposal (throwaway) tips. As compared to carbides, they almost completely resist crater wear, usually require no coolant, and have the same tool life if operated at double the cutting speed of carbides. Ceramics require more rigid tool holders and rigid machine tools to take the benefit of their capacities. Ceramics are particularly inert to steel up to its melting point. They retain their cutting ability for temperature, up to 1200°C, which enables steel, semihard steels, and CI to be machined at high speeds (up to 600 m/min) and reasonable feeds (up to 0.25 mm/rev). Higher speeds and feeds are realized when machining nonferrous metals. Ceramic is a multipurpose tool material as compared to carbide tooling because a wide range of materials (ferrous and nonferrous) can be machined at various cutting speeds by one type of ceramic material. However, ceramics are not suitable for Al, Ti, and other materials that chemically react with alumina-based ceramic.

Because ceramics have poor mechanical and thermal shock resistance, interrupted cuts and application of cutting fluids can lead to premature tool failure. It is advisable to chamfer the edge of the workpiece with a carbide or HSS tool before cutting with ceramics to avoid severe impact when touching with the ceramic tool. Being nonmetallic, ceramics are less likely to adhere to metal during cutting. Therefore, BUE is not liable to occur, and consequently high surface quality is achieved. Better chip flow on the tool face is realized owing to the low coefficient of friction; the result is reduced cutting forces and machining power when using ceramic tools. It should be considered that the tool/chip interface temperature is high because ceramic tool is a poor conductor of heat.

Experience has shown that when ceramic tools are properly applied as a replace of carbide tools, the machining time can be reduced by one-third to one-half when machining steel and by two-thirds when machining CI. Unfortunately, ceramics are more brittle than other sintered inserts; for this reason, their use is still limited in metal cutting. Of all sintered materials in current use, only about 2%–3% are ceramics, about 60%–70% cemented carbides and cermets, 25% coated carbides, and 5% titanium carbides.

15.2.5.2.6 Cermets

The so-called cermets (metallic or black ceramics) are composites of alumina Al_2O_3 and TiC. Cermets are somewhat more refractory than TiC. The addition of 20%–30% TiC promotes toughness and reduces brittleness of alumina. Therefore, performance of cermets is between that of ceramics and carbides. Compared to cemented carbides, cermets have higher hot hardness, less toughness, lower thermal conductivity, and greater thermal expansion, so thermal cracking can be a problem during interrupted cuts. For these reasons, cermets are best suited for finishing. The better surface quality imparted by cermets is due to its low level of chemical reaction with iron, and hence less craters and BUE. Moreover, cermets are applicable for the following machining duties:

- High-speed machining of hard CI (HB = 360–600 kg/mm^2) and steels of hardness up to 500 VH30, under conditions of limited shock loads
- Machining of martensitic stainless steels
- Machining of carbon and plastics

15.2.5.2.7 Diamond

1. Natural diamond: Diamond is the hardest material known to man, so it should, in theory, make a good cutting tool material. Unfortunately, this is not the case, as it is extremely brittle and the slightest impact or fluctuation in cutting force causes it to fracture. Its use is, therefore, severely restricted and is limited to high cutting speeds and uninterrupted cutting of soft materials such as Al, Al alloys, bronze, and plastics. Furthermore, it is not applicable to cut ferrous metals and Ni alloys owing to the adverse chemical reactions between diamond and these materials. As compared to other cutting tool materials, diamond is characterized by
   - Extreme hardness and high abrasion resistance
   - Shock resistance
   - Very low coefficient of friction
   - Low coefficient of thermal expansion (12% that of steel)
   - High thermal conductivity (2 times that of steel)
   - Chemical inertness up to 800°C, thereby it burns to CO_2
   - No BUE formation
   - High chemical affinity for ferrous metals and Ni alloys
   - Offering tool lives 50–100 times those of cemented carbides
   - High initial and sharpening cost

   Precautions when using natural diamond tools include
   - Special care should be taken for proper mounting.
   - Not applicable for cutting steel, CI, and Ni alloys.
   - Abrasion resistance varies by at least 10-fold depending on crystal orientation.
   - Interrupted cuts are to be avoided and negative rakes to be used to avoid edge chipping due to lack of toughness.
   - Using rigid and precise machine tools operating at high speeds.

   There are several applications of natural diamond. Because of its high thermal conductivity, less tendency to form BUE, and low coefficient of friction, natural diamond has proved to be efficient as a cutting tool material. It is mainly used for turning and boring operations. The tools are tipped with large size diamond tips weighing from 0.3 to 1.5 carat (1 carat = 0.2 g). Natural diamond is particularly used in fine machining of Al and Al alloys, Mg alloys, Cu, brass, bronze, zinc, gold, silver, platinum, rubber, plastics, babbits, and so on, at fantastic cutting speeds up to 3000 m/min, and small feeds up to 0.05 mm/rev; thereby, mirror like surfaces and high precision are produced. Moreover, natural diamonds are frequently used in truing and dressing of grinding wheels. They are also used in the powder form to produce metallic bonded grinding wheels and in lapping operations. Outside the domain of metal cutting, diamonds are used to produce drawing dies for fine wires and as penetrators for hardness tester, as well as many other applications.

2. Polycrystalline synthetic diamond: A recent development (1970s) has led to introducing the polycrystalline synthetic diamond. Its main advantage is due to the random orientation of their crystals lending improved impact strength, thus making it suitable for interrupted cutting. The polycrystalline diamond is a composite of very small synthetic diamond crystals, which are fused with metallic bond together by a high temperature and pressure process. A consolidated polycrystalline layer, usually 0.5–1 mm thick is then bonded to WC substrate to form a tool tip of about 3 mm thick. The random orientation of the diamond crystals prevents the propagation of cracks through the structure, whereas the WC substrate provides the necessary toughness for shock loads acting on the tool. The composite tip can be clamped or brazed on the tool shank, then ground, lapped and polished, and resharpened

when worn. It takes longer sharpening time than carbides. It is more expensive, costing typically 20–30 times the equivalent carbide tool.

Characteristics of polycrystalline diamonds include

- They are composed of randomly oriented particles, behaving as an isotropic material of improved toughness.
- Their edges are less sensitive to accidental damage, while maintaining exceptional wear resistance.
- Like natural diamond, they are chemically inert up to 800°C.

There are several applications of polycrystalline diamond. They are recommended for machining Al, Al alloys, Cu, Cu alloys (e.g., Cu-commutators), hypereutectic Al–Si alloys, and so on. They are applicable for operations of intermittent cutting like milling and threading. However, they are not recommended for machining steels, heat resisting steels, and Ni alloys.

15.2.5.2.8   Cubic Boron Nitride

Cubic boron nitride (CBN) was developed to overcome the problem of diamond being unsuitable for machining ferrous metals and, while not quite as hard as diamond, it is still harder than any other cutting tool material developed to date. Exactly as polycrystalline diamond, this tool is made of a bonding layer of polycrystalline CBN to a WC substrate. The carbide provides the shock resistance, whereas the CBN layer provides a very high wear resistance and strength. Two main commercially available products of CBN are BZN (of GEM) and Amborite (of De Beers). BZN is produced as laminated tool tips of about 0.5 mm thick on WC substrate, whereas Amborite is produced as a tip consisting entirely of consolidated CBN.

The greatest chemical stability of CBN for a long period at elevated temperature of 1000°C makes it possible to machine steel, CI, and Ni alloys at high cutting speeds. The greatest potential of CBN appears in machining hardened steel, HSS, chilled CI at cutting speeds of 60 m/min and feeds of 0.2–0.4 mm/rev. With correct tool geometry and negative rake angles, CBN can be employed for taking interrupted cuts on hardened steel, e.g., bars and holes with slots, and milling operations. CBN grinding wheels are used for grinding carbides, HSS, Stellites, and so on. CBN is expensive and it is about the same price as synthetic diamond. The material removal rate is several times greater than that possible with carbide tools. It is available in the form of indexable inserts.

15.2.5.2.9   Sialon

Si–Al–O–N stands for silicon nitride Si_3N_4-based materials with Al and O additions. It is produced by milling Si_3N_4, aluminum nitride, alumina Al_2O_3, and yttria. The mixture is dried, pressed to shape, and sintered to 1800°C. It is not as tough as cemented carbide; however, it is considerably tougher than alumina and thus suited for interrupted cuts. Sialon possesses high thermal conductivity and low coefficient of thermal expansion, thus providing resistance to thermal shocks as compared to alumina. Sialon is mainly used to machine aerospace alloys and Ni-based gas turbine discs at cutting speeds of 200–300 m/min. It is available as inserts of negative rakes and provided chamfers at the cutting edge to strengthen the tool. Improvement in machine tool rigidity enables Sialon tips to be used effectively.

15.2.6   Tool Wear and Tool Life

During chipping operations, cutting tools are subjected to an extremely severe rubbing action. The tools are in metal-to-metal contact with chip and workpiece under severe conditions of mechanical and thermal stresses. Such conditions undermine the edges until they cease to cut satisfactorily.

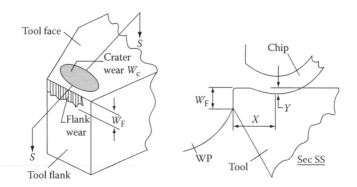

FIGURE 15.36 Formation of flank and crater wear.

15.2.6.1 Tool Wear

Tool wear could occur because of one or more of the following mechanisms

- Adhesion wear: Wear caused by the fracture of welded asperity junctions between the materials of workpiece and tool, where small fragments from the tool are torn out and carried away either by chip or workpiece.
- Abrasion wear: Occurs when hard particles on the chip underside pass over the tool face and remove tool material by chipping action.
- Diffusion wear: In metal cutting, intimate contact occurs between the tool and workpiece, while high temperatures exist. A diffusion process occurs where atoms move from tool to work, and the surface structure of the tool is thereby undermined.

Typical forms of tool wear are identified as flank wear and crater (face) wear.

1. Flank wear: It is generally attributed to sliding of the work surface against the tool flank causing adhesion and/or abrasion wear depending on the materials involved.

    Flank wear is characterized by the wear land W_F (Figure 15.36). Figure 15.37 illustrates the three different stages of flank wear. The wear-in stage (I) is due to high initial stresses caused by the sharp tool edge. The second stage (II) is of linear steady character. The third is the destructive wear stage (III), where the wear increases progressively resulting in catastrophic and uncontrollable failure due to thermal cracks in case of ceramic and

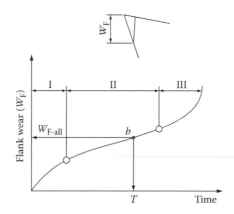

FIGURE 15.37 Flank wear and typical wearing curve.

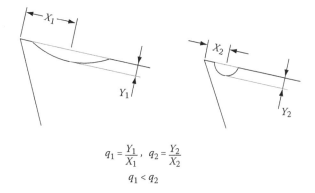

$$q_1 = \frac{Y_1}{X_1}, \quad q_2 = \frac{Y_2}{X_2}$$

$$q_1 < q_2$$

FIGURE 15.38 Crater wear and q criterion.

carbide tools or plastic deformation in case of HSS and TCS tools. In practice, complete tool blunting, stage III is not allowable because it leads to heavy tool damage and increases resharpening cost. Moreover, the machining accuracy is lost. It is, therefore, suggested that the tool should be earlier withdrawn before reaching the point b, when an allowable wear land $W_{F.all}$ is reached.

$$W_{F.all} = 60\% \ W_{F.I+II} \ \text{(carbide and ceramic tools)}$$

$$= 75\% \ W_{F.I+II} \ \text{(HSS tools)}$$

where $W_{F.I+II}$ is the wear land for useful tool life.

2. Crater (face) wear: It occurs on the tool face, at the point of impingement of the chip with the tool. It does not actually reach the cutting edge but ends beyond it. Crater takes the form of cavity, which grows with the time in width and depth (Figure 15.36). Crater wear may be characterized by crater depth Y or the factor q, which is the ratio crater depth to distance of crater center from the cutting edge X (Figure 15.38).

$$q = \frac{Y}{X} \tag{15.25}$$

q increases as Y increases and X decreases; it should not exceed a certain allowable value (0.4 for carbide tools and 0.6 for HSS tools); otherwise, tool weakening and catastrophic failure may occur. The assessment of the crater wear W_C is not so easy as flank wear W_F; this is the main reason why W_F is frequently used as a tool life criterion. Crater wear is more prominent when machining ductile metals and their alloys, whereas flank wear is prevalent in brittle materials of discontinuous chips regardless of the tool material used. Figure 15.39 visualizes schematically the conditions at which flank and crater wear may occur. High speeds and feeds call for crater wear.

15.2.6.2 Tool Life

Tool life or cutting edge durability is defined as the total time at which the tool is able to take off chips. It is the sum of actual cutting times in which the tool is operating from sharpening to economical blunting. In other words, it is the cutting time elapsed between two consecutive sharpenings. As throwaway tips are not resharpenable, their tool life does refer to the time until a particular edge is no longer usable.

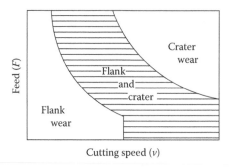

FIGURE 15.39 Predominance of flank and crater wear.

The tool life may be expressed in one of the following measures:

- Actual time to failure expressed in minutes
- Length of work cut to failure expressed in meters
- Volume of material removed to failure expressed in cubic centimeter
- Number of components produced to failure
- Cutting speed for a given time to failure expressed in meters per minute

15.2.6.2.1 Tool Life Criteria

Depending on the nature or the type of machining operation performed, the end of the tool life can be judged by one of the following criteria:

- Wear land W_F, crater wear W_C, or a combination of both.
- Limiting value of roughness of the machined surface or dimensional change of component.
- Sudden increase of cutting force or power (Schlesinger Criterion). Drawback of this criterion is that it necessitates the use of force and power dynamometers, which may not be available in the workshop.

Other criteria may be sometimes used, including chipping and cracking of cutting edge, bright band criterion, and noise criterion as applied in drilling.

15.2.6.2.2 Taylor's Tool Life Equation

The most important factor that does affect the tool life T is the cutting speed v. For the practical range of cutting speeds, the tool life decreases as the cutting speed increases. Taylor (1906) established the T–v relationship empirically when cutting various steels using HSS tools. The Taylor equation is expressed as

$$vT^n = C$$

or

$$T = \frac{C}{v}^{1/n} \qquad (15.26)$$

where v is the cutting speed (m/min), T is the tool life (min), and n, C are the Taylor exponent and Taylor constant, respectively.

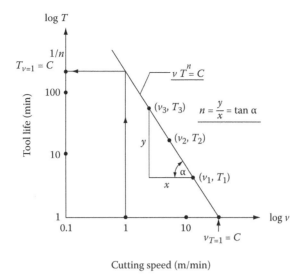

FIGURE 15.40 Taylor relationship as determined on a double logarithmic coordinate system.

Taylor obtained his relationship (Figure 15.40) by performing cutting tests at cutting speeds v_1, v_2, v_3 at otherwise constant conditions until a maximum allowable wear land $W_{F,all}$ is reached. The corresponding tool lifetimes T_1, T_2, T_3 are determined (Figure 15.41). The points (v_1,T_1), (v_2,T_2), and (v_3,T_3) are the points of the straight line relationship on the double-logarithmic system (Figure 15.40). Table 15.7 illustrates the recommended values of $W_{F,all}$ for different types of tool materials. Referring to Equation 15.26, for $T = 1$ min, then $v = C$, that means C is the cutting speed to realize a tool life of 1 min. On the other hand, for $v = 1$ m/min, then $T = C^{1/n}$ min, which means for unity cutting speed, the tool life would be $C^{1/n}$. Figure 15.40 illustrates precisely what are the constants n and C and how they are graphically determined from the logarithmic plot. The value n is mainly a function of the cutting tool material used (Table 15.8), whereas C is heavily influenced by the cutting tool/workpiece material combination, as well as tool geometry and whether or not coolant is used. Their values are derived from carefully controlled experimental tests, which are published by cutting tool manufacturers.

The most important conclusion to be drawn from Taylor's relationship is that the tool life is mainly a function of cutting speed rather than either depth of cut or feed rate. This is one of the

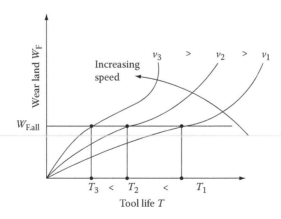

FIGURE 15.41 Wearing curves to determine the three points (v_1,T_1), (v_2,T_2), (v_3,T_3) of the Taylor plot.

TABLE 15.7
Recommended Values of $W_{F.all}$

Tool Material	$W_{F.all}$
HSS tools	1.5 mm for turning and face milling
	0.5 mm for end milling
	0.4 mm for drilling
	0.2 mm for finish turning
	0.15 mm for reaming and broaching
Carbide tools	0.8 mm for roughing
	0.4 mm for finishing
	0.15 mm for reaming and broaching
Ceramic tools	0.8 mm for roughing

Note: The allowable wear land $W_{F.all}$ is selected small for higher accuracy and better finish.

TABLE 15.8
Typical Values of Taylor Exponent n

Material to be Machined	Tool Material	n
Steels, alloy, and cast steels	Ceramics	0.4
	Carbides (P grade)	0.3
	HSS	0.15
	TCS	0.1
CI	Carbides (K grade)	0.3
Cu alloys	Carbides (K grade)	0.35
Light metals and Al alloys	Carbides (K grade)	0.45

Note: The value of n increases as the tool material becomes more refractory.

most important laws of material cutting. It explains why, when removing large amounts of material (roughing operations), feed rate and depth of cut can be increased without a major reduction in tool life, whereas if cutting speed is significantly increased, tool life shortens drastically. With finish turning, in which only small amounts of material are removed, high cutting speeds are normally used to enhance the surface quality. Some cutting tool manufacturers quote a modified version of the basic Taylor equation that includes the effects of both feed rate and depth of cut parameters.

15.2.7 Machining Economy

As it has been already discussed, the cutting speed has the greatest influence on the tool life compared to the feed rate and depth of cut, such that it greatly influences the overall economics of the machining process. First of all, the feed should be selected according to the type of cutting operation, if it is finishing or roughing. When a finish cut is to be taken, the appropriate feed will be that which provides an acceptable surface finish as specified by the designer. On the other hand, in rough cuts it is recommended to increase the feed rate rather than the cutting speed. The guiding principle in roughing is that the feed should always be set at the maximum possible. A limit of feed increase will depend on the maximum force the tool or the machine is able to withstand.

According to Taylor's equation, machining at higher speeds leads to much earlier failure of the cutting tool, whereas machining at lower speeds gives a lower production rate. Therefore, some optimum speed and optimum tool life must exist. Two criteria of optimization are frequently considered. One is the minimum cost criterion; and the other is the minimum machining time criterion. These criteria give different cost and different production rates. The minimum cost criterion gives a lower production rate, while the minimum time criterion gives higher cost. In general, these two objectives cannot be realized simultaneously. A compromise, usually somewhere in between, is selected. The optimization is performed to determine the optimum durability and the corresponding cutting speed for both criteria. For simplicity, a single-pass operation, at constant feed and depth of cut, will be considered.

15.2.7.1 Minimum Cost Criterion

The total unit machining cost C_t is the sum of three cost elements

$$C_t = C_M + C_T + C_L \tag{15.27}$$

where C_M is the cost related to the machine tool, C_T is the tool cost, and C_L is the labor cost.

If the cutting is performed at very high speed, C_L will be very low, but C_T will be very high. The opposite is true if the cutting is performed at low speed. C_M is found to be independent of the cutting speed. Accordingly, there is an optimum (economical) cutting speed v_e at which minimum unit cost is realized (Figure 15.42). The economical cutting speed v_e and the corresponding economical durability T_e can be determined by differentiating Equation 15.27 with respect to v and equating to zero.

$$\frac{dC_t}{dv} = 0$$

From which, the economical durability T_e min is obtained

$$T_e = \left(\frac{1}{n} - 1\right)\left(t_{ch} + \frac{W}{L_e(1+r)}\right) \tag{15.28}$$

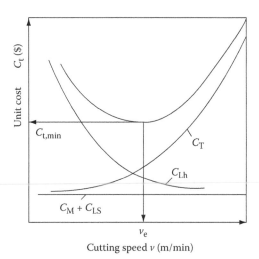

FIGURE 15.42 Cost elements and economical cutting speed.

where

n = Taylor's exponent
t_{ch} = tool exchanging time, min = const.
W = prime and sharpening cost per one tool life = const.

$$= \frac{(C_1 - C_2) + n_s C_s}{n_s + 1} \text{ for ordinary tools}$$

$$= \frac{C_1 - C_2}{n_i} \text{ for indexable throwaway tips}$$

C_1 = tool or tip prime cost ($)
C_2 = tool or tip scrap value ($)
n_s = number of tool sharpenings until scraping
n_{s+1} = number of tool lives until scraping
n_i = number of tip indexing times (for square tip, $n_i = 8$)
C_s = tool sharpening cost ($)
r = overhead ratio
L_c = labor cost/unit time

The corresponding economical cutting speed v_e (m/min) is then obtained by substituting the economical durability T_e in Taylor's equation.

Therefore,

$$v_e = C T_e^{-n} \tag{15.29}$$

v_e may also be obtained from the graphical logarithmic plot of the Taylor equation.

15.2.7.2 Minimum Machining Time Criterion

This criterion is developed by considering the total machining time t_m for one component.

$$t_m = t_h + t_s + \frac{t_{ch}}{z} \tag{15.30}$$

where

t_h = actual cutting time ($t_h \alpha v^{-1}$)
t_s = secondary or auxiliary machining time (t_s is independent of v)
z = T/t_h = number of components machined within one tool life ($z \alpha v^{1-\frac{1}{n}}$)
$\frac{t_{ch}}{z}$ = tool exchange time/component $\frac{t_{ch}}{z} \alpha v^{\frac{1}{n}-1}$

The total machining time t_m behaves similar to the total cost, because it consists of three elements. The value t_h decreases with the increase of cutting speed, t_{ch}/z increases with the cutting speed, and t_s does not depend on the speed. Accordingly, minimum time is realized at a cutting speed v_p, which is the speed for maximum productivity (Figure 15.43). The corresponding durability is the durability for maximum productivity. Both are determined by differentiating Equation 15.30 with respect to v and equating to zero.

$$\frac{dt_m}{dv} = 0$$

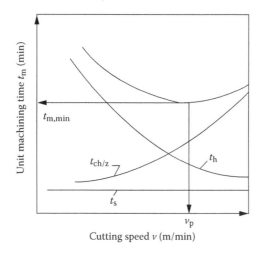

FIGURE 15.43 Elements of machining time and cutting speed for maximum productivity.

From which, the durability for maximum productivity T_p (min) can be calculated.

$$T_p = \left(\frac{1}{n} - 1\right) t_{ch} \qquad (15.31)$$

The corresponding cutting speed v_p (m/min) is then obtained by substituting T_p in Taylor's equation,

$$v_p = CT_p^{-n} \qquad (15.32)$$

v_p may also be obtained graphically from the logarithmic Taylor plot.

15.2.7.3 Comparison of the Two Criteria

In the majority of machining duties, the minimum cost criterion (economical durability T_e) is selected. However, if the work should be performed as fast as possible, then the minimum time criterion (durability for maximum productivity T_p) should be selected. The latter is advisable in special machining situations, such as in case of delay punishment, some military industries, and times of political crises. Close examination of both criteria as assessed by the durability Equations 15.28 and 15.31 leads to the following conclusions.

1. The economical durability T_e depends on
   a. Taylor exponent n, which is the slope of Taylor relationship
   b. Tool exchanging time t_{ch}
   c. The term $\dfrac{W}{L_c(1 + r)}$, expressing the tool and labor cost
2. The durability for maximum productivity T_p depends only on the two factors a and b.
3. Therefore, T_p is always less than T_e; consequently, v_p is always greater than v_e (Figure 15.44).

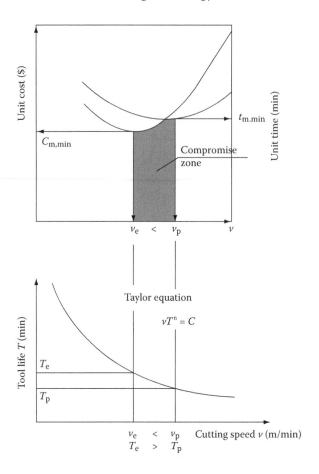

FIGURE 15.44 Comparison between minimum cost and minimum machining time criteria, and the compromise zone.

15.2.8 MACHINABILITY

Machinability is a term that was suggested for the first time in the 1920s to describe the machining properties of workpiece materials. Since that time, it is frequently used but seldom fully explained, as it has a variety of interpretations depending on the view point of the person using it. In its broadest interpretation, a material of good machinability requires lower power consumption, with high tool life, and achieving a good surface finish without damage. The relative importance of these three factors depends mainly on whether the machining is roughing or finishing. In actual production, tool life for rough cuts and surface finish for finish cuts are generally considered to be the most important criteria of machinability (Table 15.9).

TABLE 15.9
Relative Importance of Machinability Criterion in Roughing and Finishing

Order of Machinability Criterion	Rough Cut	Finish Cut
1	Tool life	Surface finish
2	Power consumption	Tool life
3	Surface finish	Power consumption

An additional machinability criterion sometimes highly considered is the chip disposal criterion. Long, thin curled ribbon chips, unless being broken up with chip breakers, can interfere with the operation leading to hazardous cutting area. This criterion is of vital importance in automatic machine tool operation. Chip formation, friction at the tool/chip interface, and BUE phenomenon are determinant to machinability. A ductile material that has a tendency to adhere to the tool face or to form BUE is likely to produce a poor finish. This has been observed to be true with such materials as low-carbon steel, pure aluminum, Cu, and stainless steel.

Mechanical and physical properties also play a role in the magnitude of energy consumption and temperatures generated during cutting. For instance, Ti is not machinable partly because of the high temperature generated owing to its poor thermal conductivity and partly because of its tendency to adhere to the cutting tool. Owing to the above described complex aspects of machinability, it is really difficult to establish quantitative relationship to evaluate the machinability of a material. For this reason, it is advisable to refer to machining recommendations that are based on extensive testing, practical experience, data collected in manufacturing manuals, and specialized handbooks.

From the foregoing, the parameters affecting machinability are

1. Work material properties and previous history
2. Tool material and geometry
3. Type of cutting operation
4. Machine tool power, rigidity, and accuracy
5. Machining conditions
6. Type and quantity of cutting fluid

Because there is no unit of machinability, it is usually assessed by comparing one material against another, one of which is taken as a reference. For roughing operations, the tool life is taken as a yardstick for ranking materials for machinability. The machinability of a reference material can be expressed in terms of cutting speed v_{60} for a tool life $T = 60$ min for a given tool material. The machinability of any other material is determined in the same way.

The relative machinability R of a material is therefore

$$R = \frac{v_{60} \text{ of the material}}{v_{60} \text{ of reference material}}$$

Table 15.10 lists the relative machinability of some common ferrous and nonferrous alloys in a descending order. The problem associated here is that if different tool materials are used to asses relative machinability, different ratings may occur. Thus, tables and data supplied should be used as guidelines.

TABLE 15.10
Relative Machinability Rating for Different Materials

Machinability Rating	Materials
Excellent rating	Mg alloys, Al alloys, Duralumin
Good rating	Zn alloys, Gunmetal, Gray CI, Brass, Free cutting steel
Fair rating	Low-carbon steel, Cast Cu, Annealed Ni, Low alloy steel
Poor rating	Ingot iron, Free cutting 18-8 stainless steel
Very poor rating	HSS, 18-8 stainless steel, Monel metal
Not machinable	White CI, Stellite, Carbides, Ceramics

15.2.9 CUTTING FLUIDS

Taylor did not confine his metal cutting experiments to study the tool life; he also discovered the benefits of flooding the cutting area with a coolant. It costs much to apply a cutting fluid (the fluid itself, pumping systems, collection and filtering systems, containers, etc.). Therefore, its use must be economically justified in the form of gains resulting due to increase of quality and productivity.

Cutting fluids are commonly applied to machining operations, chiefly to

1. Cool the cutting zone (cutting tool, workpiece, and chips). Accordingly, the tool life is increased, and the accuracy and surface quality of the workpiece is improved.
2. Lubricate the area of contact on the tool flank and tool face, thus reducing friction, which means
   - Reducing tool wear
   - Reducing cutting forces and consumed power
   - Increasing accuracy and surface quality of machined surface
3. Flush away chips and swarf.
4. Protect the workpiece and machine element from corrosion.

The cutting fluid can have interchangeable roles as coolant or a lubricant depending on temperatures encountered, cutting speed, type of machining operation, and method of its application. The cutting fluid should be flowed in a copious stream on top of the chip and tool. Some advantage is realized by directing the flow upward between the tool flank and workpiece. The flow must be uninterrupted when using carbide tools, because they may crack owing to sudden temperature changes. The lubricating action of cutting fluids is only effective at low cutting speeds (e.g., thread cutting). It is poor at 30 m/min and totally absent at 120 m/min and above. At high speeds (e.g., grinding) the fluid is not capable to reach the cutting zone, where it can lubricate. In all cases, the cutting speed is more effective than cutting fluid in improving finish.

A proper cutting fluid should be characterized by

- Good lubricating quality
- High heat absorption capacity
- High flash point
- High chemical stability
- Emits no fumes while in contact with hot surfaces
- Of less biological and environmental hazards
- Has no corrosive effect on the workpiece and machine guide ways
- Available at low price

Types of Cutting Fluids: Cutting fluids may be in the form of solids, liquids, gases, or chemical solutions. Liquid cutting fluids are widely used. These are of two general types.

1. Water-base (emulsion): These are composed of mineral oil and soft water. The thermal capacity (mass × specific heat) of water is more than double that of mineral oil, and this is one of the reasons that most coolants tend to be water-soluble oil compounds. Emulsions are generally used, where the cooling action is the most important. For this reason, emulsions are most frequently used in high-speed cutting operations (about 90% of all lubricants). Lubricating properties, however, are adequate, and with the inclusion of extreme pressure additives, one coolant is now suitable for most cutting duties. Emulsions are nonexpensive type of coolants, in which oil is mixed with water in certain proportions. A small amount of soap is added as an emulsified agent; thus, the emulsion has a milky white color. If the mixture is weak (low in oil) it may cause corrosion, and it is of low lubricating properties. Typical concentrations (oil/water) are

$$\frac{1}{40} - \frac{1}{50}, \text{ used for grinding}$$

$$\frac{1}{20}, \text{ used for turning, milling, drilling, and so on}$$

It is important to soften hard water before mixing by addition of 2 grams of common soda/1 L of water. The main causes of oil separation after mixing are the incorrect mixing procedure or the excessive water hardness.

Example

In a cutting operation, under the following conditions:

- Specific cutting power of material to be machined $k_s = 2$ W/mm³/s = 2000 $\frac{N}{mm^2}$
- Cutting speed $v = 40$ m/min
- Feed × depth of cut = $f \times a = 0.35$ mm/rev × 4 mm
- Cutting fluid: emulsion, flow rate $Q = 6$ L/min, $\rho = 0.96$ kg/L
- Specific heat $c_p = 3.56$ kJ/kg°C

Assuming that only 90% of the generated heat is conducted to coolant, calculate the temperature rise of the coolant.

Solution

$$\text{Volume removed/s} = \frac{a \times f \times v \times 1000}{60} \text{ mm}^3/\text{s}$$

$$= \frac{0.35 \times 4 \times 40 \times 1000}{60} \text{ mm}^3/\text{s}$$

$$\text{Power consumed} = \frac{k_s \times a \times f \times v \times 1000}{60}$$

$$= \frac{2 \times 4 \times 0.35 \times 40 \times 1000}{60} = 1870 \text{ W} = 1.87 \text{ kW}$$

Heat balance

$$\rho Q c_p \Delta t = 0.9 \times 1.87 \times 60$$

Therefore, the temperature rise Δt is

$$\Delta t = \frac{0.9 \times 1.87 \times 60}{0.96 \times 6 \times 3.56} = 5°C$$

The example shows that the temperature rise of coolant depends on its flow rate and specific heat. The heat absorbed by the coolant is radiated to the shop atmosphere. A large coolant container provides a large radiating area, and the larger the coolant volume stored, the greater the amount of heat that can be absorbed. In any process, where dimensional accuracy is important, cutting fluid temperature must be confined to reasonable limits; this is particularly important for cylindrical grinding of large dimensions.

2. Neat oils: These are only used when the lubricating action is the most important consideration. So they are usually confined to operations of slow cutting speeds, such as gear cutting, threading, honing, broaching, tapping, and reaming to promote superior finish of machined surfaces. For operations such as tapping and reaming, greases that melt at the tool edge may be more convenient than oils. Paraffin is sometimes used on Al alloys in preference to soluble oil because of its superior wetting properties.

Despite their much higher price, straight cutting oils find another important application in automatics, because water-based coolants, which are quite satisfactory for most turning operations, are likely to find their way to the head stock, causing oil contamination and hence serious deterioration of indexing mechanism. Straight cutting oils are blended from two basic types of oil.

- Mineral oils (e.g., paraffin) and other petroleum oils. Mineral oils are cheaper and more stable than fatty oils, and for these reasons, they are generally blended with fatty oils.
- Fatty oils are generally organic, of animal or vegetable origin (lard and whale oil, and rapeseed oil). They are expensive and emit an unpleasant odor; however, they are environmentally safe. Fatty oils possess very high lubricating properties and promote good finishes especially if lard oil is used for high tensile steels. On the other hand, they are less stable than mineral oils and may decompose if used in the natural state for a long time.

Extreme pressure (EP) additives are added, where cutting forces are practically high (tapping and broaching), or in operations that are performed at high feeds. EP additives provide a tougher, more stable form of lubrication at the chip/tool interface. These additives include sulfur, chlorine, or phosphorus compounds that react at higher temperatures in cutting zones to form metallic sulfides, chlorides, and phosphides. Most cutting oils are sulfurized by introducing chemically combined sulfur into the oil. Its main advantage is that it prevents pressure welding of chips to the tool (BUE) by forming a film on the tool/chip interface possessing antiweld properties. These oils may cause dark staining on copper-rich alloys; hence, sulfur-free grades may be used in this case. Chlorinated oils are used for the same purpose, and some cutting oils contain both sulfur and chlorine compounds.

3. Liquid Gas Coolants. Liquid gases (cryogenics) are recently used as coolants in cutting and grinding to eliminate the adverse environmental impact caused by conventional mineral oils and emulsions. Nitrogen in liquid state (–200°C) is injected through small nozzles into the cutting zone to reduce drastically its temperature, then directly evaporates. Consequently, the tool hardness and tool life are maintained, and high cutting speeds are allowed. Because the hot chips are severely quenched, they become more brittle, hence easily broken without the need of chip breakers; they are then disposed and recycled. The outcome is an improved machining economy without environmental effects.

15.2.10 THERMAL ASPECT IN METAL CUTTING

Most of the power delivered by machine tools is ultimately dissipated in the form of heat through the chip, tool, and workpiece. The adverse effects of high temperatures at the cutting edge have already been discussed, and heating of the workpiece makes accurate machining very difficult. Fortunately, about 80% of the heat generated during cutting is conveyed away by chip (Figure 15.45a). In metal cutting operations, the material is deformed elastically or plastically. A negligible proportion of mechanical power is consumed in elastic deformation with no heat development. The major proportion of power is consumed in plastic deformation and is completely converted into heat, which is mainly developed on two principal regions of plastic deformation (Figure 15.45b).

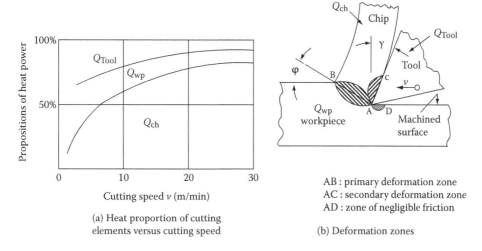

(a) Heat proportion of cutting elements versus cutting speed

AB : primary deformation zone
AC : secondary deformation zone
AD : zone of negligible friction

(b) Deformation zones

FIGURE 15.45 Heat proportions of chip, tool, and WP, and the deformation zones.

These are

- Primary deformation zone in the shear plane AB, $P_s = F_s \times v_s$
- Secondary deformation zone due to the friction existing in the tool/chip interface AC, $P_f = F_f \times v_{ch} = F_f \times v \times r$

where

P_s and P_f = powers consumed in primary and secondary deformation zones
F_s = shear force component
F_f = friction force component
v = cutting speed
v_s = velocity component in the direction of shear plane
v_{ch} = chip velocity = $v \times r$
r = chip thickness ratio

Friction in the region AD (the tool flank/workpiece interface) is neglected if sharp edged tool is used (Figure 15.45b). The total mechanical power P_c and the total power Q are correlated by the relationship.

$$Q = \frac{P_c}{J} = \frac{P_s + P_f}{J} \tag{15.33}$$

where

J = mechanical equivalent of heat
= 4270 N m/kcal

Referring to Figures 15.45 and 15.46

$$Q = Q_{wp} + Q_T + Q_{ch} \tag{15.34}$$

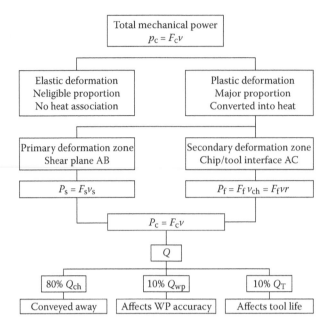

FIGURE 15.46   Mechanical power and heat energy relationships.

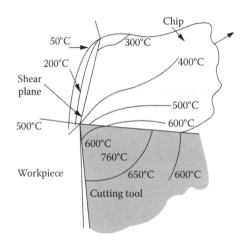

FIGURE 15.47   Temperature in cutting area by turning.

If a cutting fluid is used not only the tool life is increased, but also the workpiece accuracy and surface quality are improved. Figure 15.47 shows the temperature distribution in the cutting zone, which necessitates, in most cases, using a cutting fluid.

15.3   BASICS OF ABRASION PROCESSES

The history of abrasive machining technology begins with the primitive man, who presumably sharpened his flint knife with a piece of sandstone. Earliest records of grinding are from ancient Egypt (ca. 2000 BC). However, modern abrasive technology was only relatively recently established (since the early nineteenth century). Since that time, these processes played, in their developed form, an important role in manufacturing technology today. Developments in abrasive technology, particularly grinding with super abrasive wheels (diamond and CBN) are remarkable. The need for high accuracy machining of difficult-to-machine materials with high efficiency is making the application of abrasive technologies increasingly important.

Abrasive processes are classified according to the type of abrasive tool, which may be in the form of grinding wheels, stones, and sticks. Other processes use loose abrasives during lapping and polishing. In contrast to the machining by chipping, in the abrasive machining the individual cutting edges are randomly distributed and oriented. The chip thickness is small and not equal for all abrasive grains that are simultaneously in contact with the workpiece. Because there are many sources of friction, the energy required for removing unit volume may be up to 10 times (and even more) higher than that in machining by cutting. Most of this energy is converted into heat causing distortion of the work surface. When grinding steels, the high temperature may develop transformation to austenite followed by chilling effect; thus, martensite is formed and cracks may occur leading to reduced surface integrity. Abrasive machining is a mixture of cutting, plowing, and rubbing. The percentage of each depends on the geometry and condition of the abrasive grit. Grits with negative rakes or of rounded shapes do not form chips but instead plow or simply rub a groove in the surface.

15.3.1 GRINDING

Grinding is a metal removal abrasion process that employs a grinding wheel (GW), whose cutting elements are grains of abrasive materials of high hardness and refractoriness. It is generally among the final operations performed on manufactured products. It is not necessarily confined to small-scale material removal; it is also used for large-scale material removal operations and specifically competes economically with some machining processes such as milling and turning.

The sharp-edged hard grains are held together by bonding materials. Projecting grains (Figure 15.48) abrade layers of metal from the work in the form of very minute chips as the wheel rotates at high speeds (up to 60 m/s). Owing to the small cross-sectional area of the chip and the high cutting speed, grinding is characterized by high accuracy and good surface finish; however, it is also used in snagging. The chip formation in grinding is similar to milling; the chip has the same comma form similar to that obtained by milling. However, in grinding, not all the grains do participate equally in the metal removal as in milling. The GW has a self-sharpening characteristic. As the grains wear during grinding, they either fracture or are torn off the wheel bond, exposing new sharp grains to the work.

15.3.1.1 Grinding Wheels

Grinding wheels of all shapes are composed of carefully sized grains held together by a bonding material. Pores between the grains and the bond allow the grains to act as single-edged tools and provide chip clearance to prevent clogging of the grinding wheel. Grinding wheels are produced using appropriate grain size with the required bond, and the mixture is bonded to shape.

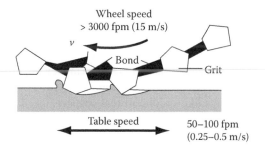

FIGURE 15.48 Cutting principle in surface grinding.

A number of variables are influencing the performance of a GW; these are

1. Abrasive materials

   Conventional abrasives: These are aluminum oxide Al_2O_3 and silicon carbide SiC. Al_2O_3 (corundum), of high hardness (Knoop = 2100) and toughness. It is mainly used for grinding metals and alloys such as steels, malleable iron, and soft bronze. SiC (carborundum), harder than Al_2O_3 (Knoop = 2500). It is more friable. It is mainly used for grinding materials of low strength like CI, Al, cemented carbides, and so on.

   Super abrasives: These are CBN (Knoop = 4500) and diamond which is the hardest of all materials (Knoop = 7500).

2. Abrasive grain size

   The size is identified by the grit number, which is a function of the sieve size. The smaller the sieve size, the higher the grit number. The sieve sizes (mesh number) are grouped into four categories

   Coarse: 10, 12, 14, 16, 20, 24
   Medium: 30, 36, 46, 56, 60
   Fine: 70, 80, 90, 100, 120, 150, 180
   Very fine: 220, 240, 280, 320, 400, 500, 600

   The choice of the grain size is governed by the nature of grinding operation and the material to be ground. Coarse and medium sizes are normally used for roughing and semifinishing operations. Fine and very fine grains are used for finishing operations, and also for making form GWs.

3. Wheel grade

   The wheel grade designates the force holding the grains. It is a measure of strength of the bond. It depends on the type and amount of the bond, the wheel structure, and the amount of abrasive grains relative to the bond. Because strength and hardness are directly related, the grade is also referred to as the hardness of the wheel.

   The grade is designated by letters, as follows:

   Very soft: A, B, C, D, E, F, G
   Soft: H, I, J, K
   Medium hard: L, M, N, O
   Hard: P, Q, R, S
   Very hard: T, U, V, W, X, Y, Z

   Soft grades are generally used for grinding hard materials, and vice versa.

4. Wheel structure

   The structure of a GW is a measure of its porosity. Some porosity is essential to provide clearances of the chips; otherwise, they would interfere with the grinding process. Loaded wheel with chips loses its cutting ability and it needs frequently to be dressed.

   Wheels of open or porous structure are used for high metal removal rates that produce rough surfaces, whereas those of dense or compact structure are used for precision grinding at low metal removal rates. Wheel structure is designated by numbers from 1 for extra dense to 15 for extra open.

5. Wheel bond

   The wheel bond holds the grains together in the wheel with just the right strength that permits each grain of the wheel cutting face to perform its work efficiently. As the grains become dull, they may be either broken, forming new cutting edges, or torn out leaving the bond. Thus, the bond acts like a tool post that supports the abrasive grains. When the amount of bond is increased, the size of the posts connecting the grains is increased, and the wheel becomes of harder grade.

   Seven standard bonds are available:

Vitrified bond (V). It is a refractory clay, which vitrifies (fuses) into glass. About 70% of GWs are made of vitrified bond, as its strength and porosity yield high stock removal rates. Moreover, it is not affected by coolants. However, it is brittle and sensitive to impact. It can withstand peripheral velocities up to 2000 m/min.

Resinoid bond (B). It is stronger and more elastic than the vitrified bond; however, it is not resistant to heat and chemicals. Resinoid-bonded wheels (also called organic wheels) can be used for rough grinding, parting off, and high-speed grinding at 3500 m/min.

Silicate bond (S). This is a silicate soda bond ($NaSiO_3$) that releases abrasives more rapidly than vitrified. It is used in large GWs for tool sharpening.

Rubber bond (R). This is the most flexible bond, as the principal constituent is natural or synthetic rubber. It is not so porous, and is widely used in thin cut-off GWs, portable snagging wheels, and centerless regulating wheels.

Shellac bond (E). It is frequently used for strong, thin wheels having some elasticity. It tends to produce high finish; thus, it is used in grinding cam shafts and mill rolls. Thin cut-off wheels may be shellac bonded.

Oxychloride bond (O). It is magnesium chloride bond. Its application is limited to certain wheels and segments used in disc grinders.

Metallic bond (M). It is made of Cu or Al alloys. It is used in diamond and CBN GWs, especially for electrochemical grinding (ECG) applications. The depth of abrasive layer may be up to 6 mm.

15.3.1.1.1 Grinding Wheel Marking

A standard marking system has been adopted by ANSI, which is implemented by all GW manufacturers today. This system involves the use of numbers and letters in the sequence indicated in Figure 15.49, which is

Abrasive type – grain size – grade – structure – bond.

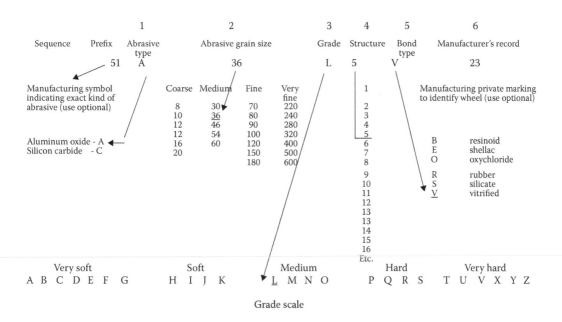

FIGURE 15.49 The marking of grinding wheels according to ANSI.

The selected wheel in Figure 15.49 is, therefore, designated as

$$(51 \text{ optional}) \ A - 36 - L - 5 - V \ (23 \text{ optional})$$

Moreover, the maximum allowable peripheral cutting speed should be printed on the grinding wheel, and must be strictly observed.

The following guidelines are generally considered when selecting a GW marking properly.

- Choose Al_2O_3 for steels, and SiC for CI, carbides, and nonferrous metals.
- Choose a hard grade for soft materials and vice versa.
- Choose a large grit for soft materials and a small grit for hard brittle materials.
- Choose a small grit for good finish and coarse grit for high MRR.
- Choose an open structure for rough cutting and a compact one for finishing.
- Choose resinoid, rubber, or shellac bonds for finish cuts, and vitrified for high MRR.
- Do not use vitrified for cutting speeds of more than 33 m/s.
- Choose softer grades for surface and internal cylindrical grinding and harder grades for external cylindrical grinding.
- Choose harder grades on nonrigid machines.
- Choose softer grades and friable abrasives for heat-resisting alloys.

15.3.1.2   Truing and Dressing, Balancing, and Safety Measures

15.3.1.2.1   Truing and Dressing of GWs

In grinding process, the sharp grains of the grinding wheel become rounded and hence lose its cutting ability. This condition is termed as grinding wheel glazing. Along with grain wear (glazing), another factor that reduces the cutting ability is the loading of voids between the grains with the chips and waste of the grinding process, resulting in a condition known as wheel loading. Loading especially occurs when grinding ductile and soft materials. A worn and loaded wheel ceases to cut. Its cutting ability can be restored by dressing or truing. Dressing may be described as a sharpening operation, which removes the worn and dull grits and embedded swarf in order to improve the cutting action.

Truing is an allied operation with the same tools to restore the correct geometrical shape of the wheel that has been lost owing to nonuniform wear. Truing makes the face of the wheel concentric and its sides plane and parallel, or it forms the wheel true for grinding special contours. It also restores the cutting ability of a worn wheel as in dressing. Dressing a wheel does not necessarily make it true: however, the distinction between truing and dressing is mostly hard to draw. There is some difference between diamond truing and crush dressing. The abrasive grit, being crystalline, tends to fracture along the most highly stressed crystallographic plane. Diamond truing tends to chip the grits along planes that make a small angle with respect to the direction of motion of the grit.

In diamond truing and dressing a single diamond (0.25–2 carat) is held in a steel holder. The grinding wheel is rotated at a normal speed and a small depth typically of 25 μm is given while moving the diamond across the face of the wheel in an automatic feed. The diamond tool is pointed in the same direction during wheel rotation to prevent gouging the wheel face. It is placed at the height of the wheel axis or 1–2 mm below it. For best results in diamond truing and dressing, the maximum rate of traverse should be 0.05–0.4 m/min, and the infeed 5–30 μm/pass, with 2 or 3 roughing, and 1 or 2 finishing passes. The lower the rates of longitudinal traverse and infeed, the smoother the active surface of the wheel will be.

Wheel truing and dressing that do not require diamonds, make use of

- Solid cemented carbide rollers
- Rollers of cemented carbide grains in a brass matrix

- Steel rollers and star-type dressers
- Abrasive wheels of black SiC with a vitrified bond, of diameter 60–150 mm and width of 20–32 mm

Wheel truing and dressing without diamond is less efficient and does not require the expensive diamonds tool. Of all the dressing tools not requiring diamonds, the abrasive wheel dressers are the most widely employed. They have a grain size from 3 to 5 steps coarser and 5 or 6 grades harder than the wheel that is to be dressed or trued. From 3 to 5 passes are made in dressing or truing; the traverse feed is 0.5–0.9 m/min and the infeed is 10–30 μm/pass. The last (finishing) passes are made without infeed and at reduced traverse speed (0.4–0.5 m/min). Ample coolant is applied in all dressing and truing methods that do not use diamond.

15.3.1.2.2 Balancing of GWs

Because of the high rotational speeds involved, GWs must never be used unless they are in good balance. A slight imbalance produces vibrations that cause waviness errors and harm the machine parts. This may cause wheel breakage, leading to serious damage and injury. Static imbalance of a GW is due to the lack of coincidence between its center of gravity and its axis of rotation. Imbalance is measured at the manufacturing plant in special balancing machines and is eliminated. The user balances grinding wheels either on a balancing stand or directly in the grinder. In the first case, and before mounting the wheel on the spindle, each wheel with its sleeve should be balanced on an arbor that is placed on the straight edges or revolving disks of a balancing stand. The wheel is balanced by shifting three balance weights in an annular groove of the wheel sleeve (or mounting flange). The wheel is rotated until it no longer stops its rotation at a specific position. Certain grinders are equipped with a mechanism for balancing the wheel during operation without stopping the wheel spindle rotation.

15.3.1.2.3 Safety Measures When Grinding

Any unsafe practice in grinding can be hazardous for operation and deserve careful attention.
   Various important aspects in this respect are

- Mounting of grinding wheels: The wheel should be correctly mounted and enclosed by a guard. Wheel bore should not be of tight fit on the sleeve.
- Wheel speed: The printed speed on the grinding wheel should not be exceeded.
- Wheel inspection: Before mounting the wheel, it should be checked for damage, cracks, and other defects. Ringing test should be performed. It is good enough for vitrified bonded wheels.
- Wheel storage: When not used, the wheels should be stored in a dry room and placed on their edges in racks.
- Wheel guards: They should always be used during grinding.
- Dust collection and health hazard: When grinding dry, provisions for extracting grinding dust should be made. Operator should wear safety devices to protect himself from abrasives and dust.
- Adequate machine power: Otherwise, the wheels slow down and develop flat spots, making the wheel run out of balance.
- Wet grinding: The wheel should not be partly immersed, as this would seriously throw the wheel out of balance.

15.3.1.3 Mechanics of the Grinding Operation

Grinding is similar to milling in that the grinding wheel may be considered as a rotating cutter with a very large number of edges (abrasives), distributed all over the periphery. The setting depth in grinding is very small compared to that of milling. Accordingly, it is expected that the chip thickness in grinding is considerably smaller, and hence the specific cutting power when grinding a given material is much greater.

Figure 15.50 shows the chip formation in case of external plunge-cylindrical grinding. The maximum chip thickness, h_g, is calculated as

$$h_g = \frac{GF}{n} \tag{15.35}$$

where

n = number of abrasives contributing in cutting the section ECF

$$= \frac{EC}{\lambda}$$

λ = effective periphery distance between two consecutive grains

Therefore,

$$h_g = \frac{GF}{EC}\lambda$$

but

$$\frac{CF}{EC} = \frac{v_w}{v_g},$$

$$GF = CF \sin(\alpha + \beta), \text{ and}$$

$$\sin(\alpha + \beta) = 2\sqrt{e}\sqrt{1/d_g \pm 1/d_w} \tag{15.36}$$

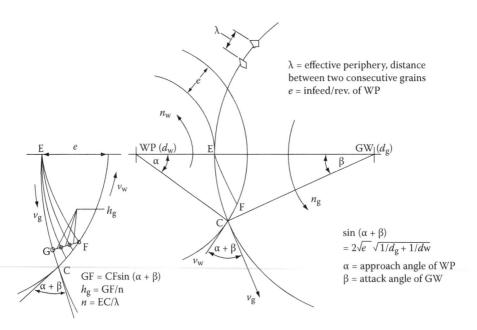

λ = effective periphery, distance between two consecutive grains
e = infeed/rev. of WP

$\sin(\alpha + \beta)$
$= 2\sqrt{e}\sqrt{1/d_g + 1/dw}$

α = approach angle of WP
β = attack angle of GW

$GF = CF\sin(\alpha + \beta)$
$h_g = GF/n$
$n = EC/\lambda$

FIGURE 15.50 Chip formation of external plunge grinding.

where

d_w = diameter of workpiece (mm)
d_g = diameter of GW (mm)
e = infeed rate of workpiece (mm/rev)
v_w = peripheral speed of workpiece (m/s)
v_g = peripheral speed of GW (m/s)

The positive sign under the root of Equation 15.36 is used for external cylindrical grinding, while the negative sign is used for internal cylindrical grinding ($d_g < d_w$). For surface grinding $d_w = \infty$.

From Equations 15.35 and 15.36, the maximum chip thickness h_g can be derived as

$$h_g = 2\lambda \frac{v_w}{v_g} \sqrt{e} \sqrt{\frac{1}{d_g} \pm \frac{1}{d_w}} \qquad (15.37)$$

Consequently, the mean chip thickness h_m is given by

$$h_m = \frac{h_g}{2} = \lambda \frac{v_w}{v_g} \sqrt{e} \sqrt{\frac{1}{d_g} \pm \frac{1}{d_w}} \qquad (15.38)$$

According to Equations 15.37 and 15.38, and for a given grinding wheel diameter and workpiece diameter, the mean chip thickness h_m and the maximum chip thickness increase linearly with the effective grain distance λ, and the linear speed of the workpiece v_w. They increase also, but not linearly, with the infeed rate e. For external cylindrical and surface grinding, the chip thicknesses decrease with increasing diameter d_g of the GW (Figure 15.51). It is noteworthy that increasing h_m and h_g calls for an increase of the acting force on individual grits of the GW and consequently for an increase of the wear rate of the GW. Consequently, increasing infeed e and work speed v_w calls for using a grinding wheel of harder grade, whereas increasing the speed v_g calls for using softer grade of GW.

15.3.1.3.1 Forces and Power in Grinding

Figure 15.52 illustrates the cutting force components in case of traverse cylindrical grinding. In this case there are three components. These are F_c (main cutting force acting tangentially in the direction of v_g), F_f (feed force component acting in the direction of longitudinal traverse feed f_{tr}), and F_r (radial, or passive force component acting perpendicular to the work rotational axis). Therefore, the resultant cutting force R is given by

$$R = \sqrt{F_c^2 + F_f^2 + F_r^2}$$

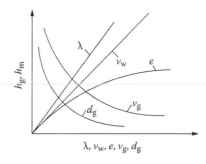

FIGURE 15.51 Effect of grinding parameters on the mean and maximum chip thickness, respectively.

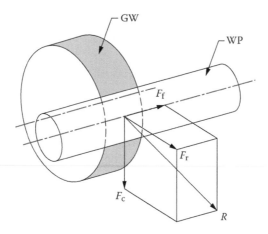

FIGURE 15.52 Cutting force components in traverse cylindrical grinding.

In case of plunge-cut cylindrical grinding, there are only two components acting, which are the main force component F_c and a radial feed component F_r acting in the direction of the radial infeed f_{in} (Figure 15.53).

15.3.1.3.2 Calculation of the Main Cutting Force F_c in Plunge-Cut Cylindrical Grinding

$$F_c = z_e h_m B f_\gamma k_{sm} \qquad (15.39)$$

where
B = width of GW
z_e = number of grains contributing in cut at the same time
= $z\beta/2\pi$

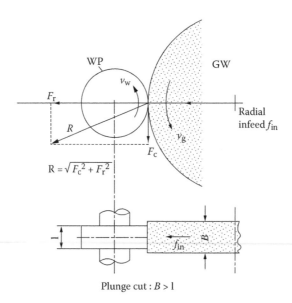

FIGURE 15.53 Force components in case of plunge-cut cylindrical grinding.

z = number of grains on the periphery of GW, pitched with λ

$\quad = \pi d_g / \lambda$

$\widehat{\beta}$ = attack angle of the GW in radians (Figure 15.50)

k_{sm} = spec. power of WP material corresponding to chip thickness h_m (Figure 15.27)

f_γ = factor considering negative rake of abrasives (ranging from 1 to 7, depending on h_m and grain size of abrasives)

$$\widehat{\beta} = \sin\beta = 2 \sqrt{\frac{e}{d_g[1 \pm (d_g/d_w)]}}$$

Then,

$$z_e = \frac{1}{\lambda} \sqrt{\frac{ed_g}{1 \pm (d_g/d_w)}} \tag{15.40}$$

Substituting the values of z_e and h_m, in Equation (15.39), the main cutting force F_c is then given by

$$F_c = \frac{v_w}{v_g} eB f_\gamma k_{sm} \tag{15.41}$$

Equation 15.41 is applicable for cylindrical (external or internal) and surface plunge grinding. In case of traverse grinding (external, internal, or surface), F_c can be calculated as

$$F_c = \frac{v_w}{v_g} e f_{tr} f_\gamma k_{sm} \tag{15.42}$$

where

e = depth of cut

f_{tr} = traverse feed/rev of WP in cylindrical grinding, or the traverse feed/stroke of WP in surface grinding. It is generally taken 0.1–0.4 the width of GW for finishing and 0.4–0.8 the width of GW for rough cuts (Figure 15.54)

The consumed power in grinding P_c can then be calculated as

$$P_c = \frac{F_c v_g}{1000} \, kW \tag{15.43}$$

F_c in (N) and v_g in (m/s).

The feed power P_f is calculated as

$$P_f = \frac{F_c v_w}{1000} \, kW \tag{15.44}$$

It is very small as compared to P_c because $\dfrac{v_w}{v_g} = \dfrac{1}{20} ..... \dfrac{1}{150}$.

The necessary motor powers for the grinding wheel and the workpiece can be calculated by considering the mechanical efficiency of the machine.

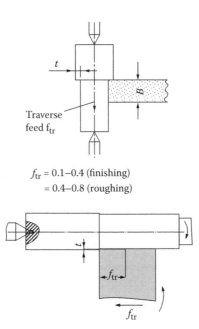

f_{tr} = 0.1–0.4 (finishing)
= 0.4–0.8 (roughing)

FIGURE 15.54 Traverse feed in roughing and finishing as related to the width of grinding wheel.

15.3.2 HONING

Honing is a controlled, low speed sizing and surface finishing process in which stock is abraded by the shearing action of a bonded abrasive honing stick. In honing, simultaneous rotating and reciprocating action of the stick (Figure 15.55a) results in a characteristic cross-hatch lay pattern (Figure 15.55b). Because honing is a low-speed operation, metal is removed without the increased temperature that accompanies grinding, and thus, any surface damage caused by heat-affected zone (HAZ) is avoided.

In addition to removing stock, honing involves the correction of errors from previous machining operations. These errors include

- Geometrical errors such as out-of-roundness, waviness, bell mouth, and reamer chatter
- Dimensional inaccuracies
- Surface character such as roughness and lay pattern

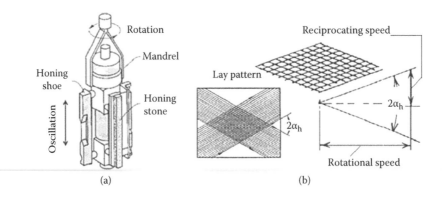

FIGURE 15.55 Honing operation: (a) honing head and (b) cross-hatched angle. (Adapted from ASM International, *Machining*, vol. 16, *Metals Handbook*, Materials Park, Ohio, 1989.)

Honing corrects all of these errors with the least possible amount of material removal; however, it cannot correct hole location or perpendicularity errors. The most frequent application of honing is the finishing of internal cylindrical holes. The hone is allowed to float by means of two universal joints so that it follows the axis of the hole (Figure 15.55). Owing to the fact that the tool floats, the honing sticks are able to exert an equal pressure on all sides of the bore regardless of machine vibration, and therefore, round and straight bores are produced. As the hone reciprocates through the bore, the pressure and the resulting penetration of grit is greatest at high spots and consequently the waviness crests are abraded, making the bore straight and round. After leveling high spots, each section of the bore receives equal abrading action. The hole axis is usually in the vertical position to eliminate gravity effects on the honing process; however, for long bores, the axis may be horizontal. Although CI and steel are the most commonly honed materials, the process can also be used for finishing materials ranging from softer metals like Al and Cu alloys to extremely hard materials like case nitrided steels or sintered carbides. The process can also be used for honing ceramics and plastics.

Bores as small as 1.5 mm diameter can be honed. The maximum bore diameter is governed by the machine and its ability to accommodate the WP. Machines powered by motors of up to 17 kW are available that can hone bores up to 1200 mm in diameter. Honing bores up to 760 mm in diameter is a common practice. Although most internal honing is done on simple, straight through holes, blind holes can also be honed. It is not feasible to hone the sides of a blind hole flush with the bottom. Bores having keyways can be honed and so can male or female splines.

15.3.2.1 Stock Removal

In honing, a general rule is to remove twice as much stock as the existing error in the WP. For example, if a cylinder is 50 μm out of round or tapered, a removal of 100 μm will be required for complete clean up. The work in preceding operations is usually planned so that the amount of stock removed in honing is minimized. Honing is performed at a rate of 32 cm³/min from soft steel tubes. For steel tubes hardened to 60 HRC, the rate is reduced to 16 cm³/min. Rough honing is employed before finish honing when large amounts of stock are to be removed and special finishes are required. Sticks containing abrasives of 80 grits or even coarser are used for rough honing to maximize the removal rate. Finish honing is accomplished by abrasives of 180–320 grits or finer.

15.3.2.2 Dimensional Accuracy and Surface Finish

Internal honing to tolerances of 2.5–25 μm is common. Surface roughness Ra of 0.25–0.4 μm can be easily obtained by rough honing, and roughness of less than 0.05 μm can be achieved and reproduced by finish honing. Figure 15.56 compares typical ranges of surface roughness obtained by honing compared to other common microfinishing processes.

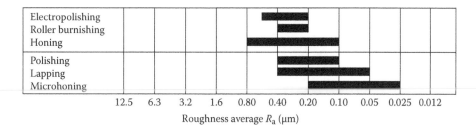

FIGURE 15.56 Average surface roughness of common microfinishing operations. (Adapted from ASM International, *Machining*, vol. 16, *Metals Handbook*, Materials Park, OH, 1989.)

15.3.3 Superfinishing (Microfinishing)

It is an abrading process that is used mainly for external surface refining or cylindrical, flat, and spherical shaped parts. It is not a dimension-changing process but is used for producing finished surfaces of superfine quality. Only a slight amount of stock is removed (2–30 μm), which represents the surface roughness. The process of honing involves two main motions, whereas superfinishing requires three or more motions. As a result, the abrasive path is random and never repeats itself. The primary distinction between honing and superfinishing is that in honing, the tool rotates, while in superfinishing the workpiece always rotates (Figure 15.57). The bonded abrasive stone, whose operating face complies with the form of the WP surface, is subjected to a very light pressure. A short, HF stroke, superimposed on a reciprocating traverse, is used for superfinishing of long lengths.

15.3.4 Lapping

The usual definition of lapping is the random rubbing of WP against a CI lapping plate (lap) (Figure 15.58), using loose abrasives carried in an appropriate vehicle (oil) to improve fit and finish. It is a low-speed, low-pressure abrading process. In general, the surface quality that can be obtained by lapping is not easily or economically obtained by any other process. Moreover, the life of the moving parts that are subjected to wear can be increased by eliminating hills and valleys.

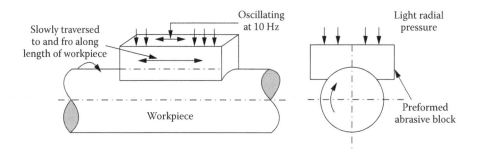

FIGURE 15.57 Principle of superfinishing process.

FIGURE 15.58 CI-lapping plate. (From Water, T. F., *Fundamentals of Manufacturing for Engineers*, CRC Press, Boca Raton, FL, 2002. With permission.)

Lapping is a final machining operation that realizes the following major objectives:

- Extreme dimensional accuracies
- Mirror-like surface quality
- Correction of minor shape imperfections
- Realizing close fit between mating parts

Lapping does not require holding devices and consequently no WP distortion occurs. Also, less heat is generated than in most other finishing operations. Therefore, metallurgical changes are totally avoided. The temperature increase of the surface is only 1°C–2°C over ambient.

Parameters that affect the lapping process are

- Lapping abrasives: Diamond is used for lapping WC and precious stones. B_4C is used for lapping dies and gauges. B_4C is more expensive than SiC and Al_2O_3 (10–25 times). SiC is intended for rapid removal and it is mainly used for lapping CI, hardened steel, and nonferrous metals. Al_2O_3 is intended for improved finish of soft steels.
- Grit size and grading: Grit size (mesh number) ranging from 100 to 1000 is used depending on the degree of finish required. Soft materials require finer grains. Abrasives increase in cost as their grading becomes closer. The use of low cost, loosely graded commercial abrasive is not recommended for reasons of economy.
- Vehicle: This prevents scoring of lapped surfaces and varies from clean water to heavy grease. It is selected to suit the work, method, and type of surface finish required. For machine lapping, an oil-base type is recommended; however, a commercial mixture of kerosene and machine oil can be used. Grease-based vehicles are recommended for lapping soft metals.

15.4 REVIEW QUESTIONS

1. What is the main difference between oblique and orthogonal cutting? Give practical examples of each.
2. Enumerate some conditions that would allow for a continuous chip to be formed in metal cutting.
3. What is the BUE? Explain how it is formed. What conditions promote its formation?
4. Derive an expression for the shear angle in orthogonal cutting in terms of rake angle γ and chip thickness ratio r.
5. Draw the Merchant force circle in orthogonal cutting; then derive equations for calculating the shear and friction forces.
6. Discuss the importance of the shear angle in metal cutting performance. What factors influence its value?
7. Sketch the various forms of wear in cutting tools.
8. What is the main parameter that affects the tool wear and tool life?
9. Numerate the different tool life criteria that are generally practiced in machining.
10. What are the factors that basically affect the tool life?
11. Define machinability. What indices are used for its evaluation?
12. Derive an expression for the economical durability and another for the durability for maximum productivity, then compare.
13. Which of the three cutting forces in oblique cutting consumes most of the power? and why?
14. What is the most important material property for metal cutting tools?
15. What is a cermet?
16. How is a CBN tool manufactured?

17. What is a chip breaker?
18. Taylor has discovered HSS. For what else is he well known?
19. Why is CBN better than diamond for machining steels?
20. What are the chief functions of cutting fluids?
21. Name and describe the three kinds of classical chip forms.
22. Name the principal cutting tool materials.
23. Explain the reasons that the same tool life may be obtained at two different cutting speeds.
24. What are the possible disadvantages of a cutting operation in which the type of chip produced is discontinuous?
25. The tool life can be almost infinite at low cutting speeds. Would you then recommend all machining be done at low speeds? Explain.
26. List alloying elements used in HSS cutting tools. Explain why they are used.
27. What precautions would you take in machining with ceramics?
28. Explain whether it is desirable to have a high or low constant in the Taylor's tool life equation.
29. Mark true [T] or false [F]:
    [   ] In metal cutting at larger rakes, the cutting power is lower.
    [   ] The chip velocity v_c is always greater than the cutting speed v.
    [   ] The specific cutting resistance k_s increases at small feeds.
    [   ] The specific cutting resistance k_s is a material constant.
    [   ] The specific cutting resistance A_s increases at larger rake.
    [   ] It is highly desirable to have higher n-exponent in Taylor-life equation.
    [   ] Crater wear occurs when machining brittle materials.
    [   ] The durability for maximum productivity is always greater than the economical durability.
    [   ] Higher machinability is ensured if the machined surface is rough.
    [   ] Ceramic tools are frequently used on planers and shapers to produce the best surface finish.
    [   ] In metal cutting the angle of shear decreases as the tool rake increases, and that improves the machinability.
    [   ] BUE mostly forms in case of discontinuous chips.
    [   ] Grinding can be followed by a milling operation.
    [   ] Open structured GWs are recommended for soft materials.
    [   ] A hard grade GW is recommended for grinding hard materials.
    [   ] The grade of a GW represents the hardness of its abrasives.
    [   ] In creep grinding, the depth of cut is removed in several machining passes.
    [   ] Truing restores the shape of GW.
    [   ] For a certain metal or alloy, the specific cutting power in grinding is considerably greater than in turning and other chipping processes.
30. What are the various abrasive processes you are familiar with?
31. Describe briefly the numbering system used for grading grinding wheels. A grinding wheel is specified as A46J8V. Give the meaning of each symbol, explaining its significance.
32. What is creep feed grinding?
33. Define dressing, truing, and balancing of a grinding wheel.
34. What is the correlation between the main and feed power in plunge cylindrical grinding?
35. What are the main factors influencing the surface quality in grinding?
36. Compare between grinding and honing.
37. Draw a neat sketch to show the kinematics of honing process.
38. Why is greater precision possible by grinding than chipping?
39. Derive a relationship for estimating the maximum chip thickness in case of external plunge-cylindrical grinding. From your derivation show how the speed and the wheel speed do affect the wear of the grinding wheel.

15.5 PROBLEMS

1. A turning operation is to be adopted under the following machining conditions:
   Cutting speed: $v = 80$ m/min
   Feed rate: $f = 0.2$ mm/rev
   Depth of cut: $a = 1.5$ mm
   External workpiece diameter: $D = 40$ mm
   Calculate
   a. The final workpiece diameter
   b. The rotational speed of the workpiece
   c. The tool feed rate expressed in m/min
   d. The material removal rate Z in mm³/min
   e. The specific cutting power k_s assuming that the measured main force component $F_c = 750$ N.
   f. The power P_f in kW assuming that the measured feed force component $F_f = 500$ N

2. In an orthogonal cutting test on mild steel, the following results are obtained:
   $t_1 = 0.25$ mm
   $t_2 = 0.75$ mm
   $b = 2.5$ mm
   $F_c = 900$ N
   $F_p = 450$ N
   Tool rake $\gamma = 0$
   Calculate
   a. Mean angle of friction β on tool face
   b. Mean shear stress τ_s

3. A bar of 100 mm diameter of SAE 1020 steel is to be turned on a lathe using a HSS tool at a speed of 30 m/min. The feed is 0.8 mm/rev, and the depth of cut is 3 mm. The main component of the cutting force is measured as 3250 N.
   What is the power needed to perform this cut and what would be the specific power?

4. A HSS tool has been used for a metal cutting operation shows a T–v Taylor relationship of $vT^{0.125} = 44.5$. Originally, 15 mins were required to remove a dull tool. A new tool holder has made it possible to reduce the time to 5 min. What increase in cutting speed does this permit to obtain the maximum rate of production from the operation?

5. For a cemented carbide tool, if $vT^{0.2} = 152$, find the economical durability and the durability for maximum productivity and their corresponding speeds assuming the following conditions:
   - Tool of replaceable clamped insert of eight cutting edges.
   - Tool insert costs $9.60 new.
   - Tip exchange time $t_{ch} = 0.5$ min.
   - Labor cost considering overhead $L_c(1 + r) = \$ 0.4$/min.

6. A shaft made of mild steel ($k_s = 1990 \times h_m^{-0.26}$) with a diameter of 55 mm is ground under the following conditions:
   Grinding wheel
   d_g: 250 mm
   b: 25 mm
   n_g: 2350 rev/min
   Grain mesh 60
   Workpiece
   n_w: 75 rev/min
   Table feed u_t: 1 m/min
   Depth of cut e: 0.01 mm

It is required to determine the following:
a. Mean chip thickness h_m
b. Mean force component F_c
c. Power of the wheel P_c and work P_f

BIBLIOGRAPHY

Arshinov, V., and Alekseev, G. 1970. *Metal Cutting Theory and Cutting Tool Design.* Moscow: Mir.

Boothroyd, G. 1981. *Fundamentals of Metal Machining and Machine Tools.* New York: McGraw-Hill.

BS 1296, BS 1886.

Degarmo, E. P., Black, J. T., and Kohser, R. A. 1997. *Materials and Processes in Manufacturing,* 8th ed. Upper Saddle River, NJ: Prentice-Hall.

El-Hofy, H. 2007. *Fundamentals of Machining Processes—Conventional and Nonconventional Processes.* Boca Raton, FL: CRC Press.

Ghenis, B., Doktor, L., and Tergan, V. 1976. *Cylindrical Grinding Practice.* Moscow: Mir Publishers.

Insasaki, I., Tönshoff, K., and Howes, T. D. 1993. Abrasive machining in the future, *Annals of the CIRP* 42(2): 723–732.

ISO/DIN 3022, 1973.

Jain, R. K. 1993. *Production Technology.* 13th ed. Delhi: Khanna.

Kalpakjian, S. 1985. *Manufacturing Processes for Engineering Materials,* 4th ed. Upper Saddle River, NJ: Prentice-Hall.

Keczmarek, J. 1976. *Principles of Machining by Cutting, Abrasion, and Erosion.* Stevenage, Hertfordshire, UK: Peter Pergrines.

Kienzle, O. and Victor, H. 1956. Zerspanungstechnische Grundlagen für die Kräftmäsige Berechnung und den Einsatz von Drehbänken, Hobelmaschinen und Bohrmaschinen. *Werkstattstechnik und Maschinen* 46: 283–288.

Shaw, M. C. 1984. *Metal Cutting Principles.* New York: Clarendon Press.

Taylor, F. W. 1906. On the Art of Cutting Metals, Trans. ASME, Vol. 28, p31.

Water, T. F. 2002. *Fundamentals of Manufacturing for Engineers.* Boca Raton, FL: CRC Press.

Youssef, H. A. 1977. *Theory of Metal Cutting-Bases of Conventional and Nonconventional Machining Processes.* Alexandria, Egypt: DAR AL-MAAREF.

Youssef, H., and El-Hofy, H. 2008. *Machining Technology—Machine Tools and Operations.* Boca Raton, FL: CRC Press.

16 Machine Tools for Traditional Machining

16.1 INTRODUCTION

Machining is usually adopted whenever the part accuracy and surface quality are of a prime importance. Machining activities form around 20% of the manufacturing activities in United States. It is carried out on machine tools that are responsible for generating the motions required for producing a given part geometry. Such machine tools form 70% of the operating production machines. They are characterized by their high production accuracy compared to metal-forming machine tools. Conventional machine tools are classified as general purpose or special purpose machine tools.

16.2 GENERAL PURPOSE MACHINE TOOLS

Typical examples of the general purpose machine tools include turning, drilling, shaping, milling, grinding, broaching, jig boring, and lapping machines. Gear cutting and thread cutting are examples of special purpose machine tools. When using the general or special purpose manual machine tools, the product accuracy and productivity depend on the operator's interference during operation. Capstan and turret lathers are typical machines that slightly reduce the operator's role during machining at higher rates and better accuracy. Semiautomatic machine tools perform automatically controlled movements, while the workpiece is hand loaded and unloaded. Fully automatic machine tools are those machines in which workpiece handling and the cutting and other auxiliary activities are performed automatically. Semiautomatic and automatic machine tools are best suited for large production lots. The general purpose machine tools can be classified according to the shape of product to those used for cutting cylindrical surfaces and those used for machining flat surfaces.

16.2.1 MACHINE TOOLS FOR CUTTING CYLINDRICAL SURFACES

16.2.1.1 Lathe Machines

Lathes are employed for turning external cylindrical, tapered, and contour surfaces, boring cylindrical and tapered holes, machining face surfaces, cutting external and internal threads, knurling, centering, drilling, counterboring, countersinking, spot facing and reaming of holes, cutting off, etc. Engine lathes provide a means for feeding the cutting tool along the axis of workpiece revolution. Metal cutting lathes may differ in size range from manual to automatic ones. The different types of lathe machines are given.

1. Universal engine lathes: These are widely employed in job and lot production, as well as for repair work. Its size varies from small bench lathe to heavy-duty lathe for machining parts weighing many tons. Figure 16.1 illustrates a typical universal engine lathe. The bed 2 carries the headstock 1 which contains the speed gearbox. The bed also mounts the tailstock 6 whose spindle usually carries the dead center. The work may be held between centers, clamped in a chuck, held in a fixture, or mounted in a face plate. If a long shaft is

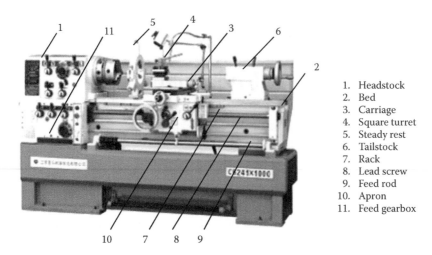

1. Headstock
2. Bed
3. Carriage
4. Square turret
5. Steady rest
6. Tailstock
7. Rack
8. Lead screw
9. Feed rod
10. Apron
11. Feed gearbox

FIGURE 16.1 A typical lathe. (Courtesy of Anhui Chizhou Household Machine Tool Co., China.)

to be machined, it is necessary to support the other end by the tail stock center. In many cases when the length of the shaft exceeds 10 times its diameter ($\ell > 10$ D), a steady rest 5 or follower rest are used to support these long shafts.

Single-edge tools are clamped in a square turret 4 mounted on the carriage 3. The types of tools such as drills, core drills, and reamers are inserted in the tail-stock spindle after removing the center. Carriage 3, to which apron 10 is secured, may be traversed along the guideways either manually or powered. The cross-slide can also be either manually or power traversed in the cross-direction. Surfaces of revolution are turned by longitudinal traverse of the carriage. Cross-slide feeds the tool in the cross-direction to perform facing, recessing, forming, knurling, and parting off operations. Power traverse of the carriage or cross-slide is obtained through the feed mechanism. Rotation is transmitted from the spindle through change gears and the quick change feed gearbox 11 to either the lead screw 8 or feed rod 9. Powered motion of the lead screw is used only for cutting threads. The carriage is traversed by hand or powered from the feed rod. Carriage feed is obtained by pinion and rack 7 fastened to the bed. The pinion may be actuated manually or powered from the feed rod. The cross-slide is powered by the feed rod through a gearing system in apron 10. Table 16.1 shows the various turning operations that can be performed on the center lathe machine. Tapered surfaces are turned by employing one of the following methods (Figure 16.2).

a. Swiveling the compound rest to the required angle α: when turning short internal and external tapers with large taper angles (Figure 16.2a).
b. Using a straight-edge broad nose tool of width that exceeds the taper being turned. The tool is cross-fed (Figure 16.2b).
c. Setting over the tailstock at an angle of taper α that should not exceed 8° and using longitudinal feed as shown in Figure 16.2c.
d. Using taper turning attachment for long tapered work. The cross-slide 1 is disengaged from the cross-feed screw and is linked through the tie 2 to the slide 3 (Figure 16.2d).

When cutting a screw thread, the tool is moved along the bed and is driven by a nut engaging with the lead screw. The lead screw is driven by a train of gears from the machine spindle (Figure 16.3). The gear train may be one of the following arrangements.

TABLE 16.1
Lathe Operations and Relevant Tools

Lathe Operations	Sketch
Cylindrical turning	
Taper turning	
Facing off a workpiece	
Necking or recessing	4. Necking or recessing
Parting off with parting-off tool	
Boring of cylindrical hole	
External threading	

(continued)

TABLE 16.1 (Continued)
Lathe Operations and Relevant Tools

Lathe Operations	Sketch
Internal threading tool	
Drilling and core drilling with a twist drill	
Forming	

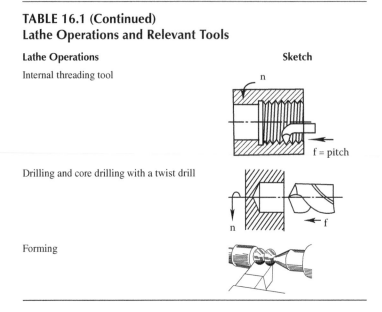

In the simple gear train, shown in Figure 16.3, the following ratio holds:

$$\frac{\text{Turns of leadscrew}}{\text{Turns of spindle}} = \frac{\text{Teeth on Driver}(A)}{\text{Teeth on Driven}(B)} \tag{16.1}$$

The intermediate gear has no effect on the ratio. It simply acts as a connection that makes the lead screw rotate in the same direction of the machine spindle.

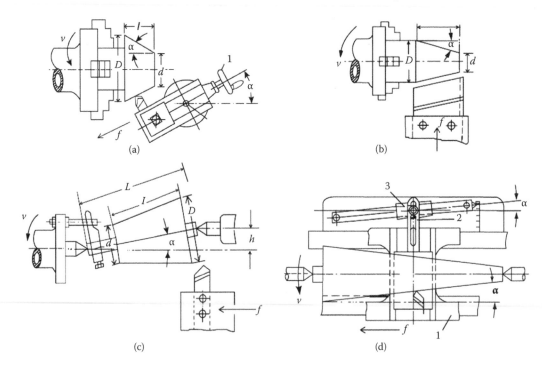

(a)

(b)

(c)

(d)

FIGURE 16.2   Methods of taper turning.

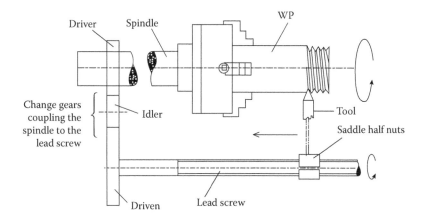

FIGURE 16.3 Diagrammatic representation of screw cutting on a lathe.

In the compound gear train, shown in Figure 16.4b, the gear ratio becomes

$$\frac{\text{Turns of leadscrew}}{\text{Turns of spindle}} = \frac{\text{Teeth on } C}{\text{Teeth on } D} \times \frac{\text{Teeth on } E}{\text{Teeth on } F} = \frac{\text{Teeth on Drivers}}{\text{Teeth on Driven}} \qquad (16.2)$$

Gears supplied with lathes, generally, range from 20 to 120 T in steps of 5 T with two 40's or two 60's and one gear 127 for British system threads. The lead screw on lathes is always single threaded of a pitch varying from 5 to 10 mm, depending on the size of the machine. For English lathes, the most common screw threads have 2, 4, or 6 threads per inch (tpi).

Example 1

Calculate suitable gear trains for cutting the following threads (Chapman, 1981):

   a. 2.5 mm pitch on a 6 mm lead screw
   b. 7/22 inch pitch, 3 start on a lathe with 2 tpi
   c. 2.5 mm pitch on a 4-tpi lead screw

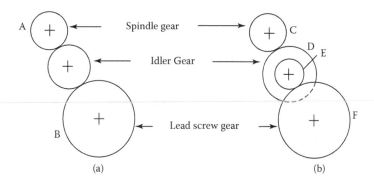

FIGURE 16.4 Gear trains for thread cutting on a lathe.

Solution

a. 2.5 mm pitch on a 6 mm lead screw

$$\frac{Drivers}{Driven} = \frac{\text{Pitch to be cut}}{\text{Pitch of lead screw}}$$

$$\frac{Drivers}{Driven} = \frac{2.5}{6} = \frac{5}{12} = \frac{25}{60}$$

25 teeth driving 60 teeth in a simple gear train

b. 7/22 inch pitch, 3 start on a lathe with 2 tpi
   Lead of the thread = 3 x 7/22 = 21/22 inch
   Pitch of lead screw = 1/2 inch

$$\frac{Drivers}{Driven} = \frac{\dfrac{21}{22}}{\dfrac{1}{2}} = \frac{42}{22} = \frac{21}{11} = \frac{3 \times 7}{2 \times 5\dfrac{1}{2}} = \frac{30}{20} \times \frac{70}{55}$$

A compound train with 30 teeth and 70 teeth as the drivers and 20 teeth and 55 teeth as the driven

c. 2.5 mm pitch on a 4-tpi lead screw
   To cut p mm pitch would require a ratio p as large as 5p/127

$$\frac{Drivers}{Driven} = \frac{5 \times 2.5}{127} \Big/ \frac{1}{4} = \frac{50}{127}$$

A simple gear train with 50 teeth driving 127 teeth

Thread chasing is the process of cutting a thread on a lathe with a chasing tool, shown in Figure 16.5. Chasing is used for the production of threads that are too large in diameter for a die head. It can be used for internal threads above 25 mm diameter. Thread chasing reduces the threading time by 50% compared to single point threading. Depending on the size of the thread, 20–50 passes may be required to complete a thread.

Workpiece fixation on an engine lathe depends, mainly, on the geometrical features of the workpiece. The workpiece can be held using one of the following methods:

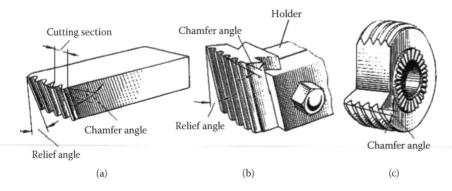

(a)   (b)   (c)

FIGURE 16.5 Thread chasers. (From Rodin, P., *Design and Production of Metal Cutting Tools*, Mir Publishers, Moscow, 1985. With permission.)

a. Between centers (Figure 16.6a) is an accurate method for clamping a long workpiece. The tailstock center may be dead center or a live center when the work is rotating at high speed. In such a case, rests are used to support long workpieces in order to prevent their deflection under the action of the cutting forces. The steady rest is mounted on the guideways of the bed, while the follower rest is mounted on the saddle of the carriage as shown in Figure 16.6b,c.

b. Clamping hollow workpieces with previously machined holes on mandrels (Figure 16.6d).

c. In a chuck that is commonly employed for holding of short work (Figure 16.3e). If the work length is considerably large relative to its diameter, supporting the free end with the tailstock dead or live center (Figure 16.6f). Chucks may be universal (Figure 16.3g)

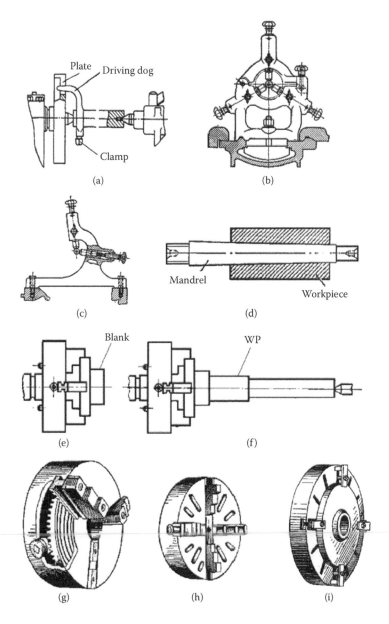

FIGURE 16.6 Clamping workpieces.

(self-centering) of three jaws, which are expanded and drawn simultaneously, to clamp circular and hexagonal rods or may be of four independent jaws that are especially useful in clamping irregular and nonsymmetrical workpieces (Figure 16.6h).

    d. Clamping large workpiece directly on a face plate (Figure 16.6i) or on a plate fixture that is attached to face plate.

2. Plain turning lathes: These do not have a lead screw. They perform all types of lathe work except threading and chasing. The absence of the lead screw substantially simplifies the kinematic features and the construction of the feed gear box.

3. Facing lathes: They are used to machine work of large diameter and short length, in single-piece production and repair jobs. These machines are, generally, used for turning external, internal, and taper surfaces, facing, boring, etc. Facing lathes have relatively small length and large diameter of face plates (up to 4 m). Sometimes, they are equipped with a tail-stock. They consist of the base plate 1, headstock 4 with face plate 5, bed 2, carriage 3, and tailstock 6 (Figure 16.7).

4. Vertical turning and boring mills: These machines are employed in machining heavy pieces of large diameters and relatively small lengths. They are used for turning and boring of cylindrical and tapered surfaces, facing, drilling, countersinking, counterboring, and reaming. The heavy work is mounted on the rotating table more conveniently and safely as compared to facing lathes. As their name implies, they are equipped with a turret head that increases the productivity of such machines.

5. Turret and capstan lathes: These are the natural development of the engine lathe. The tail-stock is replaced by an indexable multistation tool head, termed as the capstan or the turret. This head carries a selection of standard tool holders and special attachments while a square turret is also mounted on the cross-slide. Sometimes a fixed tool holder is also mounted on the back end of the cross-slide. Capstan and turret lathes bridge the gap between the manual engine lathes and the automated lathes and are most practical for batch and short runs of production. A less skilled operator is needed as compared with center lathes. No need to change tooling or move the work to another machine as many operations can be performed without the need to change tooling layout.

    In the capstan lathe (Ram type), the hexagonal turret is mounted on a slide that moves longitudinally in a *stationary saddle* (Figure 16.8a). For the turret lathe (saddle type), the turret is mounted directly on a movable saddle and furnished with both hand and power longitudinal feed (Figure 16.8b). The capstan lathe is used for bar work; the turret lathe is

1. Base plate
2. Bed
3. Carriage
4. Headstock
5. Face plate
6. Tailstock

FIGURE 16.7 Facing lathe.

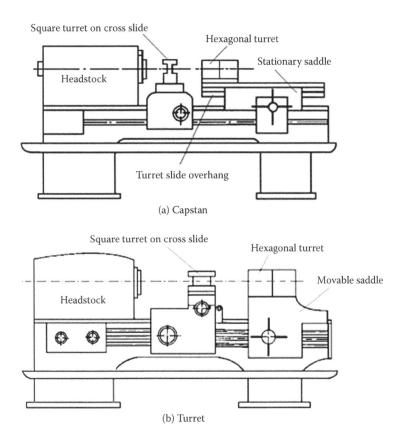

Square turret on cross slide

Hexagonal turret

Stationary saddle

Headstock

Turret slide overhang

(a) Capstan

Square turret on cross slide

Hexagonal turret

Movable saddle

Headstock

(b) Turret

FIGURE 16.8 Difference between capstan and turret lathes.

designed for machining chuck work and bar work. For both the capstan and turret lathes, the tools mounted on cross-slides are used for turning, facing, necking, knurling, and paring off. Those mounted on the turret head are used for drilling, boring, reaming, threading, recessing, etc. The accuracy and cost of the machined components mainly depend on the tool layout that differs according to the nature of the workpiece. Figure 16.9 shows a typical tool layout used for machining a threaded adaptor.

6. Turret semiautomatic lathes: These are used for the same type of work carried out by the turret lathe. They require hand loading and unloading while completing the machining cycle automatically. These machines are used when producing intermediate production lots more economically compared to turret lathes or multispindle automatics. The setup time is much lower than that of multispindle automatics. The machine has a control unit that automatically selects speeds, feeds, length of cuts, and machine functions, such as dwell, cycle stop, index, reverse, cross-slide actuation, and many others.

7. Automated lathes: Fully automatic lathes are those machines in which workpiece handling and the cutting activities are performed automatically. All movements related to the machining cycle, loading of blanks, and unloading the machined parts are performed without the operator interference. Automatics are used as a rule in mass production. Automatic lathes are designed to produce parts of complex shapes by machining the blanks or bar stock. They include the following types.

   • Turret automatic screw machine: This machine is the development of the capstan and turret lathes. It was originally designed for cutting screws and currently is used extensively for producing complex external and internal surfaces on workpieces by

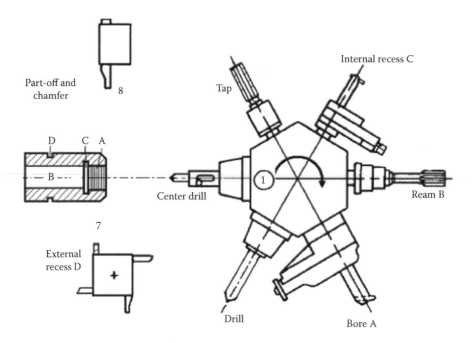

FIGURE 16.9 Tooling layout for machining a threaded adaptor. (Adapted from *Metals Handbook, vol. 16, Machining,* ASM International, Materials Park, OH, 1989.)

using several and parallel working tools. A general view of a classical automatic screw machine is shown in Figure 16.10, together with its basic elements. Figure 16.11 shows typical parts produced on turret screw automatic machines.

- Swiss-type automatic: This machine (Figure 16.12) is also called long part, sliding headstock, or bush automatic. It was originally developed by the watch making

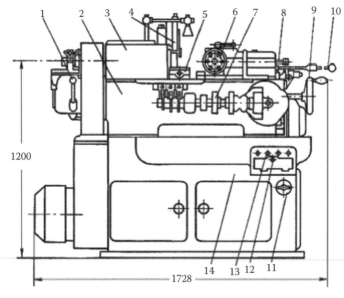

1. Lever to engage auxiliary shaft
2. Bed
3. Headstock
4. Tool slide (vertical)
5. Turret-tool slide (horizontal)
6. Turret slide
7. Main cam shaft
8. Adjustable rod for positioning turret slide with respect to spindle nose
9. Hand wheel to rotate auxiliary shaft
10. Lever to traverse turret slide
11. Rotary switches
12. Console panel for setting up spindle speeds
13. Push button controls of spindle drive
14. Base

FIGURE 16.10 General view of the automatic screw machine. (From Acherkan, N., *Machine Tool Design, vol. 1–4,* Mir Publishers, Moscow, 1969. With permission.)

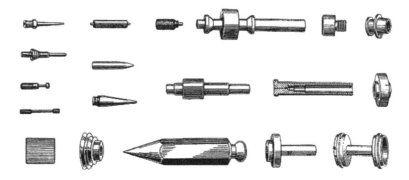

FIGURE 16.11   Typical parts produced on turret screw automatics. (From Acherkan, *Machine Tool Design, vol. 1–4*, Mir Publishers, Moscow, 1969. With permission.)

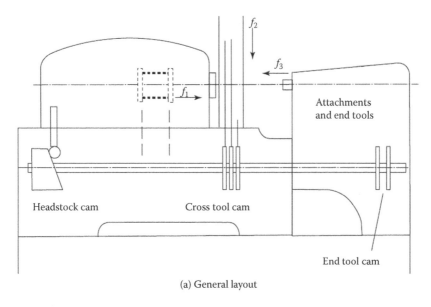

(a) General layout

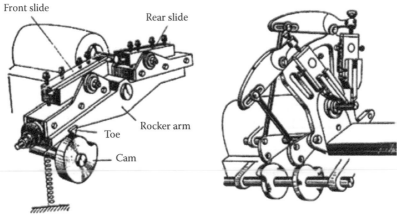

(b) Radial feed sliders

FIGURE 16.12   Swiss-type automatics. (From Boguslavsky, B. L., *Automatic and Semi-Automatic Lathes*, Mir Publishers, Moscow, 1970. With permission.)

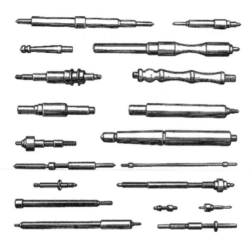

FIGURE 16.13 Typical parts produced by Swiss-type automatics.

industry of Switzerland. It is now extensively used for machining long and slender precise and complex parts as shown in Figure 16.13. A Swiss-type automatic lathe has a distinct advantage over the automatic screw machine because it is capable of producing slender parts of extremely small diameters with a high degree of accuracy, concentricity, and surface finish. Such an advantage became possible because the machining is performed using stationary or cross-fed single point tools in conjunction with longitudinal working feed of the bar stock. Turning takes place directly at the guide bushing supporting the bar stock. It is possible to turn a diameter as small as 60 μm.

- Multispindle automatics: Multispindle automatics are designed for mass production of parts from a bar stock or separate blanks. The distinguishing characteristic is that several workpieces are machined at the same time. According to the type of stock material of the workpiece, they are classified as bar- or chucking- (magazine) type automatics. Chucking machines have the same design as bar automatics, with the exception of stock feeding mechanisms. Typical parts produced by these machines are illustrated in Figure 16.14, while Figure 16.15 shows a six spindle automatic lathe.

16.2.1.2 Boring Machines

Boring machines are used for enlarging and finishing holes or other circular contours of large length-to-diameter ratio at greater dimensional accuracy and improved finish. It is used for finishing large holes in castings and forgings that are too large to be produced by drilling. The boring tools

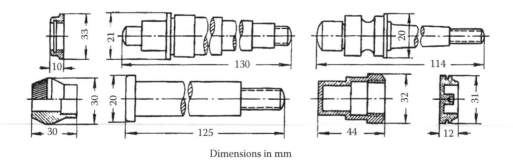

Dimensions in mm

FIGURE 16.14 Typical parts produced on multispindle automatics. (Courtesy of Wickman Group, Binley, Coventry, UK.)

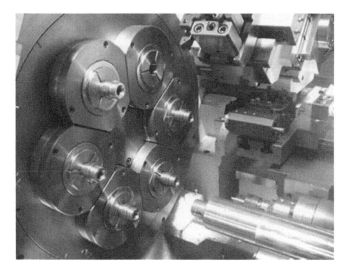

FIGURE 16.15 Multispindle automatic lathe. (Courtesy of Shimada Machinery, Japan.)

can be mounted in either a stub type bar, held in the spindle, or in a long boring bar that has its outer end supported in a bearing. Figure 16.16a to e illustrates typical boring tools that include

- A single-edge cutter mechanically secured to a boring bar
- Adjustable single-edge cutter for wear compensation
- Tool clamped in a universal boring head that is attached to the end of the boring bar
- Fixed cutter, held in a shank type boring bar
- Blade-type boring tool that provide two cutting edges for increasing the machining rate

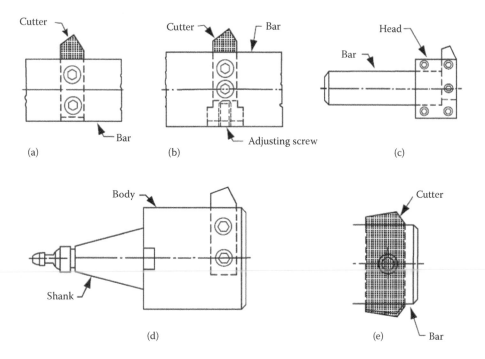

FIGURE 16.16 Typical boring tools.

1. General purpose boring machines: The boring machine is designed to machine relatively large, irregular, and bulky workpieces, which cannot be easily rotated. Hence the work-piece remains stationary and the tool rotates and may simultaneously perform a feed motion. This machine can perform boring, facing, drilling, counterboring, counterfacing, external and internal thread cutting, and milling. Horizontal boring machines are suitable for work, where several parallel bores with accurate center distances are to be produced. A typical general purpose horizontal boring machine is shown in Figure 16.17. The cutting tool is mounted either in the spindle or on the facing slide 8. The rotation of spindle and face plate 7 is the principal movement that is effected by the main motor 11 through speed gearbox housed in the headstock 9. The spindle can also be fed axially, so that drilling and boring can be done over a considerable distance without moving the work. The workpiece is installed either directly on the table 6 or in a fixture. The table is moved longitudinally or transversally on the cross-slide 5. The table and cross-slide are located on a saddle 4, which moves longitudinally on the bed 3. The headstock 9 moves vertically along the column 10 simultaneously with the spindle rest 2, which is moving vertically along the end support column 1. The spindle travels axially when boring or cutting internal thread, etc. The fac-ing slide is moved radially on the face plate to perform facing operations. The table feed and its rapid reverse are powered by the motor 12. In some setups, the work is fed toward the tool, while in other cases, the tool is fed toward the work.

2. Jig boring machines: Jig borers are extra precise vertical boring machines intended for pre-cise boring, centering, drilling, reaming, counterboring, facing, spot facing, etc. They are mainly designed for use in such tool making, jigs and fixtures, and machining of other pre-cision parts. A jig boring machine contains similar features of a vertical milling machine, except that the spindle and its bearings are constructed with very high precision, and the worktable permits extra precise movement and control. Figure 16.18 shows a typical jig borer that is of much more rigid and accurate construction than any other machine tool. The table and saddle ensure the longitudinal and cross-movements X and Y. The machine has a massive column, which supports and accurately guides the spindle housing in the vertical direction, thus achieving the third position adjustment Z. They are equipped with special devices ensuring accurate positioning of the machine operative units including precision lead screw and nut and supplemented by Vernier dials and precision scales in combination with optical read out devices, inductive transducers, and also optical and

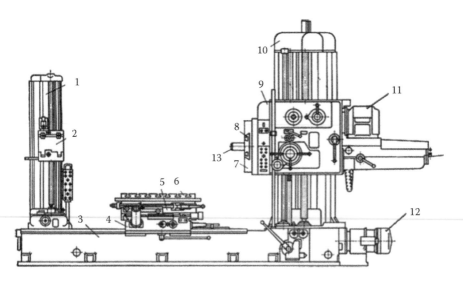

FIGURE 16.17   Horizontal boring machine.

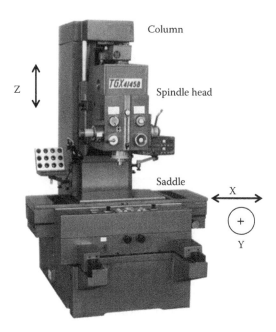

FIGURE 16.18 Jig boring machine. (Courtesy of SJMC Machine Tools, China.)

electrical measuring devices. Jig borers are installed in special environmental enclosures with temperature maintained at a level of 20°C. Currently, jig boring machines are being replaced by NC machining centers that do similar work.

16.2.1.3 Drilling Machines

Drilling machines are used to produce through or blind holes in a workpiece. They employ twist drills that are available in diameters ranging from 0.25 to 80 mm. A standard twist drill, Figure 16.19, is characterized by a geometry in which the normal rake and the velocity of the cutting edge are a function of their distance from the center of the drill. The helix angle of the twist drill is the equivalent of the rake angle of other cutting tools. The standard helix is 30°, which, together with a point angle of 118°, is suitable for drilling steel and cast iron (Figure 16.19a). Drills with a helix angle of 20°, known as low-helix drills, are available with a point of 118° for cutting brass and bronze (Figure 16.19b), and with a point of 90° for cutting plastics. Quick helix drills, with a helix

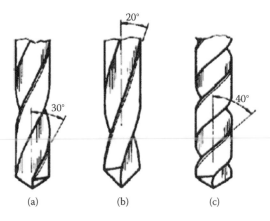

FIGURE 16.19 Helix drills of different helix angles.

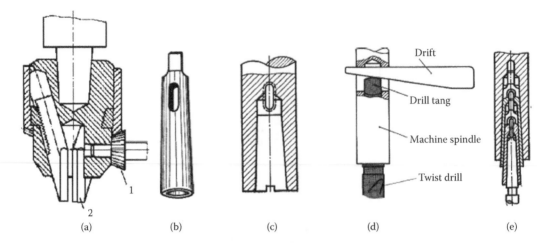

FIGURE 16.20 Holding drills in spindle socket or sleeves and drifting it out from socket or sleeve.

angle of 40° and a point of 100°, are suitable for drilling softer materials such as aluminum alloys and copper (Figure 16.19c).

For tool fixation a self-centering, three-jaw drill chuck (Figure 16.20a) is used to hold small drilling tools (up to 15 mm) using straight shanks. The chuck itself is fitted with a Morse-taper shank that fits into the spindle socket. Alternatively, a tapered sleeve, is used for holding tools with taper shanks in the spindle socket (Figure 16.20b,c). At the end of the taper shank, a tang is used to remove the tool from the spindle socket by a drift as shown in Figure 16.20d. When the cutting tool has a Morse taper smaller than that of the spindle socket, the difference is made up by using one or two tapered sleeves (Figure 16.20e).

The type of work holding device used depends on the shape and size of the workpiece, the required accuracy, and the production rate. Work is held on a drilling machine by clamping to the worktable or in a vise. Vises do not accurately locate the work and provide no means for holding cutting tools in alignment. Small workpiece can be held in a vise. Larger work is best clamped on to the worktable surface using standard tee grooves. Additionally, in case of mass production, drilling jigs are designed to hold a particular workpiece and guide the cutting tool. Figure 16.21 shows a typical plate jig that performs

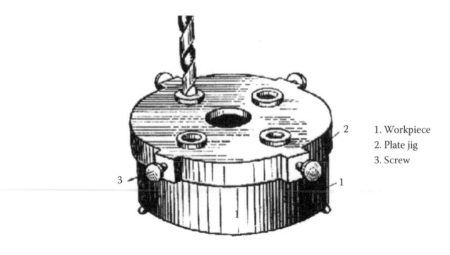

1. Workpiece
2. Plate jig
3. Screw

FIGURE 16.21 Simple plate jig.

accurate and quicker drilling than standard methods. However, larger quantities of workpieces must be required to justify the additional cost of the design and manufacture of a drilling jig.

Drilling allied operations such as core drilling, step drilling, counterboring, countersinking, spot facing, reaming, tapping, and other operations can also be performed on drilling machines as shown in Figure 16.22. In these processes, the tool shape and geometry depend on the machining process to be performed.

General purpose drilling machines are classified as

1. Bench-type sensitive drill presses: They are used for machining small holes of 0.25–12 mm diameter. Manual feeding and high rotational speed characterizes these machines, and that is why they are called sensitive.
2. Upright drill presses: These machines are used for machining holes up to 50 mm in diameter in relatively small size work. Figure 16.23 shows such a typical drilling machine. It has a wide range of spindle speeds and feeds. Therefore it is employed not only for drilling from solid but also for core drilling, reaming, and tapping operations.
3. Radial drilling machines: These machines (Figure 16.24) are especially designed for drilling, counterboring, countersinking, reaming, and tapping holes in heavy and bulky workpieces that are inconvenient or impossible to machine on the upright drilling machines. The spindle axis of the machine is made to coincide with the axis of the hole being machined by moving the spindle in a system of polar coordinate to the hole, while the work is stationary. That is achieved by swinging the radial arm 4 about the rigid column 2, raising or lowering the radial arm on the column by the arm elevating and clamping mechanism 3 to accommodate the workpiece height, and moving the spindle head 5 along the guideways of the radial arm 4. Accordingly, the tool is located at any required position on the stationary workpiece that is set either on detachable table 6 or directly on base 1. During operation the radial arm and spindle head are held in position using power-operated clamping devices.
4. Multispindle drilling machines: They are mainly used in lot production for machining workpieces requiring simultaneous drilling, reaming, and tapping of a large number of holes in different planes of the workpiece. In such a case, a single spindle drilling machine is not economical. The main types of multiple-spindle drilling machines are
   • Gang multiple-spindle drilling machines: The spindles (2–6) are arranged in a row and each spindle is driven by its own motor. The gang machine is in fact several upright drilling machines having a common base and single worktable. They are used for

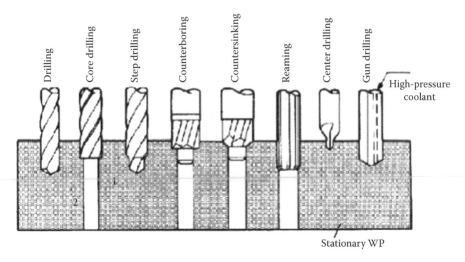

FIGURE 16.22 Drilling and drilling allied operations.

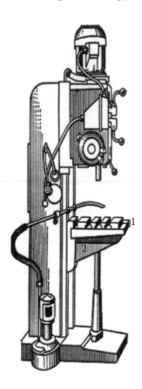

FIGURE 16.23 Typical upright drill press.

consecutive machining of different holes in one workpiece or for the machining of a single hole with different cutting tools.

- Adjustable-centers multiple-spindle vertical drilling machines: They differ from gang type in that they have common drive for all working spindles. The latter are adjusted in the spindle head for drilling holes of varying diameters at random locations on the workpiece surface. Figure 16.25 shows a typical multispindle drilling machine together with the spindle head of adjustable tool centers.

1. Base
2. Column
3. Clamping mechanism
4. Radial arm
5. Spindle head
6. Detachable table

FIGURE 16.24 Radial drilling machine. (Courtesy of Ajax Machine Tools.)

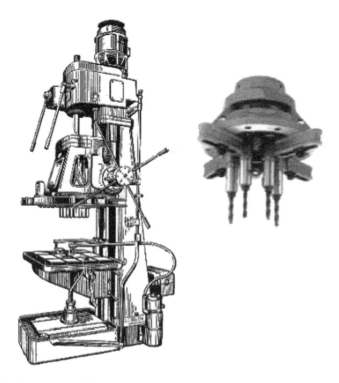

FIGURE 16.25 Multiple-spindle drilling machine and multispindle head. (Courtesy of Awanti Yantra Nirmiti, India.)

16.2.2 MACHINE TOOLS FOR CUTTING FLAT SURFACES

16.2.2.1 Shapers, Planer, and Slotters

Shapers, planers, and slotters are used for machining horizontal, vertical, and inclined flat and contoured surfaces, slots, grooves, and other recesses by means of special single-edge tools. Tools used in these processes should not be shock sensitive such as ceramics and cubic baron nitride (CBN). It is sufficient to use low cost and easily sharpened tools such as high-speed steel (HSS) and Widia. Shaper and planer tools are strongly dimensioned tools to withstand the operating impact loads. Such tools have rake angles of 5° to 10° for HSS tools and between 0 and −15° for Widia tools depending on the workpiece material. The cutting edge inclination angle is normally 10°, while a nose radius of 1–2 mm is used in case of roughing tools. Figure 16.26 shows typical cutting tools that are used for different related machining purposes.

1. Shapers: These are commonly used in single-piece and small-lot production as well as in repair shops and tool rooms. Owing to its limited stroke length, it is adapted to small jobs and best suited for surfaces comprising straight-line elements, contoured surfaces, and keyways and splines on shafts. The shaping machine is quite popular because of the short set up time, inexpensive tooling, and ease of operation. In comparison to a planer, it occupies less floor space, consumes less power, costs less, is easier to operate, and is about three times quicker in action as stroke length and inertia forces are less. Its stroke length is limited by 750 mm, as the accuracy decreases for longer strokes because of ram overhanging. Figure 16.27 shows a typical shaper where the column 1 houses the speed gearbox, the crank, and the slotted arm mechanism. The power is therefore transmitted from the motor 2 to ram 3. Ram travel is the primary reciprocating motion, while the intermittent cross-travel of the table is the feed motion. The tool head 5, carrying the clapper box and the tool

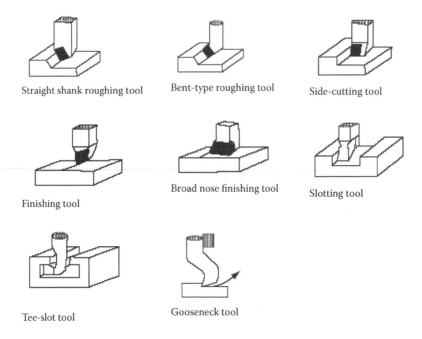

Straight shank roughing tool   Bent-type roughing tool   Side-cutting tool

Finishing tool   Broad nose finishing tool   Slotting tool

Tee-slot tool   Gooseneck tool

FIGURE 16.26   Shaper and planer tools and operations.

holder 6, is mounted at the front end of the ram and is fed manually or automatically. The slot with the clamp 7 serves to position the ram in setting up the shaper. The tool head has a tool slide and feed screw rotated by a ball crank handle 8 for raising and lowering the tool to adjust the depth of cut. A swivel motion of the tool head enables it to take angular cuts to machine inclined surfaces. The workpiece is clamped either directly on the table or is held in a machine vise. By means of ratchet and pawl mechanism 9, driven from the crank and

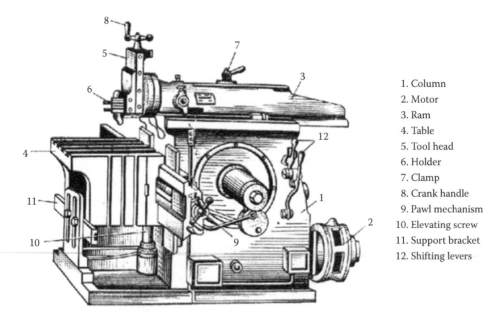

1. Column
2. Motor
3. Ram
4. Table
5. Tool head
6. Holder
7. Clamp
8. Crank handle
9. Pawl mechanism
10. Elevating screw
11. Support bracket
12. Shifting levers

FIGURE 16.27   Mechanical shaper.

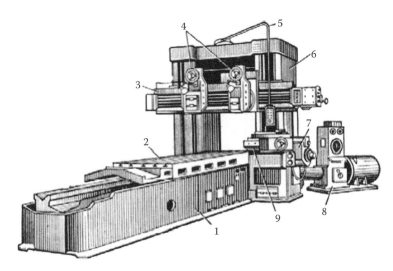

1. Bed
2. Table
3. Cross-rail
4. Upper tool heads
5. Cable
6. Housing
7. Feed gearbox
8. Motor
9. Side tool head

FIGURE 16.28 A typical double housing planer.

slotted arm mechanism, the table is fed crosswise in a horizontal plane. The table is raised or lowered by the elevating screw 10. Support bracket 11 is provided to clamp the table rigidly during operation. The number of ram strokes per minute is set by shifting levers 12.

2. Planers: These are used for machining large size workpieces because they have long table travel (1–15 m) and robust construction. Both the productivity and accuracy of planers are considerably enhanced by taking multicuts on the workpiece in a single stroke. Since it is usual to mount two tool holders on the cross-rail and one each side of the column, the setting time is of the order 5 to 6 times that of the shaper. A depth of cut up to 18 mm and a feed rate of 1.5–3 mm/stroke can be taken for roughing, while a depth of 0.25–0.5 mm may be used for finish cuts. According to Figure 16.28, table 2 carrying the workpiece reciprocates on the bed 1. The table is powered from motor 8 through a reduction gearbox and a rack-and-pinion drive. The housing 6 mounts the side tool head 9, while the cross-rail 3 is raised and lowered from a separate motor on the housings to accommodate workpieces of different heights set up on the table. The upper tool heads 4 are traversed by a lead screw (feed motion). The side tool head, is traversed vertically (feed motion) by the feed gearbox 7 to machine vertical surfaces. All tool heads operate independently. The control panel and the suspended cable 5 are shown in the same figure. The tool heads 4 may be swiveled to machine-inclined surfaces. Like all reciprocating machine tools, planers are equipped with a clapper box to raise the tools on the return stroke.

3. Slotters: These are useful for machining keyways, cutting of internal and external teeth on large gears. As illustrated in Figure 16.29, the job is generally supported on table 3, which has a rotary feed in addition to the usual table movement in cross-directions. The ram 1 travels vertically along the ways of the column 2. The ram stroke of a slotter ranges from 300 to 1800 mm. Slotters are generally very robust machines having a possibility of tilting the ram up to ±15° from vertical to permit machining of dies with relief. Ram speeds are usually from 2 to 40 m/min. Longitudinal and transverse feeds range from 0.05 to 2.5 mm/stroke. Cutting action takes place on a downward stroke.

16.2.2.2 Milling Machines

Milling is the metal removal by feeding the work past a rotating multitoothed cutter. Milling operations may be classified as peripheral (plain) milling or face (end) milling. In peripheral (horizontal)

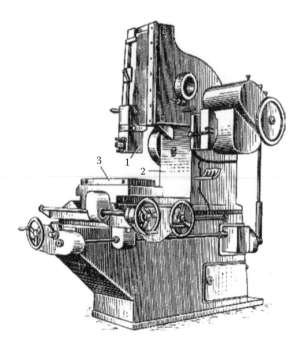

FIGURE 16.29 Slotting machine. (Courtesy of SJMC Machine Tools, China.)

milling (Figure 16.30a), the cutting occurs by the teeth arranged on the periphery of the milling cutter, and the generated surface is a plane parallel to the cutter axis. Peripheral milling is usually performed on a horizontal milling machine. In face (vertical) milling, the generated surface is at right
angle to the cutter axis (Figure 16.30b) and is usually performed on vertical milling machines. It is
more productive than plain milling. During peripheral milling, the appearance of the surface and
the type of chip formation are affected by the direction of cutter rotation with respect to the movement of the workpiece. In this regard, two types of peripheral milling are differentiable, namely,
up- and down-milling as shown in Figure 16.31.

Up-milling (conventional) is accomplished by rotating the cutter against the direction of the
feed of the workpiece (Figure 16.31a). In some metals, up-milling leads to strain hardening of the
machined surface and to chattering and excessive teeth blunting. Up-milling does not require a
backlash eliminator; it is safer in operation because the cutter does not climb on the work, loads on
teeth are acting gradually, built-up edge (BUE) fragments are absent from the machined surface,
and the milling cutter is not affected by the original sandy or scaly surfaces of the work. Down-
milling (climb) is accomplished by rotating the cutter in the direction of the work feed as shown
in Figure 16.31b. The milling cutter attempts to climb the workpiece. Chips are cut to maximum
thickness at initial engagement of cutter teeth with the work and decreases to zero at the end of its

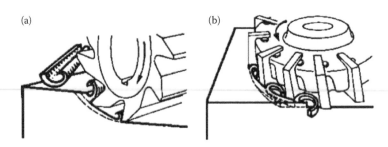

FIGURE 16.30 Peripheral and vertical milling.

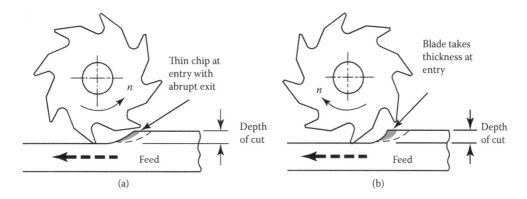

FIGURE 16.31 Up-milling and down-milling.

engagement. The cutting forces are directed downward, fixtures are simpler and less expensive, blunting is less possible, and tendencies of chattering and vibration are less, leading to improved surface finish.

Milling cutters may be provided with a hole to be mounted on the arbor of the horizontal milling machines or provided with a straight or tapered shank for mounting on the vertical or horizontal milling machines. Figure 16.32 shows the commonly used milling cutters during their operation. Plain and side milling cutters are mounted on an arbor whose taper shank is drawn up tight into the taper socket of the spindle 2 with a draw-in bolt 1 (Figure 16.33a). The outer end of the long arbor 3 is supported by an overarm support 5 in horizontal milling machines, and the cutter 4 is mounted at the required position on the arbor by a key (or without key in case of slitting saws) and is

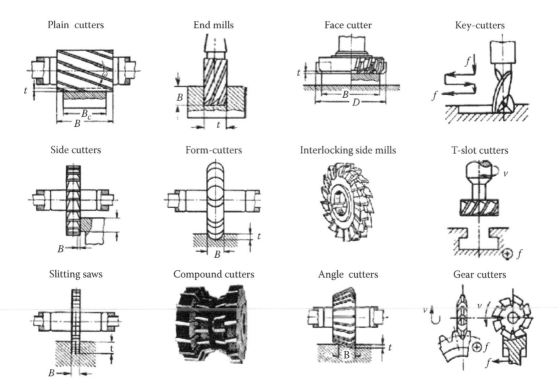

FIGURE 16.32 Different types of milling cutters during operation.

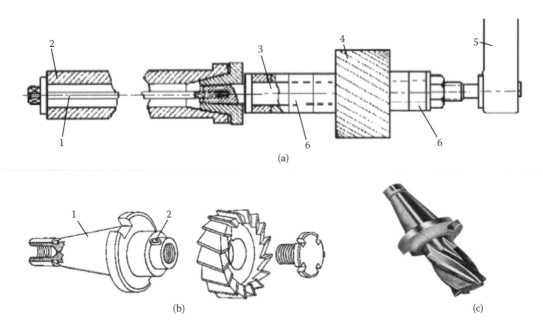

FIGURE 16.33 Cutter mounting on milling machines.

clamped between collars or spacers 6 with a large nut. The shell end mill or the face milling cutters are mounted on the stub arbors and driven by a feather key as shown in Figure 16.33b. Additionally, end mills, T-slot cutters, and other milling cutters of tapered shanks are secured with a draw-in bolt directly in the taper socket of the spindle by means of adaptors (Figure 16.33c), while straight shank cutters are held in chucks.

Small workpieces and blanks are clamped most frequently in a general purpose plain, swivel, or universal milling vise fastened to the worktable (Figure 16.34a). Large workpieces and blanks that are too large for a vise are clamped directly on the worktable, using standard fastening elements

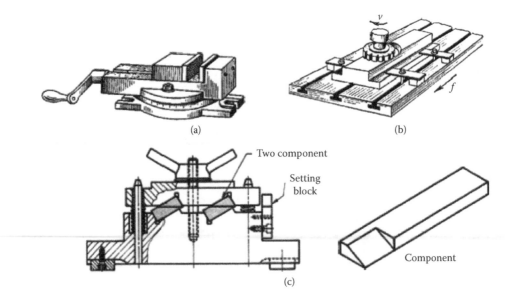

FIGURE 16.34 Workpiece fixation on milling machines.

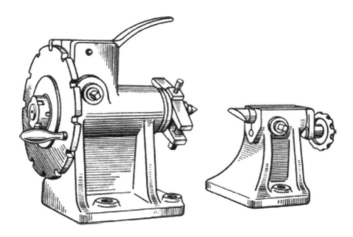

FIGURE 16.35 Universal dividing head.

such as strap clamps, support blocks, T-bolts, etc. (Figure 16.34b). For more accurate and productive work, expensive milling fixtures are frequently used as shown in Figure 16.34c.

Dividing heads (Figure 16.35) are attachments that extend the capabilities of the milling machines. They are used for milling spur and helical gears, spline shafts, twist drills, reamers, milling cutters, etc. Dividing heads are capable of indexing the workpiece through predetermined angles as well as continuously rotating the workpiece, which is set at the required helix angle during milling of helical slots and helical gears. The universal dividing head is the most widely used. Figure 16.36 illustrates an isometric view of the gearing diagram of a universal dividing head in a simple index-ing mode. Periodical turning of spindle 3 is achieved by rotating the index crank 2, which transmits the motion through a worm gearing 6/4 to the workpiece (gear ratio 1:40, i.e., one complete revolu-tion of the crank corresponds to 1/40 revolution of the workpiece). Index plate 1 having several con-centric circular rows of accurately equispaced holes serves for indexing the index crank 2 through the required angle. The workpiece is clamped in a chuck screwed on the spindle 3. It can also be clamped between two centers. The dividing head is provided with three index plates (Brown and Sharp) or two index plates (Parkinson). The plates have the following number of holes:

Brown and Sharp:
    Plate 1: 15, 16, 17, 18, 19, and 20
    Plate 2: 21, 23, 27, 29, 31, and 33
    Plate 3: 37, 39, 41, 43, 47, and 49

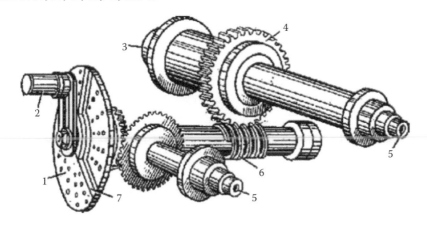

FIGURE 16.36 An isomeric gearing diagram of a universal dividing head.

Parkinson:
   Plate 1: 24, 25, 28, 30, 34, 37, 38, 39, 41, 42, and 43
   Plate 2: 46, 47, 49, 51, 53, 54, 57, 58, 59, 62, and 66

The universal dividing head can be set up for simple or differential indexing, Figure 16.37, or for milling helical slots. In simple indexing, the index plate 1 is fixed in position by a lock pin 4. The work spindle 3 is rotated through the required angle by rotating the index crank 2. For determining the number of index crank revolutions n to give the number of divisions z on the job periphery (assuming a worm/worm gear ratio of 1:40). The kinematic balance equation is given by

$$n = \frac{40}{z} \tag{16.3}$$

Example 2

It is required to determine the suitable index plates (Brown and Sharp) and the number of index crank revolutions n necessary for producing the spur gear of 37 teeth.

Solution

$$z = 37 \text{ teeth}$$

$$n = \frac{40}{37} = 1 + \frac{3}{37} \text{revolutions}$$

Then choose the plate 3, and select the hole circle 37. The crank should be rotated one complete revolution plus 3 holes out of 37.

Differential indexing is employed where an index plate with number of holes required for simple indexing is not available. In differential indexing a plunger 5 is inserted in the bore of the work spindle (Figure 16.37b) while the index plate is unlocked. The spindle drives the plate through change and bevel gears while the crank through the worm is driving the spindle. Hence the required turn of the work spindle is obtained as the sum of two turns:

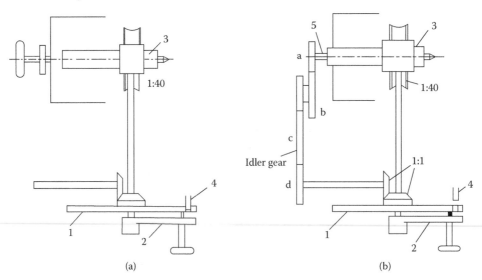

(a)   (b)

FIGURE 16.37   (a) Simple and (b) differential indexing.

1. A turn of index crank 2 relative to index plate 1
2. A turn of the index plate itself, which is driven from the work spindle through change gears $\dfrac{a}{b} \times \dfrac{c}{d}$ to provide the correction

Depending on the setup, the index plate rotates either in the same direction with the index crank or in the opposite direction. An idler gear should be used if the crank and plate move in opposite directions to each other (Figure 16.37). In order to perform a differential indexing, the following steps are to be considered

- The number of revolutions of index crank is set up in the same manner as in simple indexing, but not for the required number of divisions z. Another number z' nearest to z that makes it possible for simple indexing to be carried out.
- The error of such setup z' is compensated for by means of a respective setting up of the differential change gears a, b, c, and d (Figure 16.37). The change gears supplied to match the three plate system (Brown and Sharp) are 24(2), 28, 32, 40, 44, 48, 56, 64, 72, 86, and 100 teeth.
- The number of teeth of the change gears, a, b, c, d are determined from the corresponding kinematic balance equation

$$\frac{40}{Z'} + \frac{1}{Z}\frac{a.c}{b.d} = \frac{40}{Z}$$

From which

$$\frac{ac}{bd} = \frac{40}{Z}(Z' - Z) \tag{16.4}$$

It is more convenient to assume that $Z' > Z$ to avoid the use of an idler gear. If $Z' < Z$, then an idler gear must be used.

Example 3

Select the differential change gears and the index plate (Brown and Sharp), and determine the number of revolutions of the index crank for cutting a spur gear of $Z = 227$ teeth.

Solution

Assume $Z' = 220$ $Z' < Z$, therefore, idler is required

$$n = \frac{40}{z} = \frac{40}{220} = \frac{2}{11} = \frac{6}{33}$$

$$\frac{a}{b} \times \frac{c}{d} = \frac{40}{Z}(Z' - Z)$$

$$= \frac{2}{11}(220 - 227) = -\frac{2 \times 7}{11}$$

$$= -\frac{8}{4} \times \frac{7}{11} \qquad = -\frac{64}{32} \times \frac{28}{44}$$

$a = 64$, $b = 32$, $c = 28$, and $d = 44$ teeth with an idler gear.

Milling machines are used for machining flat surfaces, contoured surfaces, complex and irregular areas, slotting, threading, gear cutting, production of helical flutes, twist drills, spline shafts, etc. to close tolerances. The general purpose milling machines are extremely versatile and are subdivided into several types.

1. Knee-type milling machines: These machines have three Cartesian directions of the table motion. This group is further subdivided into plain horizontal, universal horizontal, vertical, and ram-head knee-type milling machines. The name "knee" has been adopted because it features a knee that mounts the work table and travels vertically along the vertical guideways of the machine column. In plain horizontal milling machines, the spindle is horizontal and the table travels in three mutually perpendicular directions. The universal horizontal milling machines (Figure 16.38) are similar in general arrangement to the plain horizontal machines. The principal difference is that the table can be swiveled about its vertical axis through ±45°, which makes it possible to mill helical grooves and helical gears. The overarm serves to hold the bearing bracket supporting the outer end of the tool arbor in horizontal machines. In contrast to horizontal milling machines, vertical type milling machines have a vertical spindle as shown in Figure 16.39.

2. Vertical bed milling machines: These machines are rigid and powerful; hence they are used for heavy duty machining of large workpieces (Figure 16.40). The spindle head containing a speed gearbox travels vertically along the guideways of the machine column and uses a separate drive motor. In some machines, the spindle head can be swiveled. The work is fixed on a compound table, which travels mutually in two perpendicular directions. The adjustment in the vertical direction is accomplished by the spindle head.

3. Planer-type milling machines: They are intended for machining horizontal, vertical, and inclined planes as well as form surfaces by means of face, plain, and form milling cutters. These machines are of single or double housing, with one or several spindles, each having

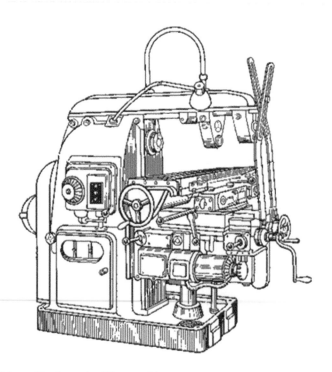

FIGURE 16.38 Universal horizontal milling machine.

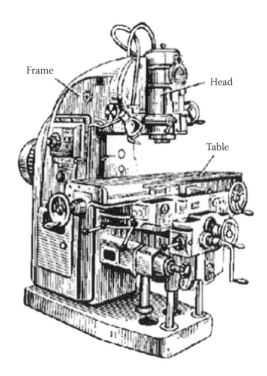

FIGURE 16.39 Vertical milling machine.

a separate drive. Figure 16.41 shows a single housing machine with two spindle heads traveling vertically and horizontally.

4. Rotary table milling machines: Rotary table machines are highly productive and are frequently used for both batch and mass production. The workpieces being machined are clamped in fixtures installed on the rotating table (Figure 16.42). The machines may

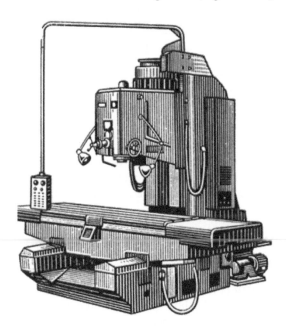

FIGURE 16.40 Vertical-bed general purpose milling machine.

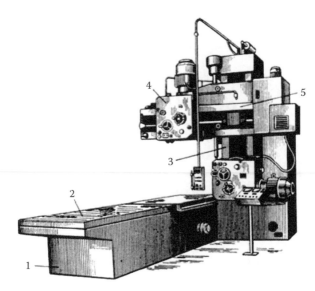

1. Bed
2. Table
3. Column
4. Spindle heads
5. Cross-arm

FIGURE 16.41   Planer-type milling machine.

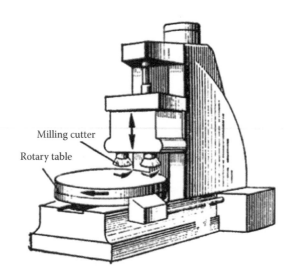

FIGURE 16.42   Rotary table milling machine.

be equipped with one or two spindle heads. When several surfaces are to be machined, the workpieces are indexed in the fixtures after each complete revolution of the table. The machining cycle provides as many table revolutions as the number of surfaces to be machined.

16.2.2.3 Broaching Machines

Broaching is a cutting process using a multitoothed tool (broach) having successive cutting edges; each protrudes to a distance further than the preceding one in the direction perpendicular to the broach length. In contrast to all other cutting processes, there is no feeding of the broach or the workpiece (Figure 16.43). The total depth of the material removed in one stroke T is the sum of rises of teeth of the broach. Broaching is generally used to bore or enlarge through holes of any cross-sectional shape, straight and helical slots, external surfaces of various shapes, and external

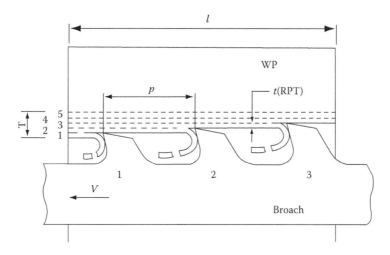

FIGURE 16.43 Broaching operation.

and internal toothed gears (Figure 16.44). Broaching, usually, produces better accuracy and surface finish than drilling, boring, or reaming operations. A tolerance grade of IT6 and a surface roughness R_a of about 0.2 μm can be easily achieved by broaching.

Figure 16.45 illustrates the terminology of a pull-type internal broach used for enlarging circular holes. Irregular shapes are produced by starting from a workpiece originally provided with drilled, bored, cored, or reamed holes. Broaches must be designed individually for a particular job. They are very expensive to manufacture and therefore can only be justified when a very large batch size (100,000 to 200,000) is to be machined. The pitch (P) of the broach is the distance between two consecutive teeth of a broach. It depends on the material of the workpiece and its mechanical properties and the rise per tooth S_z (super elevation).

$$P = 3\sqrt{S_z l x} \tag{16.5}$$

where l is the length of cut and x is the chip space number, 3–5 for brittle workpiece materials and 6–10 for ductile and soft workpiece materials. Large pitch and tooth depth are required for roughing

FIGURE 16.44 Typical parts produced by internal and external broaching. (Courtesy of Miller Broach, Capac, MI.)

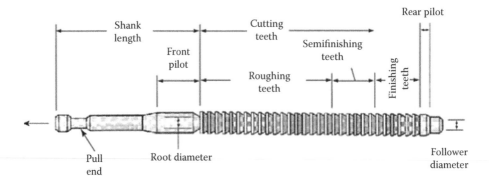

FIGURE 16.45   Solid-pull broach configuration.

teeth. For finishing teeth, the pitch is reduced to about 60% of that of roughing teeth to reduce the overall length of the broach. In order to provide better guide of the tool the pitch should not be greater than $l/2$ (l is the length to be cut).

Regarding the application of broaching force, two types of broaching are distinguished (Figure 16.46). In the pull-broaching, the broach is pulled through the hole and the main cutting force is applied to the front of the broach, subjecting its body to tension. In push-broaching, the main cutting force is applied to the rear of the broach, thus subjecting the body to compression. A push-broach is shorter than pull-broach, and its length does not usually exceed 15 times its diameter to avoid buckling.

Because the shape of the surface produced in broaching depends on the shape and arrangement of the cutting edges on the broach, broaching machines are simple in construction and operation. They have no feed mechanisms because the feed is provided by the gradual increase in the height of the broach teeth. The only cutting motion of the broaching machine is the straight-line motion of the ram. Most broaching machines have hydraulic drives that are of smooth running and safe operation.

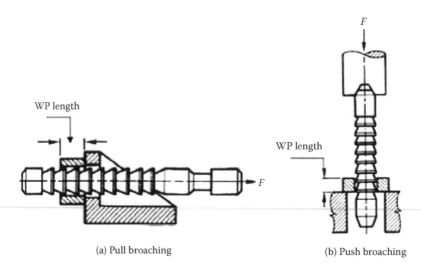

(a) Pull broaching   (b) Push broaching

FIGURE 16.46   Pull- and push-broaching.

Horizontal machines find wide acceptance, because of their long strokes and the limitation that ceiling height places on vertical machines. They are used mainly for automotive engine blocks and are seldom used for broaching small holes. The pulling capacity ranges from 2.5 to 75 ton, strokes up to 3 m, and cutting speeds limited to less than 12 m/min. Broaching that require rotating of the broach, as in rifling and spiral splines, is usually done on horizontal internal broaching machines. The surface hydraulic broaching machines are built with capacities up to 40 ton, strokes up to 4.5 m, and cutting speeds up to 30 m/min. On the other hand, the electromechanically driven horizontal surface broaching machines are available with higher capacities, stroke lengths, and cutting speeds (up to 100 ton, 9 m, and 30 m/min, respectively).

Vertical broaching machines are hydraulically driven and may be pull-up, pull-down, or push-down units. Figure 16.47 illustrates a pull-down dual ram vertical broaching machine. They are available with pulling capacities from 2 to 50 ton, strokes from 0.4 to 2.3 m, and cutting speeds up to 24 m/min. The choice between vertical and horizontal machines is determined primarily by the length of stroke required and the available floor space. Vertical machines seldom have strokes greater than 1.5 m because of ceiling limitation. Horizontal machines can have almost any stroke length; however, they require greater floor space. In the continuous horizontal surface broaching machines, the broaches are stationary and mounted in a tunnel on the top of the machine, while the work is pulled past the cutters by means of a conveyor (Figure 16.48). Fixtures are usually attached to the conveyor chain, so that the workpieces can be provided automatically by the loading chute at one end of the bed and removed at the other end. In the rotary continuous horizontal broaching machines, the broaches are also stationary, while the work is passed beneath or between them. The work is held in fixtures on a rotary table.

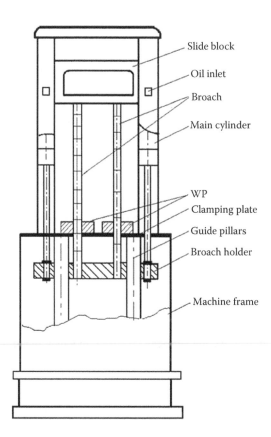

FIGURE 16.47   Dual ram vertical broaching machine.

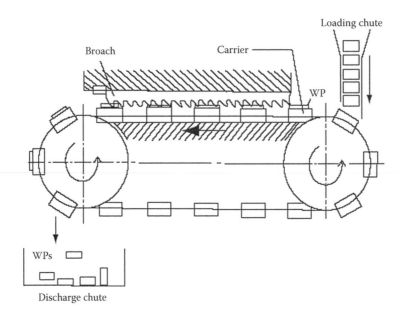

FIGURE 16.48 Continuous horizontal surface broaching machine. (Adapted from *Metals Handbook, vol. 16, Machining,* ASM International, Materials Park, OH, 1989.)

16.2.3 Grinding Machines

Grinding machines are characterized by high accuracy and good surface finish. Consequently, they are usually employed in finishing operations. The tools used on the grinding machines are the grinding wheels. A standard marking system for the grinding wheels has been adopted by the American National Standard Institute (ANSI), which is previously described in Chapter 15.

Grinding wheel shapes must permit proper contact between the wheel and the surfaces to be ground. Figure 16.49 illustrates eight standard shapes of grinding wheels. Because of the high rotation speeds involved, grinding wheels must be used after wheel balancing in order to avoid vibrations that cause the wheel breakage, serious damage, and injury. Before mounting the wheel on the spindle, the wheel with its sleeve should be balanced on an arbor that is placed on the straight edges or revolving disks for a balancing stand (Figure 16.50). The wheel is balanced by shifting three balance weights 1 in an annular groove of the wheel sleeve (or mounting flange). Proper and reliable clamping on its spindle is of a prime importance. Figure 16.51 shows different methods of wheel mounting, which depends on type and construction of the grinder and the shape and size of the grinding wheel.

Owing to the abrasion action of the grinding process, the sharp grains of the grinding wheel become rounded and hence lose their cutting ability (wheel glazing). Additionally, loading of voids between the grains with the chips reduces their cutting ability when grinding ductile and soft materials. Grinding wheel cutting ability can be restored by dressing or truing (see Chapter 15).

16.2.3.1 Surface Grinding Machines

Surface grinders are used to finish flat surfaces that are previously machined by shaping, planing, and milling. The horizontal-spindle reciprocating-table machine uses a straight-shaped wheel. The bed contains the drive mechanisms and the main table hydraulic cylinder. The table reciprocates longitudinally at a speed v_w. The grinding wheel is periodically fed laterally, after each stroke, at a rate f_1 which is less than the grinding wheel width. The wheel is fed down to provide the infeed f_2 (Figure 16.52a) after the entire surface has been ground. In the horizontal-spindle rotary table grinders, the reciprocating cross-feed motion f_1 is transmitted in these machines to either the

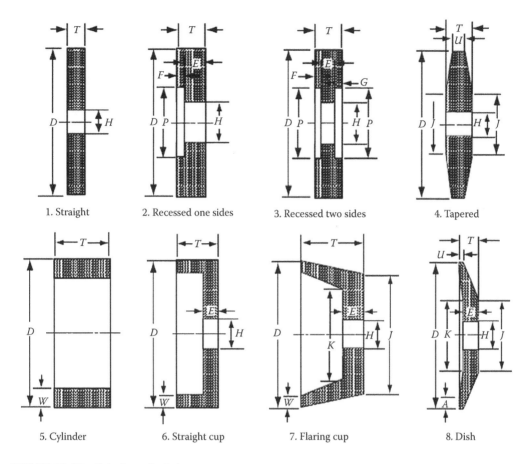

1. Straight 2. Recessed one sides 3. Recessed two sides 4. Tapered

5. Cylinder 6. Straight cup 7. Flaring cup 8. Dish

FIGURE 16.49 Grinding wheel shapes.

grinding wheel or the table unit, the feed f_2 is actuated per table revolution (Figure 16.52b). The worktable rotates at a speed v_w. In the vertical-spindle reciprocating-table grinders shown schematically in Figure 16.53a, a cup, ring, or segmented-wheel grinds the work over its full width, using the end face of the wheel in one or several strokes of the table. The tool is fed down periodically at the infeed rate f. Vertical-spindle rotary-table grinders are similar to the previous type, except that

FIGURE 16.50 Balancing grinding wheels.

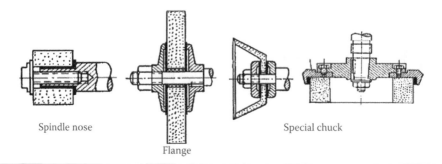

Spindle nose

Flange

Special chuck

FIGURE 16.51 Mounting grinding wheels.

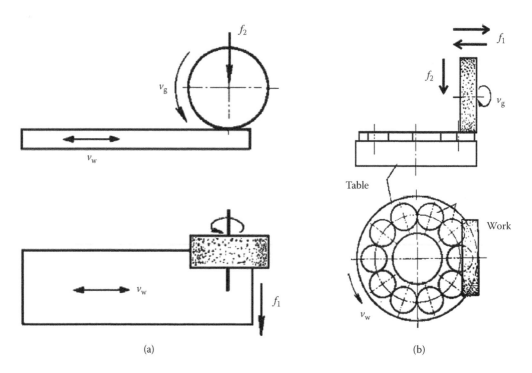

(a)   (b)

FIGURE 16.52 Horizontal spindle surface grinders.

the spindle is vertical. The configuration of these machines allows a number of pieces to be ground in one setup (Figure 16.53b).

16.2.3.2 Cylindrical Grinding Machines

External cylindrical grinding machines are used for grinding external cylindrical surfaces. As shown in Figure 16.54a, the rotating cylindrical workpiece reciprocates laterally along its axis. Universal-type grinders are provided with possibility for swiveling the workpiece and grinding wheel by swiveling the headstock. This enables steep tapers to be ground. Owing to their versatility, universal cylindrical grinders are best suited for tool-room applications. As shown in Figure 16.54b, the table assembly 1 is reciprocated using a hydraulic drive at infinitely variable speed. The stroke can be controlled by means of adjustable trip dogs 2. Infeed is provided by the movement of the wheel head 3, crosswise to the table axis. The wheel is rotating by the motor 4 against the workpiece that is driven by motor 5. Such machines are also equipped with an automatic diamond wheel truing device that dresses the wheel before grinding is started on each piece. These machines are generally

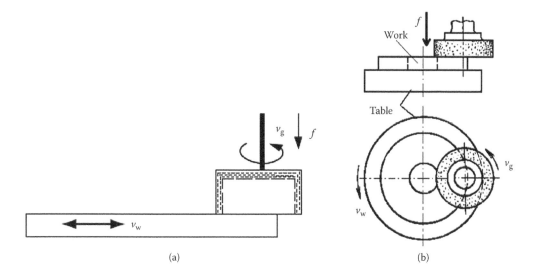

FIGURE 16.53 Vertical spindle surface grinders.

equipped with computer control thereby reducing labor cost and producing parts accurately and repetitively. In the traverse cylindrical grinding, the work rotates about its axis and also traverses longitudinally past the wheel so as to extend the grinding action over the full length of the work (Figure 16.55a). The longitudinal traverse should be about ¼ to ½ of the wheel width per revolution of the work. The depth of cut (infeed) ranges from 50 to 100 μm for rough cut and 6 to 12 μm for finish cut. The grinding allowance ranges from 125 to 250 μm for short parts and from 400 to 800 μm for long parts. In the plunge-cut cylindrical grinding, there is no traverse motion of either the wheel or the work. Because the grinding wheel extends over the entire length of the surface being ground on the work ($B > l$), the rotating wheel is continuously fed into the work at a rate of 2.5–20 μm per revolution of the work (Figure 16.55b). This method is used in form grinding of relatively short work at high output.

Internal cylindrical grinding machines are used to grind the inside diameter of bushing, bearing races, and heavy housings. It is usually of the traverse type; however, the plunge-cut technique may also be used. As shown in Figure 16.56, the grinding wheel and, consequently, the machine spindle

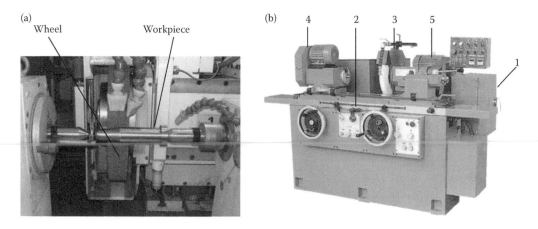

FIGURE 16.54 Cylindrical grinders. (Courtesy of Ajax Machine Tools.)

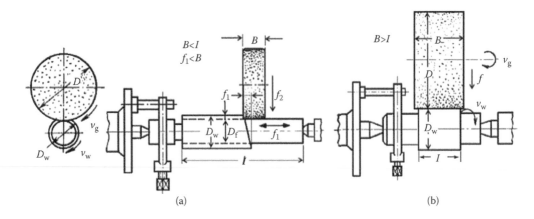

FIGURE 16.55 Cylindrical grinding methods.

are small to suit small internal holes. The rotational speed of the small grinding wheel must be very high (up to 150,000 rpm) to achieve the recommended cutting speeds. Therefore high-speed drive for the grinding wheel with special spindle mounting is required. A churching-type machine is used in grinding comparatively small workpieces where the primary cutting motion of the grinding wheel v_g and the feed motions f_1 for the wheel are encountered (Figure 16.57a). A planetary-type machine is used to grind holes in large irregular parts that are difficult to mount and rotate (Figure 16.57b). In this case, the work is stationary, while the wheel rotates, not only around its own axis v_g but also around the axis (v_w) of the hole being ground. In addition to these two motions, traverse feed f_1 and infeed f_2 are also affected.

In centerless grinding machines, the work is not supported between centers but is held against the face of the grinding wheel, supporting rest, and a regulating wheel (Figure 16.58a). The process is applicable for long slender parts. During cutting, the workpiece 1 is supported on the work rest blade 2 by the action of the grinding wheel 3. The regulating wheel 4 of infinite variable speed holds the workpiece against the horizontal force controlling its size and imparting the necessary rotational

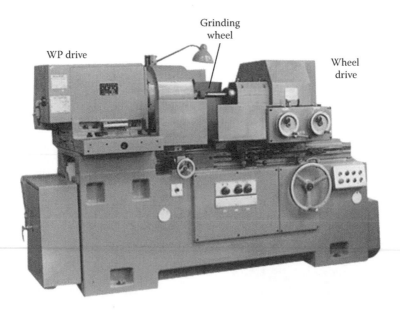

FIGURE 16.56 Internal grinder. (Courtesy of SJMC Machine Tools, China.)

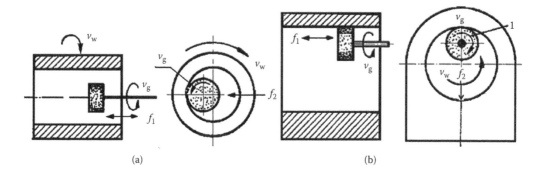

FIGURE 16.57 Internal grinding methods.

and longitudinal feeds of the workpiece. The wheels rotate clockwise, and the work driven by the regulating wheel rotates counterclockwise at a peripheral speed of 20–30 m/min. To increase friction between the work and the regulating wheel, the latter has a fine grain size of mesh number 100 to 180, rubber bond and hard grade (R or S). The grinding wheels run at a much higher speed of 2000 m/min to accomplish the cutting action. For grinding true cylindrical surface the work is set above the centers of the regulating and grinding wheels by 0.15 to 0.25 of the work diameter, but not over 10 mm to avoid chattering. Centerless grinders are now capable of wheel surface speeds of the order of 10000 m/min, using CBN abrasive wheels. The accuracy that can be obtained from centerless grinding is of the order of 2–3 μm, and with the suitable selected wheels high degrees of finishes are obtained. However, when grinding tubes, the internal and external diameters may not be concentric. Additionally, lobbing (unevenly ground surface) may occur during grinding of steel bars, whose surfaces have some high and some low spots because of hot or cold rolling.

The arrangement of grinding internal surfaces of short or long tube work is schematically illustrated in Figure 16.58b. The workpiece is supported by two steel rollers 1, 2, and a regulating wheel 5. Roller 1 is a supporting roller, and roller 2 is a pressure roller. The grinding wheel 4 and the workpiece 3 rotate in the same direction, while the regulating wheel 5 rotates in the opposite direction. The grinding wheel is generally smaller than the regulating wheel. In internal centerless grinding, because the roundness of the internal surface depends on the external surface, the latter must be ground first.

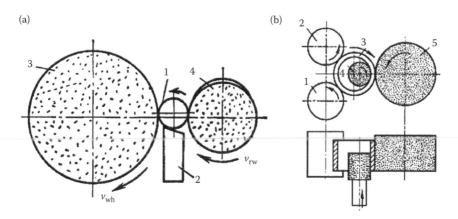

FIGURE 16.58 Centerless grinding operation.

16.2.4   Microfinishing Machines

These machines perform the final machining stage for parts that fit for the service for which it is intended. They remove a very small amount of metal and hence the surface finish obtained is specified in the range of microfinishes. Typical microfinishing machines include honing, super finishing, and lapping.

16.2.4.1   Honing Machines

The most frequent application of honing machines is the finishing of internal cylindrical holes. However, outside surfaces also can be honed. Gear teeth, valve components, and races for antifriction bearings are typical applications of external honing. The hone is allowed to float by means of two universal joints so that it follows the axis of the hole. The honing sticks are able to exert an equal pressure on all sides of the bore, regardless the machine vibrations and therefore round and straight bores are produced. The axis of the hole is usually in the vertical position to eliminate gravity effects on the honing process; however, for long parts the axis may be horizontal. Honing is characterized by rapid and economical stock removal with minimum of heat and distortion. It generates round and straight holes by correcting form errors caused by previous machining operations, and it achieves high surface quality and accuracy. Honing sticks commonly used may be vitrified, resinoid, or metallic bonded. The grit size depends on the desired rate of material removal and the degree of surface finish required. Al_2O_3 is widely used for steels while SiC is generally used for cast iron and nonferrous materials. CBN is used for all steels (soft and hard), while Ni- and Co-base super alloys, stainless steels, Br–Cu alloy, zirconium, and diamonds are used for chromium plate, carbides, ceramics, glass, cast iron, brass, bronze, and surfaces nitrided to depths greater than 30 µm. Figure 16.59 shows a vertical honing where the machine head rotates and reciprocates the floating hone to finish the internal hole of the workpiece.

16.2.4.2   Superfinishing Machines

Superfinishing machines impart three or more motions. This ensures that the abrasive path is random and never repeats itself. During superfinishing, the bonded abrasive stone, whose operating face complies with the form of the rotating workpiece surface, is subjected to very light pressure. A short, high-frequency stroke, superimposed on a reciprocating traverse is used for superfinishing of long lengths.

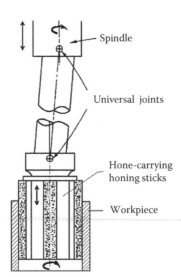

FIGURE 16.59   Vertical honing.

16.2.4.3 Lapping Machines

Lapping is the random rubbing of a workpiece against a cast iron lapping plate (lap) using loose abrasives, carried in an appropriate vehicle (oil) in order to improve fit and finish. Lapping is a final machining operation that realizes extreme dimensional accuracies, mirror-like surface quality, correction of minor shape imperfections, ensuring a close fit between mating surfaces; no workpiece distortion occurs, less heat is generated, and therefore metallurgical changes are totally avoided. Figure 16.60 shows a typical vertical lapping machine used for lapping cylindrical surfaces in production quantities. The laps are two opposed CI circular plates that are held on vertical spindles of the machine. The workpieces are retained between laps in a slotted-holder plate and are caused to rotate and slide in and out to break the pattern of motion by moving over the inside and outside edges of the laps that prevent grooving. The lower lap is usually rotated at a regulated speed and drives the workpieces. The upper one is held stationary while it is free to float so that it can adjust to the variations in workpiece size. In order to avoid damage of the surface being lapped, the holder plate or carrier is made of soft material (copper, laminate fabricate base, etc.). An alternative design is also available where both the upper and lower laps are rotating.

Figure 16.61 illustrates a dual face lapping machine, having two-bonded abrasive laps (400-grit SiC) that are rotated in opposite directions at 88 rpm. The head is air actuated in order to provide the lapping pressure to the top lap. The workpiece carrier is eccentrically mounted over the bottom lap and rotates at 7.5 rpm. The viscous cutting oil is fed to the laps during operation. The laps are dressed 2 or 3 times during an 8-hour shift.

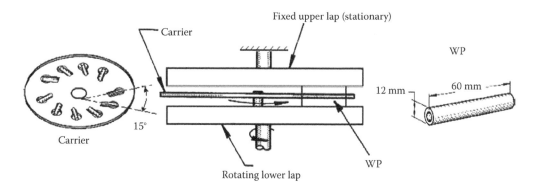

FIGURE 16.60 Vertical lapping machine used for lapping cylindrical surfaces. (Adapted from Hoffman Co., Carlisle, PA.)

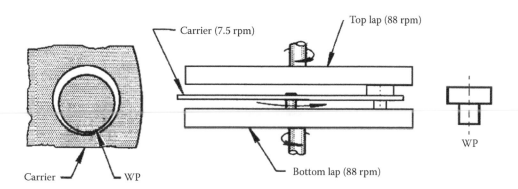

FIGURE 16.61 Dual face lapping machine for flat surfaces. (Adapted from Hoffman Co., Carlisle, PA.)

16.3   SPECIAL PURPOSE MACHINE TOOLS

16.3.1   THREAD CUTTING MACHINES

Screw threads are commonly used as fasteners to transmit power or motion and for adjustment. They should conform to some established standard in order to be interchangeable and replaceable. Cast threads may be finished by machining or left in the as-cast state. With molded or die-cast, the threads can be very nice indeed straight from the mold or die. Threads can be produced by means of taps and threading dies that cut internal and external threads, respectively. Die heads with radial, circular, and tangential chasers are available. Threads can be produced by a variety of general purpose machine tools. In this respect, coarse-pitch screw threads can be cut using thread cutting tools in an engine lathe, manual turret lathe, and automatic turret lathe. Special purpose thread cutting machines include the following.

1. Thread tapping machines: Taping machines are basic drill presses equipped with lead screws, tap holders, and reversing mechanisms. Lead screws convert the rotary motion into a linear one so that the axial motion of the tap into the hole to be threaded conforms with the pitch of the thread. Tension/compression tapping spindles and attachments provide axial float and compensate for any differences between machine feed and correct tap feed. This provides the possibility to tap different thread pitches at the same time with a single machine feed rate. Self-reversing tapping attachments eliminate the need for reversing motors for tap retraction. Tapping machines are used for small to medium production lots. The simpler modes have no lead control devices but depend on the screw action of the tap in the hole to control the feed (Figure 16.62). Multiple-spindle tapping machines (Figure 16.63) are used for high-volume production lots. They may have up to 25 spindles that are rotated by a common power source. Holes of different sizes can be tapped

FIGURE 16.62   Herbert flash tapping machine with automatic cycle. (Courtesy of Alfred Herbert, Coventry, UK.)

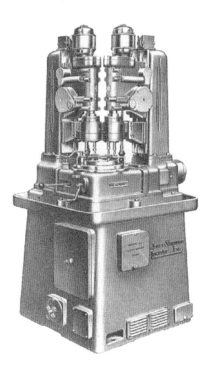

FIGURE 16.63   Jones and Shipman multiple-spindle automatic drilling and tapping machine.

simultaneously. Spindles having axial float compensate for differences between the lead of the tap and the feed of the spindle. Thus different thread pitches can be cut simultaneously on the same machine. Gang machines permit in line drilling, reaming, and tapping operations and are generally used for low production lots.

2. Die threading machines: These include drill presses that are easy to set and simple to operate. Special threading machines are available for only die threading in either cylindrical or irregular-shaped parts. Workpiece loading and unloading can be achieved manually, hopper fed, or fully automatic. These machines usually incorporate lead-control devices. Bar-type machines with collets can handle long parts to thread rods, shafts, and pipes.

3. Thread milling machines: Thread milling machines are used for cutting threads usually of too large diameter for die heads. As the milling cutter is held on a stub arbor, the length of the thread is limited to short ones. As shown in Figure 16.64, the cutter rotates at a cutting speed of 3.6 m/min, and the work rotates at the correct feeding speed. As the work rotates, the cutter is fed outward under the action of a master lead screw. A right- and left-hand thread can be machined by controlling the direction of cutter feed and workpiece rotation. Universal thread mills (Figure 16.65) have a lead screw and cut internal and external threads. Change gears permit milling of threads with leads of 0.8–1520 mm. Pick-off gears in the cutter drive provide a wide range of speeds. The cutter head on the cross-slide can be set at the proper angle for right- or left-hand thread helix angle. A single form cutter must be set at such an angle and then traverse the full length of the thread. Planetary thread mills are used to thread odd-shaped parts that are difficult to hold in a chuck. Consequently, the workpiece is held in a special fixture that does not rotate during thread cutting. The milling cutter rotates around its axis and revolves around the work.

4. Thread broaching machines: These are normal broaching machines developed for thread cutting operations. The broaches used for the application, shown in Figure 16.66 have a special form and are guided by lead screws. Such broaches are available in sizes up to

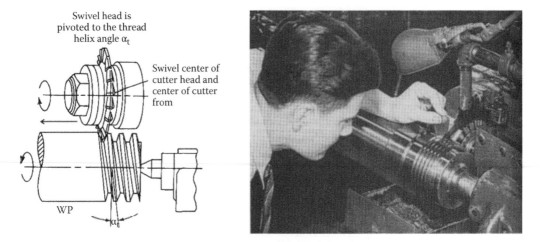

FIGURE 16.64 Thread milling. (Courtesy of Lees-Brander.)

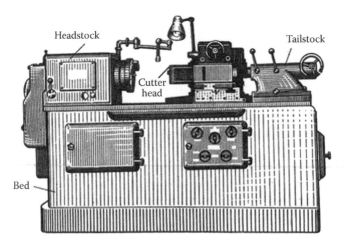

FIGURE 16.65 Semiautomatic thread milling machine. (From F. Barbashov, Thread Milling Practice, Mir Publishers, Moscow, 1984. With permission.)

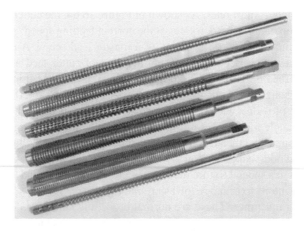

FIGURE 16.66 Threading broaches. (Courtesy of Chin Liang Broach Tool Works, Taiwan.)

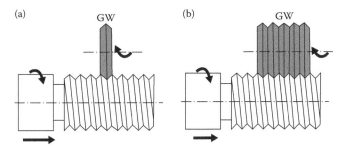

FIGURE 16.67 Thread grinding operations.

50 mm in diameter and 750 mm in length. Threads are cut by drawing up the part and
fixture against the revolving tool.

5. Thread grinding machines: Threads are ground by contact between a rotating workpiece
   and a rotating grinding wheel that has been shaped to the desired thread form. In addition
   to the rotation, there is a relative axial motion between the wheel and workpiece to match
   the pitch of the thread being ground. In the center-type thread grinding machines, the
   workpiece is held between centers or in the machine chuck. The number of passes required
   may be from 1 to 6 passes. Depending on the design of the threading wheel, single rib
   wheel traverse grinding or multirib wheel grinding are possible (Figure 16.67a,b). Multirib
   traverse grinding is more productive than single rib wheel grinding because of the higher
   material removal rate per pass. In the centerless-type thread grinding machines, screw
   threads are cut by feeding the bars between the grinding and regulating wheels in a continuous stream as shown in Figure 16.68. The wheel is positioned at a full thread depth, and
   then the work is traversed past the wheel. The first thread form on the wheel removes the
   majority of metal and therefore subjected to the most wear, whereas the following threads
   affect the finishing. A single ribbed wheel may be used for large threads.

6. Thread rolling machines: Thread rolling machines (Figure 16.69) are better suited to high-
   volume production and produce threads of diameters typically smaller than one inch. Also,
   materials with good deformation characteristics are necessary for rolling; these materials

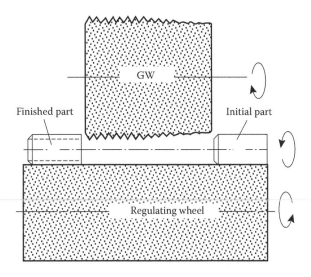

FIGURE 16.68 Traverse centerless grinding of headless screws. (Adapted from *Metals Handbook, vol. 16,
Machining,* ASM Internationals, Material Park, OH, 1989.)

FIGURE 16.69   Thread rolling machines. (Courtesy of Power Channel Manufacturers Corp., Taiwan.)

FIGURE 16.70   Roll threading dies and their operation. (Courtesy of Yieh Chen Machinery, Taiwan.)

include ductile metals. A rolled thread can often be easily recognized because the thread has a larger diameter than the blank rod from which it has been made. The process produces external rolled threads with excellent strength and surface finish. It forms major thread diameters greater than blank diameter. It is usually 3 to 5 times faster than thread cutting. When compared to a cut thread, the load capacity of the rolled thread is increased by 5%–15%. Thread rolling machines use either flat dies or circular dies. Figure 16.70 shows typical circular dies and a closer look at the rolling operation. Being a plastic forming process, the principles of thread rolling are presented in section 9.4.4.

16.3.2 GEAR CUTTING MACHINES

Gears are machine elements that transmit power and rotary motion from one shaft to another. They have the advantage over friction and belt drives in that they are positive in their action, a feature that most of the machine tools require, as exact speed ratios are sometimes essential. Thread cutting and indexing movements in gear cutting are typical examples that acquire synchronized rotary and linear movements without any slip.

16.3.2.1 Common Gear Types

Depending on the specific application, gears can be selected from the following types.

Spur gears: The most common types that transmit power or motion between parallel shafts or between a shaft and a rack. They are simple in design and measurement.

Helical gears: Used to transmit motion between parallel or crossed shafts or between a shaft and a rack by meshing teeth that lie along a helix at an angle to the shaft.

Herringbone gears: Sometimes called double helical gears. These gears transmit motion between parallel shafts. They combine the principal advantages of spur and helical gears, because two or more teeth share the load at the same time.

Worm gear sets: Used where the ratio of the speed of the driving member (worm) to the speed of the driven member (worm wheel) is large and for a compact right angle drive. They are frequently used in indexing heads of milling machines and in hobbing machines.

Crossed-axes helical gears: Operate with shafts that are nonparallel and nonintersecting.

Internal gears: May be of spur or helical tooth form.

Racks: A gear of infinite pitch circle radius. The teeth may be at right angles to the edge of the rack and mesh with a spur gear or at some other angle and engage a helical gear.

Bevel gears: Transmit rotary motion between two nonparallel shafts.

16.3.2.2 Forming and Generating Methods in Gear Cutting

Gear production by cutting involves two main principal methods; these are forming and generating processes, while gear finishing involves four operations which are shaving, grinding, lapping, and burnishing (Figure 16.71).

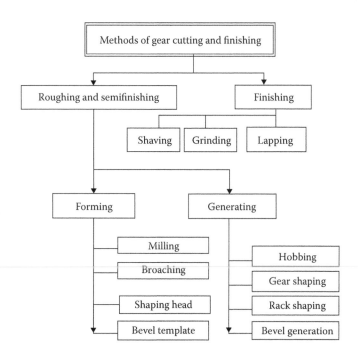

FIGURE 16.71 Methods of gear cutting and finishing.

1. Gear cutting by forming: The tooth profile is obtained by using a form cutting tool. This may be a multiple-toothed cutter used in milling, broaching machines, and shaping cutter head, or a single-edge tool form for use in a shaper and a bevel gear planer.
   a. Gear milling. The usual practice in gear milling is to mill one tooth space at a time, after which the blank is indexed to the next cutting position. Figure 16.72 shows teeth in a spur gear cut by peripheral (horizontal) milling with a disc cutter (Figure 16.72a). Similarly, end milling can also be used for cutting teeth in spur or helical gears (Figure 16.72b) and is often used for cutting coarse-pitch teeth in herringbone gears.

      In practice, gear milling is usually confined to
      - One-of-a-kind replacement gears
      - Small-lot production
      - Roughing and finishing of coarse-pitch gears
      - Finish milling of gears having special tooth forms

      As the tooth profile depends on the module, pressure angle, and number of teeth, it is theoretically necessary to have a tool with a certain profile for each gear with a different number of teeth or module. In practice, however, sets of gear tooth milling cutters, according to ASA B.9-1959, are used (8 cutters per set or for more accurate gears 15 and less frequently 26 cutters for each module of gears). Each cutter in the set is designed for cutting a limited range of numbers of teeth (Table 16.2).

      When cutting helical gears, the cutter number as obtained from Table 16.2 has to be modified because of the helix angle β_g

      $$\text{Equivalent teeth number } Z = \frac{\text{number of teeth of helical gear}}{(\cos\beta_g)^3} \qquad (16.6)$$

   b. Gear broaching: This is usually confined to cutting teeth in internal gears. However, not only internal but also external, spur, or helical gears can be broached. Figure 16.73a shows progressive broach steps in cutting an internal spur gear. Figure 16.73b shows how an external spur gear is produced using a rotating broach. In such arrangements, the blank is withdrawn for indexing to cut another space between two teeth. Broaching

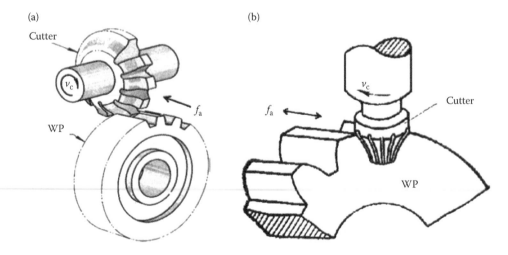

FIGURE 16.72   Spur gear cutting on milling machines.

TABLE 16.2
Gear Cutter Sets for Milling According to ASA B.9-1959

		8-Cutter Set for Spur Gears						
Cutter no.	1	2	3	4	5	6	7	8
Number of teeth	135-rack	$\dfrac{55}{134}$	$\dfrac{35}{54}$	$\dfrac{26}{34}$	$\dfrac{21}{25}$	$\dfrac{17}{20}$	$\dfrac{14}{16}$	$\dfrac{12}{13}$

		15-Cutter Set for Accurate Gears						
Cutter no.	1	1½	2	2½	3	3½	4	4½
Number of teeth	135-rack	$\dfrac{80}{134}$	$\dfrac{55}{79}$	$\dfrac{42}{54}$	$\dfrac{35}{41}$	$\dfrac{30}{34}$	$\dfrac{26}{29}$	$\dfrac{23}{25}$
Cutter no.	5	5½	6	6½	7	7½	8	
Number of teeth	$\dfrac{21}{22}$	$\dfrac{19}{20}$	$\dfrac{18}{17}$	$\dfrac{15}{16}$	14	13	12	

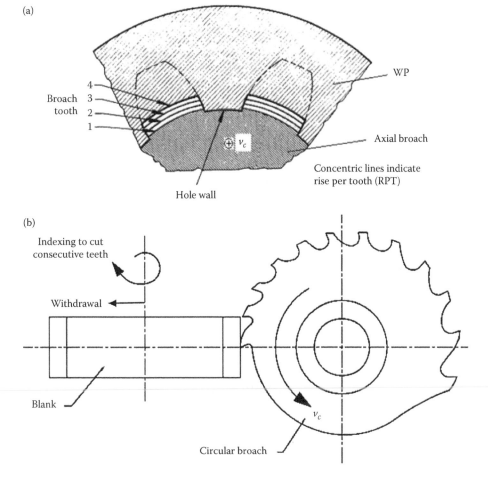

FIGURE 16.73 Gear broaching by forming.

is fast, accurate, and provides an excellent surface quality. However, the cost of tooling is high; therefore, gear broaching is best suited to large production runs.

c. Gear forming by a multiple-tool shaping head. It is a high production and accurate method of producing teeth in external and internal spur gears. This method is not applicable to helical gears. As in broaching of internal gears, all tool spaces are cut simultaneously and progressively (Figure 16.74). Prior to each cutting stroke, each tool is fed radially toward the blank by an amount equal to the prescribed infeed. All the tools are simultaneously retracted from the work on the return stroke to avoid rubbing of the tool against the machined surfaces. The gear is finished when the tools reach the full depth of cut. Cutting speeds in this process are similar to those used for broaching the same work metal using the same tool material. Machines with shaping heads are available for cutting spur gears up to 500 mm in diameter, with face width up to 150 mm.

For example, a machining time of not more than 1 min is required to produce a spur gear of 160 mm pitch diameter, face width of 30 mm, and a module of 4 mm; therefore the process is best suited to large production runs. Drawbacks of the process are the comparatively complex shaping heads and the necessity of having a separate head for each gear size and module.

2. Gear cutting by generation: This technique is based on the fact that two involute gears of the same module and pith mesh together, one is the workpiece blank and the other is the cutter. So this method makes it possible to use one cutting gear for machining gears of the same module with a varying number of teeth.

Gear generation methods are characterized by their higher accuracy and machining productivity. They comprise hobbing and gear, or rack, shaping for the manufacture of spur and helical gears, worm and worm wheels, and bevel gear generation.

a. Gear hobbing: It is a gear generation method most widely used for cutting teeth in spur gears, helical gears, worms, worm wheels, and many special forms. Hobbing machines are not applicable to cutting bevel or internal gears. The tooling cost for hobbing is lower than for broaching and multiple-tool shaping heads. For this reason, hobbing is used in low-quantity production or even for a few pieces. Compared with milling, hobbing is fast, accurate, and is therefore suitable for medium- and high-production quantities. The hob is a fluted worm of helix angle α_h with form relieved teeth that cut into the gear blank in succession. A simplified gear train of a hobbing machine is shown in Figure 16.75.

The ability to cut teeth in two or more identical gears in one setup can encourage the use of this method. A typical hobbing fixture, which is a common mandrel-type fixture for flat-face gears, is illustrated in Figure 16.76. Incorporated in the fixture is an interchangeable bottom plate in order to utilize the same fixture for various sizes of

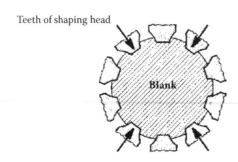

Teeth of shaping head

Blank

FIGURE 16.74 Cutting with progressive gear shaping head.

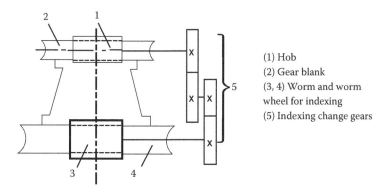

(1) Hob
(2) Gear blank
(3, 4) Worm and worm
wheel for indexing
(5) Indexing change gears

FIGURE 16.75 Elementary hobbing machine setup.

gears. Figure 16.77 illustrates the cutting action used for different types of gears. The rotary motions imparted to the blank and hob is the same as those of the worm wheel and worm gearing.

Hobbing of spur gears: The hob is set up so that the thread of the hob on the side facing the gear blank is directed vertically along the axis. This is done by setting the hob axis, at an angle α_h to the horizontal equal to the helix angle of the hob. The hob attains a continuous feed motion f_a along the axis of the gear blank as shown in Figure 16.77a.

Hobbing of helical gears: To cut helical gears, the hob is set up so that the thread of the hob facing the gear blank is directed at the helix angle of the teeth. This is done by setting the hob at an angle $\gamma = \beta_g \pm \alpha_h$, where β_g is the helix angle of the helical gear being cut and α_h is the helix angle of the hob. If the hand of helical gear and that of the hob are different, the positive sign is considered; if the hand is the same, the negative

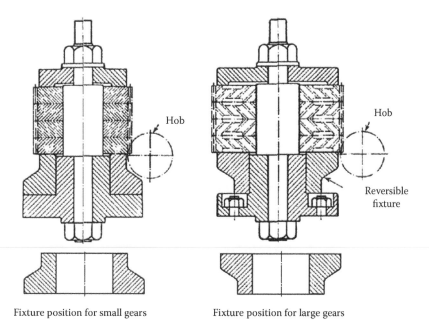

Hob

Hob

Reversible fixture

Fixture position for small gears   Fixture position for large gears

FIGURE 16.76 Interchangeable hobbing fixture for various gear sizes.

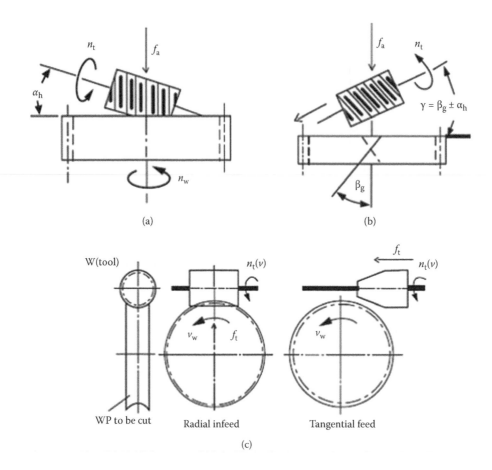

FIGURE 16.77   Cutting action for (a) spur gear, (b) helical gear, and (c) worm wheel.

sign should be used. Also, the hob attains a continuous feed motion f_a along the axis of the gear blank (Figure 16.77b). In cutting helical gears an incremental motion is imparted to the blank, with an angular velocity that would provide one full additional revolution of the blank during vertical feed of the hob through a distance equal to the lead of the helical teeth on the gear.

Hobbing of worm wheels: When cutting worm wheels, the axis of the hob is set perpendicular to the axis of the rotation of the blank. The following principal motions are shown in Figure 16.77c.

1. Principal rotary cutting motion $n_t(v)$ of the hob.
2. Continuous indexing rotary motion v_w of the gear bank.
3. Feed motion of the hob that may be either
   a. Worm wheel hobbling through radial infeed f_t. The radial infeed ceases when the full depth of cut is reached.
   b. Worm wheel hobbing through tangential feed f_t. The hob is set at the beginning to the full depth of cut and is fed tangentially into the blank.

The radial infeed method has a higher production capacity; however, a small part of the hob in the midlength is actually doing the cutting. As a result, the hob wears nonuniformly and has an unfavorable effect on the tooth profile accuracy. If high gear accuracy is required, the tangential-feed method is used.

Hobbing of worms: Hobbing produces the highest grade worm at the lowest machining cost, but it can be used only when production quantities are large enough to justify the high tooling cost. The number of flutes in a worm hob is increased to improve surface finish.

Example 4

An HSS hob of pitch diameter 70 mm is used to cut a spur gear of 48 teeth at a cutting speed of 30 m/min. The gear tooth has a face width of 90 mm and the hob is fed at a rate of 1.8 mm/rev of the blank. What is the time required to achieve the hobbing operation, provided that an approach and over travel of 36 mm is assumed?

Solution
Hob rotational speed is N

$$N = \frac{1000V}{\pi D} = \frac{1000 \times 30}{\pi \times 70} = 136.5 \text{ rev/min}$$

Axial feed u_a

$$u_a = f \times N = 1.8 \times 136.5 = 245.70 \text{ mm/min}$$

Time to achieve the cut t

$$t = \frac{90 + 36}{1.8 \times 136.5} = 0.51 \text{ min}$$

3. Gear shaping with pinion cutter: It is the most versatile of all gear cutting processes. Although shaping is most commonly used for cutting teeth in spur and helical gears, this process is also applicable to cutting herringbone teeth, internal gears (or splines), chain sprockets, elliptical gears, worm gears, and racks. Shaping cannot be used to cut bevel gears.

Figure 16.78 shows the principle of gear shaping with a pinion cutter. In this process, the cutter is mounted on a spindle that reciprocates axially as it rotates. The workpiece spindle is synchronized with the cutter spindle and rotates slowly as the tool meshes and cuts while it is being fed into the work at the end of each return (upward) stroke. The downward movement of the tool represents the principal cutting motion. To prevent the flanks of the cutter teeth from scoring the blank as the cutter is returned upward, the blank (or the cutter) is withdrawn radially in the direction of arrow X. Because tooling cost is relatively low, gear shaping is practical for any production volume. Workpiece design often prevents the use of milling cutters or hobs (e.g., cluster gears), and shaping is the most practical method for such cases (Figure 16.79). Shaping can also be applied in cutting a worm where the cutter involves no axial stroke.

Characteristics of gear shapers include:
- Gears produced by gear shapers are accurate.
- Both internal and external gears can be cut by this method.
- Production rate of gear shapers is lower than hobbers.
- Bevel and worm gears cannot be generated on gear shapers.

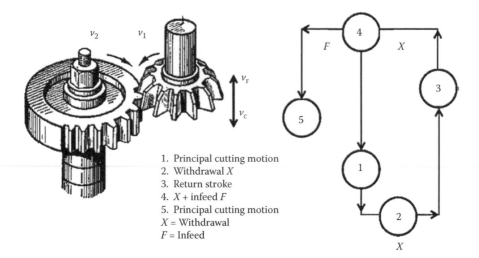

1. Principal cutting motion
2. Withdrawal X
3. Return stroke
4. X + infeed F
5. Principal cutting motion
X = Withdrawal
F = Infeed

FIGURE 16.78   Principles of gear shaping modify.

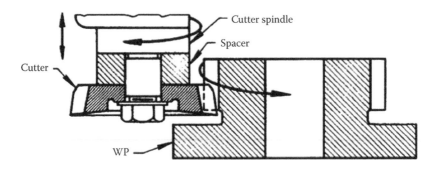

FIGURE 16.79   Shaping of cluster gear.

4. Gear shaping with rack cutter: Gear shaping is performed by a rack cutter with three to six straight teeth (Figure 16.80). The cutters reciprocate parallel to the work axis when cutting spur gears and parallel to the helix angle when cutting helical gears. In addition to the reciprocating action of the cutter, there is synchronized rotation of the gear blank with each stroke of the cutter, with a corresponding advance of the cutter in a feed movement. Rack cutters are less expensive than pinion cutters and hobs. A rack cutter is especially adapted for cutting of large gears of modules typically of 5 to 10 mm.

5. Cutting straight bevel gears by generation: The generation principle of bevel gear cutting is based on reproducing the sides of the teeth on an imaginary crown gear in space by means of the cutting edges of rotating interlocking cutters or reciprocating two tool generators. The profiles of the straight cutting edges coincide with the opposing sides of two teeth of the imaginary crown or generating gear with which the gear being cut is in mesh. The primary cutting motion, either rotation or reciprocation, is transmitted to these cutting edges.

    In this method, two interlocking disc-type cutters rotate at the same speed on axes inclined to the face of the mounting cradle and both cut in the same tooth space. The gear blank is held in a work spindle that rotates in timed relation with the cradle on which the cutters are mounted (Figure 16.81).

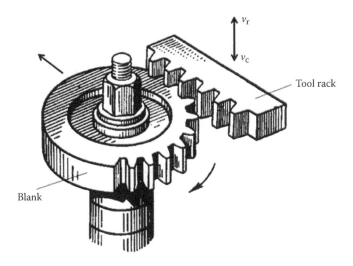

FIGURE 16.80 Principles of gear shaping using rack cutter.

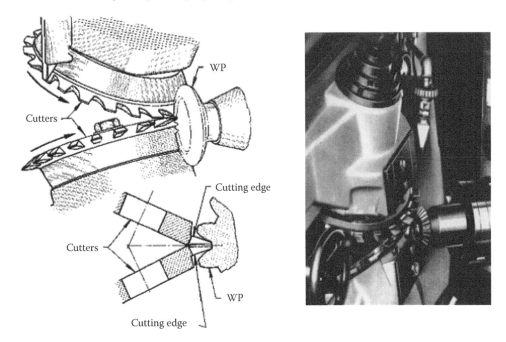

FIGURE 16.81 Bevel gear generating by interlocking cutters (Konvoid generators).

16.3.2.3 Selection of Gear Cutting Method

The following factors should be considered in the final choice of the gear cutting method:

- Size of the gear and its module
- Configuration of the workpiece to be machined
- Batch size
- Gear ratio
- Accuracy
- Cost related to the tool and the machine
- Cycle time and productivity

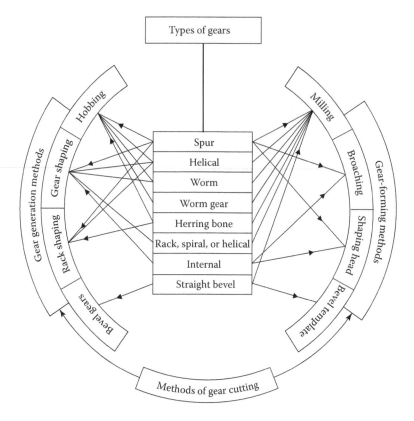

FIGURE 16.82 Selection of gear cutting methods.

TABLE 16.3
Gear Cutting Methods and Their Capabilities to Produce Different Types of Gearing

	Forming				**Generation**			
	Milling	**Broaching**	**Shaping Head**	**Bevel Template**	**Hobbing**	**Gear Shaping**	**Rack Shaping**	**Bevel Generators**
Type of Gears	Spur				Spur	Spur		
	Helical	Spur exterior			Helical	Helical	Spur	
	Worm							
	Worm gear			Straight bevel	Worm	Worm	Helical	Straight bevel
	Herringbone							
	Rack	Spur interior			Worm wheel	Rack Internal	Herringbone	
	Straight bevel				Herringbone			

Gear Cutting Method

Figure 16.82 summarizes the possibilities to produce a certain gear type by cutting. The outcome of this layout leads to the conclusion expressed by Table 16.3.

16.3.2.4 Gear Finishing Operations

Gear finishing operations are distinguished from gear cutting operations in that they are used for improving the accuracy, uniformity, and surface quality of the various gear tooth elements. Higher accuracy is necessary if the gears are required to operate quietly and at high speeds and to transmit heavy loads. Gear finishing methods include burnishing, shaving, lapping, and grinding. Unhardened teeth of gears are finished by shaving or burnishing, whereas hardened teeth are finished by grinding or lapping operations.

1. Finishing gears prior to hardening
   a. Gear shaving: This is a finishing process based on consecutively removing thin layers of chips (2–10 μm thick) from the profiles of the teeth by a tool called a gear shaving cutter. Shaving is currently the most widely used method of finishing spur and helical gear teeth following the gear cutting operation and prior to hardening the gear. It is not intended to salvage gears that have been carelessly cut, although it can correct small errors such as tooth spacing, helix angle, tooth profile, and concentricity. Shaving reduces noise level and increases load carrying capacity, surface quality, and accuracy.

      Shaving is performed with cutter and gear at crossed axes; the value of the crossed axes angle controls the finish produced to some extent. The smaller the angle, the finer will be the finish. Angles ranging from 8° to 15° are generally found more satisfactory. In this process, generally helical cutters of a helix angle 10° to 15° are used for spur gears and vice versa. In some cases, helical gears are shaved by helical cutters. The action between gears and cutter is therefore a combination of rolling and sliding. Vertical serration (0.6–1 mm deep) in the cutter teeth (Figure 16.83) takes thin hair-like chips from the profile of the gear teeth.
   b. Gear burnishing: This is another method of surface finishing for teeth of a gear. It consists essentially of rolling the work gear with burnishing gears whose teeth are very hard, smooth, and accurate. The inaccuracies and asperities of the surface of the work gear are leveled by kneading action of the material. Burnishing is of no use to gears that are to be subsequently heat treated, because it may set up stresses that will be released during heat treatment, hence leading to increased distortion, surface cracks, and peeling off the carburized and deformed surface layer.

      Three burnishing gears (spur or helical depending on the type of burnished gear) are meshed with and spaced at 120° positions around the work gear. One of the burnishing gears is the driver and the other two are idlers, which exert burnishing pressure against the work gear. The burnishing cycle starts by rotating the gears in one direction for the necessary period of time, then reversing the direction of rotation for an equal

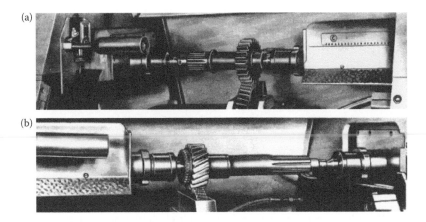

(a)

(b)

FIGURE 16.83 Gear shaving.

period of time (Figure 16.84). During burnishing, a lubricant is supplied to produce the desired surface quality and prevent abrasion.

2. Finishing gears after hardening
   a. Gear grinding: It is a specially adapted process to finish gears having considerable stock to be removed after hardening in order to obtain the most accurate and the highest-quality gears. It is also especially used in producing gear tools. The low rate of production and the high cost of gear grinding exclude the use of this method for mass production. As a rule, it is used only for finishing gears of precise machinery. Similar to gear cutting, gear grinding may be also performed by forming or generation (Figure 16.85).

      Disadvantages of gear grinding process include
      – The process is characterized by its low production capacity.
      – The scratches or ridges formed increase both wear and noise. To eliminate this defect, ground gears are frequently lapped.
      – Dimensional instability is an inherent characteristic of the gear grinding method.
      – The process necessitates the application of complex and expensive gear grinding machines tended by highly skilled operators.
   b. Gear lapping: It is a microfinishing process performed on the gear after hardening. This method is based on finishing of the gear teeth profiles, using a lapping tool (called lap) and fine-grained abrasives, with the purpose of imparting a high accuracy and fine surface finish to the gear teeth. It is, however, impossible to correct considerable errors (exceeding 30–50 μm) by lapping. Prolonged lapping associated with large allowance, besides being time-consuming, may distort the gear profile and impair the teeth accuracy. Usually, lapping is performed on special machines using three laps made of soft and fine grained cast iron, where a lapping compound (oil + fine abrasives) is applied on the tools.

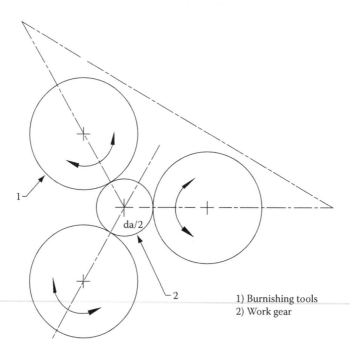

FIGURE 16.84 Burnishing operation and maximum limit of addendum diameters of burnishing tools modify.

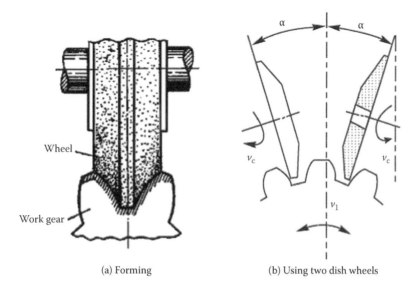

(a) Forming   (b) Using two dish wheels

FIGURE 16.85 Gear grinding by forming and generation.

16.4 REVIEW QUESTIONS

1. How does form turning differ from ordinary turning?
2. What will happen to a workpiece held between centers if the centers are not exactly in line?
3. How does a steady rest differ from a follower rest?
4. Numerate the methods of taper turning on a lathe.
5. On which diameter is the rpm of the work is based for facing cut, assuming given work and tool material?
6. Define the following: boring, broaching, counterboring, countersinking, reaming, and spot facing.
7. When a large diameter hole is to be drilled, why is a smaller diameter hole often drilled first?
8. What is unique about broaching compared to other basic machining operations?
9. Why is broaching more practically suited for mass production?
10. Can continuous broaching machines be used for broaching holes? Explain why.
11. How does the process of shaping differ from planing?
12. How does a gang drilling machine differ from a multiple-spindle drilling machine?
13. What may result when holding the work by hand during drilling?
14. Describe the relative characteristics of climb milling and up milling, mentioning the advantages of each.
15. Which type of milling (up or down) do you think uses less power, provided the same cutting conditions?
16. Explain the steps required to produce a T-slot by milling.
17. What is the basic principle of a universal dividing head?
18. The input end of a universal dividing head can be connected to the lead screw of the milling machine table. For what purpose?
19. What is the purpose of indexing plates on a universal dividing head?
20. How is a workpiece controlled in centerless grinding?
21. What are the different methods of thread cutting and grinding.
22. Show using a sketch how a thread is cut on the center lathe.

23. Compare between thread chasing and thread cutting on a lathe.
24. Compare between thread milling using disc and multiple-thread cutters.
25. Show the arrangement of thread chasers in threading die heads.
26. Show the arrangements of single- and multiple-rib traverse grinding of threads.
27. What are the advantages of gear generation by shaping?
28. Why is a heat treatment process not recommended after a gear burnishing finishing method?
29. Explain the main advantages and limitations when using a gear shaping head. Is it a forming or a generating gear production method?
30. Can a helical gear be machined on a universal milling machine?
31. Numerate methods of finishing gears before and after hardening.
32. What is the difference between a turret lathe and a capstan lathe?
33. Mark true (T) or false (F)
    [　] In turret lathes, the turret is mounted on a saddle.
    [　] Capstan lathes are characterized by higher accuracy if compared to turret lathes.
    [　] Heavier cuts can be taken by turret automatics rather than the capstan machines.
    [　] Thread cutting on a lathe can be performed using chasers.
    [　] Thread rolling is not a machining process.
    [　] Change gears are not necessary when cutting threads on the lathe machine.
    [　] Tapping blind holes is done easier and faster than through holes.
    [　] Thread grinding is recommended for hard workpieces of 37 HRC.
    [　] Broaching is one method for the production of gears by forming.
    [　] Helical gears cannot be produced by milling.
    [　] Radial, not a tangential feed, is preferred for the production of accurate worm-wheels by hobbing.

16.5 PROBLEMS

1. Calculate suitable gear trains for cutting the following threads:
    a. 11 tpi on a 4 tpi lead screw
    b. 7 threads in 10 mm on 6 mm lead screw
    c. 12 tpi on a lathe having 6 mm pitch lead screw
2. Calculate the broach length required for machining a key way of 8 mm depth and 20 mm width for 100 mm length in a steel workpiece, S_z is taken as 0.08, and chip space number is 8.
3. Calculate the indexing particulars for cutting eight flutes in a reamer blank.
4. Calculate the indexing movements and gear ratio required for cutting 57 gear teeth.
5. Calculate the time required to achieve the hobbing operation using HSS hob of pitch diameter 80 mm to cut a spur gear of 48 teeth at a cutting speed of 40 m/min and the gear toot has a face width of 94 mm. The hob is fed at a rate of 2 mm/rev of the blank and assuming the approach and over travel of 36 mm.
6. Calculate the angle of tilting the compound slide when turning a short taper of steel bar having $D = 80$ mm, $d = 60$ mm, and length of taper $l = 40$ mm (see Figure 16.2a).
7. Find the tailstock setting required for turning a taper of 85 mm diameter to 75 mm diameter over a length of 200 mm. The total length of the job is 300 mm.

BIBLIOGRAPHY

Acherkan, N. 1968. *Machine Tool Design*, vols. 1–4. Moscow: Mir Publishers.
Arshinov, V., and Alekseev, G. 1970. *Metal Cutting Theory and Cutting Tool Design*. Moscow: Mir Publishers.

Barbashov, F. 1984. *Thread Milling Practice*. Moscow: Mir Publishers.

Boguslavsky, B. L. 1970. *Automatic and Semi-Automatic Lathes*. Moscow: Mir Publishers.

Chapman, W. A. J. 1981. *Elementary Workshop Calculations*. London: Edward Arnold.

Chernov, N. 1975. *Machine Tools*. Moscow: Mir Publishers.

Davis, J. R. 1989. *Metals Handbook, vol. 16, Machining*. Material Park, OH: ASM International.

Düniß, W., Neumann, M., and Schwartz, H. 1979. *Trennen-Spanen and Abtragen*. Berlin: VEB-Verlag Technik.

Rodin, P. 1968. *Design and Production of Metal Cutting Tools*. Moscow: Mir Publishers.

Town, H. C., and Moore, H. 1980. *Manufacturing Technology—Advanced Machines and Processes*. London: Bastford Academic and Educational Ltd.

WMW. *Bevel Gear Hobbing Machine,* Technical Information, ZFTX 250x5. Berlin: WMW Export.

WMW. *Gear Cutting Practice*, Technical Information, Special Edition 12. Berlin: WMW Export.

Youssef, H., and El-Hofy, H. 2008. *Machining Technology, Machine Tools and Operations*. Boca Raton, FL: CRC Press.

17 Fundamentals of Nontraditional Machining Processes

17.1 INTRODUCTION

Engineering materials have been recently developed. Their hardness and strength are considerably increased, such that the cutting speed and the material removal rate (MRR) tend to fall when machining such materials using traditional methods like turning, milling, grinding, and so on. In many cases, it is impossible to machine hard materials to certain shapes using these traditional methods. Sometimes it is necessary to machine alloy steel components of high strength in a hardened condition. It is no longer possible to find tool materials that are sufficiently hard to cut at economical speeds, such as hardened steels, austenitic steels, Nimonic, carbides, ceramics, and fiber-reinforced composite materials. The traditional methods are unsuitable to machine such materials economically, and there is no possibility that they can be further developed to do so because most of these materials are harder than the materials available for use as cutting tools.

By utilizing the results of relevant applied research, it is now possible to process many of the engineering materials that were formerly considered to be nonmachinable using traditional methods. The newly developed machining processes are often called modern machining processes or nontraditional machining processes (NTMP). These are nontraditional in the sense that traditional cutting tools are not employed; instead, energy in its direct form is utilized. The NTMPs have specifically the following characteristics as compared to traditional processes.

- They are capable of machining a wide spectrum of metallic and nonmetallic materials irrespective of their hardness or strength.
- Complex and intricate shapes, in hard and extra-hard materials, can be readily produced with high accuracy and surface quality and commonly without burrs.
- The hardness of cutting tools is of no relevance, especially in much of NTMPs, where there is no physical contact between the work and the tool.
- Simple kinematic movements are needed in the NTM equipment, which simplifies the machine design.
- Micro and miniature holes and cavities can be readily produced by NTM.

However, it should be emphasized that

1. NTMPs cannot replace TMPs. They can be used only when they are economically justified or it is impossible to use TMPs.
2. A particular NTMP found suitable under given conditions may not be equally efficient under other conditions. A careful selection of the NTMP for a given machining job is therefore essential. The following aspects must be considered in that selection:
   - Properties of the work material and the form geometry to be machined
   - Process parameters
   - Process capabilities
   - Economic and environmental considerations

17.2 CLASSIFICATION OF NONTRADITIONAL MACHINING PROCESSES

NTMPs are generally classified according to the type of energy utilized in material removal. They are classified into three main groups (Table 17.1):

1. Mechanical processes: In these, the material removal depends on mechanical abrasion or shearing.
2. Chemical and electrochemical processes: In chemical processes, the material is removed in layers because of an ablative reaction where acids or alkalis are used as etchants. Electrochemical machining is characterized by a high removal rate. The machining action is due to anodic dissolution (AD) caused by the passage of high-density dc current in the machining cell.
3. Thermoelectric processes: In these, the metal removal rate depends on the thermal energy acting in the form of controlled and localized power pulses, leading to melting and evaporation of the work material.

An important and latest development has been realized by adopting what is called hybrid machining processes (HMPs). These are new processes produced by integrating one NTMP with one or more TMPs and NTMPs to improve the performance and promote the removal rate of the hybrid process (HP). Examples of these processes are electrochemical grinding (ECG), electrochemical honing (ECH), electrochemical ultrasonic machining (ECUSM), and abrasive water jet machining (AWJM).

17.3 JET MACHINING

Jet machining (JM) processes based on using high-energy jets to cause machining due to mechanical abrasion.

17.3.1 ABRASIVE JET MACHINING

In abrasive jet machining (AJM), a fine stream of abrasives is propelled through a special nozzle by a carrier gas (CO_2, N_2, or air) of a pressure ranging from 1 to 9 bar. Thus the abrasives attain a high speed ranging from 150 to 350 m/s, exerting impact force and causing mechanical abrasion of the workpiece (target material). The workpiece is positioned from the nozzle at a distance called the stand-off distance (SOD), or the nozzle-tip distance (NTD) as shown in Figure 17.1. In AJM, Al_2O_3 or SiC abrasives of grain size ranging from 10 to 80 μm are used. The nozzles are generally made of tungsten carbides (WC) or synthetic sapphires of diameters 0.2–2 mm. To limit the jet flaring, nozzles may have rectangular orifices ranging from 0.1 × 0.5 to 0.18 × 3 mm. The optimum jet angle is determined according to the ductility or brittleness of the WP material to be machined (Figure 17.1).

TABLE 17.1
Classification of NTMPs according to the Type of Fundamental Energy

Fundamental Energy	Removal Mechanism	NTMP
Mechanical	Abrasion, shearing	AJM, WJM, USM, MFM, AFM
Chemical	Ablative reaction (etching)	CH milling, PCM
Electrochemical	Anodic dissolution	ECM, ECT, ECD
Thermoelectric	Fusion and vaporization	EDM, LBM, EBM, IBM, PBM (PAC)

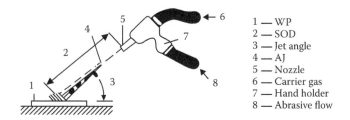

FIGURE 17.1   Abrasive jet machining (AJM). (From Düniß, W., Neumann, M., and Schwartz, H., *Trennen-Spanen und Abtragen*, Berlin, VEB-Verlag Technik, 1979. With permission.)

AJM is not considered to be a gross material removal process. Its removal rate when machining the most brittle materials, such as glass, quartz, and ceramics, is about 30 mg/min, whereas only a fraction of that value is realized when machining soft and ductile materials. Owing to the limited removal rate, and also the significant taper, AJM is not suitable for machining deep holes and cavities. However, the process is capable of producing holes and profiles in sheets of thicknesses comparable to the nozzle diameter. AJM is applicable for cutting, slitting, surface cleaning, frosting, and polishing.

Work Station of AJM: Figure 17.2 shows a typical workstation of AJM, which is connected to a gas supply (gas bottles or compressed air). The carrier gas must not flare excessively when discharged from nozzle to atmosphere. Furthermore, it should be nontoxic, cheap, available, and capable of being dried and filtered. Air is widely used owing to its availability. In small stations, CO_2 and N_2 gas bottles are commonly used. After filtering, the pressure of the compressed gas of 7–9 bars is regulated to suit the working conditions. The gas is then introduced to the mixing chamber containing the abrasives. The chamber is equipped with a vibrator providing amplitude ξ of 1–2 mm at a frequency f from 5 to 50 Hz. The abrasive flow rate is controlled through the adjustment of ξ and f. From the mixing chamber, the gas/abrasive mixture is directed to the nozzle that directs the jet onto the target or workpiece. The jet velocity of 150–350 m/s depends on the

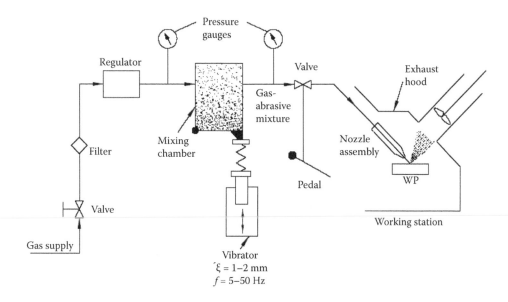

FIGURE 17.2   Typical work station of AJM.

gas pressure at the nozzle, the orifice diameter of the nozzle, and the mixing ratio. The flow rate of a typical working station is about 0.6 m³/h, which is controlled through a foot control valve. The nozzle is mounted in a special fixture and sometimes held in hand. The AJM station must be equipped by a vacuum dust collector to limit the pollution. Strict measures and precautions should be undertaken in case of machining toxic materials such as beryllium to collect produced dust and debris.

Process Capabilities: The performance of AJM in terms of MRR and accuracy is affected by the selected machining conditions. The MRR for a certain material is mainly affected by the kinetic energy of the abrasives, that is, the speed with which the abrasive is bombarding the work material. This speed depends on

- Gas pressure at nozzle
- Nozzle diameter
- Abrasive grain size
- Weight mixing ratio, β_m (abrasive flow rate/air flow rate)
- SOD

MRR attains a maximum value at a mixing ratio $\beta_m = 0.15$ and a SOD from 15 to 17 mm. The type of material to be machined and the abrasive grain size has an influence on the MRR. The latter increases with increasing grain size. Sharp-edged abrasives of irregular shape are best suited to perform the job. The limiting size of abrasive grains that permits the grain to be suspended in the carrier gas is about 80 μm. SiC and Al_2O_3 abrasives are used for cutting and slitting operations, whereas sodium bicarbonate, dolomite, and glass beads are used for cleaning, frosting, and polishing. The accuracy improves by selecting smaller SOD, which reduces the MRR. The grain size is a decisive factor for determining the surface finish.

Advantages and Disadvantages of AJM: Advantages of AJM include

- AJM is capable of producing holes and intricate shapes in hard and brittle materials.
- It is used to cut fragile materials of thin walls.
- Heat-sensitive materials such as glass and ceramics can be machined without affecting their physical properties and crystalline structure.
- AJM is safe to operate.
- It is characterized by low capital investment and low power consumption.
- AJM can be used to clean surfaces, especially in areas that are inaccessible by ordinary methods.
- The produced surfaces after cleaning by AJM are characterized by their high wear resistance.

Limitations of AJM include

- AJM is not recommended for machining soft and malleable materials.
- Abrasives cannot be reused because they lose their sharpness.
- Nozzle clogging occurs if fine grains having a diameter $d_g < 10$ μm are used.
- The process accuracy is poor owing to the flaring effect of the abrasive jet.
- Deep holes are produced at significant taper.
- Sometimes cleaning is necessary to get rid of grains sticking or penetrating to the surface.
- Excessive nozzle wear causes additional machining cost.
- The process tends to pollute the environment.

AJM has been successfully applied in the following domains:

1. Deflashing and trimming of parting lines of injection molded parts and forgings
2. Cleaning metallic molds and cavities
3. Cutting thin-sectioned fragile components made of glass, refractoriness, mica, and so on
4. Cleaning surfaces from corrosion, paints, glue, and contaminants, especially those that are inaccessible
5. Frosting interior or exterior surfaces of glass tubes and marking on glass
6. Engraving on glass using metallic or rubber masks
7. Engraving registration numbers on glass windows of cars
8. Deburring fine internal intersecting holes in plastic components
9. Deburring of surgical needles and hydraulic valves, nylon, Teflon, and derlin

17.3.2 Water Jet Machining

In water jet machining (WJM), the cutting capability of liquid jets, has been reported for a wide spectrum of target materials, including Pb, Al, Cu, Ti, steels, and granite. It is hard to believe that a jet of water can cut steel and granite. However, in scientific terms, it is explainable, as illustrated in Figure 17.3. When a stream of water is propelled at high pressure (2000–8000 bar) through a converging nozzle it gives a coherent jet of water of high speed of 600–1400 m/s at the target. The kinetic energy (KE) of the jet is converted spontaneously to high-pressure energy, inducing high stresses exceeding the flow strength of target material, causing mechanical abrasion.

Equipment of WJM: Figure 17.4a visualizes a simplified layout of the WJM equipment. It consists of the following stations:

1. Multistage filtering station: It filters the solid particles down to 0.5 μm. In this stage, it is also recommended to perform deionization and demineralization of water to allow for better performance of machine elements and extended nozzle life. After filtering, the water is mixed by polymers to obtain a coherent jet.
2. Oil pump and water high pressure–intensifier station: It consists of a hydraulic pump powered by an electric motor that provides oil at about 120 bar. Such a pressure is needed to

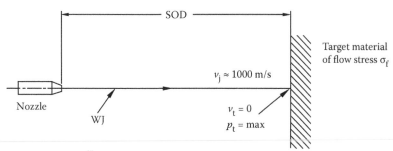

$$\text{Stagnant pressure, } p_t = \rho\frac{v_j}{2} > \sigma_f$$
$$\text{Jet velocity } \quad v_j = 2\text{--}4 \text{ Mach}$$
$$\rho = \text{water density}$$

FIGURE 17.3   Cutting principle of WJM.

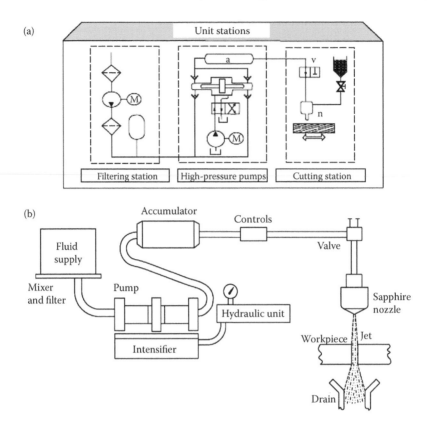

FIGURE 17.4 Simplified layout and basic elements of WJM equipment: (a) simplified layout of WJM equipment and (b) basic elements. (From König, W., *Fertigungsverfahren*, Band 3, Abtragen, VDI Verlag, Duesseldorf, 1990. With permission.)

drive a double acting plunger pump (intensifier) that pumps water from 4 bar to about 4000 bar or more. The ultrahigh-pressure water is delivered to an accumulator tank to provide the water pressure free of fluctuation and hydraulic spikes to the cutting station. During idle times the water is stored in the accumulator under pressure to be ready in any time to perform cutting.

3. Cutting station: The converging cutting nozzles (Figure 17.4b) converts the ultrahigh pressure (about 4000 bar) into a high speed of 600–1400 m/s. The nozzle provides a coherent water jet stream for optimum cutting. The jet coherency can be enhanced by adding long chain polymers such as polyethylene oxide (PEO) with a molecular weight of 4 million. Such addition provides the water higher viscosity and hence increases the coherent length up to 600 d_n, where d_n is the nozzle orifice diameter that falls between 0.1 and 0.35 mm. For optimum cutting, the SOD is selected within this range. Even if SOD is selected beyond this range, the stream is still capable of performing noncutting operations such as cleaning, polishing, and degreasing.

Nozzles are generally made from very hard materials such as WC, synthetic sapphire, or diamond. Diamond provides the longest nozzle life, whereas WC gives the lowest one. About 200 hours of operation are expected from a nozzle of synthetic sapphire, which becomes damaged by particles of dirt and the accumulation of mineral deposits if the water is not filtered and treated. High-pressure tubing transports pressurized water to the cutting nozzle. Thick tubes of diameters ranging from 6 to 14 mm and of diameter ratio 1/5–1/10 are used. For severe pressure, which may exceed the yielding stress of the tube material, shrink fit tubes should be used. To machine complex contour, the nozzle is mounted in a robot arm supplemented to the machine. Although the reaction forces in most cases do not exceed 50 N, the pressure waves due to the operation of the on–off valves impart vibration to the robot joints and consequently impair the machining accuracy. The cutting station must be equipped by a catcher, which acts as a reservoir for collecting the machining debris entrained in the jet. Moreover, it absorbs the rest energy after cutting, which is estimated by 90% of the total jet energy. It reduces the noise levels (105 dB) associated with the reduction of the water jet from mach 3 to subsonic levels.

Process Capabilities: The MRR, accuracy, and surface quality are influenced by the WP material and the machining parameters. Brittle materials fracture, while ductile ones are cut well. A material thickness ranges from 0.8 to 25 mm or more. Table 17.2 illustrates the cutting rates for different material thicknesses. The quality of cutting improves at higher pressures and lower traverse speeds. Under such conditions, greater thicknesses can be cut.

Advantages and Disadvantages of WJM: Advantages of WJM include

- Water is cheap, nontoxic, and can be easily disposed of and recirculated.
- The process requires a limited volume of water (100–200 L/h).
- No thermal degrading of the work material, as the process does not generate heat. For this reason, WJM is the process best suited for explosive environments.
- It is ideal for cutting asbestos, beryllium, and fiber-reinforced plastics (FRP), because the process provides a dustless atmosphere (environmentally safe).
- The process provides clean and sharp cuts that are free from burrs.
- Starting holes are not needed to perform the cut.
- Wetting of the WP material is minimal.
- Noise is minimized, because the power unit can be kept away from the cutting station.
- WJM is able to machine both brittle and ductile materials.
- The WP is subjected to a limited mechanical stress, because the force exerted by the jet does not generally exceed 50 N.

TABLE 17.2
Traverse Speeds and Thicknesses of Various Materials Cut by Water Jet

Material	Thickness (mm)	Traverse Speed (mm/min)
Leather	2.2	20
Vinyl chloride	3.0	0.5
Polyester	2.0	150
Kevlar	3.0	3.0
Graphite	2.3	5.0
Gypsum board	10.0	0.6
Corrugated board	7.0	200
Pulp sheet	2.0	120
Plywood	6.0	1.0

Source: Tlusty, G., 2000, *Manufacturing Processes and Equipment*, Prentice-Hall, Upper Saddle River, NJ.

Disadvantages of WJM include

- WJM is unsafe in operation if safety precautions are not strictly followed.
- The process is characterized by high production cost due to the high capital cost of the machine and the need for highly qualified operators.

Applications of WJM: WJM is used in many industrial applications comprising the following.

1. Cutting of metals and composites applied in aerospace industries.
2. Underwater cutting and ship-building industries.
3. Cutting of rocks, granite, and marble.
4. It is ideal in cutting soft materials such as wood, paper, cloth, leather, rubber, and plastics.
5. Slicing and processing of frozen foods, backed foods, and meat. In such cases, alcohol, glycerin, and cooking oils are used as alternative cutting fluids.
6. WJM is also used in cleaning, polishing, and degreasing of surfaces, removal of nuclear contaminations, cleaning of tubes and castings, surface preparation for inspection purposes, surface strengthening, and deburring.

17.3.3 Abrasive Water Jet Machining

Abrasive water jet machining (AWJM) is a hybrid process (HP) because it is an integration of AJM and WJM processes. The addition of abrasives to the water jet increases the range of materials that can be cut with a water jet and maximizes the MRR of this HP. The MRR is based therefore on using the kinetic energies of the abrasives and water in the jet. AWJM process is capable to machine both soft and hard materials at high speeds as compared with those realized by WJM. Moreover, the cuts performed by AWJM have better edge and surface qualities.

AWJM uses a comparatively lower water pressure than that of WJM (about 80%) to accelerate the AWJM. The mixing ratio of abrasive to water in the jet is about 3/7 by volume. Abrasives (garnet, sand, Al_2O_3, and so on) of a grain size 10–180 µm are often used. As previously mentioned, apart from its capability to machine soft and hard materials at very high speeds, AWJM process has the same advantages of WJM. However, it has the following limitations.

- Owing to the existence of the abrasives in the jet, there is an excessive wear in the machine and its elements.
- The process is not environmentally safe as compared to WJM.

The AWJM process has many fields of application such as

- Cutting of metallic materials such as Cu, Al, Pb, Mo, Ti, W, and steel (Figure 17.5a)
- Cutting carbides and ceramics
- Cutting concrete, marble, and granite (Figure 17.5b)
- Cutting plastics and asbestos (Figure 17.5c)
- Cutting composites such as FRP and sandwiched Ti-honeycomb without burr formation
- Cutting of acrylic and glass

In the field of machining technology, the AWJM has two promising applications that include milling of flat surfaces and turning of cylindrical shapes, assisted by AWJ. The process is also applicable

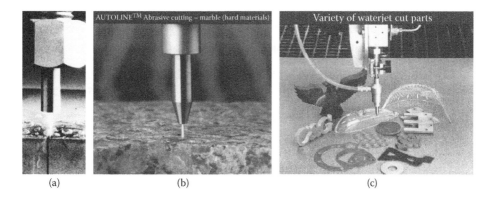

(a)   (b)   (c)

FIGURE 17.5 Cutting with AWJ: (a) 25-mm-thick carbon steel, (b) marble, and (c) plastic and asbestos. (From (a) ESAP Automation, Andover, and (b, c) Ingersoll-Rand. With permission.)

in deburring (AWJD), sharpening of grinding wheels, and surface strengthening to increase the fatigue strength.

Equipment of AWJM: The equipment of AWJM does not differ greatly from that of the basic WJM. So it is composed of the same working stations. The cutting station is provided with a jet former instead of a nozzle. In the jet former, the pressure energy of the water is first converted into kinetic energy, which, in turn, is partially converted into kinetic energy of the abrasive particles. Figure 17.6a illustrates a jet former of Ingersoll-Rand, while Figure 17.6b illustrates a sectional view of the same jet former. At the end of the high-pressure tubing an orifice is installed; it commonly consists of a hexagonal–rhomboidal sapphire Al_2O_3, a ruby or diamond having a hole of 0.08–0.8 mm inner diameter.

The high-pressure water is expelled through the orifice, and pure water jet is formed and directed into the mixing chamber. Through the interaction of the pure water jet and the surrounding air, a vacuum is created in the mixing chamber causing airflow from outside through the abrasive channels

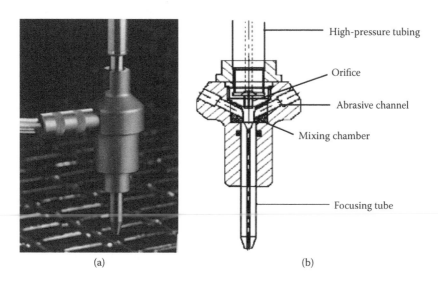

(a)   (b)

FIGURE 17.6 (a) Jet former, and (b) cross section in mixing chamber. (From Ingersoll-Rand. With permission.)

TABLE 17.3
Traverse Velocity (mm/min) When Machining Different Materials by AWJM

Materials	Material Thickness				
	6 mm	15 mm	19 mm	25 mm	50 mm
Titanium	250	150	100	50	16
Aluminum	250	150	100	50	16
Fiber reinforced plastic	500	280	130	75	25
Stainless steel	200	90	60	40	15
Glass	2000	1000	700	500	150

Source: Youssef, H. A., *Non-Traditional Machining Processes—Theory and Practice* (in Arabic), 1st ed., El-Fath Press, Alexandria, Egypt, 2005.

to the mixing chamber. In the mixing chamber, the jet loses its coherency; therefore a focusing tube (Figure 17.6b) is installed below the mixing chamber to restore the coherency of the AWJ. The resulting diameter of the AWJ is nearly equal to the focusing tube diameter.

Process Capabilities: The typical machining variables of the AWJM process include

- Water pressure
- Water nozzle diameter
- Geometry of focusing tube (length and diameter)
- Stand-off distance
- Size and type of abrasive grits
- Abrasive/water ratio
- Hardness and strength of the WP material
- Type of WP material (metallic, nonmetallic, or composite)

When machining glass by AWJ, a cutting rate of about $16–20$ mm^3/min is achieved. An AWJ cuts through 360 mm thick slabs of concrete or 76 mm thick tool steel plates at a traverse speed of 38 mm/min in a single pass. When cutting steel plates (or metallic materials) the surface roughness R_t ranges from 3.8 to 6.4 µm, while tolerances of ±130 µm are obtainable. Sand and garnet are frequently used as abrasive materials. However, garnet is preferred because it is 30% more effective than sand. A carrier liquid consisting of water with anticorrosive additives contributes to higher acceleration of abrasives with a consequent higher abrasive speed and increased MRR. The penetration depth increases with increasing water pressure and decreasing traverse velocity. The SOD has an important effect on the MRR and the achieved accuracy. It attains values between 0.5 and 5 mm. The smallest value realizes higher accuracy and smallest kerf width, whereas the largest value realizes the maximum MRR. Beyond 5 mm, the jet loses gradually its cutting capability until it reaches 50–80 mm, at which the jet is used efficiently in surface cleaning and peening.

Table 17.3 illustrates the traverse velocities when cutting different materials of different thicknesses using AWJ. Accordingly, it can be depicted that

1. Pure metals (Ti, Al) have the same machinability.
2. Glass is cut at 8–10 times faster than metals and alloys.

Surface roughness depends on the WP material, grit size, and type of abrasives. A material with a high removal rate produces large surface roughness. For this reason, fine grains are used for machining soft metals to obtain the same roughness as hard ones. Additionally, the larger the abrasive/water ratio, the higher will be the MRR. In the domain of composites, WJM process is particularly good as the cutting rates are considerably high, and it does not delaminate the layered material.

17.4 ULTRASONIC MACHINING

Ultrasonic machining (USM) is an economically viable operation, by which a hole or a cavity can be pierced in hard and brittle materials, whether electric conductive or not, using an axially oscillating tool. The tool oscillates with small amplitude of 10–50 μm at high frequencies of 18–40 kHz to avoid unnecessary noise (the audio threshold of the human ear is 16 kHz). During tool oscillation, abrasive slurry (B_4C and SiC) is continuously fed into the working gap between the oscillating tool and the stationary WP. The abrasive particles are therefore hammered by the tool into the WP surface, and consequently they abrade the WP into a conjugate image of the tool form. Moreover, the tool imposes a static force ranging from 1 N to some kilograms depending on the size of the tool tip (Figure 17.7). The static pressure is necessary to sustain the tool feed during machining. The process productivity is realized by the large number of impacts per unit time (frequency), whereas the accuracy is achieved by the small oscillation amplitude employed. The tool tip, usually made of relatively soft material, is also subjected to an abrasion action caused by the abrasives; thus it suffers from wear, which may affect the accuracy of the machined holes and cavities. Owing to the fact that the tool oscillates and moves axially, USM is not limited to the production of circular holes. The tool can be made to the shape required, and hence extremely complicated shapes can be produced in hard materials. The process is characterized by the absence of any deleterious or thermal effects on the metallic structure of the WP. Outside the machining domain, US techniques

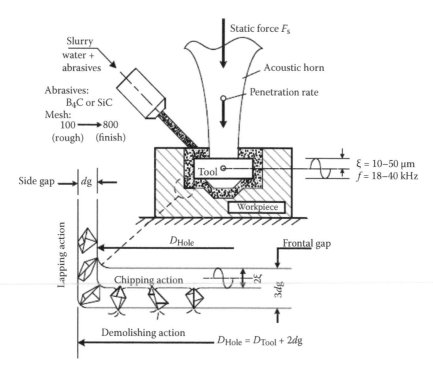

FIGURE 17.7 Characteristics of USM process.

are applied in nondestructive testing (NDT), welding, and surface cleaning, as well as diagnostic and medical applications. However, USM process is hampered by the following disadvantages.

- USM is not capable of machining holes and cavities with a lateral extension of more than 25–30 mm with a limited depth of cut.
- The tool suffers excessive frontal and side wear when machining hard materials such as steels and carbides. The side wear deteriorates the accuracy of holes and cavities.
- Every job needs a special high-cost tool, which adds to the machining cost.
- When machining through holes, the WP should be supported by a pad of machinable material to prevent breaking out.
- In case of blind holes, the designer should not allow sharp corners, because these cannot be produced by USM.
- The abrasive slurry should be regularly changed to get rid of worn abrasives, which means additional cost.

Typical applications of USM are as follows:

- Forming dies in hardened steel and sintered carbides
- Wire drawing dies, cutting nozzles for jet machining applications in sapphire, and sintered carbides
- Slicing hard brittle materials such as glass, ceramics, and carbides
- Coining and engraving applications
- Boring, sinking, blanking, and trepanning

17.4.1 USM Equipment

Figure 17.8 shows, schematically, the main elements of the equipment, which consists of the oscillating system, the tool feed mechanism, and the slurry system.

17.4.1.1 Oscillating System

The oscillating system includes the transducer contained in the acoustic head, the primary acoustic horn, and the secondary acoustic horn.

1. **Transducer and magnetostriction effect:** The transducer transforms electrical energy to mechanical energy in the form of oscillations. Magnetostrictive transducers are generally employed in USM, but piezoelectric ones may also be used. The magnetostriction effect was first discovered by Joule in 1874. According to this effect, in the presence of an applied magnetic field, ferromagnetic metals, and alloys change in length. An electric signal of US frequency f_r is fed to a coil that is wrapped around a stack made of magnetostrive material (iron–nickel alloy). This stack is made of laminates to minimize eddy current and hysteresis losses; moreover, it must be cooled to dissipate the generated heat (Figure 17.8a). The alternating magnetic field produced by the HF-ac generator causes the stack to expand and contract at the same frequency.

   To achieve the maximum magnetostriction effect, the HF-ac current i must be superimposed on an appropriate dc premagnetizing current I_p that must be exactly adjusted to attain an optimum working point. This point corresponds to the inflection point ($d^2\varepsilon/dI^2 = 0$) of the magnetostriction curve (Figure 17.8b). The premagnetizing direct current has the following functions.
   - When precisely adjusted, it provides the maximum magnetostriction effect (maximum oscillating amplitudes).
   - It prevents the frequency doubling phenomenon.

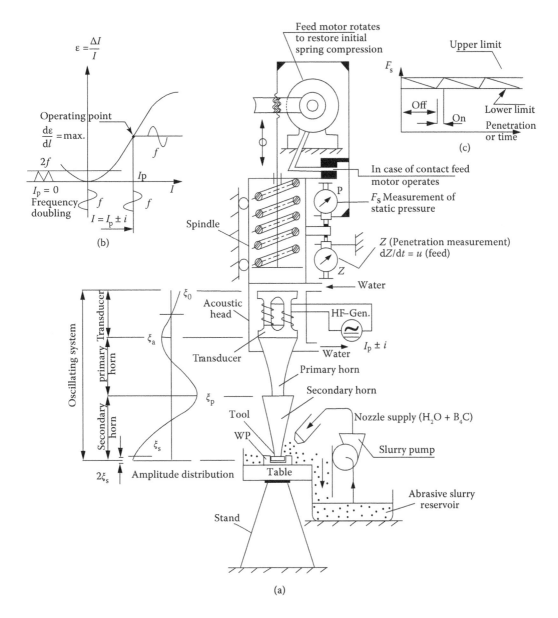

FIGURE 17.8 Schematic of vertical USM equipment. (From Youssef, H., and El-Hofy, H., *Machining Technology, Machine Tools and Operations*, CRC Press, Boca Raton, FL, 2008. With permission.)

If the frequency of the ac signal, and hence that of the magnetic field, is tuned to be the same as the natural frequency of the transducer (and the whole oscillating system), so that it will be at mechanical resonance, then the resulting oscillation amplitude become quite large and the exciting power attains its minimum value. The required resonance condition is realized if the transducer length is l, which equals to half of the wave length 1 (or a positive integer number n of it)

$$\ell = \frac{n}{2}\lambda = \frac{\lambda}{2} \quad (if\ n = 1)$$ (17.1)

and

$$\lambda = \frac{c}{f_r} = \frac{1}{f_r}\sqrt{\frac{E}{\rho}} \qquad c = \sqrt{\frac{E}{\rho}} \tag{17.2}$$

where

c = acoustic speed in magnetostrictive material (m/s)
f_r = resonant frequency (1/s)
E, ρ = Young's modulus (MPa), and density (kg/m^3) of magnetostrictive material.

Hence

$$\ell = \frac{1}{2f_r}\sqrt{\frac{E}{\rho}} \tag{17.3}$$

2. **Acoustic horns (mechanical amplifiers or concentrators):** The oscillation amplitude ξ_o as obtained from the magnetostrictive transducer does not exceed 5 μm, which is too small for effective removal rates. The amplitude at the tool should therefore be increased to practical limits of 40–50 μm by fitting one or more amplifiers into the output end of the transducer (Figure 17.8a). Moreover, the acoustic horn transmits the mechanical energy to the tool and concentrates the power on a small machining area. To attain resonance, the acoustic horns, like transducers, should be half-wavelength resonators whose terminals oscillate axially in an opposite direction relative to each other. The nodal points (points of zero amplitude $\xi_n = 0$) are a little displaced toward the upper end in the case of tapered concentrators. Figure 17.8a illustrates the amplitude distribution of the cascaded oscillating system along its longitudinal axis.

17.4.1.2   Tool Feeding Mechanism and Slurry System

The tool feeding mechanism performs the following functions:

- Brings the tool slowly to the WP
- Provides and sustains adequate static pressure during cutting
- Decreases the pressure before the end of cut to eliminate sudden fractures
- Overruns a small distance to ensure the required hole size at the exit
- Retracts the tool upward rapidly after machining

Figure 17.8c illustrates an automatic tool feeding mechanism, which operates precisely through the application of roller frictionless guides. When the oscillating system is freely suspended (no contact between the tool and the workpiece), the static pressure on the workpiece equals zero. When machining starts, the tool comes into contact with the WP and the spring in the machine spindle expands giving a measure for the static pressure. The static force is indicated by the dial gauge (F_s). As machining proceeds, the spring is compressed and the static force decreases (Figure 17.8c) until the contact switch is actuated, allowing the feed motor to rotate, and rapidly recovers the value of static force. The dial gauge (Z) indicates the tool displacement.

A centrifugal pump is used to supplement the abrasive slurry into the working zone (Figure 17.8a).

17.4.2 PROCESS CAPABILITIES

17.4.2.1 Material Removal Rate

The dominant factor involved in USM is the direct hammering of the abrasive grains, caused by the oscillating tool. Therefore the stock removal rate (MRR) depends mainly on the

- Work material
- Amplitude and frequency of tool oscillation
- Abrasive size and type
- Static pressure
- Abrasive concentration (mixing ratio) in the slurry

The efficiency of the slurry supplement in the working gap affects the MRR considerably. The conventional method of supplying the abrasive slurry is the nozzle supply system (Figure 17.7), in which the slurry is directly supplied at the oscillating tool. Pumping in or suction from the working gap through a central hole in the horn is found to be a more effective regime. The highest machining rates are realized when machining brittle materials such as glass, quartz, ceramics, and germanium, whereas the lowest machining rates are expected when machining hard and tempered steels and carbides.

USM is not applicable for soft and ductile materials, such as copper, lead, ductile steels, and plastics, which absorb energy by deformation. In practice, the oscillation amplitude is mainly selected with reference to the size of abrasive grits used. It should be selected to be approximately the same as the grit size. MRR increases with increasing oscillation amplitude (or abrasive grit size). The maximum amplitude value is governed by the maximum allowable strength of the material from which the acoustic horn is designed.

The removal rate increases with the frequency. However, the frequency is constant and exactly equal to the natural frequency of the system, and hence the frequency is not considered a factor. The specific removal rate increases with the applied static pressure. It attains a maximum value, after which it decreases with a further increase of the static pressure. Two types of abrasives are commonly used in USM, these are B_4C and SiC. B_4C is more expensive; however, it is economically recommended for USM owing to its increased cutting ability and resistance to abrasion. Moreover, grits of B_4C have less specific gravity, and hence more capability to be suspended in the slurry as compared with SiC. The maximum SRR is achieved if a slurry mixing ratio (abrasives/water) of 40% by volume is used.

17.4.2.2 Accuracy and Surface Quality

Factors affecting the accuracy and surface quality of holes and cavities produced ultrasonically are

- Work material
- Tool material and tool design
- Oscillation amplitude and grain size of abrasives
- Hole depth and machining time
- Cavitation effect

A main feature of the USM operation is that the abrasives start to cut for themselves a sideway between the tool and the WP (side gap) to move through it downward to the frontal gap, where the material removal takes place (Figure 17.7). Accordingly, the ultrasonically produced holes are somewhat larger than the tool used by a certain oversize (overcut), which approximately equals the size of the abrasive grains used. This oversize is affected more or less by the machining time, which, in turn, depends on the depth of hole, material of WP, tool design, as well as the other machining parameters. It should be emphasized that the hole accuracy does not mean the hole oversize. It means the repeatability of the oversize. Tolerances of ±25 μm can be easily obtained by USM. However, it is possible to obtain tolerances as close as ±25 μm if some provisions are taken. The wall roughness of the ultrasonically machined holes is mainly governed by the material to be machined, the oscillation amplitude, and the abrasives grain size. The surface quality deteriorates if cavitations conditions prevail. From this point of view, the use of a pumping regime is preferred over a suction one. Moreover, the rotation of the WP in case of circular holes may improve the surface quality and the hole roundness.

17.5   CHEMICAL MACHINING

Chemical machining (CHM) is the oldest nontraditional machining process, originally used for zincograph preparation. CHM depends on controlled chemical dissolution (CD) of the work material by contact with an etchant. Today, the process is mainly used to produce shallow cavities of intricate shapes in materials independent of their hardness or strength. CHM includes two main applications. These are chemical milling (CH milling), shown in Figure 17.9a, and photochemical machining (PCM), which is also called spray etching (Figure 17.9b).

17.5.1   Chemical Milling

This process has a special importance in airplane and aerospace industries, where it is used to reduce the thicknesses of plates enveloping the walls of rockets and airplanes, striving at improving stiffness to weight ratio (Figure 17.10). CH milling is used also in metal industries to thin out walls, webs, and ribs of parts that have been produced by forging, sheet metal forming, or casting. Furthermore, the process has many applications such as lamination of the decarburized layer from low-alloy steel forgings, removal of the recast layer from parts machined by electric discharge machining (EDM), removal of burrs from conventionally machined parts of complex shapes and producing burr-free printed circuit boards (PCBs). In CH milling, a special coating called maskant or resist protects areas from which the metal is not to be removed. The process consists of the following steps.

1. Preparing the WP surface by cleaning, mechanically or chemically, to provide good adhesion of the masking material.
2. Masking using a strippable mask that adheres to the surface and withstands chemical abrasion during etching.
3. Scribing of the mask using special templates to expose areas to be etched.
4. After etching, the work is rinsed and the mask is stripped.
5. The work is washed by deionized water then dried by nitrogen.

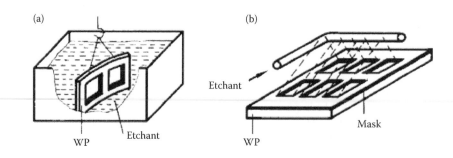

FIGURE 17.9   CHM processes: (a) CH milling and (b) PCM (spray etching).

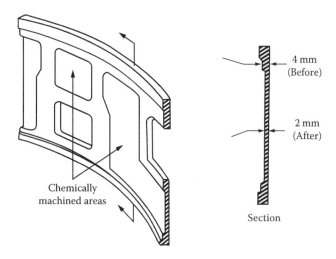

FIGURE 17.10 CH milling striving at improving stiffness to weight ratio of Al alloy plates for space vehicles. (Adapted from ASM International, *Advanced Materials and Processes*, Materials Park, OH, 1990.)

During CH milling (Figure 17.11), the etching depth is controlled by the time of immersion. The etchants used are very corrosive and therefore must be handled with adequate safety precautions. Vapors and gases produced from the chemical reaction must be controlled for health and environmental protection. A stirrer is used for the agitation of fluid. Typical reagent temperatures range from 37°C to 85°C, which should be controlled within ±5°C to attain uniform machining. Faster etching rates occur at higher etchant temperatures and concentrations.

When the mask is used, the machining action proceeds both inwardly from the mask opening and laterally beneath the mask, thus creating the etch factor (EF), which is the ratio of the undercut d_u to the depth of etch T_e (EF = d_u/T_e), as seen in Figure 17.11; this ratio must be considered when scribing the mask using templates. A typical EF of 1:1 occurs at a cut depth of 1.27 mm, while deeper cuts can reduce this ratio to 1:3.

Tooling for Chemical Milling: Tooling for CH milling is relatively inexpensive and simple. Four types of tools are required: maskants, etchants, scribing templates, and accessories.

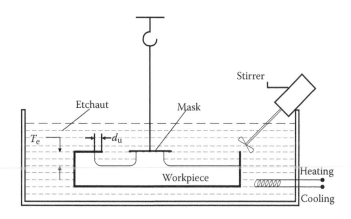

Etch factor (EF) = d_u/T_e

FIGURE 17.11 CH milling setup.

1. Maskants: Synthetic (polyvinyl chloride, polyethylene) or rubber base materials (butyl, acrylonitrile, or neoprene-rubber) are frequently used as maskants. They should possess the following properties:
   - Tough enough to withstand handling
   - Inert to the chemical reagent used
   - Able to withstand heat generated by etching
   - Adhere well to the work surface, scribe easily, and remove easily after etching
2. Etchants: Etchants are highly concentrated acidic or alkaline solutions maintained within a controlled range of chemical composition and temperature. They are capable of reacting with the WP material to produce a metallic salt that dissolves in the solution. Table 17.4 shows the machined material, the recommended etchant, its concentration and temperature, etch factor (EF), and the etch rate. When machining glass or germanium, the acidic solutions HF or HF + HNO_3 are used as etchants. When machining tungsten (W), it is recommended to use either of the following:
   - Alkaline solution, $K_3Fe(CN)_6$: NaOH = 20:3 (by volume)
   - Acidic solution. HF:HNO_3 = 30:70 (by volume)
     A suitable etchant should provide the following requirements:
   - Good surface finish of the workpiece
   - Uniformity of metal removal
   - Low cost and availability
   - Ability to regenerate, or readily neutralize and dispose of its waste products
   - Nontoxic
3. Scribing templates: These are used to define the areas for exposure to chemical dissolution (CD). The most common scribing method is to cut the mask with a sharp knife, followed by careful peeling. In such a step, the EF allowance must be included. Laser scribing of masks for CH milling of large surfaces is preferred.
4. Accessories: These include tanks, hooks, brackets, racks, and fixtures.

Advantages and Disadvantages of Chemical Milling: Advantages of chemical milling include

- Weight reduction is possible on complex contours that are difficult to machine conventionally.
- Several parts can be machined simultaneously.

TABLE 17.4
Machined Materials and Recommend Etchants in Chemical Machining

Etchant	Concentration	Temperature (K)	Etch Rate (µm/min)	Etch Factor (EF = d_u/T_e)	Metal to be Machined
$FeCl_3$	12°–18° Be[a]	320	20	1.5:1	Al alloys
HCl:HNO_3:H_2O	10:1:9	320	20–40	2:1	
$FeCl_3$	42° Be[a]	320	20	2:1	Cold rolled steel
HNO_3	10–15% (vol.)	320	40	1.5:1	
$FeCl_3$	42° Be[a]	320	40	2.5:1	Cu and Cu alloys
$CuCl_2$	35° Be[a]	325	10	3:1	
HNO_3	12–15% (vol.)	300–320	20–40	—	Magnesium
$FeCl_3$	42° Be[a]	320	10–20	(1–3):1	Nickel
$FeCl_3$	42° Be[a]	325	20	2:1	Stainless steel, tin
HNO_3	10–15°% (vol.)	320–325	20	—	Zinc

[a] Baume specific gravity scale.

- Simultaneous material removal from all surfaces improves productivity.
- No burr formation.
- No induced stresses, thus minimizing distortion of delicate parts.
- Low capital cost of equipment and minor tooling cost.
- Quick implementation of design changes.
- Less skilled operator is needed.

Disadvantages of chemical milling include

- Only shallow cuts are practical. Deep narrow cuts are difficult to produce.
- Handling and disposal of etchants can be troublesome and hazardous.
- Masking, scribing, and stripping are repetitive, time-consuming, and tedious.
- For best results, metallurgical homogeneous surfaces are required.
- Porous castings yield uneven etched surfaces.
- Welded zones frequently etch at rates that differ from the base metal.

Process Capabilities: Using fresh solutions, the etch rate ranges from 20 to 40 μm/min (Table 17.4). Etch rates are high for hard materials and low for soft ones. Generally, the high etch rate is accompanied by a low surface roughness and hence closer machining tolerances. Typically, surface roughness of 0.1–0.3 μm (R_a value), depending on the initial roughness, can be obtained. However, under special conditions, surface roughness of 0.023–0.05 μm becomes possible.

17.5.2 PHOTOCHEMICAL MACHINING

Photochemical machining (PCM) (spray etching) is a variation of CH milling, where the resistant mask is applied to the WP by photographic techniques. The two processes are quite similar because both use etchant to remove material by CD. CH milling is usually used to the three-dimensional parts originally formed by forging and casting of irregular shapes. However, PCM is a promising method for machining foils and sheets of thicknesses ranging from 0.013 to 1.5 mm to produce accurate and micro shapes. So, the PCM process becomes a realistic alternative to shearing and punching operations performed by mechanical presses in metal forming.

Additionally, a main difference between CH milling and PCM is that in CH milling, the depth of etch is controlled by the time the component is immersed in the etchant, whereas in PCM, the etch depth is controlled by the time the component is sprayed by fresh etchant though upper and lower nozzles, thus improving the performance of the PCM process by activating the etch rate and enhancing the quality. The etch rate of PCM is 5–10 times that achieved by CH milling. Of course, in PCM, highly developed expensive equipment is needed to provide high pressure/high temperature of the sprayed etchant. The PCM equipment is generally provided with a system of upper and lower nozzles, a unit for cleaning the worksheet by water then drying by hot air, a unit for measuring and controlling the etchant, and finally a unit for product inspection (Figure 17.12). In PCM, the following steps are carried out (Figure 17.13).

- The part shape that is considered as a primary image for the phototool is created by computer-aided design (CAD).
- Two photographic negatives, called artwork, are produced at the actual size of the work.
- The sheet metal is chemically cleaned, then coated with a highly sensitive photoresist. The work is allowed to dry. The photoresist adheres to the surface, protecting it during etching.

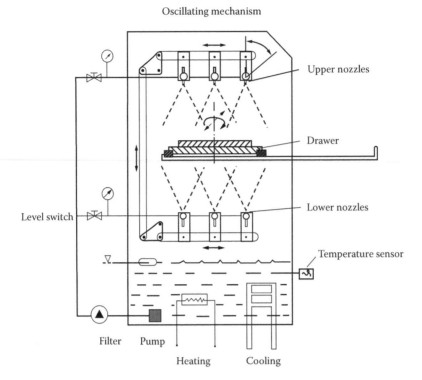

FIGURE 17.12 Schematic of PCM equipment. (From Visser A., Junker, M., and Weißinger, D., *Sprühätzen metallischer Werkstoffe*, 1st Auflage, Eugen G. Leuze Verlag, Germany, 1994. With permission.)

- After coating, the work is sandwiched between both negatives (artwork), then exposed in vacuum to an ultraviolet (UV) light. The coating is solidified in the unexposed areas and is removed from the exposed area by dissolving into developer.
- The worksheet is exposed once to a powerful water jet to remove the soft photoresist. The worksheet is then rinsed by deionized water and dried by nitrogen gas.
- The worksheet is then spray etched from the top and bottom.
- After etching, the hard photoresist is removed and the worksheet is rinsed to avoid any reactions with suspended etchant.

Applications of Photochemical Machining Process: Aluminum, copper, zinc, steels, stainless steels, lead, nickel, titanium, molybdenum, glass, germanium, carbides, ceramics, and some plastics are photochemically machined. The process also works well on springy materials, which are difficult to punch. The materials must be flat so that they can later be bent to shape and assembled into other components. Products made by PCM are generally found in the electronic, automotive, aerospace, telecommunication, computer, and other industries. Typical products such as PCB (Figure 17.14), fine screens, flat springs, and so on machined from foils are produced by PCM.

Advantages and Disadvantages of Photochemical Machining: In addition to the previously mentioned advantages of CH milling, PCM is characterized by the following.

- The accuracy and etch rate are considerably greater than those realized by EC milling.
- Because tooling is made by photographic techniques, they can be easily stored, and patterns can be reproduced easily. Lead times are correspondingly smaller.

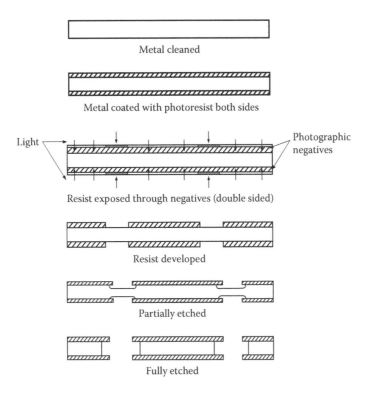

Metal cleaned

Metal coated with photoresist both sides

Light →

Photographic negatives

Resist exposed through negatives (double sided)

Resist developed

Partially etched

Fully etched

FIGURE 17.13 PCM steps. (From Youssef, H., and El-Hofy, H., *Machining Technology, Machine Tools and Operations*, CRC Press, Boca Raton, FL, 2008. With permission.)

Apart from the disadvantages of CH milling, PCM has also the following limitations:

- Requires highly skilled operator.
- Requires more expensive equipment.
- The costly machine should be protected from the corrosive action of etchants.

17.6 ELECTROCHEMICAL MACHINING

Electrochemical machining (ECM) is one of the most effective NTMPs where the metal removal is based on the anodic dissolution governed by Faraday's principle of electrolysis (1833). According to

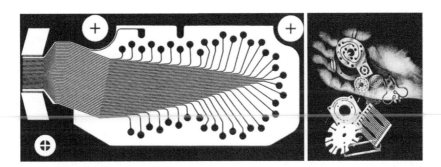

FIGURE 17.14 Typical products by PCM. (From Visser A., Junker, M., and Weißinger, D., *Sprühätzen metallischer Werkstoffe*, 1st Auflage, Eugen G. Leuze Verlag, Saulgau, Wuert, Germany, 1994, with permission, and courtesy of Chemical Corporation.)

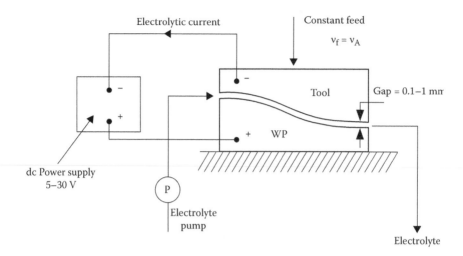

FIGURE 17.15 ECM process. (From Youssef, H., and El-Hofy, H., *Machining Technology, Machine Tools and Operations*, CRC Press, Boca Raton, FL, 2008. With permission.)

this principle, the anodic dissociation rate in the machining (depleting) cell is directly proportional to the dc electrolyzing current and the chemical equivalent ε of the anode material (ε = atomic weight/valence). In the machining cell, the WP is connected to the anode, while the tool is connected to the cathode of a dc source of 5–30 V. Both the tool and WP electrodes must be electrically conductive. They are separated by a gap of 0.1–1 mm length, into which an electrolyte (NaCl, KCl, and $NaNO_3$) is pumped rapidly to sweep away the reaction products (sludge) from the narrow machining gap. Depending on the gap thickness, a machining current of high density (20–800 A/cm^2) passes, causing a high anodic dissolution rate. The shape of the cavity formed in the WP is the female mating image of the tool shape. The tool advances axially toward the WP by means of a servomechanism at a constant feed rate, ranging from 0.5 to 10 mm/min (Figure 17.15). The anodic dissolution rate adjusts itself to match with the selected tool feed ($\nu_f = \nu_A$). This matching characteristic or the self-adjusting feature of ECM process controls and stabilizes its performance. Consequently, during machining, the gap thickness attains a constant value known as the equilibrium gap.

17.6.1 ELEMENTS OF THE ECM PROCESS

These are the tool, WP, the electrolyte.

1. Tool: The tool material should be machinable, stiff, and possess high corrosion resistance and good electric and thermal conductivity. ECM tools are usually made of copper, brass, and 316 stainless steel. Carbon steels are not recommended because of their low corrosive resistance. The tool is shaped to be not exactly a mirror image of the machined cavity. Its dimensions must be slightly different from the nominal dimensions of the cavity to allow for an overcut. The tool design must permit electrolyte flow at a rate sufficient to dissipate heat generated to eliminate boiling of the electrolyte in the interelectrode gap. To produce smooth surfaces on the WP, tool design must enable a uniform flow over the entire machining area. Ideally, flow should be laminar and free from eddies. On machining complex cavities, a good designer performs pilot tests under the same machining conditions. Accordingly, corrections to the tool form are carried out to realize the required accuracy and surface quality.
2. Workpiece: In ECM, there is no restriction on the WP whatsoever, except being able to conduct electricity. The machinability is dependent only on the chemical equivalent ε of the

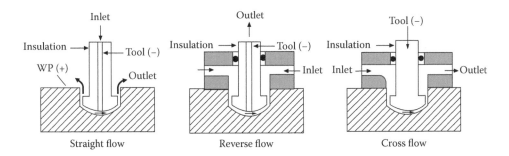

FIGURE 17.16 Methods of electrolyte feeding in ECM. (Adapted from El-Hofy, H., *Advanced Machining Processes, Nontraditional and Hybrid Processes*, McGraw-Hill, New York, 2005.)

WP. Carbon is passive in EC reactions; consequently, gray cast iron cannot be machined satisfactorily by ECM.

3. Electrolyte: Electrolytes are highly conductive solutions of inorganic salts (usually NaCl, KCl, and $NaNO_3$, or their mixtures) which meet multiple requirements. The main functions of the electrolytes are to
   - Complete the electric circuit between the tool and the WP
   - Allow desirable reactions to occur, and create conditions for anodic dissolution
   - Carry away heat generated during chemical reactions
   - Remove products of reaction (sludge) from the machining gap

Effective and efficient electrolytes should therefore have the following properties:

- High electrical conductivity to ensure high current density
- Low viscosity to ensure good flow conditions in the extremely narrow interelectrode gap
- High specific heat and thermal conductivity to be capable of removing the heat generated from the gap
- Resistance to the formation of a passive film on the WP surface
- High chemical stability
- High current efficiency and low throwing power
- Nontoxic and noncorrosive to the machine parts
- Inexpensive and available

Electrolytes also play an important role in dimensional control. Despite its lower current efficiency and its passivity, $NaNO_3$ is more preferable than NaCl when precise holes are to be machined, because it produces less hole oversizes. Moreover, $NaNO_3$ is a noncorrosive electrolyte. Several methods of supplying electrolytes to the gap are shown in Figure 17.16. The choice of the electrolyte supply method depends on the part geometry, required accuracy, and surface finish. Typical electrolyte conditions in the gap include a temperature of 90°–110°C, pressure between 10 and 20 bar, and a maximum electrolyte velocity of 25–50 m/s.

17.6.2 ECM EQUIPMENT

ECM equipment includes a dc power generator and an EC machine. EC machines equipped with power generators of current capacities ranging from 50 to 40,000 A are available on the market. The power sources supply constant voltages ranging from 5 to 30 V. They are generally characterized by a high power factor, high efficiency, and should be equipped with short-circuit protection within a small fraction of a second to prevent catastrophic short circuits across the electrodes.

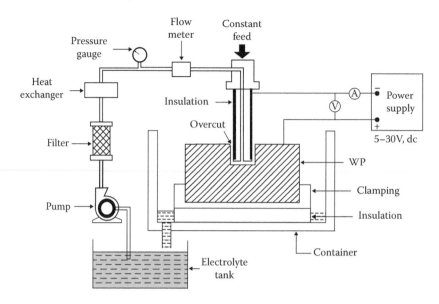

FIGURE 17.17   ECM setup. (Adapted from El-Hofy, H., *Advanced Machining Processes, Nontraditional and Hybrid Processes*, McGraw-Hill, New York, 2005.)

Figure 17.17 illustrates schematically a typical EC sinking machine. The machine must be rigid enough to withstand the hydrodynamic pressure of the electrolyte in the machining gap, which tends to separate the tool from WP. A servomechanism is necessary to control the tool movement in such a way that the material dissolution is balanced by the constant feed rate of the tool. The rate of current change is monitored and stops tool feeding when an abnormal rise in current is detected.

In contrast to conventional machine tools, EC machines are designed to stand up to corrosion attack by using nonmetallic materials. For high strength and rigidity, metals with nonmetallic coatings are recommended. To eliminate the danger of corrosion on other machinery, EC machines should be perfectly isolated in separate rooms in the workshop. The electrolyte feeding unit supplies electrolyte at a given rate, pressure, and temperature. Facilities for electrolyte filtration, temperature control, and sludge removal are also included.

17.6.3 PROCESS CAPABILITIES

In ECM, the machining rate is solely a function of the ion exchange rate, irrespective of the hardness of the work material. The process provides metal removal rates for steels in the order of 1.5–2 cm³/min/1000 A. Penetration rates up to 2.5 mm/min are routinely obtained when machining carbides and steels, either hardened or not. Table 17.5 shows the EC removal rate of most common metals, assuming an electrolyzing current of 1000 A and a current efficiency of 100%.

A well-known and unique characteristic of ECM, among all traditional and nontraditional processes, is that both the accuracy and surface quality improve when applying higher removal rates (i.e., higher current densities). A major problem of ECM is the overcut (side gap), which affects accuracy. Roughly speaking, the side gap is governed by a complex set of parameters, of which the type of electrolyte and the electrolyte flow are most crucial. A typical dimensional tolerance of ECM is ±0.13 mm; however, through proper control of the machining parameters,

TABLE 17.5
Removal Rate and Specific Removal Rate for Commonly Used Metals

Metal	ρ (g/cm³)	N Atomic Weight (g/mol)	n valance (N)	$\varepsilon = N/n$ (g)	Removal Rate (I = 1000 A, η = 100%) (g/min)	(cm³/min)	Specific RR (cm³/A min)
Aluminum (A1)	2.7	27	3	9.0	5.6	2.1	0.0021
Chromium (Cr)	7.2	52	2	26.0	16.2	2.3	0.0023
			3	17.3	10.8	1.5	0.0015
			6	8.7	5.4	0.8	0.0008
Copper (Cu)	9.0	64	1	64.0	39.5	4.4	0.0044
			2	32.0	19.7	2.2	0.0022
Gold (Au)	19.3	197	1	197.0	122.6	6.4	0.0064
			3	65.7	40.8	2.1	0.0021
Iron (Fe)	79	56	2	28.0	17.4	2.2	00022
			3	18.7	11.6	1.5	0.0015
Nickel (Ni)	8.9	59	2	29.5	16.2	2.1	0.0021
			3	19.7	12.2	1.4	0.0014
Titanium (Ti)	4.5	48	3	16.0	10.0	2.2	0.0022
			4	12.0	75.0	1.6	0.0016
Tungsten (W)	19.3	184	6	30.7	19.0	1.0	0.0010
			8	23.0	14.3	0.7	0.0007
Zinc (Zn)	7.2	65	2	32.5	200.0	20.9	0.0029

tight tolerance of ±0.025 mm can be achieved. It is difficult to machine internal radii smaller than 0.8 mm. A typical overcut of 0.5 mm and a taper of 1 μm/mm are possible. Typical surface roughness (R_a value) of ECM ranges from 0.2 to 1 μm is common, which decreases with increasing machining rate. The principal tooling cost is due to the preparation of the tool electrode, which is time-consuming and costly, requiring several cut-and-dry efforts, except for simple shapes. There is no tool wear and the process produces stress free surfaces. The capability to cut the entire cavity in one stroke makes the process very productive, but the complicated tool shape increases the process cost.

Advantages and Disadvantages of ECM: Advantages of the ECM include

- Three-dimensional surfaces with complicated profiles can be easily machined in a single operation, irrespective of the hardness and strength of the WP material.
- ECM offers a higher rate of metal removal as compared to traditional and nontraditional methods, especially when high machining currents are employed.
- There is no wear of the tool, which permits repeatable production.
- No thermal damage or HAZ.
- High surface quality and accuracy can be achieved at the highest MRR.
- The surfaces produced by ECM are burr-free and free from stresses.

Disadvantages of the ECM include

- Nonconductive materials cannot be machined.
- Inability to machine sharp interior corners or exterior edges of less than 0.2 mm radius.

- The machine and its accessories are subjected to corrosion and rust, especially when NaCl electrolyte is used; expensive electrolytes like $NaNO_3$ are less corrosive.
- The endurance limit of parts produced by ECM is lowered by about 10%–25%. In such a case, shot peening is recommended to restore the fatigue strength.
- Specific power consumption of ECM is considerably higher than required for TM.
- Cavitation channels may form, which deteriorates the surface quality.
- Pumping electrolytes at high pressures into the narrow gap gives rise to large hydrostatic forces acting on the tool and WP, which necessitates a rigid machine frame.
- The machined parts need to be cleaned and oiled immediately after machining.

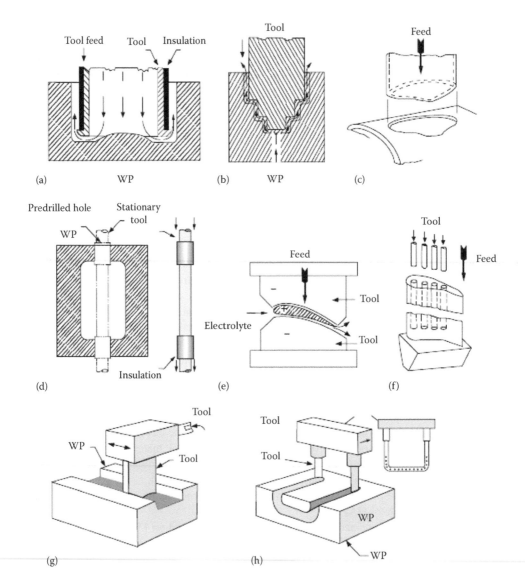

FIGURE 17.18 Typical ECM applications: (a) hole sinking with insulated tool, (b) EC sinking of stepped hole, (c) EC trepanning, (d) ECM of internal cavity by stationary tool electrode, (e) ECM of a turbine blade, (f) EC deep hole drilling, (g) EC surfacing, and (h) EC hogging. (From Youssef, H., and El-Hofy, H., *Machining Technology, Machine Tools and Operations*, CRC Press, Boca Raton, FL, 2008. With permission.)

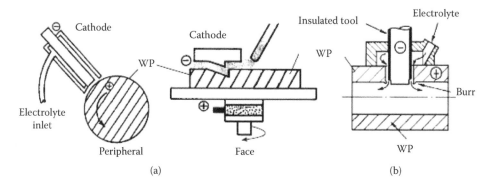

FIGURE 17.19 Other applications: (a) EC turning and (b) EC deburring.

- There is a danger of explosion if the hydrogen generated is not safely disposed.
- The tool and the WP may be damaged if arcing is initiated due to the contamination of oxides and debris in the machining gap.

Applications of ECM: ECM has been used in a wide variety of industrial applications ranging from cavity sinking to deburring. Modifications of this process are used for turning, slotting, trepanning, and profiling, in which the electrode becomes the cutting tool. Hybridization performed by integrating ECM with conventional finishing processes leads to the highly developed electrochemically assisted grinding, honing, and superfinishing processes. The process can handle a large variety of materials, limited only by their electric conductivity and not by their hardness or strength. The MRR is high, especially for difficult-to-machine alloys. Fragile parts that are otherwise not easily machinable can be shaped by ECM. The fact that there is no tool wear in this process is advantageous, as it has a positive impact on accuracy. Moreover, a large number of components can be machined without the tool having to be replaced. Hence ECM is well-suited for mass production of complex shapes in hard and extra-hard materials. The ability to machine high-strength and hardened steels has led to many cost-saving applications where other processes are impractical. In the sector of electric conductive materials, both ECM and EDM processes compete with each other. However, ECM has a special attraction owing to the absence of thermal stresses and HAZ. This characteristic is useful in the manufacturing of dies in hardened and tempered steel blocks (Figure 17.18). Figure 17.19 illustrates other applications such as EC-turning and EC-deburring.

17.7 ELECTROCHEMICAL GRINDING

Electrochemical grinding (ECG) is one of the most important HPs, in which metal is removed by a combination of EC dissolution and mechanical abrasion. The equipment used in ECG (Figure 17.20) is similar to a traditional grinding machine, except that the GW is metal bonded with diamond or borazon (CBN) abrasives. The wheel is the negative electrode that is connected to the dc supply through the spindle insulated from the machine frame, whereas the work is connected to the positive terminal of the power supply. A flow of electrolyte, usually $NaNO_3$ is provided in the direction of wheel rotation for achieving the ECM phase of the operation. The wheel rotates at a surface speed of 25–35 m/s.

The abrasives in the wheel are always nonconductive, and thus they act as an insulating spacer, maintaining a separation (electrolytic gap) of 12–80 μm between electrodes. The abrasives also mechanically remove the reaction residue from the working gap and cut chips from the WP. Therefore the removal of WP material occurs owing to the flow of current (100–300 A/cm²) through

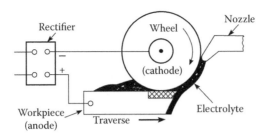

FIGURE 17.20 Electrochemical grinding.

the gap and the rate of decomposition is enhanced by the grinding action of the abrasive grains. With proper operation, typically, 95% of material removal is due to electrolytic dissolution, and only about 5% is due to the abrasion effect of the grinding wheel. Consequently, the wear of the wheel is very low, thus eliminating or considerably reducing the need for grinding wheel redressing, which reduces the sharpening costs approximately by 60%.

The lack of heat damage, distortions, burrs, and residual stresses in ECG is very advantageous, particularly when coupled with high MRR, in addition to far less wear of the grinding wheel. That is why the process has been applied most successfully in sharpening cutting tools, die inserts, and punches made of hardened high-strength steel alloys. The process offers the following specific advantages over traditional diamond wheel grinding:

- Increased MRR due to the added EC effect
- Reduced tool wear and sharpening costs
- Less risk of thermal damage and distortion
- Absence of burrs
- Reduced wheel pressure, which improves accuracy

However, ECG has the following disadvantages:

- Higher capital cost of the equipment.
- Limited to electrically conductive materials.
- Hazard due to corrosive nature of electrolyte. For that reason $NaNO_3$ of limited corrosive nature is used.
- Necessity of electrolyte filtering and disposal.

Some typical applications of ECG are sharpening carbide tools and grinding of fragile parts.

ECG machines are now available with NCs, thus improving process accuracy, repeatability, and increased productivity. In general, metal removal rates of an ECG process of the order 1 cm^3/min/100 A are realized. Surface roughness in the range of 0.2–0.6 μm can be easily obtained. Generally, the higher the hardness of metal or alloy, the better is the finish. Typical tolerances achieved by ECG are of the order of ±10 μm.

17.8 ELECTRIC DISCHARGE MACHINING

Electric discharge machining (EDM) comprises the following three alternatives.

17.8.1 Electric Discharge Sinking (Conventional EDM)

Of all the NTMPs, none has gained greater industry-wide acceptance than EDM. It is well known that when two current-conducting electrodes are allowed to touch each other, an arc is produced.

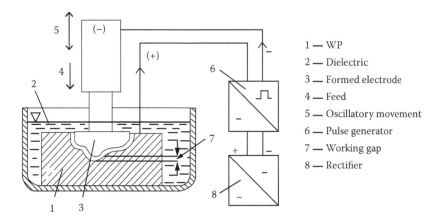

FIGURE 17.21 Concept of EDM. (From Düniß, W., Neumann, M., and Schwartz, H., 1979, *Trennen-Spanen und Abtragen*, VEB-Verlag Technik, Berlin. With permission.)

Although this phenomenon has been detected by Priestly since 1790, it was not until the 1940s that a machining process based on this principle was developed by Lazarenkos in Russia. The principle of EDM (also called spark erosion machining) is based on erosion of metals by spark discharges between a shaped tool electrode (usually negative) and the conductive WP (usually positive).

The tool and the WP are separated by a small gap of 10–500 μm. Both are submerged or flooded with electrically nonconducting dielectric fluid. When a potential difference between the tool and the WP is sufficiently high, the dielectric in the gap is partially ionized, so that a transient spark discharge ignites though the fluid, at the closest points between the electrodes. Each spark of thermal power concentration, typically 10^8 W/mm², is capable of melting or vaporizing very small amounts of metal from the WP and the tool (Figure 17.21). A part of the total energy is absorbed by the tool electrode, yielding some tool wear, which can be reduced to 1% or less if adequate machining conditions are carefully selected.

The instantaneous vaporization of the dielectric produces a high-pressure bubble that expands radially. The discharge ceases with the interruption of the current, and the metal is ejected, leaving tiny pits or craters in the WP and metal globules suspended in the dielectric (Figure 17.22).

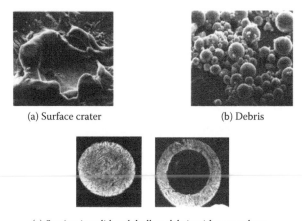

(a) Surface crater   (b) Debris

(c) Section in solid and hollow debris with trapped gas

FIGURE 17.22 SEM surface crater and debris produced by EDM. (Adapted from König, W., *Fertigungsverfahren*—Band 3, Abtragen, VDI Verlag, Duesseldorf, Germany, 1990.)

A sludge of black carbon particles is formed from hydrocarbons of the dielectric produced in the gap and expelled by the explosive energy of the discharge; it remains in suspension until removed by filtering. Immediately following the discharge, the dielectric surrounding the channel deionizes and, once again, becomes effective as an insulator. The capacitor discharge is repeated at rates between 0.5 and 500 kHz, at a voltage between 50 and 380 V and currents from 0.1 to 500 A.

17.8.1.1 Electric Discharge Sinking Machine

A typical setup of ED-sinking machine is illustrated in Figure 17.23. The gap between the electrode and the WP is critical, thus the down feed should be controlled by a servomechanism that automatically maintains a constant gap. A short circuit across the gap causes the servo to reverse the motion until proper control is restored. The WP is clamped within the tank containing the dielectric fluid that is pumped and filtered. At the beginning, the fresh supply of dielectric is clean and has a higher insulation strength than a contaminated supply. When spark discharges commence, debris are created and the dielectric insulation strength is diminished by particles. If too many particles are allowed to remain, a bridge is formed, resulting in arcing (not sparking) across the gap, which causes damage to the tool and WP. Therefore the contamination in the gap must be controlled by efficient flushing. Common flushing techniques are

- Injection flushing, in which a slight taper is produced on the sides of the cavity due to lateral discharges as debris pass up the side of the tool.
- Suction flushing, through which the side taper is avoided.
- Side flushing, in which a slight taper is produced on the side of the cavity at the outlet (downstream) of the dielectric.

17.8.1.2 Dielectric Fluids

The dielectric fluids have many important functions. They act as an insulator between tool and work. Moreover, they act as spark conductor for the plasma channel and provide a cooling and

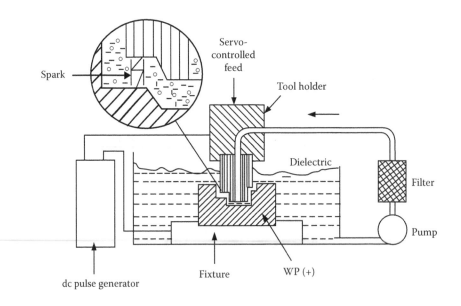

FIGURE 17.23 ED sinking setup.

flushing medium. The fluid must ionize to provide a channel for the spark and deionize quickly to become an insulator. A good dielectric has the following properties:

- Low viscosity to ensure effective flushing
- High flash point
- High latent heat
- A suitable dielectric strength, for example, 180 V/25 µm
- Rapid ionization at a potential of 40–400 V followed by rapid deionization
- Nontoxic
- Noncorrosive
- Nonexpensive

The most common dielectric fluids are hydrocarbons, although kerosene and distilled and deionized water may be used in specialized applications. Polar compounds such as glycerine water (90:10) with triethylene oil as an additive have proved to improve the MRR and decrease the tool wear as compared to kerosene.

17.8.1.3 Spark Circuits

The ED machine is equipped with a spark-generating circuit that can be controlled to provide optimum conditions for a particular application. This generator should supply voltage adequate to initiate and maintain the discharging process and provides the necessary control over the process parameters such as current intensity, frequency, and cycle times of discharge. The cycle time ranges from 2 to 1600 µs. Two main types of generators are applicable for this purpose. These are the resistance–capacitance generator (RC circuit) and the transistorized pulse generator.

1. RC circuit (relaxation): It is also called the Lazarenko circuit, which is basically a relaxation oscillator. It is a simple, reliable, rigid, and low-cost power source that is ordinarily used with copper or brass electrodes. It provides a fine surface texture of $R_a = 0.25$ µm. However, the machining rate is slow, because the time required to charge the capacitors prevents the use of high frequencies. The reversed polarity encountered in the relaxation circuit leads to an additional tool wear. In the circuit illustrated in Figure 17.24a, the capacitor C is charged from the dc power supply ($V_o = 200$–400 V) via the resistor R which determines the charging rate. Capacitor voltage V_{cap} increases exponentially (Figure 17.24b)

$$V_{cap} = V_o (1 - e^{-t/RC}) \tag{17.4}$$

where
t = charging time (s)
RC = time constant = resistance (Ω) × capacitance (F)

When V_s attains the breakdown voltage in the working gap, C discharges more quickly (t_d) across it than it can recharge C (t_c) via the resistor R, thus causing eroding the tool (wear) and WP (material removal).

$$t_d \approx 0.1 \, t_c \tag{17.5}$$

For maximum production rate,

$$V_s = 0.73 \, V_o \tag{17.6}$$

and the energy (J) per individual discharge is given by

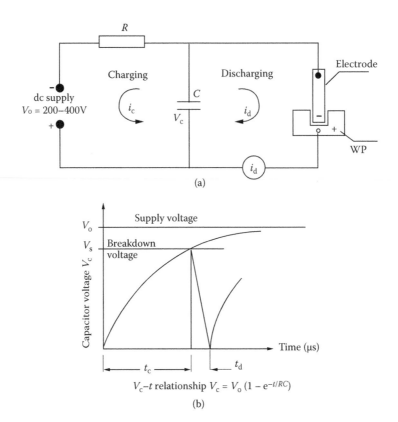

FIGURE 17.24 (a) RC circuit and (b) capacitor voltage–charging time exponential relationship. (From Youssef, H., and El-Hofy, H., *Machining Technology, Machine Tools and Operations*, CRC Press, Boca Raton, FL, 2008. With permission.)

$$E_d = \tfrac{1}{2}CV_s^2 \tag{17.7}$$

The increase of V_o, V_s, and C leads to an increase of machining rate and consequently to a poor surface quality.

2. Transistorized pulse generator circuits: The adoption of the transistorized pulse generators in the 1960s allowed the process parameters (frequency and energy of discharges) to be varied with a greater degree of control, in which charging takes only a small portion of the cycle. Furthermore, the voltage of these machines is reduced to 60–80 V range, permitting discharge that is characterized by lower current pulses of a square profile. This results in shallower and wider craters, which means better surface texture. Alternatively, when required, they provide high MRRs at the expense of surface quality by permitting high discharge currents. Moreover, this type of generator provides considerably lower electrode wear as compared to simpler *RC* circuits. In the simple form of the transistorized pulse generators, the parameters are selected and preadjusted according to the machining duty. The selected parameters remain constant, that is, not influenced by the variation of working conditions in the gap during machining.

17.8.1.4 Tool Electrodes

In ED sinking, tool electrodes are often the most expensive part of the EDM operation. Material fabrication, wear, and redressing costs must be carefully weighed to determine the best electrode

TABLE 17.6
Polarity for Most Common Electrode/WP Material Combinations

WP Material	Electrode Material		
	Graphite	Cu	Cu–W
Steel	SR	S	S
Copper	R	R	R
Cemented carbides	R	SR	SR
Al	S	S	S
Ni-base alloys	SR	S	S

Source: ASM International, *Machining, vol. 16, Metals Handbook.* Materials Park, OH, 1989.
Note: S is straight polarity (WP-positive electrode); R is reverse polarity (WP-negative electrode).

material and EDM machine setup. The ideal electrode material should have high electric conductivity, high melting point, be easy to fabricate, and strong enough to stand up EDM without deformation. Most electrodes for EDM are usually made of graphite, although brass, Cu, or Cu/W alloys may be used. These electrodes are shaped by forming, casting, and powder metallurgy, or frequently by machining. EDM tool wear is an important factor, as it affects the dimensional and form accuracy. It is related to the melting point of the tool material involved. In this regard the higher the melting point, the lower is the wear rate. Consequently, graphite electrodes have the highest wear resistance, as graphite has the highest melting point of any known material (3600°C); moreover, it is low in cost and readily fabricated. Tungsten (3400°C) and W alloys are next in melting temperature, followed by molybdenum (2600°C); however, these metals are expensive and difficult to fabricate. The tool wear can be minimized by reversing the polarity, which depends on the tool/WP combination. Table 17.6 illustrates the recommended polarity for various electrode/WP material combinations. The wear may reach a zero value during the so-called no-wear EDM process. Work material machinable by no-wear EDM can be steels, stellites, Ni-base alloys, and aluminum. However, no-wear EDM is not recommended for machining carbides.

No-wear EDM requires pulse generators and equipment capable of attaining the following conditions: reverse polarity, low pulse frequency (0.4–20 kHz), graphite and copper electrodes, high duty cycle > 90%, high intensity i_d, smooth servomechanism, $V_o < 80$ V, dielectric temperature < 40°C, dielectric with debris, no capacitance across the gap, and no inductance in series.

17.8.1.5 Process Capabilities

EDM is a slow process compared to conventional methods. It produces matte and pitted surfaces composed of small craters, which are characterized by a nondirectional, randomly distributed nature due to succession of individual sparks of the process. The metal removal rates usually range from 0.1 to 600 mm³/min. High removal rates produce a rough finish, having a molten and recast structure with poor surface integrity (SI) and low fatigue strength. The finish cuts are made at low removal rates, and the recast layer formed during rough cuts is removed later by finishing EDM operations. The MRR depends not only on the WP material but also on the machining variables, such as pulse conditions (voltage, current, and duration), electrode material and polarity, and the dielectric.

In EDM, the surface finish and MRR vary widely as a function of the spark frequency, voltage, current, and other parameters. New techniques use oscillating electrodes to provide very fine surface quality. Alternatively, bad surfaces and surface defects characterize EDM using graphite electrodes. Typically, overcut values vary from 10 to 300 µm, depending on the breakdown voltage, and the size of debris flowing in the side gap. In this respect, suction flushing is preferred, because the debris are not drawn past the side gap, and thus lateral sparking is minimized, leading to a smaller overcut and side taper. Typical taper varies from 1 to 5 μm/mm per side, depending on the machining conditions, and especially on the flushing technique used. The minimum corner radius, more or less equals the size of the overcut. Tolerance of ±50 μm can be easily achieved; however, with close control, tolerance of ±5 to ±10 μm can be obtained.

Advantages and Disadvantages of EDM: EDM is characterized by the following advantages.

- It is applicable to alloys and carbides irrespective of their hardness and toughness.
- Because there is no contact between tool and WP, very delicate work can be machined.
- The process is widely used to produce accurate cavities of intricate shapes in extra-hard materials.
- EDM electrodes can be accurately produced from machinable materials.

However, the process is hampered by the following limitations.

- It cannot be used if the WP material is a bad electric conductor.
- On machining materials, the process produces HAZ, which is characterized by hairline cracks and a thin, hard recast layer.
- EDM cannot produce sharp corners and edges.
- EDM has high specific energy.

Applications of EDM: EDM has become an indispensable process in modern manufacturing. It is used in producing die cavities for automotive body components, connecting rods, and various intricate shapes to a high degree of accuracy. About 80%–90% of EDM work is the manufacture of tool and die sets for the production of castings, forgings, stampings, and extrusions. Micromachining of holes, slots, texturing, and milling are also typical applications of the process.

17.8.2 Electrical Discharge Milling

The conventional electrical discharge (ED) sinking, discussed in the previous section, requires a preliminary phase for producing specially shaped electrodes. These electrodes are very expensive, as they are difficult to design and manufacture, and therefore they add more than 50% to the total machining cost of the product. Recently, a revolutionary breakthrough in the EDM realm has been achieved through a new ED milling technology that makes use of simple and cheap standard rotating pipe electrodes. In this process, three-dimensional cavities are machined by successive sweeps of the electrode down to the desired depth, while the NC automatically compensates by means of powerful algorithms the electrode's front wear to ensure product accuracy along the three axes. Therefore there is no need to manufacture specially shaped electrodes as in the case of conventional EDM, which means saving time and money. The theory of ED milling (also termed ED scanning) is shown in Figure 17.25. The thickness of the layer removed per path ranges from 0.1 mm to several millimeters on rough paths and from 1 to 100 μm on finish paths.

Advantages and Limitations of ED Milling: The advantages include the following.

- Design and manufacture of electrodes is totally omitted.
- Fine shapes can be readily produced.
- Electrode wear does not need to be considered.
- NC data can be directly generated from the EDM die data.
- Sharp edges and corners can be readily produced due to the excessive frontal wear.

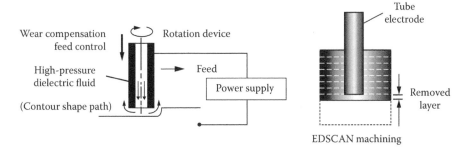

FIGURE 17.25 Theory of ED milling. (From Mitsubishi EDSCAN Technical Data, Hannover Exhibition, 1997. With permission.)

The disadvantages and limitations of ED milling are

- The removal rate may be less than that achieved by conventional EDM.
- If there is a large side taper (10° or more), it is difficult to maintain side accuracy.

Fields of Applications of ED Milling: ED milling technology is particularly applicable for machining cavities with or without taper, including three-dimensional shapes. It is used notably for making molds of parts for electrical and electronic industries, household appliances, and the automotive and aeronautical industries. Another technological breakthrough of ED milling is that the process has entered the domain of micromachining, where it is possible to produce fine and intricate shapes with sharp corners. The path of the two-dimensional shape is created by the NC system built into the machine beforehand to allow the target shape to be stored. Layered machining is then carried out by executing the NC path program several times, until the required depth is achieved, as shown in Figure 17.26. ED milling technology was announced by Mitsubishi and Charmilles at the micromachine exhibit in October 1996. Since then, this technology has gained attention in applications such as fabrication of microdies and others.

17.8.3 ELECTRIC DISCHARGE WIRE CUTTING

Electric discharge wire cutting (EDWC) is a variation of EDM that is similar to contour cutting with a band saw. A continuously moving wire travels along a prescribed path, cutting the WP, with

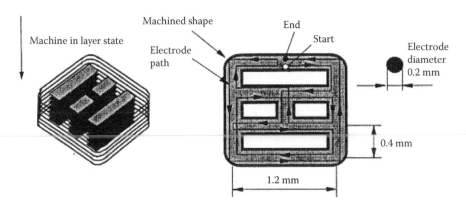

FIGURE 17.26 Micro-EDM. (From Mitsubishi EDSCAN Technical Data, Hannover Exhibition, 1997. With permission.)

discharge sparks acting like cutting teeth. The tensioned wire is used only once, traveling from a take-off spool to a take-up spool while being guided to provide an accurate narrow kerf. The horizontal movement of the worktable is numerically controlled to determine the path of the cut. EDWC is used to cut plates as thick as 300 mm and for making punches, tools, stripper plates, and extrusion dies in hard materials. It is also used to cut intricate shapes for the electronics industry.

Generators for EDWC: EDWC machines are equipped only with pulse generators where the peak current and on-time are the major variables controlling spark energy. Modern machines of EDWC are equipped with sophisticated isopulse generators, where the previously mentioned variables, along with off-time and spark frequency, can be set independently while monitoring the gap. The wire has a limited current capacity, so that the current rating rarely exceeds 30 A. The potential difference between the wire electrode and the WP is usually set between 50 and 60 V. Because wire electrode wear is of little importance, negative wire (straight polarity) is always used. As larger wire diameter can handle higher energy of sparks, therefore, it cuts at higher machining rates.

Dielectric Flushing in Electrodischarge Wire Cutting: Effective flushing in EDWC is very important. Flushing nozzles should be as close as possible to the working gap. WP of large variations in thickness are especially troublesome as they prevent effective dielectric flushing. The usual result of poor flushing is wire breakage, which may be avoided by decreasing the pulse on-time, a solution that is accompanied by a slower cutting rate. Deionized water is used almost exclusively as a dielectric. The low viscosity is ideal for the difficult flushing conditions found in EDWC. Light oils are used. However, good filtration is important. Additives are sometimes used as antirust compounds to make the dielectric slippery.

Wire Electrodes: Brass is the most commonly used wire, because it possesses high tensile strength, high electrical conductivity, and good wire drawability to close tolerances. Layered wires are also recommended, but are more expensive; however, they cut faster than brass. One example is steel/copper/graphite wire, with a steel core for tensile strength, a copper layer for electrical conductivity, and graphite on the surface for attaining high machining speeds. Zinc-coated brass, with Mo-core, is also available. Wire diameters range from 5 to 300 μm. It travels at a constant velocity ranging from 0.2 to 9 m/min.

Cutting Speed: In EDWC, the cutting speed is generally given in terms of a cross-sectional area cut per unit time. Typical examples are 18,000 mm^2/h for 50 mm thick tool steel, and 45,000 mm^2/h for 150 mm thick aluminum block. This rate indicates a linear cutting speed of 6 and 5 mm/min, respectively.

17.9 ELECTRON BEAM MACHINING

The pioneer work of electron processing is related to Steigerwald (1958), who designed a prototype of electron beam equipment that has been built by Messer-Griessheim in Germany for welding applications. This new technology has quickly spread in industry to embrace other fields of applications such as machining and surface hardening. EBM is a thermal NTMP that uses a beam of high-energy electrons focused to a very high power density on the WP surface, causing rapid melting and vaporization of its material. A high voltage, typically 120 kV, is utilized to accelerate the electrons to speeds of 50%–80% of light speed. The interaction of the electron beam with the WP produces hazardous X-rays; consequently, shielding is necessary, and the equipment should be used only by highly trained personnel. EBM can be used to machine conductive and nonconductive materials. The material properties, such as density, electric and thermal conductivity, reflectivity, and melting point, are generally not limiting factors. The greatest industrial use of EBM is the precision drilling of small holes ranging from 0.05 to 1 mm diameter to a high degree of automation and productivity.

17.9.1 EBM EQUIPMENT

A typical piece of EBM equipment (also called an electron beam gun) is shown schematically in Figure 17.27. The electrons are released from a heated Tungsten filament. A high potential voltage, typically 120 kV, is necessary to accelerate the electrons from the cathode (filament) toward the hollow anode, and the electrons continue their motion in vacuum toward the WP. A bias cup (Wehnelt electrode) between the cathode and the anode acts as a grid that controls the beam current (1–80 mA) by controlling the number of electrons. The bias cup also acts as a switch of the beam current. A magnetic lens is intended to focus the beam on a spot of diameter ranging from 12 to 25 μm, whereas deflection coils are used to deflect the beam within an angle of not more than 5° to extend the machining range. Through beam deflection, standard shapes and configurations can be produced without WP manipulation. For drilling purposes, the beam is pulsed once per hole.

Another alternative is WP manipulation which occurs when the beam is deflected, so that it moves in sequence with the part, thus allowing drilling while the part is moving. This is called an on-the-fly drilling operation. The manipulator includes a rotary axis with motion capability. The work sheet, formed as a cylinder is clamped over a tensioning drum, together with the backing material. A flying drilling operation is performed. This on-the-fly drilling process is continuously repeated. As each beam pulse starts hole drilling, the beam is deflected, while the drum rotates at constant speed. When the hole is completed, the pulse is turned off, and the beam instantaneously goes back to its original position to drill the next. A more sophisticated multiaxis manipulator (Figure 17.28) can be used for drilling holes in complicated shapes. Accordingly, the WP is held in a chuck, while the motion of the axes is computer numerically controlled. It is worth noting that the bearings must be carefully sealed to protect them against damage by metal vapor and drilling debris. The electron beam is confined with the WP in the evacuation chamber (10^{-4} torr) to prevent

- Oxidation of filament and other elements
- Collision of electrons with the massive molecules of O_2 and N_2 to eliminate loss of their kinetic energy
- Contamination of metal vapor and debris

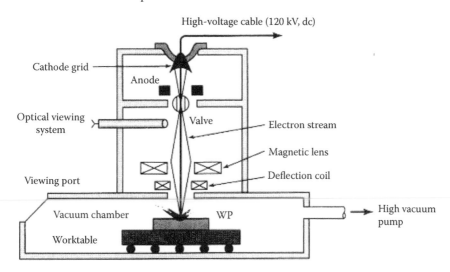

FIGURE 17.27 Electron beam gun.

(a)

(b)

FIGURE 17.28   (a) Multiaxis manipulator and (b) a product. (Courtesy of MG Industries, Steigerwald.)

17.9.2   PROCESS CAPABILITIES

The parameters that affect the performance of EBM are

- Density and thermal properties of WP (such as specific heat, thermal conductivity, and melting point)
- Accelerating voltage (50–150 kV)
- Electron beam current (0.1–40 mA)
- Power (1–150 kW)
- Pulse duration (4–60,000 μs)
- Pulse frequency (0.1–16,000 Hz)
- Minimum spot diameter (12–25 μm)
- Beam intensity 10^6–10^9 W/cm^2
- Traverse speed of the WP

Equation 17.8 expresses the traverse speed v_f, in terms of the machining parameters of the process (Kaczmarek 1976).

$$v_f = \frac{C_d}{d_f}\left(0.1\frac{P_e}{\theta_m t_1 k_t}\right)^2 \text{ m/s} \tag{17.8}$$

where
t_1 = plate thickness (m)
P_e = power of electron beam (Nm/s) = beam current × acceleration voltage = $i_b \times V_b$
d_f = beam focusing diameter (m)
k_t = thermal conductivity (N/s°C)
C_d = coefficient of thermal diffusivity = $k_t/\rho . c_1$ (m^2/s)

c_l = specific heat of WP material (N m/kg°C)

θ_m = melting point of WP material (°C)

EBM is characterized by its minor volumetric removal rate, which reaches a maximum value of 0.1 cm³/min. The volumetric removal rate increases by increasing pulse energy (power intensity), provided that the same number of pulses is used. The machinability depends on the melting point of the material to be machined. In this regard, tin and cadmium have the highest machinability, whereas W and Mo are of low machinability. In EB hole drilling, the achieved tolerance depends on the hole diameter and WP thickness, whereas the surface roughness depends mainly on the pulse energy. Depending on the working conditions, the tolerance of the drilled holes (and slits) may attain values between ±5 and ±125 μm, while the surface roughness R_a ranges from 0.2 to 6.3 μm.

Advantages and Limitations of EBM: The EBM advantages include

- Drilling of fine holes is possible at high rates (up to 4000 holes/s).
- Machining any material irrespective of its properties.
- Micromachining economically at higher speeds than that of EDM and ECM.
- Maintaining high accuracy and repeatability of ±0.1 mm for position and ±5% of the diameter of the drilled hole.
- Drilling parameters can easily be changed during machining even from row to row of holes.
- Providing a high degree of automation and productivity.
- No difficulty encountered with acute angles.

However, EBM has the following limitations:

- High capital cost of equipment
- Time loss for evacuating the machining chamber
- Presence of a thin recast layer and HAZ
- Necessity for auxiliary backing material
- Need for qualified personnel to deal with CNC programming and the X-ray hazard

Applications of Electron Beam Machining: EBM is almost exclusively used in drilling and slitting operations. Drilling is preferred when many small holes are to be made or when holes are difficult to drill because of the hole geometry or material hardness. Textile and chemical industries use EB drilling as a perforating process to produce a multitude of holes for filters and screens.

17.10 LASER BEAM MACHINING

Laser is an acronym for light amplification by stimulated emission of radiation. It is a highly collimated monochromatic and coherent light beam in the visible or invisible range. Laser beam machining (LBM) is a promising NTMP for machining any material, irrespective of its physical and mechanical properties. It is used to cut and machine both hard and soft materials, such as steels, cast alloys, refractory materials, ceramics, tungsten, titanium, nickel, Borazon (CBN), diamond, plastics, cloth, alumina, leather, woods, paper, rubber, and even glass when its surface is coated with radiation-absorbing material such as carbon. However, machining of Al, Cu, Ag, and Au is being especially problematic, as these metals are of high thermal conductivity and have the tendency to reflect the applied light. But recently, yttrium aluminum garnet (YAG) with enhanced laser focusing has been used to cut such metals after treating their surfaces by oxidizing them or increasing their surface roughness. YAG is superior more than a CO_2 laser because it emits shorter wavelengths.

Laser is a versatile tool, useful in many areas ranging from precision watch-making to heavy metal-working industrial applications. The key of laser's effectiveness lies in its ability to deliver, in some cases, a tremendous quantity of highly concentrated power, as high as 10^{12} W/cm^2. Tuning the beam makes it possible to deliver just the right intensity of power for the right amount of time to perform a specific piece of work.

It is possible to make automotive engine blocks out of aluminum with a thin cladded hard layer inside the cylinder by lasers, thus considerably reducing the engine weight. As long as the beam is not obstructed, it can be used to machine inaccessible areas. One of the laser beams main advantages is that it does not take up time for the evacuation of the machining area, as does EB. A laser can operate in transparent environments like air, gas, vacuum, and in some cases even in liquids. However, LBM is quite inefficient and cannot be considered as a mass metal removal process. A significant limitation of laser drilling is that the process does not produce round and straight holes. This can, however, be overcome by rotating the WP as the hole is being drilled. A taper of about 1/20 is encountered. HAZ is produced in LBM, and heat-treated surfaces are also affected.

High capital and operating cost, and low machining efficiency, which could be as low as 1%, prevent LBM from being competitive with other NTM techniques. Protective measures are absolutely necessary when working around laser equipment. Extreme caution should be exercised with lasers; even a low power of 1 W can cause damage to the retina of the eye. In all cases, safety goggles should be used, and unauthorized personnel should not be allowed to approach the laser working zone.

Pyrolithic and Photolithic Lasers: A laser beam is of a high power density, especially when focused to a small spot on the surface of the WP. Depending on its wavelength, the beam interacts with the WP material either pyrolithically (thermally) or photolithically. In the pyrolithic laser, the material removal occurs by melting and vaporization of the material spontaneously. This laser is used mainly in applications such as cutting, drilling, welding, and surface hardening. The removal or processing rate depends on the material being machined, its thickness, and its physical and optical properties, such as the specific heat, latent heats of melting and vaporization, and the surface reflectivity. The machinability of materials increases by decreasing the previously mentioned properties. It depends also on the beam characteristics, especially power density.

In the photolithic laser, the material removal is not removed thermally, but it is affected by the dissociation and breaking of the chemical bond between the material molecules, when its bond energy is below the photon energy of the beam. The photon energy of the beam is inversely proportional to its wavelength. The fluorine excimer laser is a beam of ultrashort wavelength ($\lambda = 157$ nm); consequently, it possesses a high photon energy of 7.43 (1 eV = 1.6×10^{-19} J), whereas the CO_2 laser is an infrared (IR) laser beam of a long wavelength ($\lambda = 10,600$ nm) that has a low photon power of 0.12 eV. It follows that an excimer laser is capable of machining plastic and Teflon photolithically, as its photon energy is greater than the chemical bond energy, which ranges from 1.8 to 7 eV for most of plastics. A CO_2 laser is not capable of machining plastics photolithically, but pyrolithically.

Industrial Lasers: Industrial lasers comprise in most cases the solid-state lasers such as neodymium yttrium aluminum garnet (Nd:YAG), neodymium glass (Nd: glass), and the ruby and gas lasers (CO_2, excimer, and He/Ne). Basically, four types prevail in metal working processes, namely, the CO_2, Nd:YAG, Nd: glass, and excimer lasers. Out of these, the CO_2 and YAG are considered the most dependable workhorses.

1. CO_2 lasers: In these lasers, the active lasing material is the CO_2 gas. However, a mixture of gases is used (CO_2:N_2:He = 0.8:1:7). Helium acts as a coolant in the gas cavity. CO_2

lasers are characterized by their long wavelength of 10,600 nm; thus, the material removal depends only on the thermal interaction with the WP. However, these lasers are bulky but economical. There are two types of CO_2 lasers.

- Axial flow: pulsed (P) of power 0.1–2 kW, pulse frequency 1–10,000 Hz, or continuous wave (CW) of power 0.15–5 kW.
- Traverse flow: only (CW) of power range 2.5–15 kW.

2. Nd:YAG lasers: This laser is a single crystal of YAG doped by 1% neodymium as an active lasing material. This laser is compact and economical, its wavelength is 1060 nm, and it can operate in either pulsed (P) or continuous wave (CW) mode. It is characterized by relatively high efficiency and high pulsating frequency and operates a simple cooling system. Its pulsating frequency ranges from 1 to 10,000 p/s and pulse energy 5–8 J/p. It has an average power output close to 1 kW.

3. Nd:glass lasers: This laser uses a glass rod doped by 2%–6% neodymium as the acting lasing material. This laser is often uneconomical, has the same wavelength as the Nd:YAG, and operates only in the (P) mode. Owing to the low thermal conductivity of glass, the pulse rate should be limited. Consequently, it is only used in drilling and welding where higher-energy output and low pulse frequency (1–2 p/s) are necessary.

4. Excimer lasers. Excimer lasers present a family of pulsed lasers (P) operating in the UV region of the spectrum. Excimer is an abbreviation of "excited dimmer." The beam is generated due to fast electrical discharges in a mixture of high-pressure dual gas, composed of one of halogen gas group (F, H, Cl) and another from the rare gas group (Kr, Ar, Xe). The wavelength of the excimer laser attains a value from 157 to 351 nm, depending on the dual gas combination. Excimer lasers have low power output, which removes the material photolithically, and has a remarkable application in the machining of plastics and micromachining applications as previously mentioned.

Industrial lasers operate either in (CW) mode or in (P) mode. Generally, CW lasers are used for welding, soldering, and surface hardening, which require an uninterrupted supply of energy for melting and phase transformation. Controlled pulse energy is desirable for cutting, drilling, marking, and so on, striving at less heat distortion and the minimum possible HAZ. Table 17.7 provides a general selection guide of industrial lasers for different applications.

17.10.1 LBM Equipment

Figure 17.29 shows schematically three important elements of LBM equipment, namely, a lasing material (solid state or gas), a pumping energy source required to excite the atoms of the lasing material to a higher energy level, and a mirror system. One of these mirrors is fully reflective, while the other one is partially transparent to provide the laser output (output mirror). It allows the radiant beam to either pass through or bounce back and forth repeatedly through the lasing material. To make the laser beam useful for machining, its power density should be increased by focusing, thus attaining power density values between 10^5 and 10^7 W/mm². The laser beam is usually delivered to the WP in the transverse excitation mode (TEM). The common mode of optical configuration is TEM_{00} mode, which is a Gaussian output beam with the lowest beam divergence, and consequently the highest power density. This mode provides the most uniform beam profile. The other widely used modes are TEM_{10}, where a broader and less intense spot is required in cutting and welding, and TEM_{04}, which is useful for machining large holes and surface hardening.

To improve the process performance, most of the LBM equipment is provided with a Q-switching facility (Q means quality factor) to amplify the power. It provides the beam, despite energy loss due to magnification, with enormous power (hundreds or thousands of its normal pulsing power), acting on extrashort pulse duration (on the order of a nanosecond). Therefore Q switching enhances

TABLE 17.7
Laser Beam Selection Guide for Different Applications

Application	Type of Laser Beam
Cutting, trepanning	
Metals	(CW) CO_2, (P)CO_2, (P)Nd:YAG
Plastics	(CW) CO_2
Ceramics	(P)CO_2
Drilling, percussion drilling	
Metals	(P) Nd: YAG, (P)CO_2
Plastics	Excimer
Marking, micromachining	
Metals	(P) Nd: YAG, (P)CO_2
Plastics	Excimer
Ceramics	Excimer
Welding, soldering	(CW) CO_2, (P)CO_2, (P)Nd:YAG
Surface hardening	(CW) CO_2

Source: Kalpakjian, S., and Schmid, S. R., *Manufacturing Processes for Engineering Materials*, 4th ed., Prentice-Hall, Upper Saddle River, NJ, 2003.

the beam capabilities regarding the removal rate and the quality of cut. The Q-switched beam is capable of evaporating the material in no time. In case of Nd:YAG and CO_2 lasers operating in the CW mode, Q switching also converts the continuous wave into a train of pulsating power. Most of the new lasers are computer-controlled to take advantage of their high-speed processing. During machining, motion can be provided to the WP, the beam, or both.

In recent LBM developments, significant progress has been made by integrating robot technology and CNC facility with lasers in a setup called a flexible machine station (Figure 17.30). A single laser beam travels to the processing locations without diffraction or loss of power, where it is divided to perform many functions simultaneously.

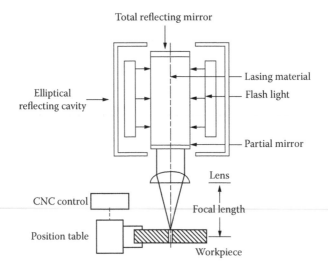

FIGURE 17.29   Schematic of LBM.

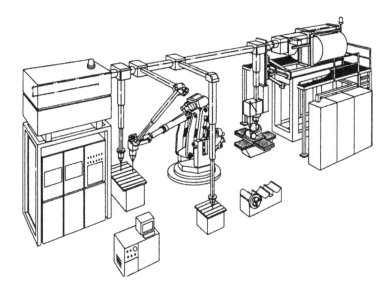

FIGURE 17.30 Flexible laser beam machine station. (From Youssef, H., and El-Hofy, H., *Machining Technology, Machine Tools and Operations*, CRC Press, Boca Raton, FL, 2008. With permission.)

Gas-Assisted Laser Cutting: In an important development, coaxial nozzle can now be supplemented with continuous jet of air, O_2, or one of the inert gases (N_2, Ar, or He). The selection of the gas depends on the type of work material, its thickness, and type of cut. Oxygen is the most commonly used assisting gas for steels and most metals. When an oxide-free surface of high-quality cut is desired, an inert gas is used. Additionally, the oxide-free edges can improve the weldability. Inert gas is used also to prevent plastics and other organic materials from charring.

17.10.2 Applications and Capabilities

Laser and gas-assisted cutting techniques are best suited for applications demanding high accuracy and for machining jobs in which the HAZ is to be as narrow as possible to avoid distortion and obtain cuts of high-quality edge finish.

Laser drilling was one of the first practical applications of laser technology. The demand for laser drilling, especially of microholes of 75 μm diameter, is increasingly emphasized. With increasing hole diameter and depth, ejected liquid metal deposits on the walls and the bottom such that perfectly cylindrical holes cannot be obtained, and that is why laser is not used for producing deep holes. In industry, laser drilling is widely used for the drilling of watch jewels, diamond drawing dies, and similar jobs where a high level of precision is not demanded.

Some special applications of LBM are

1. Machining of microholes in filter screens, carburetors, and fuel injection nozzles.
2. Machining of miniature holes of diameter 0.1–0.5 mm at rates of 1–10 hole/s.
3. Microdrilling of diamond wire drawing die (50 μm), using a Q-switched microsecond-pulse Nd:YAG laser, and nanosecond-pulse excimer laser.
4. Laser drilling of rubber cups.
5. Scribing to widths of 5–10 μm at speeds up to 12 m/min. Ultra-short wave excimer lasers now produce cuts having 0.5 μm in width.
6. Trimming of flashes from plastic parts.

7. In the aircraft turbine industry, laser drilling is used to make holes for air bleeds, air cooling, or passage of other fluids.
8. Lasers may be used for machining hard materials (white CI, Inconel), in combination with a TM process (milling or turning). The laser is directed onto a spot in front of the turning tool in laser-assisted turning (LAT).
9. CO_2 lasers have recently been used to cut and peel masks, needed for CH milling of airplane wings (laser power 75 kW, and a mask 0.4 mm thick).
10. Restoring the dynamic balance of high-speed rotors and shafts by removing infinitesimally small pieces of material during rotation on the dynamic balancing machine.

17.11   PLASMA ARC CUTTING

When a gas is heated to a high temperature on the order of 2000°C, its molecules separate out as atoms. If the gas temperature is raised above 3000°C, the electrons of some of the atoms dissociate and the gas becomes ionized, that is, to include free electrons, positive charged ions, and neutral atoms. This state of ionized gas is known as the plasma gas, which is characterized by high electrical conductivity. The plasma arc is initiated in a confined gas-filled chamber by a high frequency spark. The dc from a high voltage source sustains the arc, and the plasma stream exits from the nozzle at sonic speed.

The source of heat generation in the plasma is due to the recombination of electrons and ions into atoms and recombination of atoms into molecules. The liberated bonding energy is responsible for increased kinetic energy of atoms and molecules formed by the recombination. The temperatures associated due to the recombination can be of the order of 20,000°C–30,000°C. Such a temperature melts out and even vaporizes any work material subjected to machining or cutting. Plasma arc cutting (PAC) is a thermal NTMP that was adopted in the early 1950s as an alternative method for oxy-fuel cutting of stainless steel, aluminum, and other nonferrous metals. Recently, cutting of conductive and nonconductive materials by PAC has become much more attractive. The main attraction is that PAC is the only method that cuts faster in stainless steel than it does in mild steel.

17.11.1   Plasma Arc Cutting Systems

PAC systems operate either in a transferred arc mode or a nontransferred jet mode. In the transferred arc mode (Figure 17.31a) the arc is struck from the rear negative electrode of the plasma torch to the conductive WP (+ve electrode) causing a temperatures as high as 30,000°C. Owing to the greater efficiency of the transferred systems, they are often used in the cutting of any electrically conductive material, including those of high electrical and thermal conductivity that are resistant to oxy-fuel cutting as aluminum.

In the nontransferred jet mode (Figure 17.31b), the arc is struck with the torch itself. The plasma is emitted as a jet through the nozzle orifice, causing a temperature rise of about 16,000°C. Because the torch itself is switched as the anode, a large part of the anode heat is extracted by cooling water and therefore is not effectively used in the material removal processes. Nonconductive materials that are difficult to cut by other methods are often successfully cut by the nontransferred plasma systems. A constructional assembly of a typical transferred plasma torch is illustrated in Figure 17.32. The nozzle diameter depends on the arc current and the flow rate of the working gas. It ranges from 1.2 to 6 mm. Fine nozzles of 50 μm diameter are especially used to cut metals with a kerf width of 0.1 mm and operate at low power of 1 kW. The commonly used working gases are He, H_2, N_2, or a mixture of them. The gas flow rate ranges from 0.5 to 6 m³/h, depending on the arc power and the plate thickness. The nonconsumable electrodes are made of 2% thoriated tungsten to resist wear.

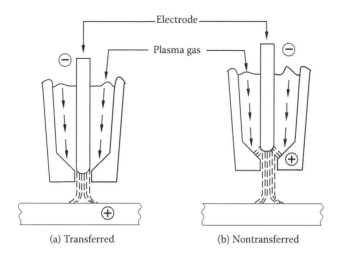

(a) Transferred   (b) Nontransferred

FIGURE 17.31 (a) Transferred and (b) nontransferred plasma torches. (Adapted from Kalpakjian, S., *Manufacturing Processes for Engineering Materials*, Addison-Wesley, Reading, MA, 1985.)

During cutting of aluminum, stainless steel, and mild steels, shielding gases are used to obtain cuts of acceptable quality. An outer gas shield of N_2 or Ar/H_2 is added around the main stream of plasma. A CO_2 shield is favorable for ferrous metals, while air or O_2 may be used for mild steel. In water-shielded plasma, N_2 is used as the main working gas, while the shield is a water curtain. It is reported that the cooling effect of water reduces the kerf width and improves the quality of cut; however, the cutting rate is not improved.

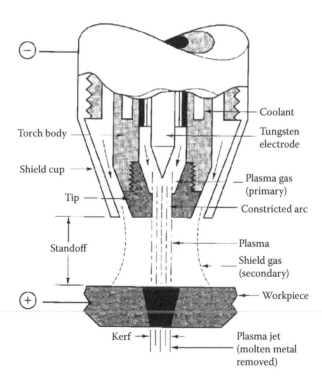

FIGURE 17.32 Cross-sectional assembly of gas shielded transferred plasma torch. (From Machinability Data Center, *Machining Data Handbook*, vol. 2, 3rd ed. Cincinnati, OH, 1980. With permission.)

17.11.2 Applications and Capabilities of PAC

PAC has many applications, some of which are

1. PAC has a special attraction in the case of profile cutting of metals such as stainless steel and aluminum that are difficult to tackle by the oxy-fuel technique.
2. Oxyacetylene flame cutting has the advantage that it cuts metals of heavier sections than PAC does. For this reason, dual operating systems (plasma/flame) are now available on the market. Dual systems have an extended application range, covering all material.
3. A plasma arc is used as a nontraditional tool, integrated with some of traditional processes such as plasma-assisted turning.
4. Underwater plasma cutting is used to reduce the plasma noise and to get rid of plasma fumes and glare. N_2 is the preferred working gas. Underwater plasma is characterized by two disadvantages, namely, the reduced cutting rate and the problem associated with immersing the cutting torch in the water.
5. PAC is also used to cut nonconductive materials such as textiles, nylon, and polypropylene with thicknesses ranging from 0.1 to 1 mm at a high traverse speed of 1000 m/min.

Advantages and Disadvantages of PAC: Advantages of PAC include

- The process provides smooth cuts, free of contaminants.
- It can cut exotic metals at high rates.
- The process has the least specific cutting energy among all NTMPs.

Disadvantages of PAC include

- Reduced accuracy and surface quality are expected.
- The process requires high power.
- It produces toxic fumes.
- Owing to high thermal effects, the WP is highly distorted and HAZ of large depth reduces the fatigue resistance.
- The plasma arc produces IR and UV radiation, which cause eye injuries (cataracts) and loss of sleep. UV radiation leads to skin cancer. Therefore gloves, goggles, and earplugs should be used.

17.12 CONCLUDING CHARACTERISTICS OF NTMPS

Most NTMPs typically have low MRRs and high specific removal rates compared to TMPs. On the other hand, they have better accuracy and surface quality. Figure 17.33 shows the range of surface roughness and dimensional tolerances of the commonest NTMPs, as based on nominal diameter of 25 mm. Tables 17.8 through 17.10 summarize the main characteristics of NTMPs dealt with in this chapter.

Table 17.8 provides a summary of these processes regarding their main features and applications, while Table 17.9 shows tooling and working media, and the typical operating parameters.

Table 17.10 illustrates typical tolerances based on lateral dimension of 25 mm and the corresponding roughness (R_a value); it provides also the MRRs, penetration rates, traverse speeds, and the specific removal rates. As shown at the end of this table, the TMPs have considerably lower specific removal rates (0.1 kW/cm³ min for turning and 0.4–1 kW/cm³ min for grinding), as compared to specific removal rates of NTMPs (ranging from 1 to 5000 kW/cm³ min), which means that the NTMPs generally necessitate a much higher power requirement.

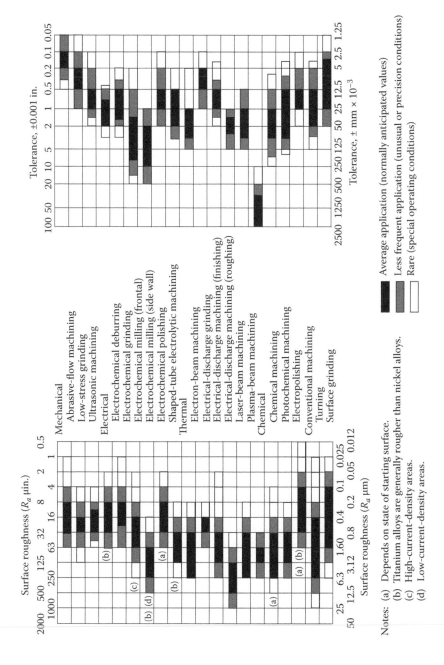

FIGURE 17.33 Surface roughness and tolerances achieved by common NTMPs. (From Machinability Data Center, *Machining Data Hand Book*, 3rd ed., Cincinnati, OH, 1980. With permission.)

TABLE 17.8
Features and Applications of NTMPs

NTMP	Features and Applications
AJM	Cutting, slotting, deburring, deflashing of brittle and hard materials; surface cleaning, glass defrosting, deburring of crossed holes in hydraulic valves, minor metal removal, environmentally pollutant.
WJM	Cutting of types of nonmetallic materials such as rocks, wood, paper, leather, meat, and frozen foods; suitable for contour cutting of soft flexible materials; no thermal damage, environmentally safe, however, noisy (catcher needed); expensive equipment.
AWJM	Cutting all materials (soft, heat resistive, hard) with high traverse speeds, granite, soft metals (Al, steel, etc.), glass, FRP, Ti alloys; large thicknesses and deep cuts, burr-free, no thermal effects, noisy (catcher needed), expensive equipment.
USM	Most effective in hard and brittle materials, conductive or nonconductive, such as Ge, ceramics, glass, carbides, hardened steels, not recommended for soft materials; accuracy and tool wear material dependent; limited tool size (up to 40 mm), so not applicable for large cavities.
CH milling	Shallow removal (<12 mm) on large flat or curved surfaces; low equipment and tooling cost; applicable for almost all materials even glass and Ge; delicate parts; applicable in aerospace and metal-working industries; no burr; no surface and thermal stresses; however, limited accuracy due to EF; requires special protection of operators; expensive etchant disposal; only suitable for low production runs.
PCM	Limited to metallic thin sheets (<2 mm); burr-free; no surface or thermal stresses; low tool cost; higher equipment cost; applicable in production of microelectronics and PCBs.
ECM	Reverse to electroplating; applicable for hard and difficult to machine electric conductive materials, highest rates of metal removal (up to 300 cm^3/min); high-power requirements; MRR current and material dependent; outstanding characteristic of ECM: highest accuracy and surface quality at highest MRR; complex shapes with deep cavities in one-pass; zero tool wear, stress and burr-free; no thermal effect; expensive tool and equipment; problematic electrolyte disposal; applicable for dies and turbine blades.
ECG	Hybrid process, applicable for grinding hard electric conductive materials using metallic-bonded GW; higher removal rate and better accuracy compared to conventional grinding; better SI; burr-free; less residual stresses; reduced GW wear compared to conventional grinding; however expensive equipment cost with corrosion hazard and problematic electrolyte disposal.
EDM	The most common of NTMPs; complex cavities in electric conductive materials regardless of their hardness, applicable in die-making in hardened steels and carbides; thermal damage due to HAZ and RL which may need to be removed; micromachining by ED milling preferred than conventional ED sinking due to reduced tool cost; expensive equipment; delicate and burr-free cuts.
EBM	Micromachining of thin sheets of any material; drilling of minute holes with high aspect ratios (100:1); vacuum needed; automation possible for multihole production; minor material removal process; consumes minor energy; HAZ and RL; expensive equipment.
LBM	All materials machinable; reflective materials difficult to machine; applicable for microholes in thin sheets; minor material removal; no vacuum needed; gas-assisted lasers applicable for plate and sheet cutting; expensive equipment; extreme caution is required.
PAC	Rapid cuts in almost all plates up to 200 mm thick; high material removal process; low accuracy and surface quality, HAZ and RL which may need to be removed; PAT for rough turning of difficult-to-machine materials.

TABLE 17.9

Tools, Working Media, and Typical Operating Parameters of NTMPs

Energy Type	MTMP		WP Material	Tool	Working Medium	Typical Operating Parameters[a]
Mechanical	Jet machining	AJM	All materials	Nozzle	Abrasives SiC, Al_2O_3	$p_n = 1\text{–}9$ atm.; $v_j = 0.6\text{–}1$ mach; SOD = 0.5–20 mm; $d_g = 10\text{–}80$ μm
		WJM		Nozzle	H_2O, oil, alcohol	$p_n = 1300\text{–}4000$ atm; $v_j = 1\text{–}3$ mach, SOD = 2.5–5 mm
		AWJM		Nozzle + Former	H_2O + Abrasives	$p_n = 1300\text{–}3500$ atm; $v_j = 1\text{–}2$ mach, SOD = 2.5–5 mm; $d_g = 90\text{–}200$ μm
	USM			Horn, tool, abrasives	Slurry	$\xi = 10\text{–}50$ μm, $f = 18\text{–}25$ kHz; $F_s = 1\text{–}50$ N; $d_g = 10\text{–}100$ μm
CH	CH milling			Mask	Etchant	$T_e = 30°\text{–}95\ °C$; suitable etchant
	PCM		Conductive materials	Photoresist	Etchant	$T_e = 40°\text{–}100°C$; $p_e = 0.2\text{–}1.5$ MPa; suitable etchant
EC	ECM			Tool	Electrolyte	$I = 40\text{–}20.000$ A, $i = 2\text{–}8$ A/mm², $E = 5\text{–}20$ V; $h = 0.1\text{–}1$ mm; $p_e = 1\text{–}10$ atm
	ECG			Metal-bonded GW	Electrolyte	$I = 20\text{–}1000$ A, $i = 1\text{–}3$ A/mm², $E = 5\text{–}15$ V, $v_g = 20\text{–}40$ m/s
Thermal	EDM			Tool	Dielectric	$I_d = 0.1\text{–}500$ A, $V_s = 50\text{–}300$ V, $h = 10\text{–}100$ μm; $t_d = 2\text{–}2000$ μs, $f_s = 1\text{–}500$ kHz, $\tau = 0.1\text{–}0.9$
	EBM		All materials	EB	Vacuum 10^{-4} torr	$V_b = 50\text{–}150$ kV, $P = 2\text{–}60$ kW, $ib = 100\text{–}1000$ μA, $f = 0.1$ Hz to 16 kHz, $t_d = 4$ μs to 60 ms
	LBM			LB	Air	$P = 2\text{–}20$ kW; SOD = 1.5 mm, $\lambda = 0.6\text{–}10.6$ μm
	PAC			Nozzle	Plasma Ar, H_2, N_2	$P = 10\text{–}200$ kW, $I_p = 50\text{–}1000$ A, SOD = 6–10 mm

[a] Here p_n = nozzle pressure; v_j = jet velocity; SOD = stand-off distance; d_g = abrasive diameter; ξ = oscillation amplitude; f = oscillation frequency; F_s = static force; T_e = electrolyte temperature; p_e = electrolyte pressure; I = electrolyzing current; i = electrolyzing current density; E = cell voltage; h = gap thickness; v_g = linear speed of GW; i_d = discharging current; f_s = sparking frequency; τ = duty cycle; V_s = supply voltage; t_d = discharge duration; V_b = accelerating voltage; P = electron beam, laser, or plasma power; λ = laser wave length; I_p = plasma current.

TABLE 17.10
Typical Tolerances, Roughness, and Removal Rates of NTMPs

Energy Type	MTMPs		WP Material	Typical Tolerance T (± μm)	Typical Roughness R_a (μm)	Material Removal			
						MRR (cm³/min)	Penetration Rate f (mm/min)	Trav. speed v_t (m/min)	Specific RR (kW/cm³ min)
Mechanical	Jet machining	AJM	All materials	—	0.1–0.8	Very low 0.015, MD	MD	MD	1000
		WJM		±100–250 ± 25 (p)	1.2–2.5	MD	MD	MD	5000
		AWJM		±125–± 500	1–1.8	MD	2–2500, MD	Up to 7.5, MD	4000
	USM			± 12.5–25 ± 5 (p) ±2.5 (r)	0.3–1.2, MD 0.2 (p) 0.1 (r)	0.05–1, MD, PD	0.3–10, MD	—	5–10
CH	CH milling			±25–80 ±12.5 (p) ± 7 (r)	0.8–6.3 0.4 (p) 0.15 (r)	0.002–250 AD, MD	0.02–0.04, MD	—	CH energy
	PCM		Conductive material	± 12.5–80 ±8 (p) ±3.5 (r)	0.8–3.2 0.4 (p) 0.2 (r)	0.006–20, AD, MD	0.06–0.2, MD, PD	—	CH energy

Category	Process	Material					
EC	ECM		±50–250 ±25 (p) ±10 (r)	1.2–6.3 0.8 (p) 0.4 (r)	0.6–300, MD, ID, AD	2.5–12.5, MD, ID	—	8
	ECG		±12.5–50 ±8 (p) ±5 (r)	0.2–0.6 0.1 (p) 0.025 (r)	0.3–15, MD, ID	MD, ID	—	4–6
Thermal	EDM		±25–40 ±12.5 (p) ±4 (r)	1.6–10 0.8 (p) 0.2 (r)	0.2–5, MD, PD	10–100, MD, PD	—	2
	EBM	All materials	±25–50 ±5 (r)	0.8–6.3 0.4 (p) 0.2 (r)	0.001–0.002, MD, PD	70, MD, PD	—	500
	LBM[a]		±25–80 ±5 (r)	0.8–6.3 0.4 (p) 0.2 (r)	0.05–0.5, MD, PD	100, MD, PD	Up to 7.5	3000
	PBM		±800–2500 ±500 (p) ±200 (r)	1.6–8 0.8 (p)	10–150, MD, PD	250, MD, PD	Up to 20	1
Conventional turning								0.1
Conventional grinding								0.4–1

Note: Here (p), possible; (r), rare; MD, material dependent; ID, current dependent; AD, area dependent; PD, power dependent.

[a] In LBM, all materials are machinable except reflecting materials.

17.13   REVIEW QUESTIONS

1. Discuss the effect of the following parameters on production accuracy and the removal rate in AJM: grain size, jet velocity, and SOD.
2. Describe at least three typical applications of AJM.
3. Using a block diagram or a line sketch, show the main components of a WJM plant.
4. Explain how material is removed in USM.
5. Explain the advantages and disadvantages of USM.
6. What are the main applications of USM?
7. Explain the effect of USM parameters on the removal rate.
8. Define the magnetostriction effect as applied in USM. What are the aims of using a pre-magnetizing dc in magnetostrictive transducers?
9. Give three examples of materials that can be machined economically by USM.
10. Mention three types of abrasive materials that are frequently used in USM.
11. What are the advantages and disadvantages of PCM?
12. What are the advantages and limitations of CHM?
13. Explain what is meant by EC deburring.
14. What measures should be considered to achieve maximum dimensional control in ECM.
15. Describe the PCM process, and list its fields of application.
16. What is the "self-adjusting feature" in ECM?
17. Explain using a neat sketch, the principle of material removal in EDM. Draw a typical relaxation circuit used for EDM power supply. Explain the main disadvantages of the relaxation circuits used in EDM.
18. What is the difference between sparking and arcing? What condition may lead to arcing in EDM?
19. Explain the term "no-wear EDM."
20. Draw a neat sketch to show the difference between injection and suction flushing as applied in EDM. What type do you recommend in the following cases? Give your reasoning.
    * Production of true cylindrical holes
    * Production of forming tools and dies
21. Define EBM.
22. State the important parameters that influence the MRR in EDM, LBM, and PAC.
23. What are the advantages and limitations of PAC?
24. List the important advantages and limitation of LBM.
25. Explain this statement: NTM should not be considered as a replacement for TM.
26. Which of the NTMPS causes thermal damage? What is the consequence of such damage?
27. Mention briefly the purpose of
    * Adding 1% of a long-chain polymer to the water in WJM.
    * Applying a premagnetizing current in addition to HF current in USM.
    * Applying suction of abrasive slurry in USM.
28. Give typical values for the following parameters:

Value	Units
Depth of HAZ in ECM	mm
Optimum breakdown voltage using a source of 150 V (assume RC circuit)	V
Surface roughness of electrochemically machined surface, R_a	μm
Penetration rate in case of USM of glass	mm/min
Gap voltage in ECM	V
Gap thickness in EDM	mm

Tool oscillation amplitude in USM	mm
Hole over size in USM	mm
Nozzle diameter in AJM	mm
SOD in WJM	mm
Oscillation frequency in USM	Hz
Gap thickness in ECM	mm
Frontal gap thickness in USM	mm
Side gap thickness in USM	mm
Tool wear in ECM	mm
Depth of HAZ in EDM	mm
Accelerating voltage of an EBM	kV
Power density in EBM	kW/mm^2

29. Mark true (T) or false (F) each of the following statements:

[   ] To produce accurate cylindrical holes by EDM, injection flushing and not suction flushing should be used.

[   ] While a relaxation oscillator is highly desirable in that it is simple and rugged, it is severely limited in metal removal capability.

[   ] EBM and EDM are thermal NT processes, which are used to machine only electric conductive materials.

[   ] It is possible to set working conditions in EDM to obtain zero or minimal tool wear.

[   ] In USM, the only function of the demagnetizing current is to avoid frequency doubling.

[   ] WJM is suitable for metallic WPs only.

[   ] In WJM, material is removed by the mechanical action of a high-velocity stream impinging on a small area, whereby its pressure exceeds the yield strength of material.

[   ] NaNO$_3$ has desirable characteristics as an electrolyte and lower corrosive than NaCl; however, it has a tendency to passivate chemical reactions.

[   ] Complex shapes are produced in glass using EDM.

[   ] The current used in EDM is an ac.

[   ] PAM produces more accurate parts than EDM.

17.14   PROBLEMS

1. Calculate the traverse speed v_f in millimeters per second for cutting tungsten carbide sheet of 2.5 mm thick, if EB equipment of 8 kW is used. The equipment is capable of focusing the beam to a diameter of 0.4 mm. The following workpiece data are given

$$\theta_m = 3400 \,°C, \qquad C_d = 8.1 \times 10^{-5} \, m^2/s, \qquad k_t = 214 \, N/s.°C$$

2. In an EDM operation using Lasarenko's relaxation generator, $V_o = 250$ V, $R = 10 \, \Omega$, $C = 3 \, \mu F$. If the cut is required to be performed under maximum removal rate condition, calculate
   a. Discharge (breakdown) voltage, V_s
   b. Charging time, t_c
   c. Cycle frequency, f_r
   d. Energy per individual discharge of capacitor, E_d

BIBLIOGRAPHY

ASM International. 1989. *Machining, vol. 16, Metals Handbook*. Materials Park, OH.

ASM International. 1990. *Advanced Materials and Processes*. Materials Park, OH.

Barash, M. M. 1962. Electric spark machining. *International Journal of Machine Tool Design and Research* 2: 281.

Benedict, G. F. 1987. *Non-Traditional Manufacturing Processes*. New York: Marcel Dekker.

Blanck. D. 1961. *Gleichmäßig keiten beim stoßläppen mit Ultraschallfrequenz*. Dissertation. Braunschweig.

Düniß, W., Neumann, M., and Schwartz, H. 1979. *Trennen-Spanen und Abtragen*. Berlin: VEB-Verlag Technik.

El-Hofy, H. 2005. *Advanced Machining Processes. Nontraditional and Hybrid Processes*. New York: McGraw-Hill.

Franz, N. C. 1972. Fluid additives for improving high velocity jet cutting. *First International Symposium of Jet Cutting Technology*, UK.

Gusseff, W. 1930. *Method and Apparatus of Electrolytic Treatment of Metals*, number 335003. British Patent.

Hoogstrate, A. M., and Van Luttervelt, C. A. 1997. Opportunities in AWJM. *Annals of CIRP* 46(2).

Ingersoll-Rand. 1996. *Product Information-Ingersoll-Rand, Wasserstrahlteclmik*. Hannover Exhibition.

Ingersoll-Rand. 1996. *Technical Data*. Hannover Exhibition.

Kaczmarek, J. 1976. *Principles of Machining by Cutting, Abrasion, and Erosion*. Stevenage, Hertfordshire: Peter Peregrines.

Kalpakjian, S. 1985. *Manufacturing Processes for Engineering Materials*. Reading, MA: Addison-Wesley.

Lehfeldt Works. 1967. *Technical Data*. Heppenheim, Germany.

Lissaman, A. J., and Martins, S. J. 1982. *Principles of Engineering Production*. London: Hodder and Soughton Educational.

Machinability Data Center. 1980. *Machining Data Handbook*, vol. 2, 3rd ed. Cincinnati, OH.

McGeough, J. A. 1988. *Advanced Methods of Machining*. London: Chapman & Hall.

Mitsubishi EDSCAN. 1997. *Technical Data*. Hannover Exhibition.

Pandey, P. C., and Shan, H. S. 1980. *Modern Machining Processes*. New Delhi: Tata McGraw-Hill.

Schmelzer, M. 1994. *Mechanisms Für Strahlerzeugung beim Wasser-Abstrahlschneiden* Dissertation. TH-Aachen.

Steigerwald, K. H. 1958. *Matecrialbearbeitung mit Elektronenstahlen. Fourth International Kongress fiir Elktronenmikroskopie*. Berlin: Springer.

Visser, A. 1968. *Werkstoftabtrag mittels Electronenstrahl*. Dissertation, T. H. Brauanschweig, Germany.

Visser, A., Junker, M., and Weißinger, D. 1994. *Sprühätzen metallischer Werkstoffe*, 1st Eugen. Württ, Germany: G. Leuze Verlag.

Youssef, H. A. 2005. *Non-Traditional Machining Processes—Theory and Practice* (in Arabic), 1st ed. Alexandria, Egypt: El-Fath Press.

Youssef, H. A. 1967. *Herstellgenauigkeit beim Stoßläppen mit Ultraschallfrequenz*. Dissertation, T. H. Braunschweig, Germany.

Youssef, H. A., and El-Hofy, H. 2008. *Machining Technology, Machine Tools and Operations*. Boca Raton, FL: CRC Press.

18 Numerical Control of Machine Tools

18.1 INTRODUCTION

In manually operated machine tools, the process starts from the part drawing. Then the operator determines the manufacturing strategy, sets up the machine, selects the proper tooling, chooses the working conditions, and manipulates the machine controls to make the part (Figure 18.1). It is accordingly clear that using the manual machines involves a considerable number of decisions that influences the quality of the final product. The level of quality obtained depends on the skill and the concentration of the operator. When batches of identical parts are required, it is preferable to use methods that do not depend on the operator skill. In this regard, automatic machines are mechanically controlled by cams and mechanisms. These machines are preferable for producing large batches; however, they are expensive and time-consuming to reset. Using numerically controlled (NC) machine tools, the control of the machine is not performed either by the operator or by cams and mechanisms. NC machines use a series of binary coded numbers that are interpreted by an electronic system. By using the NC technology, each component is an exact replica of the data contained in the NC part program, and high levels of accuracy, repeatability, and uniformity are achieved.

A numerical control system uses prerecorded information, prepared from numerical data, to control the machine tool or the manufacturing process. NC controls the machine movements and various other functions by a series of numbers that are used to drive the machine operative units through the machine control unit (MCU) as shown in Figure 18.2. Computer numerical control (CNC) is the term used when the control system includes a computer as shown in Figure 18.3. CNC controllers have powerful memories capable of storing sophisticated part programs. On some CNC machine controllers, the available memory is too small to contain the part program. In this case the program is stored in a separate computer and sent directly to the machine, one block at a time. When controlling more than one machine using the same computer and data transmission lines, the concept is called direct numerical control (DNC) (Figure 18.4). Direct numerical control, also known as distributed numerical control, is a common manufacturing term for networking CNC machine tools. DNC networking or DNC communication is always required when computer-aided manufacturing (CAM) programs are to be run on some CNC machines. The major advantage of CNC and DNC over NC is that punched tapes are not used directly to control the machine tool. Instead, all information flows from a computer that interfaces with each machine control unit.

The history and development of NC is dated back to 1952 when the first NC conventional milling machine appeared at Massachusetts Institute of Technology (MIT). In 1957, aircraft manufacturers installed a milling machine for machining complex profiles for the aircraft and aerospace industries. Drilling machines, jig borers, lathes, and other NC machine tools were soon developed. The introduction of CNC machining and turning centers to the market was mainly due to the improved NC machine parts and control units. Since the 1960s, smaller electronic components (transistors, resistors, and diodes) have increased the reliability and reduced the size and cost of NC machine tools. The development of the integrated circuits (1965) led to a further reduction of size and cost of the control units and provided the basis for the use of minicomputers for CNC and DNC control systems, which allowed a great deal of flexibility not obtainable with NC ones.

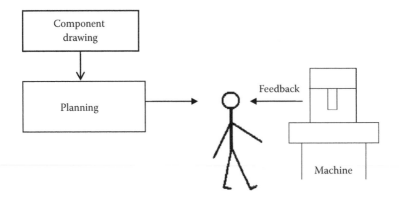

FIGURE 18.1   Manual machine tool.

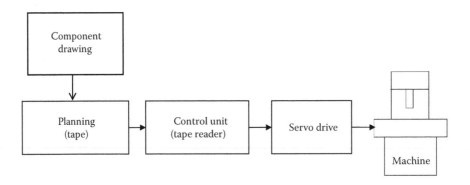

FIGURE 18.2   Conventional numerical control (NC).

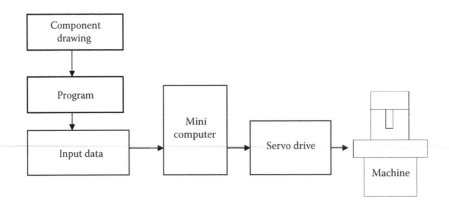

FIGURE 18.3   Computer numerical control (CNC).

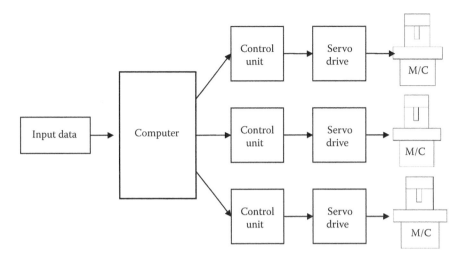

FIGURE 18.4 Direct numerical control (DNC).

18.1.1 ECONOMICS OF CNC

Figure 18.5 shows the total cost against the number of parts produced using different manufacturing methods. At a production volume of zero, the fixed cost using NC, line B, includes tape preparation and setup in addition to the cost related to the design and fabrication of holding fixtures. When using manual (conventional) machine tools, line A, the fixed cost, includes the design and fabrication of tooling, fixtures, and the machine setup, which is less than that for NC machines. Manual preparation and machine adjustments require more time than tape preparation. For special purpose and automatic machines, lines C and D, the design and fabrication of special tooling, manual setup, and adjustment of the machine tool are expensive.

With NC flexibility, the setup cost is often less and smaller lot sizes are economical. Less floor space is needed for materials in process and finished part storage. Referring to Figure 18.5, it is obvious that NC cannot compete in terms of the machining cost with the special purpose machines

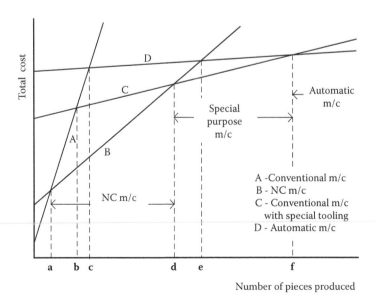

FIGURE 18.5 NC cost compared to other methods.

used for mass production. Their ultimate benefit is achieved when machining small and medium size runs between points a and d. Conventional machines with special tooling are recommended at production quantities between points d and f, while automatic machines are most suited for production quantities beyond point f. Generally, NC can be used when

- The tooling cost is high compared to the machining cost by conventional method.
- The setup time is large in conventional machining.
- Frequent changes in tooling and machine setting are required.
- Parts are produced intermittently.
- Complex shaped components are needed.
- Expensive parts are required where human errors are costly.
- Design changes are frequent.
- One hundred percent inspection is required.

18.1.2 Advantages of CNC

Computer numerically controlled machine tools have many advantages that can be summarized by

1. Greater flexibility because a wide variety of operations is performed and product design changes modifications through tape/program changes are made rapidly.
2. Elimination of templates, models, jigs, and fixtures because the control system takes over the job of locating the tools.
3. Easier setups by using more simple work holding and locating devices.
4. Reduced production time by using a wider range of speeds and feeds than conventional machine tools. Additionally, the CNC equipment moves from one operation to the next faster than the operator that reduces the total production time.
5. Greater accuracy and uniformity are possible because the same part is produced using the same stored NC program that improves parts uniformity and interchangeability and reduces scrape and rework.
6. Greater safety because the tape/program is checked out before the actual production runs, thus allowing less chance of machine damage that may cause human injuries.
7. Conversion from English to the metric system of units.

18.1.3 Disadvantages of CNC

The use of CNC has the following disadvantages.

1. It follows programmed instructions that may cause destruction if they are not properly prepared.
2. CNC cannot add any extra capability to the machine in terms of power of the original drive motor and the machine table travel.
3. CNC machines cost 5 to 10 times more than conventional machines of the same working capacity.
4. The skill required to operate is usually high.
5. CNC requires high investments in terms of wages, expensive spare parts, and special training.

18.1.4 NC System Components

As shown in Figure 18.6, the NC system consists of data input devices, machine control unit (MCU), servo drive for each axis of motion, machine tool unit, and the feedback devices. The program

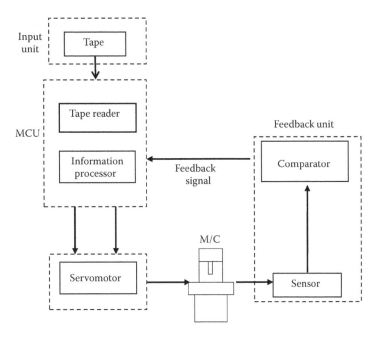

FIGURE 18.6   Main components of the CNC system.

written and stored on the tape is read by the tape reader that is a part of the MCU. The MCU translates the program and converts the instructions into an electrical signal to the servomotor that causes the machine tool movements. Such a movement is sensed and fed back to the MCU through the feedback control unit. The actual movement is compared with the input command and the servo motor operates until the error signal is zero.

18.2   NC CONCEPTS

In order to be familiar with the CNC technology, the following CNC concepts must be made clear.

18.2.1   Machine Tool Axes

In NC machine tools the standard three-dimensional axis (X, Y, Z) system is used to plan the sequence of positions and movements of the machine tool elements. There are three additional rotational axes a, b, c around X, Y, and Z, respectively. Figure 18.7 shows the right-hand rule used for machine axes and the direction of rotation around each axis. The machine axis and motion nomenclature is published according to Electronic Industry Association (EIA) standards. Figure 18.8 shows the standard axis designation for turning, horizontal, and vertical milling machines. Accordingly, Z axis is usually taken as the axis of machine spindle rotation. It has the positive value in the direction away from the workpiece surface. For drilling and milling machines, the longer machine tool slide is taken as the X axis.

18.2.2   Point Location

As shown in Figure 18.9, a quadrant is a quarter of a circle in the Cartesian coordinates. Quadrants are numbered in counterclockwise direction (ccw) from 1 to 4. In most NC machines, the work is carried out in the first quadrant. Point location is used for locating points in the X–Y plane. Zero

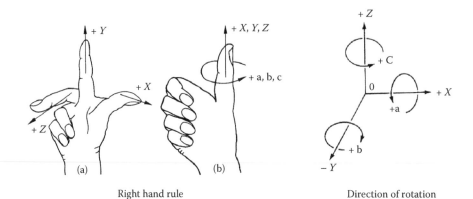

Right hand rule   Direction of rotation

FIGURE 18.7　CNC motion description.

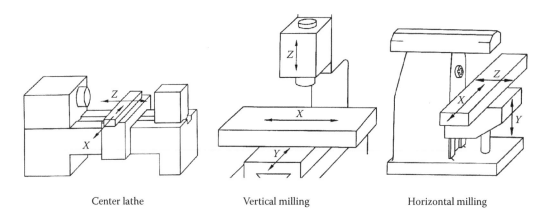

Center lathe   Vertical milling   Horizontal milling

FIGURE 18.8　Standard axes of some CNC machines.

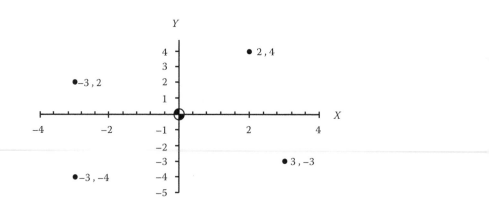

FIGURE 18.9　Point location.

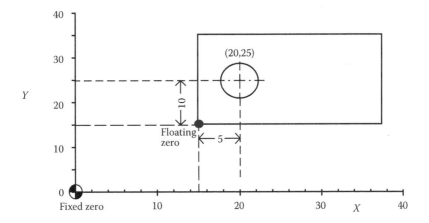

FIGURE 18.10   Fixed zero and floating zero.

point is the point where X, Y, and Z intersect and from which all coordinate dimensions are measured. This point can be either fixed by the manufacturer (fixed zero) or determined by the programmer (floating zero). According to fixed zero location, the point of $X = 0$ and $Y = 0$ is located at a specific point on the machine table and cannot be changed. Floating zero is found in some NC machine tools where the programmer selects the location of the zero point at any convenient spot on the machine table (Figure 18.10).

18.2.3   ABSOLUTE AND INCREMENTAL POSITIONING

The tool location can be described using the absolute or incremental positioning systems (Figure 18.11). In the absolute positioning, the tool locations are always defined in relation to the fixed zero point. This is easy to check and correct and programming mistakes affect only one line of the NC program. In the incremental positioning, the next tool position/location is defined with reference to the previous tool location, which is usually considered as (0, 0). In such a system, any mistake affects all the consequent programmed positions.

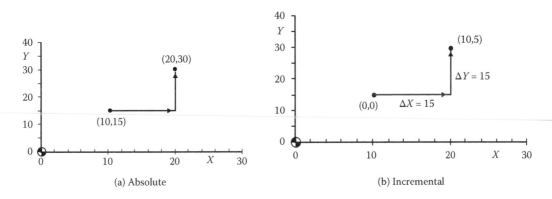

FIGURE 18.11   Absolute and incremental positioning.

18.3 MOVEMENTS IN CNC SYSTEMS

CNC control systems provide specific movements such as

1. Point-to-point, shown in Figure 18.12, where the tool performs certain operations at point one then moves to point 2 for another operation and so on. This system is suitable for drilling holes in a workpiece.
2. Straight-cut NC is used when turning successive shoulders or milling rectangular shapes as shown in Figure 18.13.
3. Contouring (continuous path) NC is used for machining contours and other complex shapes where the tool moves at a controlled feed rate in any direction in the plane described by two axes (Figure 18.14). The method by which continuous path/contouring systems move the tool from one programmed point to the next is called interpolation; in this regard, CNC control systems are supplied with linear, circular, parabolic, and cubic interpolation.
   a. Linear interpolation: The cutting tool moves at a controlled feed rate between two points in a straight line (Figure 18.15a).
   b. Circular interpolation: For machining an arc, points needed are the coordinates of the center point, and the start and finish points. It requires a code to specify the direction of the cut in addition to the desired feed rate (Figure 18.15b).

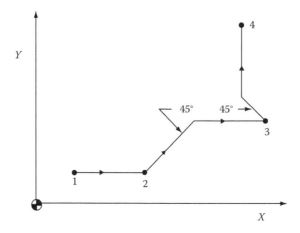

FIGURE 18.12 Point-to-point control.

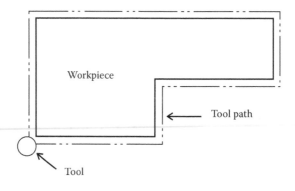

FIGURE 18.13 Straight-cut NC.

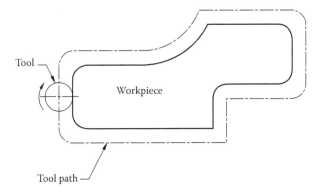

FIGURE 18.14 Contouring NC.

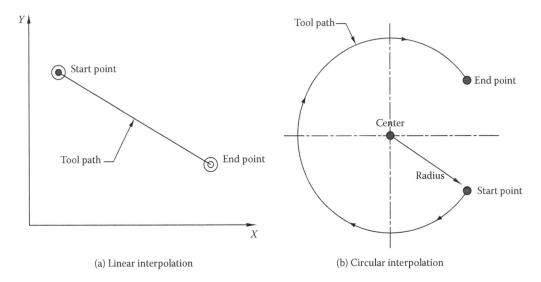

(a) Linear interpolation   (b) Circular interpolation

FIGURE 18.15 Linear and circular interpolations.

c. Parabolic interpolation: Produces parabolic tool paths with the minimum inputs and uses fewer blocks than circular interpolation.

d. Cubic interpolation: Developed with the use of computers where sophisticated cutter paths are generated with few input points.

18.4 CONTROL OF NC MACHINE TOOLS

The control of CNC machine tools comprises the spindle rotation, slide movements, tooling, work holding, and supplementary functions. The movement of these machines is controlled using the automatic control systems that include the machine control unit (MCU), the drive motors, and other equipment. The main function of the control unit is to read and interpret instructions, store information, and send signals to the machine tool to get the appropriate movement that result to the finished workpiece. The open loop control system, shown in Figure 18.16, is used in machine tools to perform specific movements without any check on whether the desired movements actually take place. Such a control system is simple and inexpensive.

In the closed loop control system shown in Figure 18.17, the actual movement is checked and then compared with the original input instructions and the difference produces an error signal that

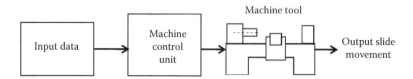

FIGURE 18.16　Open-loop control system.

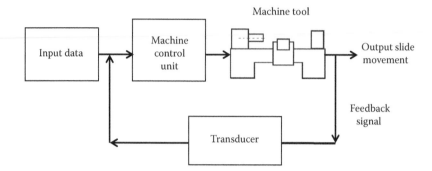

FIGURE 18.17　Closed loop control system.

is used to drive the system. In this regard, transducers are devices that are widely used with CNC machines for monitoring the following:

- Position of the work table
- Speed and angular rotation of the machine spindle
- Tool tip temperature
- Cutting power
- Oil pressure in the hydraulic and lubrication systems
- Flow rate of the coolant

In the closed loop control systems, linear transducers attached to the machine table provide the necessary feedback required for the servo motors to position the work table. Rotary transducers measure the angular displacement of a rotating member such as the lead screw or machine tool spindle. However, transducers provide, mainly, incremental data only. Encoders use more complex scale to provide the exact position of the work relative to the machine elements. They are therefore capable of providing both incremental and absolute information for the machine control system.

18.5　CNC MACHINE TOOLS

CNC machine tool components are similar to the conventional machines. They are specially designed, manufactured, and assembled to suit their special applications with respect to the high speed, high feed, and freedom from vibrations and minimum deflection due to cutting forces. This, in turn, ensures the use of high-quality components in terms of accuracy and surface quality such as

1. Machine tool structures of cast iron are widely used because they possess adequate strength and rigidity besides having a high tendency to absorb vibrations.
2. For large machines, fabricated steel structures are used for weight reduction while ensuring adequate strength and rigidity.

3. The machine tool spindles are subjected to radial and axial load that may cause deflection. Inadequate spindle support leads to dimensional inaccuracies, poor surface finish, and chatter.
4. CNC machines are fitted with recirculating ball screws that replace the normal sliding motion by the rolling motion resulting in reduction of the frictional resistance.
5. Machine tool slides must be smooth and have minimum frictional resistance and low wear, which causes dimensional inaccuracies.
6. The majority of CNC machine tools use electric rather than hydraulic motors to provide sufficient power and spindle speed for a wide range of applications.
7. The operative units that provide the feed movement are not as powerful as those used for driving the main spindles and feed motors of one kW are adequate.
8. Alternating current (ac) induction motors are generally used for coolant pumps, chip removal equipment, and driving hydraulic motors where the only control action required is on/off switching.

CNC machine tools operate at high speeds and feeds that require special attention to the tool materials, shapes, wear compensation, tool replacement, and indexing. Cutting tools of identical dimensions to the original one should be rapidly replaced owing to their wear or breakage. In this regard,

- HSS tools are used for small diameter drills, taps, reamers, end mills, and slot drills; the bulk of tooling for CNC machining involves the use of cemented carbides.
- Solid carbide tools are used when the workpiece material is difficult to machine using HSS tools. Solid carbide milling cutters of 1.5 mm diameter, small drills of 0.4 mm diameter, and reamers as small as 2.4 mm diameter are available.
- Indexable inserts of the correct cutting geometry and precise dimensions are located in special holders or cartridges. As the cutting edge becomes blunt, the insert is moved to a new position to present a new edge to the machining process.
- Automatically indexable turrets, shown in Figure 18.18, are used to accommodate cutting tools. These turrets are programmed to rotate to a new position so that a different tool can be presented at work.
- Indexable storage tool magazines are used on machining centers to store tools that are not in use (Figures 18.19 and 18.20). Chain types of 24 to 180 tools and rotary drums with 12 to 24 tools are available.

FIGURE 18.18 Indexable tool turrets.

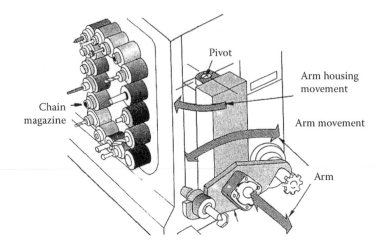

FIGURE 18.19   Chain-type magazine with automatic tool changer.

- A power tool with separate power supply rotates the tool independent of the main rotational axis of the machine tool. In such a case radial holes, flats, and cam type profiles can be machined while the workpiece is nonrotating.

The most important types of CNC machines are

1. Drilling machines: Hold, rotate, and feed the drilling tool into the workpiece. These machines are built with single spindle or multispindle and can accommodate tool turrets.
2. Milling machines: Used to machine flat surface, contours, and curved surfaces (Figure 18.21).

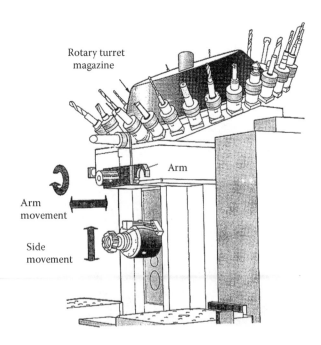

FIGURE 18.20   Rotary-type magazines with automatic tool changer.

FIGURE 18.21   Typical CNC milling machine.

3. Turning machines: Used for cutting cylindrical shapes, boring, drilling, and thread cutting. CNC lathes are equipped with either straight-cut or continuous-path control systems (Figure 18.22).
4. Machining centers: Perform milling, drilling, boring, tapping, countersinking, facing, spot facing, and profiling. They have the ability to change the cutting tools automatically.
5. Turning centers: Use tooling magazines that extends the range of their application compared to CNC turning machines.
6. Wood routers: Similar to the CNC milling machines, but they rotate at higher speeds between 13,000 and 24,000 rpm.
7. Laser cutting machines: Use a laser beam to cut materials. The workpiece traverse is numerically controlled.
8. Water jet machines: Capable of cutting materials using a jet of water or a mixture of water and abrasives at high velocity and pressure. It has several applications in mining and aerospace industries.

FIGURE 18.22   HARDINGE CNC lathe QUEST 10/65.

FIGURE 18.23   CNC WEDM machine. (Courtesy of Charmills Technologies, 560 Bond Street, Lincolnshire, IL.)

9. Plasma arc cutting (PAC) machines: Capable of multiaxis cutting of thick material, allowing opportunities for complex welding seams on CNC welding equipment that is not possible otherwise.

10. Wire electrical discharge machines: Used to cut plates as thick as 300 mm and to make punches, tools, and two-dimensional dies from hard metals that are too difficult to machine with other methods (Figure 18.23).

18.6   INPUT UNITS

Data can be input to the machine control unit using one of the following methods.

1. Manual data input (MDI) is normally used for setting the machine, writing, and editing the program.

2. Conversational manual data input (CMDI) involves the operator pressing appropriate keys on the control console in response to questions that appear in the visual display unit (VDU).

3. Punched tape can be read easily and inexpensively and is less sensitive to handling, inexpensive to purchase, and requires less equipment to make and less costly space for data storage. The combination of punched holes/bits in the tape establishes the values associated with that row. EIA RS-244-A and RS-358 (ASCII) systems are available for NC and are currently used for coding numbers on the tape as shown in Figure 18.24. It should be mentioned here that the RS-244-A coding system involves the use of odd parity where track 5 makes certain that an odd number of holes (not including the sprocket hole) appear on every row of the tape, whereas the ASCII subset uses even parity where an extra hole is added to track 8 in the tape to ensure an even number of holes in each row.

4. Magnetic tapes in the form of cassettes are widely used for transmitting data. They require special storage space and must be handled carefully.

5. Portable electronic storage units store the data, which is then connected to the MCU for data transfer.

6. Magnetic discs transfer stored data into the computer and hence into the MCU.

7. Master computers store programs on its memory and transfer them to the microcomputer of the MCU when required.

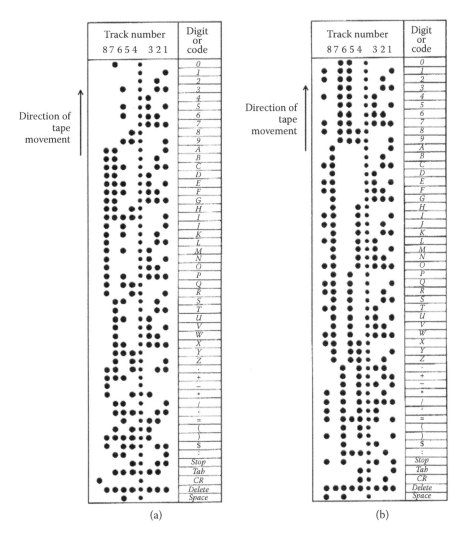

FIGURE 18.24 (a) EIA RS-244-A and (b) RS-358 (ASCII) coding systems.

18.7 CNC INSTRUCTIONS

CNC systems understand the numbers of (0) and (1), which in electrical terms corresponds to (on) or (off) when sensing pressure, magnetism, or voltage. Decimal numbers must therefore be converted to binary ones.

Example 1

Convert 327 to binary:

2	327	1	Least significant digit
2	163	1	
2	81	1	
2	40	0	
2	20	0	
2	10	0	

2	5	1	
2	2	0	
2	1	1	Most significant digit
	0		

Hence $(327)_{10} = (101000111)_2$.

Example 2

Convert (101000111) to decimal

$$(101000111)_2 = 1 \times 2^8 + 0 \times 2^7 + 1 \times 2^6 + 0 \times 2^5 + 0 \times 2^4 + 0 \times 2^3 + 1 \times 2^2 + 1 \times 2^1 + 1 \times 2^0$$
$$= 256 + 0 + 64 + 0 + 0 + 0 + 4 + 2 + 1$$
$$= (327)_{10}$$

In NC systems, the command is given to the machine control unit in blocks of data.

- *Blocks* are collection of words, arranged in a definite sequence, to form a complete NC instruction that could be understood by the machine.
- *Word* is a collection of characters used to form a part of an instruction.
- *Character* is a collection of bits which represent a letter, number, or symbol.
- *Bit* is a binary digit that has the value of 0 or 1, depending on the presence or absence of a hole in a certain row and column on the tape.

Example 3

EIA RS-244-A

Block								N001 X2.000 Y2.500 F 2.50													
Words	N001				X2.000						Y2.500						F2.50					
Characters	N	0	0	1	X	2	.	0	0	0	Y	2	.	5	0	0	F	2	.	5	0	
Bits 1	O			O		O						O	O					O	O			
(holes 2					O	O	O					O	O				O	O	O			
on the 3	O				O								O				O			O		
tracks)	O	O	O	O	O	O	O	O	O	O	O	O	O	O	O	O	O	O	O	O		
4					O							O					O					
5								O			O		O				O			O		
6		O	O		O						O	O					O	O				
7	O	O		O	O	O	O	O	O		O		O	O	O		O			O		
8	O																					

18.8 PROGRAM FORMAT

Tape format describes the general sequence and arrangement of the coded information on a punched tape. According to EIA standards, it appears as words made of individual codes written in horizontal lines according to the following formats.

18.8.1 Fixed Block Format

The fixed block format contains only numerical data arranged in sequence with all codes necessary to control the machine appearing in every block. The instructions are given in the same sequence and in every block including those unchanged from the preceding blocks. It has no word address letter to identify individual words,

001 2.000 2.500 2.50 573

18.8.2 TAB Sequential Format

The block is given in the same sequence as in the case of the fixed block format, but each word is separated by a tab character. If the word remains unchanged in the next block, the word need not to be repeated, but a tab code is required to keep the sequence of words. Because the words are written in a set order, the address letters are not required:

001 TAB2.000 TAB 2.500 TAB2.50TAB573 EOB

18.8.3 Word Address Format

Each element of information is prefixed by an alphabetical character. The alphabet acts as an address that tells the NC system what it must do with the numbers that follow each prefix. If the word remains unchanged, it needs not be repeated in the next block. Word address format is the most common format currently used.

N001 X2.000 Y2.500 F 2.50 S573 EOB

EIA Standard RS-274-A defines the various standard word addresses and describes their use as shown in Table 18.1, which includes the following.

1. Sequence number function: Identifies a block and is represented by a letter N plus three digits.
2. Preparatory functions: Instructs the machine tool for the operation to follow. These are represented by the letter G followed by two digits (see Table 18.1).

TABLE 18.1
Some Common Preparatory Codes and Functions

Code	Function
G00	Point-to-point positioning
G01	Linear interpolation
G02	Circular interpolation arc cw
G03	Circular interpolation arc ccw
G04	Dwell
G05	Hold
G08	Acceleration
G09	Deceleration
G17	x–y plane selection
G18	z–x plane selection
G19	y–z plane selection
G33	Thread cutting, constant lead
G40	Cutter compensation cancel
G41	Cutter compensation left
G42	Cutter compensation right
G80	Fixed cycle cancel
G80 to G99	Fixed cycles as selected by manufacturers

3. Dimensional data function: Represented by a symbol plus five to eight digits as shown in Table 18.2.

4. Feed rate function: Expressed by the letter F followed by three digits that may be expressed in mm/min, inch/min (ipm), or the magic three codes as follows:

Feed Rate (ipm)	Magic Three Equivalent
35.5	F535
3.55	F435
0.35	F335
0.087	F287

5. Tool selection: The information regarding the tool is given by the letter T followed by a numerical code for the tool position in the tool turret or magazine. The T word in the block specifies which tool is to be used in operation.

6. Spindle speed function: The spindle speed code may have the same value preceded by the letter S (1500 rpm will be coded as S1500). The spindle speed coding can also be performed through a chart that is supplied by the machine builder such as S10, which represents the spindle speed of 146 rpm. Similarly, the magic-three method is applied for

TABLE 18.2
EIA RS 274-A Standard Word Addresses

Code	Function
a	Angular dimension around x axis
b	Angular dimension around y axis
c	Angular dimension around z axis
d	Angular dimension around special axis or third feed function[a]
e	Angular dimension around special axis or second feed function[a]
f	Feed function
g	Preparatory function
h	Unassigned
i	Distance to arc center or thread feed parallel to x
j	Distance to arc center or thread feed parallel to y
k	Distance to arc center or thread feed parallel to z
l	Do not use
m	Miscellaneous function
n	Sequence number
o	Rewind application stop
p	Third rapid traverse dimension or tertiary motion dimension parallel to x[a]
q	Third rapid traverse dimension or tertiary motion dimension parallel to y[a]
r	Third rapid traverse dimension or tertiary motion dimension parallel to z[a]
S	Spindle speed
t	Tool function
u	Secondary motion dimension parallel to x[a]
v	Secondary motion dimension parallel to y[a]
w	Secondary motion dimension parallel to z[a]
x	Primary x motion dimension
y	Primary y motion dimension
z	Primary z motion dimension

[a] When d, e, p, q, r, u, v, and w are not used as indicated, they may be used elsewhere.

TABLE 18.3
Some Miscellanies or Auxiliary Functions and Codes

Code	Function
M00	Program stop
M01	Optional (planned stop)
M02	End of program
M03	Spindle start cw
M04	Spindle start ccw
M05	Spindle stop
M06	Tool change
M07	Coolant no. 2 on (mist)
M08	Coolant no. 1 on (flood)
M09	Coolant off
M10	Clamp
M11	Unclamp
M13	Spindle cw and coolant on
M14	Spindle ccw and coolant on
M15	Motion +
M16	Motion −
M30	End of tape
M32 to M35	Constant cutting speed

spindle speed coding. A rotating speed of 7 rpm will be coded as S470, 500 rpm as S650, and 1500 rpm will have the code of S715.

7. Miscellaneous functions: Specify auxiliary functions that do not relate to the dimensional movements of the machine. These include spindle start, spindle stop, coolant on/off, etc. They are denoted by the letter M followed by two digits as shown in Table 18.3.

18.9 FEATURES OF CNC SYSTEMS

The following are common features of CNC control systems:

1. Feed and spindle speed override: Causes deviation from the programmed feed rate or spindle speed in order to increase production rate or to reduce the tool wear. Feed rates can be varied as a percentage between 125% and 0% of the programmed rate. For the spindle speeds, 100–80% of the programmed rpm is possible.
2. Mirror imaging: Is used to machine left and right hand parts from the same tape as shown in Figure 18.25.
3. Scaling: Allows a range of components to be machined in different sizes from one set of programmed data.
4. Rotation: Enables the cutter path to be rotated by an angle and repeated if required as shown in Figure 18.26.
5. Jog: Enables the machine operator to move the machine slides manually through the machine control console.
6. Position displays: Are used for setting the machine and to inspect parts on the machine.
7. Switchable input format: Allows the operator to choose either EIA tape format (RS 244-A) or the ASCII RS-358 tape format.
8. Switchable inch/metric input: Using this feature, inch or metric dimensions can be selected.

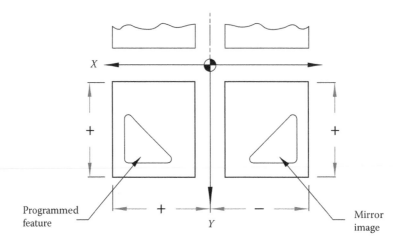

FIGURE 18.25 Mirror imaging.

9. Tool offsets: Are used in case of lathe applications where minor changes are made to the program that avoids the oversize or undersize resulting from the tool wear or from minor variations on the size and shape of the tools when they are changed.
10. Tool length compensation: The difference in tool length with respect to the presetting tool is recorded and manually entered and stored with the associated tool number (Figure 18.27).
11. Cutter diameter compensation: Allows the use of cutter diameter which is different from the diameter used in developing the original part program (Figure 18.28).
12. Operator control features: Include, on–off, manual data input (MDI), manual jog control for the machine axes, sequence number search and display, and the slide hold, which is used to inspect cut or tool condition, then restarts the cycle.
13. Canned cycles: Allow common repetitive machining patterns, like drilling, milling, threading, and turning to be done automatically with a single command; common milling cycles include drilling and peck drilling, shown in Figure 18.29, while slot milling, pocket milling, and hole milling cycles are shown in Figure 18.30. On the other hand, turning canned cycles include rough turning, finish turning, grooving, peck drilling (Figure 18.31), threading, facing, and tapping cycles.

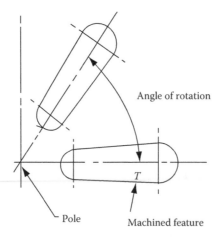

FIGURE 18.26 Rotation.

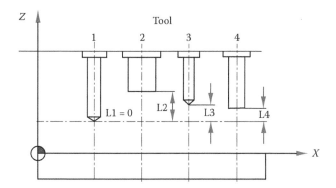

FIGURE 18.27 Tool length offset.

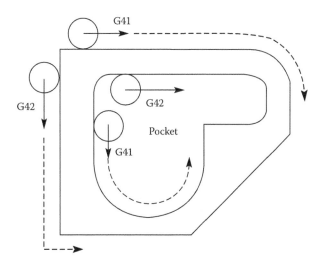

FIGURE 18.28 Cutter diameter compensation.

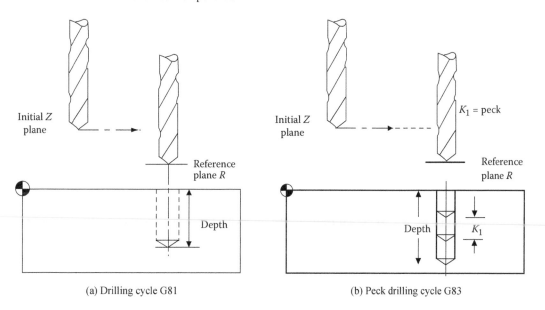

(a) Drilling cycle G81

(b) Peck drilling cycle G83

FIGURE 18.29 Canned drilling cycles.

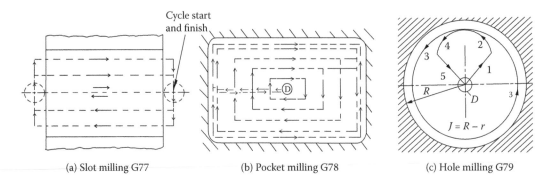

(a) Slot milling G77   (b) Pocket milling G78   (c) Hole milling G79

FIGURE 18.30 Canned milling cycles.

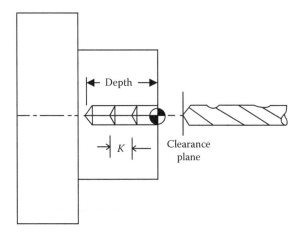

FIGURE 18.31 CNC canned peck drilling cycle G74 used on turning machines.

18.10 PART PROGRAMMING

The part program is a computer program containing a number of lines/instructions/statements called NC blocks that describe the detailed plan of machining instructions proposed for the part. Part programs can be written manually using standard words, codes, and symbols. They can be written using the automatically programmed tools (APT) language that can be converted into NC codes with the help of computers. Part programs can be directly developed using CAD/CAM systems such as Unigraphics and ProEngineer or CAM systems such as Master CAM and Surf CAM that convert the CAD file into NC codes required to drive the machine.

18.10.1 MANUAL PART PROGRAMMING

During machining using CNC machine tools, the following procedure should be followed for preparing manual part programs.

1. List the sequence of operations to be performed on the part as well as the machines intended to use.
2. Write down all the commands required to make the part, using the coded format.
3. Prepare the punched tape/program based on the coded instructions.

4. Verify, check, and correct the program/tape through a computer or by actual run using a workpiece made of foam or plastic.
5. Use the corrected tape/program for producing the part.

Various commands are used to write a part program. These include

- Positioning systems: G90 (modal) is used to specify the absolute positioning system where all coordinates are based on a single origin in the work coordinates system. G91 (modal) is used to specify the incremental positioning system where the current tool position is taken as a reference point in the tool moves.
- Units: G20 is used to specify inches as the unit of measurement, while G21 specifies millimeters as the unit of measurement.
- Absolute zero setting: G92 sets the absolute zero for the coordinates system and ties the coordinates of the machine tool to the coordinates of the program,

$$G92 \; X... \; Y... \; Z...$$

- Tool motion commands: G00 moves the tool at a rapid traverse to a specific XYZ coordinate. Takes the shortest route to reach the specified point

$$G00 \; X... \; Y... \; Z...$$

- Linear interpolation G01 moves the tool from its current position to a specific XYZ coordinate at a specified feed rate F

$$G01 \; X... \; Y... \; Z... \; F...$$

- Circular interpolations G02, G03 codes move a tool around a circular arc to a specific XYZ coordinate; it requires plane selection (Figure 18.32), arc start point, rotation direction, and arc end point. G02 is used for circular interpolation clockwise (cw) around an arc and G03 is used for circular interpolation counterclockwise (ccw) around an arc

$$G02 \; X... \; Y... \; Z... \; I... \; J... \; K...$$

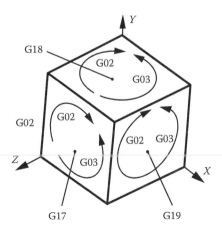

FIGURE 18.32 Plane identifier.

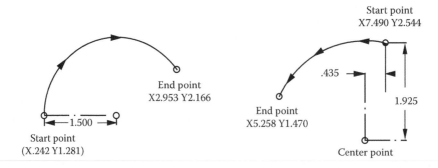

G02X2.953 Y2.166 I1.5 J0   G03X5.258 Y1.470 I-.435 J-1.952

FIGURE 18.33   Circular interpolation (milling).

I, J, and K addresses are used to specify the distances from the start of the arc to the arc center coordinates in XYZ directions as shown in Figure 18.33.

- Stop and end functions: M00 shuts down all drive motors until restarted by the machinist. M01 is used for optional stop of the program, while M02/M30 ends the program.

Example 3

Write a part program for milling, drilling, and slotting of the part shown in Figure 18.34. Use vertical feed 100 mm/min, horizontal feed: 200 mm/min and depth = 10 mm.

Solution

Point Location	X Coordinate (mm)	Y Coordinate (mm)
Tool change	−200	200
A	0	0
B	75	0
C	75	30
D	80	35
E	95	50
F	80	65
G	75	70
H	75	100
I	0	100
J	25	50
K	50	50
L	80	50

Part Program

N10	G90 G21	Absolute, metric programming
N20	G40	Cutter diameter cancel
N30	M06 T01	Change to tool 1
N40	G54X-200Y200	Workpiece preset position
N50	S300M03	Rotate spindle cw at 300 rpm
N60	G00 X-25 Y-25	Rapid positioning
N70	Z 2	
N80	Z-10	

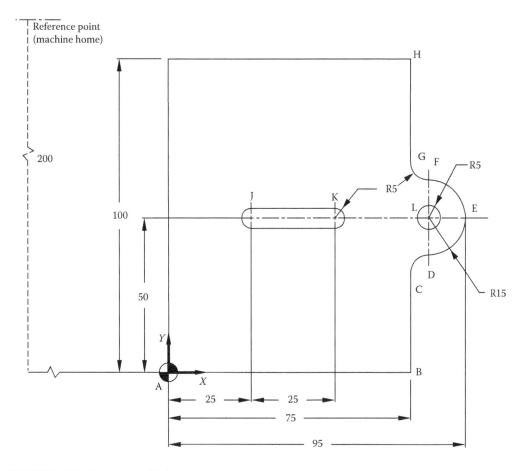

FIGURE 18.34 Part to be milled.

N90	G42X0.0Y0.0	Cutter diameter compensation right
N100	G01X75 Y0 F100	Linear interpolation to point B
N110	Y30	Linear interpolation to point C
N120	G02 X 80Y35I5J0	Circular interpolation cw to point D
N130	G03 X95Y50 I0J15	Circular interpolation ccw to point E
N140	X80Y65I-15J0	Circular interpolation ccw to point F
N150	G02 X75 Y70 I0J5	Circular interpolation cw to point G
N160	G01 X75Y100	Linear interpolation to point H
N170	X0	Linear interpolation to point I
N180	Y0	Linear interpolation to point A
N190	G40 Y-20	Cutter diameter cancel
N200	G00 Z3	Rapid positioning
N210	G00 X25Y50	Rapid positioning to point J
N220	G01 Z-10	Linear interpolation to workpiece thickness
N230	X50Y50	Mill slot to point K
N240	Z0.1	Cutter out of the slot
N250	G00 X80Y50	Rapid positioning to point L
N260	G80 Z-10	Drilling cycle
N270	G28	Home position
N280	M05	Spindle stop
N290	M30	Rewind the program

Example 4

Write down a part program to finish turn the part shown in Figure 18.35. Part to be turned. Preset the tool at X150 mm, Z150 mm.

Point Location	X Coordinate (mm)	Z Coordinate (mm)
Tool change	150	150
A	0	0
B	25	0
C	35	−25
D	35	−42.5
E	42.5	−50
F	55	−50
G	55	−70
H	60	−70

Part Program

N010	G90 G20	Absolute, metric programming
N020	G40	Cancel tool nose radius
N030	T0101	Turning tool
N040	G9 X150 Z 150	Workpiece coordinate setting
N050	S400 M03	Spindle at 400 rpm for turning
N060	G00 G42 X0.0 Z2	Rapid positioning, TNR compensation
N070	G01 X0 Z0 F10	Linear interpolation to point A
N080	X25	Linear interpolation to point B
N090	X.35 Z-25	Linear interpolation to point C
N100	Z-24.5	Linear interpolation to point D
N110	G02 X42.5 Z-50 I7.5 K0	Circular interpolation cw to pint E
N120	G01 X55	Linear interpolation to point F
N130	X55 Z-70	Linear interpolation to point G

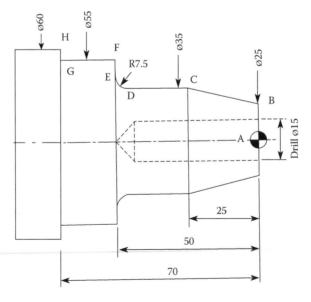

FIGURE 18.35 Turned part. (Adapted from Timings, R. L., *Manufacturing Technology*, Addison-Wesley, Reading, MA, 1998.)

N140	X60	Linear interpolation to point H
N150	G28	Home position for tool change
N160	T0404	Change to drilling tool
N170	S800 M03	Spindle at 800 rpm for drilling
N180	G00 X0 Z.2	Rapid positioning near point A
N190	G74	Drilling cycle
N200	G28	Home position
N210	M30	Program stop

18.10.2 COMPUTER-ASSISTED PART PROGRAMMING

In case of complicated point-to-point jobs and in contouring applications, writing manual part programs becomes tedious and subject to possible errors. Many part programming languages have been developed to automatically perform most of the calculations that the programmer faced to do. This, in turn, saves time and results in more accurate and more efficient part programs. The use of computer aided part programming is justified when

- The part is of complex shape.
- The part is simple, while the program required is too long.
- The CNC machine is complex, such as machining centers and four-axis CNC lathes.

The most widely used automatic programming language is the automatically programmed tools (APT). Part programs are achieved through the definition of workpiece geometry and specifying the tool path or operation sequence. Computer-assisted part programming generates the cutter positions based on APT statements and the tool path is directed to various point locations and along the surfaces of the workpiece to carry out machining. Some common APT statements are shown in Table 18.4.

TABLE 18.4
Common APT Statements

Statements	Example	Definition
Geometry	Point: P1=POINT/5.0,4.0,0.0	X, Y, and Z coordinate values
	Line : L3=LINE/P3,P4	Between two points
	Plane: PL1=PLANE/P1, P4, P5	Defined by three points
	Circle: CIRCLE/CENTER, P1, RADIUS, 5.0	Centers P1, radius is 5
Motion	FROM/TARG	From target
	FROM/-2.0,-0.20,0.0	From X, Y, and Z
	GOTO/P1	Go to point P1
	GOTO/2.0,7.0,00	Absolute
	GODELTA/2.0,7.0,0.0	Incremental
Postprocessing	COOLANT/	Coolant control (ON or OFF)
	RAPID	Rapid cutter motion
	SPINDL/	Spindle on/off, and direction of rotation
	FEDRATE/	To select feed rate
	TURRET/	To select cutter number
Auxiliary	PARTNO	Part number
	MACHIN/MILL,1	Machine used
	INTOL/.01	Inside tolerance
	OUTOL/.01	Outside tolerance
	CUTTER/20.0	Cutter diameter

1. Geometry statements describe the basic geometric elements of the part by points, lines, planes, circles, cylinders, and other mathematically defined surfaces.
2. Motion statements specify the tool path in a detailed step-by-step sequence of cutter moves that are made along the geometry elements. Motion command also include GOLFT, GORGT, GOFWD, GOBACK, GOUP, and GODOWN.
3. Postprocessing statements contain the machine instructions that are passed unchanged into the cutter location data (CLDATA) file to be dealt with by the postprocessor.
4. Auxiliary statements provide additional information such as part name and tolerances.

Figure 18.36 shows the main steps of computer-assisted part programming, which can be summarized as follows.

1. Identify the part geometry, general cutting motions, feeds, speeds, and cutter parameters.
2. Code the geometry, cutter motions, and general machine instructions into the part programming language (APT).
3. Process the source to produce a machine-independent list of movements and ancillary machine control information in a cutter location data (CLDATA) file.
4. Postprocess for CLDATA to produce the CNC code for a particular machine.
5. Transmit the CNC program to the machine and test the program.

18.10.3 CAD/CAM Part Programming

Computer-aided design (CAD) involves the use of a computer in the design process. It enables the engineer to develop, change, and interact with the graphic model for a part. During CAD stages, the engineer draws and creates the part on the screen. The computer allows the design to be viewed and tested to meet the strength requirements before manufacturing. CAD systems export the CAD part file in the form of drawing exchange file (DXF), which is essential for computer-aided manufacturing (CAM). CAD systems secure the following advantages:

1. Increase the productivity of the designers
2. Create better designs

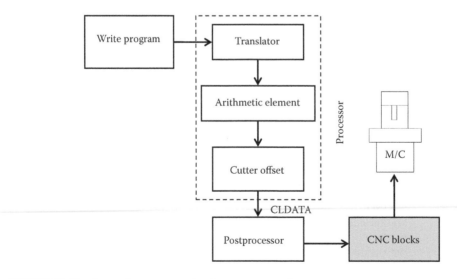

FIGURE 18.36 Steps of computer-assisted part programming.

3. Reduce redundant effort
4. Allow easy and rapid modification to be made to prints
5. Integration of engineering and manufacturing

Computer-aided manufacturing (CAM) uses computer-based software tools to assist engineers in the manufacturing of components. It allows the manufacture of parts using computer-aided design (CAD) files. CAM system works with a CAD design made in a three-dimensional environment. The CNC programmer specifies the machining operations and the CAM system will create the CNC program. CAM allows the programmer to develop a model that represents the part and the machining operations. The programmer can interact with the model graphically to make the necessary adjustments and modifications before the CNC code is generated. CAM software reads the DXF file, which contains the part geometry and levels/layers that the geometry exists on. It utilizes a job plan to assign the correct tool path for each machined layer. Programming steps using CAD/CAM are shown in Figure 18.37, which include the following steps.

1. The aspects of the part geometry that are important for machining are identified; geometry may be edited or additional geometry is added to define boundaries for the tool motion.
2. Tool geometry is defined by selecting tools from a library.
3. The desired sequence of machining operations is identified, and tool paths are defined interactively for the main machining operations.
4. The tool motion is displayed and may be edited to refine the tool motion or other details may be added for particular machining cycles or operations.
5. A cutter location data file (CLDATA) is produced from the edited tool paths.
6. The CLDATA file is post processed to CNC codes, which is then transmitted to the machine.

Using CAD/CAM systems, the CAD drawings can be changed to CNC programs. CAD/CAM approach provides the following advantages:

1. No need to encode the part geometry and the tool motion.
2. Allows the use of interactive graphics for program editing and verification.
3. Display the programmed motions of the cutter with respect to the workpiece, which allows visual verification of the program.
4. Edit interactively the tool path with the addition of the tool moves and standard cycles.
5. Incorporates the most sophisticated algorithms for generating part programs.

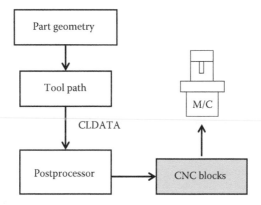

FIGURE 18.37   CAD/CAM programming sequence.

18.11   REVIEW QUESTIONS

1. Mark true (T) or false (F)
   [   ] Circular interpolation G02 and G03 are limited to 90°.
   [   ] DNC allows a computer to control one m/c only.
   [   ] G00 requires a feed rate during part programming.
   [   ] G01 performs machining of straight lines at any number of axes.
   [   ] G02 performs circular interpolation in ccw direction.
   [   ] CNC machines can be used to produce large numbers only.
   [   ] Parity check in tape code is helpful in winding the tape.
   [   ] Row 5 is used as the parity check in ASCII tape code set.
   [   ] Straight-cut NC can produce contours.
   [   ] The character is a collection of words that forms a block.
2. Explain the following using neat sketches:
   a.   Absolute and incremental programming.
   b.   Fixed zero and floating zero.
   c.   Linear and circular interpolation.
   d.   Open and closed loop control systems.
3. Show, using sketches, each of the following:
   a.   Direct numerical control (DNC).
   b.   Advantages of CAD/CAM systems.
   c.   Steps of computer-assisted part programming.
4. State the advantages and disadvantages of using CNC systems.
5. Compare between point-to-point and straight-cut NC systems.
6. Mention some applications for CNC technology.
7. Suggest machining conditions where the use of CAD/CAM system is necessary.

18.12   PROBLEMS

1. In what year was the first NC machine tool developed?
   A. 1700
   B. 1860
   C. 1952
   D. 1977
2. The binary number 10010101 is equivalent to what number in the decimal system?
   A. 159
   B. 149
   C. 139
   D. 129
3. The decimal number 45 is equivalent to
   A. 110111
   B. 101101
   C. 101001
   D. 111011
4. The term applied to the total number of holes in each row of a tape perpendicular to direction of feed is
   A. EIA
   B. Binary-coded decimal
   C. Even or odd parity
   D. ASCII subset

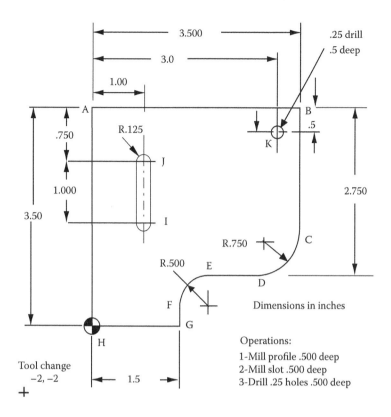

FIGURE 18.38 A part to be milled. (Modified from Senerston, J., and K. Kuran, *Numerical Control Operations and Programming*, Prentice-Hall, New York, 1997.)

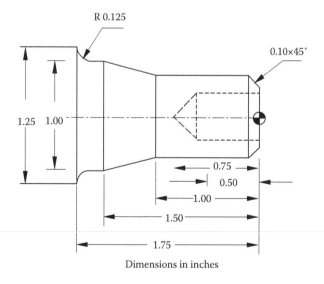

Dimensions in inches

FIGURE 18.39 Turned part. (Adapted from Senerston, J., and K. Kuran, *Numerical Control Operations and Programming*, Prentice-Hall, New York, 1997.)

5. The magic three code S675 would represent a spindle speed of
   A. 368 rpm
   B. 750 rpm
   C. 967 rpm
   D. 975 rpm

6. Write down a part program for milling slotting and drilling the part shown in Figure 18.38.
   - Preset the absolute tool reference at $X = -2''$ and $Y = -2''$.
   - Use letters A, B, C, and D to describe the tool paths.

7. Write down a part program to finish turn and center drill the part shown in Figure 18.39. Preset the tool at $X = 6''$ and $Z = 10''$.

BIBLIOGRAPHY

Gibbs, D. 1988. *An Introduction to CNC Machining*, 2nd ed. London, UK: ELBS Cassell Publishers Ltd.

Rao, P. N. 2000. *Manufacturing Technology—Metal Cutting and Machine Tools*. New Delhi: Tata McGraw-Hill.

Senerstone, J., and Kuran, K. 1997. *Numerical Control Operation and Programming*. Upper Saddle River, NJ: Prentice-Hall.

Thyer, G. E. 1988. *Computer Numerical Control of Machine Tools*, 1st ed. New York: Industrial Press.

Timings, R. L. 1998. *Manufacturing Technology*. Reading, MA: Addison-Wesley.

Waters, T. F. 2002. *Fundamentals of Manufacturing for Engineers*. Boca Raton, FL: CRC Press.

Youssef, H. and El-Hofy, H. 2008. *Machining Technology, Machine Tools and Operations*. Boca Raton, FL: CRC Press.

19 Industrial Robots and Hexapods

19.1 INTRODUCTION

Industrial robots (IRs) and hexapods are programmable multifunctional devices. Robots are designed to move materials, parts, tools, or special devices through a programmable pass to perform in a seemingly human way, whereas hexapods are intended mainly to perform a machining duty similar to numerically controlled (NC) and computer numerical control (CNC) machine tools. In hexapods, the workpiece is fixtured on a table, and three pairs of telescopic legs (struts), each equipped with its own drive, are used to maneuver a rotating cutting tool (or end effector). During the operation, the controller shortens some legs, while lengthening others, so that the tool follows a desired contour with respect to the workpiece.

Robots, hexapods, and CNC machines have a number of similar attributes. All use the same NC technology; moreover, their applications and operating modes are often planned and programmed by the same production crew. The main advantage that makes robots and hexapods cost-effective is their inherent flexibility. Robots and hexapods differ somewhat from CNC machines in that both affect higher velocity and movement in more axes of motions. While with NC only a point, namely the endpoint of the cutter, is controlled in space, the end points and orientations are manipulated with robots and hexapods. This requires more degrees of freedom, more powerful software, and more effective control algorithms. Design considerations include the coordinate system for robot and hexagon movements, and inertia effects resulting from high acceleration and deceleration of the moving components. A robot cannot of course replace a hexapod or CNC machine tool. However, it has a supportive role in industrial plants to enhance the level of automation, where it is now important equipment in flexible manufacturing cells (FMCs).

A hexapod lies midway between a CNC machine tool and industrial robots (IR), because it has the main task performed by the CNC machine, while it resembles to a certain extent the robot in that both of them are devices with several axes of freedom. CNC machines, conventional machine tools, and stacked axis robots have a serial open loop kinematic architecture, in which each axis supports the following one (Figure 19.1a). The kinematic analysis of this serial stacked axis system is easy. However, it generates cumulative errors leading to imprecision with each subsequent station. In addition, if these machines have a big frame with beds and saddles, they require a massive concrete foundation for stability and a considerable floor area.

On the other hand, hexapod is a machine based on advanced closed loop parallel kinematics representing a promising new technique of the twenty-first century that is currently receiving a lot of interest (Figure 19.1b). Its stiffness and rigidity have a positive impact on the accuracy, surface quality, and cutting tool durability. The robust light structure is self-supporting and needs minimal foundation requirements. Programming of hexapods and NC machine tools is normally performed off-line, while the majority of robots are programmed on-line with instructions being retained in the robot's electronic memory. In spite of these differences, there are similarities between the three pieces of equipment in terms of power drive technologies, feedback systems, computer control, and even some of the industrial applications. In this chapter, IRs and hexapods will be considered in detail to explore their specific features, characteristics, and industrial applications.

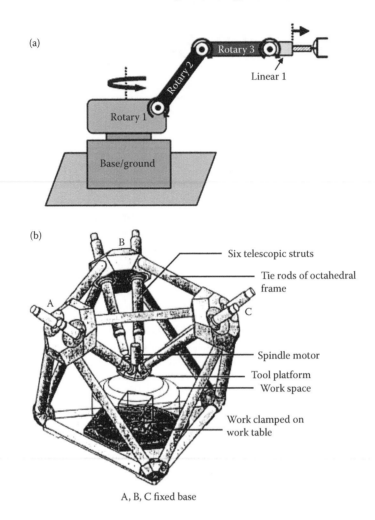

FIGURE 19.1 Constructional features of an industrial robot and a hexapod.

19.2 INDUSTRIAL ROBOTS

19.2.1 ROBOTS AND ROBOTICS

In 1962, the GMs installed their first robot, which has been used in hazardous operations, such as handling toxic materials, or hot workpieces in metal working plants and foundries. In 1968, the first intelligent robot was built, while a commercially available robot was produced in 1973. Today, there is a lot of interest of this important domain, and a separate branch of technology "robotics" has emerged. It embraces all problems of robot design, development, and applications. Industrial robots are defined by the Robot Institute of America (RIA) as "a reprogrammable multifunctional manipulators designed to move materials, parts, tools, or other specialized devices through variable programmed motions for the performance of a variety of tasks." Accordingly, industrial robot is a manipulator that has a built-in control system and is capable of stand-alone operation.

On the other hand, the ISO 8373-1994 has defined industrial robot as "An automatically controlled reprogrammable, multipurpose manipulator, programmable in three or more axes." Therefore, if a machine is programmable, capable of automatic repeat cycles, and can perform manipulations in an industrial environment, then it is an IR. The main advantage of industrial robot is that it performs the job in an exact cycle time, whereas a human being often cannot. This is an important feature in automatic manufacturing systems. Human beings can do the same work as robots, often more economically, but there are conditions under which a robot is superior. Robots can work tirelessly in dirty, hot, tiring, monotonous, unhealthy, and unsafe circumstances. They make less scrap and can lift heavy loads. They operate in foundry, die casting, plastic molding, forging, welding, machining, assembly, painting, and many other manufacturing operations. One of the main advantages of IRs is that they can be quickly reprogrammed to perform tasks that differ in sequence and character. Another point in favor of IRs is that they can be taught to do jobs that are not amenable to automation and mechanization. They pave the way to qualitatively new stage of automation, namely the development of production systems that would require minimum human attendance. Smart robots are now developed that can see, hear, and touch, and consequently make decisions. On deciding whether to use robots or not, industrial safety and operating conditions should be highly considered.

As a conclusion, reasons for using IRs include

- Relieving man of hazardous tasks and heavy loads
- Increasing productivity without sacrificing quality and product consistency
- Increasing the life of tooling, machines, and equipment
- Relieving man of repetitive and tedious work
- Ensuring labor safety
- Reducing labor requirement without loss of productivity, by offering the opportunity for multimachine manning or wholly manned promotion
- Leading the way into areas of technology where man has not entered so far

19.2.2 Major Components of an Industrial Robot

Industrial robot has generally four major components (Figure 19.2). These are the mechanical unit (manipulator), the end effector, the power unit, and the controller. This robot has six motions or six degrees of freedom (6-DOF), which provide the robot with the capability to move the end

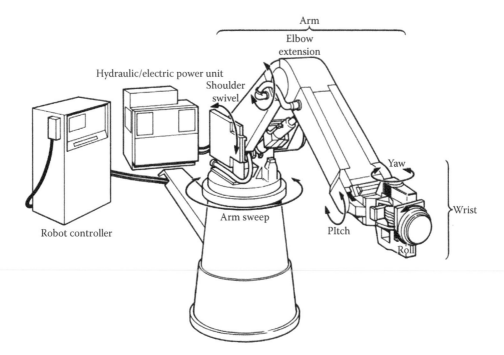

FIGURE 19.2   The major components of an industrial robot. (From Cincinnati Milacron. With permission.)

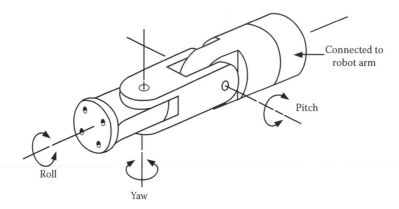

FIGURE 19.3   Robot wrist with three rotary movements: roll, pitch, and yaw.

effector with 6-DOF to ensure the versatility of movement. Not all robots are equipped to move with 6-DOF.

1. Manipulator: It comprises the arm and wrist. Each has 3-DOF (three axes of motion). The motions of the arm and wrist are similar to those of a human arm and hand. The robot movements are executed by mechanical parts like links, power joints, and transmission systems along with internal sensors housed within the manipulator. Each axis of motion is driven by an independent actuator (an electric or hydraulic motor). The wrist is equipped with an end effector and includes three rotary axes denoted by roll, pitch, and yaw as shown in Figure 19.3. The roll (twist) is a rotation in a plane perpendicular to the end of the arm, pitch (bend) is a rotation in a vertical plane, and yaw is a rotation in a horizontal plane (Figure 19.3). There are applications that require only two axes of motion in the wrist (e.g., most arc welding tasks require only two DOF, since the welding gun is a symmetrical tool). In order to reduce the weight of the wrist, its drives are to be located at the base, while the motion is transferred through chains or rigid links. Reduction of weight of the wrist increases the maximum allowable carrying capacity and reduces the moment of inertia, thus improving the dynamic performance of the arm.

2. End effector: Depending on the type of robot, the end effector (robot tool) may be equipped with a welding head, a spray gun, a power tool (drill or nut driver), measuring instrument (dial indicator), or a gripper containing on–off jaws. End effectors are generally custom-made to meet the desired handling requirements. Grippers are the most commonly used end effectors. They are provided either with two or, for more versatility, three fingers to handle a wide range of part configurations.

3. Power unit: The necessary energy is provided to the actuators and their controller. The actuators may be hydraulically, pneumatically, or electrically driven.

4. Controller: The controller acts like the brain of the robot. It performs the functions of storing and sequencing data in memory, initializing and stopping the motions of the manipulator, and interacting with the environment.

19.2.3   Types of Robot Manipulators

One of the most important characteristics of a robot is the shape of the arm and the work volume (envelope). The shape depends on the coordinate system, and the size of the work volume depends on the arm dimensions. When the end effector is attached to the wrist, the work volume exceeds the one given by the robot manufacturers, which is considered when planning for the safety of people working near the robot. In order to define the coordinate systems, a symbolic notation is often used to determine the types and number of joints, starting from the base to the end of the arm. Linear joints are designated by P and revolute joints designated R.

The various types of robot arms are illustrated in Figure 19.4.

1. Cartesian coordinate robot (notation: PPP): It consists of three Cartesian linear sliding axes (X, Y, Z) (Figure 19.4a). In this type, the manipulator, control algorithm, resolution, and accuracy are similar to those of CNC machine tools. An important feature of Cartesian robots is the equal and constant spatial resolution.
2. Cylindrical coordinate robot (notation: PRP): It consists of a horizontal arm moving in and out in a carriage moving up and down and is rotating about a vertical column (Figure 19.4b). The resolution is not constant; it depends on the distance between the column and wrist.
3. Spherical coordinate robot (notation: RRP): Its configuration is similar to the turret of a tank, consisting of a rotary base and a pivot that can be used to raise and lower a telescoping arm (Figure 19.4c). The work volume is a partial thick spherical shell.
4. Articulated or jointed—arm robot (notation: RRR): Similar in appearance to the human arm, it consists of three rigid elements, connected by two joints. The robot arm is mounted to a rotary base (Figure 19.4d). The work volume is a quasi-spherical space. Since the articulated robot accumulates the joint error at the end of its arm, its accuracy is the lowest. Due to its high speeds and excellent mechanical stability, the articulated structure is frequently used in medium size robots.
5. SCARA (Selective Compliance Assembly Robot Arm) (notation: RRR): It is a body-and-arm assembly (Figure 19.5), which is similar to the jointed-arm robot, except that the shoulder and elbow rotational axes are vertical. It is very well suited to perform insertion tasks, such as assembly in a vertical direction, requiring side adjustment to mate the two parts properly. Because of minimal orientation requirements, wrist assembly can be avoided, which means reduced cost. SCARA design results in good positioning accuracy and high operating speed.

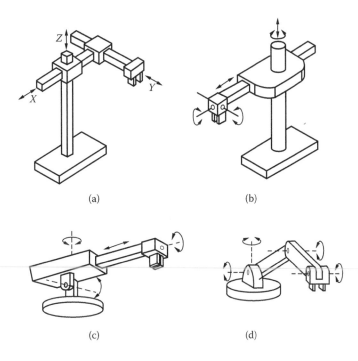

(a) (b)

(c) (d)

FIGURE 19.4 Basic types of manipulators.

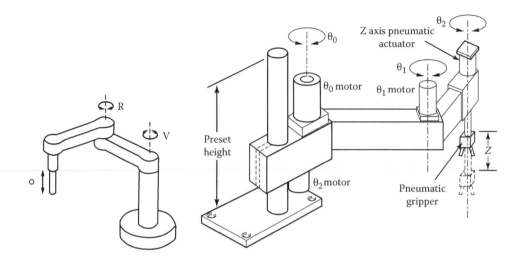

FIGURE 19.5 The SCARA manipulator.

19.2.4 DRIVE SYSTEMS (ACTUATORS)

Energy source is required to effect movement of the manipulator arm through drives. The drives of true servo-controlled robots are electric motors (stepping or dc motors), hydraulic, or pneumatic actuators, and sometimes a combination of these. Robots driven by electric motors do not possess the strength or speed of hydraulic units, but their accuracy and repeatability are generally better. Less floor space is required owing to the absence of the hydraulic power unit. Stepping motors are limited in resolution and power and thus are only suitable for small robots. Direct current motors are ideally suited for small- to medium-sized robots. They can be designed to meet a wide range of power requirements, and moreover, they are relatively inexpensive and reliable. Electric motors may connect the joints directly (direct drive) or via gears. Using such gears results in measurable backlash, especially in the case of small robots where the DC motor requires high gear ratios, so that backlash is a problem.

Hydraulic actuators or motors are well suited for large robots of high power requirements. They deliver power while being relatively small in size. The large power/weight ratio is extremely important in robotics, where extra weight tends to deteriorate the dynamic behavior. Another advantage of hydraulic actuators is that the piston itself can be used as the moving element of linear axes, thus saving the weight and cost of gearing, lead screws, and other transfer mechanisms. The cost of hydraulic actuators is not proportional to the power delivered; therefore, they are expensive for small- and medium-sized robots. They also present some problems in terms of maintenance and oil leakage.

Hydraulic actuators are preferred to electric drives in applications such as painting, where a spark could set off an explosion; however, a low internal air pressurization of the arm can prevent ingress of flammable vapors as well as contaminants. Pneumatically driven robots are technologically less sophisticated than other types. Pick-and-place tasks and other simple, high cycle-rate operations are examples reserved for pneumatically driven robots.

19.2.5 CONTROLLERS

All IRs are either servo or non–servo-controlled. Servo robots are controlled through the use of sensors that continually monitor the robot's axes and associated components for position and velocity. Non-servo robots do not have the feedback capability.

Types of robot control techniques in common use are as follows

1. Pick-and-place robots: These are the simplest and least expensive form of non-servo IRs. They are capable of picking up parts and moving them from one point to another. Hence they are ideally suited to loading and unloading machine tools as well as simple assembly operations. Their movements are limited in number, and their control systems are somewhat elementary, consisting of a series of switches or valves tripped by dogs. Movements are limited by fixed stops. Programming is performed by placing pins in a plug board connection or by push button settings of binary controls. Programming is elaborate and not expected to be done frequently. Pick-and-place robots are usually powered either pneumatically or hydraulically.

2. Point-to-point robots: These are also non–servo open loop-controlled robots, where only end points are defined, and how the arm moves to reach end point is of no concern. These robots also have the capability of moving to and from a number of points, not necessarily in the same sequence. The reprogrammability is imparted by a position servo at each joint, which secures a precise point location. Reprogramming is easily done by the operator leading the robot through its routine at first. The operator presses a button to impress the memory with the location of the end points of each movement. Some robots are programmed off line by the input of coded instructions like NC machines. An average size robot may be able to store about 100 program steps, while large sized up to 1000. Such robots are suitable for spot welding and riveting applications.

3. Continuous-path robots: IRs that move along a two-dimensional (2D) or three-dimensional (3D) path and are continually monitored by a closed loop control (similar to CNC), are called continuous path or universal robots. They utilize position, velocity, force, and torque sensors in order to estimate the deviation from the control program, and then the corrective action reduces deviation to zero. When programmed, the robot is led not only from point to point but also along the paths between points. Such servo-controlled robots have more sophisticated controls than point-to-point robots. In general appearance, they are the same as point-to-point robots; however, they have much more memory storage, and processors capable of coordinating the movements of the joints, and interpolating a given path between end points. They are capable of calculating the shortest route between two points with given constraints. Because of their complicated multiaxis movements, the programming of this type is extremely difficult. The programming is followed by the teach-in method. These robots are beneficial for operations such as spray painting, arc, and seam-resistance welding operations.

4. Intelligent or sensory robots: They are CNC servo-controlled robots, which are equipped with some form of artificial intelligence that allows them to cope with alternating situations. In addition to internal sensors, intelligent robots are provided with one or more external sensing elements, which are as follows.
   - *Visual sensing:* It requires TV camera for vision. The scanned vision information is converted into digital form and processed in the computer (image processing). Several cameras operating simultaneously may deal with complex tasks (Figure 19.6).
   - *Tactile sensing:* It requires force-sensing elements, built into the end effector. When jaws come close enough to the part, the movement is slowed, and the part is gripped with a preset force. Systems approaching human senses are also developed.
   - *Adaptive control sensing:* For example, a deburring robot may move along the edge at a high rate, while searching for burr. Increased total deflection indicates the presence of a burr, whereupon feed rate is reduced until the burr is removed.
   - *Voice sensing:* Intelligent robots are also being designed and developed to convert spoken words into operating commands, through voice sensing or voice programming. The robot controller is equipped with a speech recognition system, which analyzes the voice input and compares it with a set of stored word patterns. When a match is found between the input and the stored vocabulary word, the robot performs the action that

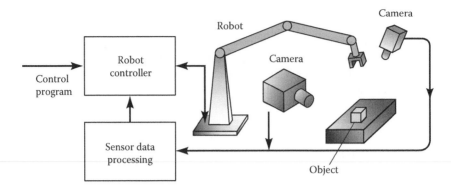

FIGURE 19.6   Intelligent industrial robot provided with visual sensor. (Adapted from Kalpakjian, S., and S. R. Schmid, *Manufacturing Processes for Engineering Materials*, Pearson Education, Inc., 2003.)

corresponds with that word. Voice sensing is beneficial in hazardous work environments as well as for performing maintenance and repair work. The robot is placed in the hazardous environment and remotely commanded to perform the repair chores by means of step-by-step instruction.

19.2.6   PROGRAMMING OF ROBOTS

The combination of six axes of motion and a non-Cartesian coordinate system in which most robots operate make the programming of a robotic system more difficult than the part programming in NC machine tools. A program consists of individual command steps that state either the position or function to be performed, along with other informational data such as speed, dwell, or delay times.

Three different teaching or programming techniques are used, namely, walk-through, lead-through, and off-line techniques.

1. Walk-through programming: The robot arm (training arm) is manually moved (walked) through the required path, with control commands inserted whenever some particular actions are desired such as switching a tool on or off, waiting for a machine tool to finish an action, and so on. The controller stores the received instructions and plays them back when the robot is later placed in the automatic mode of operation. The programmer does not have to worry about the cycle time during the walk-through. His main concern is getting the position sequence correct. With the walk-through programming, the programmer is in a potentially hazardous situation because the operational safeguarding devices are deactivated. The walk-through programming is appropriate for spray painting and arc welding robots.
2. Lead-through programming: This method of teaching uses a proprietary teach pendant (the robot's control is placed in a teach mode), which allows the programmer to lead the robot through the desired sequence of events by activating the appropriate pendant switch (button).

    Each motion is recorded into memory for future playback during the work cycle. The lead-through is popular because of its ease and convenience. Also, when using this technique, the programmer is in a potentially hazardous situation. This technique is widely used in spray painting of auto parts by robots.
3. Off-line programming: This technique involves the preparation of robot programming off-line in a similar way to NC-part programming. The computer algorithms in robots, however, are more comprehensive, because they include simultaneous control of six axes of motion rather than two or three linear axes in the NC machine tools. Off-line robot programming is accomplished on a computer terminal, interfaced with the robot (Figure 19.7).

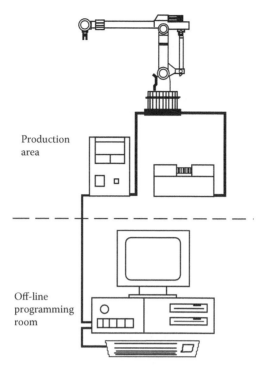

FIGURE 19.7   Off-line robot programming.

After the program has been transferred to the robot controller, either the lead- or walk-through techniques can be used for obtaining actual positional coordinate information for the robot's axes. Off-line programming is executed while the robot is still in production on the preceding job, which means higher utilization of the robotic system. Another advantage associated is the possibility of robot integration into the CAD/CAM database. In most cases the database established in CAD/CAM is used to program all robot motions and actions. Several programs may be stored and called upon to deal with a certain part. Part identification may be defined by reading a bar code applied to the part or pallet. Off-line programming is used for continuous path and sensory IRs.

19.2.7   Robot Characteristics

These characteristics determine the robot's efficiency and effectiveness in performing a given task. These include the following.

1. Number of axes: Two axes are required to reach any point in a plane. Three axes are required to reach any point in space.
2. Degree of freedom: Usually the same as the number of axes.
3. Working envelope: The region of space a robot can reach.
4. Kinematics: The actual arrangement of rigid members and joints in the robot, which determines the robot's possible motion. Classes of robot kinematics include articulated, Cartesian, polar, cylindrical, and SCARA.
5. Carrying capacity or payload: How much weight a robot can lift.
6. Speed: The characteristic where the robot can manipulate the end effector. It ranges up to a maximum of about 1.5 m/s. Almost all robots have the possibility to set the speed to the desirable

level to cope with the task. Speed may by defined in terms of the angular or linear speed of each axis or as a compound speed, i.e., the speed of the wrist when all axes are moving.

7. Acceleration: How quickly an axis can accelerate.
8. Spatial resolution: Refers to the smallest increment of motion at the wrist end that be controlled by the robot. This is mainly determined by the robots control resolution.
9. Accuracy: Shows how closely a robot can reach a commanded position. It can vary with speed and position within the working envelope and with payload. It can be improved by robot calibration. Accuracy is closely related to spatial resolution, where it would be one half the distance between two adjacent resolution points.
10. Repeatability: How well the robot returns to a programmed position. Repeatability is different from accuracy.

Some other characteristic parameters are also important. These are

- Motion control and sensing capabilities
- Programming method
- Types of actuators and their rated powers

19.2.8 Applications of Industrial Robots in Manufacturing

IRs were initially installed to replace human workers in dangerous operations and activities, such as handling hot parts from processing furnaces or in hazardous environments created when workers are subjected to long exposure of toxic materials. Today, however, IRs are installed to improve productivity and quality in manufacturing processes.

The main applications of industrial robots include the following fields:

- Material handling (loading/unloading)
- Spray painting and electrostatic coating
- Die casting
- Spot-resistance welding and riveting
- Seam-resistance and electric arc welding
- Machining (drilling, deburring, and grinding)
- Metal forming
- Inspection
- Assembly

Each of these applications requires different levels of control, which are summarized in Table 19.1.

Figures 19.8 through 19.12 illustrate some important applications of IRs. Pick-and-place robots are ideally suited to loading and unloading machine tools (Figure 19.8) and for assembly operations in mass production. A rotary (cylindrical) robot is used to perform the task. Because only a simple NC is necessary, they are the least expensive form of IRs. A point-to-point controlled robot carries a gun on a yoke to rivet or spot weld assemblies as they pass on a conveyor (Figure 19.9). Dexterity and versatile memory allow universal robots (articulated) to unload parts from two injection molders on a moving conveyor (Figure 19.10). Savings are dramatic in these labor-intensive operations. Robots work continuously without fatigue or the need for relief in hot and hostile environments. Robots can easily manipulate the hot parts in a metal working cell (Figure 19.11). It takes a hot billet out of the furnace and places it in a die on a forging press, then transfers the forging to a trimming press and the scrap to a bin.

Figure 19.12 shows a schematic of a typical FMC, in which a spherical robot is used to load and unload the parts and manipulate them through the cell. The task of the operator is reduced to loading the racks of the conveyor, while the robot picks them up, then removing the finished parts from the conveyor.

TABLE 19.1
Application of Robots and Related Types of Control

Control Type	Application
Pick and place	Loading/unloading, injection molding machines, sheet metal punches, etc.
	WP handling on machine tools
Point to point	Spot welding, riveting
	Material handling
	Simple assembly tasks
	Drilling
Continuous path	Arc welding, seam welding
	Spray painting, electrostatic coating
	Assembly
	Deburring
Synchronized control (with conveyor)	Synchronized spot welding
	Synchronized spray painting
	Synchronized loading/unloading
Sensory control (intelligent robots)	Loading from conveyors
	Complicated assembly tasks
	Inspection

Source: Koren, Y., *Computer Control of Manufacturing Systems*, McGraw-Hill, New York, 1983, with author's modifications.

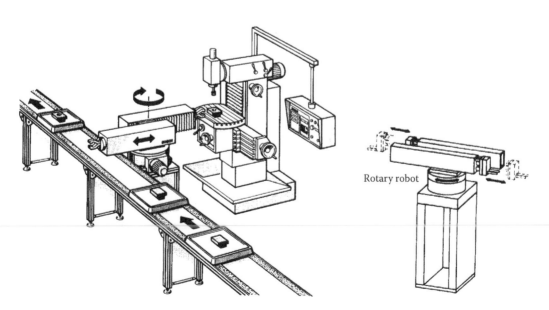

Rotary robot

FIGURE 19.8 Pick-and-place industrial robot. (Courtesy of H. Kief, Michaelstadt, Germany.)

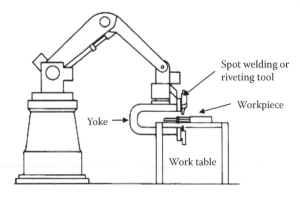

FIGURE 19.9   Point-to-point place industrial robot for riveting or spot welding.

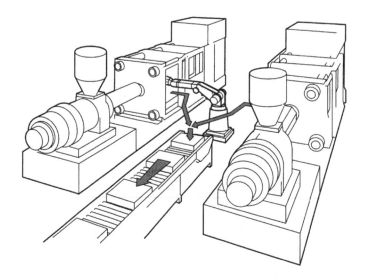

FIGURE 19.10   Universal robot for unloading parts from two-injection molder. (From Mert Corwin, Wolfsburg, Germany, 1977. With permission.)

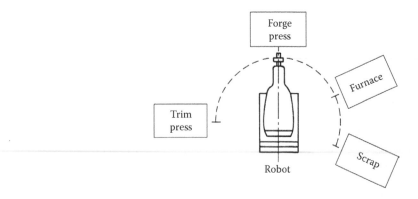

FIGURE 19.11   Industrial robot integrated with a hot working cell. (Adapted from Doyle, L. E., *Manufacturing Processes and Materials for Engineers*, Prentice-Hall, Upper Saddle River, NJ, 1985.)

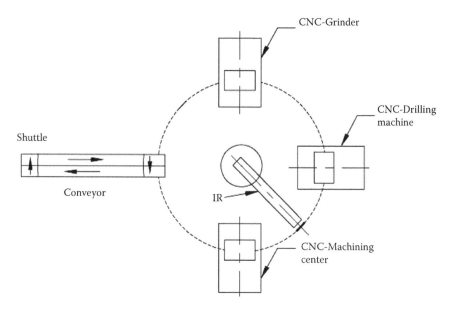

FIGURE 19.12 Industrial robot integrated with a flexible machining cell. (Adapted from Schey, J. A., *Introduction to Manufacturing Processes*, McGraw-Hill, Singapore, 2000.)

19.2.9 Robot Economics

One of the most important issues of robot economics is the reduction of payback period. The payback period when using robots for one shift ranges from 3 to 4 years, and for two shifts it ranges from 1 to 2 years, depending on the robot cost, and the type of operation performed.

The payback period P is estimated as

$$P = \frac{C_R}{L - M} \tag{19.1}$$

where
C_R = investment cost of the robot, tooling, and accessories
L = annual labor saving with overheads
M = annual maintenance, repair, power, installing, and programming cost

The payback period may even be reduced for the following reasons:

1. Increasing the productivity and improving the quality as a result of admitting robot technology
2. Material saving due to precise operation of robots, such as saving in paint cost due to less overspraying (10%–50% saving)
3. Saving of gloves, glasses, and lavatory facilities needed for saved operators
4. Eliminating the possibility of accidental injury of saved operators

Assuming a robot service life of 8 years, it follows from the above discussion ($P = 1–4$ years) that a considerable savings in production cost would be realized by introducing the robot technologies. Of course, economics is not the only reason to implement robotics. The loss of human capability and manufacturing flexibility in the system when a robot replaces a human being must be carefully evaluated.

19.2.10 Recent and Future Developments of Industrial Robots

Now the robotic arm approaches a mature state, where the speed and accuracy achieved enough values for most applications. Vision guidance is bringing a lot of flexibility to intelligent sensory robots. However, the end effectors are often simple pneumatic, which do not allow robots to easily handle different parts, in different orientations. With increasing off-line programmed applications, robot calibration is becoming more and more important to guarantee a good positioning accuracy. Other developments include downsizing arms for light industrial use such as production of small-sized parts and quality control robots. Such robots are classified as benchtop robots.

Prices of IRs are varying according to their features but are usually from 12,000 USD for entry-level models to as much as 100,000 USD or more for heavy-duty, long-reach robots. In 2006, it was reported that Japan leads the world in both stock and sales of multipurpose IRs. About 60% of the installations were articulated robots, 22% were gantry, 13% were SCARA, and 4% were cylindrical robots. The majority of installations were in the automobile sector. In 2007, the world market grew by 3% with approximately 114,000 new installed IRs. At the end of 2007, around 1 million IRs were in use. In the future, robots will be developing very rapidly. The abilities and ease of use of robots will be improved steadily for at least the next 10 years. Major developments will be expected in the following areas:

- Faster control loops leading to more accurate and faster motion control
- Simpler generation with sensors including force control and vision systems
- Enhanced structures with more than six degrees of freedom
- Advanced off-line programming techniques

In conclusion the range of possible applications for robots is rapidly expanding, and it is imperative that world-class manufacturing companies should keep an eye on developments and implement robots whenever benefits can be gained.

19.3 HEXAPODS

19.3.1 Historical Background

The earliest known hexapod machine was the car tire tester designed by Eric Gough of Dunlop in 1948. Stewart described the attributes of a simple hexapod in 1965, thus giving his name to the platform. This platform has been successfully used as a flight simulator to train pilots. Recently, the cost of computing has fallen drastically, and therefore many companies are offering hexapod-based machines for various applications that include hexapod machine tools of Geodetic in 1997. In 2006, Hitachi, Seiki, and Toyota have joined the U.S. builders Giddings Lewis, Ingersoll, and Britain's Geodetic for developing hexapod technology that started in the early 1990s. In 1990, Ingersoll announced and exhibited its own prototype of the Octahedral Hexapod. This hexapod utilized a 12-node hexapod suspended from an octahedral framework, with the spindle pointing down toward the workpiece (Figure 19.13). The Variax Hexacenter milling machine, introduced in 1994 for high-speed milling of aluminum, took its design impetus from the flight simulator with the struts crossing over with the spindle pointing downward from the platform toward the working volume, enclosed by the machine.

19.3.2 Hexapod Mechanism

The hexapod mechanism consists of a fixed upper dome platform (base) of a hexagonal shape and a moving platform of triangular shape, connected by six struts (telescopic or ball screw). Starting from the six joints on the base, each of the two struts are intersecting at three nodes on the moving

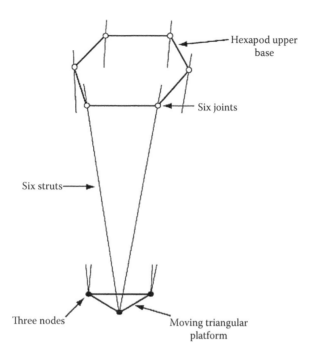

FIGURE 19.13 Hexapod mechanism. (From Youssef, H., and El-Hofy, H., *Machining Technology, Machine Tools and Operations*, CRC Press, Boca Raton, FL, 2008. With permission.)

triangular platform. The movement of the platform is actuated when the six struts change their lengths in a coordinated manner as illustrated in Figure 19.14.

a. When the six struts simultaneously expand or contract at the same feed rate, the platform moves downward or upward horizontally in *extending* movement.
b. When some struts expand and others contract and change orientation, such that the platform moves horizontally, then it is said that the hexapod exhibits *panning* movement.
c. When strut orientations and lengths are changed to achieve a specific platform inclination in space with respect to coordinate axes, then the hexapod exhibits *rotation*.
d. When all struts are of the same length and similarly rotated, such that the platform moves horizontally, then the hexapod is being *twisted*.

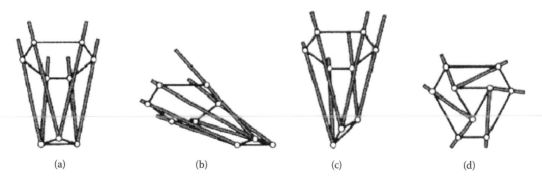

FIGURE 19.14 Typical hexapod movements: (a) extending, (b) panning, (c) rotating, and (d) twisting. (From Ingersoll Machine Tools Inc., Illinois, USA. With permission.)

Therefore, through the rotational and axial movements of struts, the moving platform is capable to reach any point on the workpiece, and the hexapod becomes an excellent universal positioning apparatus. Of course, provisions should be taken to prevent struts crossing and colliding against each other or into the platform.

19.3.3 CONSTRUCTIONAL FEATURES OF HEXAPODS

Hexapods have two main constructional features.

1. Hexapods of telescopic struts (Ingersoll System): In this design, the hexapod consists of six hydraulic struts of telescopic nature. The struts may be of circular cross section or of square shape and are free to expand or contract between a base and a platform. The platform represents the output element that gets the six degrees of freedom of the system (Figures 19.15 and 19.16). By combining an octahedral structural frame (Figure 19.1b) with the above described hexapod actuator, Ingersolls have invented the stiffest and the most possible rigid machine tool. The stiffness and rigidity have a positive impact on the product accuracy, surface quality, and cutting tool durability. The octahedron (the frame structure) consists of 12 beams of similar length that are joined at six junction points (Figure 19.1b). This robust structure is self-supporting and needs the minimal foundation requirements. The mechanism that guides the spindle, the hexapod with its telescopic struts, is attached to the top of the octahedron at top corners, A, B, C. Ingersolls have provided two versions of octahedral hexapods. These are horizontal spindle HOH-600 and vertical spindle VOH-1000 hexapods. The technical specifications of the vertical model VOH-1000 are given below:
   - Maximum feed along and traverse strut axis: 30 m/min
   - Acceleration: 0.5 (4.8 m/s^2)
   - Spindle speed: 0–20,000 rpm (stepless)
   - Maximum torque: 49 Nm
   - Maximum power: 37.5 kW
   - Tool storage magazine: 80 tools (max tool weight = 12 kg)
   - Volumetric accuracy: 20 μm
2. Hexapods of ball screw struts (Hexel and Geodetic System): Hexel and Geodetic dramatically simplified their approach by developing a bifurcated ball between pairs of struts meeting at the working platform. This development reduced the number of nodes to nine

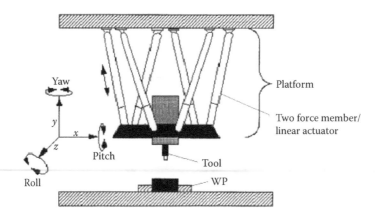

FIGURE 19.15 Hexapod of telescopic struts, Ingersoll system. (From Ingersoll Machine Tools Inc., Illinois USA. With permission.)

FIGURE 19.16 Vertical spindle hexapod VOH-1000.

(six nodes in the work cell and three on the working platform) that improves the stiffness, simplifies the control, and allows for automated calibration. This design approach strives for precision to be derived from software. Such flexibility reduces cost and time, thus leading to truly soft machine.

This type of hexapod is extremely stiff and reliable because it does not require any telescoping mechanism. It is equipped with low pressure, spherical universal joints to facilitate accurate and repeatable calibration while inherently providing excellent strut damping. The coordinated motion of struts (telescopic or ball screw) enables the moving platform (and spindle) to perform six DOF: orthogonally (X, Y, Z) and rotary (pitch, yaw, and roll). This makes the spindle very dexterous, easily accessing unusual angles and geometric features.

19.3.4 HEXAPOD ELEMENTS

The elements of a ball screw hexapod are briefly described.

1. Strut assembly: The Geodetic ball screw strut assembly is illustrated in Figure 19.17. The ball screw is connected to a movable triangular platform through a bifurcated ball and is then driven through the pivot on the dome by the sphere drive. This strut assembly displays a very high extension/contraction ratio, because the ball screw struts can pass into unlimited space behind the pivot. The pivot points (bifurcated nodes and the sphere drive joints) are hermetically sealed to retain flexibility and to provide very smooth running conditions even under shock loads. A variety of lower platforms can be used to accommodate various tool attachments.

2. Sphere drive: The sphere drive is a special mechanism that forms the heart of any ball screw strut–type hexapod. It is located in the upper base and provides the accurate positional movements of the hexapod struts. The sphere drive is a hollow ball that accommodates a high-powered, high-specification, frameless, brushless dc motor. This arrangement drives the ball screw in and out through the hollow cored servomotor. The unit is water cooled to maintain reliability and control of thermal inaccuracies. A high-resolution radial

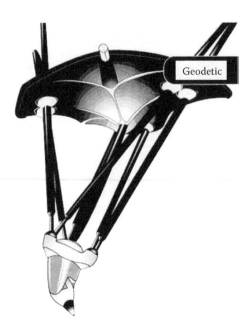

FIGURE 19.17   Geodetic ball screw strut assembly. (Adapted from Geodetic, UK, 1997.)

incremental encoder is mounted outside the unit to provide the accurate positional information required by the controller. An integral thermocouple on the motor windings feeds back data to the controller allowing thermal expansion to be compensated. All critical parts included in the sphere drives are effectively protected from foreign matter. The sphere drive (3200 rpm to provide ball screw feed up to 25 m/min) is clamped to the top plate by six bolts. A location dowel ensures precise and repeatable positioning. All sphere drive components such as ball screws, ball nuts, bearings, motors, and encoders are standardized. This makes the sphere drive mechanically simple, cost-effective, and reliable. Moreover, this drive can be easily tailored to satisfy a wide range of performances for a variety of manufacturing applications.

3. Bifurcated balls: This nodal joint allows a pair of struts to terminate at a common focal point at the lower platform (Figure 19.18). The design includes a spilt ball in a variety of socket arrangements. The hydrostatic bifurcated ball joint is the latest addition to the family of ball joints. The new design uses a system similar to that of the hydrostatic sphere drives, thus ensuring elimination of clearances. This promotes a smoother, vibration free, and well-lubricated environment leading to higher accuracy and better surface quality.

4. Spindles: Hexapods are equipped with high-speed spindles that pack enormous power into a very compact unit. Hexapod builders make full use of the conical shape design of their spindles, which reduces interference with the workpiece (Figure 19.19). Geodetic machining units are equipped with motors of power rating between 3 and 20 kW, running at a 20,000 rpm maximum spindle speed. Geodetics also produce a range of air drive spindles for special application. Hexapod spindles provide speed feedback to ensure precise speed control. They are usually water cooled. To ensure maintenance-free operation at such high speeds, they should feature self-lubricating precision ceramic roller bearings. The spindles accommodate tools up to 20 mm in diameter and 200 mm in length. High speed promotes the productivity, reduces tool wear, achieves tighter machining tolerances, and provides higher surface quality.

5. Articulated head: Its dexterity is comparable to the human arm (wrist and hand). The two-axis head incorporates the high-speed spindle and can tilt by over 90° and rotate over 540° (Figure 19.20). The mechanism configures itself to bend around obstructions in its working

FIGURE 19.18 Hydrostatic bifurcated ball joint. (Adapted from Geodetic, UK, 1997.)

FIGURE 19.19 Hexapod spindle head. (Adapted from Geodetic, UK, 1997.)

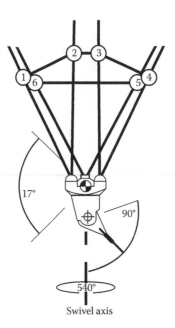

FIGURE 19.20   Dexterity of hexapod articulated head. (Adapted from Geodetic, 1997, UK.)

environment to allow the task to be approached at the optimum angle. Tilting is achieved by a geared sector with backlash elimination. Bearings of drive are protected by a powerful air curtain, which prevents contamination.

6. Upper platform: The cast iron dome is the most logical configuration to achieve the best angular coverage while maintaining maximum rigidity and stiffness. Spherical drives are sited in the dome.

7. Control system: As machine tools and robots, the hexapod requires sophisticated control, compared to conventional machine tools. In hexapods, a continuous and exact relationship of strut movements should exist to control the triangular platform movements. The contour to be followed by the cutting tool is controlled by CAD/CAM, which is software based on Cartesian coordinates X, Y, Z, and the orientation vectors A, B, C of the six struts. The location coordinates X, Y, Z, A, B, C of each strut are calculated by the on-line controller, which necessitates few milliseconds to be performed. Besides the contour geometry, the calculations of the tool movements require additional data such as contouring speed and acceleration. The contour calculations are then analyzed and tested, on-line, to make sure that the dynamic limits of the machine are not exceeded to prevent the damage of the tool on the machine. The hexapod has excessive movement possibilities that may lead to collision between its elements. The struts may touch or even cross each other. Such possibilities must be perceived and prevented by the hexapod controller. The controller software must also be compensated for inaccuracies inherent to the machine elements. The vibration, the speed profile of struts, and the positioning errors of joints and nodes due to incorrect calculations should also by compensated by the controller. Geodetic used the QNX controller, a real-time Art-to-Part that is capable of driving unlimited number of axes simultaneously by using G and M programming codes, cutter location data (CLDATA), and automatically programmed tools (APT).

The new high-performance PC-based controller provides a cost-effective approach to multiaxis machine control. A simple to use comprehensive graphical user interface guides the operator through all tedious tasks. Interaction with the machine is handled through a programmable logic Geodetic controller, QNX. Art-to-Part uses the latest forward and inverse kinematics transform algorithms. These new algorithms have been streamlined and are extremely fast. Art-to-Part is completely hardware independent and includes a tool management database, support for automatic tool change, automatic head change, palette changers, probes, and many other features. Written entirely in C++, Art-to-Part can be ported across to any platform with minimal effort.

19.3.5 HEXAPOD CHARACTERISTICS

Hexapod is a modern technology that bridges the gap between robots and NC machine tools. The objective of this hexapod is to enhance quality and reliability and to reduce the cost and overall cycle time through the dramatic departure from conventional mechanism design. As development and refinements continue, it is believed that hexapod will eventually proliferate. Hexapod provides significant benefits to the end-user because it offers many new attributes for the manufacturing processes.

The merits of hexapod are numerous.

1. Six degrees of freedom: The hexapod with its six struts provides the tool platform with six DOF. This advantage allows the spindle to reach unusual angles and to machine parts of difficult geometrical features such as turbine blades, plastic injection moulds, and other parts requiring high precision.
2. Flexibility and agility: Flexibility is the ability to react to planned changes, while agility is the ability to react to unplanned changes. Its mechanical simplicity plus its foundation independence gives the user the flexibility to quickly reconfigure with changes in production lines with the easy option of storing the machines disassembled when they are not needed. The agile strut supported spindle platform positions in all six DOF.
3. Productivity: Hexapods provide higher production rates through
   - Designing the machine to be above the worktable
   - Making use of high-speed automatic tool changer
   - Reducing the mass of moving parts (no beds or slider as in conventional machine tools) to achieve very fast acceleration/deceleration
   - Using high-speed/high-power precise spindles
4. Stiffness and rigidity: A well-constructed hexapod is characterized by its rigid frame, which does not deflect significantly under acting loads. The stiffness of Ingersoll's octahedral hexapod is about 3 to 4 times that of the five-axis CNC machining center of the same rating.
5. Precision and accuracy: It is easy to control a strut length to 2.5 μm using sophisticated software control. Some hexapods attain submicron accuracy. The parallel strut arrangement lends itself to error averaging.
6. Unique installation: By concentrating all forces of the machining process within the hexapod frame, an important advantage is offered that avoids the need for a special machine tool foundation. Design and installing a foundation for a conventional machine tool represents a substantial cost. The hexapod foundation independence may be demonstrated by using a crane to lift one hexapod corner during machining. A hexapod could function on a ship at sea, laid on its side, or suspended from the ceiling without sacrificing the precision of its performance.
7. Simplicity: Another potential of hexapods is its simplicity and ease to manufacture. The part count in a hexapod is only about 300 compared to about 1000 in conventional machine tools. The other important characteristic is that many are duplicate parts. Assembly is so easy and takes so little time that the hexapod could be sold as a kit.
8. Portability: Hexapod is characterized by its high potential for portability. It is a machine that could be taken to jobs like remote oil fields as well as different manufacturing plants.
9. High load/weight ratio: The high nominal load (power/weight) is a very important characteristic of hexapods. The cutting force acting on the moving platform is approximately

equally distributed on the six parallel struts. It means that each strut suffers only from 1/6 of the total load. Furthermore, the struts are stressed longitudinally either in tension or compression; consequently, there is no need to be designed as massive and strongly dimensioned as in conventional machines.

10. Scalability: Hexapods are scalable in size, both upward and downward to accommodate a multitude of applications ranging from microassembly and surgery to milling, drilling, turning, welding, painting, inspection, and assemsbly.

11. Dexterity: The hexapods have a complex working volume (truncated hexa cone) based on the polar sweep of struts between maximum expansion and compression and the degree of angular freedom. Dexterity extends substantially with the addition of a two-axis articulated unit.

12. Enhanced control systems: Hexapods are generally equipped with control systems and software that are capable of processing the complex algorithmic calculations necessary to command the struts. The Hexapod has calculation power faster than that of several PC's combined. Additionally, the software is capable to compensate offset data and thermal deviations.

13. Cost: After conventional CNC machine tools comes the realm of hexapods. As more hexapods are built, it is expected in the near future that their prices will be reduced to 20% or less as compared to equivalent CNC machine tools. This is because they are simple and easy to design and assemble. Fast assembly means lower inventory, less space, and lower labor costs. Six identical struts simplify the construction, providing easy and fast assembly and reducing maintenance cost. The replacement of faulty parts subjected to wear is also easy. Control and calibration is facilitated by highly efficient software. Moreover, the power consumption of a hexapod is considerably less than that of conventional CNC machines. They are capable to adapt in flexible manufacturing systems (FMS). Figure 19.21 illustrates comparison of cost, number of parts, power consumption, and time elements between hexapods and CNC machine tools.

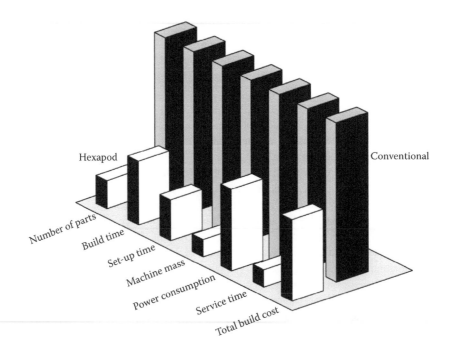

FIGURE 19.21 Comparison of total cost, number of parts, power consumption, and time elements of hexapod and conventional machine tools. (From Youssef, H. and El-Hofy, H., *Machining Technology, Machine Tools and Operations*, CRC, Boca Raton, 2008. With permission.)

However, as a new design, hexapods have still some problems that need to be tackled by further research and development. The main problems are the high coefficient of friction, long struts, dynamic thermal growth, and the cumbersome calibration. All these issues affect the hexapod accuracy and repeatability.

19.3.6 MANUFACTURING APPLICATIONS OF HEXAPODS

Hexapods are coming, and they will likely change the manufacturing paradigm. Their applications in industry include the following.

1. Machining technology: Typical applications in the domain of traditional machining (TM) include machining of press tools, mould making, turbine blades cutting, and drilling at inclined angles (Figure 19.22). Stiffness and precision are expensive to achieve in conventional multiaxis grinders. The hexapod grinder offers a cost-saving upgrade and a flexible architecture suited to precise grinding. Typical application of hexapod grinders includes tool grinding and precision grinding of ceramics. Hexapods have a multitude of applications in the domain of nontraditional machining (NTM). They provide contour machining capability, so they are equipped with lasers for cutting, welding, or hardening. Similarly, they could be set up for high-pressure water jet machining (WJM).
2. Precision assembly technology: Hexapods are used for delicate welding in automatic assembly lines and aircraft production. The small format of low cost hexapods, which plug into the back of the PC, are used for such application. Coupled with the Art-to-Part control software, these smart hexapod centers will be easy to operate.
3. Measuring technology: Hexapod could be an ideal shop-floor coordinate-measuring machine (CMM).

FIGURE 19.22 A typical hexapod machining application.

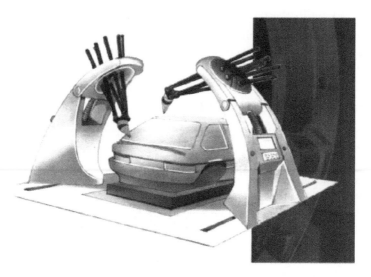

FIGURE 19.23   Hexapod car painting station.

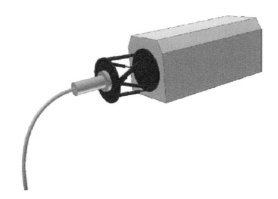

FIGURE 19.24   Fiber adjustments using a micropositioning hexapod.

4. Car-painting station: A pair of hexapods mounted on a simple structure is shown in Figure 19.23. This system is expandable for different applications such as milling where a milling tool can be used in place of a painting nozzle gun.

5. Electronic industry and fiber handling applications: Micropolishing hexapods are used in fields that require very accurate positioning such as the electronic industry, semiconductor, and fiber handling applications. Figure 19.24 shows a hexapod used for fiber alignments where it is attached to the moving platform.

19.4   REVIEW QUESTIONS

1. What are the major components of an industrial robot?
2. What is the main advantage of a robotic unmanned cell versus a human worker in a manned cell?
3. Explain tactile and voice sensing as applied to intelligent industrial robots.
4. List three types of robot control.
5. What are the advantages of a SCARA robot? For what operation is it best suited? What type of control do you suggest for SCARA?

6. What are the types of manipulators used in IRs?
7. What are the recent developments in robotics?
8. Discuss the types of actuators in industrial robots. What are their characteristics and applications?
9. What type of control do you suggest for IRs intended for the following operations: arc welding, spot welding, deburring, and loading/unloading machines?
10. Write brief notes on these methods of programming: walk-through, lead-through, and off-line.
11. Define by symbolic notation (P for translation and R for rotation) the following robot manipulators: Cylindrical, SCARA, and Articulated.
12. Using neat sketches, differentiate between telescopic and ball screw strut hexapods.
13. Draw a neat sketch to illustrate extending, panning, rotating, and twisting of a hexapod mechanism.
14. In what applications is the magnetic bifurcated ball hexapods best suited, and why?
15. Explain the following terms as applied to hexapods: flexibility, agility, scalability, dexterity, and scalability.
16. What are the basic elements of a ball screw hexapod?
17. Discuss the main applications of hexapod in manufacturing technology.
18. What is the main purpose of the Art-to-Part software as applied to hexapods?

19.5   PROBLEMS

1. For a new robot installation, the total cost of an industrial robot including tooling is $80,000. The estimated annual maintenance and programming cost is $10,000 for one shift and $12,000 for two shifts. The robot replaces one worker whose annual salary including overhead is $30,000. What would be the payback period P for one- and two-shift use?
2. Five IRs are replacing the workers in a production line. Each will cost $70,000, including accessories. The annual maintenance is estimated as $4000/robot, and the programming cost for the whole system is $12,000. What is the payback period P, as based on one shift, if the annual salary of each worker is $25,000, including overhead?

BIBLIOGRAPHY

AKIMA, 1997. *First European Conference on Advanced Kinematics for Manufacturing Applications*, Hannover, Germany.

Degarmo, E. P., Black, J. T., and Kohser, R. A. 1997. *Materials and Processes in Manufacturing*, 8th ed. Upper Saddle River, NJ: Prentice-Hall.

Doyle, L. E., Keyser, C. A., Leach, J. L., Schrader, G. F., and Singer, M. B. 1983. *Manufacturing Processes and Materials for Engineers*, 3rd ed. Upper Saddle River, NJ: Prentice-Hall.

El-Midany, T. 1994. *Computer Automated Manufacturing and Flexible Technologies*, 1st ed. Mansoura City, Egypt: Mansoura University.

Geodetic. 1997. *Hannover Exhibition, Hexapod—Breakthrough*, Technical information.

Ingersoll Co., 2000. Octahedral hexapod design promises enhanced machine performance, *Research and Data for Status Report 92-01-0034*.

Koren, Y. 1983. *Computer Control of Manufacturing Systems*, 1st ed. New York: McGraw-Hill.

Schey, J. A. 2000. *Introduction to Manufacturing Processes*, 3rd ed. New York: McGraw-Hill.

Stewart, D. 1965. A platform with 6-DOF, *Proceedings of the Institute of Mechanical Engineers*, 180: 371–386.

U.S. Department of Labor, Occupational Safety & Health Administration, n.d. *OSHA Technical Manual*. http://www.osha.gov/dts/osta/otm/otm_iv/otm_iv_4.html, Accessed 4/3/2009.

Volker Kreidler and Siemens, A. G. 1997. Hannover Exhibition, Report, Offene Objectorientierte CNC—Steuerungsarchitektur am Beispiel der Hexapod-Maschine.

Warwick Manufacturing. 1993. *Introduction to Industrial Robots*. http://html.rincondelvago.com/robots_2.html.

Waters, F. 1996. *Fundamentals of Manufacturing for Engineers*. Boca Raton, FL: CRC Press, Accessed at 4/3/2009.

Wikipedia, n.d. Industrial robot. http://en.wikipedia.org/wiki/Industrial_robot, Accessed at 3/29/2009.

Youssef, H., and El-Hofy, H. 2008. *Machining Technology, Machine Tools and Operations*. Boca Raton, FL: CRC Press.

20 Surface Technology

20.1 INTRODUCTION

Component surfaces are required to be treated for one or more of the following main reasons:

- To clean their surfaces
- To improve their surfaces, usually by smoothing and polishing
- To round up their sharp edges (deburring)
- To provide protective coating to enhance their corrosion, electrical, or heat resistance
- To provide a hard surface layer (hard facing)
- To provide decorative finish

It should be emphasized that the case hardening techniques, both diffusion and selective methods, which were discussed in Chapter 6, are designed solely to increase the surface hardening, either thermally or thermo chemically, through conventional heat treatment operations. In the present chapter, we focus on the surface treatment techniques underlined above. Figure 20.1 classifies the techniques dealt with in this chapter.

20.2 SURFACE SMOOTHING

It is the process of conditioning surfaces to achieve mainly a specific degree of smoothness, besides deburring, and removal of lightly attached inclusions or particles.

20.2.1 MECHANICAL SMOOTHING

1. Wire Brushing: It can vary from a hand operation to the use of automatic power driven rotary brushes (for large parts) to remove loose rust and scales. It is cheap and easy to use, and it leaves a dry surface. Wire brushing is adequate for many noncritical applications. For example, when cleaning prior to shop repairs, preparing for welding, and painting—in another modification—the brushes are replaced with fiber wheels that are loaded with abrasive.

2. Hand Grinding and Belt Sanding: Hand grinding is classified as a cleaning process despite the fact that significant metal removal is usually involved. It is more a fettling/deburring/smoothing/dressing activity. Welded joints need to be dressed. A hand-held portable grinder is frequently used for this purpose. Safety precautions (mask and gloves) should be provided.

3. Polishing and Buffing: Polishing is a process used to enhance the appearance of already cleaned and smoothed surfaces. In polishing, the abrasive used is extremely fine, which removes the multiplicity of fine scratches. The end result is a highly polished surface free from visible imperfections. Polishing is performed by applying fine abrasive to a smooth lint-free cloth and then rubbing it on the work surface by hand in a rotary (or reciprocating) action. The cloth polishing wheel can be charged with abrasive. Buffing is similar to polishing, however, finer abrasives (lapping rouge) are used. Buffing is recommended when an exceptionally smooth, mirror-like, finish is required. Polishing and buffing are labor-intensive processes and should only be used when absolutely necessary for technical or decorative reasons.

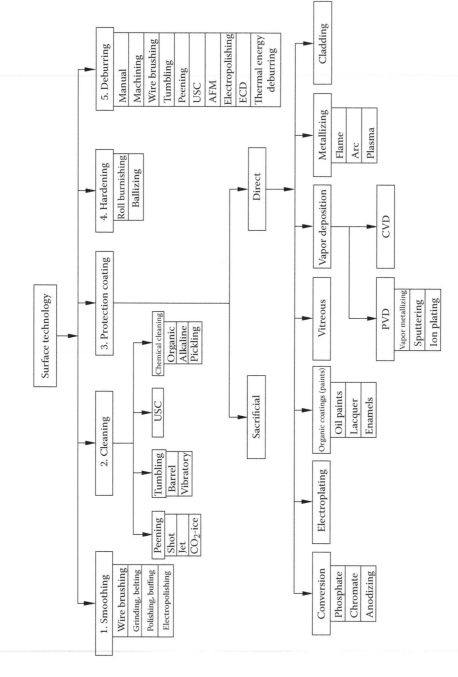

FIGURE 20.1   Classification of surface technology processes.

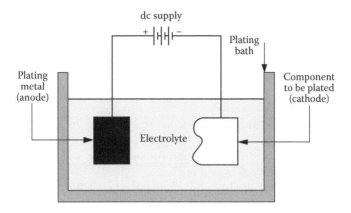

FIGURE 20.2 Electroplating cell.

20.2.2 ELECTROPOLISHING

Electropolishing is basically the reverse of electroplating (discussed later on) because material is removed from the surface rather than being deposited. The material removal obeys the well-known Faraday's law of electrolysis. In an electrolytic cell, if a small, direct current (dc) intensity is applied, a very thin layer (up to 25 μm) of the component surface is stripped away by deplating. This is achieved by reversing the polarity of a normal electroplating cell (Figure 20.2). Both the workpiece (WP) anode and a plate cathode are immersed in a suitable electrolyte such as phosphoric acid. Fortunately, the electrolyte attacks projections and peaks on the workpiece surface at a higher rate of dissolution than for the rest of the surface, thus producing a leveled surface of a mirror-like bright finish.

Electropolishing is a finishing process for machined (ground) surfaces and is particularly adapted for irregular shapes that would be difficult to buff or polish. The process is ideal for such applications as surgical instruments and internal polishing of watch cases. It has been found that bacteriological growth occurs much more slowly on electropolished surfaces than on ordinary machined surfaces owing to their extra-finished quality that makes the process attractive for the pharmaceutical, food, and brewing industries.

20.3 SURFACE CLEANING

Surface cleaning involves the removal of unwanted material contaminations (solid, semisolid, or liquid) from the part's surface, using either mechanical or chemical means. Cleaning is an important manufacturing operation for preparing the surface for consequent surface treatment operations, such as electroplating, painting, hardening, coating, and decorative activities. Surface cleaning covers a multitude of operations. The choice of a specific operation depends on the type of contaminants (soil), which may be rust, scales, chips, debris, lapping compounds, grinding abrasives, lubricants, and any environmental contaminants. There are two types of cleaning methods, mechanical and chemical. They are equally varied and range from coarse cleaning methods involving physical removal of heavy scales to extremely mild processes involving no more than degreasing and fine polishing.

20.3.1 PEEN CLEANING

Peening is the mechanical cold working of surfaces by repeated blows of impelled shot of round nose particles or jets. The highly localized impacts flatten and broaden the surface, but the broadening is restricted by the underlying material, resulting in a surface that is loaded in residual compression. Because the residual compression is subtracted from any applied tensile loading, peening tends to enhance fracture resistance and endurance limit. For this reason, most cyclic loaded components are frequently peened. Surface roughness after peening, however, in the majority of cases increases. Therefore, the objectives of peen-cleaning methods are

1. A primary reason is to clean surfaces.
2. The physical impact between components that occurs in tumbling processes is eliminated.
3. It increases the surface hardness and improves the fatigue and corrosion resistance. It improves surface wearability as in gear teeth.
4. It enhances lubricating characteristics by improving oil retentivity of surfaces.

20.3.1.1 Shot Peening

Shot peening uses compressed air containing sand, steel grit, metal shots, fine glass shots, or other forms of abrasives, which is mechanically impelled against the surface to be cleaned. Steel grit tends to clean more rapidly and generates much less dust but is more expensive. It is important to isolate the jet cleaning operation from the surrounding environment (e.g., enclosed cabinet). The component(s) to be cleaned is loaded into the working area of the jet-blaster cabinet, and the blast is directed either manually at dirty surfaces, or less commonly, the workpiece is moved around under one or more fixed jets. Manual manipulation of either the jet or the component is carried out through a rubber glove box. When the parts are large, it may be easier to bring the cleaner to the part rather than the part to the cleaner.

A common technique of the process is sand or shot blasting, where only abrasives are carried out by a high velocity blast of air of relatively low pressure (6 bar) is used. Fortunately, it is the same pressure of the air found in most factory piped air supplies. Much lower pressures are used when cleaning nonferrous metals. In sand blasting, the sand should be clean, sharp-edged silica sand. Sand blasting is limited to surfaces that can be reached by the blast; however, if the blast is properly directed, internal surfaces may be effectively cleaned. The process cannot be used when sharp edges or corners must be maintained. An advantage of jet blasting is that the surface finish achieved provides a particularly good key for subsequent surface treatment operation such as painting. Special portable units of shot peening that invariably use water mixed with chemical agents are frequently used to remove atmospheric contamination from the outside of buildings.

20.3.1.2 Jet Peening

Jet peening types include abrasive jet peening (AJP), water jet peening (WJP), and abrasive water jet peening (AWJP). These processes are efficiently used for delicate cleaning jobs. Moreover, they promote the surface hardness and improve the fatigue resistance of the peened parts. In WJP (and AWJP, where abrasives are added to water), the WJ hits the surface to be cleaned at a supersonic speed v (2–3 mach). The kinetic energy KE of the jet converts spontaneously to high-pressure energy p at the target surface ($p = \rho v^2/2$, where ρ is density of water), thus removing the contaminants and increasing the surface hardness as a result of cold forming. The fatigue resistance is also improved. Depending on the working conditions, a hardness layer thickness ranging from 20 to 80 μm is achieved. Polymers, if added to water, enhance the jet coherency and improve its peening effect. WJP is less pollutant as compared to AJP. In AJP, the cleaning and hardening effect is due to the impingement of abrasives (Al_2O_3, SiC). Similar to sand blasting, small lot blasting is commonly done in a cabinet where the operator from the outside manipulates a nozzle through safety gloves. In mass production, nozzles are mounted in fixed positions in a large enclosure. WPs are carried past the nozzles on rotating tables, conveyors, etc.

20.3.1.3 CO₂–Ice–Pellet Peening

This is an environmental friendly cleaning process that has been adopted for industrial applications since 20 years ago (Lockheed, U.S. patent, 1987). CO_2 gas is liquefied by cooling to a temperature under $-78°C$, then cooled for ice making, which is palletized in a stick or ball form (1–3 mm diameter), then transported as dry pellets to the cleaning location. The pelletization is generally performed by gas producers, which also produce the relevant cleaning equipment. Figure 20.3 illustrates schematically the basics of this peening process; the dry CO_2 pellets ($\rho = 1.3$ g/cm³) are sucked owing to the high negative pressure developed in the Laval nozzle induced by the side air flow. The ice pellets are directed from the nozzle orifice to the target to be cleaned (placed at suitable SOD), thus impinging it at a supersonic speed.

The mechanism of cleaning may be described in the following steps:

- The target is cooled due to the impingement of ice pellets.
- Differential contraction of metal and contaminants (paint) takes place.
- The contaminants become brittle and crack when subjected to stresses due to pellet impinging.
- The ice pellets melt and convert to gas spontaneously.
- The contaminants are removed with the action of high-pressure gas.

Important applications of the process include

- Cleaning of buildings from outside
- Cleaning of apparatus and machinery
- Removing old paints of ships
- Cleaning of planes and vehicles
- Cleaning of castings, forgings, and welded constructions
- Cleaning of tanks and containers of food and conservative industry
- Deburring applications

20.3.2 BARREL TUMBLING AND VIBRATORY FINISHING

These are mass-finishing processes to clean and deburr components. These processes are adaptable for batches of continuous and automatic flow. Tumbling is confined to finish outside surfaces, but vibratory finishing, under appropriate conditions does well in finishing holes and internal recesses. Vibratory finishing equipment costs about 50% more than that of tumbling. Both processes are applied for ferrous and nonferrous metals, plastics, rubber, and wood. They clean castings, forgings, and screw machine products, remove burrs, fins, skin, scales, take off paint and plating, improve

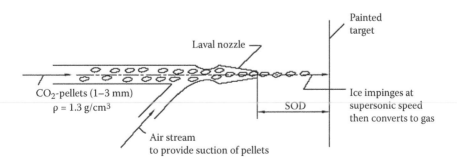

FIGURE 20.3 Schematic of shot peening by CO_2 ice pellets.

surface finish and appearance, and relieve surface strains. Some reduction in size may be experienced, but results are uniform in each lot.

1. Barrel tumbling: Tumbling is a coarse cleaning process for cleaning and deburring of castings. It involves placing castings inside a circular steel drum, which is then rotated typically at 10–15 rpm: the castings are randomly hitting one another sufficiently hard to clean and deburr sharp edges and to shake free any molding sand not removed during fettling. Loose materials such as sand, granules of granite, slag, or ceramic pellets can be added in the tumbler (Figure 20.4). Tumbling is usually done dry, but it can be also performed with an aqueous solution. To enhance cleaning, chemical compounds can be added, and rust inhibitors are sometimes provided. This process can be a very inexpensive way to clean castings. The cycle time is often long (1–10 h), and the process can be quite noisy. The so-called barrel burnishing is considered as a refined form of tumbling and is usually done wet using lubricating and cleaning agents, such as soap. The nonabrasive material used in tumbling is replaced by mesh-screened natural or artificial abrasives. The volume ratio of abrasive media to work should be about 2 : 1, so the workpieces rub against the media and not each other to produce smoother surfaces.

2. Vibratory finishing: In contrast to the barrel processing, vibratory finishing is performed in an open rubber or plastic lined bowl near filled with workpieces and media. The bowl is vibrated at a frequency ranging from 15 to 60 Hz, and with amplitude of 3–10 mm (Figure 20.5). The action makes the entire load rotate slowly in a helical path, and at the same time the whole mass is agitated, thus scoring, trimming, and finishing action takes place through the mixture. Therefore, vibratory finishing is much faster (from 2 to 15 times) than tumbling. The frequency and the amplitude are determined by the size, shape, weight, and material of the part, as well as the media and compound used. The process is less noisy, easily controlled, and automated. In addition, open bowls allow for direct observation during the process.

   A significant objective of the media is to prevent parts from impinging upon each other during operation. Fillers such as scrap punchings, leather scraps, and saw dust are often added to provide bulk and cushioning cleaning media include slag, sand, corundum, limestone and hardwood shapes, or synthetic media. These fillers containing 50–70 wt% of

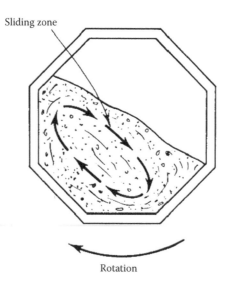

FIGURE 20.4 Barrel tumbling.

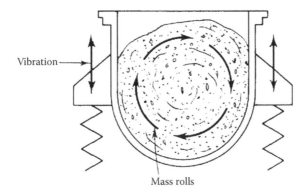

FIGURE 20.5   Vibration finishing.

abrasives, such as alumina, emery, flint, and SiC embedded in a matrix of ceramics or resin plastics. Steel media without abrasives are frequently specified for burnishing and light deburring. Compounds are added to the media and WPs. Cleaning compounds such as dilute acids and soaps are used to remove excessive soil. Corrosion inhibitors are particularly important when steel media is being used. Liquid compounds may also provide cooling to WPs and media.

20.3.3 ULTRASONIC CLEANING

Ultrasonic cleaning (USC) is only applicable for small parts when high-quality cleaning is required. This process is principally different from those so far described. The parts to be ultrasonically cleaned are placed in a wire mesh basket, which is then immersed in a liquid (often water) bath (Figure 20.6). A number of piezoelectric transducers are externally bonded to the bottom of the liquid bath. These transducers are energized by a high-frequency (HF) power source (10 W/L of tank capacity) at a high-frequency current (ranging from 40 to 100 kHz) (Figure 20.7). When the power is switched on, intense cavitation bubbles that form and implode provide the majority of cleaning

FIGURE 20.6   Ultrasonic cleaning.

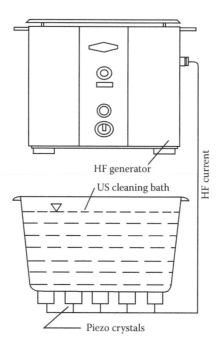

FIGURE 20.7   Ultrasonic cleaning bath with piezo-crystals bonded to the bottom, energized by high-frequency generator (10 W/L tank capacity).

action. Upon reaching a certain size (0.15 mm diameter), they implode (collapse inward). As each bubble collapses, a pressure wave is created in the fluid and propagates at a fantastic high speed and blasts away the contamination from the dirty surface (Figure 20.8). The sustained generation, growth, and implosion of huge numbers of bubbles throughout the liquid bath is referred to as the cavitation effect. USC units ranging from portable (50 W) to large size stationary units (500 W) are

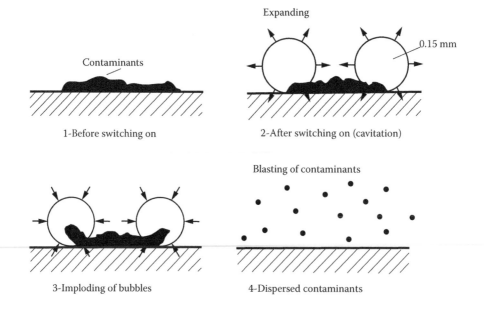

FIGURE 20.8   Stages of ultrasonic cleaning (cavitation and imploding of bubbles).

FIGURE 20.9 Delicate components cleaned ultrasonically.

FIGURE 20.10 Ultrasonic cleaning of highly contaminated valve seat (cleaning time, 3 min).

available in the market. The cleaning of this process is generally quicker and more efficient than that achievable by most methods; moreover, both intricate and delicate components can be cleaned without any risk of damage (Figure 20.9). If gross dirt, grease, and oil are removed prior to USC, excellent results can be obtained in 1–3 min (Figure 20.10). Because of the ability to use water-based solutions, USC has replaced many of the environmentally unfriendly solvent processes.

20.3.4 CHEMICAL CLEANING

Another highly effective way of surface cleaning is using chemicals for removing oil, dirt, scale, or other foreign material that may adhere to the surface of a product. Such a method is used for preparing surfaces for subsequent painting or electroplating processes. While normal care has to be exercised when using mechanical cleaning methods, the use of chemicals is always a hazardous affair, requiring special care and protection of the personnel engaged in such work. The more mechanized operation, the less the risk to staff. Chemical methods often require the disposal of contaminated solutions that are generally hazardous, toxic, or environmental unfriendly materials. Unfortunately, environmental issues have forced the almost complete demise of vapor degreasing. While the solvents used in vapor degreasing are chemically stable, have low toxicity, are nonflammable, evaporate quickly, and can be recovered for reuse, they also have been identified as ozone depleting compounds and have essentially been banned from use. Process changes to comply with added regulations can significantly shift its economics. For this reason, most manufacturers have been forced to replace with water-based processes using alkaline, neutral, or acid cleaners or processes using chlorine-free, hydrocarbon-based solutions. Some cleaning fluids are used in conjunction with electrochemical process for more effective cleaning.

The selection of a suitable cleaning method depends on

- What soils have to be removed
- How clean the surfaces have to be
- The lot size to be processed
- The part configuration
- The part material
- Desired surface finish

The methods of chemical cleaning are categorized as

1. Organic solvent cleaning: Oils and greases are the most common corrosion inhibitors used in industry. If parts are protected by these substances, they should be removed by organic solutions. These solvents are normally applied by
   a. Immersion or spraying: The common solvents in this case are kerosene, mineral spirits, chlorinated hydrocarbons, such as methylene chloride, and trichloroethylene, acetone, benzene, and various alcohols. Small parts are generally cleaned by immersion, with or without agitation, or spraying. Large parts can be cleaned by spraying or wiping.
   b. Vapor degreasing: An organic nonflammable solvent (e.g., trichloroethylene) in the bottom of a tank is heated to cause it to vaporize at 40°C–125°C. The vapor is then passed over the component surface to be degreased, but the vapor condenses on the relatively cool component, the condensate dissolves oil and grease and runs off, removing contaminants, and residues collect at the bottom. Components are then dried when they come out of the tank. The key to this process is that the surfaces onto which the solvent vapor is required to condense must always be cold, or at least well below the condensation temperature of the solvent. Small light workpieces may become hot quickly and cease to condense vapor before they are cleaned. Thin sheet materials, particularly those of high thermal conductivity (Cu and Al), face the same problem. Some parts that have exceptionally heavy soil to be cleaned may be immersed in warm or violently boiling liquid or sprayed before vapor degreasing. Volatile solvents are quite toxic and can be dangerous if used or handled improperly. Strict regulations for safety of the workers and protection of the environment must be closely followed by users.
2. Alkaline surface cleaning: Organic solvents are effective cleaning agents as long as the surface contaminants are soluble organic solvents, but if they are not, then alkaline solutions are normally used. Furthermore, adding a water-soluble detergent to the alkali creates a highly versatile cleaning medium that removes organic and most other forms of surface contamination. This explains why alkali-based cleaning is the most popular of all chemical cleaning processes. Alkaline solutions can be applied to work surfaces by immersion or spraying and are also used as active agents in steam cleaning when the parts to be cleaned are too large to be dipped. The cleaning is then followed by a water rinse to remove all residues of the cleaning solution, as well as flush away some small amounts of remaining soil. A drying operation is also required because the aqueous cleaners do not evaporate quickly, and some form of corrosion inhibitor may be required. Steel and cast iron are particularly suitable for this type of cleaning. However, because of affinity of steel for rusting, thorough drying is essential if the parts are not immediately used after water rinsing. Certain nonferrous metals such as Al, Zn, Sn, and brass should not be alkali cleaned.

   Environmental issues related to alkaline cleaning include
   - Reducing or eliminating phosphorus effluent
   - Reducing toxicity and increasing biodegradability
   - Recycling the cleaners to extend their life and reduce the volume of discard

3. Pickling and oxidizing: Pickling is just the chemical removal of surface oxides and scale from metal by acid solutions. Thus it is a chemical descaling process rather than a surface cleaning one, although it is partially ineffective if, before pickling, the component's surface is not properly degreased. The process is commonly applied to remove scales from rolled shapes, wire, sheets, heat-treated steel parts, wrought, and cast aluminum parts, etc. In some applications such as on aluminum, it is called oxidizing. It is essential that overpickling does not occur, as this results in surface pitting and, in the case of steels, hydrogen embrittlement may occur. Common pickling solutions contain sulfuric or hydrochloric acids with water. Nitric and hydrofluoric acids are used for some applications. Pickling is usually done by immersion for several minutes. In pickling, acids cannot penetrate to a surface covered with dirt; thus parts must be cleaned first. After pickling, the part must be completely neutralized by an alkaline solution and then rinsed. Any residue of acid will harm the paint or other subsequent coating.

20.4   SURFACE PROTECTION

The surface of a component frequently needs to be protected from immediate hostile environment (sacrificial protection) or to be coated to enhance its resistance against environmental attack and/or to enhance its wear resistance (direct protection).

20.4.1   Sacrificial Protection (Cathodic or Galvanic Protection)

As the name implies, the process relies on sacrificing one material to protect another (the work-piece). The work material (e.g., steel) if surrounded by a corrosive environment (atmosphere, earth, and seawater) will corrode extremely quickly. However, if it is directly or electrically connected to another sacrificial metal (Zn, Cd, and Mg) that is more reactive in the particular environment, the attacking medium will, surprisingly, leave the steel workpiece and attack the sacrificial metal. This will continue until the sacrificial metal has been eroded after which the cathodic protection is lost. Hence the steel will start to corrode, if a new supply of sacrificial metal is not provided.

This method is applied to protect the ship bodies by providing Zn pieces at some strategic points along the body length below the water level. They are periodically replaced to maintain corrosion resistance (Figure 20.11a). Another application is to protect a steel pipeline against corrosion using a magnesium sacrificial electrode (Figure 20.11b). An effective method of cathodic protection is made by using an external dc rectifier (Figure 20.11c) to protect an underground tank. In such a case the negative terminal of the rectifier is connected to the structure to be protected, while the other terminal is joined to an inert anode (often graphite), buried in the soil. High-conductivity backfill material provides good electrical contact between the anode and the surrounding soil and thus closes the current circuit. The cathodic protection is especially useful in preventing corrosion of water heaters, underground tanks, pipelines, and marine equipment. Another application based on the galvanizing process applies a layer of Zn to the surface of steel by hot dipping, which is an alternative to electroplating. In a corrosive environment and in case of surface damage, Zn is anodic and thus will cathodically protect the steel (Figure 20.11d). A common application of galvanizing is the protection of steel sheets that are used as a cheap roofing material and as a steel coating in vehicle body panel manufacture.

20.4.2   Direct Protection

20.4.2.1   Conversion Coatings (Chemical Reaction Priming)

A coating forms on the component surface as a result of chemical or electrochemical reactions, thus providing the desired properties to the surface texture. These properties range from enhanced

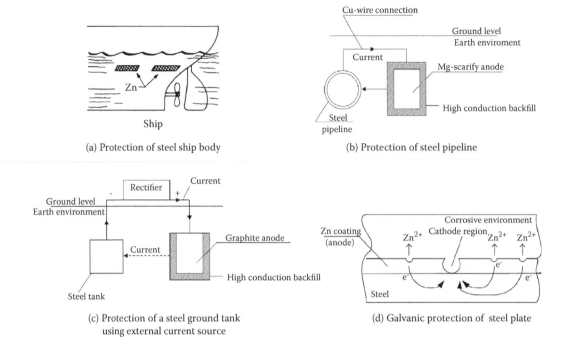

FIGURE 20.11 Types of sacrificial protection.

corrosion resistance, improved surface for subsequent painting, and lubricant retentions on sheet, wire, and tube stock to enhance the cold working properties by applying lubricants. Lubricants may not always adhere properly to workpiece surfaces, particularly when subjected to high normal and shearing stresses during forming. This condition represents a problem especially in forging, extrusion, and wire drawing of steels, stainless steels, and high-temperature alloys. In these applications, acids react with the workpiece surface, leaving a somewhat rough, spongy surface that acts as a carrier for the lubricant. After treatment, borax or lime is used to remove any excess acid from the surfaces. A liquid lubricant such as soap is then applied to the coated surface. The lubricant film adheres to the surface and cannot be scraped easily. Zn-phosphate conversion coatings are often used on carbon and low-alloy steels, while oxalate coatings are used for stainless steels and high-temperature alloys. The three principal types of conversion coatings are phosphate, chromate, and anodizing. The first two are due to chemical reactions, which are applied to both ferrous and nonferrous metals, whereas the third is due to electrochemical action, being mainly suitable for Al, Mg, Zn, Ti, and their alloys.

- Phosphating: It is a widely used process where a diluted phosphoric acid is applied to the surface (usually steel, Al, or Zn). The chemical reaction provides the surface with a corrosion-resistant spongy phosphate layer to a depth of about 50–75 µm, with no significant increase of size, as with all conversion coatings.
- Chromating: It is exactly similar to phosphating; however, a very thin amorphous film (0.1 µm) forms that resists abrasion, provides an excellent base for painting, and may be dyed to serve as a decorative finish. In chromating, acidic solutions of Cr compounds are occasionally applied with electrolytic assistance.
- Anodizing: The workpiece (commonly Al and Mg) is immersed in acid solution (sulfuric acid) and connected as anode in an electrolytic cell. Anodic coating provides a layer ranging from 2 to 250 µm thick, thus enhancing the corrosion resistance of the thin oxide film (1–2 µm) that naturally protect these metals against corrosion.

Although anodizing is used to prepare surfaces for painting, it is extensively used in the aerospace in their unpainted condition. The surfaces can also be dyed after anodizing to give an enhanced visual appearance.

20.4.2.2 Electroplating

The main objectives of the process are to produce metal coating that imparts corrosion or wear resistance and improves appearance. It also contributes to rework of worn parts by increasing their sizes and to stop off areas on steel parts from being carburized during heat treatment. Virtually all commercial metals can be electroplated, including Al, Cu, brass, steel, and zinc-based die castings. Plastics can be electroplated, provided that they are first coated with an electrically conductive material. The most common platings are Zn, Cr, Ni, Cu, tin, gold, platinum, and silver. The principle of electroplating is illustrated in Figure 20.2. The work to be plated is immersed in a water solution of salts of the metal to be plated and connected as the cathode in a direct current electrolytic cell. Table 20.1 summarizes the commonly used plating materials and their typical coating thicknesses in micrometers.

Anodes of the coating metal replenish the solution when the current is flowing and the anodic metal ions are attracted to the work to form the coating layer. The rate of deposition and the properties of the coating, such as hardness, uniformity, and porosity, depend on the proper balance among the electrolyte composition, current density, agitation, solution acidity, and temperature of electrolyte. Efforts should be directed toward the bond strength between coating and substrate. For instance, the higher the current density, the faster the metal is deposited. However, a rate above a critical value for a specific electrolyte and temperature results in a rough and spongy coating.

The electroplating process presents several difficulties of concern. One of these is that plating is not always uniformly deposited. A thicker layer builds up on edges, is thin in recesses, and almost does not exist in some corners. Therefore, part shape should be properly designed to compensate for differential material build up. Figure 20.12 illustrates good design features and those to be avoided for some plated components. Parts with irregular surface configurations can be uniformly covered if solutions of better throwing power are used. Electroplating does not hide defects in the work surface. Accordingly, the surfaces must be well finished if they are to be plated for appearance as well as corrosion resistance. Hydrogen released at cathode causes embrittlement of hardened or cold-worked steel work. Its quantity is kept to a minimum by proper control of the operation. Embrittlement can be alleviated by heating after plating.

Parts such as automobile bumpers, which are nickel or chrome plated, are commonly given an initial copper plating of 5–15 μm thick. This adheres well to and effectively covers the steel and its surface is easier to buff out than steel. Purely decorative chrome plating is ordinarily only 0.5–1 μm thick to maintain brightness over a protective Ni intercoat. This type is mainly used for domestic products and office furniture. Tin and Zinc are deposited continuously on steel sheets for further working. Zinc and cadmium are deposited on parts for corrosion resistance, but cadmium cannot be used for food applications for toxicity. Cu is deposited for corrosion protection and for making electrical circuits. Ni is extensively used for corrosion resistance. Cr of layer thickness ranging from 10 to 500 μm is deposited directly on the base metal for hard Cr coating (60–70 HRC), which may

TABLE 20.1

Commonly Used Plating Metals and Corresponding Coating Thicknesses

	Zinc	Cadmium	Tin, Tin Alloys	Precious Metals (Silver, Gold, Platinum)	Cu, Cu Alloys	Ni	Cr, Decorative	Cr Hard Layer
Coating thickness (μm)	2.5–25	5–10	0.1–2	0.5–100	10–50	8–500	0.3–0.8	75–500

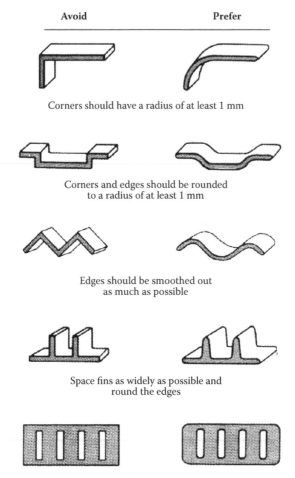

Avoid   Prefer

Corners should have a radius of at least 1 mm

Corners and edges should be rounded
to a radius of at least 1 mm

Edges should be smoothed out
as much as possible

Space fins as widely as possible and
round the edges

Eliminate sharp edges and corners of slots

FIGURE 20.12 Good and bad design features of some electroplated components. (Adapted from Metal Finishing Association, Birmingham, UK.)

need diamond polishing, but then impart wear resistance to dies and reduce their adhesion to many workpiece materials including Al and Zn. Precious metals are used for decorative purposes and, primarily in electronic devices, for corrosion protection.

The basic unit of electroplating equipment is the electrolyte tank, which is made of various materials such as lead, rubber, plastics, or tile to resist alkaline and acidic electrolytes. Large pieces are individually suspended, while small pieces are mounted on racks. A batch of pieces is put in a nonconducting perforated barrel. The electroplating process entails cleaning, washing, and rinsing in addition to the deposition process. Electroplating is commonly done manually for small lots, but labor work is saved and quality controlled better by automatic and even NC-plating machines for moderate- to large-quantity production. Figure 20.13 illustrates a return-type equipment, which is tied with a conveyor system running through a series of baths. In certain industries, specialized machines have been developed for continuously electroplating sheet metal strips (Figure 20.14). Exhausted fluids from electroplating tanks can be highly polluting. A common remedy to eliminate pollution is to resort to chemical reactions to convert the waste to nontoxic substances.

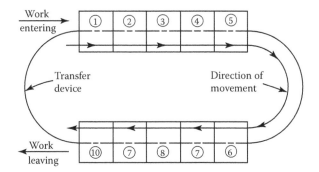

FIGURE 20.13 A return-type electroplating equipment. (Modified from Doyle, L. E. et al., *Manufacturing Processes and Materials for Engineers*, 3rd ed., Prentice Hall, Upper Saddle River, NJ, 1985.)

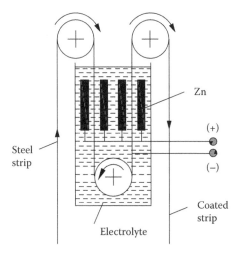

FIGURE 20.14 Electroplating of steel strip (continuous process).

20.4.2.3 Organic Coatings (Paints)

Organic coating is the topmost of many products to improve their appearance and corrosion resistance, which is found in all domestic and commercial environments. It has a wide range of properties: flexibility, durability, hardness, abrasion and corrosion resistance, color, texture, and gloss. Organic coatings can be applied on almost all materials offering unlimited glossy colors thus providing excellent decorative appearance. They are generally superior to conversion coatings and usually cost less than metallic coating and thus are preferred if they function satisfactorily.

20.4.2.3.1 Important Fields of Application

- Naval aircrafts that are subjected to humidity, rain, seawater, pollutants, aviation fuel, battery acids, dust, gravel and stones, and deicing salts.
- Aluminum structures having a dual layer consisting of epoxy primer (important for coating durability) and polyurethane topcoat. This dual layer coating has a lifetime of 6–8 years.
- Organic coatings are sometimes applied to sheets, plates, and coil stocks on continuous lines with thicknesses ranging from 0.005 to 0.2 mm. The coated sheets are used for applications such as television cabinets, house appliances, paneling, shelving, and metal furniture.

20.4.2.3.2   Classification of Organic Coatings (Paints)

Depending on the type of solvent, organic coatings may be classified as

- Oil paints
- Lacquers
- Enamels (not to be confused with vitreous enamels)

The basic constituents of paint are

- Film forming material (natural or synthetic).
- Pigment(s) to provide the opacity (hiding power), color, and surface texture (matt or glass) required.
- Solvent or carrier that enables the paint to be easily applied.
- Drying agents may be added.

1. Oil paints: Oil paint is a dispersion of metallic pigments such as white lead, in vegetable drying oil (linseed), and solvent thinner and perhaps dryers. The thinner evaporates and the oil oxidizes to form the film. Drying time depends on the oil and drying agents added, but it is relatively long. These paints are used domestically and are not suitable for engineering applications, partly because of their long drying times.

2. Lacquers: Industrial applications require coatings that are not only durable and attractive but also easy and quick to apply and offer a significant degree of surface hardness. These are the main reasons why oil-based paints have now been superseded by lacquer polymer based ones.

   A lacquer is essentially a solution of plastic resins and plasticizers with or without pigments in a solvent. When the lacquer is applied, the solvent evaporates and leaves a film that can be made quite attractive by polishing. Lacquer is relatively easy to apply and dries quickly, but the film is not as durable and resistant to some solvents as other coatings.

3. Enamels: These are synthetic types that are based largely or entirely on plastic resins and elastomers and harden by polymerization. Synthetic resins can be compounded to be equal or even superior to vitreous enamels.

   Plastic resins and elastomers are provided as liquids or powders.

   - Liquid coatings: These contain a large proportion of organic solvent that evaporates to leave a polymer film. The vapors can cause air and water pollution, requiring costly equipment to meet environmental standards. Accordingly, these coatings have been greatly curtailed. Other liquid coatings with little or no solvent content have become more popular. Special finishes are obtained using metallic pigments, with oils and dryers that give a crinkled surface, silicone additives for high temperature, plastisols, and organisols.

   - Powder coatings (powder spraying): An increasingly popular surface coating process involves the use of plastic powders as the covering medium, which is usually applied by electrostatic spraying in a similar manner of liquid painting. The powders are either thermoplastics (Nylon II, PTFE, and PVC) heated prior to application or thermosets (epoxy resin and polyester) cured after application. Powder coating can be used to coat most metal surfaces. Powder does not contain solvents, which make powder spraying much less hazardous and more environmental friendly than liquid spraying. The powder-sprayed surfaces do not suffer from paint runs. Thicker coatings can be applied in a single pass, and painting time is reduced. Powder coating requires heating the work to form the film of powder coat.

20.4.2.3.3 Preparation for Painting

The first step for preparing a surface for coating is to be thoroughly cleaned. A conversion coating, especially phosphating, is frequently applied. Sometimes one coat of paint is enough, but usually two or more layers of different coatings are applied to obtain combinations of desirable characteristics. An undercoat may be applied to obtain a good base for the finish coat that provides a desired color, luster, and appearance. Undercoatings include primers that form a bond and inhibit corrosion on the interface, while intermediate coats serve as fillers or sealers. Usually, powder coating does not need a primer.

20.4.2.3.4 Painting Methods

Painting may be applied by different methods, which are

1. *Brushing* is easy but slow and only used for domestic purposes. Roller application is suitable for mechanization of flat surfaces.
2. *Dipping* demands little equipment and is easily mechanized, but stirring is needed, and workpieces without pocket can be immersed easily. Drips and tears that collect by dropping at the bottoms can be drawn off electrostatically. Another way of collecting residual paints is to whirl the wet pieces in a centrifuge.
3. *Flow coating* is achieved by pouring paint on workpieces and recirculating the runoff. It is fast and applicable for small and medium size work.
4. *Spraying* is the most used method of industrial painting. It is fast, versatile, and uniform. It is based on the principle that a liquid stream atomizes when it exceeds a certain speed by introducing a liquid (or powder) into a high-velocity stream of compressed air released through a fine nozzle. The paint may be heated as discharged with a high velocity through the nozzle at a pressure of about 350 bar. In another method, the paint is slung off the edge of a rapidly rotating disc-shaped atomizer. The spray gun may be directed by hand, but for continuous production, it is commonly automated to spray pieces passing on a conveyor. Robots and hexapods are being utilized for automated paint spraying. Manual touch-up may be necessary at the end of an automated line; however, the cost is less than if each piece were hand sprayed completely. An advantage of automated painting is that it can be done at 40°C, which is ideal for paint but not for people. Most industrial painting is done in well-ventilated enclosures or booths, which promote cleanliness and quality control. Because of the toxicity and flammability of solvents, enclosures are necessary to help control air and water pollution, and aid the recovery of excess paint. Much paint can be lost when sprayed. This loss can be considerably reduced (5%–10%) by using an electrostatic charge on atomized liquid paint or powder (Figure 20.15). A potential of 80–150 kV

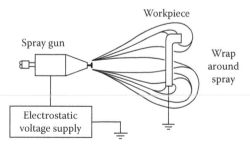

FIGURE 20.15   Electrostatic-painting gun.

charges the paint slung from a spray gun. Because of the electrostatic attraction principle involved, surfaces that are not in the line of sight of the applicator gun can still be coated with ease. A "wraparound" effect occurs to deposit the coating material with almost no loss on all exposed external and internal surfaces, although restrictions may cause variations in film thickness.

5. *Powder bed coating* is done by dipping a heated metal piece in an agitated swirling pool of fine powder suspended in a stream of air. The powder is a mixture of resin, catalyst, pigment, and stabilizer, which melts on the hot workpiece surface to form a uniform film. The workpiece is then oven heated to flow out and cure the coating. Coatings of up to 1.5 mm thick can be obtained in one immersion by powder coating, but a thickness of less than 0.25 mm is not achievable. Powder coatings are pore free, smooth, tightly held, and cover edges and corners as well. However, a uniform coat is difficult to achieve on parts of variable sections, because the heating effect is different for different masses.

20.4.2.3.5   Baking (Drying and Curing)

Paints can conventionally be finally baked in ovens heated by electricity, oil, gas, or steam. Infrared, ultraviolet rays, laser beam, and electron beam are recently used to cure organic coatings by radiation. These methods save fuel, reduce pollution, and increase productivity. Curing needs seconds instead of many minutes and even hours by conventional baking. Baking using IR lamps has become popular because the process can be easily adjusted and controlled.

20.4.2.4   Vitreous Coatings (Porcelain and Ceramic Enamels)

Metals may be coated with a variety of glass-like coatings to provide excellent corrosion, chemical, electrical, heat, and abrasion resistance. Vitreous coatings are usually classified as porcelain and ceramic enamels. The ingredients used in vitreous coatings are finely ground frit of silicates, feldspar alumina (for ceramic enamels), borax and soda ash, and various metallic oxides (for coloring), mixed with clay and water, forming a multicomponent slurry. These coatings are applied by spraying, dipping, flow coating, or brushing at room temperature. Both coatings and substrates are fired (heated) to 500°C–900°C for a few minutes depending on the coat and substrate.

Vitreous coatings are used for coating steels, cast iron, Al, Cu, bronze, stainless steels, and refractory metals. The coating thickness is usually ranging from 0.1 to 0.6 mm. These coatings are smooth, hard, lustrous, and resist high temperatures but unfortunately tend to be brittle and therefore prone to surface chipping and cracking. Porcelain enamels are frequently used as coatings for cooking utensils, cook pots, and frying pans; health service equipment; inner perforated tubes of many washing machines; containers resisting chemical attack; surfaces resisting wear and abrasion; jet engine combustion chambers; and exhausts. Ceramic enamels are coatings of alumina or zirconium oxides, which are frequently used for such applications to withstand repeated arcing. Typical applications include rocket nozzles, wear-resistant parts, hot extrusion dies, turbine blades, and similar applications. Glassing is the application of glassy coating on ceramics and earthen wares to give them decorative finishes and to make them impervious to moisture.

20.4.2.5   Vaporized Metal Coating (PVD and CVD)

In this respect, there are two major techniques that include the physical vapor deposition (PVD) and the chemical vapor deposition (CVD). Both techniques allow an effective control of coating composition, thickness, properties, and generally provide thin coatings of metals and their compounds on metallic and nonmetallic materials. PVD embraces a group of processes in which the material to be deposited is carried out physically to the substrate. These are vacuum metalizing, sputtering, and ion plating, which are carried out in some form of vacuum, in which the substrate is positioned in line of sight relative to the source (evaporant). In contrast, the CVD deposits the material through chemical reactions and generally requires higher temperatures. Hence CVD is a thermochemical process.

20.4.2.5.1 Physical Vapor Deposition

1. Vacuum metallizing (vacuum deposition): In vacuum metallizing the metal or compound (called source or evaporant) to be deposited is placed in a vacuum chamber (10^{-4} bar), along with a substrate workpiece to be coated. When the source is heated to a high temperature, its vapor pressure exceeds that of its environment and evaporates. The vapor condenses over the relatively cold workpiece, which is usually at room temperature or slightly higher, forming a thin coat (0.5–0.7 μm) (Figure 20.16). This coat is used to provide reflective surfaces of decorative appearance or imparts desirable electrical properties for many applications.

   The vacuum minimizes the atomic collision, such that atoms travel in a straight line away from the source. If an entire product is to be coated, it should be rotated in such a way that exposes all its surfaces. Uniform thin coatings can be obtained on complex shapes. Gases (O_2, N_2, or hydrocarbon gases) may be metered into the vacuum chamber to produce oxides, nitrides, or carbides by reactive operation (a transition to CVD). For better bonding, the substrate is often heated. The rate of deposition is accelerated by the application of a dc electric field, where the source is the cathode, and the substrate is the anode. Material is removed from the source in the form of negatively charged ions that are accelerated toward the substrate (cathode). The resulting depositions offer the surface oxidizing-resistant coating for high-temperature applications, electronics and optics, and decorative nature as applied for hardware and jewelry. The thin film produced usually lacks durability and often requires a top coat of transparent lacquer or resin. Dyes or pigments can be incorporated into the top coat to achieve a wider variety of metallic finishes. A thin layer of Al with a pigmented resin top coat can be applied to make the surface look like Cu, brass, bronze, or even gold.

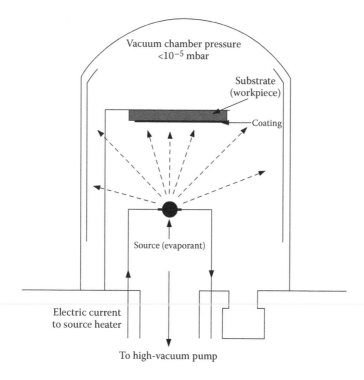

FIGURE 20.16 Vacuum metallizing, PVD. (Courtesy of Edwards High Vacuum International, Crawley. With permission.)

2. Sputtering: Sputtering offers another means of depositing a thin film onto a prepared substrate. It is a form of plasma deposition, which involves bombarding a substrate with high-energy argon ions. Both the substrate and the target are confined in a chamber that is first evacuated and then given a backfill of heavy inert gas such as argon. A high dc voltage (2–6 kV) is applied between target (cathode) and substrate (anode) to obtain the energized argon ions A_r^+. If the applied voltage exceeds the ionization energy of gas, electrons emitted at the cathode strike argon atoms, ionizing them to create plasma. The gas ions are attracted to the target with sufficient kinetic energy to knock metallic atoms from its surface. These atoms are driven to the work with enough energy to bind them to the substrate surface, where they form a film as they collect (Figure 20.17). Owing to the high KE of atom impacting the substrate and the cleanliness of the substrate (due to partial vacuum), the adhesion is considerably stronger than with other deposition techniques. For this reason, sputtering is only applicable where the adhesion strength is of prime concern. However, the main disadvantage of the process is its relatively low deposition rate. Sputtering of Cr on blades to prevent corrosion is an important field of application in the razor and knife blade industry.

   If argon is partially or totally replaced by a reactive gas such as O_2, atoms knocked from the cathode are immediately oxidized, and the oxide is deposited on the substrate in a process called reactive sputtering. In magnetron sputter deposition, the cathode is surrounded by a magnetic field that captures electrons, thus increasing their ionization efficiency and consequently increasing the rate of sputtering. Another advanced sputtering technique involves the use of radio frequency power source (RF sputtering), which is used for coating nonconductive materials such as insulators and semiconductor devices.

3. Ion plating: Ion plating is a combination of sputtering and vapor metallizing. Accordingly, the substrate is reverse sputtered in an inert gas atmosphere, which means that the work is given a negative potential and is highly bombarded by gaseous ions, thus cleaning the substrate thoroughly. The plating material (target) is then heated until atoms evaporate from its surface and are ionized and driven to plate the work surface. The evaporation

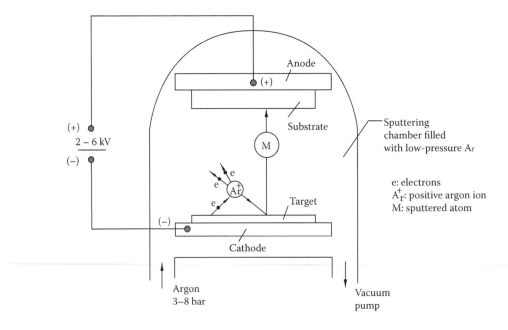

FIGURE 20.17 Sputtering principle (PVD).

rate is kept higher than the rate of sputtering from the substrate. In consequence, a dense adherent coating is formed, while the surface is continually cleaned by sputtering ions. In reactive ion plating, a reactive gas is admitted into the argon gas. The familiar gold-colored TiN coating is formed on cutting tools by evaporating titanium in an Ar–N_2 atmosphere. Coatings are smooth, dense, and have high compressive residual stresses. A more recent development of ion plating is the dual ion-beam–assisted deposition that combines PVD with bombardment by ions produced independently in an ion-assisted gun. This process provides dense coatings of good adhesion on metals, ceramics, and polymers. If reactive atmosphere is used, a compound layer such as Si_3N_4 can be deposited. Ceramic bearings and dental accessories are two important applications.

Concluding remarks concerning the PVD techniques are shown below.

1. The PVD techniques generally provide thinner films than those obtained by all other methods of surface coating. They use less material and energy and create less pollution problems.
2. In PVD, atoms travel in a straight line (line-of-sight deposition). Reentrant shapes cannot be coated.
3. Deposition rate is constant for points lying on the inside surface of a sphere; therefore, substrates are attached to rotary spherical sections.
4. Coatings up to 0.15 mm thick give good corrosion and oxidation resistance.
5. Film adhesion is good on ordinary clean surfaces (vacuum metallizing) and excellent on sputter cleaned surfaces (ion plating), which is the main advantage of the ion plating process.
6. Vacuum metallizing is best suited to coat pure metals with moderate melting points.
7. Vacuum metallizing is fast. A large batch can be coated in few seconds plus 5–15 min needed for evacuation. The equipment cost is high; however, it is justified for batch size and mass production.
8. Sputtering is only 5% as fast as vacuum metallizing; however, it is capable to deposit almost any material. Film thickness can be held to ±5% of coat thickness. Moreover, the workpiece need not be heated, so its surface integrity is not affected.
9. Workpiece is heated when ion plated; consequently, heat-sensitive materials cannot be coated by this method.
10. Ion plating has deposition rates between those of sputtering and vacuum metallizing.
11. Better adhesion and more uniform coverage are developed by sputtering as compared to vacuum metallizing.

20.4.2.5.2 Chemical Vapor Deposition

It is a thermochemical gas plating process, where pure metals, carbides, nitrides, borides, and oxides can be deposited. A substantial area of industrial application is the hard coating of tools made of HSS's, stainless steels, and cemented carbides. The primary limitation of this process is the high temperature required to promote the reactions. The substrate must be capable to withstand the process temperature. Medium temperature chemical vapor deposition (MTCVD) is preferred, because if high temperatures are used, the coating will crack upon cooling due to the large difference in coefficient of thermal expansion between substrate and coating. Furthermore, tool steels have to be reheat treated after coating. A schematic layout of the CVD reactor is visualized in Figure 20.18. The carrier gases are selected according to the desired type of coating. A mixture of hydrogen, methane, and titanium tetrachloride gases are mixed in a gas chamber (1). The reactive gases are directed by the suction pump (2) to the electric heated coating furnace (3), where the carbide tools (4) are positioned on graphite shelves (5), mounted on a stainless steel retort (6). To initiate the

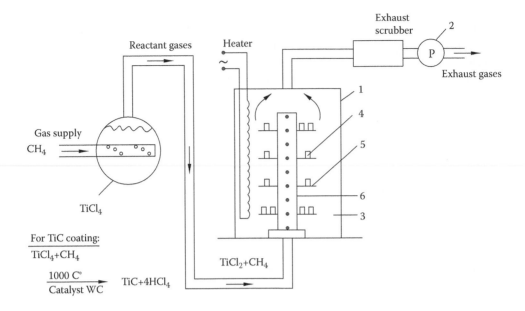

FIGURE 20.18 Schematic of CVD reactor.

reaction, the tools are heated to 950°–1050°C. The following reaction produces a thin coating (5 μm) of titanium carbide TiC, where the WC substrate acts as a catalyst.

$$TiCl_4 + CH_4 \xrightarrow{\text{Heat}} TiC + 4HCl \uparrow$$

If the methane is replaced by nitrogen, a coat of TiN is obtained according to

$$2TiCl_4 + N_2 + 4H_2 \xrightarrow{\text{Heat}} 2TiN + 8HCl \uparrow$$

Alternatively, if ammonia gas is used instead of N_2, then

$$TiCl_4 + NH_3 + \tfrac{1}{2}H_2 \xrightarrow{\text{Heat}} TiN + 4HCl \uparrow$$

For alumina coat, the reaction will be

$$2AlCl_3 + 3H_2 + 3CO_2 \rightarrow Al_2O_3 + 3CO + 6HCl \uparrow$$

Alternatively,

$$2AlCl_3 + 3H_2O \rightarrow Al_2O_3 + 6HCl\uparrow$$

SiC coat is obtained from methyl trichlorosilane,

$$CH_3SiCl_3 \rightarrow SiC + 3HCl\uparrow$$

and Si_3N_4 is obtained according to

$$3SiCl_4 + 4NH_3 \rightarrow Si_3N_4 + 12HCl\uparrow$$

Coatings of TiC, TiN, Al_2O_3, SiC, and Si_3N_4 provide extraordinarily extended tool life in harsh cutting conditions. They offer high hardness, high chemical stability, low friction, and antigalling characteristics. The above coatings may be applied singly or consecutively. Multicoatings allow a stronger bond between coating and substrate and provide good protection. A top coat of TiN effectively reduces tool crater wear considerably (2–3 times) in comparison to uncoated inserts. The control of critical variables such as temperature, pressure, gas concentration, and its flow pattern is required to secure adhesion to substrate. Adhesion must be exceptionally good for tool inserts than for most applications to avoid flaking off under severe cutting conditions.

20.4.2.6 Metal Spraying (Metallizing)

Metal spraying involves propelling a stream of fine molten metal particles at the surface of the workpiece. The stream impacts the surface at speeds up to 100 m/s; thus the particles bond themselves securely to the previously prepared surface. When solidified, the molten particles form the required new metal layer. The more metal that is sprayed onto the substrate the thicker the coating that is built up. The process is extensively used to produce metallic coatings offering specific mechanical properties such as surface hardness, corrosion resistance, wear resistance, and resistance to high temperature environments.

There are three basic methods of creating the necessary molten metal spraying:

1. Flame spraying: The spray nozzle is supplied with a constant supply of metal to be sprayed, in either wire or powder form. As it leaves the nozzle, the metal is melted in an oxy-fuel flame and the molten metal is atomized and ejected to the work surface by compressed air. It solidifies forming a dense, strongly adherent coating (Figure 20.19). The powder option enables metals that are not suitable to deliver as wire to be applied without difficulty.
2. Arc spraying: The spraying nozzle is fed with a pair of wires made from coating metal. They are brought closely at the nozzle exit such that when supplied with a high current, low voltage dc power supply, they sustain an electric arc of 4000°C, which is more than sufficient to melt any coating metal. The coating produced has greater bond strength than that achieved by flame spraying. Arc spraying only requires an electric source and a compressed air supply (Figure 20.20).

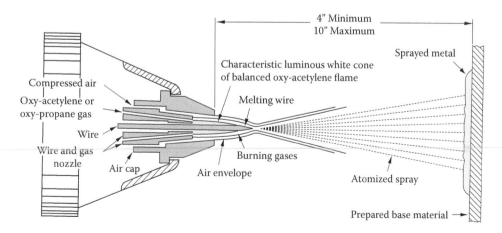

FIGURE 20.19 Flame spraying. (From Sulzer Metco Inc. With permission.)

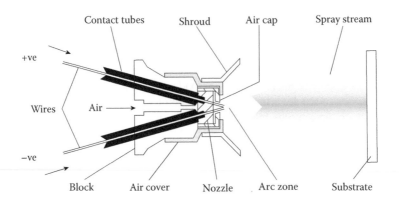

FIGURE 20.20   Arc spraying. (From Metallisation Ltd., Dudley, UK. With permission.)

3. Plasma spraying: It is the most important type. An ionized plasma arc is established using a gas mixture (Ar/H$_2$ or N$_2$/H$_2$). An extremely high temperature (10,000°C–15,000°C) is generated as the plasma jet emerges from the nozzle. This permits vaporization of any metal. It can also spray ceramics and other materials possessing high vaporizing temperatures. No compressed air is required to atomize the powder, as it is already in a vaporized state when it leaves the plasma arc zone and is somewhat akin to conditions encountered in vapor deposition (Figure 20.21). Plasma spraying provides a coating of exceptional uniformity and excellent metallurgical integrity.

Whatever spraying method is used, there must be a traverse motion of the nozzle relative to the workpiece if a uniform coating is to be produced. This is achieved manually for portable units or by rigidly mounting of the nozzle on a center lathe and traversing the metallizing unit along the slow rotating component. The sprayed thickness practically ranges from 0.1 to 10 mm. The noise level generated can be high (over 100 dB), such that ear protection is essential. Like in all coating methods, appropriate surface preparation is a must. In metal spraying, a rough surface is preferred, which is usually achieved by either jet basting or shot peening. Such surfaces provide an excellent base onto which the molten spray can adhere. The sprayed surfaces tend to be slightly porous that provide an attractive finish for journal bearings to retain lubricants. Typical applications of

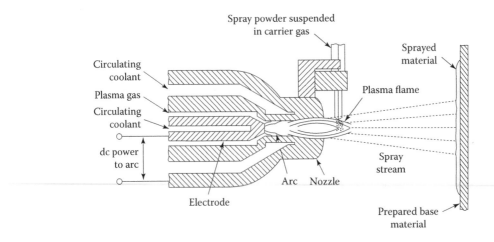

FIGURE 20.21   Plasma spraying. (From Sulzer Metco Inc. With permission.)

metallizing include steel structures and storage tanks that are sprayed with Zn or Al up to 0.25 mm in thickness.

20.4.2.7   Cladding (Clad or Diffusion Bonding)

Cladding is the bonding of metals with a thin heated layer of corrosion resistant or conductive metal by applying pressure with rolls or other means. A typical process is Al clad in which a corrosion resistant layer of Al alloy is cladded over Al. Another application is stainless steel clad over steel, reducing the need for using a full Cr, Ni alloy base metal (Figure 20.22). Steel wires are clad with Cu using extrusion dies. Here the surface layer imparts electrical conductivity, while the core provides strength and rigidity. Cladding may be performed on either one or both sides by pressing the sandwich together under high pressure, and a temperature approximately half of the melting point of the base metal for typically 1 hour. This combination of pressure, temperature, and time is resulting in a condition at interface(s) known as diffusion bonding. When sufficient transfer of atoms at the interface has occurred—hence the time factor is important—a strong and permanent bond is produced. Providing adequate pressure and a sufficiently slow traverse speed, it is possible to diffusion bond clad strip as a continuous process. Surface preparation ensuring thorough cleaning and degreasing is vital for successful bonding. When cladding only one side, it is necessary to ensure that the cladding material is facing the correct way in service. Also, special measures should be considered such that the cut edges do not present a source of subsequent surface failure.

20.5   ROLL BURNISHING AND BALLIZING

Roll burnishing and ballizing is a surface hardening process that involves rolling or rubbing smooth hard objects under considerable pressure over minute surface irregularities that are previously produced by machining or other operations. It is also used to improve the size and finish of internal and external cylindrical and conical surfaces. Because surfaces are cold worked and in residual compression, they possess improved wear and fatigue resistance. Also, the surfaces produced have high corrosion resistance due to their improved surface quality. Soft and hard surfaces can be burnished. The boss and tapered parts of holes are roller burnished, as shown in Figure 20.23a, thus improving the surface quality by smoothing and leveling scratches, tool marks, and pitting. In ballizing (Figure 20.23b), a slightly over dimensioned hard ball (D) is axially pushed (forced) into the

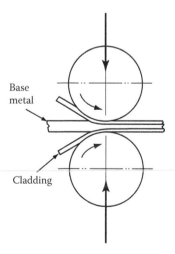

FIGURE 20.22   Cladding.

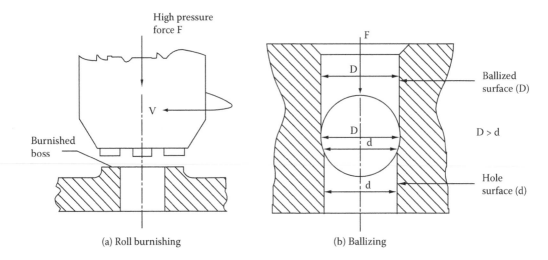

FIGURE 20.23   (a) Roll burnishing and (b) ballizing.

hole (D) through its entire length. Burnishing and ballizing are applicable for smoothing hydraulic components, valves, sealing journals, shafts, plungers, and fillets.

20.6   DEBURRING

Burrs are thin ridged projections on surfaces that develop along the edge of a component from previous processing such as machining, punching, grinding, casting, forging, etc. They harm the assembly operation, causing jamming, misalignment, short circuiting, and safety hazards and injuries to personnel. Burrs may be reduced by providing chamfers to sharp-edged surfaces.

A number of processes are commonly used for burr removal. These are

- Manually by files and scrapers during bench working
- Mechanically by cutting and grinding
- Wire brushing and abrasive belting
- Barrel and vibratory tumbling
- Shot peening
- Ultrasonic cleaning
- Abrasive jet machining (AJM), water jet machining (WJM)
- Electropolishing
- Electrochemical deburring (ECDB)
- Thermal energy deburring

Some of the above mentioned methods were covered in this chapter, while others were discussed in previous chapters. Thermal energy deburring will be briefly described here.

Thermal Energy Deburring: In thermal energy deburring, the part is loaded into a chamber, which is injected with combustible gas mixture. When the gas is ignited, the short-duration wave front heats the small burrs violently (3500°C). The burrs are vaporized instantly in less than 20 ms, while the remainder material remains cold (<200°C). A disadvantage of the process is that a thin recast layer and heat affected zone (HAZ) form. Moreover, the process is not applicable for thin and slender parts because they are liable to distortion caused by excessive heating. The process does not buff or polish the work surface as in other deburring processes.

20.7   REVIEW QUESTIONS

1. Define reactive sputtering.
2. What is the main advantage of sputtering PVD process?
3. What is the primary limitation of CVD?
4. Why is MTCVD generally preferred than CVD?
5. Why may surface treatment of manufactured products be necessary? Give examples.
6. Give some applications of mechanical surface treatment.
7. Explain how roller burnishing induces residual stresses on the surface of the workpiece.
8. Describe vapor degreasing and state its advantages.
9. Define pickling. What precautions must be taken with it?
10. Define conversion coatings.
11. What are organic coatings? Why are they popular?
12. What are the benefits of electrostatic painting, and how is it done?
13. Describe the vacuum metallizing process and its uses. Why is it only suitable for mass production?
14. What is cladding?
15. How is vitreous coating applied? What is its advantage?

20.8   PROBLEMS

1. An automobile bumper has an area of 1.4 m². It receives one of the following listed electroplates and a cell voltage of 6 V.

Material	Thickness (μm)	Allow. Current Density (A/m²)	Atomic Weight (A_w)	Valence n	Specific Weight γ (g/cm³)	Current Efficiency ζ%
1. Cu plate	12	130	64	1	8.9	50
2. Ni plate	12.5	270	59	2	8.8	95
3. Cr plate	0.5	1600	52	6	6.9	15
4. Cd plate	12.5	215	112	2	8.6	90

How much plate material and electrical energy are consumed for each bumper, and how long should the plating time be for the used allowable current densities?

2. Instead of plating Cu and Ni on a bumper, the same requirements can be met by a single Ni plate of 25 μm thick. In either case, a final Cr plate of 0.5 μm must be applied. Cu costs $1.7/kg, and Ni costs $5/kg. Electricity costs 30 cent./kW h. Which of the two alternatives do you recommend from the economical point of view?

BIBLIOGRAPHY

Degarmo, E. P., Black, J. T., and Kohser, R. A. 1997. *Materials and Processes in Manufacturing*, 8th ed. Upper Saddle River, NJ: Prentice-Hall.

Doyle, L. E., Keyser, C. A., Leach, J. L., Schrader, G. E., and Singer, M. B. 1985. *Manufacturing Processes and Materials for Engineers*, 3rd ed. Upper Saddle River, NJ: Prentice-Hall.

Kalpakjian, S. 1985. *Manufacturing Processes for Engineering Materials*, 4th ed. Upper Saddle River, NJ: Prentice-Hall.

Waters, F. 2000. *Fundamentals of Manufacturing for Engineers*. London: UCL Press.

21 Joining Processes

21.1 INTRODUCTION

Manufactured products are made by assembling a number of components that may need to be joined together either permanently or semipermanently, where they can be taken apart for maintenance or repair purposes. Additionally, permanent joints are used for parts that do not require disassembly as in the case of car bodies, ships, and bridges. Manufacturing engineers should be aware of the different joining techniques, their principles, and limitations that ensure proper joint design and manufacture for a wide range of industrial applications. Figure 21.1 shows the different joining methods. In order to establish semipermanent joints, screws and bolts are commonly used. This chapter introduces the methods of establishing permanent joints.

Each method of joining has its own attributes, and a number of aspects should be evaluated. In this regard, strength, ease of manufacture, cost, permanency, corrosion resistance, and appearance, depending on the specific application, should be carefully considered. Permanent joining methods are classified according to Figure 21.2. The main classes are as follows.

1. Liquid state (fusion) welding, which depends on the principles of metal casting and heat treatment.
2. Solid-state welding which is based on the adhesion and deformation processes.
3. Liquid solid joining processes are achieved through solidification and adhesion.
4. Mechanical joining, which depends on metal-forming processes.

21.2 FUSION WELDING

Fusion welding involves melting the parts to be joined in the localized area where they meet. The molten metal within the joint area resolidifies to form a metallurgical uniform joint. For each fusion welding process, the provision of a suitable source of heat is of major importance. There is a need to protect the weld pool against atmospheric contamination. The heat source should be capable of generating localized fusion in a controlled manner and operate at a temperature significantly higher than the melting point of the metal to be joined. The heat should also be concentrated at a small area in order to restrict the welding pool. It should be controlled so that conditions can be set and maintained constant during the welding operation. Table 21.1 shows the different heat sources used in a wide range of fusion welding processes. Parameters involved in the effective melting of the parent metal during welding involve metal thickness, joint type, heat input, thermal conditions, the temperature of the parent metal before welding, and the melting point.

In order to achieve melting of the weld joint, the rate of heat supply to the part should be greater than the rate at which it flows into the parent metal, which depends on the thermal conditions. Figure 21.3 shows the main requirements of the welding process.

The term weldability has been defined by the American Welding Society (AWS) as "The capacity of metal to be welded under the fabrication conditions imposed into a specific, suitably designed structure and to perform satisfactorily in intended service." A metal of good weldability must be welded readily so as to perform satisfactorily in the fabricated structure and also must not require expensive or complicated procedures in order to produce sound joints. Weldability influences weld quality and determines which welding process to use. Thus, weldability is a measure of how easy it

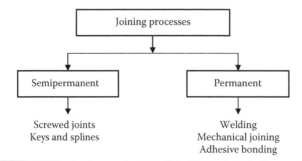

FIGURE 21.1 Joining methods.

is to obtain crack free welds, achieve adequate mechanical properties, and produce welds that are resistant to service degradation.

The weldability of steels is inversely proportional to its hardenability, which measures the ease of forming martensite during heat treatment. The hardenability of steel depends on its chemical composition. Greater percentage of carbon and other alloying elements results in a higher harden-

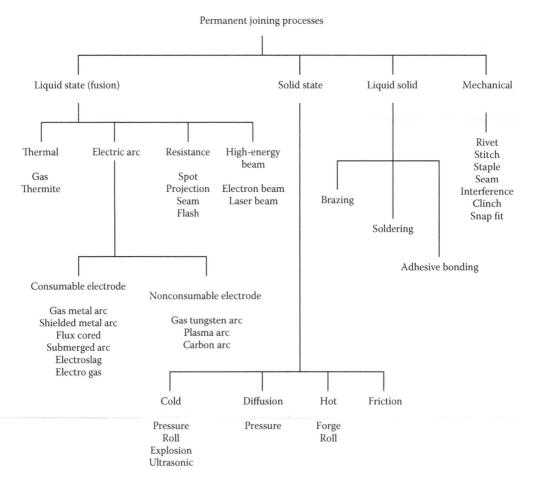

FIGURE 21.2 Classification of welding processes.

TABLE 21.1
Heat Sources for Fusion Welding Processes

Welding Operation	Source of Heat
Gas welding	Burning gases
Electric arc welding	Electric arc
Laser beam welding	Focused beam of light
Electron beam welding	Focused beam of electrons
Plasma arc welding	Ionized gases through arc
Resistance welding	Electrical resistance
Thermite welding	Chemical reaction

ability and thus a lower weldability. The equivalent carbon content is used to compare the relative weldabilities of different alloys by comparing their properties to plain carbon steel.

The weldability of any metal can be changed by altering the physical, chemical, thermal, and metallurgical properties. This includes welding procedure, shielding atmosphere, flux material, filler material, and heat treatment of metal to be welded. Carbon steel, low alloy cast steel, cast iron, and stainless steel are arranged in a descending order with respect to their weldability.

There are four types of welding joints: butt, T, corner, and lap (Figure 21.4). Butt joints are formed by welding the end surfaces or edges of the joint. The load is transmitted along the common axis. The preparation of the joint depends on the thickness of the metal being welded. Flanging is used for metal below 3 mm thick. The height of the flange should be twice the thickness of the metal. Square butt joints with no special edge preparation are suitable for thickness from 3 to 8 mm in thickness. Single V edge preparation is applied for metal from 14 to 16 mm thick. If the metal is thicker than 16 mm, a double V is used, while U joints are used for metal over 20 mm. Lap joints are produced by fillet welds where the two members should overlap by 3 to 5 times their thickness.

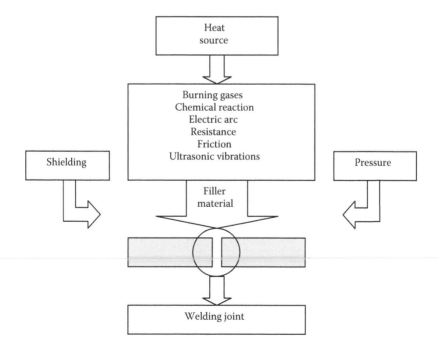

FIGURE 21.3 Welding requirements.

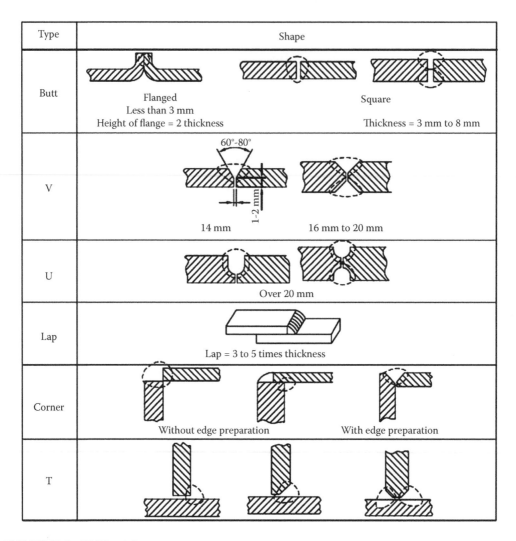

FIGURE 21.4 Welding joints.

The corner joints are welded with or without edge preparation. The T joint is probably the most commonly used in welding one element to another at an angle of 90°.

Edge preparation is an important aspect for obtaining sound welds. Figure 21.5 shows typical edge preparation for butt joints. The choice of a joint preparation depends on many factors

- Type of welding process
- Type of work
- Welding position
- Access for arc and electrode
- Volume of deposited weld metal
- Cost of edge preparation
- Shrinkage and distortion

Figure 21.6 shows the different welding positions. Flat welding provides the best conditions for the welder to control the weld pool because it enables high currents to be used and leads to a

Joint type	Joint shape	A°	G (mm)	RF (mm)	B°	R (mm)	t_1/t_2
Single V	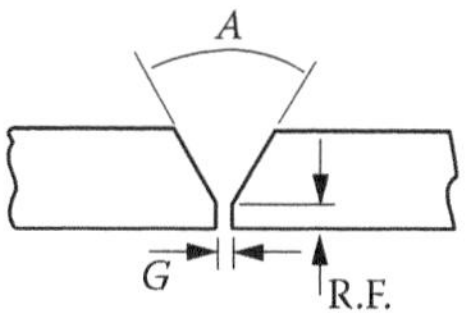	60–80	1–3	1–3			
Double V (symmetrical)	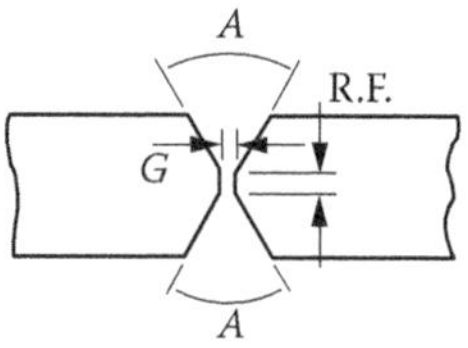	60	0–3	0–3			
Double V (asymmetrical)	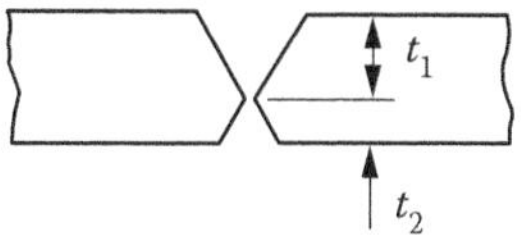						2 : 1
Single U	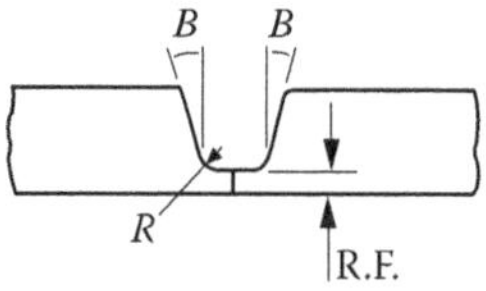			5	10–15	5	
Double U (symmetrical)	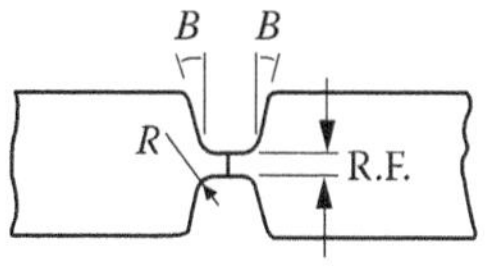			5	10–15	5	

FIGURE 21.5 Typical edge preparation for butt joints.

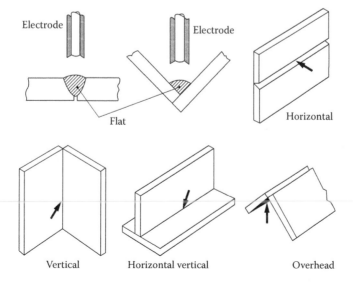

FIGURE 21.6 Welding positions.

faster welding process. In addition to flat welding, there are three positions that can be identified: horizontal, vertical, and overhead. There is a subdivision of the horizontal position, known as the horizontal–vertical, which is used for T joints. During these arrangements, the metal tends to run out of the joint under the effect of gravity. Under such circumstances, the welder controls the weld by lowering the heat input to reduce the fluidity and produce a small weld pool that solidifies before it has time to run out of the joint. While in flat joint, the maximum current can reach 350 A; in overhead welding it should not exceed 160 A.

21.2.1 GAS WELDING

Gas welding (Figure 21.7) is a popular joining method that uses a gas flame as a source of heat, which is produced by burning a combustible gas, such as acetylene mixed with oxygen. Other gases, including propane, natural gas, and hydrogen, can also be used as the heat source, particularly for welding aluminum and lower melting point metals. The process is widely used in maintenance and repair work because of the ease in transporting oxygen and fuel cylinders.

During welding a temperature of 3150°C is generated in the flame. Figure 21.8 shows the different zones of the welding flame together with the temperature distribution along the flame. Accordingly, three distinct zones can be identified: the inner luminous cone 1, adjacent to the torch, the reducing zone surrounding the luminous cone 2, called the welding zone, and the external oxidizing zone 3

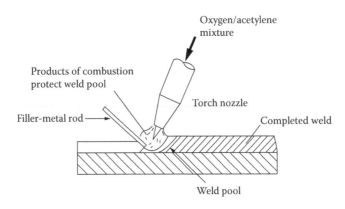

FIGURE 21.7   Oxyacetylene welding.

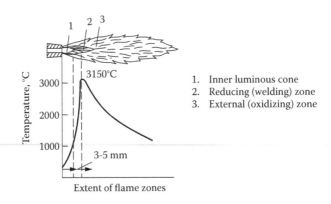

FIGURE 21.8   Gas welding flame temperature.

that forms the external envelop of the flame. In the inner white zone, two thirds of heat is generated and the highest temperature is obtained through

$$2C_2H_2 + 2O_2 \rightarrow 4CO + 2H_2$$

Complete combustion occurs in the outer cone by

$$4CO + 2O_2 \rightarrow 4CO_2$$

$$2H_2 + O_2 \rightarrow 2H_2O$$

Depending on the ratio of mixing C_2H_2 and O_2, three types of gas flame are commonly used for oxy-gas welding. By controlling such a ratio, neutral, carburizing, and oxidizing flames are created (Figure 21.9).

Neutral flame: The most common flame for oxyacetylene welding is the neutral flame where the ratio of oxygen to acetylene is kept at 1 : 1. Neutral flames do not oxidize or add carbon to the metal during welding. This flame acts like the inert gases that protect the weld from the atmosphere. It is widely used for welding low carbon steel, copper, cast iron, and aluminum alloys.

Oxidizing flame: When the ratio of oxygen to acetylene becomes 1.2 : 1, the flame becomes oxidizing. This flame is suitable for welding brass and bronze.

Carburizing flame: In this case, the ratio of acetylene to oxygen is 1.2 : 1. The carburizing flame always shows distinct colors: the inner cone is bluish white, the intermediate cone is white, the outer envelope flame is light blue, and the feather at the tip of the inner cone is greenish. Carburizing flames are not used for welding low carbon steels because they add carbon that causes embrittlement and hardness increase of the weld joint. Carburizing flames are, therefore, ideal for welding high carbon steels, cast iron, and nonferrous metals where the additional carbon poses no problems and the flame adds more heat to the metal because of its size.

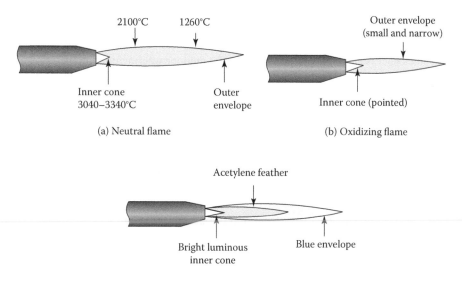

FIGURE 21.9   Gas welding flames.

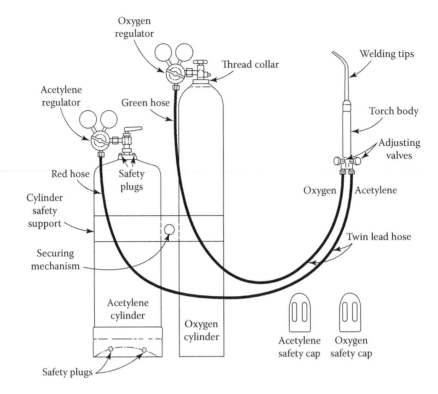

FIGURE 21.10   Gas welding system.

21.2.1.1   Oxyacetylene Welding Equipment

Oxyacetylene welding equipment (Figure 21.10) usually consist of a cylinder of acetylene gas, a cylinder of oxygen, two regulators, two lengths of hose with fittings, and a welding torch with tips. In addition to the basic equipment mentioned, auxiliary equipment that consists of tip cleaners, cylinder trucks, clamps, and holding jigs are used. Safety tools, including goggles, hand shields, gloves, leather aprons, sleeves, and leggings, are essential. A portable oxyacetylene unit is an advantage when it becomes necessary to move the equipment for outdoor welding operations.

The oxy-gas welding torch mixes oxygen and acetylene gas in the proper proportions and controls the amount of the mixture burned at the welding tip. Torches have two needle valves: one for adjusting the oxygen flow and the other for adjusting the acetylene gas flow. Other basic parts include a handle (body), two tubes (one for oxygen and another for fuel), a mixing head, and a tip. Welding tips are made from a special copper alloy and are available in different sizes to handle a wide range of applications and plate thicknesses. A typical welding torch is shown in Figure 21.11.

A filler metal in the form of wire or rod is used in gas welding to supply filler metal to the joint. Filler rods for welding steel are often copper coated to protect them from corrosion during storage. Most rods are furnished in 900 mm lengths and a wide variety of diameters, ranging from 0.8 to 9 mm. Rods for welding cast iron vary from 300 to 600 mm in length and are frequently square rather than round. The filler wire diameter for a given job is selected according to the thickness of the metal to be joined.

In certain gas welding applications, metal fusion is hindered by the oxides of the base metal, which have a higher melting point than the metal itself. These oxides become entrapped in the solidifying metal instead of flowing from the welding zone, thus leading to poor weld quality. This problem can be encountered by adding certain fluxes that react chemically with these oxides and

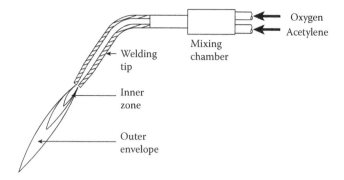

FIGURE 21.11 Gas welding torch.

form a fusible slag. Such a slag floats at the top of the welding puddle and does not interfere with the deposition of the filler metal. Fluxes also protect the molten puddle from absorption of oxygen into the molten weld pool that results in a poor weld joint.

Fluxes are available in several forms such as dry powder, paste, or coating on the welding rod. The use of such fluxes is essential for welding cast iron, brass, bronze, stainless steel, and aluminum. Fluxes are not recommended for gas welding of carbon steel because the formed oxides are normally lighter and float to the surface of the weld in the form of a scale.

21.2.1.2 Oxy-Gas Welding Techniques

Oxyacetylene welding adopts either the forehand (rightward) or the backhand (leftward) method. The deciding factor in determining the method used is the relative position of the torch, the welding rod during welding, and the direction of welding. The best method to use depends on the type of joint, joint position, and the need for heat control on the parts to be welded.

21.2.1.2.1 Backhand (Rightward) Welding

In the backhand welding technique, shown in Figure 21.12a, the torch tip precedes the rod in the direction of welding and the flame points back at the molten puddle and completed weld. The welding tip makes an angle of about 60° with the joint being welded. The end of the welding rod is placed between the torch tip and the molten pool. The backhand method is recommended for welding materials of more than 3 mm thickness. A narrower V at the joint with included angle of 60° is provided for a good joint. The backhand method requires less welding rod than the forehand method.

21.2.1.2.2 Forehand (Leftward) Welding

In this method (Figure 21.12b), the rod is kept ahead of the flame in the direction of welding. The flame is pointed in the direction of travel, and the tip is made at an angle of about 45° to the working surfaces. The flame preheats the edges under welding just ahead of the molten pool. The rod is moved in the same direction of the tip. Additionally, the torch tip and the welding rod are moved

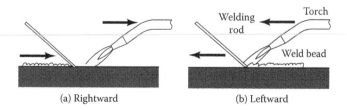

(a) Rightward   (b) Leftward

FIGURE 21.12 Gas welding techniques.

back and forth in opposite, semicircular paths, in order to ensure that the heat is evenly distributed. As the flame passes the welding rod, it melts a short length of the rod and adds it to the welding pool. The motion of the torch distributes the molten metal evenly to both edges of the joint and to the molten puddle. Forehand welding is used in all positions for welding sheet metals and plates up to 3 mm thick where better control of a small welding pool and a smoother weld are achieved. The forehand technique is not recommended for welding heavy plate owing to the lack of base metal penetration.

Oxyacetylene welding has the following advantages.

1. Easily controlled.
2. Suitable for thin sheets.
3. Low cost of welding equipment.
4. Portable welding equipment, allowing the process to be used outdoors and for repair work.
5. Most metals can be welded by this method.
6. By changing the nozzle, the torch can be used for heating, welding, brazing, and cutting operations.

The process has several limitations because it is slower than arc welding while the heat-affected zone and distortion is larger than arc welding. Moreover, the gases are expensive and require safety precautions for handling.

21.2.2 Thermit Welding

Thermit welding (TW) is used to join metals using a superheated liquid filler metal from a chemical reaction between a metal oxide and aluminum or other reducing agent. The heat for welding is obtained from an exothermic reaction or chemical change between iron oxide and aluminum (Thermit mixture) as follows:

$$8Al + 3Fe_3O_4 \rightarrow 9Fe + 4Al_2O_3 + Heat$$

The exothermic reaction is relatively slow, nonexplosive, and requires 20 to 30 s. The temperature resulting from this reaction is approximately 2482°C, which is nearly twice that of the melting point of steel. The amount of Thermit provides sufficient metal to produce the weld. Generally, the amount of steel produced by the reaction is approximately one half the original quantity of Thermit material by weight and one third by volume.

As shown in Figure 21.13, the parts to be welded are aligned with a gap between them. The superheated steel runs into a mold that is built around the parts to be welded. The molten metal fills the cavity between the parts being welded by gravity. Because its temperature is almost twice that of the melting temperature of the base metal, melting occurs at the edges of the joint and fuses with the molten steel from the crucible. As the molten metal solidifies, coalescence occurs, and the weld

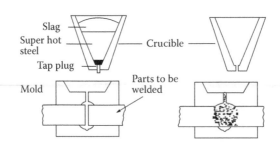

FIGURE 21.13 Thermit welding.

is complete. If the parts to be welded are large, preheating within the mold cavity may be necessary to bring the parts to welding temperature and to dry out the mold.

Thermit welding is applicable in the repair of heavy parts such as trucks, motor castings, and connecting rods, as well as welding pipes that can be welded in place. Thermit welded joints have mechanical properties that approach those of forged steel. The need for molds, however, hinders its application.

Thermit welding can also be used for joining of copper and aluminum conductors for the electrical industry. In such a case, the exothermic reaction is a reduction of copper oxide by aluminum, which produces molten superheated copper that flows into the mold, melts the ends of the parts to be welded, and as the metal cools, a solid homogeneous weld results. When welding nonferrous materials, molds are made of graphite, and the parts to be joined must be extremely clean by applying a flux to the joint prior to welding.

Among the advantages of thermit welding is that the deposited weld metal is homogeneous and its quality is relatively high. Distortion is minimized because the weld is accomplished in one pass and cooling is uniform across the entire weld cross section. Welds can be made in almost any position as long as the cavity has vertical sides. No external power source is required, and very large heavy sections may be joined.

The process has several limitations because only ferrous (steel, chromium, nickel, copper, and aluminum) parts may be welded and the rate of welding is low. The high temperature causes distortion and changes in grain structure to the weld region. The welded joint may contain hydrogen gas and slag contaminations.

21.2.3 ELECTRIC ARC WELDING

The electric arc welding process uses an electric power supply to create and maintain an electric arc between an electrode and the base material to melt the edges of the welding point. Arc welding uses either direct (dc) or alternating (ac) current and consumable or nonconsumable electrodes. The welding region is sometimes protected by shielding fluxes or an inert gas. A filler material is sometimes used in case of using nonconsumable electrodes.

Polarity: Polarity is the direction of the current flow in a circuit, as shown in Figure 21.14. Direct current welding machines uses either a straight polarity or a reversed polarity. In straight polarity (Figure 21.14a), the electrode is negative while the workpiece is positive; electrons flow from the electrode to the workpiece. In reversed polarity (Figure 21.14b), the electrode is made positive and the workpiece is negative; the electrons flow from the workpiece to the electrode. During straight polarity, the majority of the heat is directed toward the workpiece which causes deeper penetration in the workpiece (Figure 21.15a). In reversed polarity, the heat is concentrated on the electrode, thus leading to a larger penetration rate from the welding electrode (Figure 21.15b). Intermediate

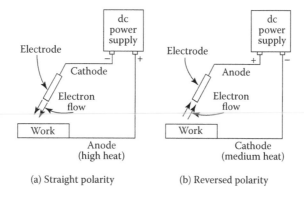

(a) Straight polarity   (b) Reversed polarity

FIGURE 21.14 Electrode polarity in arc welding.

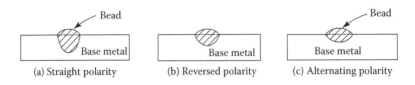

(a) Straight polarity   (b) Reversed polarity   (c) Alternating polarity

FIGURE 21.15   Welding polarity.

condition occurs when using ac current due to the heat balance between the welding electrode and the workpiece (Figure 21.15c).

During overhead welding the filler metal is forced to fall by gravity. Using reverse polarity, less heat is concentrated at the workpiece, which allows the filler metal to cool faster, giving it greater holding power. Cast-iron arc welding uses reverse polarity that provides rapid deposition of the molten metal from the electrode and prevents overheating of the base metal.

In general, straight polarity is used for welding mild steel using bare or lightly coated electrodes where the majority of heat is developed at the positive workpiece. However, when heavy-coated electrodes are used, the released gases cause the greatest heat to be produced on the negative side. Reverse polarity is used in the welding of nonferrous metals, such as aluminum, bronze, Monel, and nickel.

Direct current welding causes the arc to wander during welding (arc blow) in corners on heavy metal or when using large-coated electrodes. It generates a magnetic field that deviates the arc from its intended path, causing excessive spatter, porosity, and incomplete fusion.

Electrodes: Bare, fluxed, and heavy-coated electrodes are commonly used in arc welding. Bare electrodes have limited applications because, during welding, they are subjected to oxygen and nitrogen, which forms nonmetallic constituents that decrease the strength and ductility of the joint. These undesirable oxides are eliminated using flux coated electrodes. The type of flux coating depends on the weld metal composition. The choice of the correct electrode for a particular job is vital for producing a sound weld. According to the AWS standards, the electrode is marked by a number consisting of a prefix letter, which indicates the following characteristics (Figure 21.16):

- Method of manufacture
- Type of flux covering
- The suitable welding position

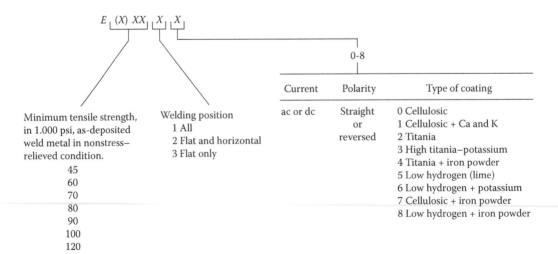

FIGURE 21.16   Coding of welding electrodes.

- Welding current condition
- Characteristics of deposited metal
- Suffix letter to indicate that deeper penetration is possible

Arc welding equipment: Depending on the application, ac or dc machines are used in arc welding. Direct current welding is mostly used for heavy work and for sites where electricity is not available, while ac transformers are used where the supply mains are available. The transformer steps down the voltage to the normal welding open gap voltage of 80–100 V. Alternating current ac welding has the advantage of high speed, low power cost, silent operation, easier to operate and maintain, and the absence of arc blow that normally occur during dc welding of magnetic materials. Generally, arc welding equipment includes the following:

- An ac or dc machine
- Electrode and its holder
- Chipping hammer
- Wire brush
- Cables and connectors
- Safety goggles
- Helmet
- Aprons, sleeves, hand gloves, etc.

Direct current dc generators are either an engine driven electric generator or a transformer and silicon rectifiers. Alternating current power sources require a transformer; however, additional circuits are provided to establish the desired voltage/current relationship. These are distinguished as constant current or constant voltage characteristics. Using the constant current characteristics (Figure 21.17), as arc length increases, the voltage rises and the current decreases. This type is suitable for manual shielded metal arc welding (SMAW). At a current level of zero the open circuit voltage is held between 70 and 80. Figure 21.18 shows the power source characteristics together with the

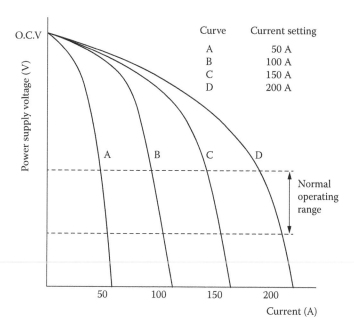

FIGURE 21.17 Constant current characteristics. (From Gourd, L. M., *Principles of Welding Technology*, Edward Arnold Publishers, London, 1982. With permission.)

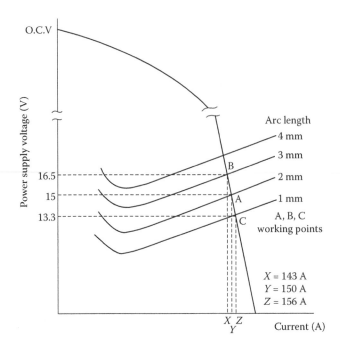

FIGURE 21.18 Changes in arc length maintains constant current during welding. (From Gourd, L. M., *Principles of Welding Technology*, Edward Arnold Publishers, London, 1982. With permission.)

arc characteristics for four different arc lengths. The working point is found in the intersection of the power characteristics with the arc characteristics. If the welder is manually guiding the electrode over the work, by a few millimeters, the voltage will change by about 20%, while the current changes will be around 8%. This means that the deposition rate remains fairly unchanged in spite of the unsteady hand movements of the welder.

Measuring heat input in arc welding: During arc welding, almost all the electrical energy is converted to heat. A small proportion is used in the generation of the arc glow and ultraviolet radiation given by the arc. Assuming that the arc current is I and voltage is V, the power input is given by

$$\text{Power input} = I \times V \text{ (kW)}$$

$$\text{Heat input to the arc} = I \times V \text{ (kJ/s)}$$

If the welding speed is V_w (mm/s),

$$\text{The heat input (J/mm)} = I \times V \times 60/V_w$$

Example 1

For a welding current of 60 A, voltage 14 V, and a welding speed 120 mm/min, calculate the heat input per millimeter of weld length.

Solution

$$\text{Power input} = 60 \times 14 = 840 \text{ W}$$

$$\text{Heat input} = 840 \qquad \text{J/s}$$

For the welding speed 120 mm/min

$$\text{Heat input (J/mm)} = 60 \times 14 \times 60/120 = 420 \text{ J/mm}$$

Arc welding is performed using different methods that depend on the use of electrodes and the shielding method.

21.2.3.1   Shielded Metal Arc Welding

Shielded metal arc welding (SMAW), also known as manual metal arc welding (MMAW), is shown in Figure 21.19. Electric current is used to strike an arc between the base metal and a consumable electrode that acts as a filler material. The electrode, made of steel, is covered with a flux that protects the weld area from oxidation and contamination by producing a shielding CO_2 gas. The process is versatile and uses relatively inexpensive equipment that makes it suitable for job shop type of work. It is commonly used in general construction, shipbuilding, and pipelines as well as for maintenance and repair work. It can be used in remote areas where portable fuel-powered generators can be used as the power supply. In case of multiple-pass welding, the slag should be removed before another pass is applied. In general, the operator factor or the percentage of operator's time spent laying weld is approximately 25%. However, SMAW remains one of the most flexible techniques that have advantages in areas of restricted access.

SMAW is often used to weld carbon steel, low and high alloy steel, stainless steel, cast iron, and ductile iron. It can also be used for welding nickel, copper, and their alloys and, rarely, for aluminum. The thickness of the material being welded is bounded on by the skill of the welder and starts from 1.5 mm. With proper joint preparation and use of multiple passes, materials of virtually unlimited thicknesses can be joined. Furthermore, depending on the electrode used and the skill of the welder, SMAW can be used in any welding position.

The choice of electrode for SMAW depends on a number of factors, including the weld material, welding position, and the desired weld properties. The electrode is coated by a flux, which generates gases as it decomposes to prevent weld contamination, introduces deoxidizers to purify the weld, causes weld-protecting slag to form, improves the arc stability, and provides alloying elements to improve the weld quality. The composition of the electrode core is generally similar to that of the base material. However, a slight difference in alloy composition can strongly impact the properties of the resulting weld properties. Electrode fluxes consist of a number of different compounds, including rutile, calcium fluoride, cellulose, and iron powder.

The welding technique utilized depends on the electrode, the composition of the workpiece, and the position of the joint being welded. The choice of electrode and welding position also determine the welding speed. Flat welds require the least operator skill and can be done with electrodes that

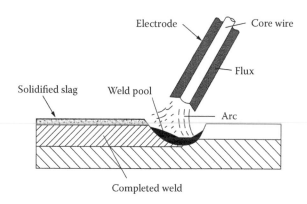

FIGURE 21.19   Shielded metal arc welding (SMAW).

melt quickly but solidify slowly, thus permitting higher welding speeds. Sloped, vertical, or upside-down welding positions require more operator skill and often necessitate the use of an electrode that solidifies quickly to prevent the molten metal from flowing out of the weld pool. However, this generally means that the electrode melts less quickly, thus increasing the time required to lay the weld.

Shielded metal arc welding equipment typically consists of a constant current welding power supply and an electrode, with an electrode holder, a work clamp, and welding cables (also known as welding leads) connecting the two.

The power supply used in SMAW has constant current output, even if the arc distance and voltage change. This is important because most applications of SMAW are manual. Maintaining a suitably steady arc distance is difficult if a constant voltage power source is used instead because it causes heat variations that make welding more difficult. However, for performing complicated welds the welder can vary the arc length to cause minor fluctuations in current.

The preferred polarity of the SMAW system primarily depends on the electrode being used and the desired properties of the weld. Reverse polarity causes heat to build up on the electrode, increasing its melting rate and decreasing the depth of the weld. Direct polarity increases the weld penetration because electrons bombarding the workpiece are associated with heat. With alternating current, the polarity changes 100 times per second, which creates an even heat distribution and provides a balance between electrode melting rate and penetration.

21.2.3.2 Flux Cored Arc Welding

In flux cored arc welding (FCAW), the electrode used is tubular in shape and is filled with a flux (Figure 21.20). This type of electrode ensures more stable arc, improved weld quality, and enhanced mechanical properties of the weld metal. It is convenient to use carbon steel for the tubular part of the electrode and to pack the core with the alloying elements. Tubular electrodes can be provided in long coiled lengths of 0.5–4 mm in diameter. Small diameter electrodes make it relatively easy to weld parts out of position. The flux chemistry enables welding of different base metals. The process is economical for welding a variety of joints, mainly of steels, stainless steels, and nickel alloys. Major advantages of fluxed electrodes are

- Ease of control of the weld metal characteristics by adding alloying elements to the flux.
- Welding in all positions.
- The absence of shielding gases makes it suitable for outdoor welding/windy conditions.
- The high deposition rate makes it suitable for automotive applications.
- Less precleaning of metal is required.
- Weld metal is protected until the flux is chipped away, which adds metallurgical benefits.
- Low operator skill is required.
- The process is easy to automate and is adapted to flexible manufacturing systems and robotics.

In addition to the incomplete fusion between base metals, slag inclusion, nonmetallic inclusions, and cracks in the welds, there are further disadvantages including irregular wire feed, gas porosity, and the high cost of wire compared to gas metal arc welding (GMAW).

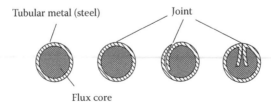

FIGURE 21.20 Flux cored electrodes.

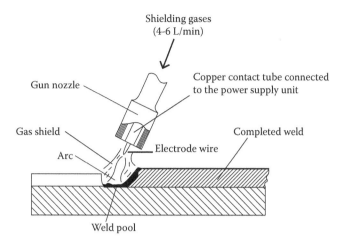

FIGURE 21.21 Gas metal arc welding (GMAW).

21.2.3.3 Gas Metal Arc Welding

In the gas metal arc welding (GMAW) process, a consumable wire electrode is fed continuously through the welding gun at a controlled speed. The consumable wire electrode maintains the arc and provides the filler metal to the joint. GMAW welding (Figure 21.21) is a one-handed operation that does not require the high degree of skill normally required in gas tungsten arc welding (GTAW). The weld area is shielded by argon, helium, carbon dioxide, or various other gas mixtures that flow at a rate of 4–6 L/min. Table 21.2 shows the shielding gases for GMAW of different meals and alloys. Accordingly, helium is used for highly conductive materials like copper, while argon is preferred for thin metals. For welding steel the addition of CO_2 stabilizes the arc, promotes metal transfer, and minimizes spatter. In some cases, flux is used in GMAW to improve the quality of weld. This flux is added as a fine powder mixed with the shielding gas or as a fine coating to the metal electrode. In order to prevent oxidation of the molten weld puddle, deoxiders can be added to the electrode metal itself. In GMAW, the high rate of metal deposition and the high speed of welding ensure the minimum distortion and a narrow heat affected zone. GMAW is, therefore, recommended for welding thicker material.

The constant voltage characteristic is used in GMAW to permit a wide choice of wire feed rates. This requires a wide variation of burn-off rates and needs a wide variation of the current while keeping the voltage and the arc length fairly constant, which makes it preferred for semiautomatic and automatic welding processes.

TABLE 21.2
Shielding Gases for GWAW

Metal to Be Welded	Shielding Gas
Aluminum and alloys	Pure argon
Nickel and alloys	Pure argon
Copper	Argon + helium
Stainless steel	Argon + 5% oxygen
Low carbon steel and carbon manganese steel	Carbon dioxide or argon + 20% CO_2 or argon + 5% O_2
Steel with 1%–2% chromium	Argon + 20% CO_2 or argon + 5%O_2
Steel with more than 2% chromium	Argon + 5%O_2

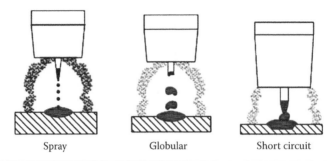

Spray   Globular   Short circuit

FIGURE 21.22   Metal transfer in gas metal arc welding.

During GMAW the metal is transferred to the weld joint by spray transfer, globular transfer, or short-circuiting transfer, as shown in Figure 21.22. The type of metal transfer depends on the arc voltage, current setting, electrode wire size, and shielding gas used.

Spray transfer: Spray transfer is a high-current range method that produces a rapid disposition of weld metal. It is effective for welding heavy-gauge metals because it produces deep weld penetration. The use of argon or a mixture of argon and oxygen is necessary for spray transfer.

Globular transfer: Carbon dioxide–rich gases are used and globules propelled by forces of the electric arc transfer the metal causing a considerable spatter. Globular transfer occurs when the welding current is low; consequently, few drops of molten metal are transferred per second, whereas many small drops are transferred with a higher current setting. GMAW is, therefore, used for thin materials where the heat input is low.

Short-circuit transfer: Metal is transferred in the form of individual droplets at a rate of over 50 per second. The electrode tip touches the molten weld pool and causes short circuits. In most cases, it is used at current levels below 200 A and wire of less than 1.0 mm diameter, which produces weld puddles that are small and easily controllable. This mode of metal transfer is suitable for welding in vertical and overhead positions. The shielding gas is a mixture of 75% carbon dioxide and 25% argon. The carbon dioxide provides greater heat and thus higher welding speeds, while the argon controls the weld spatter. This method is suitable for welding thin sheets and small sections of less than 6 mm. Pulsed arc systems are currently used for welding thinner ferrous and nonferrous metals.

In GMAW, there is no flux and no slag to remove, the process is rapid, versatile, and economical. It does not require high operator skill and can be automated and lends readily to flexible manufacturing systems and robotics. However, the disadvantages of GMAW include the following.

* Welding equipment is more complex, expensive, less portable, and used indoors.
* There is difficulty welding in small corners and hard to reach places.
* Weld joint properties are affected by the high cooling rate.

21.2.3.4   Submerged Arc Welding

Submerged arc welding (SAW) (Figure 21.23) requires a continuously fed consumable solid or tubular electrode. The molten weld and the arc zone are protected from atmospheric contamination by being submerged under a blanket of granular fusible flux consisting of lime, silica, manganese oxide, calcium fluoride, and other compounds. The molten flux provides a current path between the electrode and the weld joint. It covers the molten metal, prevents spatter and sparks, and suppresses the intense ultraviolet radiation and fumes generated from the welding operation. The solidified flux is removed while the granular unused flux is reused again.

SAW is normally operated in the automatic mode; however, semiautomatic systems with pressurized or gravity flux feed delivery are available. The process is normally limited to the flat or horizontal-fillet welding positions. The process is performed with ac or dc power supplies. However,

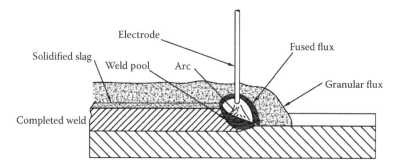

FIGURE 21.23 Submerged arc welding (SAW).

dc gives better control of weld shape, penetration, and welding speed. Alternating current ac machines minimize arc blow and give intermediate penetration between straight and reverse polarities (Figure 21.15c). Currents ranging from 300 to 5000 A ensure high rate of metal transfer (45 kg/h) and welding speeds (5 m/min) for thin sheet of steel. The process is used for welding carbon steels, low alloy steels, stainless steels, and nickel-based alloys. SAW is used for the manufacture of large boilers, vessels, heavy marine components, ship and barrage buildings, pipes, beams, and girders.

The advantages of SAW include the following:

• Deep weld penetration.
• Sound welds can be readily made with proper process control.
• Minimal welding fume, spatter, and arc light are avoided.
• Welding indoors and outdoors.

The process has the following limitations:

• Used for ferrous and some nickel-based alloys
• Used in flat welding positions for long straight seams or rotated pipes or vessels
• Requires a flux handling system and post weld slag removal
• Presents a health and safety issue caused by slag and flux fine particles

21.2.3.5 Underwater Welding

Recently, offshore structures such as oil drilling rigs, pipelines, and platforms are being installed on a large scale. Some of these structures will experience failures during normal usage and unpredicted storms, and collisions. Their repair requires the use of underwater welding that can be classified as wet welding and dry welding. In wet welding the operation is performed underwater and is directly exposed to the wet environment. A special electrode is used and welding is carried out manually just as one does in open air welding. The increased freedom of movement makes wet welding the most effective, efficient, and economical method. Welding power supply is located on the surface with connection to the diver/welder maintained via cables and hoses. In SMAW wet welding, a dc power supply is used with negative work polarity in order to avoid its possible electrolysis. Wet underwater welding has now been widely used for many years in the repair of offshore platforms. The benefits of wet welding are

1. The versatility and low cost.
2. The speed with which the operation is carried out.
3. Welding difficult to reach portions of offshore structures.
4. No enclosures are needed.

In dry (habitat) welding, a dry chamber is created near the area to be welded and the welder does the job by staying inside the chamber. This method produces high-quality weld joints that meet X-ray and code requirements. The gas tungsten arc welding (GTAW) process is performed under water in the dry mode of operation.

21.2.3.6 Electroslag Welding

Electroslag welding (ESW) (Figure 21.24) is a highly productive, single pass welding process for thick materials of 25–300 mm in a vertical position. It eliminates the need for multiple passes and joint preparation. An electric arc is initially struck by wire that is fed into the desired weld location, and then a flux is added until the molten slag reaches the tip of the electrode that extinguishes the arc. The wire is then continually fed, melted, and transmitted through the molten slag to cause welding.

Electroslag welding is used to join low carbon steel plates and/or sections that are very thick. This process uses dc power supplies of current about 600 A and a voltage from 40 to 50 V or ac power supply. Benefits of electroslag welding include

- High metal deposition rates of 15 to 20 kg/h.
- Ability to weld thick materials.
- Single pass is sufficient for electroslag welding.
- Joint preparation and material handling are minimized.
- Filler metal utilization is high.
- Safe and clean.
- No arc flash.
- Low weld distortion.
- Lends itself to mechanization, thus reducing the requirement for skilled manual welders.

Electrogas welding is a variation of ESW where an inert gas is used for shielding, while a flux cored wire is automatically fed to the molten weld pocket and an electric arc is continuously maintained between the wire electrode and the weld puddle.

21.2.3.7 Gas Tungsten Arc Welding

Gas tungsten arc welding (GTAW), shown in Figure 21.25, or as sometimes called tungsten inert gas (TIG) welding, uses a nonconsumable tungsten electrode to produce the weld joint. The weld

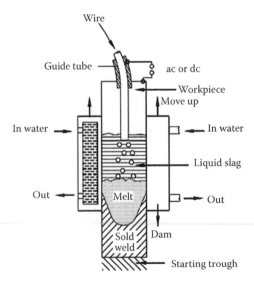

FIGURE 21.24   Electroslag welding.

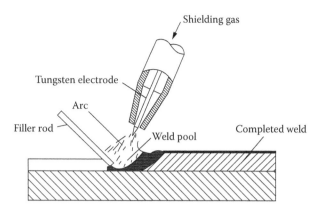

FIGURE 21.25 Gas tungsten arc welding (GTAW).

area is protected from atmospheric contamination by an inert gas such as argon and a filler metal is normally used. A constant-volt welding power supply, shown in Figure 21.26, produces energy which is conducted across the arc through a column of highly ionized plasma channel. Depending on the material of the joint, GTAW uses different polarities as follows:

Direct polarity: Mild steel, stainless steel, copper, and titanium.
Reverse polarity: Aluminum and heavily oxidized aluminum castings.
Alternating current: Aluminum and magnesium have oxide coatings that are removed by the current alternation at high frequency.

Figure 21.27 compares the different electrode polarities used in GTAW. This welding method is the most commonly used to produce high-quality, high-strength welding of thin sections made of stainless steel, aluminum, magnesium, and copper alloys with notable exceptions being lead and zinc. Its application involving carbon steels is limited, because GMAW is a more economical welding technique. Furthermore, GTAW can be performed in a variety of other-than-flat positions, depending

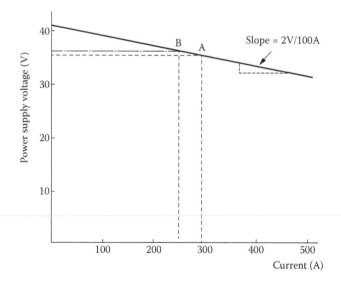

FIGURE 21.26 Constant voltage characteristics. (From Gourd, L. M., *Principles of Welding Technology*, Edward Arnold Publishers Ltd., London, 1982. With permission.)

Current type	DCEN	DCEP	Alternating current (balanced)
Electrode polarity	Negative	Positive	
Electron flow			
Penetration characteristics			
Oxide cleaning	No	Yes	Yes, (once every half cycle)
Heat balance in the arc	70% at work end	30% at work end	50% at work end
	30% at electrode end	70% at electrode end	50% at electrode end
Penetration	Deep, narrow	Shallow, wide	Medium
Electrode capacity	Excellent	Poor	Good
	(3.2 mm, 400 A)	(6.4 mm, 120 A)	(3.2 mm, 225 A)

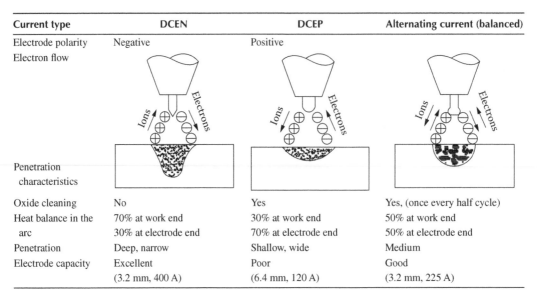

FIGURE 21.27　Comparison between GTAW electrode polarities.

on the skill of the welder and the materials being welded. Pulsed gas tungsten arc welding reduces the heat input to the metal. This results in a smaller heat affected zone, less brittle, less distortion, and more ductile weld. GTAW is comparatively more complex, difficult to control, and significantly slower than GMAW.

The shape of welding electrode and the type of shielding gas affect the process performance. In this regard, Figure 21.28 shows the different electrode configurations. The impact of shielding gas used in GTAW is shown in Table 21.3.

GTAW has several advantages

- Concentrated arc resulting in a narrow heat-affected zone
- No slag, sparks, or spatter
- Little smoke or fumes
- Welds more metals and alloys than any other process
- Good for welding thin material

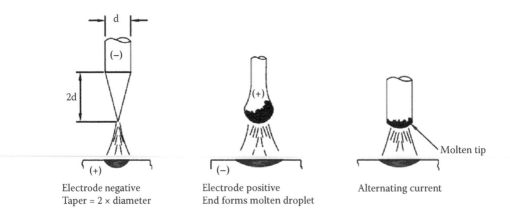

FIGURE 21.28　Tungsten electrodes.

TABLE 21.3
Shielding Gases for GTAW

Shielding Gas	Effect on GTAW
Argon	Smooth and quiet arc
	Superior shield
Helium	Wider and deeper penetration puddle
	Expensive
Argon–H_2	Increases fluidity and makes low thermal
	conductivity metals to be welded faster and easier
Argon–CO_2	Used for carbon steel

The process disadvantages include

- Smaller filler metal deposition rates and slower welding speeds
- Requires high operator skill
- Generates brighter UV rays
- Higher equipment costs than other welding processes

21.2.3.8 Atomic Hydrogen Welding

The atomic hydrogen arc welding (AHW) is maintained between two tungsten electrodes in an atmosphere of hydrogen. As the hydrogen enters the arc, its molecules are broken into atoms and recombines to hydrogen molecules outside the arc. This reaction is accompanied by intense heat that attains a temperature of about 4200°C. In addition, hydrogen provides proper shielding that protects the electrodes and molten metal from oxidation. A filler metal is added to the welding joint in the form of a welding rod. Atomic hydrogen welding is used to weld difficult-to-weld metals, such as chrome, nickel, molybdenum steels, Inconel, Monel, and stainless steel. Its main application is in the field of tool and die repair.

21.2.3.9 Carbon Arc Welding

Carbon arc welding (CAW) has been replaced by gas tungsten arc welding (GTAW) in many applications. In carbon arc welding, shown in Figure 21.29, the arc is drawn between a single carbon electrode and the workpiece and produces extreme temperatures of 3000°C. Generally, no shielding is required while a filler metal may be used. CAW may adopt two carbon electrodes, made of pure graphite, that do not erode away as quickly as the single carbon electrode, although such an arrangement is more expensive. In the single electrode arc welding, shielding can be made by using inert gas, or gas mixture, or from the combustion of a flux fed into the arc.

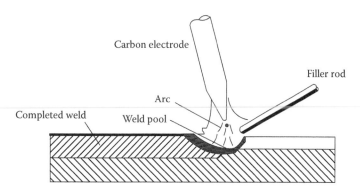

FIGURE 21.29 Carbon arc welding (CAW).

In the twin carbon electrode welding, the arc is drawn between two carbon electrodes. Shielding is not used, while a filler metal may be used. The welding joint is not a part of the welding electric circuit; therefore, the welding torch may be moved from one workpiece to another without extinguishing the arc. The advantages of carbon arc welding include low cost of equipment and welding operation, low level of operator skill, the process can be easily automated, and low distortion of workpiece. Limitations of carbon arc welding include the unstable quality of the weld (porosity) and contamination of weld material with carbides.

21.2.3.10 Plasma Arc Welding

Plasma arc welding (PAW) utilizes the heat generated by a constricted arc struck between a tungsten nonconsumable electrode and either the workpiece in case of transferred arc welding (Figure 21.30a) or the water-cooled constricting nozzle during the nontransferred arc welding process (Figure 21.30b). The temperature generated reaches 10,000 to 14,000°K. The metal to be welded is melted by the intense heat of the arc and fuses together. Auxiliary inert shielding gas or a mixture of inert gases is normally used. By forcing the plasma gas and arc through a constricted orifice, the torch delivers a high concentration of heat to a small area of the welding zone. The plasma gas (helium or argon) flows at 1.5–15 L/min around the tungsten electrode and subsequently forms the core of the plasma arc. The shielding gas that flows at 10–30 L/min provides protection for the molten pool. Transferred arc produces plasma jet of high-energy density, suitable for high-speed welding and cutting of ceramics, steels, aluminum alloys, copper alloys, titanium alloys, and nickel alloys. In the nontransferred plasma welding torch, the arc is initiated between the tungsten electrode and the nozzle tip. In such a case, the workpiece is not a part of the electric circuit, the plasma arc torch that can be moved from one workpiece to other without extinguishing the arc. Nontransferred arc process produces plasma of relatively low energy density suitable for welding of various metals.

Advantages of plasma arc welding include

- Requires less operator skill
- High welding rate
- No contamination with tungsten occurs
- Requires lower heat input and less filler metal
- Does not need edge preparation
- Minimized possibility of human error

Plasma arc welding requires expensive equipment that forms a major process limitation.

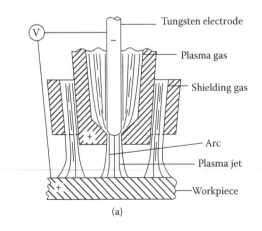 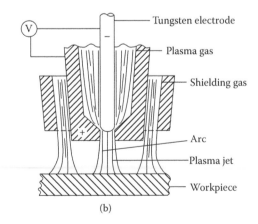

(a)   (b)

FIGURE 21.30 Plasma arc welding (PAW).

Example 2

Compare between GTAW and PAW under the following conditions:

GTAW: 125 A, 12 V, 26 cm/min
PAW: 75 A, 18 V, 34 cm/min

Solution

$$\text{Heat input} = \frac{V \times I \ (A) \times 60}{\text{Welding speed } V_w \ (mm/min)} \quad J/mm$$

For GTAW

$$\text{Heat input} = \frac{12 \times 125 \times 60}{260} = 346 \quad J/mm$$

For PAW:

$$\text{Heat input} = \frac{18 \times 75 \times 60}{340} = 238 \quad J/min$$

In addition to the fact that a higher welding speed is possible by PAW, the lower heat input provides less distortion, less stress in welded components, and a lower risk of damaging any heat sensitive parts adjacent to the weld joint.

21.2.4 RESISTANCE WELDING

In resistance welding, coalescence is produced by the heat obtained from resistance of the workpiece to electric current flow and by the application of pressure. The process is limited to sheet metals, not highly conductive. A filler metal and fluxes are not employed. The main factors involved in making a good resistance weld include the welding current, the welding pressure, and the time of current flow through the work. The welding pressure is applied before, during, and after the current flow. It forces the heated parts together so that coalescence occurs. It is an advantage to shorten the welding time and minimize the heat losses. The mechanical pressure that forces the parts together refines the grain structure of the weld. During resistance spot welding, shown in Figure 21.31, high current at a low voltage flows through the circuit in accordance with Ohm's law,

$$I = V/R$$

$$\text{Heat energy } (H_e) = I \times V \times t$$

$$= I^2 \times R \times t$$

For practical reasons, a factor of heat losses K_1 is considered. Therefore, the actual resistance welding formula is given by

$$H_e = I^2 \times R \times t \times K_1$$

where
V = Voltage in volts
I = Current in A

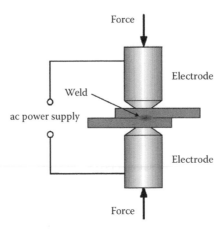

FIGURE 21.31   Resistance spot welding.

R = Total resistance of the work in Ω
t = Time of current flow in s
K_1 = Heat losses through radiation and conduction

Resistance welding is used in mass production where long production runs and consistent conditions should be maintained. The process is used for manufacturing a variety of products made of thinner gauge metals. The automotive and appliance industries are the major users of the resistance welding. There are also applications in the steel industry where it is used for manufacturing pipes, tubes, and smaller structural sections.

Resistance welding has the following advantages.

- It produces a high volume of work at high speeds.
- It does not require filler materials.
- Resistance welds are reproducible.
- High-quality welds are normal.
- Resistance welding operations can be automated.
- Weld quality does not depend on operator skill.

21.2.4.1   Resistance Spot Welding

Resistance spot welding (RSW) produces welding in one spot by the heat obtained from resistance to electric current through the work parts held together under pressure by electrodes as shown in Figure 21.31. Figure 21.32 shows the current–pressure timing for RSW where the squeeze time, weld time, and forge time are interrelated. The size and shape of the individually formed welds (called nuggets) are limited by the size and contour of the electrodes. The equipment for resistance spot welding is relatively simple and inexpensive, and the process is used in mass production for the automotive industry. In such applications, multiple spot welding machines carrying electrode tips are used.

21.2.4.2   Resistance Projection Welding

Resistance projection welding (RPW) is a resistance welding process where the resulting welds are localized at predetermined points made by projections, embossments, or intersections. As shown in Figure 21.33, localization of heating is obtained by a projection or embossment on one or both of the parts being welded. The major advantage of projection welding is that the electrode life is increased

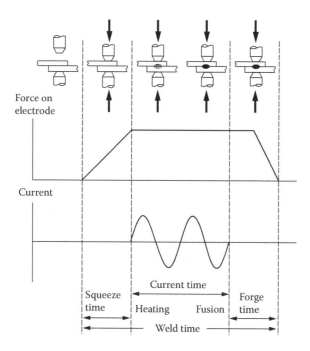

FIGURE 21.32 Resistance spot welding current–pressure timing for RSW.

because larger contact surfaces are used. A common use of projection welding is for joining special nuts and studs that have projections in the form of points or lines to sheet metals and small plates for the assembly purposes.

21.2.4.3 Resistance Seam Welding

Resistance seam welding (RSEW) is a welding process where a series of overlapping resistance spot welds are progressively made along the joint using rotating electrodes. When the spots are not overlapped, it is known as the roll resistance spot welding. As shown in Figure 21.34, the electrodes are powered wheels, while the pressure is applied in the same manner of the spot welding. Water cooling is not provided internally, and therefore, the weld area is flooded with cooling water to keep the electrode wheels cool. High conductive metals are not recommended to be welded by seam welding (RSEW).

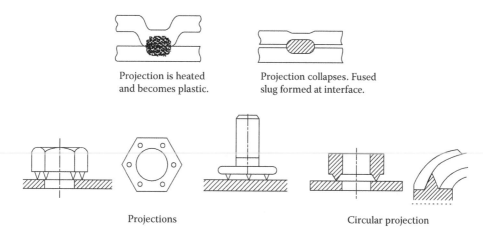

Projection is heated and becomes plastic.

Projection collapses. Fused slug formed at interface.

Projections

Circular projection

FIGURE 21.33 Projection welding (RPW).

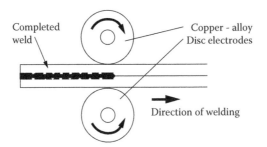

FIGURE 21.34 Resistance seam welding (RSEW).

In resistance seam welding, a rather complex control system that involves the travel speed as well as the sequence of current flow to provide for overlapping welds is required. The welding speed, the spots per unit length, and the timing schedule are dependent on each other. Welding schedules provide the pressure, the current, the speed, and the size of the electrode wheels. As examples, this process is used for making flange welds and watertight joints for tanks. Another variation of seam welding is the mash seam welding where the lap is narrow and the electrode wheel is twice as wide as used for standard seam welding.

21.2.4.4 Flash Welding

Flash welding (FW) (Figure 21.35) is a resistance welding that produces coalescence simultaneously over the entire area of abutting surfaces. The heat is obtained from the resistance to electric current between the two surfaces and by the application of pressure after heating is completed. During the welding operation, there is an intense flashing arc and heating of the metal on the abutting surfaces. After a predetermined time, the two pieces are forced together and welding occurs at the interface.

21.2.5 High-Energy Beam Welding

This group of welding processes utilizes a beam of accelerated electrons that bombard the work-piece to generate the heat required for fusion (electron beam welding) or a focused beam of light (laser beam welding).

21.2.5.1 Electron Beam Welding

In electron beam welding (EBW) coalescence of metals is made by the heat obtained from a concentrated beam of high-velocity electrons bombarding the surfaces to be joined. Virtually all of the kinetic energy is transformed into heat upon impact, which is sufficient to cause complete fusion and vaporization. No flux or shielding gas is required. In the electron beam system, shown in Figure 21.36, the workpiece is housed in a vacuum chamber.

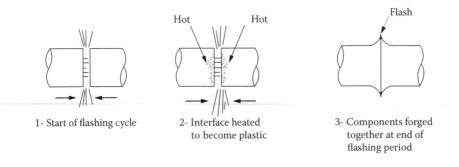

1- Start of flashing cycle   2- Interface heated   3- Components forged
                                 to become plastic   together at end of
                                                             flashing period

FIGURE 21.35 Steps of flash welding (FW).

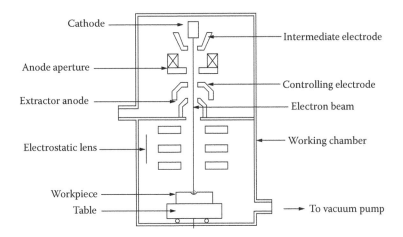

FIGURE 21.36   Electron beam welding (EBW).

One of the major advantages of electron beam welding is the tremendous penetration that occurs when the highly accelerated electron hits the base metal. It penetrates slightly below the surface and at that point releases the bulk of its kinetic energy, which turns to heat energy. The addition of the heat brings about a substantial temperature increase at the point of impact. The width of the penetration pattern is extremely narrow. The depth-to-width ratio can exceed 20 : 1. As the power density is increased, penetration is increased. This results in narrow almost parallel weld with very small distortion and small width of heat-affected zone. There is no possibility of contamination by atmospheric gases. Any metal can be welded including refractory metals such as molybdenum, zirconium, stainless steel, and tungsten for automotive and aerospace industries. One of the disadvantages of the electron beam process is the high capital cost.

21.2.5.2   Laser Beam Welding

Laser beam welding (LBW) produces welding of materials by the heat obtained from a concentrated coherent light beam impinging upon the surfaces to be joined as shown in Figure 21.37. Laser

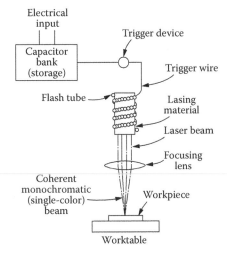

FIGURE 21.37   Laser beam welding.

beam welding has high power density (on the order of 1 MW/cm^2) resulting in small heat-affected zones and high heating and cooling rates. The spot size of the laser varies between 0.2 and 1.3 mm. The depth of penetration is proportional to the amount of power supplied and depends on the location of the focal point. Penetration is maximized when the focal point is slightly below the surface of the welding joint. A continuous or pulsed laser beam may be used depending on the application. Milliseconds long pulses are used to weld thin materials such as razor blades, while continuous laser systems are employed for deep welds. Further details about the types of laser are presented in Section 17.10.

LBW is a versatile process, capable of welding carbon steels, high strength low alloy (HSLA) steels, stainless steel, aluminum, and titanium. Owing to the high cooling rates, cracking is a concern when welding high-carbon steels. The high capital and maintenance costs of equipment is another limitation. The weld quality is high, similar to that of electron beam welding. The traverse speed of welding is proportional to the amount of power supplied and depends on the type and thickness of the workpiece. Compared to EBW, laser beam can be transmitted through air and, therefore, does not require a vacuum. The process is easily automated with robotic machinery, X-rays are not generated, and LBW produces higher-quality welds. Laser welding has the following advantages.

- High welding speeds are possible.
- No filler metals are necessary.
- No slag or spatter occurs.
- The process is automated and can be manipulated by robotics.
- The process can be used in open air and transmitted over long distances with a minimal loss of power.
- Low total thermal input causes narrow heat-affected zone and low distortion.
- Welds dissimilar metals.
- Extremely accurate and do not require secondary finishing operations.

21.3   SOLID-STATE WELDING

Solid-state welding produces coalescence at temperatures below the melting point of the materials being joined without the addition of filler metal. In solid-state welding time, temperature, and pressure individually or in combination produce welding without significant melting of the base metals. In some processes the time element is extremely short, in the microsecond range or up to a few seconds while in other cases, the time is extended to several hours. As the temperature increases, the welding time is usually reduced. Because the base metal does not melt and form a nugget, the metals being joined retain their original properties without the heat-affected zone problems associated with fusion welding. When dissimilar metals are joined, their thermal expansion and conductivity is of much less importance with solid-state welding than with the arc welding processes. Solid-state welding depends on surface deformation, surface oxide layer and oil films, recrystallization and grain growth at the interface, and diffusion. Solid-state welding techniques include the types discussed in the following sections.

21.3.1   COLD WELDING

Cold welding (Figure 21.38) is accomplished using extremely high pressures on extremely clean interfacing materials at room temperature to produce coalescence of metals with substantial deformation at the weld. The high pressure is obtained using either simple hand tools when thin materials are being joined or presses when cold welding of heavier sections is required. A satisfactory weld may be obtained with either an impact blow or a slow squeeze. The process is readily adaptable for joining ductile metals such as pure aluminum and other nonferrous metals.

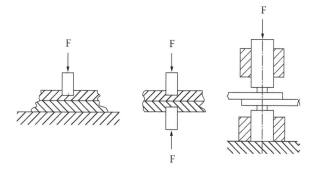

FIGURE 21.38 Cold welding.

21.3.2 DIFFUSION WELDING

In diffusion welding (DW), joining is produced by the application of pressure at elevated temperatures slightly over half the normal melting temperature of the metal. Heating is usually accomplished by induction, resistance, or furnace. Close tolerance joint preparation is required, and a vacuum or inert atmosphere is used. The process is extensively used for joining dissimilar metals. Diffusion brazing occurs when a layer of filler material is placed between the surfaces of the parts being joined. DW is used for joining refractory metals for aerospace and aircraft industries at temperatures that do not affect their metallurgical properties.

21.3.3 EXPLOSION WELDING

In explosion welding (EW) (Figure 21.39), coalescence is achieved by high-velocity movement of the parts to be joined using a controlled detonation. Explosion welding creates a strong weld between almost all metals as well as dissimilar ones that are not weldable by the arc welding processes. The strength of the weld joint is equal to or greater than the strength of the weaker of the two metals joined. The process is widely used in a cladding of base metals with thinner alloys, joining of tube-to-tube sheets used for heat exchangers. It is self-contained and portable and can be achieved quickly over large areas.

21.3.4 FORGE WELDING

The forge welding (FOW), shown in Figure 21.40, is an old welding technique that produces coalescence of metals by heating them in a forge below the melting temperature and applying pressure or blows sufficient to cause permanent deformation at the interface. Normal practice was to apply flux to the interface. The physical properties of the forge weld depend on the operator skill, weldability of metal, fuel used in the forge, the atmospheric conditions, the amount of flux used, and the time

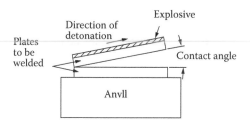

FIGURE 21.39 Explosive welding (EW).

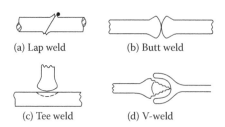

(a) Lap weld   (b) Butt weld

(c) Tee weld   (d) V-weld

FIGURE 21.40   Forge welding (FW).

of application. This process is not widely used as a production method as it is a very costly and slow process.

21.3.5 FRICTION WELDING

In friction welding (FRW), coalescence of materials is achieved by the heat obtained from the mechanically induced sliding motion between rubbing surfaces under pressure. There are three variations of the friction welding process, namely, the rotating, linear, and friction stir welding processes. In the rotating friction welding, shown in Figure 21.41, one part is held stationary while the other is rotated at constant rotational speed. The two parts are brought in contact under specific pressure for a specific period of time. The actual operation of the machine is automatic and can be accurately controlled when speed, pressure, and time are closely regulated. When a suitable high temperature has been reached, the rotational motion ceases, additional pressure is applied, and coalescence occurs.

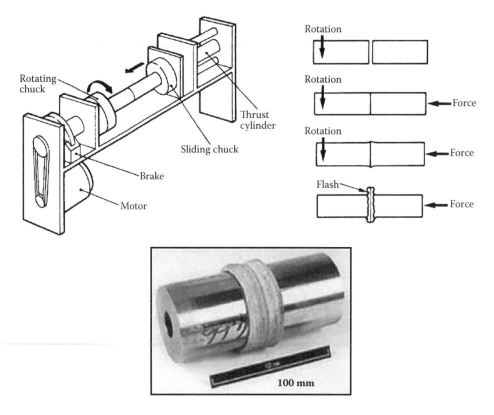

FIGURE 21.41   Rotary friction welding.

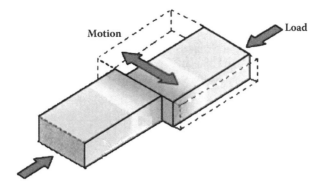

FIGURE 21.42 Linear friction welding.

Rotary friction welding is commonly used for joining carbon steel vehicle axles and subaxles. The process is also used to weld suspension rods, steering columns, gear box forks, and drive shafts.

In the linear friction welding, a linear relative motion across the interface is used to join two parts (Figure 21.42). The process is widely used in automotive applications, such as the fabrication of brake discs, wheel rims, and engine parts. Friction stir welding (Figure 21.43) uses a nonconsumable rotating tool, pushed into the materials to be welded. The central pin, or probe, followed by the shoulder, is brought into contact with the two parts to be joined. The rotation of the tool heats up and plasticizes the joint materials. As the tool moves along the joint line, material from the front of the tool is swept around this plasticized annulus to the rear, so eliminating the interface. Friction stir welding finds many applications that include

1. Shipbuilding and marine industry: Helicopter landing platforms, marine and transport structures, and refrigeration plants
2. Aerospace industry: Wings, cryogenic fuel tanks for space vehicles, aviation fuel tanks, and military rockets
3. Railway industry: Railway tankers, goods wagons, high speed trains, underground carriages, and trams
4. Land transportation: Mobile cranes, fuel tankers, caravans, airfield transportation vehicles, motorcycles, and cycle frames
5. Construction industry: Aluminum bridges, window frames, aluminum pipelines, heat exchangers, and air conditioners
6. Electrical industry: Electric motor housing and electrical connectors

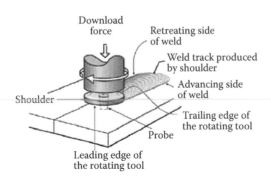

FIGURE 21.43 Friction stir welding.

Among the advantages of friction welding is the ability to produce high-quality welds without porosity in a short cycle time. No filler metal is required, and flux is not used. The process is capable of welding most of the common metals as well as many combinations of dissimilar metals. It is environment friendly because no fumes or spatter are generated, and there is no arc. Another major advantage is that, by avoiding the creation of a molten pool that shrinks significantly on resolidification, the distortion after welding and the residual stresses are low. Friction welding requires relatively expensive apparatus.

21.3.6   HOT PRESSURE WELDING

In hot pressure welding (HPW), joining occurs at the interface between the parts to be joined using pressure and heat. The process is accompanied by noticeable deformation that removes the surface oxide film and increases the areas of clean metal. Welding is accomplished by diffusion across the interface and coalescence of the joint surface. This operation is normally carried on in closed chambers where vacuum or shielding gases are used. It is mainly used in aerospace industry.

21.3.7   ROLL WELDING

Roll welding (ROW) process joins metals by heating the metals to be joined and applying pressure using rolls that cause deformation at the faying surfaces (Figure 21.44). Coalescence occurs at the interface between the two parts through diffusion. The process is used for cladding of mild steel or stainless steel, as well as for welding bimetallic materials for the instrument industry.

21.3.8   ULTRASONIC WELDING

In ultrasonic welding (USW), joining is produced by the local application of high-frequency vibratory energy as the work parts are held together under pressure. Welding occurs when the ultrasonic tip or electrode, which is clamped against the workpiece, oscillates in a plane parallel to the weld interface (Figure 21.45). The combined pressure and oscillating forces introduce dynamic stresses in the base metal that produce minute deformations and create a moderate temperature rise in the base metal at the weld zone. Incidentally, coalescence across the interface produces the weld joint. The resulting temperature is not raised to the melting point, and therefore, there is no weld nugget similar to that which occurs during resistance welding. The ultrasonic oscillation also cleans the weld area by breaking up the oxide films and removing them. Most ductile metals as well as many combinations of dissimilar metals can also be welded. The process is, however, restricted to relatively thin materials normally in the foil or extremely thin gauge thickness. USW is used extensively in the electronics, aerospace, and instrument industries in addition to producing sealing packages and containers. Other process advantages include

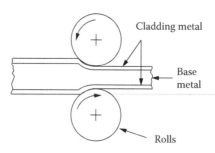

FIGURE 21.44   Roll welding (RW).

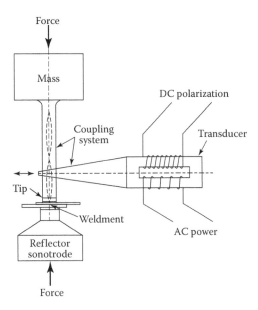

FIGURE 21.45 Ultrasonic welding (USW).

- Low pressure in the order 2–3 kg/cm² used.
- No grain growth, no gas absorption, and no porosity.
- Minimum embrittlement.
- Environment friendly process.
- Improved product uniformity and joint appearance.
- Fast bonding with low energy consumption.

21.4 SOLID–LIQUID STATE WELDING

In welding processes, edges of the workpieces are either fused (with or without a filler metal) or pressed to each other without any filler material. Solid–liquid state welding joins two parts without melting them but through a fused filler metal. In this regard, brazing and soldering are typical examples.

21.4.1 BRAZING

Brazing is a method of joining two metal workpieces using a filler material at a temperature above its melting point (>450°C) but below the melting point of either of the materials being joined. The flow of the molten filler material into the gap between the two metal workpieces is driven by the capillary force. Capillary effect is achieved through proper surface cleaning and the use of a flux for wetting. The flux is applied onto the metal surface by brushing, dipping, or spraying. It melts during the heating stage and spreads over the joint area, wets it (low surface tension), and protects the surface from oxidation. It also cleans the surface by dissolving the metal oxides. When the filler material cools down, it solidifies and forms a stronger metallurgical joint than the parent materials that are not fused in the process. Figure 21.46 shows typical brazed joints. Brazing filler materials depend on the material to be welded as shown in Table 21.4.

Brazing utilizes different methods for heating that can be summarized by the following.

1. Torch brazing utilizes a heat of the flame from a torch that mixes a fuel gas with oxygen or air in the proper ratio and flow rate.
2. Furnace brazing uses a furnace for heating the workpieces.

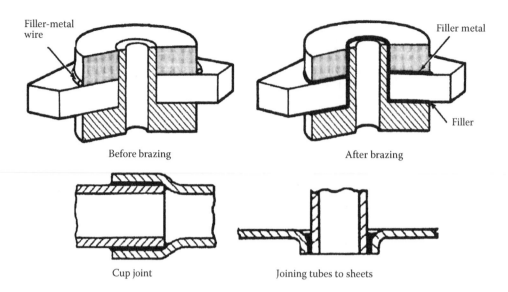

Before brazing　　　　　　　　After brazing

Cup joint　　　　　Joining tubes to sheets

FIGURE 21.46　Brazed joint.

3. Vacuum brazing is a furnace brazing in which heating is performed in vacuum.
4. Induction brazing utilizes alternating electromagnetic field of high frequency for heating the workpieces together with the flux, while the filler metal is placed in the joint region.
5. Resistance brazing uses a heat generated by an electric current flowing through the workpieces.
6. Dip brazing is a brazing method, in which the workpiece together with the filler metal are immersed into a bath with a molten salt. Hence the filler material melts and flows into the joint.
7. Infrared brazing utilizes the heat of a high power infrared lamp.

Advantages of brazing include the following.

- Dissimilar materials can be joined.
- Easily automated process.
- Various materials can be joined.
- Low thermal distortions and residual stresses in the joint parts.

TABLE 21.4
Brazing Filler Alloys and Their Applications

Filler Alloy	Filler Composition	Applications
Copper	BCuP-2 (Cu-7P), BCuP-4 (Cu-6Ag-7P)	Copper alloys, steels, nickel alloys
Aluminum	Al-4Cu-10Si, Al-12Si, Al-4Cu-10Si-10Zn, 4043 (Al-5.2Si), 4045 (Al-10Si)	Aluminum alloys
Magnesium	BMg-1 (Mg-9Al-2Zn), BMg-2 (Mg-12Al-5Zn)	Magnesium alloys
Nickel	BNi-1 (Ni-14Cr-4Si-3.4B-0.75C), BNi-2 (Ni-7Cr-4.5Si-3.1B-3Fe), BNi-3 (Ni-4.5Si-3.1B)	Nickel alloys, cobalt alloys, stainless steels
Silver	BAg-4 (40Ag-30Cu-28Zn-2Ni), BAg-5 (45Ag-30Cu-25Zn), BAg-6 (50Ag-34Cu-16Zn), BAg-7 (56Ag-22Cu-17Zn-5Sn)	Used for most of metals and alloys except aluminum and magnesium alloys

- Microstructure is not affected by heat.
- Moderate skill of the operator is required.
- Thin wall parts can be joined.

Disadvantages of brazing are as follows.

- Careful removal of the flux residuals is required in order to prevent corrosion.
- Fluxes and filler materials may contain toxic components.
- Large sections cannot be joined.
- No gas shielding that causes porosity of the joint is needed.
- Relatively expensive filler materials are used.

21.4.2 SOLDERING

Soldering is a method of joining two metal pieces using a third filler metal (solder) at a relatively low temperature, which is above the melting point of the solder but below the melting point of either of the materials being joined. Flow of the molten solder into the gap between the welded parts is driven by the capillary force. When the solder cools down, it solidifies and forms a joint. The difference between soldering and brazing is in the melting point of the filler alloy. While solders melt at temperatures below 450°C, brazing filler materials melt at temperatures above this point. Figure 21.47 shows different soldered joints. Generally, soldering joints have relatively low tensile strength of about 70 MPa. For this purpose, a variety of solders are available such as

1. Tin–lead solders such as the eutectic alloy of 63% tin and 37% lead are used for most copper-joining applications. Tin–lead alloys are cheap and have a low melting point. The process requires simple equipment and low operator skill; however, these alloys are toxic due to lead.
2. No lead solders are widely used for environmental reasons. Most of these solders are not eutectic formulations, and therefore, reliable joints are difficult to achieve.
3. Special alloys are available with properties such as higher strength, better electrical conductivity, and higher corrosion resistance.

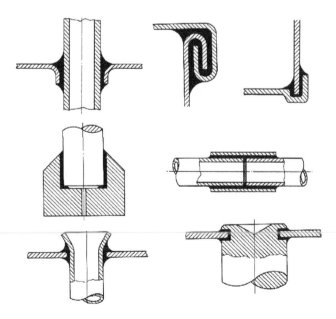

FIGURE 21.47 Soldered joint designs.

Soldering flux is added during the heating stage, which melts and spreads over the joint area, prevents the formation of metal oxides, and acts as a wetting agent. Traditionally, soldering fluxes require postprocess removal due to their chemical activity that causes erosion of the base material and produces unreliable joints. Soldering methods are as follows.

1. Hand soldering: This includes iron soldering that utilizes a heat generated by a soldering iron and torch soldering, which uses the heat of a flame supplied from a torch. The torch flame is directed to the workpiece with a flux applied on their surfaces.
2. Wave soldering: This method uses a tank full with a molten solder. The solder is pumped, and its flow forms a wave of a predetermined height. The method is used for soldering through-hole components on printed circuit boards.
3. Reflow soldering: In this method a solder paste of a mixture of solder and flux particles is applied onto the surface of the parts to be joined and then heated to a temperature above the melting point of the solder. The joint forms when the solder cools down and solidifies.

Soldering is inexpensive, easy to use, and a valuable method for making prototypes and experimental devices. Other applications include the following.

- Assembling electronic components on printed circuit boards (PCB).
- Permanent connections between copper pipes in plumbing systems.
- Joints in sheet-metal objects such as food cans, roof flashing, drain gutters, and automobile radiators.
- Jewelry and small mechanical parts are often assembled by soldering.

The advantages of soldering are as follows.

- Low temperature and hence low power is required.
- No thermal, metallurgical distortions, or residual stresses in the joint parts.
- Easily automated process.
- Wide variety of materials including dissimilar ones can be joined.
- Thin wall parts may be joined.
- Moderate skill of the operator is required.

Soldering has some limitations such as

- Careful removal of the flux residuals is required in order to prevent corrosion.
- Large sections cannot be joined.
- Fluxes may contain toxic components.
- Soldering joints cannot be used in high-temperature applications.
- Low strength of joints.

21.4.3 ADHESIVE BONDING

Adhesive bonding (AB) is a modern assembly technique where two similar or dissimilar materials (metals, plastics, composites, etc.) are joined using an adhesive that is placed between the faying surfaces and solidifies to produce an adhesive bonded joint. Adhesive bonding of metal-to-metal applications accounts for less than 2% of the total metal joining applications. The bonding of metals to nonmetals, especially plastics, is the major use of adhesive bonding. Adhesive joints are designed according to the loads carried by the joint. A properly designed adhesive joint provides for adherend failure rather than adhesive one. Surface preparation is required to remove oils and greases, increase the adherend surface energy, and provide a surface with enough roughness. In some cases, a primer is applied to the adherend surface before applying the adhesive. Generally, the adhesive requires a set time for solidification while the adherend are kept in place. The adhesive selected must be compatible with each material involved. If the materials expand and contract at different rates, a flexible bond is required. Several characteristics to consider when selecting an adhesive are

- Chemical compatibility with the parts to be joined
- Aesthetics of the finished joint
- Expansion or contraction with temperature changes
- Brittleness, rigidity, and flexibility
- Durability, service life, and strength
- Suitability for food contact

Main types of adhesives include the following.

- Natural adhesives: These are made from inorganic minerals or biological sources such as vegetable matter, starch (dextrin), natural resins, or animal skin. Natural adhesives are often referred to as bioadhesives.
- Synthetic adhesives: Elastomers, thermoplastic, and thermosetting adhesives are examples of synthetic adhesives.
- Drying adhesives: These adhesives are a mixture of polymers dissolved in a solvent. As the solvent evaporates, the adhesive hardens. White glue and rubber cements are members of the drying adhesive family. These adhesives are weak and are, therefore, used for household applications.
- Contact adhesives: These are applied to both surfaces and allowed some time to dry before the two surfaces are pushed together. Natural rubber and polychloroprene (Neoprene) are common contact adhesives used for bonding Formica to a wooden counter, and in footwear, when attaching an outersole to an upper.
- Hot (thermoplastic) adhesives: These are also known as hot melt adhesives. These adhesives are thermoplastics that are applied hot and allowed to harden as they cool.
- Reactive adhesives: It works either by chemical bonding with the surface material or by in situ hardening as the two reactant chemicals complete a polymerization reaction. These include two-part epoxy, peroxide, silane, metallic cross-links, or isocyanate. Such adhesives are frequently used to prevent loosening of bolts and screws in rapidly moving assemblies of automobile engines.
- Ultraviolet light curing adhesives: UV light curing adhesives have rapid curing time and strong bond strength. Light curing adhesives can bond dissimilar substrates and withstand high temperatures. UV curing adhesives find applications in electronics, telecommunications, medical, aerospace, glass, and optical industries.
- Pressure-sensitive adhesives: These form a bond by the application of light pressure. They are designed for either permanent or removable joining applications.

Adhesives are used in a wide range of applications including electronics, automotive, aircraft, furniture construction, and plywood manufacture. Other applications include paper binding, carton sealing (hot-melt adhesives), and envelope sealing. In medicine, adhesives are used as tissue sealants during surgeries and transdermal drug delivery systems. Plastics, elastomers, and certain metals such as aluminum and titanium can be more reliably joined with adhesives than with other methods. The main advantages provided by adhesive bonding are

- Clean-looking high-strength joints.
- Different materials with dissimilar thicknesses can be assembled.
- Easy to automate.

- Enhanced attenuation and absorption capabilities.
- Lighter structures.
- Uniform stress distribution.
- Welding dissimilar metals having different coefficients of thermal expansion or thermal conductivities.

There are also some limitations for adhesive bonding.

- Change in the mechanical properties of the adhesive with time.
- Curing time and temperature must be optimized.
- Assessing the durability of the bonded structure is difficult.
- Inspection without destructive testing methods is difficult.
- Disassembly is difficult.
- Heat resistance is limited.
- Surface pretreatment is usually required.

21.5 WELDING OF PLASTICS

The plastic welding is confined to the thermoplastic polymers because these materials can be softened by heat. Thermosetting polymers once hardened cannot be softened again after heating. The heat required for welding thermoplastic polymers is less than that required for metals. This heat can be generated using mechanical movements during ultrasonic welding, friction welding, and vibration welding or by using an external source during hot plate welding, hot gas welding, and laser beam welding. Full details about the joining of plastics are explained in Section 13.8.12.

21.6 METALLURGY OF WELDED JOINTS

The heat generated during welding causes a heat flow that raises the temperature of the parent metal adjacent to the fusion boundary. Therefore, a number of metallurgical changes occur in this region that may affect the mechanical properties of the joint by causing a loss of the tensile and impact strength or the formation of cracks. The structure of the parent metal that has been raised in temperature depends on its composition and the heating and cooling cycles (Figure 21.48). The boundary of this heat-affected zone (HAZ) depends on the lowest temperature that causes metallurgical changes. The heat-affected zone is subjected to temperatures above recrystallization. HAZ is, therefore, softened, and consequently the tensile strength and hardness fall below that of parent metal.

21.7 WELDING DEFECTS

There is a host of possible welding defects (Figure 21.49) that are caused by the improper welding procedure. Defects usually encountered include incomplete fusion, porosity, inclusions, cracking, and undercutting.

1. Lack of fusion: In order to achieve a good-quality joint, the fusion zone must extend through the full thickness of the sheets being joined. Lack of fusion occurs owing to little heat input and/or rapid welding speed. This condition forms a problem during welding of thin sheets as a higher level of skill is needed to keep balance between the heat input and torch traverse speed.
2. Porosity: This occurs when gases are trapped in the solidifying weld metal. It can be avoided by storing all the welding materials in dry conditions and carefully cleaning and degreasing the work prior to welding.

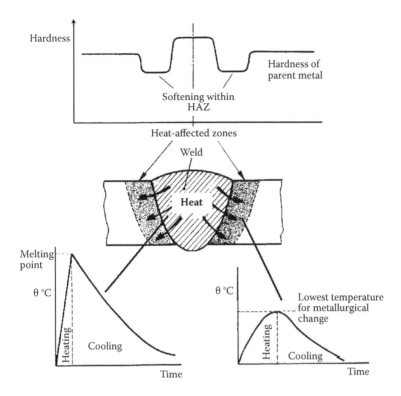

FIGURE 21.48 HAZ in welding. (Adapted from Gourd, L. M., *Principles of Welding Technology*, Edward Arnold Publishers, London, 1982.)

3. Inclusions: These occur when several runs are made along a V joint in a thick plate using flux-cored or flux-coated rods. In such a case, the slag covering a welding run is not totally removed before the following run.
4. Cracking: This occurs owing to thermal shrinkage or to a combination of strain accompanying phase changes and thermal shrinkage. In order to prevent such a problem, controlled

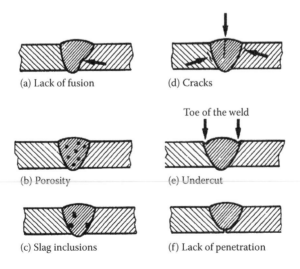

FIGURE 21.49 Welding defects.

heating in stages is needed before welding, while a slow controlled post cooling in stages is also required after welding

5. Undercutting: In this case, the thickness of one (or both) of the sheets is reduced at the toe of the weld. This causes stress concentration that reduces the strength of the welded joint.

6. Lack of penetration: This occurs when the weld metal fails to penetrate the joint. It allows a natural stress riser from which a crack may propagate.

21.8 WELDING QUALITY CONTROL

Tests are used to determine the quality and soundness of the welded joints. The type of test adopted depends on the requirements of the welds and the availability of the testing equipment. In this regard, nondestructive and destructive testing methods are used.

21.8.1 NONDESTRUCTIVE TESTING

Nondestructive testing (NDT) does not destroy or impair the usefulness of a welded item. These tests disclose all of the common internal and surface defects that occur when improper welding procedures are used. NDT methods are easier to implement than the destructive ones, especially when working on large and expensive items.

NDT methods include the following.

1. Visual and optical testing (VT): The most basic NDT method is visual examination. Visual examiners follow procedures that range from simply looking at a part to see if surface imperfections are visible, to using computer-controlled camera systems to automatically recognize and measure features of a welded joint.

2. Magnetic particle testing (MT): This method is performed by inducing a magnetic field in a ferromagnetic material and then dusting the surface with iron particles (either dry or suspended in liquid). Surface and near-surface flaws distort the magnetic field in such a way that the iron particles are attracted and concentrated, thus producing a visible indication of surface or subsurface defects or a crack in the material.

3. Radiography testing (RT): RT involves the use of a penetrating X-ray to examine defects and internal features. Radiation is directed through a part and onto film or other media. Hence the resulting shadowgraph shows the internal features and soundness of the joint. The darker areas in the radiograph represent internal voids in the component. This test is used to inspect almost any material for surface and subsurface defects.

4. Penetrant testing (PT): Liquid penetrant is used to detect surface defects that are similar to those revealed by magnetic particle inspection. While magnetic particles inspection reveal subsurface defects, liquid penetrant inspection reveals those defects that are open to the surface. After thoroughly cleaning and drying the surface, it is coated with the liquid penetrant. The penetrant is allowed to soak into all the cracks, crevices, or other defects that are open to the surface. Any excess penetrant from the surface is removed, and the test surface is allowed to dry. A developer in the form of powder or liquid is applied for a minimum of 7 minutes before starting the inspection that clearly shows the surface cracks. PT is used to locate cracks, porosity, and other defects that break the surface of a material and have enough volume to trap and hold the penetrant material.

5. Ultrasonic testing (UT): Ultrasonic inspection uses high-frequency waves to locate and measure defects. It can be used in both ferrous and nonferrous materials to locate very fine surface and subsurface cracks as well as other types of defects in metals and plastics. In ultrasonic testing, high-frequency sound waves are transmitted into a material to detect imperfections or to locate changes in material properties. The most

common ultrasonic testing technique is pulse echo, whereby a sound wave is introduced into a test object and reflections (echoes) from internal imperfections are returned to a receiver.

6. Eddy current testing (ET): Eddy currents are generated in a conductive material by a changing magnetic field. The strength of these eddy currents can be measured. Material defects cause interruptions to the flow of the eddy currents that alert the inspector to the presence of a defect. The process is used to detect surface and near-surface flaws in conductive materials such as the metals.

Each NDT method has its own set of advantages and disadvantages, and therefore, individual methods are better suited for a particular application than others. Table 21.5 provides some guidance in the selection of NDT methods for common detection of defects.

TABLE 21.5
Nondestructive Testing Methods

				Inspection Method				
Flaw Type	Flaw Shape	Visual	Liquid Penetrant	Magnetic Particle (A)	Ultrasonic Straight Beam	Ultrasonic Angle Beam	Eddy Current (B)	X-ray
Linear surface crack		L	H	H	L	M	H	L
Volumetric surface crack		H	H	H	H	H	H	H
Surface crack linear, normal				M	L	M	H	L
Surface crack linear, parallel					H	H		
Near surface volumetric				M	H	H	H	H
Subsurface crack linear, normal					L	M		L
Subsurface crack linear, parallel					H	H		L
Subsurface volumetric					H	H		H

Source: Brian F. Larson, Center for Nondestructive Evaluation, Iowa State University, Ames, Iowa 50011, USA. With permission.

Note: L, Not well; M, Fairly well; H, Ideal application; A, Ferromagnetic materials only; B, Conductive materials only.

21.8.2 DESTRUCTIVE TESTING

In destructive testing, sample portions of the welded structures are subjected to loads until they actually fail. The failed pieces are then studied and compared to known standards to determine the quality of the weld. The most common methods of destructive testing are known as free bend, guided bend, nick-break, impact, fillet welded joint, etching, and tensile testing. The primary disadvantage of destructive testing is that an actual section of weldments must be destroyed to evaluate the weld. This type of testing is usually used in the certification process of the welder.

21.9 MECHANICAL JOINING

Several methods are used for establishing a permanent joint by mechanical means. Mechanical fasteners using auxiliary elements should be chosen so that both fastener and the components to be joined are compatible as far as corrosion and recycling aspects are concerned. The parts that come in contact with each other must have similar electrochemical potentials. Mechanical joining includes the following.

1. Riveting: Riveting is a method of high-quality joining in the aerospace industry. Riveting joins materials that are otherwise difficult or impossible to spot weld as well as similar or dissimilar materials. Figure 21.50 shows different types of rivets used for producing permanent joints.
   a. Solid rivets are one-piece joining elements in which the rivet shaft is plastically formed into the closing head. Such rivets are used for components that are accessible from both sides (Figure 21.50a).
   b. Huck bolts (screw rivets) are used for highly stressed rivet joints. Because screw rivets are made of high strength materials that cannot be formed easily during assembly, a closing collet (self-locking nut) is fixed on to the rivet (Figure 21.50b).
   c. Blind rivets consist of one or more elements and require only accessibility from one size. Generally, blind rivets consist of a hollow shaft and a pull-stem (mandrel) that serves as a tool for forming the closing head (Figure 21.50c).
   d. Self-piercing riveting requires no predrilled hole in the sheets to be joined, thus eliminating the need for aligning prepared parts and then placing these correctly in the rivet setting equipment. A punch and die are used to complete the joining operation in a single step. It is mainly used for the automotive industry.

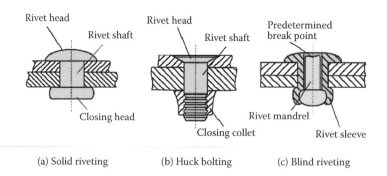

(a) Solid riveting   (b) Huck bolting   (c) Blind riveting

FIGURE 21.50 Rivets.

2. Clinching: Clinching uses a punch that forms the two materials to be joined into a die. A button is formed on the underside that provides an interlock between the two sheets as shown in Figure 21.51.
3. Stitching or stapling: This method is used to join thin sheets. It avoids the need for preliminary drilling and is commonly used to fasten metal sheets to wooden backing (Figure 21.52a).
4. Seams: Seams are produced by a sequence of bends on tight radii. Lock seams are made impermeable without or with a filler material such as soldering adhesives or polymeric seals. Seams can be found in radiator tubes and beverage cans (Figure 21.52b).
5. Mechanical interface: It uses plastic deformation such as bending of lanced protrusions and crimping (Figure 21.53a, b).
6. Snap fit joint: This relies on the elastic spring back of a cantilevered element (Figure 21.53c).
7. Shrinkage: Shrinkage is used to join a sleeve into a core of a circular section. Such shrinkage is attainable by heating either the sleeve or the core or by pressing the two parts that maintain a small angle.

Table 21.6 presents general comparison to the joining processes.

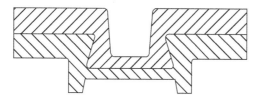

FIGURE 21.51   Clinch riveting.

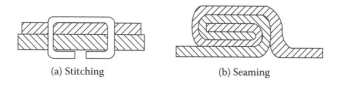

(a) Stitching   (b) Seaming

FIGURE 21.52   Stitching and seaming.

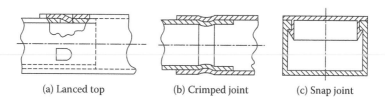

(a) Lanced top   (b) Crimped joint   (c) Snap joint

FIGURE 21.53   Lanced, crimped, and snap joints.

TABLE 21.6
Summary of Joining Processes

Joining Process	Category	Heat Source								Filer Metal	Shielding	Pressure
		Burning Gases	Chemical Reaction	Arc	Beam	Resistance	Ultrasonic Vibrations	Friction	Other			
Oxyacetylene	Fusion	X								X		
Thermit			X							X		
Shielded metal arc				X						X	X	
Flux cored arc				X						X	X	
Gas metal arc				X						X	X	
Submerged arc				X						X	X	
Underwater welding				X						X	X	
Electroslag				X						X	X	
Gas tungsten arc				X						X	X	
Carbon arc				X						X	X	
Atomic hydrogen				X						X	X	
Plasma arc				X						X	X	
Resistance spot welding						X						X
Resistance seam welding						X						X
Projection welding						X						X
Flash welding						X						X
Electron beam					X							
Laser beam					X							
Cold welding	Solid state											X
Diffusion welding									X			X
Explosion welding									X			X
Forge welding									X			X
Friction welding								X				X
Hot pressure welding									X			X
Roll welding												X
Ultrasonic welding							X					X
Brazing	Solid–liquid state									X		
Soldering										X		
Adhesive bonding										X		
Mechanical joining			Depends on metal-forming principles										X

21.10 REVIEW QUESTIONS

1. Explain what is meant by arc blow.
2. Compare straight and reversed polarity in arc welding.
3. What are the advantages of using ac arc welding?
4. Compare transferred and nontransferred plasma arc welding.
5. Explain the behavior of constant current power supplies.
6. Show how a constant voltage power supply is used for GTAW welding.
7. Compare laser and electron beam welding operations.
8. Draw a sketch showing the different defects in a welding joint.
9. What are the advantages of SAW?
10. Using sketches, explain the modes of metal transfer during GMAW.
11. What are the different methods of brazing?
12. State the advantages of brazing.
13. Explain what is meant by Thermit welding and Thermit mixture composition.
14. Using sketches, show the different flames and their applications in oxyacetylene welding.
15. Show the different techniques adopted during gas welding.
16. What is forge welding? What conditions affect the weld quality?
17. Using sketches, show the lap, butt, corner, and edge joints.
18. Compare shielded and unshielded arc welding.
19. What is carbon electrode arc welding?
20. What are the advantages and disadvantages of ac and dc arc welding?
21. Show the principal components of a SAW welding system.
22. Using sketches, compare GTAW and GMAW.
23. Explain the principles of atomic hydrogen welding.
24. What are the principles and applications of underwater welding?
25. Explain the brazing operation. Why does it need flux?
26. Compare welding, brazing, and soldering operations.
27. Compare submerged arc welding and electroslag welding processes.
28. Show the principles of resistance spot welding.
29. Compare resistance seam and resistance projection welding.
30. Explain how resistance flash welding is used to produce a butt joint.
31. Explain the procedure for spot welding.
32. Show the different types of friction welding processes.
33. What are the advantages of friction welding?
34. Explain what is meant by soldering.
35. Compare roll welding and explosive welding.
36. What is the difference between cold welding and diffusion welding?
37. Explain what is meant by adhesive bonding.
38. What are the main applications of adhesive bonding.
39. List the different types of adhesives used in adhesive bonding.
40. List the different methods used in joining of plastic materials.
41. Show how defects can be inspected using nondestructive techniques.
42. What are the different methods used for producing permanent mechanical joints?

21.11 PROBLEMS

1. During arc welding, if the current used is 80 A, arc voltage is 20 V, and the welding speed is 96 mm/min, calculate
   - Energy input to the arc
   - Heat input in J/mm

2. During arc welding using a current of 70 A, voltage 18 V, calculate the welding speed if the heat input to the arc is 400 J/mm. How much heat in kilocalories is required for seam welding of 100 mm length?

3. Calculate the welding speed during arc welding using 100 A, and 20 V if the heat input is 600 J/mm. If the thickness is doubled and the welding speed is reduced to 50%, calculate the heat input required in the new case.

4. During arc welding a seam deposited during 10 min using an arc current of 200 A and voltage of 20 V, calculate
   - Amount of kW hours consumed
   - Amount of heat used in kcal

5. Calculate the amount of energy (J) and the heat required (kcal) during resistance spot welding under the following conditions:
   - Welding cycle: 3 on, 2 off
   - Power supply: 50 Hz
   - Welding current : 12,000 A
   - Resistance between electrodes: 100 $\mu\Omega$

6. Calculate the number of welds per minute, welding speed, electrode rotational speed, and the amount of energy required during resistance seam welding under the following conditions:
   - Electrode diameter: 250 mm
   - Welding current : 10,000 A
   - Welding rate: 4 welds/cm
   - Welding cycle 3 on, 2 off
   - Power supply: 50 Hz
   - Effective resistance between electrodes 100 $\mu\Omega$

BIBLIOGRAPHY

Gourd, L. M. 1982. *Principles of Welding Technology*. London: Edward Arnold.
Jain, R. K. 1993. *Production Technology*. Delhi: Khanna.
Little, R. L. 1973. *Welding and Welding Technology*. New York: McGraw-Hill.
Polukhin, P. 1964. *Metal Process Engineering*. Moscow: Mir Publishers.
Schey, J. A. 2000. *Introduction to Manufacturing Processes*. New York: McGraw-Hill.

22 Advanced Manufacturing Techniques

22.1 INTRODUCTION

Advanced manufacturing methods play an important role in the development and prosperity of nations. The need for modern methods that achieve the highest level of accuracy, minimum part size, high production rate, minimum material utilization, and the lowest energy consumption and manufacturing cost became the focus of many scientists and researchers. In order to achieve such goals, near net shape processes are used to produce parts very close to the final dimensions, thus reducing the need for traditional machining and finishing operations with subsequent reduction in energy, materials, and manufacturing cost.

The need for small size parts currently used in modern mechanical and electronic devices requires special manufacturing techniques. In this respect, micromanufacturing techniques produce ever decreasing sizes with tightly specified dimensions and accuracies. The need for further reduction in the size of parts led to the introduction of nanotechnology, which is currently used for the production of parts and systems at the molecular scale used in manufacturing parts, medicine, and electronic devices, where semiconductor manufacturing plays the major role in the production of integrated circuits and computer systems.

22.2 NEAR NET SHAPE MANUFACTURING

Near net shape is a manufacturing technique that produces items very close to their final (net) shape. It reduces the need for traditional finishing operations and eliminates about two thirds of the production costs. Near net shape manufacturing includes the following processes.

22.2.1 Metal Injection Molding

Metal injection molding (MIM) combines plastic injection molding with the strength and integrity of machine-pressed, small, complex metal parts. It combines fine metal powders with plastic binders that are injected into a mold. The binders are removed with solvents and thermal processes. The resultant metal parts are sintered at temperatures suitable to bind the particles without melting the metal. The final products are up to 98% as dense as wrought iron. They are used in the medical, dental, firearms, aerospace, and automotive industries. Complex features are possible through injection molding, including internal and external threads. Tolerances of ± 0.3 μm can usually be held without secondary processes.

22.2.2 Rapid Prototyping

Rapid prototyping (RP) techniques, referred to as solid free form fabrication (SFF), produce object sizes ranging from microscopic to entire buildings. Materials used range from paper and plastic to metals and ceramics. RP is used to produce complex net/near net shaped parts in materials that are hard to machine conventionally, such as difficult-to-machine metals, ceramics, and composites. Other applications include toys, aerospace, and advanced medicines. The process is fully automated based on computer-aided design (CAD) models, requires the minimum or no human intervention,

TABLE 22.1
Rapid Prototyping Methods

Prototyping Technologies	Base Materials	Condition
Steriolithography (SLA)	Photopolymer	Liquid
Liquid thermal polymerization (LTP)	Thermo setter	
Fused deposition modeling (FDM)	Thermoplastics, eutectic metals	
Selective laser sintering (SLS)	Thermoplastics, metals, powders	Powder
3DP Printing	Various materials	Powder
Laminated object manufacturing (LOM)	Paper	Solid

and produces accurate prototypes in a short time at minimum cost. Table 22.1 shows the different methods of rapid prototyping.

22.2.2.1 Steriolithography

Steriolithography (SLA) employs a layer-by-layer manufacturing based on photopolymerization that causes solidification of a liquid by the impact of a laser light on the upper surface of the liquid. Solidification extends to a few tenths of a millimeter below the surface and is restricted to a pattern that corresponds to the part's cross section. Once the layer is completely formed, it is lowered by a small distance and a second layer of the liquid traced right on the top of the first and so on to form a complete, three-dimensional (3D) object as shown in Figure 22.1. Objects having overhangs or undercuts are usually supported by extra support structures. At the end of the process, the part is removed from the liquid and the supports are cut off. The process is fully automated, and it produces high accuracy of ±0.1 mm and a good surface finish. Moreover, semitransparent materials can be processed for optical clarity. However, the limited range of materials used, high cost, and the need for post curing operation hinders its application.

22.2.2.2 Liquid Thermal Polymerization

Liquid thermal polymerization (LTP) uses a thermo setter that solidifies by heat dissipation. The system employs two jets for the plastic object and the wax-like support materials. The liquids are fed to the individual jetting heads, which squirt tiny droplets of the materials as they move in the X–Y coordinates to form a layer of the object as shown in Figure 22.2. As the material solidifies and hardens, the milling head makes the layer of uniform thickness, and cut particles are collected. The process is repeated to form the entire part. At the end of the process, the wax support

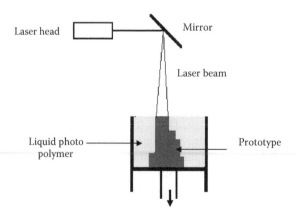

FIGURE 22.1 Principles of steriolithography (SLA).

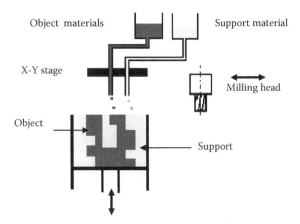

FIGURE 22.2 Liquid thermal polymerization (LTP).

material is either melted or dissolved away. LTP produces extremely fine resolution and surface finish. However, for large components, the process is very slow, the milling head creates noise, and part materials are limited.

22.2.2.3   Fused Deposition Modeling

In fused deposition modeling (FDM) the material is applied directly in a semiliquid state from a computer numerical controlled (CNC) extruder head. FDM builds up parts layer by layer using the thermoplastic filament having 0.003 mm diameter. In the apparatus shown in Figure 22.3, the nozzle is guided using CNC in the required X, Y, and Z coordinates to form each layer. The plastic material hardens immediately and bonds to the layer below. Several materials are used including acrylonitrile butadiene styrene (ABS), thermoplastic polyester-based elastomer (E20), and investment casting wax. The process is quiet, nontoxic, and office friendly; it produces strong parts, is cost-effective, does not need part clean up, and can produce multicolored parts using colored ABS. However, FDM is slow when making large cross-sectional areas and requires additional support materials. It produces relatively low accuracy and a poor surface finish. It is not suitable for producing complicated parts, and a limited number of materials can be used.

22.2.2.4   Selective Laser Sintering

Selective laser sintering (SLS) is based on sintering metallic or nonmetallic powders selectively into individual objects. As shown in Figure 22.4, a thin layer of fusible powder is laid down and heated

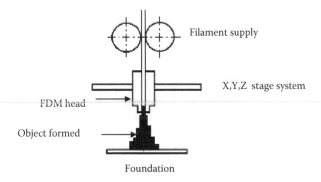

FIGURE 22.3 Fused deposition modeling (FDM).

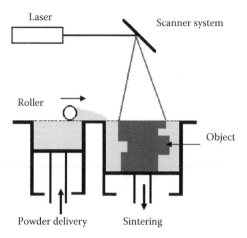

FIGURE 22.4   Selective laser sintering (SLS).

to just below its melting point. A laser beam, guided on the basis of 3D-CAD of the part, sinters and fuses the desired pattern of the first layer of the powder. This fused layer descends, the roller spreads out another layer of powder, and the process repeats. SLS uses materials including metal, plastic, ceramics, wax, nylon, elastomers, and polycarbonate. No post curing processes and no additional supports are required. However, it produces a rough surface finish; the time needed to complete an object is about 8–10 h, and toxic gases are generated when fusing plastic materials.

22.2.2.5   Three-Dimensional Printing

In three-dimensional printing (3DP), shown in Figure 22.5, the roller distributes and compresses the powder material at the top of the fabrication chamber. The multichannel jetting head deposits a liquid adhesive in a two-dimensional pattern onto the layer of the powder, thus bonding the areas where the adhesive is deposited to form a layer of the object. The piston moves down by the thickness of a layer, and the process is repeated until the entire object is formed. The object is elevated, and the extra powder is brushed away leaving the object. 3DP is a high-speed method used for producing color parts. However, part resolution, surface finish, and the limited materials available form a major drawback of the process.

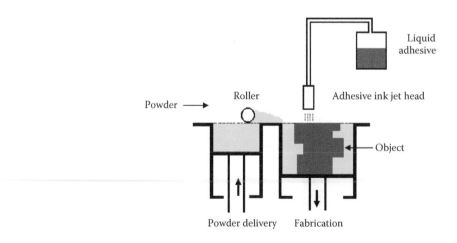

FIGURE 22.5   Three-dimensional printing (3DP).

22.2.2.6 Laminated Object Modeling

In laminated object modeling (LOM) (Figure 22.6), profiles of object cross sections are cut from paper using a laser beam. The paper is unwound from a feed roll onto the stack and first bonded to the previous layer using a heated roller, which melts a plastic coating on the bottom side of the paper. The profile is then traced by an optical laser driving system that is mounted to an X–Y stage. In LOM, the surface finish, accuracy, and stability of paper objects are not as good as for materials used with other RP methods. However, material cost is very low and objects have the look and feel of wood that can be worked and finished in the same manner to form patterns for sand casting operations.

22.2.2.7 Spray Forming (Casting)

Spray forming is used for casting near net shape metal components of homogeneous microstructures by the deposition of semisolid sprayed droplets onto a shaped substrate. The molten metal exits the induction furnace as a thin free-falling stream that is broken up into droplets by an annular array of gas jets. These droplets proceed downward, accelerated by the gas jets to deposit onto a substrate in the semisolid condition. Deposition continues to build up a spray-formed metal part.

22.2.2.8 Superplastic Forming

The superplastic forming process is mainly used for sheet metal-forming operations. It is based on the theory of superplasticity, which means that a material can elongate beyond 100% of its original size. The material used in this process must have fine grain size and be heated to promote superplasticity. The process forms large and complex parts for aerospace applications and the automotive industry in a single operation. The finished products have high precision, fine surface finish, and do not suffer from spring back or residual stresses. The forming rate is slow, and it is therefore recommended for low production volumes. Further details of superplasticity are presented in Section 8.9 and on superplastic forming that is presented in Section 10.9.

22.2.2.9 Gelcasting

The gelcasting process uses a slip of ceramic powders in a solution of organic monomers that is poured into a mold. The mixture is polymerized to form a strong, cross-linked solvent–polymer gel filled with the ceramic powder. The wet body is then dried and fired. Gelcasting produces large, complex-shaped parts that are either too complicated or too expensive to be manufactured by other processes. The process is simple, economical, and uses conventional casting equipments.

22.2.2.10 Hot Isostatic Pressing

Hot isostatic pressing (HIP) subjects a component to both elevated temperature and isostatic gas pressure in a high-pressure containment vessel containing an inert gas, so that the material does not

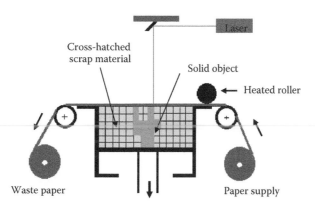

FIGURE 22.6 Laminated object modeling (LOM).

chemically react. The simultaneous application of heat and pressure eliminates internal voids and microporosity. Primary applications of HIP include the reduction of microshrinkage, the consolidation of powder metals and ceramic composites, as well as the fabrication of metal matrix composites. Further details of HIP are presented in Section 12.6.2.

22.3 MICROFABRICATION TECHNOLOGY

Micro in microfabrication indicates micrometer and represents the range of 1–999 µm. Owing to the high capability of machining processes in terms of accuracy and surface quality, compared to other manufacturing processes, microfabrication is restricted to micromachining operations. Their applications include semiconductor devices, compact electrical circuits, and integrated circuit packages containing devices of microdimensions.

22.3.1 MICROCUTTING OPERATIONS

In conventional micromachining, the tool material must be stronger than the workpiece. Turning of micropins, drilling of microholes, milling for microgrooves, and cutting micro 3D shapes are typical examples. Diamond microturning is mainly used in the optical and electronic industries. The process produces high profile accuracy, good surface finish, and low subsurface damage in semiconductors, magnetic read–write heads, and optical components. Diamond micromachining is used for producing spherical molds for plastic ophthalmic lenses, medical instruments, reflecting hybrid lenses, aluminum alloy automotive pistons, and aluminum alloy substrate drums for photocopying machines. Additionally, microdrilling is capable of fabricating holes several tens of micrometers in size. Conventionally, microdrilling operations are used to produce microholes in the circuit board, fuel injection nozzles for the automobile industry, and miniaturized medical tools for inspection and surgery. Furthermore, grooves, cavities, and 3D convex shapes may be fabricated by micromilling operations.

22.3.2 MICROFINISHING OPERATIONS

Diamond microgrinding is used for finishing ceramics using a grinding wheel speed of 30–60 m/s, workpiece speed of 0.1–1.0 m/min, depth of cut 1–10 µm, specific removal rate 0.05–0.2 mm^3/(mm s), and the total power is less than 1 kW. The accurate dimensions and tight tolerances achieved by this method depend on stiffness of the grinder and motion control system between the grinder and the workpiece. Additionally, microsuperfinishing is characterized by achieving low stock removal rate and a mirror-like surface finish. Microsuperfinishing of bearings having initial surface roughness of 0.1 µm using SiC grits gave a surface finish of 0.025 µm Ra in 6 s. Microlapping is used for finishing ceramics, silicon, quartz wafers, and germanium crystals. Typical lapped parts include integrated circuit (IC) devices, turbine engine blades, and glass lenses. Microlapping is also used to produce flat parts such as gauge blocks and sealing surfaces and to finish steel and ceramic balls used in bearing technology.

22.3.3 NONCONVENTIONAL MICROMACHINING

Micro ultrasonic machining (µUSM) is achieved using very fine grains, smaller vibration amplitudes, and smaller static force during sinking and contouring operations. Therefore, tools for µUSM should have diameters as small as a few micrometers to 1 mm, grain size from 0.2 to 20 µm, vibration amplitudes from 0.1 to 20 µm, and forces from 0.1 mN to 1 N. Microholes of 5 µm diameters in quartz, glass, and silicon have been produced using tungsten carbide (WC) tools (Figure 22.7). Applications of the process include electronic, aerospace, biomedicine, and surgery.

Micro electrodischarge machining (µEDM) die sinking applications include, ink jet nozzles for jet color printers, gasoline injector spray nozzles, liquid and gas microfilters, high aspect ratio holes and

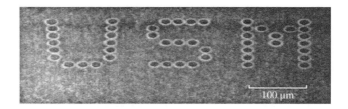

FIGURE 22.7 Forty-eight holes (22 μm diameter) produced in silicon workpiece, using a sintered diamond tool (20 μm diameter), 0.8 μm amplitude, static load 0.5 mN. (From Masuzawa, T., and Tonshof, H. K., *Annals of CIRP*, 46(2), 821–828, 1997. With permission.)

slots, and square cornered cavities. Micro-EDM systems can generate 5-μm hole, superfine nozzles such as the fuel injection nozzles for diesel engines (Figure 22.8). Preparing electrodes for cutting complex 3D microshapes is costly, time-consuming, and requires difficult design and fabrication procedures. Moreover, because the material is eroded from the workpiece as well as from electrodes, 3D microshapes cannot be accurately cut (Figure 22.9a). Alternatively, these microshapes can be machined layer by layer along the Z axis using simple rotating electrodes, which are driven according to specific CNC codes as shown in Figure 22.9b. Currently, commercial wire electrodischarge grinding (WEDG) machines can fabricate cylinders, rods, and other convex shapes of size around 10 μm (Figure 22.10). Various microtools for micro-EDM can be fabricated using this method.

Excimer laser offers high-precision machining without the formation of a resolidified layer and a heat affected zone at the machined surface. The femtosecond (FS) laser has also used laser micromachining where the pulse duration is as short as tens of femtoseconds and the peak power reaches terawatt order. Laser beam micromachining finds applications in drilling holes of 20–60 μm diameters, cutting diamond knife blades for eye surgery, and microstructuring of fine surfaces in cornea shaping for myopia correction. Other applications include texturing and structuring, scribing silicon transistor wafers, and dynamic balancing of gyrocomponents.

Electrochemical micromachining (EMM) is carried out using maskless or through mask material removal. Capillary drilling and electrolytic jet EMM are typical examples of maskless electrochemical micromachining (Figure 22.11). Capillary drilling produces high aspect ratios by moving the tool at a constant feed toward the workpiece. Masked EMM include photochemical machining (PCM), which is used for mass production of microelectronic packages, microstructures, sensors, and microelectronic mechanical systems. Figure 22.12 shows one- and two-sided through-mask EMM applications.

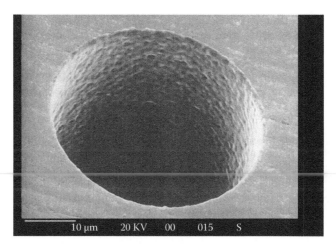

FIGURE 22.8 Inkjet nozzle fabricated by micro-EDM die sinking. (From McGeough, J., *Micromachining of Engineering Materials*, Marcel Dekker, New York, 2002. With permission.)

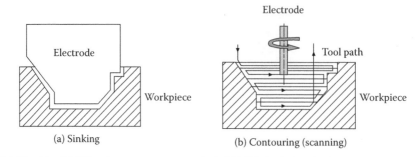

FIGURE 22.9   Micromachining by EDM.

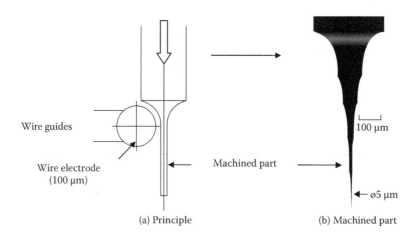

FIGURE 22.10   Micromachining by WEDG.

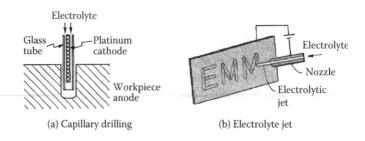

FIGURE 22.11   Maskless electrochemical micromachining (EMM). (From McGeough, J., *Micromachining of Engineering Materials*, Marcel Dekker, New York, 2002. With permission.)

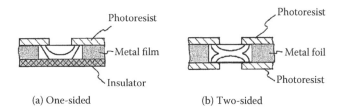

(a) One-sided (b) Two-sided

FIGURE 22.12 Through-mask electrochemical micromachining (EMM) applications. (From McGeough, J., *Micromachining of Engineering Materials*, Marcel Dekker, New York, 2002. With permission.)

22.4 NANOTECHNOLOGY

A nanometer is one-billionth of a meter, which is roughly the width of three or four atoms. The average human hair is about 25.000 nm wide. The term nanotechnology was first developed by Eric Drexler in 1981. The idea behind nanotechnology or manipulating atoms to build things was first proposed by Richard Feynman in 1959. Nanotechnology is the science of manufacturing materials and machines at the nanometer or atomic/molecular scale. It is also called molecular manufacturing because parts can be made by moving atoms and molecules into the desired arrangements until the required product is finished. It is a clean and cheap production method for the finest computer processors and food. In nanomanufacturing, two approaches are adopted to produce parts:

1. Top-down approach: In the top-down approach an existing solid is gradually reduced in size using some external radiation and/or chemical reaction. Lithography: starts with an existing extended structure and reduces it, using a radiation of photon, electrons, and ions and/or chemical solution. The basic steps of photolithography are shown in Figure 22.13. Photolithography can reach down to 80 nm in resolution, while deep UV to less than 50 nm and electron radiation to 1 nm.

2. Bottom-up approach: In the bottom-up approach, thin film deposition, lithographic patterning, and etching are used to produce nanowires, nanotubes, and nanopowders of the desired material and size/morphology with high yield. It is a completely new approach to device architectures, down to microscale. In nanotechnology the nanostructures are built atom by atom from scratch using one of the following techniques:

   • Epitaxial method: Epitaxy was developed in the 1970s by growing crystals of atomic layer on a substrate using molecular beam epitaxy (MBE) and liquid phase epitaxy (LPE). Using such methods, the formed layers can be made arbitrarily thin. The elements forming the epitaxial layers are evaporated in diffusion cells at the appropriate temperature and deposited onto a substrate. Controlling the temperature and the deposition time leads to an accurate control of the layer thicknesses down to single monolayer thicknesses.

   • Chemical vapor deposition: Chemical vapor deposition (CVD) uses chemical precursors to deposit nanostructures on a substrate. The heat or plasma breaks the precursors into reactive radicals that diffuse and adsorb to the substrate. Surface chemical reactions lead to the deposition of the solid.

   • Pulsed laser vaporization: Pulsed laser vaporization (PLV) is an example of a bottom-up method where a pulsed laser ablates a target creating a vapor. These atoms are then carried by an inert gas (Ar) to a region where they deposit and form the desired nanostructure. The most common growth mechanism is vapor–liquid–solid (VLS), whereby the vapor diffuses into the Au nanoparticle where it forms a liquid, and then supersaturates out of the catalyst to form the solid nanowire.

   • Electrochemical deposition: Electrochemical deposition (ED) in nanoporous template leads to the formation of metallic nanowires. Accordingly, Cu or Au film is sputtered

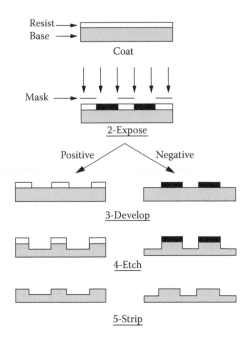

FIGURE 22.13 Top-down approach using photolithography.

on one side of the template. The metal is deposited from solution into the template pores to form the nanowires.
• Wet-chemical method: Colloidal metallic nanoparticles of Au are commonly made using this technique.

22.4.1 APPLICATIONS OF NANOTECHNOLOGY

There are many applications of nanotechnology. A few of them are shown in Figure 22.14.

22.4.1.1 Electronics
Nanotechnology increases the capabilities of electronic devices and reduces their weight and power consumption by

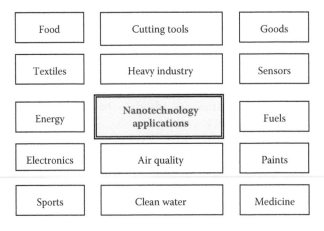

FIGURE 22.14 Nanotechnology applications.

- Improving display screens of electronic devices by reducing power consumption and decreasing their weight and thickness.
- Increasing the density of memory chips.
- Nanoscale integrated circuit can be produced.
- Nanosized magnetic rings are used to make magnetoresistive random access memory.
- Nanowire electrodes are used to produce thin, flat panel, and more flexible displays.
- Reducing the size of transistors used in integrated circuits.

22.4.1.2 Heavy Industry

The use of nanotechnology in heavy industry will cover the following areas.

- Aerospace: Lighter and stronger materials will be used for aircraft manufacturing thus leading to increase of their performance. Nanotechnology reduces the size of equipment, thereby decreasing the fuel-consumption required.
- Construction: Nanotechnology has the potential to make construction faster, cheaper, safer, and more varied. Automation of nanotechnology construction can allow for the creation of structures from advanced homes to massive skyscrapers much more quickly and at much lower cost.
- Refineries: Using nanotechnology refineries in producing materials such as steel and aluminum will remove any impurities in the materials they create.
- Vehicle manufacturers: Lighter and stronger materials will be useful for producing vehicles that are both faster and safer. Combustion engines will also benefit from parts that have higher strength and wear resistance.

22.4.1.3 Foods

Nanotechnology is affecting several aspects of food science because it will influence the taste, safety, and the health benefits the food delivers. It can also be applied in the production, processing, safety, and packaging of food.

22.4.1.4 Household

The most prominent application of nanotechnology in the household is the self-cleaning or easy-to-clean surfaces on ceramics and glasses. Nanoceramic particles have improved the smoothness and heat resistance of common household equipment such as flat irons.

22.4.1.5 Optics

For optics, nanotechnology offers scratch resistant coatings based on nanocomposites. These coatings are transparent, ultrathin, and well-suited for daily use.

22.4.1.6 Textiles

Nanofibers are used to make water- and stain-repellent or wrinkle-free clothes by attaching molecular structures to cotton fibers. Nanotextiles can therefore be washed less frequently and at lower temperatures. Nanomaterials can be used for making battle suits that withstand blast waves or incorporate sensors to detect or respond to chemical and biological weapons.

22.4.1.7 Medicine

Nanoparticles can be used to deliver drugs, heat, light, or other substances to specific cells in the human body. Nanoparticles at the size of molecules that deliver drugs directly to diseased cells in the human body, reducing the damage caused by what chemotherapy does to a patient's healthy cells. Nanorobots can be programmed to repair specific diseased cells in a similar way to antibodies in our natural healing processes.

22.4.1.8 Energy

Nanotechnology can reduce the cost of catalysts used in fuel cells to produce hydrogen ions from fuels such as methanol. It improves the efficiency of membranes used in fuel cells to separate hydrogen ions from other gases such as oxygen. Nanotechnology solar cells can be manufactured at lower cost than conventional solar cells. Batteries made of nanomaterials can work for decades and can be recharged significantly faster than conventional ones. Additionally, fuels such as diesel and gasoline can also be produced, economically, from low-grade raw materials.

22.4.1.9 Air Quality

Nanotechnology can improve the performance of catalysts used to transform vapors escaping from cars or industrial plants into harmless gases. Catalysts made from nanoparticles have a greater surface area to interact with the reacting chemicals better than catalysts made from larger particles. Nanostructured membranes, on the other hand, are developed to separate carbon dioxide from industrial plant exhaust streams.

22.4.1.10 Clean Water

Nanotechnology improves water quality by the removal of industrial wastes. In this regard, nanoparticles are used to remove salt or metals from water using electrodes composed of nanosized fibers, which, in turn, reduce the cost and energy required for turning salt water into drinking water. Additionally, filters of nanometers in diameter are currently developed to remove virus cells from water.

22.4.1.11 Chemical Sensors

Nanotechnology-based sensors can detect very small amounts of chemical vapors. In this respect, carbon nanotubes, zinc oxide nanowires, or palladium nanoparticles can be used in the nanotechnology-based sensors where a few gas molecules are sufficient to change the electrical properties of the sensing elements.

22.4.1.12 Sporting Goods

A high-performance ski wax produces a hard and fast-sliding surface, where the ultrathin coating lasts much longer than conventional waxing systems. A racket with carbon nanotubes is also introduced, thus leading to an increased torsion and flex resistance. The rackets are more rigid and provide more power than current carbon rackets. Long-lasting tennis balls are made by coating the inner core with clay polymer nanocomposites.

22.4.1.13 Paints

Incorporating nanoparticles in paints could improve their performance by making them lighter. Thinner paint coatings are used on aircraft. This reduces their weight by reducing the solvent content, which is beneficial to the environment. Paints that change their color in response to changes in temperature or chemical environment and those with reduced infrared absorptivity are also available.

22.4.1.14 Lubricants

Nanospheres of inorganic materials can be used as lubricants, thus acting as nanosized ball bearings. These nanoparticles, dispersed in a conventional liquid lubricant, reduce friction between metal surfaces, particularly at high normal loads of high-performance engines and drivers. Such a lubricant is effective when the metal surfaces are not highly smooth, which reduces the machining cost.

22.4.1.15 Cutting Tools

Nanotechnology is used to create multiple nanoscale layers of thin elemental coatings with special crystalline structures on cutting tools. Each thin nanocoating layer is optimized to a size of merely 3–4 nm in thickness and reaching half the hardness of diamond. The process involves the use of multilayer nanocoating on cemented carbide tools by physical vapor deposition (PVD). Nanocoated tools have longer tool life, which leads to machining cost reduction in the automotive and aircraft industries.

22.5 SEMICONDUCTOR DEVICE FABRICATION

Semiconductor device fabrication is used to create chips and integrated circuits for electronic devices. It consists of a series of steps that deposit special layers of materials on the wafers, in precise sequence, amounts, and patterns. Accordingly, photographic and chemical processing techniques gradually generate electronic circuits on a wafer made of semiconducting material such as silicon. Semiconductor manufacturing consists of the following steps:

1. Production of silicon wafers from pure silicon ingots
2. Fabrication of integrated circuits onto these wafers
3. Assembly of every integrated circuit on the wafer into a finished product
4. Testing and packaging of the finished products

For each layer of the integrated circuit, a pattern is made that is optically reduced in size to make an optical mask. In order to transform the pattern into the wafer, the surface of the wafer is oxidized and a photosensitive resist is deposited and exposed through the optical mask. Unwanted portions of the photoresist are then dissolved and the exposed oxide film is etched away. Such an oxide mask controls the diffusion of doping elements into the substrate. The sequence is repeated several times until the device is constructed. Metallic conducting paths are then deposited on the surface of the circuit that is provided with connections to the real world. These major steps are shown in Figure 22.15.

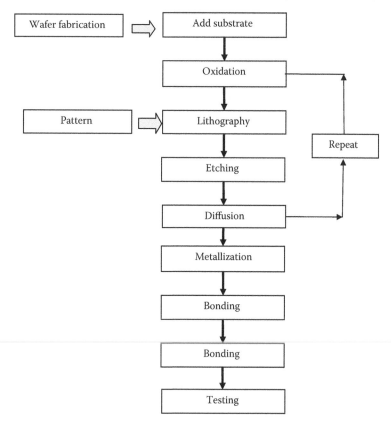

FIGURE 22.15 Fabrication sequence of integrated circuits.

22.5.1 WAFER FABRICATION

Wafer fabrication consists of a long series of mask/etch and mask/deposition steps until the circuit is completed. Semiconductor device processing steps fall into four general categories: deposition, removal, patterning, and modification of electrical properties as shown in Table 22.2. The process of wafer fabrication is a series of 16–24 loops that puts down a layer on the device. Each loop comprises some or all of the major steps of photolithography, etch, strip, diffusion, ion implantation, deposition, and chemical mechanical planarization. The detailed steps of wafer manufacturing include the following steps.

1. Wafer production starts by the crystal growing process where a single seed crystal of silicon is immersed in a bath of molten silicon, which forms a cylindrical ingot of pure silicon of a diameter larger than required. The ingot is then ground down to the required diameter and sliced into individual wafers that are finely polished to meet the surface flatness and thickness specifications (Figure 22.16).
2. A photographic process generates the fine featured patterns for each layer of the integrated circuit. Each layer of the chip is defined by a specific mask (reticle) and there are 16–24 mask layers in each IC.
3. A thin layer of pure silicon is grown on the raw wafer using the epitaxial growth (epilayer).
4. The epilayer is exposed to high temperature to form a silicon dioxide layer (Figure 22.17).
5. Photolithography is performed for a layer of photoresist covering the oxide layer using the proper mask.
6. The wafer with patterned photoresist is then etched to remove the oxide where there is no pattern and then the photoresist is removed.
7. Dopant molecules are implanted vertically into the surface of the silicon using a high-energy ion beam. These regions are now doped with negative ions, creating n-type source and drain regions of the transistor in a p-type silicon base.
8. Depositing the gate oxide of silicon nitride film via a chemical vapor deposition (CVD) process. The gate itself is either made of polysilicon or a metal that is deposited using physical vapor deposition (PVD) known as sputtering.
9. Deep field oxides are grown into the silicon in order to electrically isolate each transistor from its adjacent partners.

TABLE 22.2
Semiconductor Device Fabrication

Process	Methods
Deposition: Grows, coats, or transfers a material onto the wafer	Physical vapor deposition (PVD)
	Chemical vapor deposition (CVD)
	Electrochemical deposition (ECD)
	Molecular beam epitaxy (MBE)
	Atomic layer deposition (ALD)
Removal: Removes material from the wafer either in bulk mode or selectively	Wet etching
	Dry etching
	Chemical–mechanical planarization (CMP)
Patterning: Shapes or alters the shape of the deposited materials	Lithography (SLA)
Modification of electrical properties: Dopes transistor sources and drains	Ion implantation
	Rapid thermal anneal (RTA)
	Ultraviolet light processing (UVP)

FIGURE 22.16   Wafer slicing.

10. Dielectric isolation oxides are deposited in layers to insulate the transistors from the inter-connecting layers which will be built above.
11. Using reticles and photolithography, contact areas in the silicon dioxide are unmasked so that they can be etched down to the silicon and polysilicon areas of the transistor's source, drain, and gate regions. These holes (vias) are essentially chemically drilled holes that expose the contacts to the three terminals of the transistor.
12. Metallization process that is followed by mechanical planarization.
13. Add the interconnect layer.

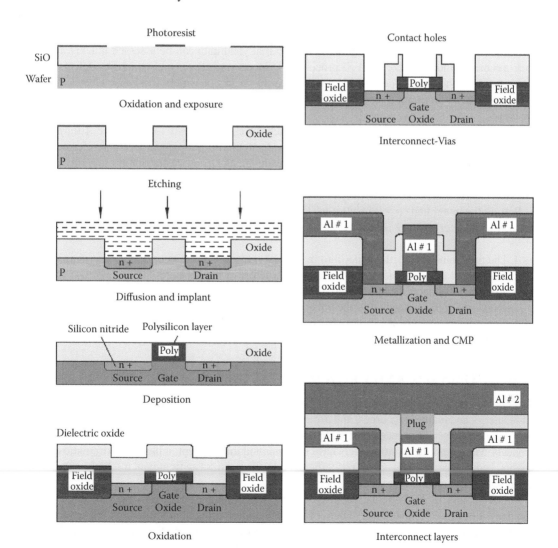

FIGURE 22.17   Wafer fabrication steps.

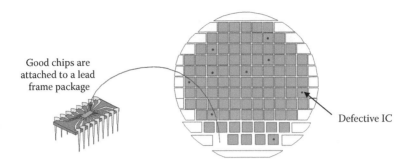

FIGURE 22.18   Die cut and assembly. (From Sorenson, C. T., NSF/SRC Engineering Research Center for Environmentally Benign Semiconductor Manufacturing, Arizona, 1999. With permission.)

22.5.2   Testing, Assembly, and Packaging

Measurement of wafer flatness, film thickness, electrical properties, and critical dimensions are essential. Wafer test is performed near the Fab, so that yield data can be assessed quickly and fed back to correct and optimize the Fab processes. Inspection observes and quantifies defects using scanning electron microscopes (SEM). Defects on the masks or wafers can cause electrical short circuits between aluminum lines or breaks in aluminum traces that are fatal to the functionality of the chip.

Semiconductor devices are subjected to a variety of electrical tests to determine if they function properly. The proportion of devices on the wafer found to perform properly is referred to as the yield. Testing the chips on the wafer is made using an electronic tester that presses tiny probes against the chip. The machine marks each bad chip with a drop of dye. Prior to shipment to the assembly devices, they are electrically tested using automated systems.

Putting the integrated circuit inside a package to make it reliable and convenient to use is known as a semiconductor package assembly, which consists of the steps shown in Figure 22.18

1. Die preparation: Cuts the wafer into individual integrated circuits or dice
2. Die attach: Attaches the die to the support structure of the package
3. Bonding: Connects the circuit to the electrical extremities of the package, thereby allowing the circuit to be connected to the outside world
4. Encapsulation: Provides a body to the package of the circuit for physical and chemical protection

22.6   SUSTAINABLE AND GREEN MANUFACTURING

Sustainable/green manufacturing is the production of parts that use nonpolluting processes, conserve energy and natural resources, and are economically sound and safe for employees, communities, and consumers. It reduces cost, enhances process efficiency, and develops new eco-friendly products. Many leading manufacturers have incorporated this concept; it has led to dramatic energy savings, reduced carbon emissions, cost efficient recycling and waste treatment methods, and a wide range of new clean technologies and products.

Green manufacturing benefits the environment, impacting the consumer, shareholders, and the company. Insurance companies give better rates to manufacturing companies, and the government is offering tax breaks for companies that have gone green. Machinery that is Earth friendly goes green by using wind and solar energy, which saves the company thousands of dollars. Green manufacturing provides help to the community through the use of renewable energy sources. It produces a better air quality, a healthier environment, more jobs, and saves money for manufacturing companies. Steps of going green include

1. Find all wastes and emissions; look at the harmful ones in order to decide what to fix.
2. Look at the products of the company to determine what can be recycled at the facility to save money.
3. Place bins around the facility where employees can easily access them to toss products that can be recycled.
4. Offer incentives to staff that regularly practice recycling.

22.7 REVIEW QUESTIONS

1. State the main applications of diamond microturning operation.
2. What are the advantages of microgrinding and micromilling processes?
3. State the major applications of microgrinding.
4. Mention some applications for micro-EDM.
5. What are the main applications of micro-ECM?
6. Explain what is meant by near net shaping.
7. List advantages of rapid prototyping techniques.
8. Give an example for a rapid prototyping method based on liquid material, powders, and solid materials.
9. Explain what is meant by spray forming and hot isostatic pressing.
10. Explain what is meant by microfabrication and nanofabrication.
11. Differentiate between bottom-up and top-down nanofabrication.
12. Explain how photolithography is used to build nanoshapes.
13. List some applications of nanotechnology in the following areas: heavy industry, textile, medicine, and electronics.
14. Using a line sketch, show the main steps of manufacturing ICs.
15. Explain the following terms: wafer fabrication and green manufacturing.

BIBLIOGRAPHY

El-Hofy, H. 2005. *Advanced Machining Processes, Non-traditional and Hybrid Processes*. New York: McGraw-Hill.

Kalpakjian, S. 1997. *Manufacturing Process for Engineering Materials*. Reading, MA: Addison Wesley.

Masuzawa, T., and Tonshof, H. K. 1997. Three-dimensional micro machining by machine tools. *Annals of CIRP* 46(2): 821–828.

McGeough, J. 1988. *Advanced Methods of Machining*. London: Chapman & Hall.

McGeough, J. 2002. *Micromachining of Engineering Materials*. New York: Marcel Dekker.

Saxhs, E. et al. 1993. Three-dimensional printing: The physics and implications of additive manufacturing. *Annals of CIRP* 40(1): 257–260.

Schey, J. A. 2000. *Introduction to Manufacturing Processes*, 3rd ed. New York: McGraw-Hill.

Wright, P. K. 2001. *21st Century Manufacturing*. Upper Saddle River, NJ: Prentice-Hall.

Youssef, H., and El-Hofy, H. 2008. *Machining Technology, Machine Tools and Operations*. Boca Raton, FL: CRC Press.

23 Materials, Processes, and Design for Manufacturing

23.1 INTRODUCTION

Selection of an optimal manufacturing process for a desired product requires the consideration of the workpiece material and shape. The production quantity, size, economic aspects, and environmental aspects should also be taken into consideration. A manufacturing process should ensure product performance, reliability, quality, reduced manufacturing costs, and product success in the market. Moreover, design considerations cover the technical feasibility of design decisions, while the process quality is concerned with tolerance capability associated with characteristic dimensions. This chapter provides a criterion for the selection of product materials and manufacturing processes that have the capability to satisfy the engineering design specifications, application, and quality requirements. It presents the principles of design for manufacturing and assembly.

Manufacturing high-quality products at the lowest possible cost requires an understanding of the complex relationships among product design, material selection, and manufacturing processes. Product quality has always been a major consideration in manufacturing. Accordingly, a high-quality product is generally considered to have the following characteristics:

- Satisfies the need and expectations of the customer
- Functions reliably over its intended life
- Has pleasing aesthetics
- Provides a high level of safety
- Is easy to install
- Less expensive maintenance and future improvements

A wide variety of the available materials and manufacturing processes exists. Producing high-quality products by selecting the proper materials and the best manufacturing process while minimizing the manufacturing costs has become a major challenge. Selecting the right manufacturing process is not easy owing to the several existing alternatives that cover certain product requirements and specifications. In this regard, computer programs, databases, and expert systems play an important role in facilitating the selection of an optimal process that depends on how far the process characteristics are able to meet the requirements of the desired product characteristics as shown in Figure 23.1. Figure 23.2 shows that product design for manufacturing, material selection, and process selection are interrelated. Design modification improves the product performance, considers the advantages of new materials, and makes the manufacturing processes and assembly easier. Figure 23.3 shows the general strategy for process and material selection.

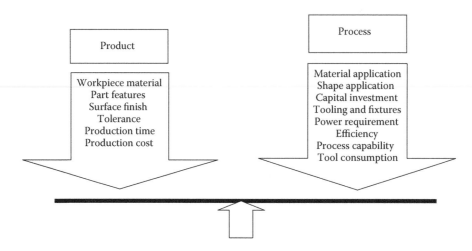

FIGURE 23.1   Relationships between product and process characteristics.

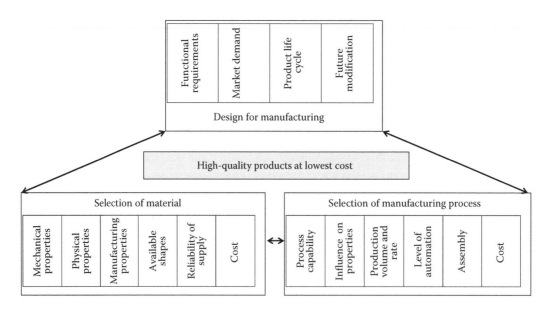

FIGURE 23.2   Production of high-quality products.

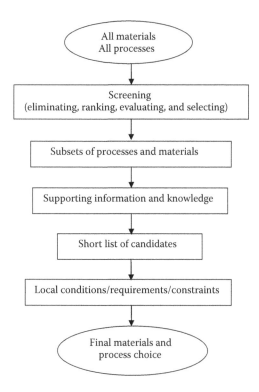

FIGURE 23.3 Process and material selection.

23.2 FUNCTION, MATERIAL, PROCESS, AND SHAPE INTERACTION

Selection of part material is currently made using computer software and expert systems that determine the appropriate material for a particular application, depending on the following aspects.

- Shape of material: Materials are the physical and chemical substances the product is made off. Materials are generally available in many forms that include cast, extruded, forged, bar, plate, sheet, foil, rod, tube, pipe, wire, or powder. The shape and size of the commercially available material should be economically justified in order to avoid additional processing. Material characteristics such as surface quality, tolerances, and straightness can also reduce the need for additional manufacturing processes. Each manufacturing process produces parts that have their own shape, surface finish, and tolerance characteristics. For example, hot formed parts have coarser surface finish and wider tolerances than cold formed ones. Extrusions have smaller cross-sectional tolerances than roll formed parts. Additionally, turned parts have rougher surfaces than ground ones.
- Manufacturing properties: These properties include castability, weldability, formability, machinability, and hardenability by heat treatment. Such properties are crucial toward the proper selection of a material for a given manufacturing process.
- Reliability of supply: This is affected by geopolitical factors, strikes, shortages, and reluctance of suppliers to produce a particular shape, quality, or quantity.
- Cost of material and processing: The cost of material depends on the material itself, its shape, size, and condition. In this regard, thin wires are more expensive than round bars made of the same material and metal foils are more expensive than metal plates. During manufacturing by machining, sheet metal forming, or forging, the cost of material should consider the value of the scrap in order to obtain the net materials cost. The value of the scrap ranges between 10% and 15% of the original cost of the material.

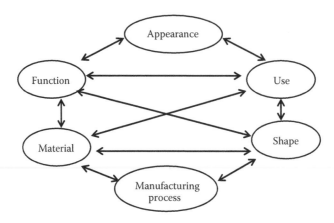

FIGURE 23.4 Interaction between product characteristics, material, and manufacturing processes.

The relationship between product appearance, product function, use of materials, shape in terms of texture and finishing affect the choice of the manufacturing process as shown in Figure 23.4. Product appearance reflects how the user's senses react on the appearance as well as the associations it elicits in the user. Product function covers what you can do with the product and how you can achieve. The use of a product is related to the designed interaction the user can have with the product.

23.3 MANUFACTURING PROCESS CAPABILITIES

Each manufacturing process has its particular advantages and limitations. For example, casting and powder injection molding, generally, produce more complex shapes than what forging can produce. Forgings have better strength and toughness compared to castings and powder metallurgy products. The minimum section size and dimensions that can be produced satisfactorily depend on the process selected. For example, thin sections can be produced by cold rolling but would be difficult or impossible to produce by casting, forging, or powder metallurgy. The shape of a product is an important factor that is used for measuring the manufacturing process capabilities. Some products can be fabricated from several individual components and then assembled with fasteners, brazing, welding, and adhesive bonding. For another product, one piece manufacturing may be more economical because of the reduced assembly cost. Moreover, single piece manufactured parts are recommended for high-rigidity requirements. The variety in the required product characteristics, material, size, and the required shape leads to select the proper manufacturing processes to meet the product requirements.

23.4 PROCESS SELECTION FACTORS

Choosing the optimum manufacturing process to produce a certain product is related to many factors such as product material, dimensional and geometrical features, surface finish, production quantity and rate, and the economic and environmental aspects. Manufacturing process selection factors include the following (Figure 23.5).

- Product shape: Product shape represents the geometry including details such as tolerance and finishing. The shape of a part depends on its function. Because not all manufacturing processes are equally suitable to produce a given part, designers often change the part shape, without affecting its main function to become easier to produce by a certain manufacturing

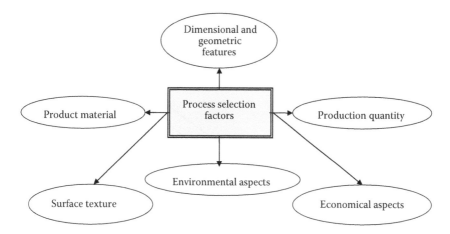

FIGURE 23.5 Factors affecting manufacturing process selection.

method. Complex shapes require more machine tool motions and complex control systems in many axes such as the case of CNC machines. Computer-aided design/computer-aided manufacturing (CAD/CAM) is currently used to link the design phase to manufacturing in order to facilitate the production and assembly with minimum complexity. Fixing the part shape in the design stage may exclude the most economical manufacturing processes. Once the optimum process is identified, the part shape is optimized for that particular process.

- Size: The maximum size that can be made by any manufacturing process is limited by the availability of large size equipment and the process conditions. Thinner, smaller, and larger parts may well be made but under special conditions at extra cost.
- Product material: Product materials have different manufacturing characteristics such as castability, forgeability, workability, weldability, and machinability. A material that is castable may be difficult to work using machining, grinding, or finishing operations, which may be required for achieving acceptable surface finish, dimensional accuracy, and quality of geometrical features. Materials also have different responses to manufacturing conditions to which they are subjected.
- Dimensional and geometric features: On the basis of each process capability, the selection of an optimum process depends on dimensional and geometric features of the product such as part size, which includes length, width, thickness, depth, and diameter; part shape; or geometry. Dimensional tolerance is defined as the permissible or acceptable variation in the dimensions of a part and affects both the product design and the manufacturing process selection.
- Surface texture: The surface texture is classified into surface roughness, waviness, as well as lays and flaws. When producing any component, it is necessary to satisfy the surface technological requirements in terms of high product accuracy, a good surface finish, and a minimum of drawbacks that may arise as a result of the manufacturing process. The nature of the surface layer has a strong influence on the mechanical properties of the part. Primary metal-forming and casting processes produce surfaces that require further machining operations to control their dimensions at closer tolerances. The surface roughness is considered an important factor in contact-to-contact surfaces and for functional properties such as wear resistance and fatigue strength of a part. According to the surface roughness required by the design specifications, the optimum manufacturing method can be selected. Figure 23.6 shows the different tolerances and surface roughness achieved by different manufacturing methods. Each manufacturing process is capable of producing certain surface finish and tolerance range without extra cost. It is necessary to specify

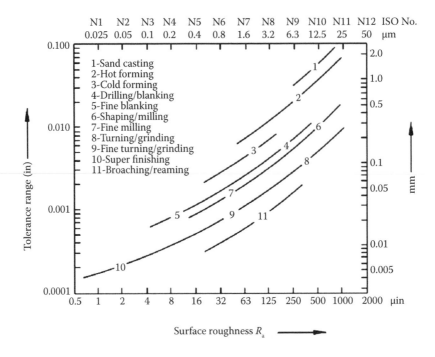

FIGURE 23.6 Manufacturing processes capabilities in terms of roughness and tolerances. (Adapted from Schey, J. A., *Introduction to Manufacturing Processes*, 3rd ed. McGraw-Hill, New York, 2000.)

both maximum and minimum surface roughness values for proper functioning and ease of manufacturing the part. Additionally, the specified tolerances also should be within the range obtained by the selected manufacturing process so as to avoid further finishing operations and rise in cost. Table 23.1 shows the approximate relative cost for achieving certain tolerances and surface finishes. Therefore the use of unnecessarily tight tolerance and fine surface finish specifications is a major source of excessive manufacturing costs.

TABLE 23.1
Approximate Relative Cost for Machining Tolerances and Surface Finishes

Machining Process	Tolerance		Roughness (R_a)	
	± mm	Relative Cost	µm	Relative Cost
Rough machining	0.77	100	6.25	100
Standard. machining	0.13	190	3.12	200
Fine machining (rough grinding)	0.03	320	1.56	440
Very fine machining (ordinary grinding)	0.01	600	0.8	720
Fine grinding, shaving, honing	0.005	1100	0.4	1400
Very fine grinding, shaving, honing, lapping	0.003	1900	0.2	2400
Lapping, burnishing, super honing, polishing	0.001	3500	0.18	4500

- Production quantity: The production quantity plays an important role in the selection of any manufacturing process. The rate of production is defined as the number of pieces produced per unit time. For example, bearings, bolts, tires, and automobiles are produced in large quantities, so the production quantity/lot size varies widely. On the other hand, ships, diesel engines, and jet engines are produced in limited quantities. The effect of product quantity on the selected manufacturing process required to produce a low carbon steel pump gear is shown in Table 23.2. It is accordingly clear that, for the same production unit, the relative cost by machining is 2.6 that of cold extrusion, which produces stronger parts at low material waste and high production rate.
- Economic aspects: The economic aspects of manufacturing processes lead to the ranking of the processes in order to obtain the most economical one. In this regard, the total cost of a product consists of many items such as cost of material, tooling, and fixed, direct, and indirect labor costs.
  1. Tooling cost is involved in making tools, dies, molds, patterns, special jigs, and fixtures necessary for manufacturing a product. The tooling cost for die casting is higher than that for sand casting. However, this high cost can be justified by the large production volume made in die casting.
  2. Fixed cost includes the cost that the company will pay whether or not it makes a particular product. It is not related to the production volume.
  3. Labor cost includes all labor from floor-to-floor time. The direct labor cost is calculated from the labor rate by the time that the worker spends in producing the part. Indirect labor costs are paid for those servicing the manufacturing operations such as repair, maintenance, quality control, research, sales, and the cost of office staff. Such costs do not contribute directly to the production of the finished part and are usually called overheads.

The method of manufacturing a component depends on the production volume required. Small batches are commonly made on general purpose machines, which are versatile and capable of producing different shapes and sizes. Under such conditions, the direct labor costs are higher. For large quantities (medium batches), CNC machines or jigs and fixtures are used, leading to the reduction of labor cost. For larger volumes, the labor costs can further be reduced by using machining centers, flexible manufacturing systems (FMS), or special purpose machine tools.

TABLE 23.2
Component Cost for Different Manufacturing Processes

Part Required (Pump Gear)	Manufacturing Method	
	Machining	Cold Extrusion
	High waste	Low waste
	Low/medium production rate	High production rate
	Poor strength	High strength
	Relative cost index: 2.6	Relative cost index: 1

Material: low carbon steel
Number: 5000 piece

Source: Swift, K. G., and Booker, J. D., *Process Selection from Design to Manufacture*, Butterworth-Heinemann, Oxford, 2003.

- Environmental aspects: Manufacturing processes generate solid, liquid, or gaseous by-products that present hazards for workers, machine, and the surrounding environment. The adverse effect of manufacturing processes leads to rank them according to their environmental impacts by determining an index that explains their net hazardous effect on human, machine, and environment.

23.5　MANUFACTURING PROCESS SELECTION

Process information maps (PRIMAs) and elimination and ranking strategy are fundamental methods used for selecting any manufacturing process.

23.5.1　Process Information Maps

PRIMAs present the knowledge and data including material suitability, design considerations, quality issues, economics, and process fundamentals and its functional characteristics so that an overall understanding is achieved. Such data enable the selection of manufacturing processes that have the capability to satisfy the engineering needs of a product manufacturing. Within the standard format PRIMA is an outline of the process, how it works, and under what conditions it functions best. There is also a summary of what it can do, the limitations and opportunities it presents, and, finally, an overview of quality considerations including process capability charts for relating tolerances to characteristic dimensions.

PRIMAs focus on the identification of candidate processes based on strategic criteria such as material, process technology, and production quantity. Having identified the possible targets, the data in the PRIMAs are used to do the main work of selection. Each PRIMA is divided into seven categories that cover the characteristics and capabilities of the process

1. Process description: An explanation of the fundamentals of the process together with a diagrammatic representation of its operation and a finished part.
2. Materials: A description of the materials currently suitable for the given process.
3. Process variations: A description of any variations of the basic process and any special points related to these variations.
4. Economic considerations: A list of several important points including production rate, minimum production quantity, tooling costs, labor costs, lead times, and any other points that may be of specific relevance to the process.
5. Typical applications: A list of components that have been successfully manufactured using the process.
6. Design aspects: Any points, opportunities, or limitations that are relevant to the design of the part as well as standard information on minimum section size range and general configuration.
7. Quality issues: Standard information includes a process capability chart, typical surface roughness detail, and common process faults.

A flowchart of PRIMA selection is shown in Figure 23.7. The manufacturing process PRIMA selection matrix (Table 23.3) has been devised on the basis of material type and production quantity per annum. The justification for basing the matrix on material and production quantity is that it combines technological and economic issues of prime importance. Each manufacturing process has been assigned an identification code rather than using process names, as shown at the bottom of the table. There may be just one or a dozen processes at each node in the selection matrix representing the possible candidates for the final process selection.

Many manufacturing processes are only viable for low-volume production owing to the time and labor involved, while others require expensive equipment and are therefore unsuitable for low

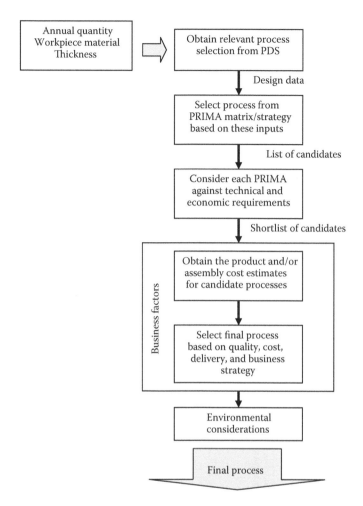

FIGURE 23.7 PRIMA selection flowchart.

production volumes. By considering production quantities in the early stages, the process that will prove to be the most economical later in the development process can be identified and selected. The PRIMA focuses attention on processes that are most appropriate based on material and production quantity. Therefore PRIMAs guide the manufacturing engineer toward the final process selection.

23.5.2 ELIMINATION AND RANKING STRATEGY

Elimination and ranking strategy (EARS) (Figure 23.8) uses a combination of elimination and ranking steps for the manufacturing processes selected for a given product. In order to obtain the optimum manufacturing process, the following levels are considered.

- First level: Employs an elimination strategy, using material-based classification of manufacturing processes where the specified nonapplicable processes are eliminated from the initial list of processes.
- Second level: Uses an elimination strategy based on shape generation capabilities of the manufacturing processes; nonapplicable processes are eliminated from the initial list of manufacturing processes. The manufacturing processes common to the short lists of first level and second level are retained for the next level.

TABLE 23.3
PRIMA Selection Matrix

Material Quantity	Carbon Steel	Tool and Alloy	Copper and Alloys	Aluminum and Alloys	Nickel and Alloys	Titanium and Alloys	Refractory Metals
1–100	[1.5][1.6] [1.7][4M]	[1.5][1.7][2.10][3A] [4.1][4.5][4.6]	[1.5][1.7] [2.10][3M] [4.1]	[1.5][1.7][2.7][2.10] [3M][4.5]	[1.5][1.7][2.10][3M] [4.1][4.5][4.6]	[1.1][1.6][1.7][4.10] [3M][4.1][4.5][4.6] [4.7]	[4.5]
100–1000	[1.2][1.5][1.6][4m] [4.3][4.4]	[1.1][1.2][1.7][3M] [4.1][4.3][4.4][4.5] [4.6][4.7]	[1.2][1.5][1.7][1.8][2.5] [2.10][3M][4.1][4.3] [4.4]	[1.2][1.5][1.7][1.8] 2.7] [2.10][3M][4.3][4.4] [4.5]	[1.2][1.5][1.7][2.10] [3M][4.1][4.3][4.4] [4.5]	[1/1][1.6][2.7][2.10] [3M][4.1][4.3][4.4] [4.5][4.6][4.7]	[4.5]
10^3–10^4	[1.2][1.3][1.5][1.6] [1.7][2.11] [4A] [4.2]	[1.2][1.5][1.7][2.1] [2.4][2.11][3A][4.2] [4.3][4.4][4.5]	[1.2][1.3][1.5][1.8][2.1] [2.3][2.10][2.11][3A] [4.2][4.3][4.4]	[1.2][1.3][1.5][1.8][2.1] [2.3][2.7][2.10][2.11] [3A][4.3][4.4][4.5]	[1.2][1.3][1.5][1.7] [2.1][2.3][2.10][2.11] [3A][4.2][4.3][4.4] [4.5]	[2.1][2.7][2.10][2.11] [3A][4.2][4.3][4.4] [4.5]	[4.5]
10^4–10^5	[1.2][1.3] [2.11][4A]	[1.9][2.1][2.3][2.4] [2.5][2.11][2.12] [3A][4.2][4.5]	[1.2][1.4][1.9][2.1][2.3] [2.4][2.5][2.11][2.12]	[1.2][1.3][1.4][1.9][2.1] [2.3][2.4][2.5][2.11] [2.12][3A][4.5]	[2.1][2.3][2.4][2.5] [2.11][2.12][3A][4.2] [4.3]	[2.1][2.4][2.11][2.12] [3A][4.2][4.5]	[2.5]
>10^5	[1.2][1.3] [2.11][4A]	[1.9][2.1][2.2][2.3] [2.4][2.5][2.12]	[1.2][1.9][2.1][2.2][2.4] [2.5][2.7][2.8][2.11] [2.12][3A]	[1.2][1.3][1.4][1.9][2.1] [2.2][2.3][2.4][2.5][2.8] [2.12][3A]	[2.2][2.3][3A]	[3A]	[2.5]
all	[1.1]	[1.1][1.6][2.6] [2.8][2.9]	[1.1][1.5][2.6][2.8] [2.9][4.5]	[1.1][1.6][2.5][2.8][2.9]	[1.1][1.6][2.6][2.8] [2.9]	[2.8][2.9]	[1.6]

Casting Processes		Forming Processes		Machining Processes	NTM Processes
1.1 Sand casting	1.6 Investment casting	2.1 Closed die forging	2.7 Super plastic forming	3A Automatic machining	4.1 Electrical discharge machining, EDM
1.2 Shell molding	1.7 Ceramic mold casting	2.2 Rolling	2.8 Sheet metal shearing	3M Manual machining	4.2 Electrochemical machining, ECM
1.3 Gravity die casting	1.8 Plaster mold casting	2.3 Drawing	2.9 Sheet metal forming		4.3 Electron beam machining, EBM
1.4 Pressure die casting	1.9 Squeeze casting	2.4 Cold forming	2.10 Spinning		4.4 Laser beam machining, LBM
1.5 Centrifugal casting		2.5 Cold heading	2.11 Powder metallurgy		4.5 Chemical machining, CHM
		2.6 Swaging	2.12 Continuous extrusion		4.6 Ultrasonic machining, USM
					4.7 Abrasive jet machining, AJM

Source: Swift, K. G., and Booker, J. D., *Process Selection from Design to Manufacture*, Butterworth-Heinemann, Oxford, 2003.

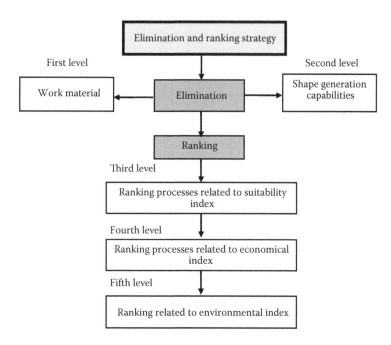

FIGURE 23.8   Elimination and ranking strategy for manufacturing process selection.

- Third level: A suitability index based on the operational requirements of the desired application is computed for each of the short listed manufacturing processes using the database of their capabilities.
- Fourth level: Required when there is more than one manufacturing process having the same suitability index based on the operational requirements. In such a case, the economic suitability index of the manufacturing processes is used to make the final choice.
- Fifth level: If the manufacturing processes have the same suitability index and the same economic index their environmental index is used to make the final choice.

23.6   DESIGN FOR MANUFACTURING

Because of the highly competitive nature of manufacturing processes, the question of finding ways to reduce cost is ever present. A good starting point for cost reduction is in the design stage of the product. The design engineer should always keep in mind the possible alternatives available in making his design. Unfortunately, designers often consider that their job is to design the product for performance, appearance, and reliability and that it is the manufacturing engineer's job to produce whatever has been designed. Of course, there is often a natural reluctance to change a proven design for the sake of a reduction in manufacturing cost. As a subject, design for manufacturing hardly exists as compared with design for strength.

Manufacturing cost is the key factor of the economic success of a product. The number of units sold and the sales price depend on the product quality. Successful design is ensured through high product quality while minimizing the manufacturing cost. Design for manufacturing (DFM) is one method of achieving this goal. Effective DFM practice leads to low manufacturing costs without sacrificing product quality. Figure 23.9 shows the outline of design for manufacturing. The following principles aid the designers to make products that can be produced at minimum cost.

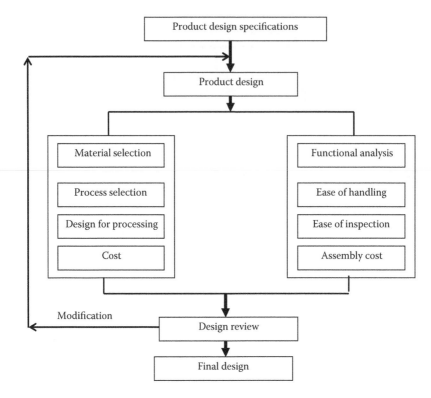

FIGURE 23.9   Outline of design for manufacturing.

1. Simplicity of the product: The minimum number of parts, the least intricate shape, the fewest precision adjustments, and the shortest production sequence.
2. Standard material and components: This enables benefits of mass production and simplifies inventory management, avoids tooling and equipment investment, and speeds up the manufacturing cycle.
3. Standard design of the product: When several similar products are to be produced, specify the same materials, parts, and subassemblies for each as much as possible.
4. Specify liberal tolerances: The higher costs of tight tolerance arise because of
   a. Extra machining operations such as grinding, honing, or lapping after primary machining operations
   b. Higher tooling cost
   c. Longer operating cycles
   d. Higher scrap and rework costs
   e. The need for more skilled and highly trained workers
   f. Higher materials cost
   g. High investment for precision equipment

A number of general rules have been developed to aid designers when thinking about the manufacture of the product. Accordingly, it is recommended to consider the following.

1. Simplify the design by reducing the number of parts required.
2. Design for low-labor cost operations wherever possible.
3. Avoid generalized statements on drawings that may be difficult for the production personnel to interpret.

4. Dimensions should be made from specific points or surfaces on the part itself.
5. Once the functional requirements are met, designers should strive for minimum weight.
6. Dimensions should be made from one datum line rather than from a variety of points to simplify tooling and gauging; avoid overlap of tolerances.
7. Design to use general purpose tooling rather than special ones.
8. Avoid sharp corners for ease of production and avoidance of stress concentration on the part.
9. Design a part so that many operations can be performed.
10. Space holes in machined parts so that they can be made in one operation without tooling weakness.
11. Whenever possible, cast, molded, or powder–metal parts should be designed without stepped parting line and with uniform wall thickness.

Design for manufacturing includes several aspects such as design for casting, forging, machining, joining, and design for assembly.

23.6.1 DESIGN FOR CASTING

Depending on the mold-making method, sand casting is economical for producing parts at all quantity levels. Few pieces can be made with minimum level of mechanization at the lowest cost, while highly automated systems are required for large volumes. Regarding the economic issue related to sand casting, it is important to consider the following.

1. For short runs, wooden patterns are inexpensive.
2. High production runs require more expensive metal patterns to withstand the wear of repeated use.
3. Handling small orders costs 10 times that of large ones.
4. Light sectioned castings cost 4 times per unit weight than the same weight in massive parts.

Design considerations for sand casting include the following points.

- Consider the shrinkage allowances of cast metal as it cools and solidifies.
- Use the parting line on a flat plane (Figure 23.10).
- Allow some degree of taper (draft) to avoid tearing the mold.
- Design the part to facilitate machining and finishing operations.
- Add risers in heavy sections to supply metal during solidification (Figure 23.11).
- Eliminate undercuts to avoid mold failure (Figure 23.12).
- Avoid heavy sections at rib intersections to eliminate hot spots (Figure 23.13).
- Minimize the number of ribs intersecting at one point (Figure 23.14).

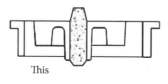

This

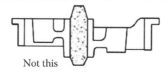

Not this

FIGURE 23.10 Use the parting line on a flat plane.

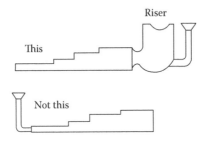

FIGURE 23.11   Add risers in heavy sections.

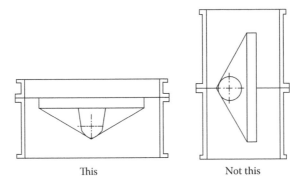

FIGURE 23.12   Eliminate undercuts.

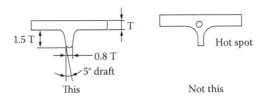

FIGURE 23.13   Avoid heavy sections where the rib intersects to eliminate hot spots.

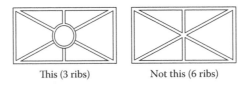

FIGURE 23.14   Reduce the number of reinforcing ribs.

- Use a cored hole or a circular web to bring a number of ribs or members together (Figure 23.15).
- Use round corners to reduce the severity of hot spots (Figure 23.16).
- Avoid acute angles at sharp corners (Figure 23.17).
- Keep the intersection of any two walls at right angles (Figure 23.18).
- Adopt the normal minimum wall thickness for various metals (>6 mm).
- Sections and walls should be uniform in thickness.

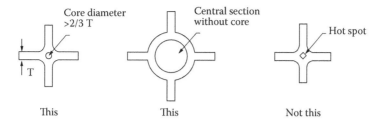

FIGURE 23.15 Use a cored hole or a circular web to bring a number of ribs or members together.

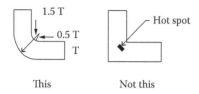

FIGURE 23.16 Use round corners to reduce the severity of hot spots.

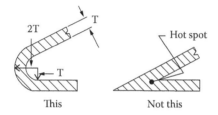

FIGURE 23.17 Avoid acute angles at sharp corners.

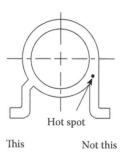

FIGURE 23.18 Keep intersection at right angles.

- Avoid abrupt changes in sections by using fillets and tapers (Figure 23.19).
- Interior walls should be 20% thinner than the outside members.
- Add lightener holes to reduce weight at low stressed areas.
- Avoid small cored holes (13–25 mm is recommended).
- Minimize material thickness at bosses (Figure 23.20).
- Minimize the need of cores by eliminating undercuts.
- Keep balance between section size of the rim spokes and hub of gears, pulleys, and wheels (Figure 23.21).
- Consider an odd number of curved wheel spokes to dissipate the casting stresses.

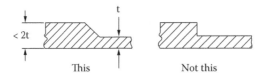

FIGURE 23.19   Avoid uprupt changes in sections.

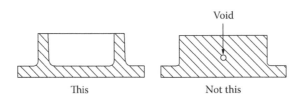

FIGURE 23.20   Minimize material thickness at bosses.

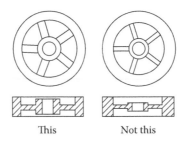

FIGURE 23.21   Keep balance between section size of the rim spokes and hub of gears, pulleys, and wheels.

23.6.2   DESIGN FOR SHEET METAL FORMING

The most common processes of sheet metal forming are blanking, piercing, and bending where the following design recommendations apply

- Design for ease of blanking (Figure 23.22).
- Tolerance of pierced holes is attained for 25% of its length (Figure 23.23).
- Avoid sharp corners to stop tearing the material (Figure 23.24).
- Shear and forming operations should have a minimum height (Figure 23.25).
- Position holes and slots away from bends (Figures 23.26 and 23.27).
- Avoid narrow web bulging by providing an ear in the blank or include the hole as a notch (Figure 23.28).
- Keep offset of bends to stop metal tear (Figure 23.29).
- Use separated straight flanges whenever possible (Figure 23.30).
- Allow a cutout when bending flanges (Figure 23.31).

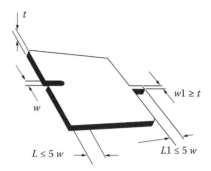

FIGURE 23.22   Design for ease of blanking.

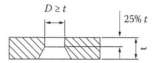

FIGURE 23.23   Tolerance of pierced holes is attained for 25% of its length.

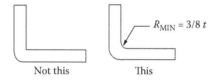

FIGURE 23.24   Avoid sharp corners.

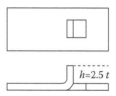

FIGURE 23.25   Shear and form height should be minimum of 2.5 thickness.

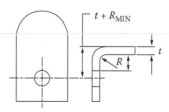

FIGURE 23.26   Position holes away from bends.

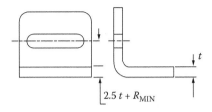

FIGURE 23.27   Position openings away from bends.

FIGURE 23.28   Avoid narrow web.

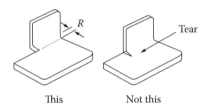

FIGURE 23.29   Keep offset of bends to stop metal tear.

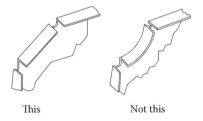

FIGURE 23.30   Use separated straight flanges.

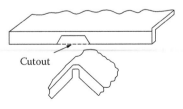

FIGURE 23.31   Allow a cutout when bending flanges.

23.6.3 DESIGN FOR DIE FORGING

Die forging is used to produce large numbers of parts. The designer should consider the following recommendations.

- Parting surface should be along a single plane if possible.
- All features should be oriented so that they can be formed in impressions moving in opposite directions (Figure 23.32).
- Avoid undercuts and holes oriented other than in the direction of forging.
- Keep forging loads balanced and eliminate side loads on the machine members. Figure 23.33 shows an unbalanced condition with a die counter lock, while Figure 23.34 shows forging rotation to balance the lateral load and avoid the need for counter lock.
- Avoid sharp exterior corners that require high forging pressures to fill the corresponding die features (Tables 23.4 and 23.5).
- Avoid sharp interior corners (fillets) that cause difficulties in metal flow and may require one or more dies to attain (Table 23.5).
- Allow maximum draft angles consistent with function, assembly, and weight constraints (Table 23.6).
- Avoid high or narrow ribs that make the material flow difficult.

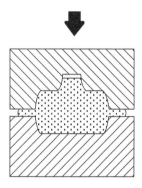

FIGURE 23.32 Forging is performed in opposite directions.

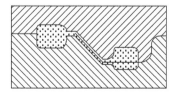

FIGURE 23.33 Forging generates a side thrust in the die requiring the counter lock to prevent lateral shift of the die.

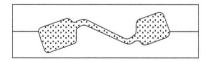

FIGURE 23.34 Forging rotation that balance the lateral loads and eliminate the counter lock.

TABLE 23.4
Minimum Fillet and Corner Radii for Forgings with 25 mm High Ribs

Alloy	Fillet Radius (mm)	Corner Radius (mm)
Carbon steel	6	1.5
Stainless steel	5	2.5
Titanium alloys	10	3
Iron base heat resistant alloys	8	3

TABLE 23.5
Effect of Protrusion Height on the Minimum Corner and Fillet Radius

Protrusion Height (mm)	Corner Radius (mm)	Fillet Radius (mm)
12.5	1.5	5
25	3	6.25
50	5	10
100	6.25	10
400	22	50

TABLE 23.6
Recommended Draft Angles

Material	Draft Angle (°)
Aluminum	0–2
Brass	0–3
Carbon steel	5–7
Stainless steel	5–8

23.6.4 DESIGN FOR MACHINING

Machining is recommended if surface finish, flatness, roundness, circularity, parallelism, or close fit is involved. Additionally, if the part is in motion or fits precisely with another part, machining operations will be employed. Machined parts can be as small as miniature screws, shafts, and gears. They can be as large as huge turbines, turbine housings, and valves found in hydroelectric power stations. Machined components are made from ferrous and nonferrous materials. However, plastics, rubber, carbon, graphite, and ceramics are also employed. Designers should follow these recommendations.

- Avoid machining operations if the surface or the feature required can be produced by casting or forming.
- Specify the most liberal surface finish and dimensional tolerances consistent with the function of the surface in order to avoid costly grinding, lapping, and other finishing operations (Figure 23.35).
- Design the part for ease of fixation and secured clamping during the machining operation.

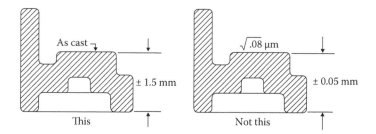

FIGURE 23.35 Avoid tolerances that necessitate machining if as-cast, as-forged, or as-formed dimensions and surface finishes are satisfactory for the parts function. (From Youssef, H., and El-Hofy, H., *Machining Technology–Machine Tools and Operations*, CRC Press, Boca Raton, FL, 2008. With permission.)

- Avoid sharp corners and sharp points in cutting tools to avoid their breakage.
- Use stock dimensions whenever possible (Figure 23.36).
- Avoid interrupted cuts during single-point machining operations.
- Design parts that are rigid enough to withstand clamping and cutting forces.
- Avoid tapers and contours to simplify tooling and setups.
- Reduce the number and the size of shoulders as they require extra materials and operations.
- Avoid undercuts because they involve more operations and special ground tools.
- Substitute a stamping operation for the machined component (Figure 23.37).
- Avoid the use of hardened or difficult-to-machine materials unless their functional properties are required.
- For thin and flat parts that require machining, allow sufficient stock for rough and finish operations.
- Put the machined surfaces in one plane.
- Provide access room for cutters, bushings, and fixture elements.
- Design parts so that standard cutters can be used (Figure 23.38).
- Avoid the use of parting lines or draft surfaces for clamping and locating.
- Avoid projections and shoulders that interfere with the cutter movement.
- Provide relief space for burr formation and furnish means for easy burr removal.

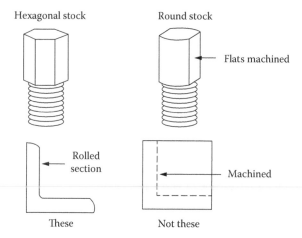

FIGURE 23.36 Use stock dimensions and minimize the machining allowance. (From Youssef, H., and El-Hofy, H., *Machining Technology–Machine Tools and Operations*, CRC Press, Boca Raton, FL, 2008. With permission.)

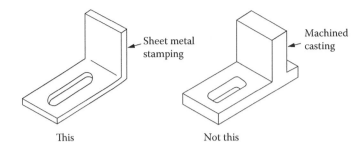

FIGURE 23.37 Metal formed parts are better than machined castings. (From Youssef, H., and El-Hofy, H., *Machining Technology–Machine Tools and Operations*, CRC Press, Boca Raton, FL, 2008. With permission.)

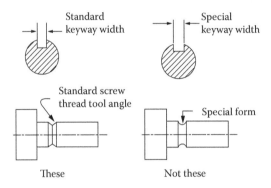

FIGURE 23.38 Design parts to be machined by standard tools. (From Youssef, H., and El-Hofy, H., *Machining Technology–Machine Tools and Operations*, CRC Press, Boca Raton, FL, 2008. With permission.)

23.6.5 Design for Welding

Design for welding emphasizes how to design a joint so that components can be produced most efficiently and without defects. This involves the selection and application of good design practices based on understanding welding aspects such as accessibility, quality, productivity, and the overall cost.

1. Fusion welding: The general design recommendations for fusion welding include the following.
   - Ensure minimum amount of weld by using simple and straight contours.
   - Balance the weld around the neutral axis of the part.
   - Ensure symmetry of parts to be welded along weld line to minimize distortion.
   - Provide access to the joint area for vision, electrodes, filler rods, cleaning, etc.
   - Provide sufficient edge distance to avoid welds meeting at the end of runs.
   - Allow provision for the escape of gases and vapors.
   - Follow design recommendations regarding joint type, welding position, minimum thickness, maximum sheet thickness, multiple weld runs, unequal thickness, and design complexity level (Table 23.7).
2. Spot welding: Design recommendations for spot welding include the following.
   - Thickness of the parts to be welded should have the ratio between 1:1 and 3:1.
   - The minimum distance between welding positions is 10 times the part thickness.

TABLE 23.7
Design Aspects for Fusion Welding Processes

Design Aspect	Gas Welding	MIG/GMAW	TIG/GTAW	MMA/SMAW	SAW	PAW
Joint type	But, lap, fillet, edge	But, lap, fillet, edge	But, lap, fillet, edge	But, lap, fillet, edge	But, fillet	But, lap, fillet, edge
Welding position	All	Vertical/overhead	Horizontal	All	Horizontal	Horizontal
Minimum thickness	C steel = 0.5 mm CI = 3 mm	0.5 mm	0.2 mm	1.5 mm, CI = 6 mm	5 mm	0.05 mm
Maximum thickness	C steel/CI = 30 mm	Refractory alloy = 6 mm Others = 80 mm	Al/Ti alloys = 15 mm Others = 36 mm	200 mm	Steels = 500 mm Ni alloy = 20 mm	Al = 3 mm Copper/refra matr = 6 mm Steels = 10 mm Ti alloys = 13 mm Ni = 15 mm
Multiple weld runs	$\geq$4 mm	$\geq$5 mm	$\geq$5 mm	$\geq$10 mm	$\geq$40 mm	$\geq$10 mm
Unequal thickness	Possible	Possible	Difficult	Difficult	Difficult	Difficult
Complexity level	Moderate	Possible	High	Possible	Limited	High

- The minimum center of weld to edge distance is twice the weld diameter.
- The minimum weld to form distance equals the bend radius plus the weld diameter.
- Allow adequate access for spot welding of small flanges in U channels.
- Flat surfaces are easier to spot weld.
- Multiple bends impose access restrictions and need special fixtures.
- For minimum setups and maximum throughput, choose the same spot weld size.
- Avoid plating of spot welded assemblies where plating salts may be trapped, thus requiring special cleaning or causing potential long-term corrosion problems.

23.6.6 Design for Assembly

In the design for assembly, products are made for the ease of assembly in mind. A product containing fewer parts takes less time to assemble, thereby reducing the assembly costs. Similarly, parts that are easier to grasp, move, orient, and insert, reduce assembly time and costs. Therefore the major cost benefit of the application of design for assembly is achieved through the reduction of the number of parts in an assembly. Assembly methods are generally divided into three major groups

1. Manual assembly: Workers manually assemble the product or its components using hand tools. It is the most flexible and adaptable method. However, there is usually an upper limit to the production volume and the high labor costs.
2. Fixed (hard automation) assembly: It uses machines and feeders that assemble a specific product. This machinery requires a large capital investment that is justified by the large production volume. This kind of assembly is sometimes called Detroit-type assembly.
3. Soft automation (robotic) assembly: Incorporates the use of a single robot or a multistation robotic assembly cell with all activities simultaneously controlled and coordinated by a programmable logic controller (PLC) or a computer. Although this type of assembly requires large capital costs, its flexibility helps offset the expenses across many different products.

Design for assembly guidelines includes the following aspects.

1. Simplify the design and reduce the number of parts: As the number of parts increase, the total cost of purchasing, stocking, fabricating, and assembling the product goes up. Additionally, automation becomes difficult and more expensive.
2. Standardize and use common parts and material: This minimizes the amount of inventory in the system, standardizes handling and assembly operations, reduces costs, and leads to higher quality.
3. Design for ease of fabrication. Select processes compatible with the materials and production volumes and simplify part features to avoid extra processing effort and/or more complex tooling.
4. Design within process capabilities and avoid unneeded surface finish requirements: Avoid unnecessarily tight tolerances that are beyond the capability of the manufacturing processes. Tolerances on connected parts will "stack-up," making maintenance of overall product difficult. Surface finish requirements is also established based on standard practices.
5. Mistake-proof product design and assembly (poka-yoke): Components should be designed so that they can only be assembled in one way. Notches, asymmetrical holes, and stops can be used to mistake-proof the assembly process.
6. Design for parts orientation and handling: Basic principles to facilitate parts handling and orienting are as follows.
   a. Parts must orient themselves when fed into the assembly process.
   b. Avoid parts that can become tangled, wedged, or disoriented.
   c. Avoid holes, tabs, and designed closed parts.

    d. Incorporate symmetry around both axes of insertion wherever possible.

    e. Provide an external feature or guide surface to correctly orient the part.

    f. Guide surfaces should be provided to facilitate insertion.

    g. Parts should be designed with surfaces so that they can be easily grasped.

    h. Minimize thin, flat parts that are more difficult to pick up.

    i. Avoid parts with sharp edges, burrs, or points.

    j. Avoid parts that can be easily damaged or broken.

    k. Avoid heavy parts that increase worker fatigue, risk of worker injury, and slow down the assembly process.

7. Minimize flexible parts and interconnections.

    a. Avoid flexible and flimsy parts such as belts, gaskets, tubing, cables, and wire harnesses.

    b. Use plug-in boards and backplanes to minimize wire harnesses.

    c. Consider fool-proofing electrical connectors by using unique connectors to avoid connectors being wrongly connected.

    d. Partition of the product to minimize interconnections between modules and co-locate related modules to minimize routing of interconnections.

8. Design for ease of assembly

    a. Part features should be provided such as chamfers and tapers.

    b. Enable assembly to begin with a base component with a large relative mass and a low center of gravity upon which other parts are added.

    c. Assembly should proceed vertically with other parts added on top and positioned with the aid of gravity.

    d. Assembly that is automated is more uniform, more reliable, and of a higher quality.

9. Design for efficient joining and fastening

    a. Threaded fasteners (screws, bolts, nuts, and washers) are time-consuming to assemble and difficult to automate.

    b. Standardize fasteners to minimize variety and use self-threading screws and captured washers.

    c. Consider the use of integral attachment methods (snap-fit).

    d. Match fastening techniques to materials, product functional requirements, and disassembly/servicing requirements.

10. Design modular products: This modular or building block design minimizes the number of part or assembly variants. Modules can be manufactured and tested before final assembly.

11. Design for automated production: The product must be designed in a way that it can be handled more with automation. For hard automation, the following points are important.

- Reduce the number of different components.
- Use self-aligning and self-locating features.
- Avoid screws/bolts.
- Use the largest and most rigid part as the assembly base and fixture. Assembly should be performed in a layered, bottom-up manner.
- Use standard components and materials.
- Avoid tangling or nesting parts.
- Avoid flexible and fragile parts.
- Avoid parts that require orientation.
- Use parts that can be fed automatically.
- Design parts with a low center of gravity.

For flexible (robotic) assembly, the designer should consider the following points.

- Design parts to utilize standard gripper and avoid gripper/tool change.
- Use self-locating parts.

- Avoid the need to secure or clamp parts.
- Use a minimum number of parts or standard parts for minimum of feeding bowls, etc.
- Use closed parts (no projections, holes, or slots) to avoid tangling.
- Consider the potential for multiaxis assembly to speed the assembly cycle time, and use preoriented parts.

23.6.7 Design for Environment

Design for environment (DFE) is a product design approach adopted for reducing the impact of products on the environment during their manufacture through the use of highly polluting processes and the consumption of large quantities of raw materials. Products can also have an adverse effect through the consumption of large amounts of energy during manufacturing, use, and disposal. The entire product life cycle from manufacture through the use and disposal is shown in Figure 23.39. During this life cycle, many events create pollution and many opportunities for recycling, remanufacturing, and reuse, thus reducing the environmental impact. Manufacturing products that impact the environment less become a market advantage. In order to reduce the environmental impacts during the product life cycle time, the following design approaches can be adopted:

- Design to minimize material usage.
- Design for disassembly.
- Design for recyclability.
- Design for remanufacturing.
- Design to minimize hazardous materials.
- Design for energy efficiency.

There are three major elements of design for the environment

1. Design for environmental manufacturing involves the following considerations.
   - Use nontoxic processes and materials.
   - Minimize the energy utilization.
   - Minimize emissions.
   - Minimize waste, scrap, and by-products.
2. Design for environmental packaging involves the following aspects.
   - Minimize the packaging materials.
   - Adopt reusable pallets, totes, and packaging.
   - Use recyclable and biodegradable packaging materials.
3. Design for disposal and recyclability through the following activities:
   - Reuse components and assemblies.
   - Select material that can be reused.
   - Use the minimum number of materials that can be easily identified and separated.

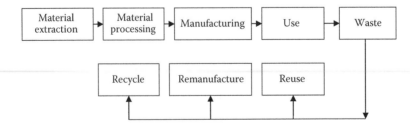

FIGURE 23.39   Stages of a product life cycle.

- Design for easy disassembly and avoid the use of adhesives.
- Limit contaminants, additives, coatings, metal plating of plastics, etc.

23.7   REVIEW QUESTIONS

1. Explain what is meant by the manufacturing process capability.
2. State the main factors that affect the selection of a manufacturing process.
3. What are the main methods used for manufacturing process selection?
4. Explain the method of elimination and ranking strategy.
5. What are the main design recommendations for sand casting and forging?
6. State the main design recommendations for fusion welding.
7. List some design recommendations for sheet metal-forming processes.
8. What are the different methods of assembly?
9. What are the recommendations needed for automatic and manual assembly methods?
10. Explain the following terms: design for environment and product life cycle.

BIBLIOGRAPHY

Cogun, C. 1994. Computer-aided preliminary selection of non-traditional machining processes. *Journal of Machine Tools Manufacture* 34(3): 315–326.

Degarmo, E. P., Black, J. T., and Kohser, R. A. 1997. *Materials and Processes in Manufacturing*, 8th ed. Upper Saddle River, NJ: Prentice-Hall.

El-Hofy, H. 2005. *Advanced Machining Processes—Nontraditional and Hybrid Processes*. New York: McGraw-Hill.

Jain, N. K., and Jain, V. K. 2003. *Process Selection Methodology for Advanced Machining Processes*, vol. 2, no. 1, pp. 5–45. Hackensack, NJ: World Scientific.

Kalpakjian, S., and Schmid, S. R. 2003. *Manufacturing Processes for Engineering Materials*, 4th ed. Upper Saddle River, NJ: Pearson Education.

Metcut Research Association. 1980. *Machining Data Handbook*, vol. 2, 3rd ed. Newport, KY: Zimmerman and Son.

Otto, K. N., and Wood, K. L. 2001. *Product Design Techniques in Reverse Engineering and New Product Development*. Upper Saddle River, NJ: Prentice-Hall.

Schey, J. A. 2000. *Introduction to Manufacturing Processes*, 3rd ed. New York: McGraw-Hill.

Swift, K. G., and Booker, J. D. 2003. *Process Selection from Design to Manufacture*, 2nd ed. Oxford: Butterworth-Heinemann.

Youssef, H., and El-Hofy, H. 2008. *Machining Technology–Machine Tools and Operations*. Boca Raton, FL: CRC Press.

Zha, X. F. 2005. A Web-based advisory system for process and material selection in concurrent product design for a manufacturing environment. *International Journal of Advanced Manufacturing Technology* 25: 233–243.

24 Quality Control

24.1 INTRODUCTION

Quality control (QC) is a procedure or set of procedures intended to ensure that a manufactured product adheres to a defined set of quality criteria or meets the requirements of the customer. It involves sampling inspection of the product during its manufacture by means of a planned system of patrol inspection of the production process where a fixed number of parts are examined by the inspector at each site visit. The sample size varies from 2 to 10 in each case of quality control by variables. The sampling inspection results are plotted on the quality control chart that is kept adjacent to the machine performing the manufacturing operation to be controlled. The control chart provides a visual judgment for significant deviations from standard quality. By taking appropriate action as indicated by the control chart, any tendencies for defective part production can be corrected before they are produced.

Quality control is different from the normal inspection of the finished products that result mainly in throwing out defectives. On the other hand, quality control is mainly designed to prevent these defectives. The use of quality control charts focuses the attention on the manufacturing process rather than the resulting product. Basically, quality control involves evaluating whether or not the product, activity, process, or service is satisfactory. By contrast, quality assurance ensures that a product or service is produced in the right way.

24.2 STATISTICAL QUALITY CONTROL

Variations in quality characteristic are inevitable and can be attributed to either natural or assignable causes. Variations due to natural causes inherent in the manufacturing process are difficult to identify, uneconomical to eliminate, and follow statistical laws. On the other hand, assignable variations are due to individual causes that can be identified and eliminated. The objective of controlling a process or a product is achieved by

1. Restricting the causes of variations in the quality characteristic due to natural causes
2. Detecting and eliminating the assignable causes of variations

Events of a random nature that are without any particular trend or pattern are occurring during manufacturing. These cause variability in production that was first monitored by Eli Whitney date. Owing to the large number of materials and process variables involved, modern statistical concepts relevant to manufacturing engineering were developed in the early 1900s. In order to understand the basics of statistical quality control (SQC), the following terms are introduced.

- Sample size: The number of parts to be inspected in a sample whose properties are studied to gain information about the whole population.
- Random sampling: Taking a sample from a population or lot in which each item has an equal chance of being included in the sample.
- Population: The totality of individual parts from which samples are taken.
- Lot size: A subset that represents the population.
- Variable measure: A product characteristic that is measured on a continuous scale such as length, weight, temperature, or time.

- Attribute: A product characteristic such as color, surface texture, cleanliness, or perhaps smell or taste; can be evaluated quickly with a discrete response such as good or bad, acceptable or not, yes or no.

24.2.1 STATISTICAL PRINCIPLES

A frequency distribution or a histogram is a graphical tool that is frequently used for quality control purposes. Accordingly, two principal analytical methods for describing a collection of data are used. The measure of central tendency of a distribution mean describes the central position of the data and how it is built up in the center. It is the most commonly used measure of central tendency that is used for further statistical measurement of dispersion and designing limits for control charts.

For ungrouped data, the mean $\bar{X}$ is the sum of values in the distribution divided by the total number of values n. Hence

$$\bar{X} = \frac{\sum_{i=1}^{n} X_i}{n}$$

where X_i are the values of the ith observation.

For grouped data, the mean $\bar{X}$ can be calculated from

$$\bar{X} = \frac{\sum_{j=1}^{m_c} f_j X_j}{n}$$

where

    f_j = frequency of observations in the jth cell
    X_j = midpoint of the jth cell
    m_c = number of cells

The measure of process dispersion describes how the data are spread out or scattered on each side of the central value (mean). In this respect, the range of a series R is taken as the difference between the largest X_{high} and the smallest X_{low} values of observations as follows:

$$R = X_{high} - X_{low}$$

However, the standard deviation (σ) is the most common measure of process dispersion that can be calculated, for ungrouped data

$$\sigma = \sqrt{\frac{\sum_{i=1}^{n} (X_i - \bar{X})^2}{n}}$$

where

    X_i = values of the ith number in the series
    $\bar{X}$ = mean of series
    n = number of observations

TABLE 24.1
Diameter Measurements of 100 Workpieces

Work Diameter, mm	Midpoint X	Frequency, f	fX	fX²
40.70–40.74	40.72	1	40.72	1658.1184
40.74–40.78	40.76	2	81.52	3322.7552
40.78–40.82	40.80	6	244.8	9987.8400
40.82–40.86	40.84	13	530.92	21682.7728
40.86–40.90	40.88	18	735.84	30081.1392
40.90–40.94	40.92	21	859.32	35163.3744
40.94–40.98	40.96	16	655.36	26843.5456
40.98–41.02	41.00	12	492	20172.0000
41.02–41.06	41.04	6	246.24	10105.6896
41.06–41.10	41.08	3	123.24	5062.6992
41.10–48.14	41.12	2	82.24	3381.7088
Sum		100	4092.2	167461.6428

When the data are grouped into a frequency distribution, σ can be calculated from

$$\sigma = \sqrt{\dfrac{\sum\limits_{j=1}^{m_c} f_j X_j^2}{n} - \left[\dfrac{\sum\limits_{j=1}^{m_c} f_j X_j}{n}\right]^2}$$

where

X_j = midpoint of the jth cell
f_j = frequency of the jth cell
m_c = number of cells

Random variations in manufacturing processes are shown by measuring the diameter in mm of 100 turned parts (Table 24.1). The histogram (Figure 24.1) and the frequency polygon (Figure 24.2)

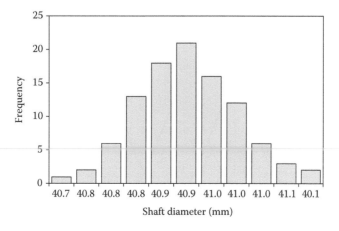

FIGURE 24.1   Histogram for data of shaft diameters.

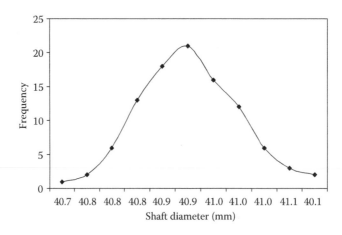

FIGURE 24.2 Frequency distribution of shaft diameters.

present some of the characteristics of the turning process variability where the produced size is subject to random variation. It can be seen that the frequency polygon is similar to the normal or Gaussian curve shown in Figure 24.3. The properties of such a curve are important in quality control because 99.73% of the area under the curve represent the total number of parts considered to lie between ±3σ, 95.46% of the parts are included between ±2σ, and 68.26% of the parts are included between ±σ. Figure 24.4 shows three normal curves with different mean values $\bar{X}$ and the same standard deviation. Figure 24.5 shows three normal curves with the same mean but different standard deviations σ. Accordingly, the larger the standard deviation, the more the dispersion the data and vice versa.

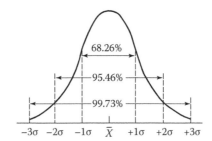

FIGURE 24.3 Properties of a Gaussian normal curve.

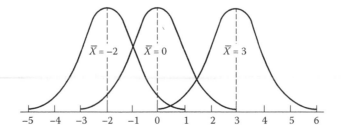

FIGURE 24.4 Changes in process average having the same dispersion.

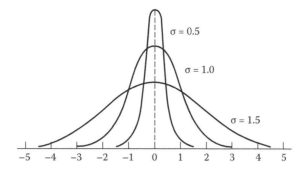

FIGURE 24.5   Changes for process dispersion having the same mean.

Example 1

Calculate the mean $\bar{X}$ and standard deviation σ for the data shown in Table 24.1.

Solution

$$\bar{X} = \frac{\sum_{j=1}^{m_c} f_j X_j}{n} = \frac{4092.2}{100} = 40.92\,\text{mm}$$

$$\sigma = \sqrt{\frac{\sum_{j=1}^{m_c} f_j X_j^2}{n} - \left(\frac{\sum_{j=1}^{m_c} f_j X_j}{n}\right)^2} = \sqrt{1674.616 - 40.92^2} = 0.412\,\text{mm}$$

Process variability = $\pm 3\sigma = \pm 3 \times 0.412 = \pm 1.236$ mm

If a tolerance of less than $\pm 3\sigma$ is taken, there will be a probability of some scrap being produced as soon as the setting changes owing to assignable causes. Generally, if the machine is set to produce a given size, changes in the product quality may arise owing to changes in the mean size of the work diameter caused by tool wear or due to changes in something fundamental in the machine. Under such conditions, a scrap will be produced unless a corrective action is taken. Control charts are thus essential for controlling the process average using $\bar{X}$ chart or controlling the process variability through range (R) chart. Historically, control charts have been used to monitor the quality of manufacturing processes, products, and services.

24.2.2   STATISTICAL PROCESS CONTROL

The technique that uses control charts to see if any part of a production process is not functioning properly and could cause poor quality is called statistical process control (SPC). In this regard, control charts visually show if a sample is within the statistical control limits that represent the upper control limits are the upper and lower bands of the control chart. Once the control charts are established for a process, it is then monitored to indicate when it is out of control. SPC is also used to measure the capabilities of the manufacturing process as well as the characteristics of the machines used.

24.2.3 Control Charts

Control charts are continuous graphical records of product quality that provide a visual method of indicating variations in product quality that cannot be allowed to continue. By plotting successive findings of the process inspection on such a chart, the presence of assignable causes of variations is immediately revealed and a corrective action is taken to eliminate them before defective products are made. As these assignable causes are eliminated, the process becomes under control and produces parts whose variations in quality are entirely of a random nature. Attaining such a goal makes it possible to predict, in advance, the percentage of the future product that falls within any specified limits or specifications. There are two main types of control charts, control charts for variables and those for attributes.

Variable control charts are widely used for continuous quantities that can be measured, such as weight, length, temperature, hardness, strength, or volume. These include the mean (X bar) and the range (R bar) charts. The control limits are generally set according to statistical control formulas designed to keep the actual production within the acceptable ±3σ range.

For the X bar chart, the mean of each sample (four to five parts) is computed and plotted on the chart. In the R chart, the range taken as the difference between the smallest and largest values in the sample reflects the process variability. Figure 24.6 shows a typical X bar chart where points above the upper control limit are due to assignable causes that should be avoided to keep the product quality characteristic within the control limits.

Example 2

Table 24.2 shows the surface roughness of turned parts in micrometers. Using a rational subgroup of four, the inspector obtained 25 subgroups of data during 5 days of work. Draw the X- and R-bar charts.

Solution
The central value of $\overline{X}$ is given by

$$\overline{\overline{X}} = \frac{\sum_{j=1}^{m_c} \overline{X}_j}{m_c} = \frac{47.95}{25} = 1.918 \ \text{m}$$

$\overline{\overline{X}}$ = average of the subgroup means
$\overline{X}_j$ = mean of jth subgroup
m_c = number of subgroups

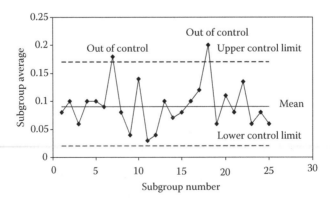

FIGURE 24.6   Typical X bar control chart.

TABLE 24.2
Data of Surface Roughness of Machined Parts

Subgroup Number	Measurements (μm)				Average $\bar{X}$ (μm)	Range (μm)
	X1	X2	X3	X4		
1	1.9	1.93	1.95	2.05	1.96	0.15
2	1.76	1.81	1.81	1.83	1.80	0.07
3	1.8	1.87	1.95	1.97	1.90	0.17
4	1.77	1.83	1.87	1.9	1.84	0.13
5	1.93	1.95	2.03	2.05	1.99	0.12
6	1.76	1.88	1.95	1.97	1.89	0.21
7	1.87	2	2	2.03	1.98	0.16
8	1.91	1.92	1.94	1.97	1.94	0.06
9	1.9	1.91	1.95	2.01	1.94	0.11
10	1.79	1.91	1.93	1.94	1.89	0.15
11	1.9	1.97	2	2.06	1.98	0.16
12	1.8	1.82	1.89	1.91	1.86	0.11
13	1.75	1.83	1.92	1.95	1.86	0.2
14	1.87	1.9	1.98	2	1.94	0.13
15	1.9	1.95	1.95	1.97	1.94	0.07
16	1.82	1.99	2.01	2.06	1.97	0.24
17	1.9	1.95	1.95	2	1.95	0.1
18	1.81	1.9	1.94	1.97	1.91	0.16
19	1.87	1.89	1.98	2.01	1.94	0.14
20	1.72	1.78	1.96	2	1.87	0.28
21	1.87	1.89	1.91	1.91	1.90	0.04
22	1.76	1.8	1.91	2.06	1.88	0.3
23	1.95	1.96	1.97	2	1.97	0.05
24	1.92	1.94	1.97	1.99	1.96	0.07
25	1.85	1.9	1.9	1.92	1.89	0.07
Average					$\bar{\bar{X}} = 1.918$	$\bar{R}_m = 0.138$

The upper control limit $UCL_{\bar{X}}$ for $\bar{X}$ chart is given by

$$UCL_{\bar{X}} = \bar{\bar{X}} + A_2\bar{R}_m$$

A_2 is a factor of quality control limits that can be taken from Table 24.3. Accordingly, for $n = 4$, $A_2 = 0.729$.

$$UCL_{\bar{X}} = 1.918 + (0.729)(0.138) = 2.0168 \quad m$$

TABLE 24.3
Factors for Control Limits

No. of Observations in Samples, n	d_2	A_2	D_3	D_4
2	1.128	1.880	0	3.269
3	1.693	1.023	0	2.575
4	2.059	0.729	0	2.282
5	2.326	0.577	0	2.115
6	2.534	0.483	0	2.004
8	2.847	0.373	0.136	1.864
10	3.078	0.308	0.223	1.777
15	3.472	0.223	0.348	1.625
20	3.735	0.180	0.414	1.586
25	3.931	0.153	0.459	1.541

The lower control limit is $LCL_{\bar{X}}$

$$LCL_{\bar{X}} = \bar{\bar{X}} - A_2 \bar{R}_m$$

$$LCL_{\bar{X}} = 1.918 - (0.729)(0.138) = 1.817398 \quad m$$

$\bar{R}_m$ = mean or average of subgroup ranges

$$\bar{R}_m = \frac{\sum_{j=1}^{m_c} R_j}{m_c}$$

Control limits for R chart are calculated as follows.
The upper control limit UCL_R is

$$UCL_R = D_4 \bar{R}_m = (2.282)(0.138) = 0.314916 \quad m$$

The lower control limit is LCL_R

$$LCL_R = D_3 \bar{R}_m = (0)(0.138) = 0 \quad m$$

$\bar{R}_m$ = mean or average of subgroup ranges
R_j = range of the jth subgroup

The factors for quality control limits, d_2, D_3, and D_4, are taken from Table 24.3. Hence, $d_2 = 2.059$, $D_3 = 0$, and $D_4 = 2.282$.

$$\bar{R}_m = \frac{\sum_{j=1}^{m_c} R_j}{m_c} = \frac{3.45}{25} - = 0.138 \quad m$$

The standard deviation of the data can be calculated from

$$\sigma = \frac{\bar{R}_m}{d_2} = \frac{0.138}{2.058} = 0.067$$

As can be seen in Figure 24.7 the subgroup 2 is out of control state due to a part of normal variation. However, the process is statistically under control. The R chart in Figure 24.8 shows that the mean of ranges is stable, while the process dispersion increases and stays within the control limits of the chart.

24.2.4 CONTROL LIMITS AND SPECIFICATIONS

The upper and lower control limits are established from the averages of subgroups, while specifications (tolerances) are the permissible variations in the product quality characteristic of the part that are set for individual values by the manufacturing engineer so that the part meets a particular function. As shown in Figure 24.9, the location of the upper and lower specifications (tolerance) is optional and does not depend on the spread of the process. On the other hand, the control limits, process spread, distribution of averages, and distribution of individuals are interdependent because

$$\sigma_{\bar{X}} = \frac{\sigma}{\sqrt{n}}$$

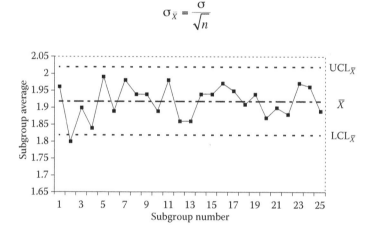

FIGURE 24.7 X bar chart.

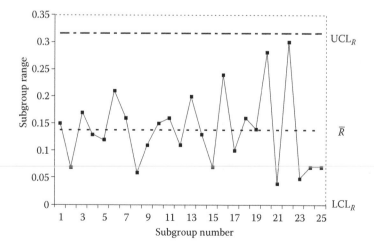

FIGURE 24.8 R chart.

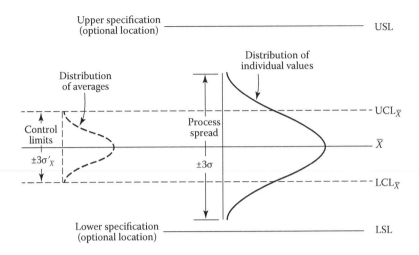

FIGURE 24.9   Relationship between limits, specifications, and distributions.

where

n = Subgroup size
$\sigma_{\bar{X}}$ = Standard deviation of averages
σ = Standard deviation of the individual values

24.2.5 PROCESS CAPABILITY

A manufacturing process combines tools, materials, methods, and people in order to produce a measurable output. A measure of the process capability is attained when meeting customer requirements by comparing the process limits to the specified tolerance limits. The process capability index C_{pk} measures the variability of a process 6σ and compares it with a proposed upper specification limit (USL) and a lower specification limit (LSL) as shown in Figure 24.9.

$$C_{pk} = \min \; \frac{(\bar{X} - \text{LSL})}{3\sigma} \; \text{or} \; \frac{(\text{USL} - \bar{X})}{3\sigma}$$

where

$\bar{X}$ = Mean of the process
σ = Standard deviation of the process

The process capability index indicates if the process mean $\bar{X}$ has shifted away from the design target and in which direction it has shifted. If C_{pk} is greater than 1.00, then the process is capable of meeting design specifications. If it is less than 1.00, then the process mean has moved closer to one of the upper or lower design specification limits and will therefore generate defects.

The process capability ratio (C_p) measures the capability of a process to meet design specifications. It is defined as the ratio of the range of the design specifications (the tolerance range) to the range of process variation, which is typically ±3σ.

$$C_p = \frac{(\text{USL} - \text{LSL})}{6\sigma}$$

Hence, if C_p is less than 1.0, the process range 6σ is greater than the tolerance range (USL − LSL). In such a case, the process is not capable of producing within the design specifications all the

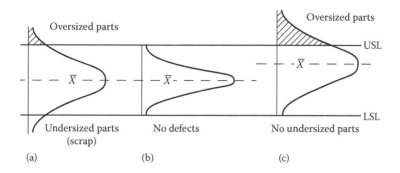

FIGURE 24.10 Changes in average and dispersion when $6\sigma <$ USL – LSL.

time. In case of Figure 24.10a, the process is in control, but defects will be produced. The process is therefore not capable of manufacturing products that will meet specifications. Defective parts can be eliminated by reducing the process dispersion as 6σ, shown in Figure 24.10b. Moving the process average as in Figure 24.10c gives the possibility of producing oversized parts that can be reworked and avoids the possible undersized or scraped occurring in the lower end of Figure 24.10a. Furthermore, as shown in Figure 24.11, the spread of the process 6σ or process variability is equal to the difference between specifications (USL – LSL). In case of Figure 24.11a, the process is centered and capable. The shift of average shown in Figure 24.11b or the increase in the process dispersion 6σ in Figure 24.11c causes individual values to exceed specifications. Under such cases, assignable causes must be corrected as soon as they occur to avoid defective production.

A process capability ratio C_p rather than 1.0 (Figure 24.12a) indicates that the process is capable of meeting specifications. In such a case, the difference between specification limits is greater than the process spread 6σ and no defective parts will be produced due to the shift in either average $\bar{X}$ or dispersion 6σ as shown in Figures 24.12b and 24.12c, respectively. Under these conditions, frequent machine adjustments or search for assignable causes are not necessary, and the control chart may be discontinued. When C_{pk} equals C_p, the process mean is centered on the design (nominal) target.

Machine capability: Because a dispersion of $\pm 3\sigma$ is expected in manufacturing, it is usual to compare 6σ to the tolerance to express the machine capability (MC) as follows:

$$\text{MC} = \frac{6\sigma}{(\text{USL} - \text{LSL})} \times 100\%$$

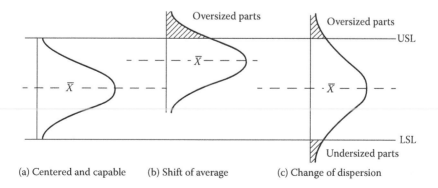

FIGURE 24.11 Changes in average and dispersion when $6\sigma =$ USL – LSL.

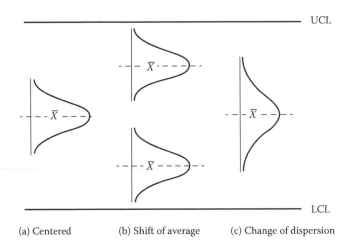

(a) Centered   (b) Shift of average   (c) Change of dispersion

FIGURE 24.12   Changes in average and dispersion when $6\sigma < $ USL – LSL.

If MC is greater than 100%, the machine capability is poor and defective parts will be produced. At MC = 100% the machine is just capable, and when MC < 100%, the machine is capable to produce parts within the specified limits.

24.2.6 ACCEPTANCE SAMPLING AND CONTROL

Acceptance sampling consists of taking a few random samples from a lot and inspecting them for the purpose of judging whether the entire lot is acceptable or should be rejected or reworked if economically feasible. It was developed during World War II and is widely used as a valuable technique for inspecting high production rates where 100% inspection is expensive. Figure 24.13 shows a sample of the operating characteristic curve, which gives the probability of acceptance. For a given sample, if the percentage of nonconforming parts exceeds the predetermined and limiting percentage, the entire lot is rejected or reworked. As the number of samples taken from the lot increases, the probability of rejecting the lot will be higher. The acceptance quality level (AQL) is the level at which there is 95% probability of acceptance of the lot while 5% will be rejected by the con-

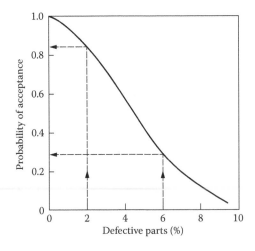

FIGURE 24.13   Acceptance sampling.

sumer. Acceptance sampling requires less time and fewer inspections than other sampling methods. However, automated inspection makes 100% inspection feasible and economical.

24.3 TOTAL QUALITY CONTROL

The total quality control (TQC) is a system approach where the management and workers make an integrated effort to manufacture high-quality parts consistently. Total quality management (TQM) is a management approach for an organization, centered on quality, based on the participation of all its members and aiming at long-term success through customer satisfaction, and benefits to all members of the organization and society. In total quality management, employees use SPC to see if their process is in control. By continuously monitoring the production process and making improvements, the employee contributes to the TQM goal of continuous improvement and production of few or no defects. Through the quality circle, regular meetings by groups of workers are held to discuss how to improve and maintain the quality level through the manufacturing stages. Comprehensive training of workers on statistical data analysis identifying the causes of poor quality and taking immediate actions to correct the situation is essential.

24.4 THE ISO 9000 STANDARD

The ISO 9000 standard on quality system provides a sound base for TQM (Figure 24.14). Accordingly, in addition to the need for quality control and standards, TQM must be management-led and the company should be dedicated to continuous improvement. Team work participation and total involvement of each employee is essential. The standards of ISO 9000 detail 20 requirements for an organization's quality management system in the following areas:

- Management responsibility
- Quality system
- Order entry
- Design control
- Document and data control
- Purchasing
- Control of customer supplied products
- Product identification and tractability

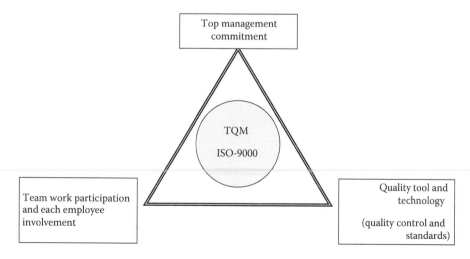

FIGURE 24.14 TQM bases of success.

- Process control
- Inspection and testing control
- Inspection and test status
- Control of nonconforming products
- Corrective and preventive action
- Handling, storage, packaging, and delivery
- Control of quality records
- Internal quality audits
- Training
- Servicing
- Statistical techniques

ISO advantages include

- Optimized company structure and operational integration
- Improved awareness of company objectives
- Improved communications and quality of information
- Clearly defined responsibilities and authorities
- Improved tractability to root causes of quality problems
- Improved utilization of time and materials
- Formalized systems ensure consistent quality and punctuality
- Documented system provides useful references and training tools
- Fewer rejects, therefore, less repeated work and warranty cost
- Improved control and relations
- Improved quality and reduced number of audits
- Improved records in case of litigation

24.5 DIMENSIONAL CONTROL

Engineering metrology is concerned with the measurements of product dimensions such as length, thickness, diameter, taper, and flatness. Measurements can be made while the part is produced (in process) or after the part is finished. Dimensional control is performed by checking that the measured dimensions comply with the specified dimensional specifications.

24.5.1 INTERCHANGEABILITY

Interchangeability is the manufacturing system in which every part of every batch will assemble freely with one of the mating parts produced without any fitting or modification to the parts. Also, if different firms are producing parts to the same standard specifications, then the firms separate products must be interchangeable. Interchangeability reduces the assembly time and facilitates required for the replacement of worn or defective components.

The principle of interchangeable manufacturing was developed throughout the nineteenth century based on the use of templates, jigs, and fixtures by semiskilled labor. In order to achieve interchangeability, appropriate component tolerances must be specified to suit the type of fit required. Moreover, the selected manufacturing process must be capable of producing parts within the required tolerance. Finally, the system of quality control must ensure that only components within the specifications are accepted for use. Interchangeable manufacturing is possible if the tolerance that is allowed on each dimension is established and the type of fit between the mating parts is determined.

24.5.2 TOLERANCE

The nominal (basic) size is the dimension of a part specified in the part drawing and assigned on the bases of design considerations. It is difficult to obtain such a nominal size during manufacturing operations using machine tools due to the presence of assignable and random errors. Dimensional tolerances are therefore the permissible or acceptable variations in dimensions (height, width, depth, diameter, angles) of a part. Because it is impossible to manufacture two parts that have precisely the same dimensions, tolerances are unavoidable. Obtaining close tolerances is economically undesirable because they substantially increase the product cost.

Tolerances are important for parts that need to be assembled or mated together. Free functional surfaces do not need close tolerance control. Figure 24.15 shows the ISO system for basic size, deviation, and tolerance on a shaft or a hole that have minimum and maximum diameters, respectively. It is accordingly clear that tolerance is the difference between the maximum and minimum limits of size of a part, i.e., the total permissible variation of a size. In this respect, the nominal size (zero line) serves as the origin from which the deviations to either side are specified. The upper deviation is the difference between the maximum limit of size and the nominal size. Additionally, the lower deviation is the difference between the minimum limit and the nominal size. The actual size lies between the maximum and minimum limits and all dimensions that are within the tolerance zone are accepted. Tolerances on dimensions can be specified as a unilateral tolerance when the total tolerance as it relates to the basic dimension, varies in one direction such as $30^{+0}_{-0.01}$ mm. The nominal size 30 mm is allowed to change between 30 and 29.99 mm. In the bilateral tolerance, the total tolerance varies in both directions in a uniform fashion such as 30 ± 0.02 mm or $30^{+0.02}_{-0.02}$ mm. In such a case, the dimension varies from 30.02 to 29.98 mm. In the bilateral tolerance the variation can be different if it is written as $30^{+0.05}_{-0.01}$, where the dimensions vary from 30.05 to 29.90 mm.

Limit tolerance shows the actual upper and lower values rather than using bilateral or unilateral dimensions. In this case, the dimension is written as $^{30.05}_{29.90}$. Furthermore, when the nominal size falls outside the tolerance, it can be written as $30^{-0.05}_{-0.15}$, which means that the nominal size 30 mm varies between 29.85 and 29.95 mm.

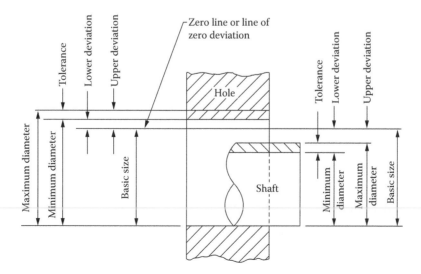

FIGURE 24.15 Principal terms of engineering assembly.

Example 3

The following limits are specified to give a clearance between a hole and shaft of 25 mm nominal diameter.

For a hole having $25^{+0.020}_{0.0}$ mm
For the shaft $25^{-0.005}_{-0.018}$ mm

Calculate

1. The basic size
2. Hole and shaft tolerance
3. Hole and shaft limits
4. Maximum and minimum clearances of the assembly

Solution

Basic size = 25 mm
Hole tolerance = 0.020 mm
Shaft tolerance = 0.013 mm
Hole high limit = 25.020 mm
Hole low limit = 25.000 mm
Shaft high limit = 24.995 mm
Shaft low limit = 24.982 mm
Maximum clearance = 25.020 − 24.982 = 0.038 mm
Minimum clearance = 25 − 24.995 = 0.005 mm

According to the ISO system of limits and fits a range of tolerance and deviations are specified, for any given size, with reference to the zero line. The position of tolerance zone with respect to the zero line is indicated by a letter symbol. Capital letters A, B, C, and D are used for holes and small letters a, b, c, and d are used for shafts (Figure 24.16). The amount of tolerance (tolerance grade) is a function of the basic size. For each of the holes, A, B, and C and shafts a, b, and c, 18 standard tolerance grades IT01, IT0 . . . IT16 are identified. Table 24.4 shows the possible tolerance grade that can be achieved with different manufacturing processes.

Table 24.5 shows that the amount of tolerance for the different grades IT01 to IT16 can be calculated on the basis of the fundamental tolerance value i which is given by

$$i = 0.45\sqrt[3]{D} + 0.001D$$

For tolerance grades IT01 and IT0, the amount of tolerance is taken as

$$IT01 = 0.3 + 0.008D$$
$$IT0 = 0.5 + 0.125D$$

where D is the dimensions (basic size) in millimeters.

The tolerance size is defined by the basic size followed by a symbol composed of a letter, representing the position of tolerance zone as shown in Figure 24.16, and a number such as 35H8.

The fundamental deviation defines the position of the tolerance zone in relation to the zero line. As can be seen in Figure 24.16, for shafts a to h the deviation is below zero line, and for j to zc it is above the zero line. For holes A to G, the lower deviation is above zero line, and for J to ZC it is below the zero line. Upper deviation of shafts is denoted by es and lower deviation by ei. For holes,

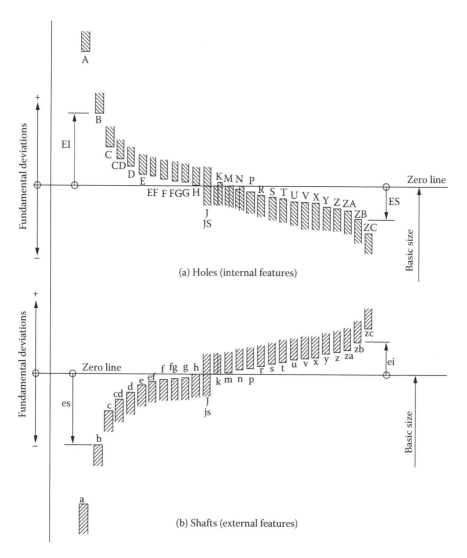

FIGURE 24.16   Position of the various tolerance zones for a given diameter in the ISO system.

the corresponding deviations are ES and EI, respectively. For each letter symbol, the magnitude and sign for one of the two deviations (upper deviation es or lower deviation ei) are given by the formulae listed in Table 24.6. It estimates the fundamental deviation corresponding to the limit, which is close to the zero line (Figure 24.17). The other deviations can be calculated using the absolute value of the standard tolerance grade IT as follows:

$$ei = es - IT$$

$$es = ei + IT$$

For holes the following algebraic relationships are used

$$EI = ES - IT$$

$$ES = EI + IT$$

TABLE 24.4
Machining Process Associated with ISO Tolerance Grade

Manufacturing Operation	IT 2	IT 3	IT 4	IT 5	IT 6	IT 7	IT 8	IT 9	IT 10	IT 11	IT 12	IT 13	IT 14	IT 15	IT 16
Lapping	▓	▓	▓												
Honing	▓	▓	▓												
Superfinishing	▓	▓	▓	▓											
Cylindrical grinding		▓	▓	▓											
Diamond turning			▓	▓	▓										
Plan grinding			▓	▓	▓										
Broaching				▓	▓	▓									
Reaming				▓	▓	▓	▓								
Boring and turning					▓	▓	▓	▓	▓	▓	▓	▓			
Sawing							▓	▓	▓	▓					
Milling							▓	▓	▓	▓					
Planing and shaping								▓	▓	▓					
Extruding							▓	▓	▓						
Cold rolling and drawing								▓	▓	▓					
Drilling								▓	▓	▓	▓				
Die casting									▓	▓	▓				
Forging										▓	▓	▓	▓		
Sand casting											▓	▓	▓	▓	
Hot rolling and flame cutting												▓	▓	▓	▓

TABLE 24.5
Tolerance for Different Grades

Tolerance Grade	Tolerance (μm)
IT 01	—
IT 0	—
IT 1	$1.2i$
IT 2	$2i$
IT 3	$3i$
IT 4	$5i$
IT 5	$7i$
IT 6	$10i$
IT 7	$16i$
IT 8	$25i$
IT 9	$40i$
IT 10	$64i$
IT 11	$100i$
IT 12	$160i$
IT 13	$250i$
IT 14	$400i$
IT 15	$640i$
IT 16	$1000i$

TABLE 24.6
Formulae for Fundamental Deviations for Shafts of Sizes up to 500 mm

Upper Deviation (es) (µm)		Lower Deviation (ei) (µm)	
Shaft Designation	(*D* in mm)	Shaft Designation	(*D* in mm)
a	$=-(265 + 1.3D)$ for $D \le 120$	j	No formula
	$=-3.5D$ for $D > 120$		
b	$=-(140 + 0.85D)$ for $D \le 160$	js	$= \text{IT}/2$
	$=-1.8D$ for $D > 160$		
c	$=-52D^{0.2}$ for $D \le 40$	k4 to k7	
	$=-(9\ 5 + 0.8D)\ D > 40$		$= +0.6\sqrt[3]{D}$
cd	Geometric mean of values for c, d	k for grade ≤ 3 and ≥ 7	$=0$
d	$=-16D^{0.44}$	m	$=+(\text{IT7}-\text{IT6})$
		n	$=+5D^{0.34}$
		p	$=+\text{IT7}+ 0$ to 5
		r	Geometric mean of values for p,s
e	$=-11D^{0.41}$	s	$=+\text{IT8}+ 1$ to 4 for $D \le 50$
			$=+\text{IT7}$ to $+0.4D$ for $D > 50$
ef	Geometric mean of values for e, f	t	$=+\text{IT7} + 0.63D$
f	$=-5.5D^{0.41}$	u	$=+\text{IT7} + D$
fg	Geometric mean of values for f, g	v	$=+\text{IT7} + 1.25D$
g	$=-2.5D^{0.34}$	x	$=+\text{IT7} + 1.6D$
		y	$=+\text{IT7} + 2D$
h	$=0$	z	$=+\text{IT7} + 2.5D$
		za	$=+\text{IT8} + 3+ 3.15D$
		zb	$=+\text{IT9} + 4D$
		zc	$=+\text{IT10} + 5D$

Note: For s: the two deviations are $= \pm\text{IT}/2$.

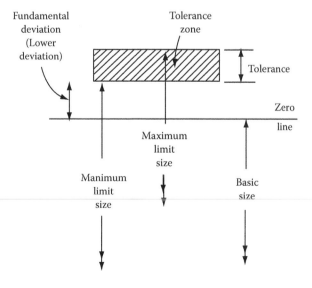

FIGURE 24.17 Fundamental deviation.

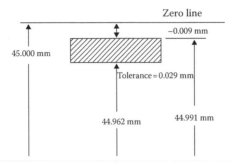

FIGURE 24.18 Deposition of tolerance and deviation for the shaft 45g7.

Example 4

Calculate the amount of tolerance and deviation for the shaft 45g7.

Solution
This given shaft has a nominal size 45 mm, positional tolerance described by the letter g, and a tolerance grade IT7. From Table 24.5 and for IT7, the amount of tolerance is 16i, where i is the fundamental tolerance;

$$i = 0.45\sqrt[3]{D} + 0.001D = 1.8004 \quad \text{m}$$

The amount of tolerance is 16i = 16 × 1.8 = 29 µm
     According to Table 24.6, the upper deviation can be calculated as follows:

$$\text{Upper deviation} = -2.5D^{0.34} = -9 \text{ µm}$$

Figure 24.18 shows the tolerance position and magnitude for 45g7 shaft.

24.5.3 FITS

A large number of combinations of holes and shafts having different deviations and tolerances are available. Manufacturing applications may require different types of fit ranging from tight fit to light running clearance one used for bearings (Figure 24.19).

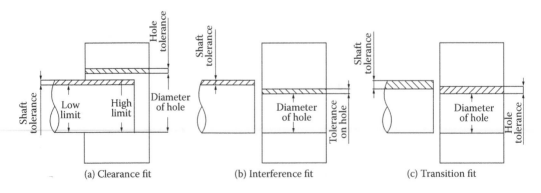

FIGURE 24.19 Typical fits in engineering assembly.

1. Clearance fit: The largest permitted shaft diameter is smaller than the diameter of the smallest hole.
2. Interference fit: The minimum permitted diameter of the shaft is larger than the maximum diameter of the hole.
3. Transition fit: The diameter of the largest hole is greater than that of the smallest shaft, but the smallest hole is smaller than the largest shaft.

Fits between mating parts may be obtained using one of the following standard systems (Figure 24.20).

- Shaft basis system: For a given nominal size a series of fits is arranged using a standard shaft and varying the limits on the hole.
- Hole basis system: For a given nominal size, the limits on the hole are kept constant, and a series of fits are obtained by only varying the limits on the shaft.

The hole basis system is commonly used because holes are more difficult to produce to a given size and are more difficult to inspect. The H series (lower limit at nominal, 0.00) is typically used and standard tooling (e.g., H7 reamers) and gauges are common for this standard. A fit is usually indicated by the basic size common to both components followed by a symbol corresponding to each component; the hole is quoted first such as $38H_8f_7$. Typical fits that can be obtained in a hole based system are depicted in Table 24.7.

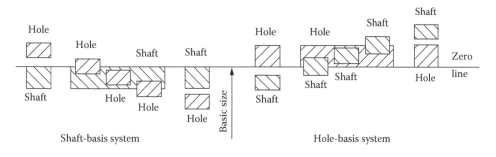

FIGURE 24.20   Different fits using the shaft and hole bases systems.

TABLE 24.7
Typical Fits in a Hole-Based System

Type of Fit	Shaft Tolerance	Hole Tolerance			
		H7	H8	H9	H11
Clearance	c11				▓
	d10			▓	
	e9			▓	
	f7		▓		
	g6	▓			
	h6	▓			
Transition	k6	▓			
	n6	▓			
Interference	p6	▓			
	s6	▓			

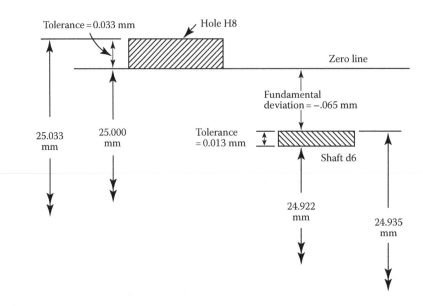

FIGURE 24.21 Disposition of tolerances and deviations.

Example 5

Calculate the limits of tolerance and allowance for a 25 mm shaft and hole pair designated by H8d6.

Solution

For the nominal size 25 mm, the fundamental tolerance i can be calculated from

$$i = 0.45\sqrt[3]{25} + 0.001 \times 25 = 1.33 \text{ m}$$

For IT8 (Table 24.5), the amount of tolerance 25 $i = 1.339 \times 25 = 33$ µm
For the hole H the fundamental deviation = 0 µm
Hole limits are 25.000 and 25.033 mm
For shaft quality IT6, the amount of tolerance 10 $i = 1.33 \times 10 = 13$ µm
The fundamental deviation = $-16 \times D^{0.44} = -65$ µm
The shaft limits are

25.000 – 0.065 = 24.935 mm
25.000 – (0.013 + 0.065) = 24.922 mm

The disposition of tolerance and deviations is shown in Figure 24.21.
Such a fit will lead to a maximum clearance = 25.033 – 24.922 = 0.111 mm
The minimum clearance = 25.000 – 24.935 = 0.065 mm.

24.5.4 GEOMETRIC DIMENSIONING AND TOLERANCING

Separate tolerances should be specified for geometric features in addition to linear tolerances. Dimensional tolerances represent the maximum possible variation of a form or position of feature. Geometric features of a component such as form, profile, orientation, location, runout, straightness, and flatness must be considered. It is a universal language of symbols that allow a design engineer to precisely and logically describe part features in a way they can be accurately manufactured and inspected. The purpose of these symbols is to form a common language that describes the

TABLE 24.8
Geometric Characteristic Symbols

Individual Feature	Tolerance	Characteristics	Symbol
	Form	Straightness	—
		Flatness	▱
		Roundness or circularity	○
		Cylindricity	⌭
Individual or related feature	Profile	Line	⌒
		Profile	⌓
Related feature	Orientation	Angularity	∠
		Perpendicularity	⊥
		Parallelism	∥
	Location	Concentricity	◎
		Position	⊕
	Runout	Circular runout	↗
		Total Runout	⤨
Notes	⌀	Ⓜ	Ⓛ	Ⓢ
	Diameter	Maximum metal condition	Least material condition	Regardless of feature size

size, form, orientation, and location of part features. The use of such symbols on drawings has the advantages of providing a uniform meaning, no misunderstanding, compact and quickly and easy drawn. Table 24.8 shows the various symbols used to specify the required geometric characteristics of dimensioned drawings. The principles of geometric tolerancing have shown the importance of maximum metal condition, which refers to the condition of a hole or shaft when the maximum amount of material is left on, i.e., high limit shaft and low limit hole. The least material condition is the reverse of the maximum metal condition. Regardless of feature size indicates that tolerances apply to a geometric feature for any size it may be. Geometric dimensioning and tolerancing is expressed in the feature control frame, shown in Figure 24.22, which is like a basic sentence that

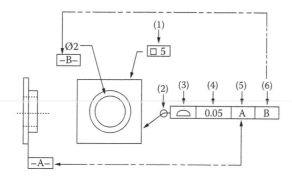

FIGURE 24.22 Geometric dimensioning and tolerancing.

can be read from left to right. For example, the feature control frame illustrated would read: The 5 mm square shape (1) is controlled with an all-around (2) profile tolerance (3) of 0.05 mm (4), in relationship to primary datum A (5) and secondary datum B (6).

24.6   MEASURING QUALITY CHARACTERISTICS

In order to measure that the quality or fitness for purpose of a component or assembly has been achieved or the design and manufacturing standards have been established measurements must be made. On the other hand, gauging is the acceptability of a given dimension that indicates whether or not the dimension lies within acceptable limits. Measurements are used for quality control purposes of some product characteristics such as length, angles, squareness, flatness, roundness, and surface roughness. Checking the part dimensions to see if it complies with the specified dimension is called inspection. In this regard, accuracy represents the degree of agreement between measured dimensions and the true value, while precision is the degree of repeatability of measurements (Figure 24.23). Measurements can be done after the part has been manufactured (post process inspection) or while the part is being produced (on-line inspection). Automated methods of inspection are currently used in industry. Direct reading instruments yield the absolute dimensions whereas a comparator compares the dimensions of the component against some known standards.

24.7   MEASURING TOOLS AND EQUIPMENT

The following factors should be considered when selecting inspection equipment.

1. Gauge capacity: The device should be at least 10 times more precise than the tolerance to be measured.
2. Linearity: Refers to the calibration accuracy of the device over its full range.
3. Repeat accuracy/precision: How repeatable is the device in taking the same reading.
4. Stability: How well the device keeps its calibration over a period of time.
5. Magnification: The amplification of the output portion of the device over the actual input dimension.
6. Resolution (sensitivity): The minimum unit of scale or dimension the device can detect.
7. Type of measurement information desired.
8. The range of size the device can handle.
9. The environment.
10. The cost of device.
11. Speed of measurement.

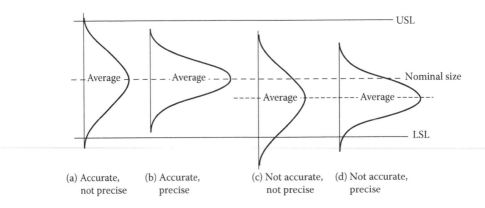

(a) Accurate, not precise   (b) Accurate, precise   (c) Not accurate, not precise   (d) Not accurate, precise

FIGURE 24.23   Dispersion of measured quality characteristics.

24.7.1 LIMIT GAUGES

The dimension of manufactured components can be measured at the required degree of accuracy using measuring instruments. This system can be used for precise work and where the actual dimensions are needed for the purpose of acceptance. Such a method is costly and time-consuming. As an alternative, properly designed limit gauge, based on Taylor's principle, can check both linear and geometric features simultaneously by a semiskilled operator. Taylor's principle, founded in 1905, is the key to the design of limit gauges. It states that

- The GO gauge checks the maximum metal condition and should check as many dimensions as possible.
- The NOT GO gauge checks the least metal condition and should only check one dimension.

Thus a separate NOT GO gauge is required for each individual dimension. In order to show this principle, consider the rectangular hole shown in Figure 24.24. The GO gauge is used to ensure that the maximum material condition is not exceeded and that metal does not encroach into the minimum allowable hole space. It should be made to the maximum allowable metal condition dimensions, due allowance being made for wear. If the NOT GO gauge was made to gauge both dimensions of least metal conditions (maximum hole size) a condition would arise where the width of the hole is within specified limit, but the length is oversize (Figure 24.25). Such a gauge will not enter the hole, and therefore the work is accepted although the length is outside the specified limits. If separate NOT GO gauges are used for the two dimensions, the width gauge would have accepted the work, but it would have been rejected by a separate length gauge.

In any measuring situation, the accuracy of the measuring equipment needs to be 10 times as good as the tolerance on the workpiece it is designed to measure. This means that the tolerance on a gauge is approximately 10% of the tolerance on the work to the nearest 0.001 mm unit. Proper positioning of the gauge tolerance relative to work tolerance limits does not allow the gauge to accept defective work. Figures 24.26 and 24.27 show the deposition of tolerance for both plug and ring gauges respectively. The following types of limit gauges are widely used.

24.7.1.1 Plug Gauges

Figure 24.28a shows the solid and renewable double ended plug gauges used for checking holes of small diameters of 1 to 6 mm. Plug gauges may be double ended with cylindrical pins 15–18 mm long that are secured to a plastic handle. For gauging larger diameters from 3 to 50 mm, double and

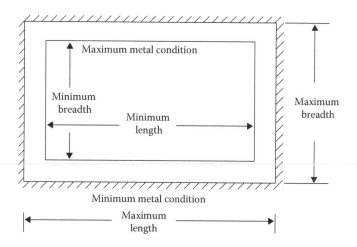

FIGURE 24.24 Tolerance zone on a rectangular hole.

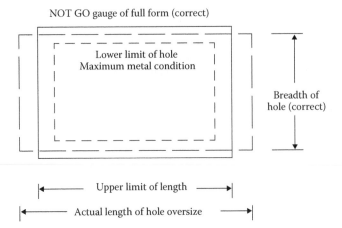

FIGURE 24.25   Rectangular hole oversize in length. A full form NOT GO will not reject such a hole.

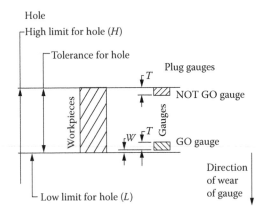

FIGURE 24.26   Deposition of tolerances in plain plug gauges.

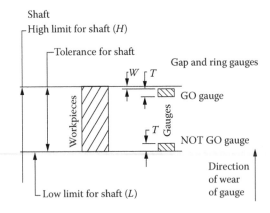

FIGURE 24.27   Deposition of tolerances in plain gap and ring gauges.

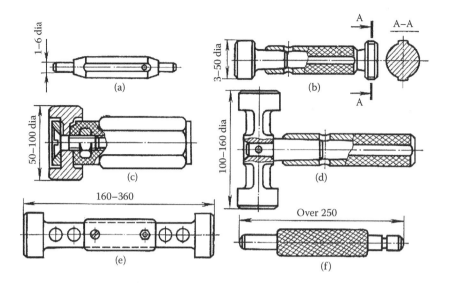

FIGURE 24.28 Plug gauges. (From Vladimirov, V., *Measuring and Cutting Tools*, Mir Publishers, Moscow, 1978. With permission.)

single ended plug gauges are used. In plugs gauges the NOT GO has a working length 1/3 that of GO plug working length. The GO element is a full form plug gauge, while the NOT GO element is a bar type of form relived (Figure 24.28b). For checking holes from 50 to 100 mm in diameter single end GO and NOT GO plug gauges that are lighter in weight are used (Figure 24.28c). Segmental type plug gauges are used for checking holes 100–160 mm in diameter (Figure 24.28d). A tapered steel shank with steel or plastic handle is pressed into the plug. Segmental type GO and NOT GO with plastic grip pads is shown in Figure 24.28e, which is used for holes from 160 to 360 mm in diameter. Internal caliper gauges with spherical precise surfaces on the ends (Figure 24.28f) are used for diameters ranging from 250 to 1000 mm.

24.7.1.2 Snap Gauges

Gauges for shafts and external threads are either of ring type or gap gauge form. Ring gauges are of nonadjustable type and separate gauges are used for GO and NOT GO gauging. Gap gauges (Figure 24.29a–d) are known as snap gauges and gauge calipers. Gap gauges for plain work may be of a solid or adjustable type. Sheet-steel double end gauges, shown in Figure 24.29a, are used for checking sizes from 1 to 70 mm. The GO gauge surface is made longer than the NOT GO and both surfaces are divided by a groove (Figure 24.29b). The modulated gap gauges (Figure 24.29c) are provided with stiffening ribs and wider measuring faces that makes them more rigid and wear resistant during checking diameters up to 170 mm. Adjustable gap gauges are used for diameters ranging from 100 to 325 mm (Figure 24.29d). In order to reduce their weight they are made with holes and moreover can be reset to compensate for wear and suit each new application. However, they are expensive and heavier than plug gauges.

24.7.2 Dimensional Measurements

Several methods and tools are used to measure dimensions. These include the following.

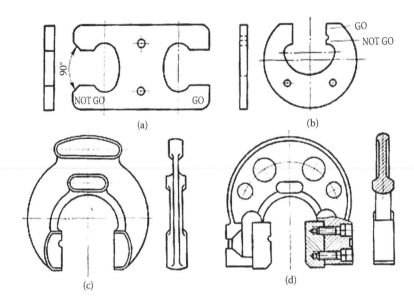

FIGURE 24.29 Gap gauges. (From Vladimirov, V., *Measuring and Cutting Tools*, Mir Publishers, Moscow, 1978. With permission.)

- Rule: A rule is used to measure dimensions without any limits or covering note as to a particular fit. The degree of accuracy by which the part is made when measured by a rule depends on the quality of the rule and the skill of the user. Rules have markings of millimeters on one face and English units on the other (Figure 24.30).
- Calipers: Calipers, shown in Figure 24.31, are for external or internal use. The inside calipers are used to measure the internal size of an object. Some calipers require manual adjustment prior to fitting, while other calipers have an adjusting screw that permits careful adjustment without its removal from the workpiece. Outside calipers are used to measure the external size of a large diameter pipe.
- Vernier caliper: A Vernier scale is the name given to any scale making difference between two scales for obtaining small difference. As shown in Figure 24.32, Vernier calipers measure internal and external dimensions using the jaws 1, 2. Depth measurements can be made by probe 3 attached to the movable head and slides along the center of the body. This probe is slender and measures deep grooves that are difficult to measure by other measuring tools. The Vernier scales 4 and 5 may include both metric and English units. The retainer 8 is used to block the movable part to allow the easy reading of a measurement. Depending

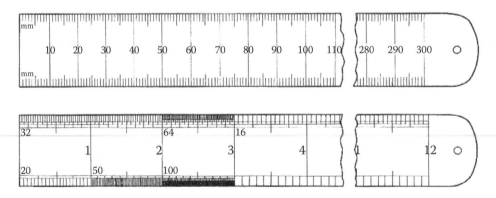

FIGURE 24.30 Rule.

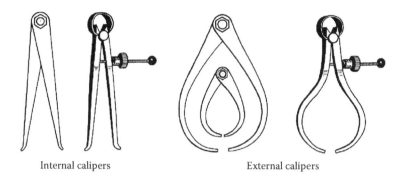

Internal calipers   External calipers

FIGURE 24.31 Different calipers.

on the way the top and Vernier scales are related, different sensitivities are available. The Vernier scale shown in Figure 24.33a considers 12 mm from the top scale that have been divided into 25 divisions on the sliding scale. Hence one division on the sliding scale represents 12/25 = 0.48 mm. Because the top division is 0.5 mm, the sensitivity of the instrument and the accuracy of readings equals to 0.5–0.4 8 = 0.02 mm. The reading of the Vernier shown in Figure 24.33b will be =28.5 + 5 × 0.02 = 28.60 mm. The Vernier scale shown in Figure 24.34a has 1 inch on the top scale, divided into 10 parts. Each part is further divided into 4, giving 1/40 or 0.025 inch. On the lower scale 6/10 is divided into 25 parts.

$$\text{Then each part will give } \frac{6}{10} \times \frac{1}{25} = \frac{24}{1000} \text{ inch}$$

$$\text{The instrument sensitivity} = \frac{25}{1000} - \frac{24}{1000} = \frac{1}{1000} \text{ inch}$$

Accordingly, the reading of the Vernier caliper, shown in Figure 24.34 is

$$\text{Reading} = 5 + 0.1 + \frac{1}{40} + 11x \frac{1}{1000} = 5.136 \text{ inch}$$

- Dial caliper: In this instrument a small gear rack drives a pointer on a circular dial. The pointer rotates once every inch, tenth of an inch, or 1 millimeter, allowing for a direct reading without reading a Vernier scale. However, one still needs to add the basic inches or tens of millimeters read from the slide of the caliper. The slide of a dial caliper can usually be

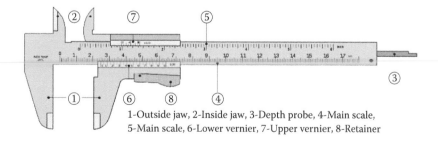

1-Outside jaw, 2-Inside jaw, 3-Depth probe, 4-Main scale,
5-Main scale, 6-Lower vernier, 7-Upper vernier, 8-Retainer

FIGURE 24.32 Parts of a Vernier caliper.

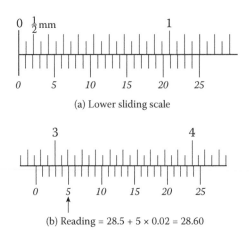

(a) Lower sliding scale

(b) Reading = 28.5 + 5 × 0.02 = 28.60

FIGURE 24.33   Metric calipers.

locked at any setting using a small lever or screw; this allows simple GO/NOT GO checks of part sizes.

- Digital caliper: This is a refined popular caliper that replaces the analog dial with an electronic digital display (Figure 24.35). Some digital calipers can be switched between metric and English units. They also provide zero display at any point along the slide that allows for differential measurements as with the dial caliper. Digital calipers offer a serial data output to be interfaced with a personal computer. In such a case, measurements can be taken and instantly stored in a spreadsheet or a software, significantly decreasing the time taken to record a series of measurements. The advantages of digital calipers include
  - Easy reading of digital display
  - One-hand operation
  - Low power consumption
  - Positive locking screw against slide
  - Zero setting at any position
  - No Vernier scale, rack, and pinion and accordingly no backlash
- Vernier height caliper: A Vernier height gauge is a variation in which a single measuring jaw is provided to measure height from the base of the height gauge as shown in Figure 24.36. The upper and lower surfaces of the measuring jaws are parallel to the base so that it can be used for measurements over or under the surface. The measuring jaw can also be

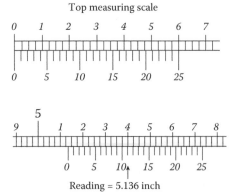

Reading = 5.136 inch

FIGURE 24.34   English calipers.

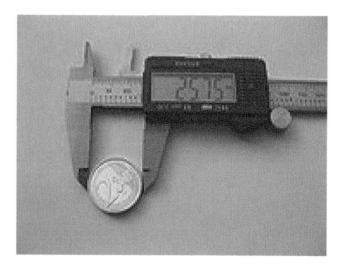

FIGURE 24.35   Digital caliper.

replaced by a scribing attachment to scribe a line at a certain height from a datum surface at the surface plate.
- External micrometers: External micrometers are end-to-end measuring devices that are used for more accurate measurements; the auxiliary scale is made on the thimble that rotate about the main scale as shown in Figure 24.37. In case of metric measurements, the smallest division on the main scale is 0.5 mm, while the thimble is divided into 50 divisions. The least count is therefore 0.5/50 = 0.01 mm. For English micrometers the screw on the spindle and on the barrel is 40 threads per inch and the thimble is marked round into 25 equal divisions. The least division is therefore = (1/40)(1/25) = 0.001 inch. The accuracy of measurement using a micrometer depends on the skill of the user as well as the accuracy of the micrometer screw. Figure 24.38 shows the metric micrometer at the left, the reading

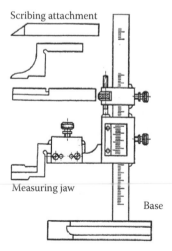

FIGURE 24.36   Vernier height gauges. (From Vladimirov, V., *Measuring and Cutting Tools*, Mir Publishers, Moscow, 1978. With permission.)

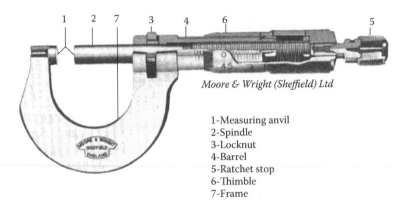

Moore & Wright (Sheffield) Ltd

1-Measuring anvil
2-Spindle
3-Locknut
4-Barrel
5-Ratchet stop
6-Thimble
7-Frame

FIGURE 24.37   External micrometer.

is 11 + 34 × 1/100 = 11.34 mm. For the English micrometer, shown in Figure 24.38b, the reading is constructed as follows:

$$\text{Reading} = 0.2 + 2x\frac{1}{40} + 22x\frac{1}{1000} = 0.272 \text{ inch}$$

During measurements, the ratchet must be used so that readings are obtained under the same pressure and therefore will be consistent. Digital micrometers (Figure 24.39) offer easy reading with both English and metric measurements.

- Depth micrometers: The depth micrometer, shown in Figure 24.40, consists of a micrometer measuring head 1 together with a number of extension rods 3. The rods are marked with their representative measuring range and are usually perpendicular to the base 2 at any setting. The scale of the depth micrometer gives the reverse readings compared to the external and internal micrometers.
- Internal micrometers: Used for the accurate measurement of the bore of holes. It has a similar head to the ordinary external micrometer and a set of extension rods to increase their range. The internal micrometer, shown in Figure 24.41, has a typical range between 50 and 210 mm with a scale range of 20 mm. It is difficult to adjust internal micrometers inside holes and slots. Moreover, they cannot be used for measuring small holes.

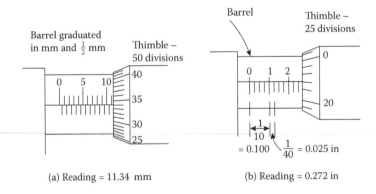

(a) Reading = 11.34 mm

(b) Reading = 0.272 in

FIGURE 24.38   Reading external micrometers.

FIGURE 24.39 Mitutoyo digital micrometer.

- Slip gauges: Slip gauges are gauge blocks having the thickness representing the dimension of the block and cross section of 32 mm × 9 mm. They are used for calibration of other measuring instruments such as Vernier scales and micrometers. The measuring surfaces are hardened and finished to a high degree with respect to flatness and accuracy. Slip gauges are available with grades of accuracies depending on the accuracy of measurement required. A typical set of 88 pieces is shown in Table 24.9. Blocks of different thickness can be wrung together and the overall length of any number of blocks represents the sum of their individual lengths. In order to assemble a dimension of 82.425 mm the following combination is used:

1.005

1.420

+80.000

82.425

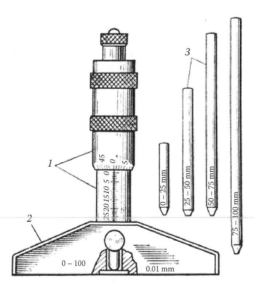

1-Micrometer head
2-Base
3-Extension rods

FIGURE 24.40 Depth micrometer. (From Vladimirov, V., *Measuring and Cutting Tools*, Mir Publishers, Moscow, 1978. With permission.)

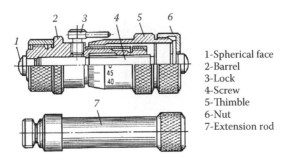

1-Spherical face
2-Barrel
3-Lock
4-Screw
5-Thimble
6-Nut
7-Extension rod

FIGURE 24.41   Internal micrometer. (From Vladimirov, V., *Measuring and Cutting Tools*, Mir Publishers, Moscow, 1978. With permission.)

- Comparators: Comparators are generally used for quick checking of the linear dimensions. They are set for a given dimension and provide the deviation from the set value and therefore cannot be used as absolute measuring devices. According to the principle used for obtaining a suitable degree of magnification of the indicating device relative to change in measured dimensions, mechanical, optical, electrical, and pneumatic comparators are available.
  a. Mechanical comparators employ mechanical methods for magnifying the small movement of the measuring stylus, which is the difference between the standard and the actual dimensions being checked. They have a robust design and do not depend on the use of a power supply. However, the wear of their moving parts affects the measuring accuracy. The simplest mechanical comparator is the dial indicator, shown in Figure 24.42. It converts linear displacement into rotation and amplifies it with a gear train to increase sensitivity up to 1 μm. It is used as a comparator by mounting it on a stand at any suitable height. Some gauges have built-in electrical contacts that activate signal lights, making them suitable for GO and NOT GO gauges. The use of the dial indicator as a comparator is shown in Figure 24.43. If the length of block gauges at position a is 25.000 mm, and the reading of the dial indicator is +0.025 mm, the height of the workpiece will be 25.025 mm. The most common error in setting the measuring head of a comparator is the cosine error (Figure 24.44). It occurs when the axis of the measuring stylus is inclined to the datum surface from which the measurement is made. Depending on the angle θ, the measured length L is larger than the required distance

TABLE 24.9
Metric Slip Gauges Set

Slip Gauge Size or Range (mm)	Increment	Number of Pieces
1.005	—	1
1.001 to 1.009	0.001	9
1.010 to 1.490	0.010	49
0.500 to 9.500	0.500	19
10 to 100	10.000	10
Total number of blocks		88

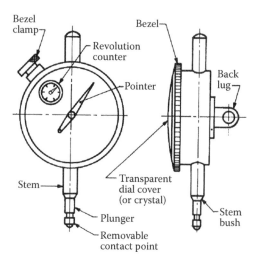

FIGURE 24.42 Dial indicator.

D because $\cos\theta = \dfrac{D}{L}$. As an example, if the angle θ is $10°$ when measuring 25 mm, the

length L will be $L = \dfrac{D}{\cos\theta} = \dfrac{25}{\cos 10} = 25.385$ mm and the error is $25.385 - 25.000 = 0.385$ mm.

b. Electrical comparators have a no moving parts and hence can retain their accuracy over long periods. Higher magnifications are possible compared to mechanical ones. Electrical comparators are available to read up to 0.0001 mm with magnification ranging between 1000 and 18,000.

c. Optical comparators have a high degree of accuracy and suffer less wear than mechanical ones. The accuracy of measurement is limited to 0.001 mm. They have their own light source that heats up the instrument and causes inaccuracies. Optical flat works on the principles of light interference and can be used to check flatness of surfaces by monitoring the fringe pattern as shown in Figure 24.45. When the monochromatic light is directed to the surface at an angle, the optical flat splits it into 2 beams appearing as light and dark beams to the naked eye. The number of fringes depends on the distance between the bottom of the optical flat and the top of the specimen. This method can be used to observe surface texture and detect surface scratches.

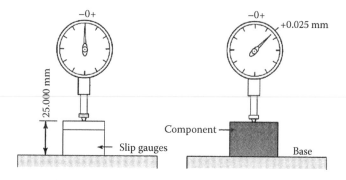

FIGURE 24.43 Principles of comparators.

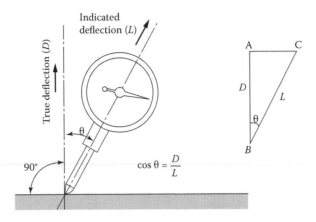

FIGURE 24.44   Cosine error.

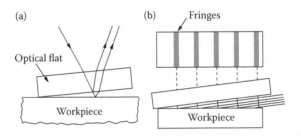

FIGURE 24.45   Use of interferometry for measuring flatness.

d. Pneumatic compotators are cheap, independent of contact pressure, free from mechanical wear, and simple to operate. Their accuracy is affected by the temperature and humidity changes, as well as the roughness of measured surfaces. They can be used for internal and external diameters, thickness measurements, and checking concentricity of circular parts and the depth of blind holes.

24.7.3 ANGULAR MEASUREMENTS

Several methods and tools of different levels of accuracy are adopted for angular measurements.

- Vernier protractor: This is the simplest instrument used for measuring angles between two faces of a component. It consists of a base plate attached to the main frame and an adjustable blade that is attached to a circular plate containing Vernier scale. The adjustable scale rotates freely about the center of the main scale engraved on the body of the instrument that can be locked in any position (Figure 24.46). The main scale on the protractor is divided up into degrees from 0 to 90° each way. The Vernier scale is divided up so that 12 of its divisions occupy the same space as 23° on the main scale. Hence one Vernier division $= \dfrac{23}{12} = 1\dfrac{11}{12}$ i.e., 1/12 degree or 5 min less than 2°. The instrument allows settings to 5 min of an angle to be obtained. However, optical bevel protractors can be used to measure angles to about 2 minutes of an angle. The same figure also illustrates a typical Vernier protractor reading.

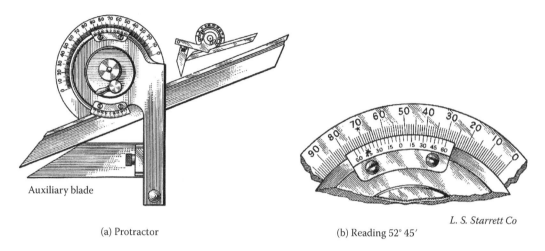

Auxiliary blade

(a) Protractor   (b) Reading 52° 45′

L. S. Starrett Co

FIGURE 24.46 Vernier protractor and reading.

- Sine bar: The sine bar uses linear measurements to define angles to a high degree of accuracy. As shown in Figure 24.47, the angle θ is given by

$$\text{sine } \theta = \frac{\text{Height of slip gauges}}{\text{Center distance of roller axes}} = \frac{H_s}{L}$$

The center distance of the contact roller axes is the nominal length L of the sine bar which is usually 250 mm. The height of the slip gauges is adjusted until the dial indicator reads zero at each end of the component. Alternatively, if the sine bar is set at a specified angle θ, the movement of the dial indicator can detect an error along the length either positive or negative. Accurate measurement of angles to 1 s can be made by angle gauge blocks. These come in sets of 16 blocks that can be assembled in the desired combinations. Rotary indexing tables having a suitable numerical control are also used to measure angles to ±0.001°.

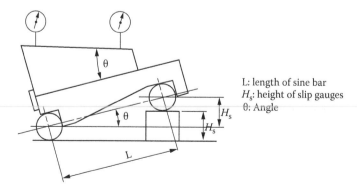

L: length of sine bar
H_s: height of slip gauges
θ: Angle

FIGURE 24.47 Sine bar.

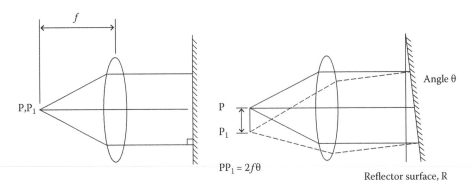

FIGURE 24.48 Measuring small angles using autocollimator.

Example 6

Calculate the setting of 250 mm sine bar to measure an angle of 45° 30′

Solution

$$\text{sine } 45.5 = \frac{H_s}{250}$$

$$H_s = 250 \times 0.713 = 178.313 \text{ mm}$$

- Autocollimator: Autocollimator measures small angular displacement from which it is possible to determine straightness, flatness, and alignment. The principle of autocollimator is shown in Figure 24.48. Accordingly, if the light source P is situated at the focus of the lens, the reflectivity surface is exactly at right angle to the path of the light, then point P_1 will coincide with P. If the reflecting surface is tilted at an angle θ, the image of point P will be displaced by an amount $PP_1 = 2f\theta$, where f is the focal length of the lens used. By measuring the distance PP_1 the angle θ can be determined. Autocollimators can measure angles up to 0.1 s. However, the range of measurement is small, and hence they are more useful when used as comparators.

24.7.4 GEOMETRICAL MEASUREMENTS

The following geometrical characteristics of manufactured products are essential to measure:

- Straightness: Straightness is measured using straight edges or with dial indicators shown in Figure 24.49. Autocollimators resembling a telescope with a light beam that bounces back from the object are also used for measuring small angular deviations on a flat surface.
- Squareness: In addition to the simple method of measuring squareness using try squares, it can be accurately evaluated using the squareness comparator shown in Figure 24.50. The dial test indicator (DTI) is set to zero with the comparator in contact with try square. When the comparator is brought into contact with the component being checked, any error is indicated as a positive or negative reading on the DTI. Hence the angular error can be calculated from

$$\tan\theta = \frac{X}{L}$$

where L is the distance between the two centers.

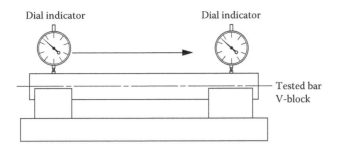

FIGURE 24.49 Measuring straightness.

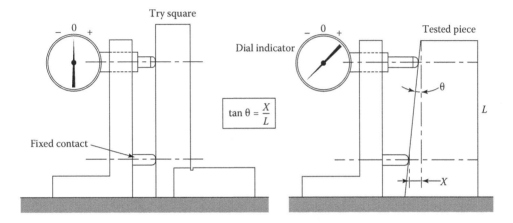

FIGURE 24.50 Measuring squareness.

- Roundness: Roundness error can be measured by putting the part on a V-block or between centers and rotating the part with a dial gauge touching the surface as shown in Figure 24.51. The maximum deviation for one revolution is called the total indicator reading (TIR). The second method of roundness measurement is done by using a platform that rotates the part against a probe as shown in Figure 24.52.

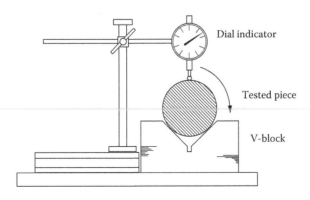

FIGURE 24.51 Measuring roundness error.

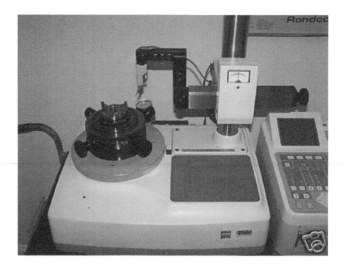

FIGURE 24.52 Zeiss Rondcom 30B roundness tester. (From Metrología Industrial, Carl Zeiss Industrielle Messtechnik GmbH. With permission.)

24.8 COORDINATE-MEASURING MACHINE

The coordinate-measuring machine (CMM), shown in Figure 24.53, is used for measuring the geometrical characteristics of an object. This machine may be manually or computer controlled. CMM machines use digital readout, air bearings, computer control, and granite table to achieve accuracies in the order of 0.0004 inch over a span of 10–30 inch. The typical CMM is composed of three axes,

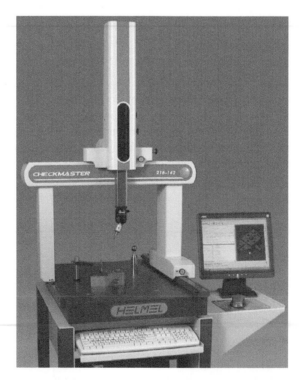

FIGURE 24.53 Coordinate measuring machine. (From Phoenix RB Model 112-102 DCC, MSI-Viking Gage, Duncan, SC, 29334. With permission.)

an X, Y, and Z. Measurements are defined by mechanical, optical, laser, or white optical probes attached to the third moving axis of this machine. Each axis has a scale system that indicates the location of that axis. The machine reads the input from the probe, as directed by the operator or the programmer. The machine then uses the X, Y, and Z coordinates of each of these points to determine size and position.

CMM is also a device used in manufacturing and assembly processes to test a part or assembly against the design intent. By precisely recording the X, Y, and Z coordinates of the target, points are generated that can then be analyzed via regression algorithms for the construction of features. CMM uses computer routines to give the best fit to feature measurements and provide the geometric tolerances such as straightness, flatness, roundness, conicity, cylindericity, squareness, parallelism, and angularity. Common applications for coordinate measuring machines include dimensional measurement, profile measurement, angularity or orientation, depth mapping, digitizing or imaging, and shaft measurement.

24.9 SURFACE MEASUREMENTS

Manufactured parts have certain surface texture that describes geometric irregularities in terms of surface roughness, waviness, lay, and flaws as described in Figure 24.54. Accordingly,

1. Surface roughness consists of the fine irregularities of the surface texture including feed marks, generated by the machining process.
2. Waviness consists of the more widely spaced component of surface texture that may occur due to the machine or part deflection, vibration, or chatter.
3. Lay is the direction of the predominant surface pattern.
4. Flaws are surface interruptions such as cracks, scratches, and ridges.

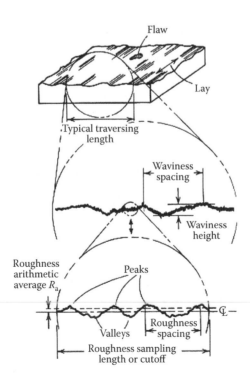

FIGURE 24.54 Surface characteristics. (From *Surface Roughness, Waviness, and Lay*, ANSI/ASME B 46.1-1985, American Society of Mechanical Engineers. With permission.)

Surface roughness can be described using the following measures:

1. The arithmetic average (R_a or CLA)

$$R_a = \frac{1}{L} \int_{x=0}^{x=L} |y|\, dx$$

   where L is the sampling length and y is ordinate of the profile from the centerline shown in Figure 24.55a.

2. The root-mean-square roughness (R_q)

$$R_q = \left[\frac{1}{L} \int_{x=0}^{x=L} y^2\, dx \right]^{1/2}$$

   This can be approximated by the following equation (Figure 24.55b).

$$R_q = \sqrt{\frac{\left(y_1 - Y_M\right)^2 + \left(y_2 - Y_M\right)^2 + \ldots \ldots \left(y_N - Y_M\right)^2}{N}}$$

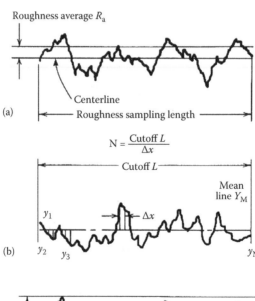

$$N = \frac{\text{Cutoff } L}{\Delta x}$$

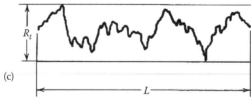

FIGURE 24.55   Commonly used surface roughness symbols: (a) average roughness R_a, (b) root-mean-square roughness (R_q), (c) maximum peak-to-valley roughness height (R_t or R_{max}). (From *Surface Texture–Surface Roughness, Waviness, and Lay*, ANSI/ASME B 46.1-1985. With permission.)

3. Maximum peak-to-valley roughness (R_t or R_{max}): The distance between two lines parallel to the mean line that contacts the extreme upper and lower points on the profile within the roughness sampling length (Figure 24.55c).

Enhanced surface texture specifications are essential in order to improve fatigue strength, corrosion resistance, appearance, and sealing. Typical applications of such surfaces include antifriction and journal bearings, food preparation devices, parts operating at corrosive environment, and sealing surfaces.

The quality of the surface finish affects the functional properties of the manufactured parts as follows.

1. Wear resistance: Larger macro irregularities result in nonuniform wear of different sections of the surface where the projected areas of the surface are worn first. In the case of surface waviness, surface crests are worn out first. Similarly, surface ridges and microirregularities are subjected to elastic deformation and may be crushed or sheared by the forces between the sliding parts.
2. Fatigue strength: Metal fatigue takes place in the areas of the deepest scratches and undercuts caused by the manufactured operation. The valleys between the ridges of the machined surface may become the focus of the concentration of internal stresses. Cracks and microcracks may also enhance the failure of the manufactured parts.
3. Corrosion resistance: The resistance of the surface against the corrosive action of a liquid, gas, water, and acids depends on its surface finish. The higher the quality of surface finish, the less the area of contact with the corrosive medium and the better the corrosion resistance. The corrosive action acts more intensively on the surface valleys between the ridges of microirregularities. The deeper the valleys, the more destructive will be the corrosive action that will be directed toward the depth of the metal.
4. Strength of interference: The strength of an interference fit between two mating parts depends on the height of microirregularities left after the machining process.

Figure 24.56 shows the American National Standard Institute (*ANSI*) Y14.36 (1978) standard symbols used for describing part drawings or specifications. Table 24.10 shows the symbols used to

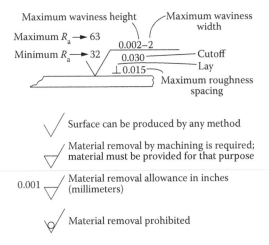

FIGURE 24.56 Surface texture symbols for drawings or specifications. (From *Surface Texture Symbols*, ANSI Y14.36 1978, American Society of Mechanical Engineers.)

TABLE 24.10
Symbols Used to Define Lay and Its Direction

Symbol	Meaning	Example	Operation
—	Lay approximately parallel to the line representing the surface to which the symbol is applied		Shaping Vertical milling
⊥	Lay perpendicular to the line representing the surface to which the symbol is applied		Horizontal milling
X	Lay angular in both directions to the line representing the surface to which the symbol is applied		Honing
M	Lay multidirectional		Grinding
C	Lay approximately circular relative to the center to which the symbol is applied		Face turning
R	Lay approximately radial relative to the center to which the symbol is applied		Lapping
P	Lay particulate, nondirectional or protuberant		ECM, EDM, LBM

define surface lay and its direction. Accordingly, a variety of lays can be machined that ranges from parallel, perpendicular, angular, circular, multidirectional, and radial ones. The same table also suggests a typical machining process for each produced lay. Figure 24.57 shows the surface roughness produced by common manufacturing methods.

Surface blocks, shown in Figure 24.58, are specimens having surfaces of a known roughness value R_a that represents a particular machining operation or other process. These specimens are used to give guidance on the feel and appearance of the particular machining process and roughness grade. The set consists of 30 comparison specimens, covering six commonly used machining methods (6 turned, 6 end-milled, 6 horizontally milled, 6 surface-ground, 3 lapped, and 3 reamed/drilled). A label gives the R_a values of each specimen in both metric and imperial units. The device shown in Figure 24.59 is used to measure surface roughness using a diamond stylus that is moved at a constant rate across the surface, perpendicular to the lay pattern. The rise and fall of the stylus is electronically detected, amplified, and recorded on a chart. The signal can also be processed electronically to produce the roughness values R_a or R_t on a scale. Microtopographers make a series of parallel traces on the surface to provide two-dimensional profile maps for the manufactured surfaces. The resolution of these devices depends on the diameter of the tip of the stylus. Scanning electron microscopes (SEM) are also used when the geometric features begin to approach the magnitude of the tip of the stylus. Laser-based instruments are currently used to measure surface roughness, thus providing the advantage of not contacting the surface to obtain a reading. The laser method of roughness measurements is fast and can measure roughness of small specimens.

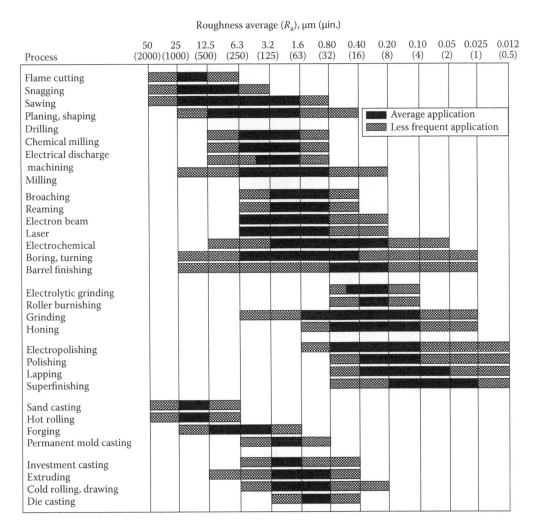

FIGURE 24.57 Surface roughness produced by common production methods: 1, surface texture. (From *Surface Roughness, Waviness, and Lay*, ANSI/ASME B 46.1-1985, American Society of Mechanical Engineers. With permission.)

24.10 NONDESTRUCTIVE TESTING AND INSPECTION

Nondestructive testing (NDT), also referred to as nondestructive inspection (NDI), is a family of inspection methods that provides information regarding the condition of the inspected components without destroying them. It is a major tool of quality control that reveals the presence of flaws and is integrated in quality programs of the aerospace, automotive, defense, pipe line, power generation, preventative maintenance, pulp and paper, refinery, and shipbuilding industries. Nondestructive evaluation (NDE) is used to describe measurements that are more quantitative in nature. It locates a defect and measures its size, shape, and orientation. NDE may be used to determine material properties, such as fracture toughness, formability, and other physical characteristics. NDT methods are described in section 21.8.1.

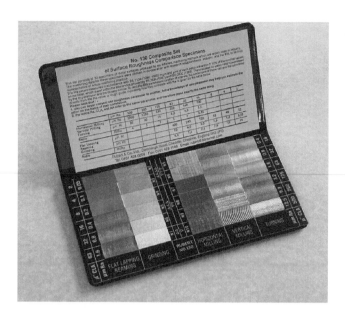

FIGURE 24.58 A set of 30 surface blocks. (From Rubert & Co., Acru Works, Cheadle, Cheshire, UK. With permission.)

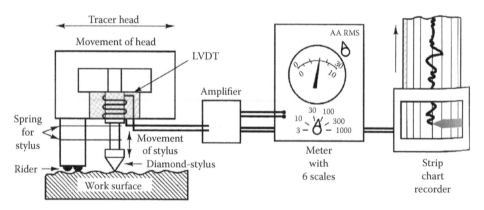

FIGURE 24.59 Measuring surface roughness. (From Degarmo, E. P., Black, J. T., and Kohser, R. A., *Materials and Processes in Manufacturing*, 8th ed. Prentice-Hall, Upper Saddle River, NJ, 1997.)

24.11 DESTRUCTIVE TESTING

Destructive testing is carried out to the specimen's failure, in order to understand a specimen's structural performance or material behavior under different loads. These tests are generally much easier to conduct, yield more information, and are easier to interpret than nondestructive testing. Destructive testing is most suitable and economic for mass production. In such a case, the cost of destroying a small number of specimens is negligible. Typical destructive tests include stress, crash, hardness, and metallographic tests.

24.12 REVIEW QUESTIONS

1. What is the difference between quality control and quality assurance?
2. What is the difference between tolerance and allowance?
3. Explain the difference between accuracy and precision.

4. What factors should be considered in selecting measuring equipment?
5. Explain how squareness, flatness, and straightness are measured.
6. Compare the use of the Vernier caliper and the micrometer.
7. What is the coordinate-measuring machine?
8. Explain what is meant by TQM and ISO 9000.
9. What are the different types of comparators?
10. Compare the use of direct measurement tools and comparators.
11. Explain Taylor's principle of designing gauge.
12. Draw the tolerance for GO and NOT GO gauges.
13. Compare hole and shaft basis systems.
14. What is meant by geometrical tolerances?
15. Explain how surface roughness affects the product characteristics.
16. Show how surface roughness is measured.
17. Compare average roughness and maximum peak-to-valley roughness.
18. Show by neat sketches the different types of gap and block gauges.
19. Explain how interferometry method is used for measuring flatness.
20. What are the advantages of nondestructive testing?
21. Explain what is meant by process capability.
22. What are the common reasons for sampling inspection rather than 100% inspection?
23. What do you know about the term cosine error? Give an example.

24.13 PROBLEMS

1. A control chart for $\bar{X}$ and $\bar{R}$ is used to control an important part dimension. The subgroup size is 5. The values of $\bar{X}$ and $\bar{R}$ are computed for each subgroup and the values of $\bullet$ $\bar{X}$ and $\bullet$ $\bar{R}$ after 25 subgroup are 614.8 mm and 120.0 mm, respectively. Determine the control limits for $\bar{X}$.
2. During a quality control inspection of manufactured parts, the following data are available: Number of subgroups = 25, subgroup size = 25, $\bar{\bar{X}}$ = 394.536 mm, and $\bar{R}$ = 21.52 mm
   • Determine the control limits for $\bar{X}$ and $\bar{R}$.
   • If the specification limits for each item are 400 ± 30 mm, compare the process capability with the specifications.
3. For the hole basis system of 38 mm diameter, specify the tolerance for shaft and hole for a close running fit.
4. Specify the type of fit for 25H7k6.
5. Choose the suitable block gauges to assemble the following dimensions: (a) 62.31 mm and (b) 36.685 mm.
6. Using the set of block gauges shown in Table 24.9, make up a set of slip gauges to check a gauge measuring 2.1758 inch.
7. Calculate the setting of a 200 mm sine bar to measure an angle of 36° 45'.
8. Calculate the setting of a sine bar 250 mm to check the angle of a taper 1:20.
9. Draw the Vernier reading for the following values: (a) 22.32 mm and (b) 1.859 inch.
10. Draw the external micrometer reading for the following values: (a) 7.79 mm and (b) 0.472 inch.

BIBLIOGRAPHY

Besterfield, D. H. 2008. *Quality Control*. Upper Saddle River, NJ: Prentice-Hall.
Brooker, K. 1984. British Standard Institution in Association with Hutchinson, *Manual of British Standards in Engineering Metrology*. London: Hutchinson.

Chapman, W. A. J. 1983. *Workshop Technology*, Part 1, 5th ed. London: Edward Arnold.

DeGarmo, E. P., Black, J. T., and Kohser, R. A. 1997. *Materials and Processes in Manufacturing,* 8th ed. Upper Saddle River, NJ: Prentice-Hall.

Galyer, J., and Shotbolt, C. 1981. *Metrology for Engineers*. London: Cassell.

Haslehurst, M. 1981. *Manufacturing Technology*, 3rd ed. London: Hodder and Stoughton.

Jain, R. K. 2002. *Engineering Metrology*. Delhi, India: Khanna Publishers.

Kalpakjian, S., and Schmid, S. 2005. *Manufacturing Processes for Engineering Materials*, 5th ed. Reading, MA: Addison-Wesley.

Prichard, R. T. 1976. *General Course Workshop Processes and Materials*. London: Hodder and Stoughton.

Rao, P. N. 2004. *Manufacturing Technology: Metal Cutting and Machine Tools*. New Delhi, India: Tata McGraw-Hill.

Timings, R. L. 1998. *Manufacturing Technology*, vol. 1. Reading, MA: Addison-Wesley.

Vladimirov, V. 1978. *Measuring and Cutting Tools*. Moscow: Mir Publishers.

25 Automation in Manufacturing Technology

25.1 INTRODUCTION

Since the early days of the eighteenth century, the first industrial revolution had been started when attempts were made to substitute the muscle power by mechanical energy. Machine tools such as boring machines, lathes, drill presses, copying lathes, turret lathes, and milling machines were introduced for the production of goods. Geared and automatic lathes were then introduced in the 1900s. Mass production techniques and mechanized transfer machines were developed between 1920 and 1940. These systems had fixed mechanisms and were designed to produce specific products. Such developments were best represented in the automobile industry, characterized by high production rates at low cost. Since that time, the productivity, which is defined as the use of all resources such as materials, energy, capital, labor, and technology became a major concern. Production may be defined as the output per labor hour, which is basically a measure of operating efficiency.

The world now is passing through the second industrial revolution with the fantastic advances occurring continuously in the fields of electronics and computer technology. The computer is substituting for the human brain in controlling machines and industrial processes. A major breakthrough in automation was the invention of the first digital electronic computer (1943), followed by the first prototype of a numerical controlled machine tool (1952). Since this historic development, rapid progress has been made in automating most aspects of manufacturing. This includes the introduction of computers to enhance automation using computerized numerical control (CNC), adaptive control (AC), industrial robots, computer-integrated manufacturing systems (CIMSs), and computer-aided design, engineering, and manufacturing (CAD/CAE/CAM) (Figure 25.1).

This chapter emphasizes the importance of the flexibility in machines, equipment, tooling, and production operations in order to be able to respond to market and ensure on-time delivery of high-quality products to attain customer's satisfaction. Important developments during the past three decades had a major impact on modern manufacturing; these include group technology (GT), cellular manufacturing, flexible manufacturing systems (FMS), and just-in-time (JIT) production.

25.2 MECHANIZATION VERSUS AUTOMATION

The definitions of mechanization and automation are based on the order of automation illustrated in Figure 25.1. Whenever a machine replaces a human function, it is considered to have taken an order level of automation. Therefore the manual (zero automation level A0), in which no human attribute was mechanized, covers the Stone Age and Iron Age. The first Industrial Revolution was tied to the development of powered machine tools, dating from 1775 (level of automation A1) and reached its peak in the 1940s (level of automation A2), where most manufacturing operations were carried out on traditional machinery such as lathes, milling machines, boring machines, and planers, all of which required skilled operators. In A2 level of automation the machine tools became single-cycle, self-feeding machines of the mass production era, displaying dexterity as they stop automatically. They still exist today in great numbers in many factories and industrial activities. Each time a different product was manufactured, the machine had to be retooled. Furthermore, new products with complex shapes required tedious work from the operator to set the proper processing parameters. Mechanization refers to A1 and A2 levels of automation, including semiautomatic machines (Figure

Development	Date	Level of Automation	
Stone and Iron Age	<1775	Ao	Muscle Power
First Industrial Revolution	(1775)	A1	Mechanization
Wilkenson boring mill	1775	A1	Mechanization
Maudsley's lathe	1794	A1	Mechanization
Whitney's milling machine	1818	A1	Mechanization
Nasymth's drilling press	1840	A1	Mechanization
First powered shaper	1851	A1	Mechanization
Whitney's lathe	1865	A1	Mechanization
Self-feeding machine tools, single cycle, mass production lathes, planers, drills, boring machines, millers, grinders. Copy and turret lathes	1880	A2	Mechanization
Gear shapers	1910	A2	Mechanization

NB: Automation levels A2 through A6 are still used in today's factories.

Level	Category	Date	Development
A3	Automation	1918	Automatic machine tools
A3	Automation	1926	Transfer machines
A3	Automation	(1943)	Second Industrial Revolution
A3	Automation	1943	Term of automation. Invention of first digital computer
A3	Automation	1952	First NC machine
A3	Automation	1960	First robot
A4	True Automation	1970	FMC, FMS, GT, NC, CNC, DNC
A5	True Automation	1980	IMS, CAD, CAM, AI. AC, industrial robots, hexapods
A6	True Automation	1990	Lean, JIT
A6	True Automation	2000	Unmanned manufacturing FOF

FIGURE 25.1 Mechanization and automation of manufacturing systems.

25.1). The mechanization may be defined as the use of various mechanical, hydraulic, pneumatic, or electrical devices to run the manufacturing process. A good example of this is the automatic feed on a conventional machine tool. In mechanized systems, the operator directly controls the particular process and checks each step of machine operation.

The next step after the mechanization was the automation, derived from the Greek word auto-mate (self-acting); this word was first used in 1943 by the U.S. automobile industry to indicate automatic handling and processing of parts in production machines. Automation also involves the replacement of human effort by a mechanized action, but it also incorporates to a greater or lesser extent an element of process control. It is therefore difficult to imagine any automated process that does not include mechanized elements. The automation as known today begins with the A3 level of automation requiring the machine to be diligent (repeat cycle automatically). These machines are open-looped of no feedback, controlled either internally by a cam or externally programmed with a computer. A3 level includes robots, NC machines without feedback, automatics, and transfer machines. Automatic machines of all types existed long before the term "automation" was conceived. Some people consider automatic machines of all kinds (bar, chucking, Swiss type, and so on) within the realm of automation. However, the discussion here is more devoted to machines in integrated manufacturing systems (IMSs).

The A4 level of automation represents the true automation, where the human judgment is replaced by a capability in the machine to measure and compare results with a desired value. One characteristic, common to all automated processes, is that they are closed-loop systems, in which feedback provides the necessary data for effective automatic process control. Therefore, truly automated systems transform the operator's role into a supervision task. An automatic grinder that checks the part diameter and automatically repositions the grinding wheel to compensate for its wear would be an example of the A4 level. The first NC machine tool was developed in 1952. It had positional feedback control and is generally considered the first A4-level machine tool.

By 1958 the first NC machining center (a compilation of milling, drilling, tapping, and boring) was being marketed by Kearney and Trecker. It could automatically change tools to provide greater flexibility. Within 10 years (1968), NC machine tools became CNC machines, having their own microprocessor and could be programmed directly. CNC machines are still A4-level machines. The A5 level, typified by adaptive control (AC), emulates human evaluation to optimize the process. In this level of automation, a computer is needed to store formulae or mathematical equations that describe how this process or system behaves and what aspect of process is to be optimized. The A6 level of automation reflects the beginning of artificial intelligence (AI), in which the control software is doped with subroutines that permit some learning and thinking capabilities applying the neural network (NNW), thus trying to relate cause to effect. In today's factory, levels A2 through A5 are found with occasional A6 levels (Figure 25.1).

25.3 AUTOMATION AND PRODUCTION QUANTITY

Production quantity is crucial in determining the type of automation required to produce parts economically. The related equipment is selected from the knowledge of inherent capabilities and limitations dictated by the production rate and quantity. The choice depends on cost factors and break-even charts constructed for this purpose. Figure 25.2 illustrates types of production methods and their related characteristics. There is no precise quantity for break-even points between different categories of production size. However, the numerical values shown in Figure 25.2 represent an approximate indication.

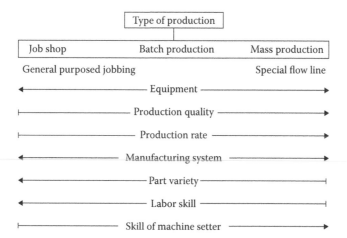

FIGURE 25.2 Types of production methods and related characteristics.

The main categories of production may be classified as

1. Jobbing production (1–20 pieces): Stand-alone general-purpose machines with manual control, requiring the smallest capital outlay are used for this purpose. Their operation is labor intensive. Labor costs do not drop significantly with increasing batch size (Figure 25.3); thus, such machines are best suited to one-off and small batch or jobbing production. The operator may be a highly skilled artisan, or in the case of repetitive production, may be a semiskilled operator. The equipment provides high part flexibility (variety). Turret and capstan lathes are preferred over manually controlled machines for batch sizes greater than those indicated by point A (Figure 25.3).

2. Batch production (10–5000 pieces): Stand-alone NC, CNC, or machining centers are most suitable for small batch production, although, with the trend toward the increasing use of friendly programming devices and with the application of group technology (GT), batch sizes involving 100–5000 may be economically machined. Once the workpiece is clamped on the CNC machine tool and the reference point is established, machining proceeds with great accuracy and repeatability. Nonproductive setup time is particularly negligible. Therefore, CNC can become economical even for small lots that are widely separated in time (Figure 25.3). The operator may again be highly skilled, this time with some programming knowledge; alternatively, the programs may be provided to the machine by a part programmer who may be working from the database of a CAD/CAM system. In this case a semiskilled operator performs machine supervision and service functions. FMS may be economically adapted for batch sizes exceeding those indicated by point B (Figure 25.3). An indication of the place of the FMS in the realm of production is illustrated in the chart of Figure 25.4.

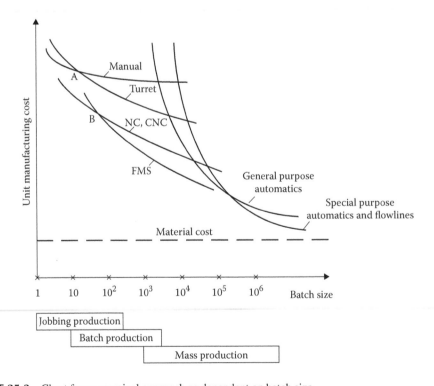

FIGURE 25.3   Chart for economical approach as dependent on batch size.

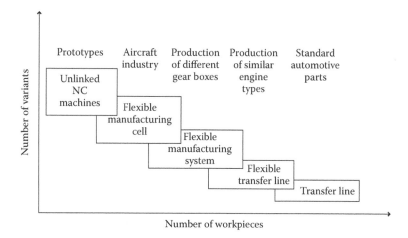

FIGURE 25.4 FMS in the realm of manufacturing systems.

3. Mass production (1000–1,000,000 pieces): In large batch and mass production, flexible lines or automatics (programmable) are most economical, while special purpose (hard programmed) transfer lines or automatics are limited to the mass production of standard parts (Figures 25.3 and 25.4). In both cases, special purpose machinery (dedicated machines) is equipped for transferring materials and parts (flow lines). Although machines and specialized tooling are expensive, both labor skills required and labor costs are relatively low. However, these equipment and manufacturing systems are generally adapted for a specific type of product and hence they lack flexibility. The changeover from one product to another is very costly.

25.4 NECESSITY FOR INTRODUCING AUTOMATION

A sound strategy would be to automate as many areas as technically possible. However, this may contravene the most common automation rule, which is, "only automate processes if it makes economic sense to do so." In this respect, there are many examples of automation being introduced for sale when it would have been economically much more sensible to have continued to use manual labor. However, because capital costs of implementing automation are considerable, it is essential to carry out a detailed analysis of both current and proposed automation methods to ensure that the most effective system is adapted.

The manufacturing situation today made the mass production of any component economically possible; however, industry in many cases demands variety in products in small lots. The economical production methods suitable for smaller lots should be followed. Further higher accuracies are required at lower cost. To meet these requirements, there is a rapidly growing need for improved communication and feedback between manufacturing and design processes by integrating them into a single system capable of being optimized as a whole.

The use of computerized-integrated manufacturing (CIM) is the answer to meet these requirements and objectives. Computers therefore have an important role to play, especially in job shop and batch production manufacturing plants, which constitute an important domain of the overall manufacturing activity.

The traditional job shop and batch manufacturing suffers from drawbacks such as

- Low equipment utilization
- Long lead times

- Inflexibility to market needs
- High inventory
- Dependence on highly skilled operators
- Poor quality control
- Increased indirect cost

It is estimated that in traditional batch production, only 5%–10% time is utilized on machines and the rest is spent in moving and waiting activities. Out of the total time on the machine, only 30% is a machining time, the rest being for positioning, gauging, and idling. These shortcomings can be overcome through the use of

- Material handling equipment
- Feedback systems and continuous flow process
- Computers for process control and data collection, planning, and decision-making to support manufacturing activities

It must be understood that a computer cannot change the basic metal working processes; it can only influence their control and their sequences so that the down time is reduced to a minimum. It provides the quick reflexes, flexibility, and speed to meet the desired needs. The computer enables the detailed analysis and the accessibility to accurate data necessary for the integration of the manufacturing system (MS). For a plant to produce diversified products of best quality at enhanced productivity and lower prices, it would be essential that all elements of manufacturing (design, machining, assembly, quality assurance, management, and material handling) be computer integrated, individually, or collectively.

The main reasons for automation may be summarized as

1. Labor costs are dramatically cut, although technically more skilled labors are normally required.
2. People are reluctant to work in certain types of environment (e.g., injurious to health), and automation may then be the only option.
3. By dramatically reducing human error, conformance to specification, and consistency to quality, greatly improves. This, in turn, results in less scrap and minimizes the risk of late delivery.
4. With implementing customer-initiated manufacture (JIT), the ability to predict precise manufacturing times is essential. Furthermore, with flexible programmable automation the lead time (time from design to completion) is dramatically reduced, resulting in financial savings due to reduced levels of work in progress.
5. All processes are guaranteed to be carried out at the programmed optimum conditions and cannot be changed without proper authority. Thus production management has real-time control over the manufacturing process at all times.
6. Certain computer-designed parts are so complex that they can only be made on equally complex computer-controlled equipment (CAD). Modern gas turbine blades and aircraft wing profiles are good examples.

25.5 MANUFACTURING SYSTEMS

The manufacturing system (MS) takes inputs and produces products for the customer (Figure 25.5). MS is a complete set of elements that include machines, people, material handling, equipment, and tooling. The system output may be consumer goods/services to user or inputs to some other processes. The materials are processed within the system and gain value as they are passed from

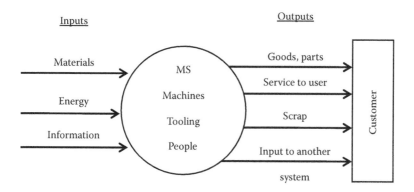

FIGURE 25.5   Manufacturing system. (From Youssef, H., and El-Hofy, H., *Machining Technology, Machine Tools and Operations*, CRC Press, Boca Raton, FL, 2008. With permission.)

one machine to another. Manufacturing systems are very interactive and dynamic; they should be designed and integrated for low cost, superior quality, and on-time delivery. Each company designing its MS will have many differences resulting from discrepancies in subsystem combinations, people, product design, and materials. Effective design of MS can easily be managed and organized to make sure that machines, tools, people, etc., are utilized to the greatest possible extent. The machines should be running at moderate speeds and feeds rather than operating at extremely fast but with interrupted machining rates.

Manufacturing systems differ in structure or design according to the production quantity. They may be classified into

1. *Job shop*, in which a variety of products is manufactured, in small lot sizes, often one a kind. It is commonly done to specific customer order. Because the plant must perform a wide variety of manufacturing processes, general purpose production equipment is required. In the job shop, workers must have relatively high skill levels to perform a range of work assignments. Job shop products include space vehicles, aircraft machine tools, and special tools and equipment. Figure 25.6 shows the functional or process layout of the job shop. Forklifts and hand carts are used for material handling.

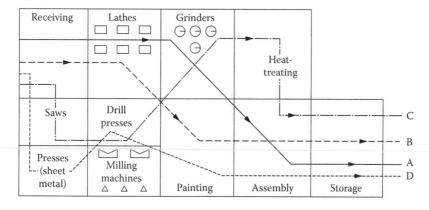

FIGURE 25.6   Job shop manufacturing system: functional or process layout. (Adapted from Degarmo, E. P., Black, J. T. and Kohser, R. A., *Materials and Processes in Manufacturing*, 8th ed. Prentice-Hall, Upper Saddle River, NJ, 1997.)

2. *Flow shop*, which has a product-oriented layout composed mainly of flow line. When the volume gets very large as in the assembly lines, it is called mass production. Specialized equipment, dedicated to the manufacture of a particular product, is used. This system may be designed to produce a particular product or a family of products, using special purpose machines of high investment rather than general purpose ones. Figure 25.7 shows an automated production flow line consisting of a number of machines or stations arranged according to a certain configuration linked to each other by conveyors, to direct the part from one machine to another. The manual skill level tends to be lower than that in a job shop. An important requirement of flow lines is that their individual operations must be balanced. This implies that each operation should take the same time to complete. In reality, different operations have different cycle times. To avoid always waiting for the slowest operation to be completed, it is usual to adjust the speeds of all other operations, so that they equal the speed of the slowest operation, or effectively to have the time required by the slowest operation by duplicating it on the production line. Failure of just one operation would rapidly cause stoppage of the line. To minimize this risk, a buffer stock is held at points of particular vulnerability (Figure 25.7). Automated production flow lines are adapted for production at reduced labor cost, increased production rate, and reduced inventory cost. In this respect, there are two types of automated production lines. Layouts are differentiated, straight (Figure 25.7a), and U-shaped (Figure 25.7b). The choice of either type depends on the number of machining stations and the available area in the manufacturing plant. More stations are consumed in the straight type, whereas the second type does not require large areas.

Figure 25.8 illustrates small transfer units that are employed when a relatively small number of operations (3–10) are to be performed. The table movement may be continuous or intermittent. The circular configuration (Figure 25.8a) has the advantage of being compact and permitting the workpieces to be loaded and unloaded at a single station without

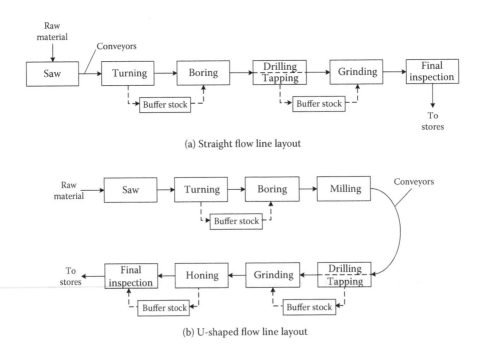

(a) Straight flow line layout

(b) U-shaped flow line layout

FIGURE 25.7 Flow line (product), layout.

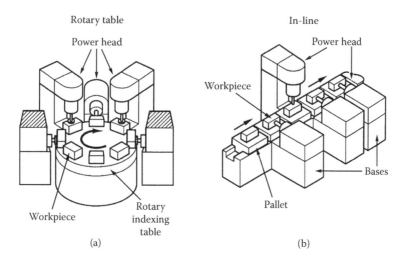

Rotary table   In-line

Power head   Power head

Workpiece

Rotary indexing table   Bases

Workpiece   Pallet

(a)   (b)

FIGURE 25.8   (a) Rotary and (b) line transfer machines. (Courtesy of Heald Machine Company, USA.)

having to interrupt machining. The mega system of Ford, illustrated in Figure 25.9, is one of the largest flow lines in the world for producing cylinder heads. A number of lines are working concurrently, and their outputs fed to an end product assembly line area. This system is capable of producing 100 cylinder heads per hour. A transfer machine of Greenlee Brothers with 10 stations for processing 1000 connecting rods and caps per hour (eight lines at a time), is illustrated in Figure 25.10.

3. *Project shops (fixed position layout)*, in which the product must remain in a fixed position or location during manufacturing because of its size or weight. Materials, machines, and labor involved in the fabrication are brought to the site. Construction jobs such as

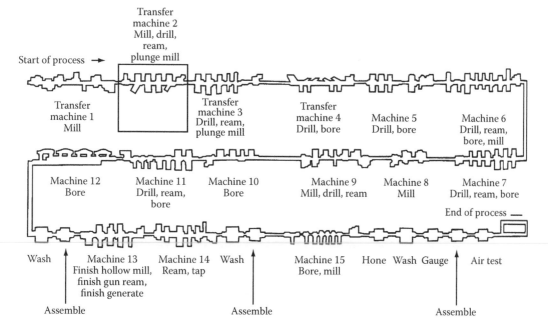

Transfer machine 2
Mill, drill, ream, plunge mill

Start of process →

Transfer machine 1
Mill

Transfer machine 3
Drill, ream, plunge mill

Transfer machine 4
Drill, bore

Machine 5
Drill, bore

Machine 6
Drill, ream, bore, mill

Machine 12
Bore

Machine 11
Drill, ream, bore

Machine 10
Bore

Machine 9
Mill, drill, ream

Machine 8
Mill

Machine 7
Drill, ream, bore

End of process —

Wash   Machine 13   Machine 14   Wash   Machine 15   Hone Wash Gauge   Air test
           Finish hollow mill, Ream, tap   Bore, mill
           finish gun ream,
           finish generate

Assemble   Assemble   Assemble

FIGURE 25.9   Ford Mega flow line.

FIGURE 25.10   Ten-station Greenlee Brothers flow line for processing 1000 connecting rods per hour.

buildings, bridges, and dams (civil engineering projects), locomotives, aircraft assembly, and shipbuilding (Figure 25.11) use the fixed position layout. When the job is completed, equipment is removed from the site.

4. *Continuous process* finds application in oil refineries, chemical processing plants, and food processing. This system is sometimes called flow production. If a complex single product (e.g., television and canning operation) is to be fabricated, continuous processes are usually

FIGURE 25.11   Project shop (fixed-position layout): assembly of a ship's hull. (Courtesy of Vosper Thorneycroft, Ltd., Southampton, UK. With permission.)

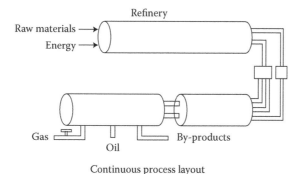

Continuous process layout

FIGURE 25.12   Continuous process layout of a refinery. (From Degarmo, E. P., Black, J. T., and Kohser, R. A., *Materials and Processes in Manufacturing*, 8th ed., Prentice-Hall, Upper Saddle River, NJ, 1997.)

easy to control, efficient, and the simplest systems; however, they are the least flexible ones (Figure 25.12).

5. *Cellular manufacturing* is intended for producing parts one at a time in a flexible design. The cell capacity (cycle time) can be altered quickly to respond to the rapid changes in market demand, thus allowing more product variety in smaller quantities that are highly desirable. Figure 25.13 illustrates an example of a FMC, comprising two machining centers, an automated part inspection, and a serving robot. Flexible cells are typically manned, but unmanned cells are beginning to emerge with a robot replacing the worker (Figure 25.13). Workpieces are placed on the rotary feeder and are loaded and unloaded one by one to the CNC machine by the robot. Machining is performed according to NC command data (cutting information) stored in advance in the CNC machine tool memory. For best results, it is not only recommended to automate the local environment of a machine tool but also to automate the global environment which comprises the following activities:
   * Management and provision of resources
   * Preparation and transportation of workpieces
   * Supply and evolutions of production data
   * Inspection of workpieces and machine tools

The flexibility is acquired when a computer is used to control the above mentioned global environment of production.

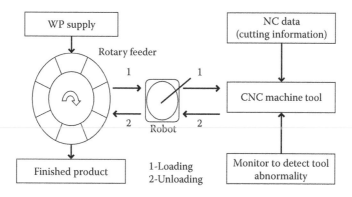

FIGURE 25.13   Unmanned manufacturing cell. (From Youssef, H., and El-Hofy, H., *Machining Technology, Machine Tools and Operations*, CRC Press, Taylor and Francis, Boca Raton, FL, 2008. With permission.)

FMC offers the following advantages:

- Flexibility for varied and small-quantity production
- Manned and unmanned operation
- Automatic operation of numerous processes
- Simple setup
- Easy operation
- Integration to FMS
- High operation economy
- Immediate stop if necessary
- Operation reliability depending on an adequate supply of NC command data

25.6 FLEXIBLE MANUFACTURING SYSTEMS

In earlier times, automation of manufacturing processes was limited to mass production (fixed automation), which could be feasible only for a large number of parts. Another field of automation is the flexible manufacturing system (FMS), which is a high automated manufacturing system comprising a collection of production devices, logically organized under a host computer and physically connected by a central transporting system. It has been developed to provide some of the economics of mass production to small batch manufacturing.

The main advantage of a FMS is the high flexibility in terms of the small effort and short time required to manufacture a new product and is therefore denoted as a flexible manufacturing system. It is an alternative that fits in between the manual job shop and hard automated transfer lines (Figure 25.14). FMS is best suited for applications that involve an intermediate level of flexibility and productivity. FMS can be regarded as a system that combines the benefits of two other systems: (1) the highly productive, but inflexible, transfer lines, and (2) the shop production that fabricates a

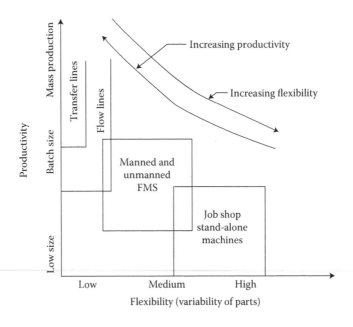

FIGURE 25.14 Flexibility against production rate of different manufacturing systems (intermediate flexibility/productivity of FMS).

large variety of products on stand-alone machines but is inefficient. In FMS, the time required for changeover to a different part is very short (1 min), thus making the quick response to product and market demand variations a major benefit of FMS.

25.6.1 ELEMENTS OF FMS

The basic elements of a FMS are

- Workstations
- Automated handling and transport of materials and parts
- Control systems

1. Workstations: They are arranged to achieve the greatest efficiency in production, with an orderly flow of materials, and products through the system. The type of a workstation depends on the type of production. For machining operations, they usually consist of a variety of three- to five-axis machining centers, CNC machines (milling, drilling, and grinding). They also include other equipment, for automated inspection (including CMMs), assembly and cleaning. For sheet metal forming, punching, shearing, and forging, the workstations of a FMS incorporate furnaces, trimming presses, heat treatment facilities, and cleaning equipment. Figure 25.15 illustrates a forging cell with two robots for material handling.

Machine tools may be equipped either with turret or automatic tool changer for supplying desired tools for machining. A turret has shorter cycle time and is preferred when turning small components; indexing tools with a turret is the fastest method. For larger components with longer cycle times, an automatic tool changer is required where a bigger magazine can be used.

2. Automated Handling and Transport: They are very important for the system flexibility. Material handling is controlled by a central computer and performed by automated vehicles, conveyors, and various transfer mechanisms. Industrial robot is best used for serving several machines in a production system (Chapter 19). A robot should not be located in front of a machine tool as it prevents manual control and supervision. However, the automatic changing of workpieces is best

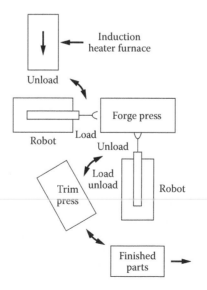

FIGURE 25.15   Workstation of a FMS.

satisfied by using handling equipment that is built integral with the machine tool. Such equipment is called computerized part changer (CPC), which is installed physically separated from the machine tool to eliminate its vibration when the parts are being handled. It consists of portal, shuttle carriage, vertical slide, and a gripping unit. The jaws of the gripping unit can clamp external and internal parts. The gripping unit axis can be positioned in four different angular positions: 0°, 20°, 180°, and 270°. Both the rotary motion and gripping action are hydraulically actuated.

3. Control Systems: The computer control of FMS is the main brain that includes various software and hardware. This subsystem controls the machinery and equipment in workstations and transportation of materials in various stages. It also stores data and provides communication terminals that display the data visually. Because FMS involves a major capital investment, efficient machine utilization is essential; machines must not stand idle. Consequently, proper scheduling and process planning are crucial. Scheduling for FMS is dynamic, unlike that in job shops, where a relatively rigid schedule is followed. Dynamic scheduling is capable of responding to quick changes in product type and thus is responsive to real-time decisions.

In brief, FMS is an integrated system of computer-controlled machine tools and other workstations with an automated flow of information, workpieces, tools, etc. The control of FMS is achieved by computer-implemented algorithms, which make all operational decisions. This system is so arranged that the automated production of a group of complex workpieces in any lot size, particularly small and medium batch, is possible (Figure 25.15). The FMS is usually planned by simulated techniques in which a model is drawn. This model is an idealized representation of the components, internal relations, and characteristics a of real-life system. The analysis of model behavior shows ways for improving the system by carrying out necessary changes. FMS offers the advantages of reduction of part cost, reduction of throughput times, and increasing flexibility toward changes of the product mix, reduced inventory and lead times, and increased productivity. This system usually incorporates features like adaptive controls, tool breakage detectors, and tool life monitoring systems.

25.6.2 Features and Characteristics of FMS

Features and characteristics of FMS are summarized as follows.

- FMS offers immerging cost and quality benefits for most engineering sectors requiring batch production.
- The batch production using conventional machine tools necessitates a minimum number of similar components to be produced economically. There is no batch size limitation in case of FMS; consequently there is no need to lock up money in extensive stocks of finished parts. The work-in-progress is reduced considerably and the inventory cost is, therefore, eliminated.
- It is possible to produce at random all varieties of products planned by a firm. FMS has the capability to quickly respond to any design changes or market demands in the product.
- FMS are usually equipped with robots and/or handling equipment. Software is developed to integrate CNC and the handling systems. All necessary tools can be stored in a magazine.
- All part programs of different models are stored in the system memory. The system identifies the model program to be produced.
- FMS can be conceived in multiples of 15- to 20-min operation. If a certain operation takes longer time, the multiples of similar machines can be installed in the automation line.
- Extensive use of touch triggers is made to minimize the operator intervention in the line.
- Industrial robots are used for material handling (loading and unloading), inspection activities, and assembly operations.

The use of FMSs is hampered by the following limitations:

- High programming cost
- Smaller degree of sophistication of fabrication and assembly processes
- Availability of reliable feedback devices for tool wear and breakage

25.6.3 New Developments in FMS Technology

These developments greatly and dramatically boost the capabilities of FMS due to

- Computerized tool setting station
- Establishing tool information like tool lengths, tool offsets, and so on by linear variable displacement transducer (LVDT)
- Automatic tool changer
- Monitoring of tool life
- Providing tool breakage detectors
- Providing compensation system and AC
- Increased use of robots and material handling systems
- Application of laser and fiber optics technology to check bore diameters and part surface location
- Spindle probes to check workpiece features like bore diameter and hole pattern location
- Improved software
- Fault analysis (vision system for on-line quality control)
- Swarf and coolant control
- Computerized simulation to establish effectiveness and programming facilities

25.7 COMPUTER INTEGRATED MANUFACTURING

CIM is a recent technology being tried and developed since the 1990s. It comprises a combination of software and hardware for product design, production planning and control, production management, and soon in an integrated manner. It is a methodology and a goal rather than merely an assemblage of equipment and computers. CIM effectiveness greatly depends on the use of a large-scale integrated communications system, involving computers, machines, and their controls.

As with traditional manufacturing approaches, the purpose of CIM is to transform product designs and materials into sellable goods at a minimum cost in the shortest possible time. The CIM begins with the design of a product (CAD) and ends with the manufacture of that product (CAM). With CIM, the usual split between CAD and CAM is supposed to be eliminated.

CIM differs from the traditional job shop in the role that the computer plays in the manufacturing process. CIM systems are basically a network of computer systems tied together by a single integrated database (DB). Using the information in the DB, a CIM system can direct manufacturing activities, record results, and maintain accurate data. Therefore, CIM is the computerization of design, manufacturing, distribution, and financial functions into one coherent system. CIM is therefore an attempt to integrate the many diverse elements of discrete parts manufacturing into one continuous process-like stream. It would result in increased manufacturing productivity and quality and reduced production cost. It employs FMS, which saves the manufacturer from replacing equipment each time a new part has to be fabricated. The current equipment can be adapted to produce a new part, as long as it is in the same product family, with programmable software and some retooling. Thus, this system has the ability to switch from one component to another with no downtime for change over. It requires NC lathes, machining centers, punch presses, and so on, which have the ability to be readily incorporated into a multimachine cell or a fully integrated manufacturing system.

Multispindle CNC machines with greater horsepower, stiffness, and wider speed ranges are important for CIM. Automatic tool changers (to change the tool in the spindle and to renew dulled tools) are a must. A robot for handling workpieces is another important machine tool peripheral essential for the CIMs. CIM is a very powerful concept and has the potential to achieve large benefits; however, it is not easy to implement. Like any powerful and complex tool, it can be dangerous and costly if not properly implemented.

The main tasks involved in CIM could be separated into four blocks (Figure 25.16)

1. Product design, for which an interactive computer-aided design (CAD) system allows drawing, analysis, and design to be performed. The computer graphics are useful to get the data out of the designers mind to be ready for interaction (Figure 25.17).
2. Manufacturing planning, where the computer-aided process planning (CAPP) helps to establish optimum manufacturing routines and processing steps, sequences, and schedules.
3. Manufacturing execution, in which computer-aided manufacturing (CAM) identifies manufacturing problems and opportunities. Intelligence in the form of microprocessors is used to control machines and material handling and collect data controlling the current shop floor (Figure 25.17).

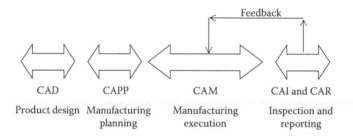

FIGURE 25.16   Main tasks of CIM. (From Youssef, H., and El-Hofy, H., *Machining Technology, Machine Tools and Operations*, CRC Press, Boca Raton, FL, 2008. With permission.)

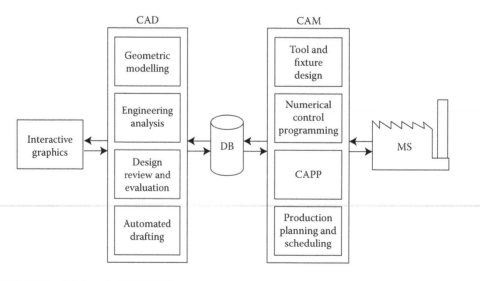

FIGURE 25.17   Database to CAD/CAM.

4. Computer-aided inspection (CAI) and computer aided reporting (CAR) so as to provide a feedback control loop (Figure 25.16).

Computer integration of these four tasks provides the most current and accurate information about the manufacturing, thereby permitting better and tighter control and enhancing the overall quality and efficiency of the entire system. Improved communication among the above activities results in enhancing productivity and accuracy if the designer considers limitations and manufacturing problems and vice versa. The availability of current data permits instantaneous updating of production-control data, which, in turn, permits better planning and more effective scheduling. All machines are fully utilized, handling time reduced, and parts move more efficiently through the production processes. Workers become more productive and do not have to waste time in coordination and searching of previous data. Figure 25.17 shows how CAD/CAM database is related to the design and manufacturing. Included in the CAM is a CAPP module, which acts as an interface between CAD and CAM.

CIM has the advantage of possessing intelligence to maximize the process performance, provided that all the parameters can be measured in real time. It has been established that in metal cutting, every parameter related to machining can be determined if the shear angle is known. It is possible to feed into the computer all the data to compute shear strain, strain rate, flow stress, coefficient of friction at the chip–tool interface, unit horse power and the total power being consumed, cutting temperature, and so on.

The intelligent machine tools, confined in the CIM system, lead to the following merits:

- Increased accuracy
- Reduced scrap
- Reduction in manned operation
- Increased predictability of machine tool operation
- Less skill for setting up and operation on the machine
- Reduced machine down time
- Reduced production cost
- Reduced setup time
- Reduced tooling due to better operation planning
- Increased range of materials and part geometries
- Increased quantity and quality of information exchange between the machine control and part designer and between human and machine tool control.

CIM technology offers the following advantages:

- High rates of production with high precision
- Remarkable flexibility for producing diverse components in the same setup
- Easy and quick manipulation of software
- Uninterrupted production with less supervision or work handling
- Economical production even in case of moderate batch sizes
- Drastic reduction of lead times
- Drastic changes in product design
- Integrating and fine tuning of all factory functions such as raw and semifinished materials flow, tooling, metal cutting, inspection, and so on

25.7.1 Computer-Aided Design

A major element of CIMs is the CAD system that involves any type of design activity that makes use of a computer to create, develop, analyze, optimize, and modify an engineering design.

The design-related tasks performed by a CAD system are (Figure 25.17)

- Geometric modeling
- Engineering analysis
- Design review and evaluation
- Automated drafting

Geometric modeling is the most important phase of the design process, through which the designer constructs the graphical image of the object on the display monitor. Engineering analysis may involve finite element analysis for stress calculations, or heat transfer computations, or the use of differential equations to describe the dynamic behavior of the system. General purpose programs are mainly used to perform these analyses. Design review and evolution are techniques used to check the accuracy of the CAD design. Other features of these techniques are checking and animation, which enhance the designer's visualization of the mechanism and help to eliminate interference. Automated drafting involves generation of hard copy of engineering drawings from the CAD database. It is also capable to perform dimensioning automatically, generating cross-hatching, scaling of drawings, developing sectional views, enlarging view for details, performing rotation of parts, and performing transformations such as oblique, isometric, and perspective views.

25.7.2 COMPUTER-AIDED PROCESS PLANNING

CAPP uses computers to determine how a part is to be made. If group technology (GT) is used, parts are grouped into part families. For each part family, a standard process plan is established, which is stored in computer files and then retrieved for new parts that belong to that particular family. For a manufacturing operation to be efficient, all its diverse activities must be planned and coordinated; this task has traditionally been done by process planners. Process planning involves selecting methods of production, tooling, fixtures, machinery, sequence of operations, standard processing time for each operation, and methods of assembly. When performed manually, this task is labor intensive, time-consuming, and also relies heavily on the process planner experience. CAPP is an essential adjunct to CAD/CAM; although it requires extensive software and good coordination with CAD/CAM as well as other aspects of integrated manufacturing systems. CAPP is a powerful tool for effective planning and scheduling operations. It is particularly effective in small volume, high variety parts production requiring forming, machining, and assembly operations. Process planning activities are a subsystem of CAM (Figure 25.17). Several functions can be performed using these activities such as capacity planning for plants to meet production schedules, inventory control, and purchasing.

Advantages of CAPP systems include the following.

- The standardization of process plans improves productivity, reduces lead times and costs of planning, and improves the consistency of product quality and reliability.
- Process plans make use of GT to retrieve plans to produce new parts.
- Process plans can be modified to suit specific needs.
- Neater and more legible routing sheets can be prepared more quickly.
- Many other functions, such as cost estimation and work standards, can be incorporated into CAPP.

25.7.3 COMPUTER-AIDED MANUFACTURING

CAM involves the use of computers to assist in all phases of manufacturing a product including process and production planning, scheduling, manufacturing, quality control, and management. CAM is another major element of CIM. Because of the increased benefits, CAD and CAM are integrated into CAD/CAM systems. This integration allows the transfer of information from the design stage to the planning stage for manufacturing of the product, without the need to manually reenter data on part geometry. The database developed during CAD is stored and then processed further by CAM into the necessary data and instructions for operating and controlling production machinery and material-handling equipment and for performing automated testing and inspection. An important feature of CAD/CAM integration in machining is the capability to describe the cutting tool path for various operations such as NC turning, milling, and drilling. The programs are computer generated that can be modified by the programmer to optimize the tool path, and to visually check for possible tool collisions with clamps or fixtures or for other interferences. The tool path can be modified at any time to accommodate other shapes to be machined.

The tasks performed by a CAM system (Figure 25.17) are

- Numerical control or CNC scheduling
- Production planning and scheduling
- Tool and fixture design
- Computer-aided process planning

Numerical control can use special computer languages. Today, APT and COMPACT II are the two most common language-based computer-assisted programming systems used in industry. These systems take the CAD data and adapt them to the particular machine control unit to make the part (see Chapter 18).

25.8   INTEGRATED MANUFACTURING PRODUCTION SYSTEM-LEAN PRODUCTION

Regardless of all that has been written about CIM, this technology is not widespread. What is called *lean production* appears to be more important to the future rather than CIM. It is evident that unless a company first adopts the approach of lean production, the conversion to CIM is likely to fail. In this approach, the functions of the production system such as production control, inventory control, quality control, and machine tool maintenance are first to be integrated. Lean production has been developed and practiced in Japan by Toyota (Toyota system) and also in the United States instead of CIM. Integration of the production system functions into the MS requires commitment from top-level management and communication with everyone, particularly manufacturing. Total employee and union participation is absolutely necessary, but it is not usually the union leadership or the production workers who raise barriers to integrated manufacturing production system (IMPS). It is those in middle management who have the most to lose in this systems-level change. In this respect, the following should be emphasized.

1. All levels in the plant, from the production workers to the president, must be educated in IMPS's philosophy and concepts.
2. Top management must be totally committed to this venture and everyone involved must be motivated.
3. Everyone in the plant must understand that cost, not price, determines the profit. Customer wants low cost, superior quality, and on time delivery.
4. Everyone must be committed to the elimination of waste to reduce cost; this is fundamental for getting a lean production.

Many Japanese and American industries are now undertaking massive educational programs to instruct their employees about quality control, machine maintenance, setup reduction, and basics of lean production. This knowledge is the key toward improving productivity. This system represents the model for the third industrial revolution.

25.8.1 Steps for Implementing IMPS (Lean Production)

Many companies have implemented IMPS (lean production) by converting a factory from a job shop flow MS to an IMPS. For this purpose, the following steps are to be followed.

1. Build foundation by forming U-shaped cells: To replace the production job shop, restructure and reorganize the FMS, composed of cells which fabricate families of parts. Creating cells is the first step in designing MS's in which production and inventory control as well as quality control are integrated.
2. Rapid exchange of tooling and dies (RETAD): Everyone on the plant floor must be informed to reduce setup time by using single-minute exchange of die (SMED), which is critical to reducing lot size.
3. Integrating quality control: A multifunctional worker can do more than operate machines. This worker is also an inspector who understands process capability, quality control, and process improvement. In lean production, every worker has the responsibility and the authority to make the product right the first time and every time, and the authority to stop the process when something is wrong. The integration of quality control into the MS considerably reduces defects while eliminating inspection. Cells provide for integration of quality control.
4. Integrating preventive maintenance: To ensure machine reliability, installing an integrated preventive maintenance program is an important task.
5. Leveling and balancing final assembly: That is done by producing a mix of final assembly products in small lots. Balancing means that each process, cell, and subassembly has essentially the same cycle time as the final assembly.
6. Linking cells: Integration of production control is realized by linking the cells, subassemblies, and final assembly elements, utilizing *Kanban*. Kanban is a visual control system that is only good for lean production, linking its cells to each other by a pull system of production inventory control. The need for a route sheet is eliminated. All linked cells, processes, subassemblies, and final assemblies start and stop together in a synchronized manner.
7. Integrating the inventory control: The inventory levels are directly controlled by the people on the floor through the control of Kanban. This is the integration of the inventory control to reduce work in process (WIP). The minimum level of WIP is determined by the percent of defectives, the reliability of the equipment, the setup time, and the transport distance to the next cell.
8. Integrating the suppliers: This is educating and encouraging suppliers (vendors) to develop their own lean production system for superior quality, low cost, and rapid on-time delivery. They should deliver material to the customer when needed, where needed, without inspection.
9. Automating and robotizing: This is to solve problems by converting manned to unmanned cells, initiated by the need to solve problems in quality, reliability, and eliminating bottlenecks.
10. Computerizing the whole production system: By restructuring the MS into just-in-time (JIT) manufacturing, the system and the critical functions are well integrated. It is expedient to restructure the rest of the company. It is basically restructuring the production system to be as waste free and efficient as the MS.

25.8.2 JIT Production

JIT production, as based on Kanban, is achieved by eliminating waste of materials, machines, capital, manpower, and inventory throughout the manufacturing system (Kalpakjian and Schmid 2003). The JIT philosophy is summarized as

Produce and deliver finished goods just-in-time to be sold, subassemblies just-in-time to be assembled into finished goods, fabricated parts just-in-time to go into finished good, fabricated parts just-in-time to go into subassemblies, and purchased materials just-in-time to be transformed into fabricated parts.

To be more specific, JIT seeks to achieve the following goals:

- Zero defects
- Zero setup time
- Zero inventories
- Zero handling
- Zero breakdowns
- Zero lead time
- Lot size of one

To achieve these goals, all elements of excess should be eliminated. Large safety stocks, long lead times, long setting times, large queues at machines, high scrap and rework levels, machine breakdowns, and so on should also be eliminated. Therefore, JIT is not a simple off-the-shelf solution to all manufacturing problems. If JIT is realized in the firm, the unnecessary inventories will be completely avoided, making stores or warehouses unnecessary, inventory cost will be diminished, and the ratio of capital turnover will therefore be increased. Consequently, JIT is sometimes called *zero inventories*, *material as need*, *stockless production*, and *demand scheduling*.

In traditional manufacturing, the parts are made in batches placed in inventory and used whenever necessary. This approach is known as *just-in-case* (JIC) or *push system*, meaning the parts are made according to a schedule and are kept in inventory to be used if and when they are needed. In contrast, JIT manufacturing is *pull system* meaning (as previously mentioned) that parts are produced to order and the production is matched with demand for the final assembly of products.

In JIT, parts are inspected by the worker and used within a short period of time. Accordingly, the worker maintains continuous production control, identifies immediately defective parts, and produces quality products. Implementation of JIT requires that all manufacturing aspects be continuously reviewed and monitored so that all operations and resources that do not have value are eliminated.

The basic aim of JIT is to produce the kind of units needed in the quantities needed at the time needed. Then the system should depend on smoothing (leveling) of the manufacturing system, so that it is necessary to eliminate fluctuation in the final assembly, which is called leveling or balancing the final assembly. The object of JIT is to make the same amount of produced part every day. Balancing is making the output from the cells equal to the necessary demand for the parts downstream. In summary, small lot sizes, made possible by setup reduction within the FMCs, single unit conveyance within the cells, and standard cycle times are the keys to accomplish a smoothed manufacturing system.

There are several advantages to JIT:

- Low inventory cost
- Fast detection of defects in production and, hence, low scrap loss
- Reduced need for inspection and reworking of parts
- Production of high-quality parts at low cost

Implementation of JIT, as compared to FMS realizes the following:

- Reductions of 20%–40% in production cost
- Reductions of 60%–80% in inventory

- Reductions up to 90% in rejections
- Reduction up to 90% in lead times
- Reduction up to 50% in scrap, rework, and warranty cost
- Increase of 30%–50% in direct labor productivity
- Increase of 60% in indirect labor productivity

25.9 ADAPTIVE CONTROL

AC machine tools are a logical extension of CNC systems (A5 level of automation). Accordingly, the part programmer sets the processing parameters on the basis of the existing knowledge of the workpiece material and various data on the particular manufacturing process. In CNC machines, these parameters are held constant during a particular process cycle. On the other hand, the AC system is capable of automatic adjustments during processing, through closed loop feedback control. It is therefore readily appreciated that this approach is basically a feedback control system. A schematic diagram of a typical AC configuration for a machine tool is shown in Figure 25.18. Accordingly, AC represents a process control that operates in addition to the CNC position or servo control system.

In manufacturing, several AC systems or strategies are distinguished

1. *Adaptive control with optimization* (ACO), in which an economic index of performance is used to optimize the process using on-line measurements. This strategy may involve maximizing material removal rate or improving surface quality.
2. *Adaptive control with constraints* (ACC), in which the process is controlled using online measurements to maintain a particular process constraint (force, power, and temperature).
3. Referring to Figure 25.19, if the cutting force and hence the torque increases excessively, the AC system changes the cutting speed or the feed rate (cutter travel), to lower the cutting force to an acceptable level. Without AC or without direct intervention of the operator (in case of conventional machining), high cutting forces may cause the tools to chip or break, or the workpiece to deflect or distort excessively. As a result, the accuracy and surface finish deteriorate.
4. *Geometric adaptive control* (GAC), in which the process is controlled using on-line measurements to maintain desired dimensional accuracy or surface finish (Figure 25.20).

Response time must be short for AC to be effective. Currently, all AC systems are based on either ACC or GAC, because the development and proper implementation of ACO is complex. The ACC is well suited for rough cutting, whereas GAC is used for finishing operations.

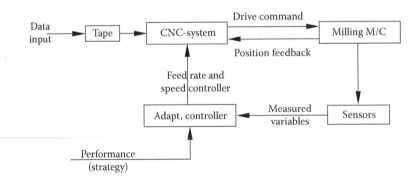

FIGURE 25.18 Typical adaptive control configuration for a machine tool. (Adapted from Koren, J., *Computer Control of Manufacturing Systems*, 1st ed., McGraw-Hill, New York, 1983.)

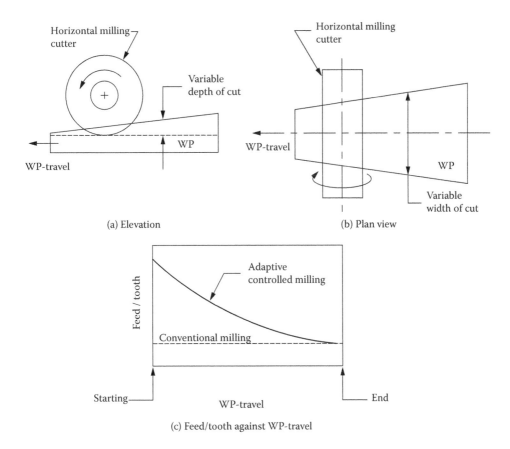

(a) Elevation

(b) Plan view

(c) Feed/tooth against WP-travel

FIGURE 25.19   Adaptive control in milling.

Integration of AC into CAD/CAM/CIM systems is an important issue in the future development of AC systems, as well as their role in the CIM hierarchy. Such integration is extremely important for unmanned manufacturing and will require additional research to extend the current understanding of AC systems. Important issues include the interface between CAD and CAM and the application of expert systems and other methods from artificial intelligence to AC systems, as well as process monitoring and diagnostics.

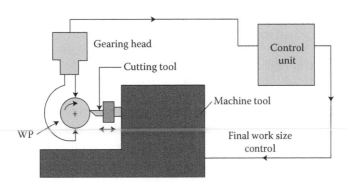

FIGURE 25.20   In-process inspection of workpiece diameter in turning. (From Kalpakjian, S., and Schmid, S. R., *Manufacturing Processes for Engineering Material*, 4th ed., Pearson Education, Upper Saddle River, NJ, 2003.)

25.10   SMART MANUFACTURING AND ARTIFICIAL INTELLIGENCE

Artificial intelligence (AI) is the basic tool for smart manufacturing (SM). AI is an area of computer science concerning systems that exhibit some characteristics which are usually associated with intelligence in human behavior, such as learning, reasoning, problem solving, and understanding of the language. Its goal is to simulate such human endeavors on the computer and represents a technique for solving problems in a better way than is available with conventional computer programs (CCP). A CCP typically relies on algorithmic solutions, in which a finite number of explicit steps produce the solution of a specific problem. These algorithms work fine for scientific or engineering calculations that are numeric in nature to produce satisfactory answers. In contrast, AI uses heuristic rule of thumb search. It should be understood that exhaustive search can only be for relatively simple, well-defined problems. For complex and uncertain problems, exhaustive search routines become impractical.

CCP is difficult to modify, but AI is usually easy to modify, update, and enlarge. In CCP, information and control are integrated, while in AI the control structure is usually separate from domain knowledge. CCP is often primarily numeric, but AI is concerned primarily with processing symbolic information in which some meaning other than a numeric value is attached to a symbol. The symbols in AI may represent a concept about a process or a condition related to it; the AI programs manipulate the relationships among such symbols and arrive at logical conclusions from these relationships. AI has a major impact on all steps of the manufacturing cycles, including design, automation, production planning, scheduling, and the overall economics of manufacturing operations. AI programs find also application in diagnosis, monitoring, analysis, interpretation learning, consultation, instruction, conceptualization, prediction, debugging, and repair. AI packages costing on the order of a few thousand dollars have been developed, many of which can now be used on personal computers for application of both office and shop floors. AI application in manufacturing, generally, encompasses expert systems, natural language, machine vision, artificial neural networks, and fuzzy logic.

25.10.1   EXPERT SYSTEMS

An Expert System (ES), also called a knowledge-based system (KBS), is generally defined as an intelligent computer program that has the capability to solve real-life problems using KB and inference procedures. The goal of ES is to develop the capability to conduct an intellectually demanding task in the way that a human expert would. Expert systems use a KB containing facts, data, definitions, and assumptions. They also have the capability to follow a heuristic approach, that is, to make good judgments on the bases of discovery and revelation and to make high probability guesses, just as a human expert would. The KB is expressed in computer codes, usually, in the form of if–then rules and can generate a series of questions; the mechanism for using these rules to solve problem is called an inference engine. ES can also communicate with other computer software packages. To construct ES for solving complex design and manufacturing problems encountered, the following is needed:

- A great deal of knowledge
- A mechanism for manipulating the knowledge to create solutions

Because the difficulty involved in modeling the many years of experience of a team of experts and the complex inductive reasoning and decision-making capabilities of humans, including the capability to learn from mistakes, the development of KBSs require much time and effort. ES operates on a real-time basis, and its short reaction times provide rapid responses to problems. The programming languages most commonly used are C, LISP, and PROLG. A significant development is ES software shells or environments (frame work). These shells are essentially ES outlines that allow a person to write specific applications to suit special needs. Writing these programs requires considerable experience and time. Several ESs have been developed to be used for

- Problem diagnosis in machines and equipment
- Modeling and simulation of production facilities
- CAD, process planning, and production scheduling
- Management of company's manufacturing strategy

25.10.2 Machine Vision

In a system that incorporates machine vision, computers and software implementing AI are combined with cameras and other optical sensors. These machines then perform such operations as inspecting, identifying, and sorting parts and guiding intelligent robots (Figure 25.21); in other words, operations that would otherwise require human involvement and intervention.

25.10.3 Artificial Neural Networks

Artificial Neural Networks (ANNs) are used in such applications as noise reduction in telephones, speech recognition, and process control in manufacturing. For example, they can be used to predict the surface finish of machined workpieces on the basis of input parameters such as cutting force, torque, acoustic emission, and spindle acceleration. However, this field is still under development.

25.10.4 Natural Language Systems

These systems allow a user to obtain information by entering English language commands in the form of simple typed questions. Natural language software shells are used in the scheduling of material flow in manufacturing and analyzing information in DBs. Major progress is made to develop computer software that will have speech-synthesis and voice-recognition capabilities, thus eliminating the need to type commands on the keyboard.

25.10.5 Fuzzy Logic (Fuzzy Models)

It is an element that has important application in control systems and pattern recognition. It is based on the observation that people can make good decisions on the basis of nonnumeric information. *Fuzzy models* are mathematical means of representing vagueness and imprecise information (hence

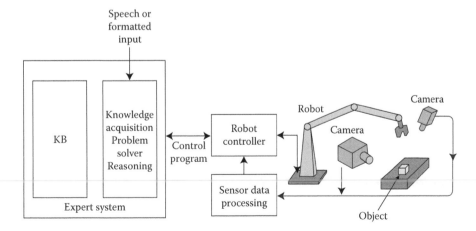

FIGURE 25.21 Expert system as applied to an industrial robot guided by machine vision. (From Kalpakjian, S., and Schmid, S. R., *Manufacturing Processes for Engineering Material*, 4th ed., Pearson Education, Upper Saddle River, NJ, 2003.)

the term fuzzy). These models have the capability to recognize, manipulate, interpret, and use data and information that are vague or lack precision. Fuzzy logic methods involve reasoning and decision-making at a level higher than ANNs. Typical concepts used in fuzzy logic are *few, more or less, small, medium, extreme,* and *almost all.*

Fuzzy logic technologies and devices have been developed and successfully applied in areas such as intelligent robotics, motion control, image processing and machine vision, machine learning, and design of intelligent systems. Some applications of fuzzy logic include automatic transmission of Lexus cars, automatic washing machines, and helicopters that obey vocal commands.

25.11 FACTORY OF FUTURE

The trend toward the automated factory seems unavoidable in modern industries. The integration of many new techniques adopted in manufacturing systems, such as CNC, machining centers, FMS, robots, material handling equipment, automatic warehouses, together with CAD, CAM, CAPP, and GT software, have made the factory of the future (FOF) closer to a reality. All manufacturing, material handling, assembly, and inspection will be done by computer-controlled machinery and equipment. Similarly, activities such as the processing of incoming orders, production planning and scheduling, cost accounting and various decision-making processes performed by management will also be done automatically by computers. The role of human beings will be confined to activities such as overall supervision; preventive maintenance, and upgrading of machines and equipment; receiving supplies of materials and shipping of finished products; provision of security for the plant facilities; programming upgrading and monitoring of computer programs; and maintaining and upgrading of computer hardware.

The reliability of machines, equipment, control systems, power supplies, and communications networks is crucial to full factory automation. Without human intervention, a local or general breakdown in even one of these components can cripple production. The computer-integrated factory of the future must be capable of automatically rerouting materials and production flow to other machines and to the control of other computers in case of such emergencies.

An important consideration in fully automating a factory is the nature and extent of its impact on employment. Although forecasts indicate that there will be a decline in the number of machine tool operators and tool and die workers, there will be major increase in the number of people working in service occupations such as computer technicians and maintenance electricians. Thus the generally high skilled, manual-effort labor force traditionally required in manufacturing will be trained or retrained to work in such activities as computer programming, information processing, implementation of CAD/CAM, and similar tasks. The development of more user-friendly computer software is making these tasks much easier. In this respect, it should be recognized that the designation of *world class,* like *quality,* is not a *fixed target* for a manufacturing country or company to reach but rather a *moving target,* rising to higher and higher levels as time passes. Manufacturing organizations must be aware of this moving target and plan and execute their programs accordingly.

25.12 CONCLUDING REMARKS RELATED TO AUTOMATED MANUFACTURING

1. Installations of FMS are very capital intensive; consequently, a thorough cost benefit analysis must be conducted before a final decision is made. This analysis should include
   - Capital cost, energy, materials, and labor.
   - Market analysis for which the products are to be produced.
   - Anticipated fluctuations in market demand and product type.
   - Time and effort required for installing and debugging the system. An FMS can take 2–5 years to install and at least 6 months to debug.

2. Although FMS requires few, if any, machine operators, the personnel in charge of the total operation must be trained and highly skilled. These include manufacturing engineers, computer programmers, and maintenance engineers.

3. The most effective FMS applications have been in medium volume, high variety batch production (50,000 units/year). In contrast, high volume, low variety parts production is best attained from transfer machines.

4. CIMS have become the most important means of improving productivity, responding to changing market demands, and better controlling manufacturing operations and management functions. Regardless of all written about CIM, this technology is not as widespread as that of lean production.

5. CAM is often integrated with CAD to transfer information from the design stage to planning stage and to production; i.e., CAD/CAM bridges the gap from design to production.

6. Advances in manufacturing operations such as CAPP, computer simulation of manufacturing processes, GT, cellular manufacturing, FMSs, and JIT manufacturing contribute significantly to improved productivity.

7. Significant advances have been made in material handling, particularly with the implementation of industrial robots and automated conveyors.

8. FOF appears to be theoretically possible. However, there are important issues to be considered regarding its impact on employment.

9. In FOF many of the functions of production systems are integrated into the manufacturing system. This requires that the job shop MS is replaced with a linked IMPS. The functions of production control, inventory control, quality control, and machine tool maintenance are the first to integrate.

25.13   REVIEW QUESTIONS

1. Describe the difference between mechanization and automation. Give some typical examples of each.
2. Explain the difference between hard and soft automation. Why they are named as such?
3. Describe the principles and purpose of adaptive control. Give some applications in manufacturing that you think can be implemented.
4. Differentiate between ACO, ACC, and GAC.
5. What are the benefits and limitations of FMS?
6. Draw a sketch to show the idea of
   - A manufacturing system
   - A flow-line manufacturing cell
   - Unmanned FMC
   - GAC for turning operation
7. What are the components of a manufacturing system?
8. Define CIM in your own words. List the benefits of a CIM system.
9. Describe the principles of FMS. Why does it require a major capital investment?
10. What are the benefits of JIT production? Why it is called a pull system? What is a push system?
11. Differentiate between JIT and JIC production.
12. When a car factory is working at full capacity, it may produce a car every 2 or 3 min. Clearly, it is not possible to build either a complete engine or a gearbox in such a short time. How do you think that this problem is overcome?
13. What is meant by the term FOF? Explain why humans will still be needed in the factory of the future.

14. What is Kanban? Why was it developed?
15. Describe the elements of AI.
16. Explain the principles of CAM, CAPP, and CIM to an older worker in a manufacturing facility who is not familiar with computers.
17. What is lean production? List and explain the main steps toward lean production.
18. What are FMS? How do they differ from transfer lines?

BIBLIOGRAPHY

ASM International. 1989. *Machining*, vol. 16, *Metals Handbook*, 9th ed. Materials Park, OH.

Degarmo, E. P., Black, J. T., and Kohser, R. A. 1997. *Materials and Processes in Manufacturing,* 8th ed. Upper Saddle River, NJ: Prentice-Hall.

El-Midany, T. T. 1994. *Computer Automated Manufacturing and Flexible Technologies,* 1st ed. Mansoura City, Egypt: El Mansoura University.

Jain, R. K. 1993. *Production Technology*, 13th ed. Delhi, India: Khanna.

Kalpakjian, S., and Schmid, S. R. 2003. *Manufacturing Processes for Engineering Material,* 4th ed. Upper Saddle River, NJ: Pearson Education.

Koren, J. 1983. *Computer Control of Manufacturing Systems,* 1st ed. New York: McGraw-Hill.

McMahon, C., and Browne, J. 1989. *CAD/CAM—Principles, Practice and Manufacturing Management,* 2nd ed. Reading, MA: Addison-Wesley.

Youssef, H., and El-Hofy, H. 2008. *Machining Technology, Machine Tools and Operations*. Boca Raton, FL: CRC Press.

26 Health and Safety Aspects in Manufacturing

26.1 INTRODUCTION

Manufacturing industry transforms raw materials into finished or semifinished products through different operations. Such operations often rely on the integration of numerous components that frequently results in the generation of toxic or hazardous wastes. A *hazard* is any source of potential damage, harm, or adverse health effects under certain conditions at work. Basically, a hazard can cause harm or adverse effects to individuals (health effects) or to the organizations in the form of property or equipment losses. Manufacturing hazards affect people, machines, ecological systems, and other inhabitants of the environment. For example, dust, radiation, temperature, and other environmental factors are hazardous to people and equipment. It is therefore necessary to protect workers within a plant from the dangerous environmental conditions and the general public from the unsafe conditions created by the manufacturing operation or products of the plant.

A *risk* is the chance or probability that a person will be harmed or experience an adverse health effect if exposed to a hazard. It may also apply to situations with property or equipment loss. The presence of toxic substances in air, water, or soil often results from inefficiencies in the production process. The presence of these substances can present a health risk to human or ecological systems. Factors that influence the degree of risk include how long a person is exposed to the hazardous condition, the way the person is exposed (breathing in a vapor, skin contact), and the severity of the conditions of exposure.

Figure 26.1 shows the different hazards and risks associated with manufacturing operations. Manufacturing hazards (Figure 26.2) can be classified by category as follows:

- Chemical: Depends on the physical, chemical, and toxic properties of the chemical substances
- Ergonomic: Because of repetitive movements, improper set up of workstation, etc.
- Physical: Radiation, magnetic fields, and pressure extremes
- Psychosocial: Stress, violence, etc.
- Environmental: air, water, and soil pollution
- Safety: Slipping/tripping, inappropriate machine guarding, equipment malfunctions, or breakdowns

Risk analysis establishes a priority ranking of hazards based on the magnitude of risk that they pose to human or ecological systems. It is a collection of methods that evaluate the probability of an adverse effect of an agent, industrial process, technology, or natural process. The adverse effect is usually related to human health (death or disease), but it may cause an economic loss that requires economic risk analysis.

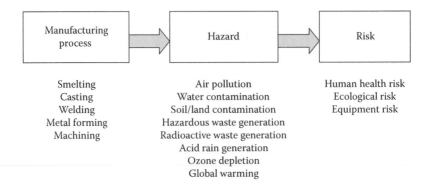

Smelting · Air pollution · Human health risk
Casting · Water contamination · Ecological risk
Welding · Soil/land contamination · Equipment risk
Metal forming · Hazardous waste generation
Machining · Radioactive waste generation
· Acid rain generation
· Ozone depletion
· Global warming

FIGURE 26.1 Hazards and risks of manufacturing operations.

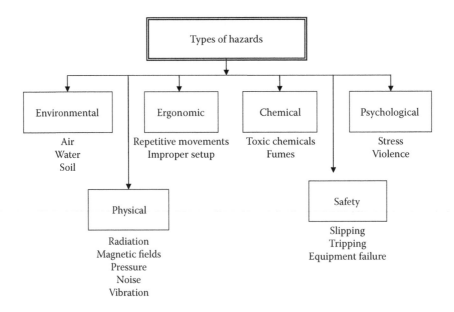

FIGURE 26.2 Classification of manufacturing hazards.

26.2 HEALTH AND SAFETY AT WORK

Health and safety act at work provides a comprehensive integrated system of law dealing with the health, safety, and welfare of workers and the general public as affected by work activity. The act has six main provisions that include the following:

1. To overhaul completely and modernize the existing law dealing with safety, health, and welfare at work
2. To put general duties on employers ranging from providing and maintaining a safe place of work to consulting on safety matters with employees
3. To create a health and safety commission
4. To reorganize and unify the various governmental inspectors
5. To provide powers and penalties for the enforcement of safety laws
6. To establish new methods of accident prevention and new ways of operating future safety regulations

The act places responsibility for safe working equally on the employers, the employee, and the manufacturers and suppliers of goods and equipment as follows:

The general duties of employers include

1. Safe plant and system at work
2. Safe use, handling, and transport at storage of substances and articles
3. Provision of information instruction, training, and supervision
4. Safe place of work, access, and egress
5. Safe working environment with adequate welfare facilities
6. Written safety policy together with organizational and other arrangement (if more than 4 employees)

Duties of employees include

1. To make responsible care for health and safety of themselves and others affected by their acts or emissions.
2. To cooperate with the employer and others to enable them to fulfill their legal obligations.

Manufacturers and suppliers, those persons who design, manufacture, import, or supply any article or substance for use at work, must ensure so far as is reasonably practicable that

1. They are safe and without risk to health when properly used according to manufacturer's instructions.
2. Carry out such tests or examinations as are necessary for the performance of their duties.
3. Carry out any necessary research to discover, eliminate, or minimize any risk to health or safety.
4. The installer or erector must ensure that nothing about the way in which the article is installed or erected makes it unsafe or a risk to health.

26.3 SOURCES OF MANUFACTURING HAZARDS

Manufacturing processes cause several hazards through the following sources.

26.3.1 HAZARDS DUE TO MANUAL HANDLING

Typical hazards of manual handling, shown in Figure 26.3, include

- Lifting a load that is too heavy or too cumbersome resulting in back injury.
- Poor posture during lifting or poor lifting technique resulting in back injury.
- Dropping a load resulting in foot injury.
- Lifting sharp edged or hot loads resulting in hand injuries.

26.3.2 HAZARDS DUE TO HAND-HELD TOOLS

Hand and power tools are commonly used in industry to perform tasks that otherwise would be difficult or impossible. However, these simple tools can be hazardous and have the potential for causing severe injuries when used or maintained improperly. Special attention toward hand and power tool safety is necessary in order to reduce or eliminate such hazards. In addition to the electrical hazards, the following are physical hazards associated with such tools (see Figure 26.4):

- Mechanical entanglement in rotating spindle or sanding discs.
- Waste material flying out of the cutting area causing eye hazard.

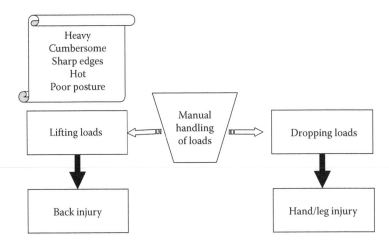

FIGURE 26.3   Common hazards due to manual handling of loads.

- Coming into contact with the cutting tools
- Manual handling problem with a risk of injury if the tool is heavy or very powerful
- Hand and arm vibration
- Tripping hazard from trailing cables or power supplies
- Explosion risk with petrol driven tools or when used near flammable liquid or gas
- High noise levels with routers and saws
- Dust level

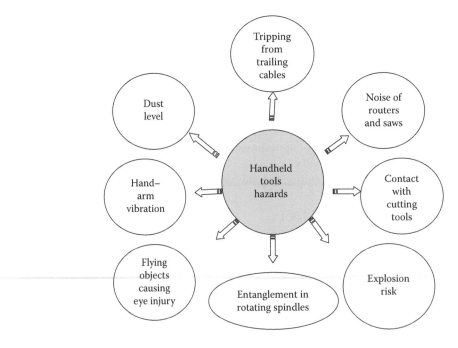

FIGURE 26.4   Common physical hazards of hand-held tools.

26.3.3   MECHANICAL MACHINING

As shown in Figure 26.5, a person may be injured at machinery as a result of the following hazards.

- Crushing through being trapped between a moving part of a machine and a fixed structure
- Shearing that traps typically a hand or a finger between moving and fixed parts of the machine
- Cutting or severing through contact with a cutting edge such as a band saw or rotating cutting disc
- Entanglement with the machine that grips loose clothing, hair, or working material
- Drawing-in or trapping between running gear wheels or rollers or between belts and pulley drives
- Impact when a moving part directly strikes a person
- A stabbing or puncture through ejection of particles from a machine or a sharp operating component
- Friction or abrasion during contact with a grinding wheel (GW) or sanding machine
- Straining and spraining (muscular skeletal disorders) are often due to poor postures, repetitive movement, and high forces exerted that may result from defective hand tools or tools not designed for the purpose for which they are used

26.3.4   ELECTRICAL HAZARDS

Engineers, electricians, and other professionals working with electricity are exposed to electrical hazards. Electrical current flows through the human body and causes a slight tingling sensation or a shock that results in muscular contractions or may knock the victim away from the electrical circuit.

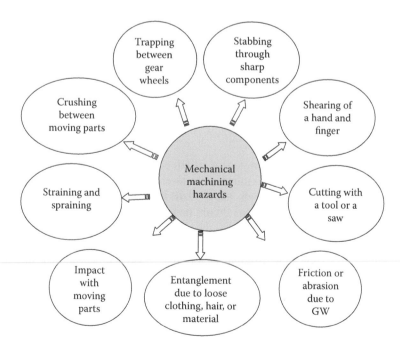

FIGURE 26.5   Common hazards of mechanical machining.

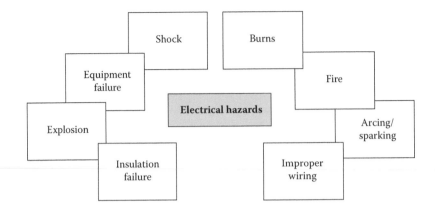

FIGURE 26.6   Electrical hazards.

In severe cases, heart muscle rhythms are disrupted and death may result. The principle hazards associated with electricity, shown in Figure 26.6, include the following:

- Electric shock
- Electric burns
- Electrical fire and explosions
- Arcing and sparking
- Equipment failure
- Insulation failure
- Improper wiring

Under these circumstances, the following electrical safety measures are considered.

- Keep electricians familiar with the electrical safety orders.
- Report as soon as possible any obvious hazard in connection with electrical equipment or lines.
- Make preliminary inspections and/or appropriate tests to determine the conditions of electrical equipment before use.
- Maintain or adjust electrical equipment regularly.
- Make sure that all portable electrical tools and equipment are grounded.

26.3.5   Noise Hazards

During manufacturing operations, vibration of different frequencies is generated and produces noise. It has been proposed that the maximum noise level regarded as safe and tolerable for an 8-h exposure is 85 dB. Exposure to noise levels below 80 dB is considered safe. However, for prolonged exposure times, personal protective devices should be used. Noise levels that exceed 90 dB in any work area cause damage of the inner ear so that the ability to hear is lost over a long period of time. As a result, progressive loss of communication, socialization, and responsiveness to the environment occurs. Figure 26.7 shows the different factors that affect the hearing loss due to excessive noise levels.

Noise protective measures include the following steps.

1. Keep noise levels below 85 dB.
2. Educate and periodically train employees to safe levels of noise exposures and its effect on their health.

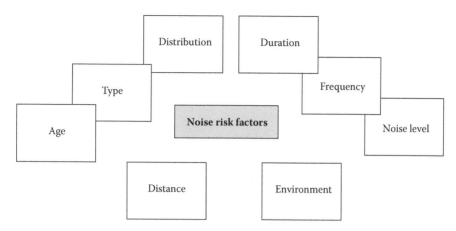

FIGURE 26.7 Critical noise risk factors.

3. Measure, record, and analyze noise levels.
4. Use engineering and administrative controls that reduce excessive noise levels.
5. Use the approved hearing protective equipment.
6. Isolate noisy machinery.
7. Apply periodic audiometric testing to employees who are exposed to noise levels above 85 dB.

26.3.6 VIBRATION HAZARDS

Vibration hazards are normally associated with noise hazards. There are two types of vibrations that are generated during manufacturing operations

1. Whole-body vibrations that occur when the labor operates heavy equipment
2. Hand and arm vibrations (HAV) that cause motion sickness and spine injury

The problem of vibrations can be solved by

1. Use of low vibration tools.
2. Limit employee exposure time to vibrations. Depending on the vibration level, the National Institute of Occupational Safety and Health (NIOSH) recommends 4 hr/day and 2 days/week.
3. Change employee work habits through
   • Wearing thick gloves that absorb vibrations
   • Taking periodic breaks
   • Using vibration absorbing mats
   • Keeping tools properly maintained
   • Keeping warm

26.3.7 CHEMICAL HAZARDS

Most machining processes use chemicals and liquids in the form of coolants, lubricants, etchants, electrolytes, abrasive slurry, dielectric liquids, gases, and anticorrosive additives. These chemicals can be transported by a variety of agents and in a variety of forms, shown in Table 26.1. Hazardous substances cause illness to people at work. These are classified according to their severity and type

TABLE 26.1
Forms of Chemical Agents and Their Hazard Effect

Form	Description	Hazard Effect
Dust	Solid particles heavier than air, suspended in it for some time (0.4 μm (fine) to 10 μm (coarse).	Fine dusts are hazardous because they are respirable, penetrate deep in the lungs, and stay there causing lung disease.
Gas	Substances at temperatures above their boiling point. Steam is a gaseous form of water. Common gases include carbon monoxide and carbon dioxide.	Absorbed into blood stream.
Vapor	Substances very close to their boiling temperature.	If inhaled, it enters blood stream causing short-term effect (dizziness) or long-term effect (brain damage).
Liquid	Substances at temperatures between freezing (solid) and boiling temperatures.	Irritation and skin burn.
Mist	Exist near boiling temperature but are closer to the liquid phase.	Produces similar effects to vapors where it penetrates the skin or is ingested.
Fume	Collection of very small metallic particles (0.4–1.0 μm) that are respirable.	Lung damage.
Aerosols	Liquid or solid particles that are so small that they can remain suspended in air long enough to be transported over a distance.	Lung damage.

of hazard that they may present to the workers. The most common types are summarized in Table 26.2. The effect of these hazards may be acute or chronic.

Acute effects are of short duration and appear fairly rapidly, usually during or after a single or short-term exposure to a hazardous substance.

Chronic effects develop over a period of time that may extend to many years. Chronic health effects are caused by prolonged or repeated exposures to hazardous substances resulting from the manufacturing processes. Such effects may result in a gradual, latent, and often irreversible illness, which may remain undiagnosed for many years. During that period the individual worker may experience symptoms. Cancers and mental diseases fall into chronic category.

Regulations under Occupational Safety and Health Administration (OSHA), Environmental Protection Association (EPA), and Department of Transport (DOT) provide the users of chemicals a user-friendly guide for assessing and minimizing the Environmental, Health, and Safety (EHS)

TABLE 26.2
Effects of Hazardous Substances

Effect	Description
Irritant	Noncorrosive substance that causes skin (dermatitis) or lung bronchial inflammation after repeated contact). Many chemicals used in machining processes are irritants.
Corrosive	Substances that attack normally by burning living tissues.
Harmful	Substances if swallowed, inhaled, or penetrates the skin may pose health risks.
Toxic	Substances that impede or prevent the function of one or more organs within the body such as kidney, lungs, and heart. Lead, mercury, pesticides, and carbon monoxide are toxic substances.
Carcinogenic	Substances that promote abnormal development of body cells to become cancers. Asbestos, hard wood dust, creosote, and some mineral oils are carcinogenic.

impacts during the selection, use, and disposal of these chemicals. Chemical exposure hazards can be avoided by adopting the following measures.

- Train employees on the safe handling practices of hazardous chemicals.
- Raise employee awareness of the potential hazards involving various chemicals.
- Keep employee exposure to chemicals within acceptable levels.
- Provide eyewash fountains and safety showers in areas where chemicals are used.
- Label containers and piping systems as to their contents.
- Use personal protective clothing and equipment.
- Keep flammable or toxic chemicals in closed containers when not in use.
- Establish standard operating procedures for cleaning up chemical spills.

26.3.8 FIRE HAZARDS

Fire hazards are conditions that favor fire development or growth. Three elements are requested to start and sustain a fire: oxygen, fuel, and heat. Combustion is the process by which fire converts fuel and oxygen to energy, usually in the form of heat. Explosion is a very rapid fire. Figure 26.8 shows the common sources of fire hazards that can be eliminated through the following measures:

- Having a fire prevention plan.
- Establishing practices and procedures to control potential fire hazards.
- Keeping employees aware of the fire hazards to which they are exposed.
- Having a certified fire alarm system.
- Inspecting interior standpipes and valves regularly.
- Keep fire doors and shutters in good operating condition.
- Periodically checking and maintaining the automatic sprinkler system.

26.3.9 HAZARDS DUE TO WORKPLACE STRESS

A lack of safety measures can cause stress at work. Workers need to feel safe from hazards such as the extreme temperature, pressure, electricity, fire, explosives, toxic materials, ionizing radiation, noise, and dangerous machinery (Figure 26.9). Stress-related problems include mood disturbance, psychological distress, sleep disturbance, upset stomach, headache, and problems in relationships with family and friends. The effects of job stress on chronic diseases is more difficult to ascertain

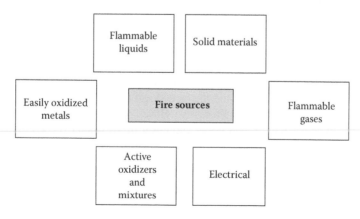

FIGURE 26.8 Sources of fire hazards.

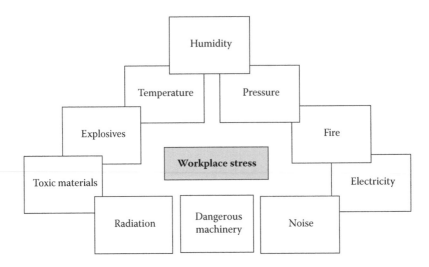

FIGURE 26.9 Sources of workplace stresses.

because chronic diseases develop over relatively long periods of time and are influenced by many factors other than workplace stress. Nonetheless, there is some evidence that stress plays a role in the development of several types of chronic health problems including cardiovascular disease, musculoskeletal disorders, and psychological disorders. In order to reduce the effect of stress due to hazards, workers should feel that their managers are committed to safety issues and their company has efficient safety and health programs.

26.4 PERSONAL PROTECTIVE EQUIPMENT

As a supplementary protection where the safety of workers cannot be ensured by other means, such as eliminating the hazard or controlling/minimizing, a suitable and sufficient personal protective equipment (PPE), depending on the type of work and risks should be used; these protective equipment include the following.

- Head protection: Helmets intended for head protection are subjected to a test for resistance to splashes of molten metal and heavy blows. The helmet should be as light as possible, be flexible, and should not irritate or injure the user.
- Face and eye protection: Face shields or eye protectors are used to protect against flying particles, fumes, dust, and chemical hazards. Face shields should be used in furnace operations and other hot work involving exposure to high-temperature radiation sources. Protection is also necessary against sparks or flying hot objects. Under these conditions, face protectors of the helmet type and the face-shield type are preferred.
- Upper and lower limb protection: Protective gloves or gauntlets, and suitable protective clothing, should be worn to protect upper and lower limbs when exposed to heat radiation or while handling hot, hazardous, or other substances that might cause injury to the skin. Burns of the lower limbs from molten metal sparks or corrosive chemicals may occur. Safety footwear and other leg protection should also be used.
- Respiratory protective equipment: Respirators, appropriate to the hazard and risk in question, should be used to protect the worker. Different sizes and models should be available to accommodate a broad range of facial types.

- Hearing protection: Hearing protectors should be comfortable, and the users should be trained to use them properly. Special attention should be paid to the possibility of increased risk of accidents because of the use of hearing protectors.
- Protection from falls: Workers should be provided with and trained to use appropriate fall protection equipment such as harnesses and lifelines.
- Work clothing: Workers should wear the appropriate protective clothing provided by the employer. Their selection should take into account the adequacy of the design and the fit of the clothing, the environment in which it will be worn, and the special requirements of workers exposed to molten metal and associated hazards.

26.5 HAZARDS OF MANUFACTURING OPERATIONS

Manufacturing operations change raw materials into final finished or semifinished products, using a variety of processes that include smelting operations used for the production of ingots, billets, and blooms. These products are further changed into sheets, slabs, rods, sections, tubes, and wires using metal-forming operations. Welding is then used to weld produced sections in order to fabricate machine elements and structures. Casting operations are used to produce complex shaped parts by melting and pouring in molds mostly made from sand. Welded and cast products are also finished through a series of machining operations that raise the accuracy and surface quality of parts. The following section explains the hazards associated with the common manufacturing operations that affect the health and safety of people at work.

26.5.1 MELTING OF METALS

Iron and steel industry exposes workers to a wide range of hazards or workplace activities that cause injury, ill health, diseases, or death. These hazards can be classified as environmental, physical, chemical, and safety hazards. The choice and implementation of specific measures for preventing workplace injury and ill health of the workforce depend on the principal hazards and the anticipated injuries and diseases, ill health, and incidents. Melting industry causes significant effects on environmental media such as air, water, and soil. Figure 26.10 shows the various hazards associated with metal melting operations which include the following.

1. Air pollution: Blast furnaces emit dust containing metals (Pb, Cd, and Hg) in pollutants such as sulfur dioxides, nitrogen dioxides, and volatile organic compounds including benzene and dioxins. Dust emissions reach 1.5 kg/t of steel produced. The implementation of the best techniques should limit emissions to 1 kg/t. Hazardous air pollutants (HAP) include the following.
   - *Lead:* The lead is a cumulative poison which is ingested in food and water, as well as being inhaled. Exposure to lead causes brain damage characterized by mental incompetence and highly aggressive behavior.
   - *Nitrogen dioxide:* Exposure to NO_2 results in cough and irritation of the respiratory tract. Continued exposure causes an abnormal accumulation of fluid in the lung.
   - *Sulfur oxides:* Sulfur oxides include sulfur dioxide, sulfur trioxides, and their acids have harmful effects on the health.
2. Water pollution: Cooling water and water used to treat gases are contaminated by various chemicals, hydrocarbons, and heavy metals. Toxic metals are individual metals and metal compounds that negatively affect people's health because they build up in the biological system and become a significant health hazard.
   Prevention and protection measures for the environment are
   - Dust reduction by using suitable filters.
   - The injection of lime and activated carbon to reduce the quantity of metals, dioxins, and sulfur oxides contained in dust.

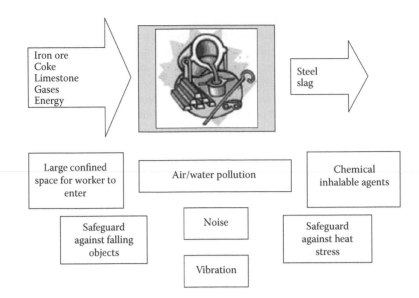

FIGURE 26.10 Melting hazards.

- Reducing nitrogen dioxides by recirculating flue gases.
- Protection of the subsoil and groundwater sources.
- Monitoring of the environment (air, groundwater, etc.).

3. Noise: Exposure to high noise levels cause noise-induced hearing loss, interfere with communication, and results in nervous fatigue with an increased risk of occupational injury.

4. Vibration: Exposure of workers to hazardous vibration is mainly caused by whole-body vibration, when the body is supported on a surface that is vibrating. Hand-transmitted vibration is caused by various processes in which vibrating tools or workpieces are grasped or pushed by the hands or fingers.

5. Heat stress: Melting operations involving high air temperatures, radiant heat sources, high humidity, and direct physical contact with hot objects, have a high potential for inducing heat stress to employees.

6. Chemical hazards: The production of molten metals involves the consumption and generation of a variety of inhalable agents including, gases, vapors, dusts, fumes, smokes, and aerosols that comprise toxicological hazards. Exposure to asbestos causes diseases of the respiratory and digestive tracts through inhalation or ingestion.

7. Confined space: A confined space is one that is large enough for the worker to enter, has limited or restricted means of entrance or exit, and is not designed for continuous employee occupancy. Examples of temporary occupancy might entail a person performing repairs on a furnace or servicing a fuel tank. Employers should be especially vigilant about all OSHA hazards that may exist in a confined space, in particular the build-up of toxic or flammable gases, poor air quality, oxygen displacement, and engulfment.

8. Falling objects: Falling materials form a dangerous hazard. The employer should take all the necessary steps to prevent materials or objects falling. Employers should ensure that areas are kept clean, in good working order, and well maintained. Also, they should ensure that covered walkways or alternative safeguards are used.

9. Handling molten metal, dross, or slag: Burns may occur at many points in the melting industries during tapping from molten metal or slag from spills, spatters, or eruptions of hot metal from ladles or vessels during processing, pouring, or transporting.

26.5.2 SAND CASTING

Sand casting is used to produce complex shaped parts by melting and pouring metals and alloys into sand molds. The process generates many hazards during sand preparation, mold and core making, melting, pouring, casting knock out, cleaning, and finishing stages. Figure 26.11 shows the possible hazards associated with sand casting that include the following.

1. Molten metal: Molten metal forms serious hazard in metal casting applications where the risk of hot metal splashes may increase when
   - Charging a furnace with impure or moist scrap metals and alloys.
   - Using damp tools and molds that touch the molten metal.
   - Pouring the molten metal from ladles into molds.

   Molten metals also emit electromagnetic radiation in the furnace and near pouring areas. Foundry workers are primarily endangered to infrared (IR) and ultraviolet (UV) radiation that may cause eye damage. Splashes of the molten metal and the radiant heat during melting and pouring processes may result in serious burns on the body, while sparks generated from molten metal may also affect the eyes. There are several measures that can be adopted to prevent or minimize the risk associated with the handling of molten metal in foundry shops.
   a. Use barriers to protect workers against the splashes of molten metal and electromagnetic radiation.
   b. Keep all the combustible materials and volatile liquids away from the melting and pouring areas.
   c. Make sure that the molten metal does not come in contact with water or other contaminants.
   d. Restrain the unauthorized access by using barriers to the furnace and pouring areas.
   e. Restrain the workers and other personnel from wearing synthetic clothing in the pouring area.
   f. Ensure proper use and maintenance of personal protective equipment, which include
      - Heat-resistant protective clothing
      - Eye protection with side shields
      - Special UV and infrared glasses

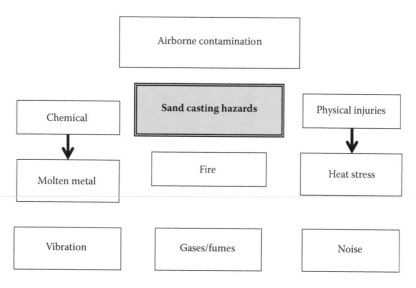

FIGURE 26.11 Sand casting hazards.

2. Airborne contamination: Substantial concentrations of airborne contaminants can be found in various aspects of foundry operations (Figure 26.12). These contaminants can be found in several areas including pattern, core, and mold making, fettling, and sand mixing regions. Dust is particulate matter produced from solids and dispersed into the air owing to the movement, loading, cleaning, and handling of metal, wood, and sand. Fumes are also airborne solid particles that are formed as a result of the condensation of molten metal in cool air. The different types of dust and fumes that expose workers to various health risks in foundries are wood dust, metal dust, and silica dust. Airborne contaminants are generally released from the

- Sand reclamation, preparation, and mixing
- Mold and core including core baking and mold drying from additives, binders, and catalysts
- Scrap handling and preparation using heat and solvent degreasers
- Melting procedure
- Treatment and inoculation of molten metal prior to pouring
- Cooling of castings that causes decomposition of organic binders
- Fettling
- Casting knockout and shake-out processes

Concentration of silica dust, coal dust, metal fragments, and other airborne contaminant should be controlled using less hazardous sands and local exhaust ventilation at the mixing stage of dry sand. The general rules that can be followed to reduce/minimize airborne contaminants are as follows.

- Use chromite sand instead of silica sand.
- Adopt wet or vacuum processes instead of compressed air.
- Set canopy hoods near the doors of furnace and the tapping outlets to seize contaminants and direct them through an emission regulation system.
- Monitor the carbon monoxide levels in the area of operation.

3. Gases and vapors: Gaseous contaminants may also arise as a result of a chemical reaction or the breakdown of complex chemical materials. Some of the gases that are found in

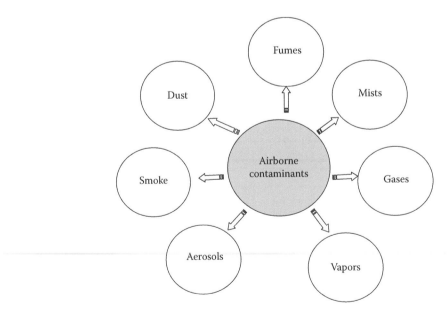

FIGURE 26.12   Common airborne contaminants.

a foundry are carbon monoxide, ammonia, carbon dioxide, methane, hydrogen chloride, hydrogen sulphide, nitrogen, sulfur dioxide, and ozone. Vapors are the gaseous form of substances, which generally exist in the form of liquids or solid state at room temperature and pressure. These vapors are produced by the natural evaporation, heating, or spraying of organic solvents that are used in foundry processes.

Gases and vapors are mostly invisible; however, some may have a strong and characteristic smell, which may give the hint of their presence in the foundry. Instead, in some cases, the gases may have no such smell and may induce health problems at extremely low concentrations. Some other gases show their presence through coughing, respiratory irritation, asthma, acidic taste, and watering of the eyes. Minimizing exposure to gases and vapors include the following steps:
- Regular monitoring of the foundry air
- Automatic alarms for the dangerous carbon monoxide levels
- Exhaust ventilation
- Increased workers awareness regarding the dangers, recognition, and treatment of poisons gases
- Biological monitoring of carbon monoxide using blood sampling or in exhaled air at the end of a working shift

4. Noise and vibrations: Noise and vibrations are generated during the following stages of the sand casting process
   - Pattern making
   - Sand ramming
   - Casting knockout and shaking
   - Cutting and removal of risers and feeders
   - Cleaning and finishing

5. Heat stress: Foundry workers experience heat load, which is determined by the time spent at each workstation, the intensity of work, the clothing worn, and the immediate workstation environment including air circulation. If the heat load is sufficiently severe, effects on health and performance will occur. These range from decreased concentration to painful cramps, fainting, heat exhaustion, and heatstroke that require immediate medical attention. Heat stress can also aggravate the effects of exposure to noise and carbon monoxide. Special clothing should be worn for protection against the heat radiating from the heat sources as well as contact with molten metal.

6. Chemicals: Specific chemicals including borax, boric acid, and other casting fluxes like ammonium chloride or sodium chloride are used. Zinc is also added to the melt a moment before casting to degas it. Some alloys may contain hazardous metals like cadmium, beryllium, arsenic, and antimony. The torch flame used for mold drying and cutting operations produces chemical by-products. Using nonprecious scrap metals produces toxic fumes when melted. Scrap metals that are coated with lead- or cadmium-based paints produce poisonous gases. Mineral oils and lubricants, used in machinery, are another source of chemical materials used in foundries. Working in such conditions, metal fume fever causes flu-like symptoms: fever, chills, and aches occur usually 2–6 h after exposure.

7. Fire: When working with gases, torch systems, hot metals, and electric and gas systems, there is an increased chance of having a fire/explosion hazard. Flammable materials and liquids should therefore be kept away from any source of heat or spark. Fire alarms and fire extinguishers together with a sound fire plan should also be available.

8. Physical injuries: Serious burns may result from splashes of molten metal. Frequent, unprotected viewing of white-hot metals in furnaces and pouring areas may cause eye cataracts. Additional eye injuries may arise from molten metal or fragments of metal. An understanding,

TABLE 26.3
Hazards Associated with Sand Casting Operation

Hazard Type	Environmental			Physical						
Hazard Source	**Water Pollution**	**Air Pollution**	**Soil Pollution**	**Noise/ Vibration**	**Dust**	**Heat**	**Moisture**	**Stress/Fatigue**	**UV Radiation**	**Electric Shock**
Foundry Steps										
Pattern making		X		X	X					
Mold preparation		X		X	X	X				X
Metal preparation		X	X		X					
Metal melting	X	X	X		X	X	X		X	
Casting	X	X	X		X	X			X	
Casting removal		X		X	X			X		X
Cleaning/finishing		X		X	X					X
Heat treatment		X				X	X			
Risk	Ill health	Lung cancer	Flora/fauna damage	Hearing loss	Silicosis	Heat stress	General fatigue	Exhaustion	Skin cancer	Severe injury

Hazard Type	Chemical				Mechanical/ Manual Handling			Safety					
Hazard Source	**Fumes**	**Chemicals/ Solvents**	**Toxic Waste**	**Heavy Metal Contamination**	**Steam**	**Sharp Edges**	**Manual Handling**	**Explosions**	**Equipment Failure**	**Hot Material**	**Molten Metal**	**Fire**	**Sparks**
Foundry Steps													
Pattern making	X	X					X						
Mold preparation	X	X					X						
Metal preparation		X	X	X		X	X			X			X
Metal melting	X	X			X	X	X	X	X		X	X	
Casting	X				X		X					X	
Casting removal	X					X	X						
Cleaning/finishing									X				
Heat treatment	X							X	X				
Risk	Lung cancer/ fever	Tissue burns		Lung cancer		Injury		Burns	Injuries		Skin burns		

appreciation, and application of the following prevention and control measures can minimize the risk of physical injury in the foundry shop.

- Check the working area to ensure good housekeeping practices.
- Display the operating instructions for each furnace.
- Use suitable protective clothing and equipment.
- Install barriers or other suitable shields against molten metal splashes.
- Provide safety blankets and automatic emergency showers to extinguish burning clothing.

Table 26.3 summarizes the various hazards and risks of foundry operations.

26.5.3 METAL-FORMING OPERATIONS

Metal forming is mainly used to produce semifinished pars and pressed forms from hot/cold metal blanks by the application of force to form the workpiece. In hot forming the billets are heated to a temperature that is dependent on the material being formed. For example, steel is heated between 1150° and 1260°C, brass between 530° and 815°C, and aluminum between 440° and 490°C. During hot forming and other presswork operations the operator may be exposed to hazards (Figure 26.13) that include the following.

1. Gases and fumes: Furnaces used for hot forming operations expel large amounts of carbon monoxide and other highly toxic fumes and gases that affect the human health.
2. Noise: Hammer forges, rolling mills, and presses generate loud noise because of the multiple impacts required to form each part from metal stock. The operators are therefore exposed to frequent loud impact noise because they are within 3–15 feet of the source of high noise levels. Over a typical 10-h shift, workers are routinely exposed to time-weighted average (TWA) exposures of 110–115 dB, which far exceeds the OSHA 8-h TWA permissible exposure limit (PEL) for a noise level of 85 dB. Noise is a hearing hazard that requires hearing protectors such as ear plugs, canal caps, and ear muffs to protect the user from hearing loss.
3. Chemicals: Potential health hazards that exist in metal-forming operations include oil, degreasing solvents, cleaning solvents, flammables, and fumes that are irritants and cause health hazards. Apron, overalls, and smocks are used to protect the operator from physical injury and other expected health hazards. Gloves are essential when hands are in contact

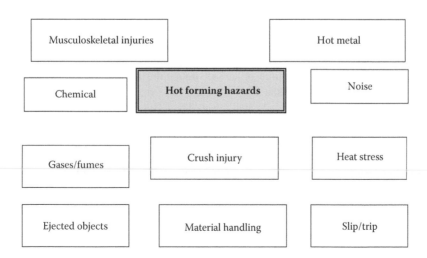

FIGURE 26.13 Hot forming hazards.

with sharp edges and rough surfaces. Safety glasses are essential in all metal-forming operations because the presence of flying metal particles is always possible.

4. Crush injuries: Repeated, alternating motions of machine parts or formed parts is a major cause of crush injuries, usually involving a person's fingers and/or hands. To eliminate or reduce the possibility of this hazard, consider the following control measures:
   - Effective inspection and maintenance programs.
   - Risk assessment plans.
   - Guards of substantial construction to the rear and side of the presses.
   - Marking, identifying, and locating emergency stops so that they are readily seen and conveniently accessed by all operator positions.
   - Shrouding mechanical foot pedal or foot switches to prevent accidental activation.
   - Developing and reviewing a system for the safe operation of the presses.
   - Instructing, informing, and training employees about the system of safe work.
   - Providing oil swabs or hand die lube spray devices or scale removers.

5. Ejected objects: Rapidly moving parts of machinery cause injuries to the operator's eyes and other parts of the body as a result of flying hot scale or other debris being ejected during the metal forming process. To reduce or eliminate the possibility of this type of hazard
   - Conduct a risk assessment of the hazard.
   - Provide a barrier to stop flying hot scale from striking the operator.
   - Provide personal protective equipment such as eye protection, leather gloves, and protective clothing.

6. Sprain and strain: Known as muscular skeletal disorders that are often due to poor postures, repetitive movement, and high forces exerted. These injuries may also result from defective hand tools or tools that are not designed for the purpose for which they are used.

7. Slip and trip: Trips or falls occur because of handling large, hot, or awkwardly shaped objects. Foot and leg injuries occur because of falling components or tools. To eliminate or reduce the possibility of this hazard, consider the following measures.
   - Arrange the layout of the work area so as to eliminate slip and trip hazards.
   - Install barriers to prevent the operator from hazards associated with tripping in front of the machines.
   - Use tools of sufficient length to keep the operator clear of the danger areas.
   - Provide personal protective equipment such as safety footwear with steel toe guards.

8. Hot metal: The potential risk exists for employees when they receive burns from exposure to hot surfaces. Careless handling of hot objects can cause thermal burns. Heating furnaces produce infrared radiation, which forms a hazard to an eye and skin. Additionally, the large amount of heat from furnaces causes heat stress. To prevent the possibility of this type of hazard, provide guarding, where practicable, for the product input and output channels of the hot press and use personal protective equipment.

9. Material handling: Another source of serious injury during metal-forming operations is handling the scrap. Chips and shavings present sharp edges and abrasive surfaces. Burrs and splinters of metal often pierce heavy gloves. The following hazards are associated with material handling in the metal-forming industry.
   - Nips occur during handling of the material during the entire process where moving equipment carries the part or form of material.
   - Pinches result when the worker is caught between the material and the equipment.
   - Sharp edges are present in scrap, tools, presses, and dies.
   - Rough surfaces or abrasive surfaces are present in scrap handling.

Machine guarding is provided to protect the operator and other employees in the machine area from ingoing nip points, rotating parts, and flying chips. Table 26.4 shows the various sources of hazards and risk associated with a variety of metal-forming operations.

TABLE 26.4
Hazards Associated with Metal-Forming Operation

Hazard Source	Water Pollution	Air Pollution	Noise	Smoke	Heat	UV Radiation	Musculoskeletal	Electric Shock	Vibrations
Hazard type	Environmental		Physical						
Forming process									
Forging	X	X	X	X	X	X		X	X
Rolling	X	X	X	X	X	X		X	X
Extrusion	X		X		X	X		X	X
Press work	X		X				X	X	X
Risk	Ill health	Cancer	Hearing loss	Lung irritation	Heat stress	Eye damage	Fatigue	Injury	Exhaustion

Hazard Source	Toxic Fumes	Chemicals	Heavy Metal Contamination	Toxic Gases	Steam	Lubricants	Flying Hot Metal	Mechanical/Manual Handling	Crushing	Equipment Failure	Pinches	Rough Surfaces	Sharp Edges
Hazard type	Chemical						Safety						
Forming process													
Forging	X	X	X	X	X	X	X	X	X	X	X	X	X
Rolling	X	X	X	X	X	X	X	X	X	X	X	X	X
Extrusion	X	X	X	X		X	X	X	X	X	X	X	X
Press work						X		X	x	x	x	x	x
Risk	Lung cancer	Skin damage	Lung cancer			Eye irritation	Burns	Injury			Severe injury		

26.5.4 Machining Operations

Machining processes generate solid, liquid, or gaseous by-products that present hazards for workers, machines, and the environment.

26.5.4.1 Traditional Machining

Metal-cutting operation removes the machining allowance from a workpiece in the form of chips. Drawbacks of such methods are noise, leakage, and flying chips. Heat and wastes of machining operations cause hazards and high risk of injuries. Therefore, safety precautions must be considered to reduce the hazards of the machining process (Figure 26.14).

1. Noise/vibrations: During machining, vibrations and noise components are generated. Noise levels of 85 dB are the maximum noise level regarded as safe and tolerable for an 8-hour exposure. When noise levels exceed 90 dB, hearing damage is liable to occur, and therefore ear plugs must be worn.
2. Flying chips: Flying chips form a major hazard and risk to the operator as they fly from the machine during the cutting process. Flying particles such as metal chips may result in eye or skin injuries or irritation. Grinding, cutting, and drilling of metal and wood generate airborne particles that affect the respiratory system. Under such circumstances, it is always recommended to wear safety glasses, goggles, or shields and use proper ventilation.
3. Cutting fluids: Cutting fluids contain many chemical additives that can lead to skin and respiratory diseases and an increased danger of cancer. This is mainly caused by the constituents and additives of the cutting fluids as well as the reaction products and particles generated during the machining process. Unfortunately, spoiled or contaminated cutting fluids are the most common wastes from the machining process that are considered hazardous wastes to the environment due to their oil content, chemical additives, chips, and dust. During machining, the high cutting speeds (>3500 m/min) produce high temperature in the machining zone that causes vaporization of the fluids and metal particles. These emissions enter the atmosphere, thus forming a complex mixture of vapors and fumes due to elements of the workpiece, cutting tool, and cutting fluids. Cutting fluids have negative health effects on the operators that appear as dermatological, respiratory, and pulmonary effects. Exposure to mists caused by the cutting fluids raises workers' susceptibility to respiratory problems that depends on the level of chemicals and particles contained in generated mists.

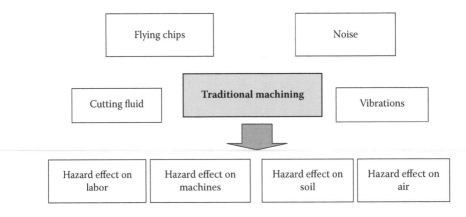

FIGURE 26.14 Traditional machining hazards.

26.5.4.2 Nontraditional Machining Processes

Nontraditional machining (NTM) processes remove materials using mechanical, thermal, chemical, or electrochemical actions. They generate solid, liquid, or gaseous by-products that cause hazards for the workers, environment, and the equipment. The level of the hazards depends on the machining method used.

1. Chemical machining: Chemical machining (CHM) uses chemical acids and alkalis to remove materials from metallic and nonmetallic workpieces. CHM induce severe effects on the surrounding environment, creates difficulties in handling and storage, and causes a damaging effect on the surrounding materials. The acidity of an etchant dissolved in water is commonly measured by the pH number. Solutions with pH values of less than 7 are described as being acidic. Acid deposition attacks steel structures and fades paint on machine tools. During CHM, exposure to hafnium (Hf) occurs through inhalation, ingestion, and eye or skin contact. The level of hazard depends on the properties of the acid/alkalis used, its concentration, and the time of exposure. The hazards of CHM are shown in Figure 26.15. These include the following.
   - Health effects on labor (Figure 26.16) in the form of irritation, corrosive injuries and burns, rapid, severe, and often irreversible damage of the eyes in addition to risk of larynx and lung cancer.
   - Steel structures forming machine tool elements are often affected by localized attack and uniform corrosion. In this regard, covering materials (e.g., paints) have been used to protect steel structures from rusting. Poor surface preparation is the prime cause of protective coating failure (see Figure 26.17).
   - Improper disposal of CHM etchants changes the level of acidity and alkalinity affect the *flora* and *fauna* in water. The change of pH from 7 (neutral water) has an adverse effect on aquatic life (Figure 26.18). As an example, at pH 6 crustaceans and mollusks start to disappear and moss increases, while at pH 5.5, salmon, trout, and white fish start to die and salamander eggs fail to hatch. Acidity of pH 4 has a lethal effect on crickets and frogs. Some alkalis such as ammonia also have an acute toxic effect on fish.

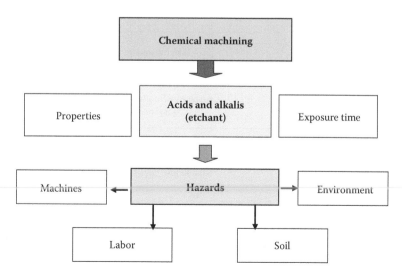

FIGURE 26.15 CHM hazards.

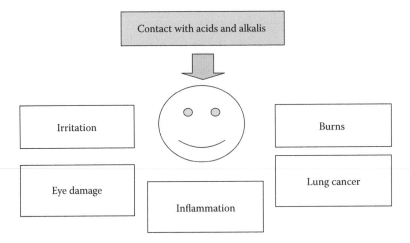

FIGURE 26.16 CHM labor impacts.

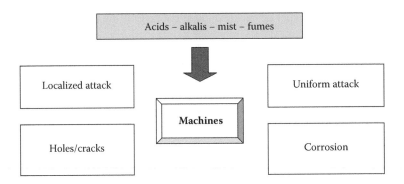

FIGURE 26.17 CHM machine impacts.

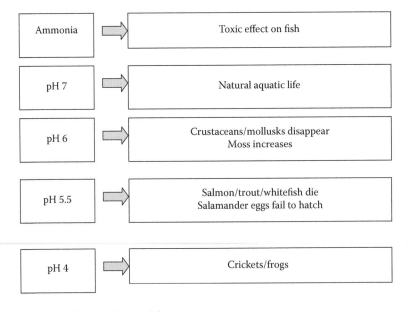

FIGURE 26.18 CHM effects on flora and fauna.

- Soil is contaminated when, owing to acidity, it has a pH value of 4–5 and heavily contaminated when the pH is 2–4. When soil has a pH value of 9–10, it is contaminated owing to alkalinity, and at pH 10–12 it is heavily contaminated. Aerosols of solid nitrogen and sulfuric oxides or liquid corrosive substances are air pollutants. They form corrosive gases and acid fumes that combine with water to form acid rain that damage plants (Figure 26.19).

2. Electrochemical machining: Electrochemical machining (ECM) is known as an environmental polluting process. Hazards of ECM are shown in Figure 26.20. Explosive hydrogen gas is generated during the electrolyzing process by the following reaction:

$$2H_2O + 2e^- \Rightarrow H_2 + 2OH^-$$

$$Fe + 2H_2O \Rightarrow Fe(OH)_2 + H_2$$

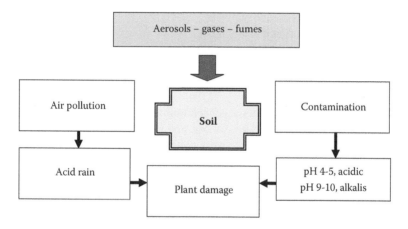

FIGURE 26.19 CHM soil impacts.

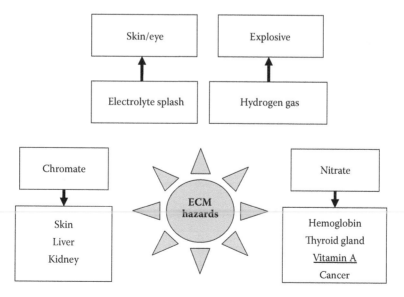

FIGURE 26.20 ECM hazards.

Local exhaust must be provided to prevent the hydrogen gas from reaching its lower flammability limit and to remove the mists from the workers breathing zone. Skin contact with the electrolyte contents causes irritation and therefore must be controlled by good work practices. The door/cover should be interlocked with the electrolyte supply system. In such a case, machining process should not be started when the interlock of the safeguarding device is opened. Exposure to trivalent chromium compounds occurs through inhalation, ingestion, and eye or skin contact affects the human skin, liver, and kidneys and causes contact dermatitis. Nitrates and nitrites are known to cause reactions with hemoglobin in blood, decreased functioning of the thyroid gland, shortages of vitamin A, and fashioning of nitrosamines, which commonly causes cancer. Effective methods of controlling worker exposures to nitrogen and trivalent chromium include the use of process enclosure, exhaust ventilation, and personal protective equipment.

3. Electrodischarge machining: During electrodischarge machining (EDM) the work material is removed by a series of sparks that occur in the dielectric liquid filling the gap between the tool electrode and workpiece. EDM has several hazard sources (Figure 26.21)
   - Hazardous smoke, vapors, and aerosols
   - Decomposition products and heavy metals
   - Hydrocarbon dielectrics
   - Sharp-edge metallic particles
   - Possible fire hazard and explosions
   - Electromagnetic radiation

In EDM, the total aerosols and vapor concentration exceed the limits of 5 mg/m^3 if no protective measures are taken. Fumes, vapors, and aerosols depend on the EDM material removal process, the dielectric, and the work material (Figure 26.22).

- EDM die sinking generates more fumes and aerosols than wire EDM.
- Material composition may contain toxic or health attacking substances such as nickel.
- Lower viscosity dielectric liquids produce fewer fumes and vapors.
- The level of the dielectric over the erosion spot condenses and absorbs a considerable part of the vapor and fumes in the dielectric itself (80 mm is recommended).

During EDM using mineral oils or organic dielectric fluids generates hazardous fumes such as polycyclic aromatic hydrocarbons (PAH), benzene, vapor of mineral oil, mineral aerosols, and other products are generated by dissociation of oil and its additives. Hydrocarbon dielectrics generate the same vapors and aerosols except PAH and benzene. For water-based solvents normally used in wire EDM, carbon monoxide, nitrous oxide, ozone, and harmful aerosols are formed.

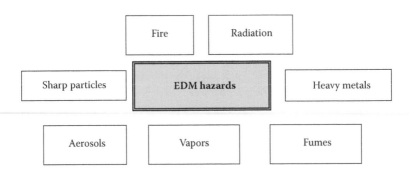

FIGURE 26.21　EDM hazards.

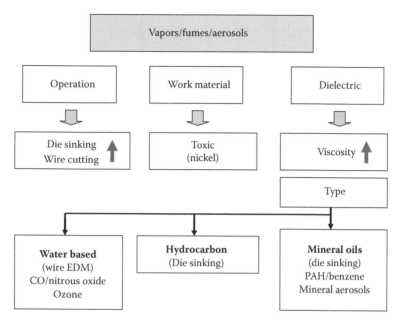

FIGURE 26.22 EDM hazardous fumes.

Because of erosion of the workpiece and tool electrodes, inorganic vapors of tungsten carbide, titanium carbide, chromium, nickel, molybdenum, and barium are released and condensate in air. The rising smoke carries additional organic components from the dielectric liquid. Additionally, the erosion slurry contains eroded workpiece and tool material particles and solid decomposition products of the dielectric.

In order to reduce the possible hazards of machining by EDM, the following measures should be strictly followed.

- Reduce air pollution using suitable filters.
- Incorporate dielectric cleaning and disposal systems.
- Keep the temperature of the working media at 15°C below its flashing point.
- Reduce the emitted electromagnetic radiation by proper shielding of the machine.
- Reduce the possibility of fire hazard.
- Use level sensors for the dielectric level.
- Avoid dielectrics with flashing point of less than 65°C.
- Raise the operator awareness regarding the risk of high voltage used in EDM.

4. Laser beam machining: During machining by lasers, the material is heated and partly vaporized thus producing hazardous materials of gases or aerosols. During laser beam machining (LBM) of thermoplastics, 99% of the particles generated have a diameter less than 10 μm and more than 90% are smaller than 1 μm. Most particles are in the range of 0.03–0.5 μm. Alloy steels emit 4–5 times more aerosols than carbon steel. Almost all particles generated by LBM have a diameter in the range of 0.04–0.35 μm. The hazard potential is influenced by power density, wave length, exposure time of radiation and if LB is visible or invisible. Figure 26.23 shows the possible hazards by laser beam machining. Repeated or even a single exposure to certain laser wavelengths causes skin damage of varying degrees more than in other parts of the body. Eye injuries are caused by thermal or photochemical mechanisms that occur when a laser beam interacts with the eye. As the beam enters the eye, its energy is then concentrated by the lens of the eye about 100,000

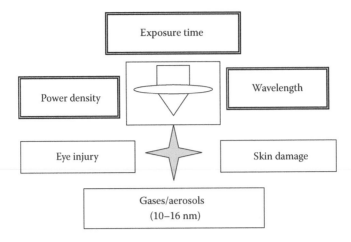

FIGURE 26.23   LBM hazards.

times at the retina, thus causing damage. Laser beam machining protective measures are as follows.

- Follow the American National Standards Institute (ANSI), ANSI Z136.1 that provides the recommendations for the safe use of lasers in typical industrial environments.
- Determine the nominal hazard zone (NHZ) where the level of direct reflected or scattered radiation exceeds the level of the maximum permissible exposure (MPE).
- Determine beam path controls unless they are totally enclosed, interlocked, and there is no beam access during normal system operation.
- Fix a laser-controlled area during periods of service.
- Use protective housings and enclosures that cover the equipment or the beam path.
- Use interlocks on the protective housings so that if they are removed, the beam is shut off.
- Use beam stops that provide safe termination of the beam path.
- Fix labels and signs that give notice of lasers operating in a given area.
- Use barriers or curtains, clothing, or eyewear.
- Train workers on rules and regulations that are designed to minimize the risk of laser beam exposure.

5. Ultrasonic machining: Ultrasonic machining (USM) is the removal of hard and brittle materials using an axially oscillating tool at ultrasonic frequency of 18–20 kHz. During tool oscillation, the abrasive slurry of B_4C, Al_2O_3 or SiC, 100–800 grit, is continuously fed into the machining zone, between a soft tool (brass or steel) and the hard and brittle workpiece. The process finds applications when machining ceramics, glass, and carbides. However, it has many environmental and health hazards (Figure 26.24) that include the electromagnetic field, ultrasonic wave, and abrasive slurry. The effects of the electromagnetic field (emf) on the health of individuals are of concern to some people. Individuals are advised to stay away from the emf sources, the strength of which drops quickly just a few feet from the source. Ultrasonic frequencies have been found to produce sound in the audible range from 96 to 105 dB, which is far above the permissible level. Excessive noise levels can be reduced by using enclosures made of thin layers of common materials that give adequate acoustic insulation at ultrasonic frequencies. Alternatively, ear protectors can be provided and used.

The ultrasound may affect hearing and produce other health effects that include noise-induced temporary threshold shift, noise-induced permanent threshold shift, acoustic

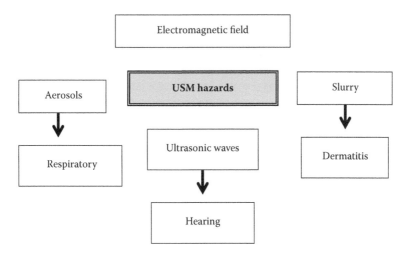

FIGURE 26.24 USM hazards.

trauma, and tinnitus. Other communication and performance effects may include isolation, annoyance, difficulty concentrating, absenteeism, and accidents. Additionally, it may also cause stress, muscle tension, ulcers, increased blood pressure, and hypertension. Mist is generated by the spray from the slurry application. Small droplets may be suspended in the air for several hours or days possibly in the workers breathing zones. Inhaled particles of diameters of less than 10 μm deposit in the respiratory system, thus causing chronic bronchitis, asthma, and even laryngeal cancer. The OSHA standard for airborne particulate (largely due to fluid mist) is 5 mg/m^3, and the United Auto Workers (UAW) has proposed a reduction in the standard to 0.5 mg/m^3. Mist can, however, be reduced by adding antimisting compounds to the abrasive slurry and using a mist collector that prevents mist from entering plant air.

6. Abrasive jet machining: The prime hazard in abrasive jet machining (AJM) is the silica dust. Particles of other cut toxic materials such as lead, mercury, arsenic, zinc, and cadmium may constitute a significant hazard (Figure 26.25). Workers are provided with appropriate respiratory protective equipment against atmospheres containing substances that are harmful if breathed. Risk factors with dust include the type and size of particulate,

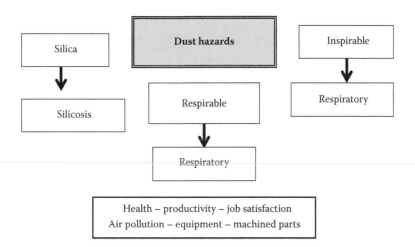

FIGURE 26.25 AJM dust hazards.

its biological effect in the breathing zone, and the duration of exposure. Silica dust is a major hazard in AJM, generated when the abrasive medium or the cut material contains silica. The major risk from silica dust is the silicosis, which causes stiffening and scarring of the lung. Symptoms include shortness of breath, coughing, chest pain, and an increased risk of lung cancer. Measures to control silica dust are made by ensuring that the process is completely isolated. Operators must wear personal protective equipment, and dust must be collected in an appropriate manner using dust collection systems. Hazardous impacts associated with AJM include

- The worker's productivity is decreased with possible high physical/mental fatigue, low job satisfaction, and high error rates.
- The abrasives are imbedded in the work surface thus leading to low product quality.
- Dusty air makes problems in other machines in the same station.

Table 26.5 shows the various sources of hazards and risk associated with a variety of machining operations.

26.5.5 Welding

Manufacturing industry employs many welding techniques that may utilize gases, electricity, friction, explosion, pressure, and chemicals to join parts together. Arc welding is the most popular welding process used in metal industry. This section describes the possible hazards of arc welding (Figure 26.26) and summarizes the hazards generated by some other welding techniques. In general, arc welding constitutes the following hazards:

1. Gases and fumes: Welding smoke is a mixture of very fine particles (fumes) and gases. It contains many substances, such as chromium, nickel, arsenic, asbestos, manganese, silica, beryllium, cadmium, nitrogen oxides, phosgene, and acrogenic. Fluorine compounds, carbon monoxide, cobalt, copper, lead, ozone, selenium, and zinc can be extremely toxic. Generally, welding fumes and gases come from
   - Base material being welded
   - Coatings and paints on the metal being welded
   - Filler materials
   - Flux coatings covering the electrode
   - Shielding gases
   - Chemical reactions occurring during welding by the action of ultraviolet light from the arc, heat process, and consumables used
   - Cleaning and degreasing materials

   Welding smokes form a serious health hazard because exposure to metal fumes such as zinc, magnesium, copper, and copper oxide causes metal fume fever. Welding smokes irritate the eyes, nose, chest, and respiratory tract and cause coughing, wheezing, shortness of breath, bronchitis, inflammation, and fluid in the lungs. Regarding shielding gases and fumes, the following hazards may occur:
   - Inhaling pure argon causes loss of consciousness in seconds.
   - Exposure to the manganese causes serious damage to the brain and nervous system.
   - Breathing iron oxide irritates nasal passages, throat, and lungs.
   - Fumes containing nickel and chromium (stainless steel welding) may cause cancer.
   - Lead harms nervous system as well as kidneys.
   - Welding electrodes that release asbestos dust causes lung cancer.

   The effect of welding smoke and fumes can be minimized, however, by keeping the head out of the fumes, avoiding breathing fumes and gases caused by the flame, and using proper ventilation systems.

TABLE 26.5
Hazards Associated by Different Machining Processes

Hazard type	Environmental			Physical							
Hazard Source	Water Pollution	Air Pollution	Soil Pollution	Noise	Vibration	Radiation	Dust	Magnetic Field	Musculoskeletal	Electric Shock	Slurry
Machining process											
Metal cutting	X	X	X	X	X				X	X	
ECM	X	X	X							X	
CHM	X	X	X							X	
EDM	X	X	X			X				X	
LBM		X	X	X	X	X				X	
USM	X	X		X	X	X	X	X		X	X
AJM		X		X	X		X			X	
Risk	Ill health	Cancer	Flora/fauna	Hearing loss	Exhaustion	Fatigue	Lung disease	Exhaustion	Fatigue	Death	Allergy

Hazard type	Chemical					Mechanical/ Manual Handling		Safety			
Hazard Source	Gases	Vapors	Liquids	Mists	Fume	Flying Chips	Manual Handling	Explosions	Equipment Failure	Sharp Edges	Fire
Machining process											
Metal cutting	X	X	X	X	X	X	X		X	X	
ECM	X	X	X	X	X		X	X	X		X
CHM	X	X	X	X	X		X	X	X		X
EDM	X	X	X	X	X		X	X	X	X	X
LBM	X	X		X	X		X		X		X
USM			X		X		X		X		
AJM							X		X		
Risk	Lung cancer	Lung cancer	Skin burns	Lung disease	Lung disease	Injury	Injury	Burns	Injury	Injury	Burns

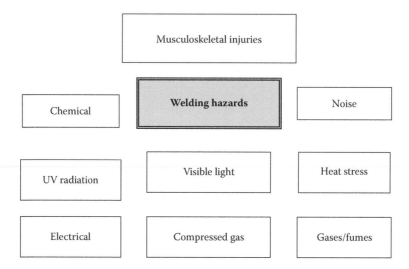

FIGURE 26.26 Welding hazards.

2. Visible light and ultraviolet and infrared radiation: The intense light associated with arc welding can cause damage to the eye retina. Infrared radiation may damage the cornea and result in the formation of cataracts, while the ultraviolet (UV) light reacts with oxygen and nitrogen in the air to form ozone and nitrogen oxides that are deadly at high doses and causes irritation of the nose and throat and serious lung disease. It can also react with chlorinated hydrocarbon solvents to form phosgene that may be deadly. Invisible UV, from the arc, causes arc eye that usually occur many hours after exposure. Exposure to ultraviolet light also causes skin burns and increases the welder's risk of skin cancer. Protective measures against these hazards include wearing protective clothing, gloves, helmet with protective filter, and safety spectacles.

3. Heat stress: The intense heat of welding and sparks can cause burns. Contact with hot slag, metal chips, sparks, and hot electrodes can also cause eye injuries. Excessive exposure to heat can result in heat stress or heat stroke. Welders should be aware of the symptoms such as fatigue, dizziness, loss of appetite, nausea, abdominal pain, and irritability. Ventilation, shielding, rest breaks, and drinking plenty of cool water protect workers against heat-related hazards.

4. Musculoskeletal injuries: Welders have a high prevalence of musculoskeletal complaints including back injuries, shoulder pain, tendonitis, reduced muscle strength, carpal tunnel syndrome, white finger, and knee joint diseases. These problems can be prevented by using proper lifting techniques, avoiding working in one position for long periods of time, working at a comfortable height, using a foot rest when standing for long periods, locating tools and materials conveniently, and minimizing vibrations.

5. Noise: Exposure to loud noise can permanently damage welders' hearing. Noise also causes stress, increased blood pressure, and may contribute to heart disease. Working in a noisy environment for long periods of time make workers tired, nervous, and irritable. If the average noise exposure exceeds 85 dB for over 8 h, hearing protection and annual hearing tests should be provided.

6. Electrical: Although welding generally uses low voltage, there is still a danger of electric shock. Dry gloves should always be worn to protect against electric shock. The workpiece being welded and the frame of all electrically powered machines must be grounded. The insulation on electrode holders and electrical cables should be kept dry and in a good

TABLE 26.6
Hazards Associated with Welding Operation

Hazard Source	Water Pollution	Air Pollution	Noise	Smoke	Heat	Visible Light	UV Radiation	Musculoskeletal	Electric Shock	Magnetic Fields
Hazard type	Environmental		Physical							
Welding process										
Gas		X		X	X			X		
Manual arc		X	X	X	X	X	X	X	X	X
GMAW/GTAW		X	X	X	X	X	X	X	X	X
Plasma arc		X		X		X	X	X	X	X
Submerged arc	X		X	X	X		X		X	X
Resistance	X		X		X				X	
Friction			X		X					X
Ultrasonic			X		X				X	
Thermit						X				
Soldering/brazing				X	X			X		
Laser beam			X		X					
Risk	Ill health	Cancer	Hearing loss	Lung irritation	Heat stress	Eye strain	Eye damage	Strain /sprain	Injury	Exhaustion

Hazard Source	Toxic Fumes	Chemicals	Heavy Metal Contamination	Toxic Gases	Ozone	Fluxes	Flying Hot Metal	Mechanical/ Manual Handling	Explosions	Equipment Failure	Molten Metal	Fire	Sparks
Hazard type	Chemical						Safety						
Welding process													
Gas	X	X	X	X		X	X	X	X	X	X	X	
Manual arc	X		X	X	X	X	X	X		X	X	X	X
GMAW/GTAW	X		X	X	X		X	X	X	X	X	X	X
Plasma arc	X		X	X	X		X	X	X	X	X	X	X
Submerged arc	X		X	X	X		X	X	X	X	X	X	X
Resistance								X		X		X	
Friction			X				X	X		X		X	X
Ultrasonic								X		X			
Thermit	X	X	X	X			X	X	X	X	X	X	
Soldering/brazing	X	X	X			X	X	X	X	X	X	X	X
Laser beam	X	X	X					X		X			
Risk	Lung cancer	Tissue damage	Lung cancer		Eye irritation		Burns	Injury	Burns	Severe injury		Burns	

condition. Electrodes should not be changed with bare hands, wet gloves, or while standing on wet floors.

7. Fires and explosions: The intense heat and sparks produced by arc welding, or the welding flame, can cause fire or explosion if combustible or flammable materials are in the area. For this reason, it is recommended to
   a. Avoid welding containers of a flammable material unless thoroughly cleaned
   b. Clear the area of sawdust, wood shavings, straw, lacquer thinners, petrol, paints, etc.
   c. Choose a sealed concrete floor rather than a gapped wooden floor as a welding bay
   d. Wear clean, fire-resistant, and protective clothing

8. Compressed gases: Hazards associated with compressed gases include fires, explosions, toxic gas exposures, and the physical hazards associated with high-pressure systems. In order to control these hazards, special storage, use, and handling precautions are necessary. Compressed gases should be used with adequate ventilation and a leak detection system. These are some recommendations to follow when using compressed gases during welding.
   • Cylinders must be stored upright.
   • Cylinders should have caps or regulators designed for the gas in use.
   • Pressure relief valves and all lines should be checked before and during welding.
   • Blowpipes, hoses, and fittings must be kept in good condition, cleaned, and regularly checked.

Table 26.6 summarizes the hazards and health associated risk by some welding processes.

26.6   REVIEW QUESTIONS

1. Explain what is meant by a hazard, risk, and risk analysis.
2. In order to ensure safety at work, explain the role of each of the following: Employers, employees, and manufacturers and suppliers.
3. What are the common causes of hazards associated with manual handling of loads?
4. Explain the main hazards and safety measures when using
   • Hand-held tools
   • Mechanical machinery
5. State the main types of vibrations occurring in manufacturing areas.
6. What are the safety measures for reducing the risk of noise exposure?
7. What are the sources of fire hazards? How are they controlled?
8. Explain how the working conditions cause workplace stress to the workers in manufacturing plants.
9. Using line diagrams, show the hazards associated with casting, welding, forming, and metal cutting operations.
10. What are the safety measures followed during melting operations, laser beam machining, and abrasive jet machining?
11. Explain how airborne contamination can be reduced during foundry operations?
12. Show the main hazards associated with chemical and electrochemical machining operations.

BIBLIOGRAPHY

Byrne, G., and Scholta, E. 1993. Environmentally clean machining processes—A strategic approach, *CIRP Annals* 42(1): 471–474.

Davis, M. L., and Cornwell, D. A. 2006. *Introduction to Environmental Engineering,* 4th ed. New York: McGraw-Hill.

El-Hofy, H., and Youssef, H. 2009. Environmental hazards of nontraditional machining. In *Proceedings of the 4th IASME/WSEAS International Conference on Energy and Environment*, pp. 140–145. New York: Cambridge University.

Geotsch, D. L. 2003. *Occupational Safety and Health for Technologists, Engineers, and Managers*, 6th ed. Upper Saddle River, NJ: Pearson Education.

Hughes, P., and Ferrett, E. 2005. *Introduction to Health and Safety at Work*, 2nd ed. New York: Elsevier.

Youssef, H., and El-Hofy, H. 2008. *Machining Technology, Machine Tools and Operations*. Boca Raton, FL: CRC Press.

Appendix A: List of Symbols

Symbol	Description	Units
A	Cross-sectional area of the gating system	mm^2
a	Depth of cut	mm
A_0	Initial area	
A_2	Constant	
A_{bcc}	Lattice parameter of bcc	Å
A_c	Area of cut	mm^2
A_f	Final area	
A_f	Fracture area	m^2
a_{fcc}	Lattice parameter of fcc	Å
a_g	Grain size	m (μm)
$a_{hcp, c}$	Lattice parameters of hcp	Å
A_i	Instantaneous area	mm^2
A_s	Surface area of particle	$μm^2$
A_w	Atomic weight	g
B	Width of grinding wheel	mm
B_1	Force due to bending and unbending	N or kN
C	Clearance on one side of the punch and die	mm
C	Constant	—
C_1	Tool initial cost	$
c_1	Specific heat	Nm/kg.°C
C_2	Tool scrap value	$
Cap	Capacitance	Farads
Cd	Coefficient of thermal diffusivity	m^2/s
C_L	Unit cost related to labor	$
CLA	The arithmetic average roughness	μm
C_M	Unit cost related to machine	$
C_p	Specific heat	kJ/kg.°C
C_p	Process capability ratio	%
C_{pk}	Process capability index	
C_s	Tool sharpening cost	$
C_T	Unit cost related to tool	$
C_t	Total unit cost	$
D	Diameter/nominal size	mm
d	Instantaneous diameter	
D_0	Outer diameter of the tube for EMF	mm
d_0	Initial diameter	mm
D_3	Constant	
D_4	Constant	
D_b	Blank diameter	mm
D_d	Die diameter	mm
d_f	Final diameters	
df	Beam focusing diameter	m (μm)
d_g	Diameter of grinding wheel	mm
D_i	Instantaneous diameter	m^2
D_o	Original diameter	m
D_p	Punch diameter	mm

Symbol	Description	Units
DP	Degree of polymerization	—
d_w	Diameter of WP in grinding	m
$d\varepsilon$	*Plastic strain increment*	—
E	Young's modulus	N/m^2 (pascal)
e	Feed rate in grinding	mm/rev of WP
Ec	Chemical energy released in the detonation	J
Ed	Discharge energy	J
E_f, E_m, E_c	Moduli of elasticity (stiffness) of fiber, matrix, and composite material, respectively	MPa or Gpa
Ei	Lower deviation of shafts	μm
EI	Lower deviation of holes	μm
e_s	Specific energy of the explosive charge	
Es	Upper deviation of shafts	μm
ES	Upper deviation of holes	μm
F	Force	N or kN
F	Feed rate	mm/rev
F_c	Main cutting force	N
F_f	Friction force (or feed force)	N
f_j	Frequency of observations in the jth cell	
F_n	Component normal to tool face	N
F_p	Component normal to shear plane	N
F_r	Radial force component	N
fr	Resonant frequency	Hz (1/sec)
F_s	Shear force	N
F_t	Thrust force	N
f_{tr}	Traverse feed in grinding	mm/rev
f_γ	Factor considering negative rake of abrasives	
G	Gravitational acceleration = 9.81	m/s^2
G	Modulus of rigidity	MPa
h	Height	mm
H	Hydrostatic head	mm
H	Chip thickness	mm
h_0	Initial height	mm
H_e	Heat energy	J
h_f	Final height	
h_g	Maximum chip thickness in grinding	mm
h_i	Instantaneous height	
h_m	Mean chip thickness in grinding	mm
H_s	Height of slip gauges	mm
I	Current	A
I	Fundamental tolerance	μm
τ_i	Shear stress of the interface between matrix and fiber	MPa
i_b	Electron beam current	A
J	Mechanical equivalent of heat	N m/kcal
K	Strength coefficient	MPa
k	Thermal conductivity	
K_h	Pressure multiplying factor for hammer forging	
Kl	Coefficient of heat loss	—
K_p	Pressure multiplying factor for press forging	
K_s	Springback factor	
k_s	Spec. cutting power (spec. cutting resistance)	N/mm^2
K_{sf}	Shape factor (Index)	

Symbol	Description	Units
k_{sm}	Mean specific cutting power of WP	N/mm^2
kt	Thermal conductivity	N/s °C
l	Length of transducer or acoustic horn	cm
l_i	Instantaneous length	m
l_o	Original length	m
l_f	Fracture length	m
L	Length	mm
l/d	Aspect ratio of the fiber	—
L_b	Bend allowance	mm
L_c	Labor cost/min	$
LSL	Lower specification limit	
m	Strain rate sensitivity	—
m_c	Number of cells/subgroups	
MC	Machine capability	%
m_p	Mass electroplated	g
MW	Molecular weight	
n	Exponent/valence/observations	—
N	Rotational speed	rpm
n_i	Number of tip indexings	—
n_s	Number of tool sharpenings	—
p	Pressure	MPa
P	Deformation power	kW
F_p	The hold down force in deep drawing	N or kN
P_c	Main power	kW
Pe	Power of EB	J/s (W)
P_f	Feed power	kW
P_m	Magnetic pressure in EMF	MPa
P_{mot}	Motor power	kW
P_{pk}	Process performance index	
Pt	Stagnant pressure	N/m^2
q	Crater wear factor	
Q	Volume flow rate	mm^3/s
Q	Flow rate (or heat power)	m^3/s or (kcal/sec)
R	Bend radius/extrusion ratio/range	mm
r	Atomic radius/percentage reduction in area	Å, %
R	Resultant cutting force/electric resistance	N, Ω
R_a	The arithmetic average roughness	μm
R_{max}	Maximum peak-to-valley roughness	μm
R_q	The root mean square roughness	μm
R_t	Maximum peak-to-valley roughness	μm
S	Standoff distance	mm
t	Thickness/time	mm, s
T	Time/torque/tool life	s - kN m, min
t_1	Undeformed chip or plate thickness	mm
t_2	Chip thickness	mm
t_c	Charging time	s
t_{ch}	Tool exchanging time	min
t_d	Discharging time	s
T_e	Economical durability	min
t_h	Actual cutting time	min
T_m	Melting point	°C

Symbol	Description	Units
t_m	Total machining time	min
T_{mf}	Mold filling time	s
T_p	Pouring temperature	°C
T_{pr}	Durability for maximum productivity	Min
t_s	Solidification/secondary time	s
U_r	Modulus of resilience	N/m^2 (pascal)
USL	Upper specification limit	
V	Volume	mm^3
V	Cutting speed	m/min
V_a	Arc volt	volt
V_c	Charge voltage	volt
V_c	Volume of composite	mm^3
v_c	Chip velocity	m/min
V_{ca}	Capacitor voltage	volt
v_e	Economical speed	m/min
v_f	Volume fraction of fibers	
V_f	Volume of fibers	mm^3
vf	Traverse speed	m/s
v_g	Linear speed of GW	m/min(m/sec)
vj	Jet velocity	m/s
v_m	Volume fraction of matrix	
Vo	Supply voltage	V
V_p	Volume of particle	μm
v_p	Speed for maximum productivity	m/min
VRR	Material removal rate	mm^3/min-m^3/min
v_s	Velocity along shear plane	m/min
V_s	Sparking voltage	V
v_v	Volume fraction of voids	
V_w	Linear speed of WP	m/min (m/sec)
v_w	Welding speed	cm/min
W	Blank width in bending	Mm
W, dw	Plastic work	MPa
w_0	Initial width of rectangular specimens	mm
w_f	Final width of rectangular specimens	
$W_{F,all}$	Allowable flank wear	mm
X_{high}	Largest value	
X_i	Values of the ith observation	
X_j	Midpoint of the jth cell	
X_{low}	Smallest value	
Y_f	Flow stress of the metallic material	MPa
Y_f'	Flow stress under plane strain $(= \dfrac{2}{\sqrt{3}} Y_f)$	MPa
Z	Number of components/tool life	—
z	Peripheral grains on grinding wheel	
z_e	Number of grains contributing in cut	

Appendix B: Greek Letters

Symbol	Description	Units
σ_s	Ultimate shear strength	MPa
σ_y	Yield stress or flow stress of the material	MPa
μ	Coefficient of friction	
ε	Strain	—
σ	Normal stress component	MPa
τ	Shear stress component	MPa
γ	Shear strain	—
ν	Poisson's ratio	—
$\sigma_x, \sigma_y, \sigma_z$	Normal stress components in Cartesian coordinate system x y z	MPa
σ_m	Mean or hydrostatic stress	MPa
$\dot{\varepsilon}, \dot{\varepsilon}_m$	Strain rate, and mean strain rate	s^{-1}
σ_y	The yield strength of the blank material	MPa
η	The efficiency of energy transfer	
τ	Shear stress	MPa
γ	Shear strain	—
ν	Poisson's ratio	—
λ	Distance between two consecutive grains of GW	mm
$\bar{\beta}$	Attack angle of WP	Radians
β	Attack angle of WP	deg
$\sigma_{\bar{x}}$	Standard deviation of averages	
σ	Standard deviation of individual values	
θ	Angle	deg/rad
UCL_R	Upper control level for R chart	
LCL_R	Lower control level for R chart	
$\bar{X}$	Mean	
$\bar{\bar{X}}$	Average of the subgroup means	
$UCL_{\bar{x}}$	Upper control level for $\bar{X}$-chart	
$LCL_{\bar{x}}$	Lower control level for $\bar{X}$ chart	
$\bar{R}_m$	Mean or average of subgroup ranges	
R_j	Range of the jth subgroup	
$\bar{R}$	Average normal anisotropy of sheet metals	—
ε	Strain	m/m
σ	Stress	N/m^2 (pascal)
α	Clearance angle	deg
β	Wedge angle (or friction angle)	deg
γ	Rake angle	deg
δ	Cutting angle	deg
ε	Nose angle	deg
λ	Angle of cutting edge inclination	deg
ρ	Density	g/cm^3
α	Attack angle of GP	deg
σ_{ind}	Endurance limit	N/m^2 (pascal)
μ_m	Mean coefficient of friction on tool face	—
η_m	Mechanical efficiency of machine tool	
Δ_t	Temperature rise	°C (K)

Symbol	Description	Units
ε_{tr}	True strain	m/m
σ_{tr}	True stress	N/m^2 (pascal)
ε_y	Yield strain	m/m
σ_y	Yield stress (elastic limit)	N/m^2 (pascal)
$\tau_{xy}, \tau_{yz}, \tau_{zx}$	Shear stress components	MPa
pa	Average pressure	MPa
p_{max}	Maximum pressure	
α	Angle	Radian
βm	Mixing ratio	
γ	Specific weight	g/cm^3
ΔR	Planer anisotropy of sheet metals	—
$\varepsilon_1, \varepsilon_2, \varepsilon_3$	Principal strain components	
$\varepsilon_r, \varepsilon_\theta, \varepsilon_z$	Radial, circumferential and axial strains in cylindrical coordinates (axisymmetric problems)	
$\varepsilon_x, \varepsilon_y, \varepsilon_z$	Normal strain components in Cartesian coordinates	
ζ	Current efficiency	%
θ	Volumetric strain	—
θ_m	Melting point	°C
λ	Wave length	cm (µm)
μ	Coefficient of friction	
μ	Coefficient of linear expansion	10^{-6}/°C
ρ	Density	g/cm^3
σ'	Stress deviator	MPa
$\sigma_1, \sigma_2, \sigma_3$	Principal stress components	MPa
σ_f	Tensile strength of fiber	MPa
σ_f	Flow stress	N/m^2
σ_f, σ_b	Front tension and back tension in flat rolling	MPa
$\sigma_r, \sigma_\theta, \sigma_z$	Radial, circumferential and axial stresses in cylindrical coordinates (axisymmetric problems)	MPa
$\sigma_x, \sigma_y, \sigma_z$	Normal stress components in Cartesian coordinates	MPa
τ_{max}	Maximum shear stress	MPa
χ	Setting angle (angle of approach)	Deg

Appendix C: List of Acronyms

Abbreviation	Meaning
AB	Adhesive bonding
ABS	Acrylonitrile butadiene styrene
ac	Alternating current
AC	Adaptive control
ACC	Adaptive control with constraints
ACO	Adaptive control with optimization
ACP	Aluminum composite panel
AD	Anodic dissolution
AE	Acoustic emission testing
AFM	Abrasive flow machining
AFNOR	Association Francaise de Normalisation
AHW	Atomic hydrogen welding
AI	Artificial intelligence
AISI	American Iron and Steel Institute
AJM	Abrasive jet machining
ALD	Atomic layer deposition
AMS	Aerospace materials specification
AMU	Atomic mass unit
ANN	Artificial neural network
ANSI	American National Standards Institute
APF	Atomic packing factor
APT	Automatically programmed tools
AQL	Acceptance quality level
ASA	American standard association
ASCII	American standard code for information interchange
ASM	American Society of Metals
ASTM	American Society of Testing Materials
AWJM	Abrasive water jet machining
AWS	American Welding Society
BMG	Bulk metallic glass
BRA	Back rake angle
BS	Breaking strength
BUE	Built-up edge
BZN	Borazon
C	Cutting
C/C	Carbon/carbon composite
CA	Cellulose acetate
CAD	Computer-aided design
CAE	Computer-aided engineering
CAI	Computer-aided inspection
CAM	Computer-aided manufacturing
CAPP	Computer-aided process planning
CAR	Computer-aided reporting
CAW	Carbon arc welding
CBN	Cubic boron nitride
CCP	Conventional computer program

Abbreviation	Meaning
ccw	Counterclockwise
CD	Chemical dissolution
CE	Concurrent engineering
CHM	Chemical machining
CH-Milling	Chemical milling
CI	Cast iron
CIM	Computer-integrated manufacturing
CIP	Cold isostatic pressing
CLDATA	Cutter location data
CMC	Ceramic matrix composite
CMDI	Conversational manual data input
CMM	Coordinate measuring machine
CMP	Chemical-mechanical planarization
CNC	Computer numerical control
CR	Chloroprene rubber
CSP	Chip-scale package
CVD	Chemical vapor deposition
CW	Continuous wave mode
cw	Clockwise
DB	Diffusion bonding/database
dBA	Decibels
dc	Direct current
DCC	Direct computer control
DCEP	Direct current electrode positive
DCEN	Direct current electrode negative
DFE	Design for environment
DFM	Design for manufacturing
DW	Diffusion welding
DIN	Deutsche Institut fuer Normung
DNC	Direct numerical control
DOF	Degree of freedom
DOT	Department of Transport
DP	Degree of polymerization
DTI	Dial test indicator
DXF	Drawing exchange file
EARS	Elimination and ranking strategy
EB	Electron beam
EBH	Electron beam hardening
EBM	Electron beam machining
EBW	Electron beam welding
ECD	Electrochemical deposition
ECDB	Electrochemical deburring
ECEA	End cutting edge angle
ECG	Electrochemical grinding
ECM	Electrochemical machining
ECUSM	Electrochemical-assisted ultrasonic machining
EDM	Electrodischarge machining
ED-Milling	Electrodischarge milling
EDWC	Electrodischarge wire cutting
EHF	Electrohydraulic forming
EHS	Environmental health and safety

Abbreviation	Meaning
EIA	Electronic Industry Association
EMF	Electromagnetic forming
emf	Electromagnetic field
EMM	Electrochemical micromachining
EPA	Environmental protection association
EPDM rubber	Ethylene-propylene-diene monomer(M-class) rubber
EPS	Expanded polystyrene
ERA	End relief angle
ES	Expert system
ESW	Electro slag welding
ET	Eddy current testing
EW	Explosion welding
Excimer	Excited dimmer
FCAW	Flux cored arc welding
FDM	Fused deposition modeling
FGM	Functionally graded materials
FL	Fuzzy logic
FLD	Forming limit diagrams
FMC	Flexible manufacturing cell
FMS	Flexible manufacturing system
FOF	Factory of future
FOW	Forge welding
FRW	Friction welding
FS	Femtosecond
FW	Flash welding
GAC	Geometric adaptive control
GFRRP	Glass fiber reinforced plastic
GH	Grind-hardening
GMAW	Gas metal arc welding
GOST	Russian Gosstandart
GT	Group technology
GTAW	Gas tungsten arc
GW	Grinding wheel
HAP	Hazardous air pollutants
HAV	Hand and arm vibrations
HAZ	Heat affected zone
HDPE	High-density polyethylene
HERF	High energy rate forming
Hf	Hafnium
HIP	Hot isostatic pressing
HIPS	High-impact polystyrene
HMP	Hybrid machining process
HP	Hybrid process
HPDL	High pressure decorative laminate
HPW	Hot pressure welding
HRC	Rockwell hardness C
HSLA	High strength low alloy
HSS	High speed steel
HT	Heat treatment
HVF	High velocity forming
IBM	Ion beam machining

Abbreviation	Meaning
IC	Integrated circuit
IMPS	Integrated manufacturing production system
IR	Infrared
ISO	International Organization for Standardization
IT	Standard tolerance
JIT	Just in time
JIS	Japanese Industrial Standard
JM	Jet machining
LASER	Light amplification by stimulated emission of radiation
LAT	Laser-assisted turning
LB	Laser beam
LBH	Laser beam hardening
LBM	Laser beam machining
LBW	Laser beam welding
LCM	Liquid composite molding
LDPE	Low density polyethylene
LDR	Limit drawing ratio
LMC	Least material condition
LOM	Laminated object manufacturing
LPE	Liquid phase epitaxy
LT	Leak testing
LTP	Liquid thermal polymerization
LVDT	Linear variable displacement transducer
M/C	Machine
MBE	Molecular beam epitaxy
MC	Machine capability
MCD	Machine control data
MCU	Machine control unit
MDF	Medium-density fiberboard
MDI	Manual data input
MF	Melamine-formaldehyde
MFM	Magnetic field machining
MIG	Metal inert gas
MIM	Metal injection molding
MIT	Massachusetts Institute of Technology
MMA	Manual metal arc welding
MMC	Metal matrix composite
MOSFET	Metal oxide substrate field effect transistor
MPE	Maximum permissible exposure
MRP	Material requirement planning
MRR	Metal (Material) removal rate
MS	Manufacturing system
MT	Magnetic particle testing
MTCVD	Medium temperature chemical vapor deposition
MW	Molecular weight
NASA	National Aeronautics and Space Administration
NBR	Nitrile-butadiene rubber
NC	Numerical control
Nd:glass	Neodymium:glass
Nd:YAG	Neodymium:yttrium aluminum garnet
NDE	Nondestructive evaluation

Abbreviation	Meaning
NDI	Nondestructive inspection
NDT	Nondestructive testing
NEMS	Nanoelectromechanical system
NHZ	Nominal hazard zone
NITINOL	Nickel titanium naval ordnance laboratories
NOISH	National Institute of Occupational Safety and Health
NR	Natural rubber
NTD	Nozzle tip distance
NTM	Nontraditional machining
OAW	Oxyacetylene welding
OSHA	Occupational Safety and Health Administration
P	Pulse mode
PA	Polyamides
PAC	Plasma arc cutting
PAH	Polycyclic aromatic hydrocarbons
PAN	Polyacrylonitrile
PAT	Plasma-assisted turning
PAW	Plasma arc welding
PBE	Polymer-bonded explosives
PBM	Plasma beam machining
PC	Polycarbonate
PCB	Printed circuit boards
PCM	Photochemical machining
PDMS	Polydimethylsiloxane
PDS	Product design specifications
PE	Polyethylene
PEL	Permissible exposure limit
PET	Polyethylene-terephthalate
PF	Phenol-pormaldehyde
PIB	Polyisobutylene
PIM	Powder injection molding
PLC	Programmable logic controller
PLV	Pulsed laser vaporization
PM	Powder metallurgy
PMC	Polymer matrix composite
PMMA	Polymethyl-methacrylate
POM	Polyoxymethylene
PP	Polypropylene
PPE	Personal protective equipment
PRIMA	Process information map
PS	Polystyrene
PT	Penetrant testing
PTFE	Polytetrafluoroethylene
PVB	Polyvinyl butyral
PVC	Polyvinylchloride
PVD	Physical vapor deposition
QC	Quality control
RBSC	Reaction bonded silicon carbide
RFS	Regardless of feature size
RIA	Robot Institute of America
RIM	Reaction injection molding

Abbreviation	Meaning
ROW	Roll welding
RP	Rapid prototyping
RPW	Projection welding
RSEW	Resistance-seam welding
RSW	Resistance-spot welding
RT	Radiographic testing
RTA	Rapid thermal anneal
SAE	Society of Automotive Engineers
SAW	Submerged arc welding
SBR	Styrene-butadiene rubber
SBS	Styrene-butadiene-styrene block copolymer
SCARA	Selective compliance assembly robot arm
SCEA	Side-cutting edge angle
SEM	Scanning electron microscope
SFF	Solid free form fabrication
SI	Surface integrity
SIALON	Si-Al-O-N
SLA	Stereolithography
SLS	Selective laser sintering
SM	Smart manufacturing
SMA	Shape memory alloy
SMAW	Shielded metal arc welding
SOD	Stand-off-distance
Sol-Gel	Solution conversion into gel
SPC	Statistical process control
SPF	Superplastic forming
SQC	Statistical quality control
SRA	Side rake angle
SRFA	Side relief angle
SRR	Stock removal rate
SS	Sweden standards
St	Steel
T	Tolerance
TCS	Tool carbon steel
TEM	Transverse excitation mode
TIG	Tungsten inert gas
TIR	Total indicator reading
TM	Traditional machining
TNR	Tool nose radius
TP	Thermoplastics
TPM	Total production maintenance
TPU	Thermoplastic polyurethanes
TQC	Total quality control
TQM	Total quality management
TS	Tensile strength
TW	Thermit welding
TWA	Time weighted average
UAW	United Auto Workers
UF	Urea formaldehyde
UHMWPE	Ultrahigh molecular weight polyethylene
USC	Ultrasonic cleaning

Abbreviation	Meaning
USM	Ultrasonic machining
USW	Ultrasonic welding
UT	Ultrasonic testing
UTS	Ultimate tensile strength
UV	Ultraviolet
UVP	Ultraviolet light processing
VDU	Visual display unit
VH	Vickers hardness
VLS	Vapor-liquid-solid
VT	Visual and optical testing
W	Wear
WEDM	Wire electrodischarge machining
WJM	Water jet machining
WP	Workpiece
XPS	Extruded polystyrene
YP	Yield point
YS	Yield strength
μ-USM	Microultrasonic machining
3DP	Three-dimensional printing

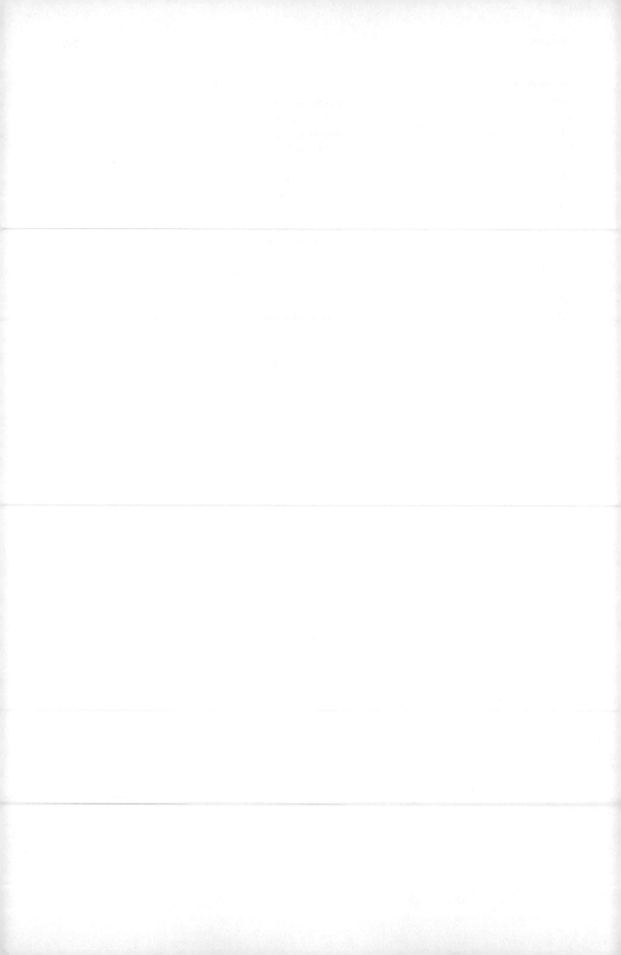

Index

Page numbers followed by *f* and *t* indicates figures and tables, respectively.